MODEL PREDICTIVE CONTROL OF WIND ENERGY CONVERSION SYSTEMS

IEEE Press
445 Hoes Lane
Piscataway, NJ 08854

MODEL PREDICTIVE CONTROL OF WIND ENERGY CONVERSION SYSTEMS

VENKATA YARAMASU
BIN WU

IEEE PRESS SERIES ON POWER ENGINEERING

IEEE PRESS

WILEY

Published by John Wiley & Sons, Inc., Hoboken, New Jersey.
Published simultaneously in Canada.

Library of Congress Cataloging-in-Publication Data is available.

ISBN: 978-1-118-98858-9

10 9 8 7 6 5 4 3 2 1

FSC www.fsc.org MIX Paper from responsible sources FSC® C013604

To my parents Subbayamma and Veeraiah

V. Y.

To my parents Lingen and Peifang and wife Janice

B. W.

CONTENTS

ABOUT THE AUTHORS

Venkata Yaramasu was born in Karumanchi, Guntur, India. He received his B.Tech degree in electrical and electronics engineering from Jawaharlal Nehru Technological University, Hyderabad, India, in 2005, an M.E. degree in electrical engineering with specialization in power electronics from S. G. S. Institute of Technology and Science, Indore, India, in 2008, and Ph.D. degree in electrical engineering from Ryerson University, Toronto, Canada, in 2014. During 2014–2015, he worked as a Postdoctoral Research Fellow at the Laboratory for Electric Drive Applications and Research (LEDAR) and Center for Urban Energy (CUE), Ryerson University. He is currently working as an Assistant Professor of Electrical Engineering in the School of Informatics, Computing, and Cyber Systems (SICCS), Northern Arizona University, Flagstaff, Arizona, USA. His research interests include renewable energy, high-power converters, variable-speed drives, electric vehicles, power quality, energy storage, smartgrid, and model predictive control.

Dr. Yaramasu worked closely with Rockwell Automation, Toronto Hydro, Hydro One, Natural Sciences and Engineering Research Council of Canada (NSERC), Wind Energy Strategic Network (WESNet) and Connect Canada, and completed 8 industrial projects in Power Electronics, Electric Drives, and Renewable Energy. He has published more than 50 peer-reviewed technical papers including 22 journal papers. He has produced more than 10 technical reports for the industry.

Dr. Yaramasu received six Best Student Paper Awards and two first prizes in National Level Technical Quiz Competitions during his undergraduate studies in India. He is a recipient of a Second Prize Paper Award from the IEEE Journal of Emerging and Selected Topics in Power Electronics (JESTPE) in 2015. During his Ph.D. studies at Ryerson University, he received Best Poster Awards from the Electrical and Computer Engineering (ECE) Department and Faculty of Engineering and Architectural Science (FEAS) in 2013, Student Research Awards from the Toronto Hydro, Hydro One, and Connect Canada in 2010, 2012, and 2013, Research Excellence Awards from the ECE Department in 2012, 2013, and 2014, a Best Poster Award at the NSERC–WESNet Annual Meeting 2010, and a Best Teaching Assistant Award from the FEAS in 2010.

Bin Wu graduated from Donghua University, Shanghai, China in 1978, and received his M.A.Sc. and Ph.D. degrees in electrical and computer engineering from the University of Toronto, Canada in 1989 and 1993, respectively. After being with Rockwell Automation Canada from 1992 to 1993, he joined Ryerson University, where he is currently a Professor in the Department of Electrical and Computer Engineering and a Senior NSERC/Rockwell Industrial Research Chair (IRC) in Power Electronics and Electric Drives.

Dr. Wu has published more than 350 peer-reviewed technical papers and two Wiley-IEEE Press books, and he holds more than 30 issued and pending patents in power electronics, adjustable-speed drives, and renewable energy systems. Dr. Wu is the founder of the Laboratory for Electric Drive Applications and Research (LEDAR), which has been recognized as the most advanced research facility of its kind in a Canadian university.

Dr. Wu has worked closely with Canadian companies and assisted them in achieving technical and commercial success through research and innovation. He has authored/coauthored more than 200 technical reports. Some of his inventions and patents have been adopted by industry and implemented in the production line, resulting in significant economic benefits.

Dr. Wu received the Gold Medal of the Governor General of Canada in 1993, the Premier's Research Excellence Award in 2001, the NSERC Synergy Award for Innovation in 2002, the Ryerson Distinguished Scholar Award in 2003, the Ryerson FEAS Research Excellence Award in 2007, the Ryerson YSGS Outstanding Contribution to Graduate Education Award and the Professional Engineers Ontario (PEO) Engineering Excellence Medal in 2014. He is a fellow of Institute of Electrical and Electronics Engineers (IEEE), Engineering Institute of Canada (EIC), and Canadian Academy of Engineering (CAE). Dr. Wu is a Registered Professional Engineer in the Province of Ontario, Canada.

PREFACE

Due to depleting fossil fuels and growing environmental concerns on global warming, electricity production from renewable energy sources has received increasing attention in recent years. Among the renewable energy sources, wind energy is rapidly becoming mainstream and competitive with conventional sources of energy. Wind energy installed capacity has increased exponentially over the past three decades and has become a real alternative to boost renewable energy penetration into the energy mix. The wind energy industry has experienced considerable technological advancements in terms of aerodynamic design, mechanical systems, electric generators, power electronic converters, control theory, and power system integration. Electric generators, power electronic converters, and control theory are the three important elements to enable the safe, reliable, and high-performance operation of wind energy conversion systems (WECS) while complying with the stringent grid code requirements.

In recent years, with the technological advancements in digital signal processors, the model predictive control (MPC) strategy has emerged as a simple and promising digital control tool in power electronics, variable-speed motor drives, and energy conversion systems. The MPC is a nonlinear control method and provides an approach that is better suited for controlling power converters in WECS. This method also mitigates several technical and operational disadvantages associated with classical control techniques, particularly during the low-switching frequency operation needed by the megawatt (MW)-level energy conversion systems. The MPC is attractive for controlling fast varying electrical variables because of its simple and intuitive concept, digital controller friendliness, finite number of optimizations, elimination of proportional-integral (PI) controllers, pulse-width-modulator-free structure, fast dynamic response, good steady-state performance during all operating conditions, capability to compensate perturbations and dead times of power conversion system, ease in incorporating nonlinearities and limitations in the design, and improved treatment of multivariable control problems with decoupling. Over the past decade, several books on power converters, wind energy systems, and MPC have been published. However, books that provide comprehensive analysis by combining these three important subject matters are unavailable. The proposed book deals with the MPC of power converters employed in a wide variety of variable-speed WECS. This book will not only fill the gap in the book market, but will also provoke further studies in the academe and industry for applications to other power electronic converters, adjustable-speed motor drives, and renewable energy conversion systems.

This book covers a wide range of topics on power converters, wind energy conversion, and MPC from the electrical engineering aspect. The contents of this book includes an overview of wind energy system configurations, power converters for variable-speed WECS, digital control techniques, MPC, modeling of power converters and wind generators for MPC design. Other topics include the mapping of continuous-time models to discrete-time models by various exact, approximate, and quasi-exact discretization methods, modeling and control of wind turbine grid-side two-level and multilevel voltage source converters. The authors also focus on the MPC of several power converter configurations for full variable-speed permanent magnet synchronous generator (PMSG)-based WECS, squirrel-cage induction generator (SCIG)-based WECS, and semi-variable-speed doubly fed induction generator (DFIG)-based WECS. By reflecting the latest technologies in the field, this book is a valuable reference for academic researchers, practicing engineers, and other professionals. It can also be used as a textbook or reference book for graduate-level and advanced undergraduate courses.

ORGANIZATION OF BOOK CONTENTS

This book contains 3 parts with 12 chapters, as illustrated in Figure 0.1. The flow of contents and interconnection between chapters is also described.

Part I is composed of four introductory-level chapters related to the power conversion and digital control of WECS. **Chapter 1** provides an overview of high-power WECS, grid code requirements (e.g., fault-ride through and reactive power generation for wind power grid integration), electric generators, power electronics, wind energy configurations, WECS control, maximum power point tracking (MPPT) techniques, and finite control-set MPC (FCS-MPC). **Chapter 2** reviews the state-of-the-art generator–converter configurations for variable-speed (Type 3 and 4) WECS in four categories with low-voltage (LV) and medium-voltage (MV) subcategories: (1) back-to-back (BTB) connected power converters, (2) passive generator-side (PGS) power converters, (3) converters for six-phase and open-winding generators, and (4) power converters without an intermediate DC link (matrix converters). **Chapter 3** presents an overview of state-of-the-art digital control techniques including hysteresis control, linear control with PI regulator and modulation stage, sliding mode control, fuzzy logic control, artificial neural network-based control, deadbeat predictive control, and FCS-MPC by considering a case study of load current control. **Chapter 4** discusses the fundamentals of MPC including its operating principle, step-by-step design procedure, cost function flexibility, weighting factor selection, control delay compensation, reference variable extrapolation, and minimum sampling time selection.

Part II contains three intermediate-level chapters on the modeling of power converters and wind generators to assist the implementation of MPC strategies for variable-speed WECS. **Chapter 5** addresses the modeling of power converters employed in variable-speed WECS. The relationship between the input and output variables of a 2L-VSC, three-level (3L) neutral-point clamped converter, 3L flying capacitor converter, current source converter, direct matrix converter, and indirect matrix converter is formulated in terms of converter switching states. **Chapter 6** deals with the state-space continuous-time modeling of PMSG, SCIG, and DFIG used in Type 3 and 4 WECS. **Chapter 7** introduces sampled-data models for MPC design. The exact discretization by the zero-order hold method, approximate discretization by forward Euler, backward Euler, and bilinear transformation, and quasi-exact discretization by matrix factorization and truncated Taylor series are discussed in detail with the help of example problems.

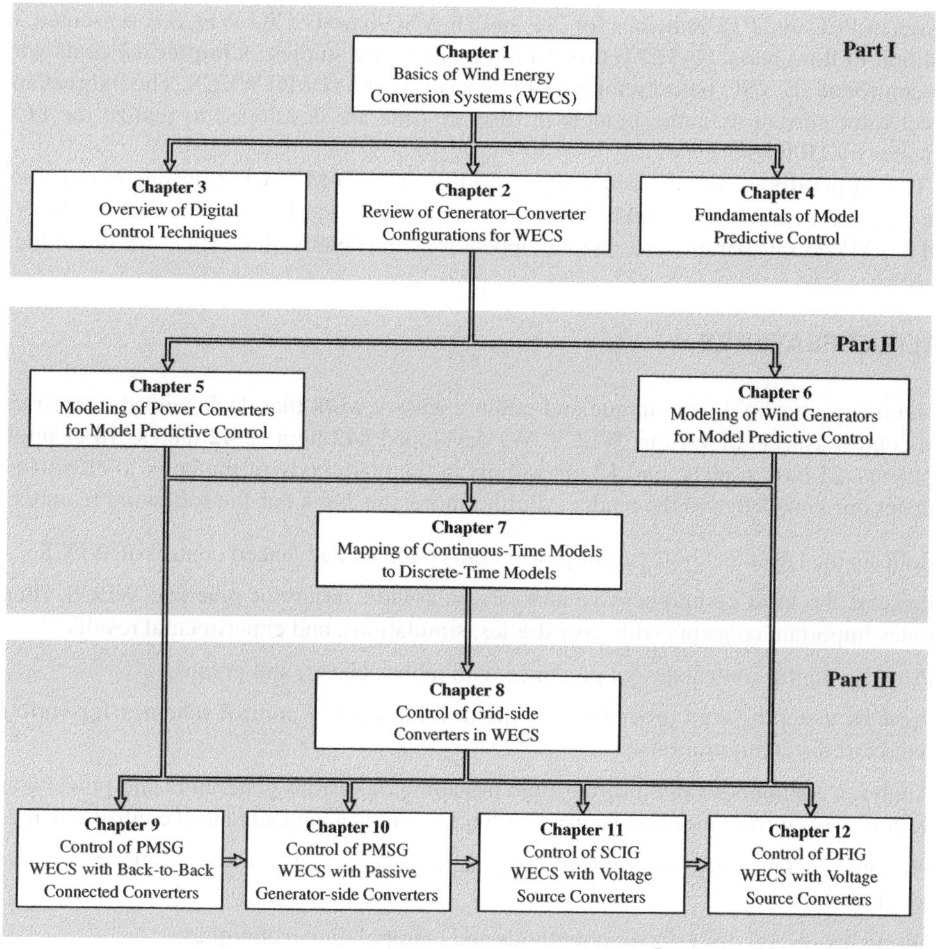

Figure 0.1 Organization of parts and chapters in this book.

Part III is composed of five advanced-level chapters with the MPC of variable-speed WECS. **Chapter 8** presents the MPC strategy for grid-side converters (GSCs) for high-power LV and MV WECS. The synchronous- and stationary-frame predictive current control (PCC) and predictive power control (PPC) for 2L and 3L GSCs are analyzed in detail in terms of design steps, control algorithms, and case studies. This chapter serves as a basic building block for the complete digital control systems in Chapters 9 to 12. **Chapter 9** discusses the MPC strategy for BTB-connected 2L- and 3L-VSC-based PMSG WECS. The zero d-axis current control and maximum torque per ampere control of surface mount and inset PMSGs are outlined. The calculation of reference control variables based on optimal tip-speed ratio and optimal torque MPPT algorithms are discussed. The PCC and predictive torque control (PTC) schemes are analyzed in detail. **Chapter 10** provides a comprehensive analysis on the PGS converters employed in LV and MV PMSG WECS. The control requirements such as MPPT, balancing of DC-link capacitors voltage, regulation of net DC-bus voltage, and grid reactive power control are analyzed with 2L and 3L boost converters, as well as PCC and PPC schemes. **Chapter 11** emphasizes the predictive control of BTB VSC for variable-speed SCIG-based LV and MV WECS. A step-by-step design of PCC and PTC schemes for 2L- and 3L-VSC-based SCIG WECS is presented, in addition to flowcharts, MATLAB programming, and case studies. **Chapter 12** deals with the control of 2L-VSC-based semi-variable-speed ($\pm 30\%$) DFIG WECS. The indirect and direct rotor current dynamic models in discrete time are developed to realize the PCC schemes for DFIG.

The **Appendix** covers the following: 750-kW and 3.0-MW wind turbine parameters; low-, medium-, and high-speed PMSG parameters; SCIG and DFIG parameters; 750-kVA and 3.0-MVA GSC parameters; chapter appendices and details of the MATLAB demo files.

SALIENT FEATURES

In summary, this book is a unique and comprehensive work that deals with the electrical and control aspects of various WECS. We developed **242** figures, **42** tables, **18** example problems, **21** case studies, and **12** algorithms in the main body of the book to effectively transfer our knowledge to the readers. Furthermore, this book has the following features:

- Reflects the latest technologies in power conversion and advanced control of WECS.
- Presents the most comprehensive analysis on a wide variety of practical WECS. Illustrates important concepts with case studies, simulations, and experimental results.
- Provides digital control design guidance with tables, charts, and graphs.
- Presents a step-by-step procedure for the development of control schemes for various wind turbine configurations.
- Analyzes continuous- and discrete-time modeling of various generators and power converters. This information can be used with other classical and advanced control schemes.
- Presents a detailed analysis on weighting factor selection, which will fill the gap in current literature.
- Discusses several discretization methods and extrapolation techniques.
- Discusses real-time implementation issues such as the configuration of prototype converters, selection of sampling time, and compensation of control delay.
- Presents useful material for other power electronic applications such as adjustable-speed motor drives, power quality conditioners, electric vehicles, photovoltaic energy systems, distributed generation, and high-voltage direct current transmission.
- Provides *S-Function Builder* programming in MATLAB environment to implement various MPC strategies.
- Supplies MATLAB demo files for quick-start on MPC design and simulation.
- Serves as a valuable reference for academic researchers, practicing engineers, and other professionals.
- Provides adequate technical background for graduate- and undergraduate-level teaching.

VENKATA YARAMASU

BIN WU

April 2016

ACKNOWLEDGMENTS

This book has been written as a result of many years of our research on wind energy and model predictive control. The manuscript took almost two and half years to complete. We would like to thank our colleagues and friends who have supported and helped us in this endeavor. The references section of each chapter also acknowledges the research work of various scholars who have been passionate about this research area.

We would like to acknowledge the valuable support of Mr. Apparao Dekka, Ph.D. student in Laboratory for Electric Drive Applications and Research at the Ryerson University, in developing the figures for Chapters 7, 8, 11, and 12. His help with the MATLAB simulations for Chapters 11 and 12 is highly appreciated. We are indebted to Mr. Tomasz Sidełko, double degree Master's student with Wroclaw University and Ryerson University, for devoting significant amount of time to develop high-quality artwork and figures during the initial phase of this book. We wish to thank Drs. Jose Rodriguez, Marco Rivera, Samir Kouro, Salvador Alepuz, and Paresh C. Sen for the collaborative works on wind energy and model predictive control.

We are also very thankful to the reviewers for painstakingly reviewing our book proposal and final manuscript and providing us constructive comments to improve the quality and readability of the book. We very much appreciate Mr. Samkruth Aluru, Master's student at Northern Arizona University and Dr. Sebastian Rivera, Postdoctoral Fellow at University of Toronto for proofreading the final manuscript.

Our special thanks to the Wiley/IEEE press editors Mary Hatcher, Brady Chin, and Tim Pletscher for their help in the preparation of this manuscript. We are also thankful to the Wiley production team members for guiding us to prepare camera-ready copy of this manuscript. We express our intense gratitude to our families and friends who supported and inspired us in every possible way to see the completion of this work.

V. Y.

B. W.

This book is developed as an outgrowth of my Ph.D. dissertation entitled "Predictive Control of Multilevel Converters for Megawatt Wind Energy Conversion Systems." I am thankful to Drs. Dewei Xu and Amir Yazdani at Ryerson University for their encouragement in transforming my Ph.D. dissertation into a book. I express my sincere gratitude to my former Ph.D. dissertation supervisor, Dr. Bin Wu, for his guidance in writing this manuscript. This book would not have been envisaged without his help.

V. Y.

ACRONYMS

2L	Two-Level (Converter)
3L	Three-Level (Converter)
1S	Single-Stage (Gearbox)
2S	Two-Stage (Gearbox)
3S	Three-Stage (Gearbox)
3ϕ	Three-Phase (Generator/Grid)
AC	Alternating Current
ADC	Analog-to-Digital Conversion
ANPC	Active Neutral-Point Clamped
BTB	Back-to-Back
CHB	Cascaded H-Bridge
CMV	Common-Mode Voltage
COE	Cost of Energy
CSC	Current Source Converter
CT	Continuous-Time
DAC	Digital-to-Analog Conversion
DBPC	Deadbeat Predictive Control
DC	Direct Current
DCC	Diode-Clamped Converter
DD	Direct-Driven
DFIG	Doubly Fed Induction Generator
DFOC	Direct Field-Oriented Control
DG	Distributed Generation
DMC	Direct Matrix Converter
DPC	Direct Power Control
DSO	Distribution System Operator

DSP	Digital Signal Processor
DT	Discrete Time
DTC	Direct Torque Control
E+P	Estimation + Prediction
FC	Flying Capacitor
FCS-MPC	Finite Control-Set Model Predictive Control
FFT	Fast Fourier Transform
FOC	Field-Oriented Control
FPGA	Field-Programmable Gate Array
FRT	Fault Ride-Through
FSWT	Fixed-Speed Wind Turbine
GCT	Gate-Controlled Thyristor
GSC	Grid-side Converter
GSF	Generator Signal Feedback
GW	Gigawatt
HAWT	Horizontal-Axis Wind Turbine
HTS	High-Temperature Superconducting
HVAC	High-Voltage Alternating Current
HVDC	High-Voltage Direct Current
HVRT	High-Voltage Ride-Through
IFOC	Indirect Field-Oriented Control
IG	Induction Generator
IGBT	Insulated Gate Bipolar Transistor
IGCT	Integrated Gate-Commutated Thyristor
IM	Induction Machine
IMC	Indirect Matrix Converter
IPMSG	Interior (Inset) Permanent Magnet Synchronous Generator
LPF	Low-Pass Filter
LTI	Linear Time-Invariant
LTV	Linear Time-Variant
LV	Low Voltage
LVRT	Low-Voltage Ride-Through
MIPS	Million Instructions Per Second
MMC	Modular Multilevel Converter
MMMC	Multi-Modular Matrix Converter
MPC	Model Predictive Control
MPP	Maximum Power Point

MPPT	Maximum Power Point Tracking
MSC	Machine-side Converter
MTPA	Maximum Torque Per Ampere
MV	Medium Voltage
MW	Megawatt
NPC	Neutral-Point Clamped
OT	Optimal Torque
OTSR	Optimal Tip-Speed Ratio
PCC	Point of Common Coupling
PCC	Predictive Current Control
PDPC	Predictive Direct Power Control
PDVC	Predictive DC-Bus Voltage Control
PF	Power Factor
PGS	Passive Generator-side
P+P	Prediction + Prediction
P+P^2	Prediction + Double Prediction
PPC	Predictive Power Control
PSC	Predictive Speed Control
PSF	Power Signal Feedback
PTC	Predictive Torque Control
PVC	Predictive Voltage Control
PI	Proportional-Integral
PLL	Phase-Locked Loop
PMSG	Permanent Magnet Synchronous Generator
PWM	Pulse Width Modulation
RL	Resistive-Inductive
RMS	Root Mean Square
RPG	Reactive Power Generation
RSC	Rotor-side Converter
SCIG	Squirrel Cage Induction Generator
SFOC	Stator Flux-Oriented Control
SG	Synchronous Generator
SGCT	Symmetrical Gate-Controlled Thyristor
SHE	Selective Harmonic Elimination
SPMSG	Surface-Mount Permanent Magnet Synchronous Generator
SRF	Synchronous Reference Frame
STATCOM	Static Synchronous Compensator

SVC	Static VAR Compensator
SVM	Space Vector Modulation
SVOC	Stator Voltage-Oriented Control
TDD	Total Demand Distortion
THD	Total Harmonic Distortion
TSO	Transmission System Operator
TSR	Tip-Speed Ratio
UPF	Unity Power Factor
VAWT	Vertical-Axis Wind Turbine
VOC	Voltage-Oriented Control
VSC	Voltage Source Converter
VSI	Voltage Source Inverter
VSR	Voltage Source Rectifier
VSWT	Variable-Speed Wind Turbine
WECS	Wind Energy Conversion Systems
WF	Wind Farm
WFCP	Wind Farm Collection Point
WRIG	Wound Rotor Induction Generator
WRSG	Wound Rotor Synchronous Generator
WT	Wind Turbine
WTPC	Wind Turbine Power Curves
ZDC	Zero d-axis Current
ZOH	Zero-Order Hold
ZVRT	Zero-Voltage Ride-Through

SYMBOLS

Notation

P, Q	Active and reactive power
R, L, C	Resistance, inductance, and capacitance
v, i, ψ	Voltage, current, and flux linkage (*peak* values)
V, I, Ψ	Voltage, current, and flux linkage (*rms* values)
s	Switching signal

Superscripts for variable x

x^{op}	Optimal value of variable
x^{p}	Predicted variable
$\widehat{x}$	Estimated variable
x^{*}	Reference variable
$\widehat{x}^{*}$	Extrapolated reference variable

Subscripts for variable x

x_a, x_b, x_c	Natural (abc) frame variables
$x_{\scriptscriptstyle B}$	Base value
x_d, x_q	Synchronous (dq) frame variables
x_{dc}	DC/DC boost converter related variable
x_g	Grid-related variable
x_i	Inverter-related variable
x_m	Mechanical system variable
x_{op}	Optimal value of variable
x_r	Rectifier/rotor related variable
$x_{\scriptscriptstyle R}$	Rated value
x_s	Generator (source in wind energy) related variable
$x_{\scriptscriptstyle T}$	Turbine-related variable
x_w	Wind-related variable
x_{WF}	Wind farm-related variable
x_α, x_β	Stationary ($\alpha\beta$) frame variables

Significant Symbols

$\mathfrak{a}$	Fortescue operator = $e^{j\frac{2\pi}{3}}$
$\mathbf{A}$	State matrix in continuous time
A_T	WT rotor swept area
$\mathbf{B}$	Input or control matrix in continuous time
B_m	WT and generator combined viscous friction
$\mathbf{C}$	Output matrix in continuous time
C_{dc}	DC-link capacitance
C_{in}	Input DC-link capacitance of boost converter
C_p	Power coefficient of WT rotor blades
C_p^{op}	Optimal power coefficient of WT rotor blades
D	Boost converter duty ratio
e_i	Mean value of current tracking error
f_g	Grid frequency
f_s	Generator stator frequency
f_{sl}	Generator slip frequency
f_{sw}	Converter switching frequency
$f_{sw,dc}$	Boost converter switching frequency
$f_{sw,i}$	Inverter switching frequency
$f_{sw,r}$	Rectifier switching frequency
g	Cost function for model predictive control
g_{cm}	Cost function for common-mode voltage mitigation
g_{dc}	Cost function for balancing of DC-link capacitors voltage
g_i	Cost function for current control
g_{id}	Cost function for d-axis current control
g_{iq}	Cost function for q-axis current control
$g_{i\alpha}$	Cost function for α-axis current control
$g_{i\beta}$	Cost function for β-axis current control
g_{op}	Optimal cost function value
g_{PQ}	Cost function for active and reactive power control
g_{sw}	Cost function for converter switching frequency minimization
$g_{sw,r}$	Cost function for rectifier switching frequency minimization
$g_{sw,i}$	Cost function for inverter switching frequency minimization
$g_{T\psi}$	Cost function for electromagnetic torque and stator flux control
$\mathbf{H}$	Output matrix in discrete time
H_m	WT and generator combined inertia constant
$\mathbf{I}$	Unity matrix
i_{ag}, i_{bg}, i_{cg}	Grid *peak* currents in abc-frame
I_{ag}, I_{bg}, I_{cg}	Grid *rms* currents in abc-frame
i_{ar}, i_{br}, i_{cr}	Generator *peak* rotor currents in abc-frame
I_{ar}, I_{br}, I_{cr}	Generator *rms* rotor currents in abc-frame

i_{as}, i_{bs}, i_{cs}	Generator *peak* stator currents in *abc*-frame
I_{as}, I_{bs}, I_{cs}	Generator *rms* stator currents in *abc*-frame
I^*_B	Reference base current (*rms*)
I^*_{Bg}	Reference grid base current (*rms*)
I^*_{Bs}	Reference generator base current (*rms*)
i_{C1}, i_{C2}	Currents flowing through DC-link capacitors C_1 and C_2
i_{dc}	DC branch current
i_{dg}, i_{qg}	Grid *peak* currents in *dq*-frame
i_{dr}, i_{qr}	Generator *peak* rotor currents in *dq*-frame
i_{ds}, i_{qs}	Generator *peak* stator currents in *dq*-frame
i_g	Grid *peak* current
I_g	Grid *rms* current
i_m	Generator *peak* magnetizing current
i_P, i_Z, i_N	Positive, mid-point, and negative DC branch currents
i_{Pi}, i_{Zi}, i_{Ni}	Inverter-side positive, mid-point, and negative DC branch currents
i_{Pr}, i_{Zr}, i_{Nr}	Rectifier-side positive, mid-point, and negative DC branch currents
i_r	Generator *peak* rotor current
I_r	Generator *rms* rotor current
i_s	Generator *peak* stator current
I_s	Generator *rms* stator current
$i_{\alpha g}, i_{\beta g}$	Grid *peak* currents in $\alpha\beta$-frame
$i_{\alpha r}, i_{\beta r}$	Generator *peak* rotor currents in $\alpha\beta$-frame
$i_{\alpha s}, i_{\beta s}$	Generator *peak* stator currents in $\alpha\beta$-frame
j	Switching state combination number
J_m	Wind turbine and wind generator combined moment of inertia
j_{op}	Optimal switching state combination number
k	Present sampling instant in discrete-time frame
$k-1$	Past sampling instant in discrete-time frame
$k+1$	Future sampling instant in discrete-time frame
k_i	Integral gain of PI controller
k_p	Proportional gain of PI controller
k_r	Generator rotor coupling coefficient
k_s	Generator stator coupling coefficient
K_T	Generator electromagnetic torque constant
L	Load inductance
L_{dc}	Inductance of boost converter DC choke
L_{ds}, L_{qs}	Generator *dq*-axes stator winding inductances
L_i	Inductance of inverter-side harmonic filter
L_{ls}, L_{lr}	Generator stator and rotor winding leakage inductances
L_m	Generator magnetizing inductance
L_s, L_r	Generator stator and rotor winding self-inductances

N	Negative DC bus
n_m	Generator mechanical speed (rpm)
N_p	Prediction horizon
n_{sw}	Number of switching changes in a converter
n_T	WT rotational speed (rpm)
P	Positive DC bus
$P_{cu,s}$	Stator winding copper losses
P_{dc}	DC-link power
P_g	Grid active power
P_g^*	Grid active power reference
P_{in}^*	Reference input DC power
P_m	Generator mechanical input power
P_p	Number of pole pairs in a generator
P_o	Overall WECS output active power
P_r	Generator rotor output active power
P_s	Generator stator output active power
P_s^*	Generator stator active power reference
$P_{s,R}$	Generator rated stator output power
P_T	WT output power
pu	Per unit
$\widehat{\text{pu}}$	Peak per unit
P_w	Kinetic power in wind
Q_g	Grid reactive power
Q_g^*	Grid reactive power reference
Q_o	Overall WECS output reactive power
Q_r	Generator rotor output reactive power
Q_s	Generator stator output reactive power
Q_s^*	Generator stator output reactive power reference
R	Load resistance
r_{dc}	Internal resistance of DC inductor L_{dc}
r_{gb}	Gearbox ratio
r_i	Internal resistance of harmonic filter L_i
R_s, R_r	Generator stator and rotor winding resistances
r_T	WT rotor blade radius
r_σ	Generator equivalent resistance
$\mathbf{s}$	Switching state combination for a power converter
s	Generator slip (when not used as a subscript) or Laplace operator
s_a, s_b, s_c	Phase-*abc* switching signals for top switch in a two-level converter
$\overline{s}_a$, $\overline{s}_b$, $\overline{s}_c$	Phase-*abc* switching signals for bottom switch in a two-level converter
$\mathbf{S_a}$, $\mathbf{S_b}$, $\mathbf{S_c}$	Phase-*abc* switching vectors
s_{aj}, s_{bj}, s_{cj}	Phase-*abc* switching signals, $j \in \{1, 2\}$

s_{anj},s_{bnj},s_{cnj}	Phase-*abc* switching signals with respect to neutral, $j \in \{1, 2\}$
s_{di}, s_{qi}	Inverter *dq*-axes switching signals
s_{dj}, s_{qj}	*dq*-axes switching signals, $j \in \{1, 2\}$
s_{dr}, $_{qr}$	Rectifier *dq*-axes switching signals
s_f	Switching signal corresponding to the grid fault
S_g	Grid apparent power
S_o	Overall WECS output apparent power
S_s	Generator stator apparent power
s_x	Switching signal of phase-x top switch in a two-level converter, $x \in \{a, b, c\}$
$\overline{s}_x$	Switching signal of phase-x bottom switch in a two-level converter, $x \in \{a, b, c\}$
$s_{\alpha i}$, $s_{\beta i}$	Inverter $\alpha\beta$-axes switching signals
$s_{\alpha j}$, $s_{\beta j}$	$\alpha\beta$-axes switching signals, $j \in \{1, 2\}$
$s_{\alpha r}$, $s_{\beta r}$	Rectifier $\alpha\beta$-axes switching signals
$\mathbf{T}$	Transformation matrix
$\mathbf{T}_{abc/abcn}$	*abc*-frame to *abcn*-frame transformation matrix
$\mathbf{T}_{abc/dq}$	*abc*-frame to *dq*-frame transformation matrix
$\mathbf{T}_{abc/\alpha\beta}$	*abc*-frame to $\alpha\beta$-frame transformation matrix
$\mathbf{T}_{dq/abc}$	*dq*-frame to *abc*-frame transformation matrix
$\mathbf{T}_{dq/\alpha\beta}$	*dq*-frame to $\alpha\beta$-frame transformation matrix
T_e	Generator electromagnetic torque
T_e^*	Generator electromagnetic torque reference
T_m	Generator mechanical input torque
T_s	Digital controller sampling time
T_T	WT output torque
$\mathbf{T}_{\alpha\beta/abc}$	$\alpha\beta$-frame to *abc*-frame transformation matrix
$\mathbf{T}_{\alpha\beta/dq}$	$\alpha\beta$-frame to *dq*-frame transformation matrix
$\mathbf{u}$	Input vector in continuous and discrete time
$\mathbf{u}^{\mathbf{op}}$	Optimal control input in continuous and discrete time
$\mathbf{v}$	Converter voltage vector
$\mathbf{v}^{\mathbf{op}}$	Optimal converter voltage vector
v_{ab}	Converter output phase-a to phase-b voltage
v_{ab1}	Converter output phase-a to phase-b fundamental voltage
v_{ag}, v_{bg}, v_{cg}	Grid *peak* voltages in *abc*-frame
V_{ag}, V_{bg}, V_{cg}	Grid *rms* voltages in *abc*-frame
v_{an}, v_{bn}, v_{cn}	Converter output voltages with respect to neutral n
v_{aN}, v_{bN}, v_{cN}	Converter output voltages with respect to negative DC-bus N
v_{ar}, v_{br}, v_{cr}	Generator *peak* rotor voltages in *abc*-frame
V_{ar}, V_{br}, V_{cr}	Generator *rms* rotor voltages in *abc*-frame
v_{as}, v_{bs}, v_{cs}	Generator *peak* stator voltages in *abc*-frame
V_{as}, V_{bs}, V_{cs}	Generator *rms* stator voltages in *abc*-frame
v_{aZ}, v_{bZ}, v_{cZ}	Converter output voltages with respect to DC-bus mid-point Z
v_{C1}, v_{C2}	Voltage across DC-link capacitors C_1 and C_2

v_{cm}	Common-mode voltage
v_C^*	Reference DC mid-point voltage (= 0)
v_{dc}	Net DC-bus voltage (voltage between positive and negative DC terminals)
v_{dg}, v_{qg}	Grid *peak* voltages in dq-frame
v_{di}, v_{qi}	Inverter output voltages in dq-frame
v_{dr}, v_{qr}	Generator *peak* rotor voltages in dq-frame
v_{dr}, v_{qr}	Rectifier output voltages in dq-frame
v_{ds}, v_{qs}	Generator *peak* stator voltages in dq-frame
v_g	Grid *peak* phase voltage
V_g	Grid *rms* phase voltage
$V_{g,ll}$	Grid *rms* line-to-line voltage
$v_{i,ab}$	Inverter output phase-a to phase-b voltage
$v_{i,ab1}$	Inverter output phase-a to phase-b fundamental voltage
v_{in}	Boost converter input voltage
v_P, v_Z, v_N	Positive, mid-point and negative DC-bus voltages
v_{nN}	Voltage between grid/generator neutral n and negative DC-bus N
v_{ns}	Generator neutral voltage
v_{nZ}	Voltage between grid/generator neutral n and DC-bus mid-point Z
v_r	Generator *peak* rotor phase voltage
V_r	Generator *rms* rotor phase voltage
$V_{r,ll}$	Generator *rms* rotor line-to-line voltage
$v_{r,ab}$	Rectifier output phase-a to phase-b voltage
$v_{r,ab1}$	Rectifier output phase-a to phase-b fundamental voltage
v_s	Generator *peak* stator phase voltage
V_s	Generator *rms* stator phase voltage
$V_{s,ll}$	Generator *rms* stator line-to-line voltage
v_w	Wind speed
v_{xN}	Converter phase-x output voltage with respect to negative DC-bus N
$v_{\alpha g}$, $v_{\beta g}$	Grid *peak* voltages in $\alpha\beta$-frame
$v_{\alpha i}$, $v_{\beta i}$	Inverter output voltages in $\alpha\beta$-frame
$v_{\alpha r}$, $v_{\beta r}$	Generator *peak* rotor voltages in $\alpha\beta$-frame
$v_{\alpha r}$, $v_{\beta r}$	Rectifier output voltages in $\alpha\beta$-frame
$v_{\alpha s}$, $v_{\beta s}$	Generator *peak* stator voltages in $\alpha\beta$-frame
$\mathbf{w}$	Disturbance input matrix in continuous time
$\mathbf{x}$	State vector in continuous and discrete time
Z	DC-bus mid-point
β	Pitch angle of WT rotor blades
$\mathbf{\Gamma_b}$	Input or control matrix in discrete time
$\mathbf{\Gamma_w}$	Disturbance input matrix in discrete time
δ	Bandwidth of hysteresis controller (relay)
θ	Phase angle of space vector

θ_e	Generator electrical angle
θ_g	Grid voltage angle
θ_m	Generator mechanical rotor position angle
θ_o	Initial phase angle
θ_r	Generator electrical rotor position angle
θ_s	Generator voltage angle
θ_{sl}	Generator slip angle
θ_T	Generator electromagnetic torque angle
λ_{cm}	Weighting factor for common-mode voltage mitigation
λ_{dc}	Weighting factor for balancing of DC-link capacitors voltage
λ_i	Weighting factor for current control
λ_{id}	Weighting factor for d-axis current control
λ_{iq}	Weighting factor for q-axis current control
$\lambda_{i\alpha}$	Weighting factor for α-axis current control
$\lambda_{i\beta}$	Weighting factor for β-axis current control
λ_P	Weighting factor for active power control
λ_Q	Weighting factor for reactive power control
λ_{sw}	Weighting factor for switching frequency minimization
λ_T	WT tip-speed ratio
λ_T^{op}	Optimal WT tip-speed ratio
λ_{Te}	Weighting factor for electromagnetic torque control
λ_ψ	Weighting factor for stator flux control
ρ	Air density
σ	Generator leakage coefficient
τ_σ	Generator stator transient time constant
τ_r	Generator rotor time constant
$\mathbf{\Phi}$	State matrix in discrete time
φ_g	Grid power factor angle (angle between grid voltage and current)
φ_s	Generator power factor angle (angle between stator voltage and current)
ψ_r	Peak value of rotor flux linkage
Ψ_r	RMS value of rotor flux linkage
ψ_s	Peak value of stator flux linkage
Ψ_s	RMS value of stator flux linkage
ω	Angular frequency of space vector
ω_g	Grid angular frequency
ω_k	Angular frequency of arbitrary reference frame
ω_m	Generator mechanical rotor angular speed
$\omega_{m,R}$	Generator rated mechanical rotor angular speed
ω_r	Generator rotor electrical angular speed
ω_s	Generator stator electrical angular frequency
ω_{sl}	Generator angular slip frequency
ω_T	Angular speed of WT blades

Vectors

$\mathbf{i}$	Current vector in natural frame = $[i_a \; i_b \; i_c]^T$
$\mathbf{i_C}$	DC-link current vector = $[i_{C1} \; i_{C2}]^T$
$\mathbf{i_{dq}}$	Current vector in synchronous frame = $[i_d \; i_q]^T$
$\mathbf{i}_{\alpha\beta}$	Current vector in stationary frame = $[i_\alpha \; i_\beta]^T$
$\mathbf{i_g}$	Grid current vector in natural frame = $[i_{ag} \; i_{bg} \; i_{cg}]^T$
$\mathbf{i_{g,dq}}$	Grid current vector in synchronous frame = $[i_{dg} \; i_{qg}]^T$
$\mathbf{i}_{\mathbf{g},\alpha\beta}$	Grid current vector in stationary frame = $[i_{\alpha g} \; i_{\beta g}]^T$
$\mathbf{i_i}$	Inverter current vector in natural frame = $[i_{ai} \; i_{bi} \; i_{ci}]^T$
$\mathbf{i_r}$	Rectifier/rotor current vector in natural frame = $[i_{ar} \; i_{br} \; i_{cr}]^T$
$\mathbf{i_{r,dq}}$	Rotor current vector in synchronous frame = $[i_{dr} \; i_{qr}]^T$
$\mathbf{i}_{\mathbf{r},\alpha\beta}$	Rotor current vector in stationary frame = $[i_{\alpha r} \; i_{\beta r}]^T$
$\mathbf{i_s}$	Generator current vector in natural frame = $[i_{as} \; i_{bs} \; i_{cs}]^T$
$\mathbf{i_{s,dq}}$	Generator current vector in synchronous frame = $[i_{ds} \; i_{qs}]^T$
$\mathbf{i}_{\mathbf{s},\alpha\beta}$	Generator current vector in stationary frame = $[i_{\alpha s} \; i_{\beta s}]^T$
$\mathbf{v}$	Voltage vector in natural frame = $[v_{an} \; v_{bn} \; v_{cn}]^T$
$\mathbf{v_C}$	DC-link voltage vector = $[v_{C1} \; v_{C2}]^T$
$\mathbf{v_g}$	Grid voltage vector in natural frame = $[v_{ag} \; v_{bg} \; v_{cg}]^T$
$\mathbf{v_{g,dq}}$	Grid voltage vector in synchronous frame = $[v_{dg} \; v_{qg}]^T$
$\mathbf{v}_{\mathbf{g},\alpha\beta}$	Grid voltage vector in stationary frame = $[v_{\alpha g} \; v_{\beta g}]^T$
$\mathbf{v_i}$	Inverter voltage vector in natural frame = $[v_{ai} \; v_{bi} \; v_{ci}]^T$
$\mathbf{v_r}$	Rectifier/rotor voltage vector in natural frame = $[v_{ar} \; v_{br} \; v_{cr}]^T$
$\mathbf{v_{r,dq}}$	Rotor voltage vector in synchronous frame = $[v_{dr} \; v_{qr}]^T$
$\mathbf{v}_{\mathbf{r},\alpha\beta}$	Rotor voltage vector in stationary frame = $[v_{\alpha r} \; v_{\beta r}]^T$
$\mathbf{v_s}$	Generator stator voltage vector in natural frame = $[v_{as} \; v_{bs} \; v_{cs}]^T$
$\mathbf{v_{s,dq}}$	Generator stator voltage vector in synchronous frame = $[v_{ds} \; v_{qs}]^T$
$\mathbf{v}_{\mathbf{s},\alpha\beta}$	Generator stator voltage vector in stationary frame = $[v_{\alpha s} \; v_{\beta s}]^T$
$\mathbf{s_{dc}}$	Boost converter state combination = $[s_{dc1} \; s_{dc2}]^T$
$\mathbf{s_i}$	Inverter switching state combination in natural frame = $[s_{ai} \; s_{bi} \; s_{ci}]^T$
$\mathbf{s_r}$	Rectifier switching state combination in natural frame = $[s_{ar} \; s_{br} \; s_{cr}]^T$
$\boldsymbol{\psi}_\mathbf{r}$	Rotor flux linkage vector in natural frame = $[\psi_{ar} \; \psi_{br} \; \psi_{cr}]^T$
$\boldsymbol{\psi}_\mathbf{r,dq}$	Rotor flux linkage vector in synchronous frame = $[\psi_{dr} \; \psi_{qr}]^T$
$\boldsymbol{\psi}_{\mathbf{r},\alpha\beta}$	Rotor flux linkage vector in stationary frame = $[\psi_{\alpha r} \; \psi_{\beta r}]^T$
$\boldsymbol{\psi}_\mathbf{s}$	Stator flux linkage vector in natural frame = $[\psi_{as} \; \psi_{bs} \; \psi_{cs}]^T$
$\boldsymbol{\psi}_\mathbf{s,dq}$	Stator flux linkage vector in synchronous frame = $[\psi_{ds} \; \psi_{qs}]^T$
$\boldsymbol{\psi}_{\mathbf{s},\alpha\beta}$	Stator flux linkage vector in stationary frame = $[\psi_{\alpha s} \; \psi_{\beta s}]^T$

PART I

PRELIMINARIES

CHAPTER 1

BASICS OF WIND ENERGY CONVERSION SYSTEMS (WECS)

1.1 INTRODUCTION

Renewable energy sources such as solar, wind, hydro, geothermal, tidal, and wave have emerged as a new paradigm to fulfill the energy needs of our civilization. In contrast to fossil fuels, renewable energy sources are clean, abundant, naturally replenished, available over wide geographical areas, and have less or no impact on the environment. Renewable energy sources are primarily used for electricity generation, heating, transportation fuels, and rural energy supply. Electricity production from renewable energy sources has come under growing attention in recent decades. Approximately 22% of global electricity consumption is compensated by all types of renewable energy sources. Driven by technological innovations, cost reduction, government incentive programs, and public demand for clean energy, wind energy is increasingly becoming mainstream, competing with not only other renewable energy sources but also with conventional fossil fuel-based power generation units [1]. At the end of 2014, global cumulative wind power capacity reached 370 gigawatts (GWs), which accounts for approximately 4% of the world's net electricity production [2–4].

Wind energy has been harnessed by mankind for millennia to carry ships across oceans, pump water, and grind grain. The conversion of wind kinetic energy to electric energy started during the 1880s with an automated wind turbine (WT) equipped with a 12-kilowatt (kW) direct-current (DC) generator. To generate electricity from WTs more efficiently and reliably, many improvements have been made in the design of the mechanical and electrical apparatus of WTs. The WT expertise has reached an adequate maturity level by

Model Predictive Control of Wind Energy Conversion Systems, First Edition. Venkata Yaramasu and Bin Wu.

Companion Website: www.wiley.com/go/yaramasu_wu_mpc_wecs

the 1980s, leading to the commissioning of the first 50-kW utility-scale WTs. Over the past 35 years, the size of WTs has gradually increased and has currently reached a massive level of 10 megawatts (MWs). Due to the rapid integration of wind power into the electric grid, many concerns have emerged on the stable and secure operation of existing electric power systems. Grid code requirements have been updated and enforced in many countries on the grid connection of large-scale WTs and wind farms (WFs).

Power electronic converters have been used in commercial WTs since the beginning of grid-connected operation; this technology has significantly evolved over the years [1, 5]. Various combinations of wind generators and power converters have also been developed in commercial WTs to achieve fixed-speed, semi-variable-speed, and full-variable-speed operations. Fixed-speed WT (FSWT) technology, which uses a power converter for the startup function (soft-start), is considered obsolete. Variable-speed WTs (VSWTs) process the electric output power of a generator through a power converter and offer enhanced wind energy conversion efficiency, power quality, and compatibility with grid codes. To fulfill various technical, operational, and grid code requirements, several generator–converter configurations have been developed for commercial WTs.

In addition to the power converter equipment, control system development is important in the safe, successful, and efficient operation of VSWTs. The electrical control system is used to control wind generators and power converters such that maximum energy is extracted from the wind and feeds the energy to the utility grid with high power quality. Electrical control systems are commonly implemented by digital control platforms such as microcontroller (μC), digital signal processor (DSP), or field programmable gate array (FPGA). With the evolution of digital control expertise, the realization of advanced and high-performance control algorithms is now possible. The finite control-set model predictive control (FCS-MPC) is a new breed of digital control technique for power converters and electric generators [6–8]. FCS-MPC opens the doors for controlling WT electric power conversion systems in an easier and more intuitive manner than traditional hysteresis and linear control techniques while offering optimal control performance. This book deals with the FCS-MPC (simply called model predictive control or MPC from here onwards) of power electronic converters in variable-speed WECS.

Chapter Overview

- An overview of wind energy technology is given in Section 1.2 with respect to the installed wind power capacity, wind kinetic energy to electric energy conversion theory, and various classes of WT technologies.
- In Section 1.3, the major mechanical, electrical, and control-related components of the grid-connected high-power WECS are presented. The grid code requirements for large-scale WECS are examined in Section 1.4.
- Fixed speed, semi-variable-speed, and full-variable-speed (Types 1 to 5) WECS are analyzed and compared in Section 1.5. An overview of the broad range of power converters employed in Types 3 and 4 WECS are presented in Section 1.6.
- Section 1.7 presents the control of WECS including pitch control, maximum power point tracking (MPPT) control, grid integration, and taxonomy of wind generator and power converter control. Section 1.8 outlines the main features and challenges of FCS-MPC. In Section 1.9, the linear and model predictive control schemes for variables-speed WECS are presented to demonstrate the simplicity of MPC over traditional control.
- The concluding remarks of this chapter are given in Section 1.10.

1.2 WIND ENERGY PRELIMINARIES

In this section, wind energy preliminaries are discussed briefly in terms of installed wind power capacity, wind energy installations by various countries, wind kinetic energy to electric energy conversion, and evolution of major WT technologies since the 1980s.

1.2.1 Installed Wind Power Capacity

Wind energy is one of the fastest growing renewable energy sources and continues to flourish each year in many countries. The cumulative and annual installed wind power capacities worldwide are shown in Figures 1.1 and 1.2, respectively. The top 10 countries in cumulative and annual installed wind power capacities are shown in Figures 1.3 and 1.4, respectively. Data are presented collectively according to the Global Wind Energy Council 2015 report [2], REN21 renewable global status report 2014 [3], and BTM wind report summary 2015 [4]. The key observations in the installed wind power capacities are summarized below.

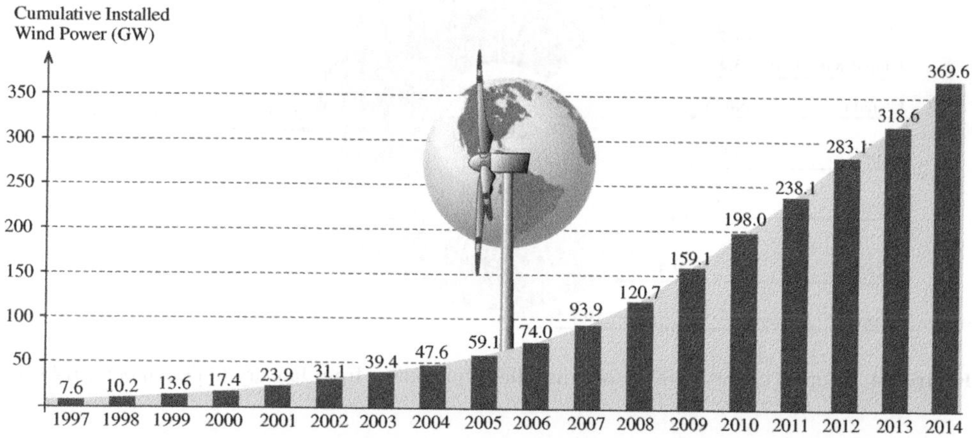

Figure 1.1 Global cumulative installed wind power from 1997 to 2014 [2].

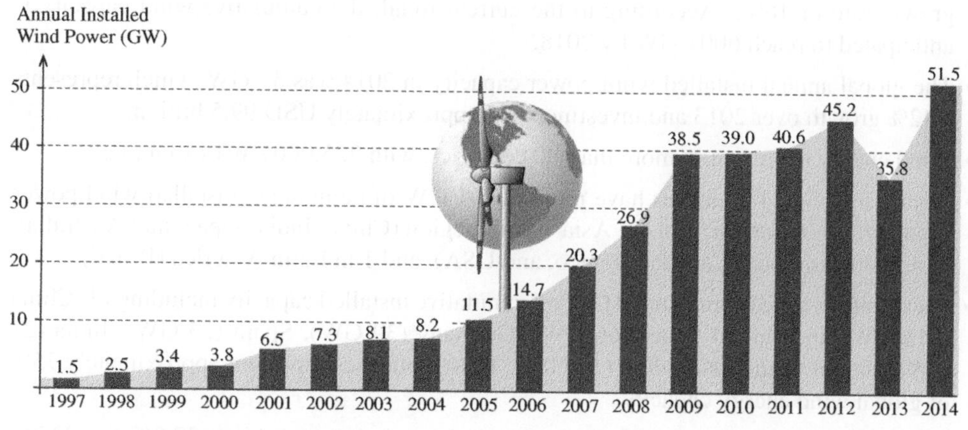

Figure 1.2 Global annual installed wind power from 1997 to 2014 [2].

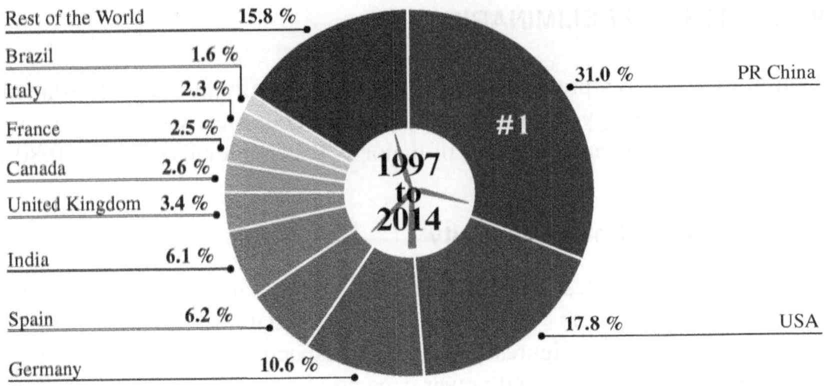

Figure 1.3 Top 10 countries in the cumulative installed wind power by December 2014 [2].

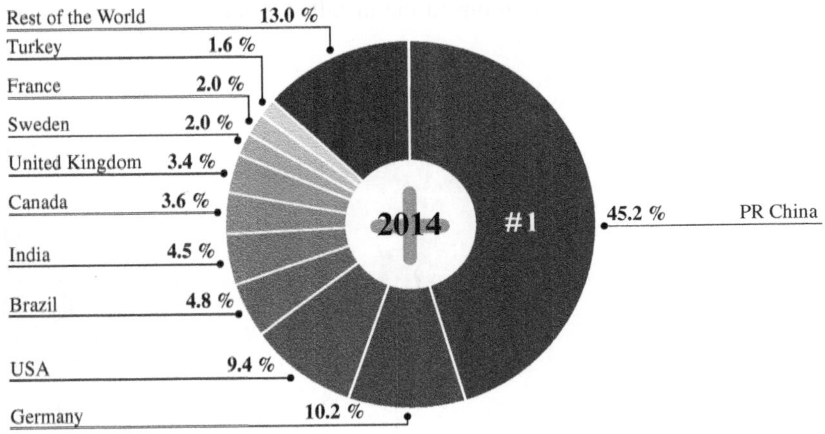

Figure 1.4 Top 10 countries in the new installed wind power from January to December 2014 [2].

- As highlighted in Figure 1.1, the cumulative installed wind power capacity increased exponentially from 7600 MW in 1997 to 369.6 GW by 2014. The data show a cumulative growth rate of 16%. According to the current trend, the cumulative wind capacity is anticipated to reach 600+ GW by 2018.
- The global annual installed wind power capacity in 2014 was 51 GW, which represents a 42% growth over 2013 and investments of approximately USD 99.5 billion.
- Wind energy is present in more than 80 countries, with 225,000+ WTs spinning.
- Approximately 24 countries have more than 1 GW of cumulative installed wind power capacity: 16 in Europe, 4 in the Asia-Pacific region (China, India, Japan, and Australia), 3 in North America (Canada, Mexico, and USA), and 1 in Latin America (Brazil).
- Six countries have more than 10 GW of cumulative installed capacity including PR China (115 GW), the United States (66 GW), Germany (39 GW), Spain (23 GW), India (23 GW), and the United Kingdom (12 GW). These countries represent approximately 75% of global wind energy capacity.
- Asia and Europe are key players in the wind energy industry and host 37.3% and 23.7% of global wind power capacity, respectively.

- Wind power accounts for approximately 4% of the world's net electricity consumption. Wind power has the highest penetration in some countries. For example, 39% of all electricity needs in Denmark is covered by wind power plants.
- The Chinese wind market maintained its top position in both global cumulative and annual installed power by doubling its capacity from 62 GW in 2011 to 115 GW by the end of 2014. PR China has added over 23 GW of new capacity in 2014, which accounts for 45% of the global annual installed capacity and is a record-high number.
- The United States is the second largest country in global cumulative installed capacity. This country added 4.8 GW of new capacity in 2014. Germany and Spain are the third and fourth top countries in wind power, respectively.
- India is the fifth largest country with a cumulative installed wind power capacity of 22.5 GW, which accounts for 6.9% of its electricity consumption. The United Kingdom holds the sixth position with 3.4% installed wind power capacity.
- The cumulative installed wind power capacity in Canada steadily increased from 137 MW in 2000 to 9.7 GW by the end of 2014, thus making it the seventh largest country in wind power. In 2014, 1871 MW of new wind capacity came online in Canada. Currently, 5% of electricity in Canada is covered by wind energy and has ambitious plans to supply 20% of net electricity from wind power by 2025.
- The last three positions in the top 10 lists are occupied by France, Italy, and Brazil.

1.2.2 Wind Kinetic Energy to Electric Energy Conversion

As illustrated in Figure 1.5, WTs convert kinetic energy into electric energy by using various mechanical and electrical components. This section discusses the concept of converting wind kinetic energy to electric energy.

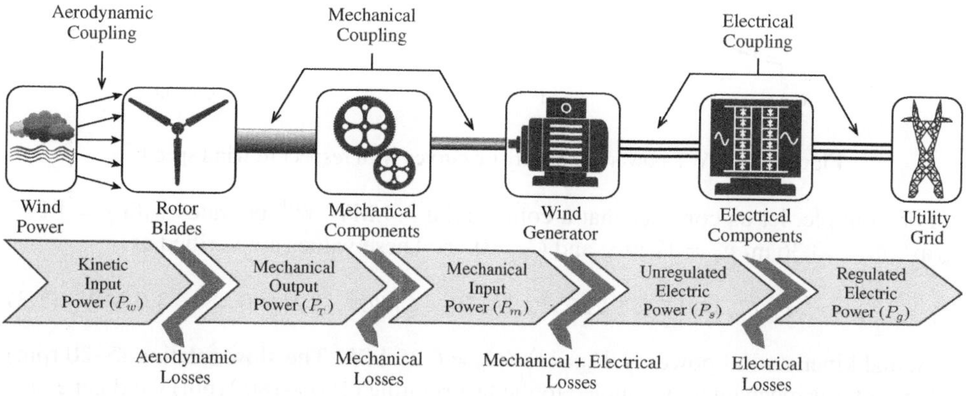

Figure 1.5 Block diagram of wind kinetic energy to electric energy conversion system.

Kinetic energy is first converted into mechanical energy by the rotor blades. The wind kinetic power P_w flowing through an imaginary area A_T at a speed v_w is

$$P_w = \frac{1}{2}\,\rho\,A_T\,v_w^3, \qquad A_T = \pi\,r_T^2 \tag{1.1}$$

where ρ is air density (kg/m^3), A_T is the rotor swept area (m^2), r_T is the blade radius (m), and v_w is the wind speed (m/s). Air density is a function of altitude, temperature, and humidity. At sea level and at 15°C, air has a typical density of 1.225 kg/m^3.

According to the theory of German scientist *Albert Betz*, mechanical power P_T extracted from the wind kinetic power P_w is given by the following:

$$P_T = P_w \times C_p = \frac{1}{2} \rho A_T v_w^3 C_p \tag{1.2}$$

where C_p represents the power coefficient of rotor blades. The P_T extraction increases in proportion to the C_p value. According to *Betz*, the theoretical or maximum value of C_p is 16/27 or 0.593. For the new generation of high-power WTs, the C_p value ranges between 0.32 and 0.52. The C_p value for various commercial WTs can be found in [9].

The P_w and P_T curves of a WT, with respect to the wind speed v_w, are plotted in Figure 1.6. The value of P_T is always lower than the P_w value because of aerodynamic power losses. The typical cut-in, rated, and cut-out wind-speed values are in the range of 3–5 m/s, 10–15 m/s, and 25–30 m/s, respectively. WTs produce negligible power when wind speed is below the cut-in value; thus, such turbines are usually kept in parking mode. To ensure safety, the turbine is shut down and kept in parking mode when wind speed is above the cut-out value or during emergency conditions. For wind-speed values between cut-in and rated, the P_T curve maintains a cubic relationship with respect to v_w. When wind speed is between the rated and cut-out value, the turbine output power P_T is regulated to its highest threshold (rated) value by the aerodynamic power control.

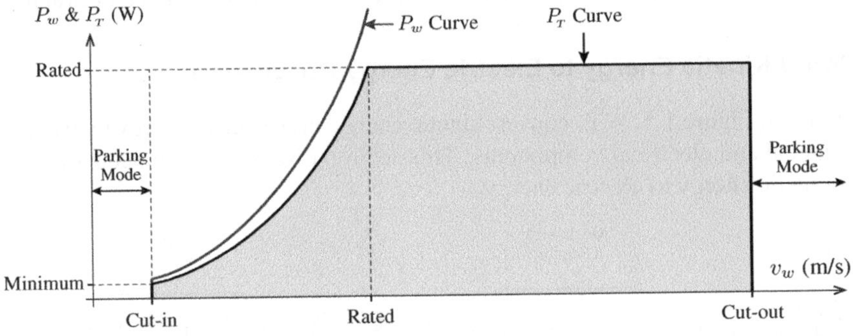

Figure 1.6 WT power characteristic curves with respect to wind speed.

For example, let us consider that a commercial 3.0 MW WT operates with $\rho = 1.225$ kg/m^3, $r_T = 43.36$ m, $v_w = 12$ m/s, and $C_p = 0.48$. These values are verified as

$$P_T = \frac{1}{2} \rho \pi r_T^2 v_w^3 C_p = 0.5 \times 1.225 \times 3.14159 \times 43.36^2 \times 12^3 \times 0.48 = 3.0 \text{ MW}. \tag{1.3}$$

Actual kinetic wind power is $P_w = P_T / C_p = 6.25$ MW. The slow rotating (5–20 rpm) turbine rotor is mechanically coupled to the fast rotating (1500–1800 rpm) wind generator through the main shaft and drivetrain. Losses in the mechanical components represent the difference between P_T and P_m. The mechanical input power P_m is converted into electric power P_s by the wind generator.

The generator output power P_s is considered unregulated because the generator output voltage and frequency varies according to the wind speed. To feed electric power P_g to the utility grid with fixed voltage and frequency, while complying with the grid code requirements, a power converter is used. To connect the WTs to the utility transmission network, the output voltage of the power converter is boosted usually from 690 V to 33 kV by using a step-up transformer. A more detailed discussion about the mechanical and electrical components of the WT is given in Section 1.3.

1.2.3 Classification of Wind Energy Technologies

As shown in Figure 1.7, in the present wind energy industry, major WT technologies are classified according to various criteria/factors [10]. The most prominent classification factors are based on the following: (1) WT electric output power rating (low, medium, and high power), (2) aerodynamic power regulation scheme during high-wind-speed conditions (stall and pitch control), (3) alignment of wind generator shaft with respect to the ground (vertical and horizontal axis), (4) type of application to feed the turbine electric output power (standalone and grid-connected), (5) wind generator operating speed with respect to the varying wind speeds (fixed and variable speed), (6) location for erection of WTs (onshore and offshore), (7) type of mechanical coupling between the turbine and generator shaft (geared and gearless), and (8) wind-speed velocities (low, medium, and high) affecting the WT. This section provides a brief overview of the various types of predominant wind energy technologies. The other insignificant classification factors are ignored in the analysis.

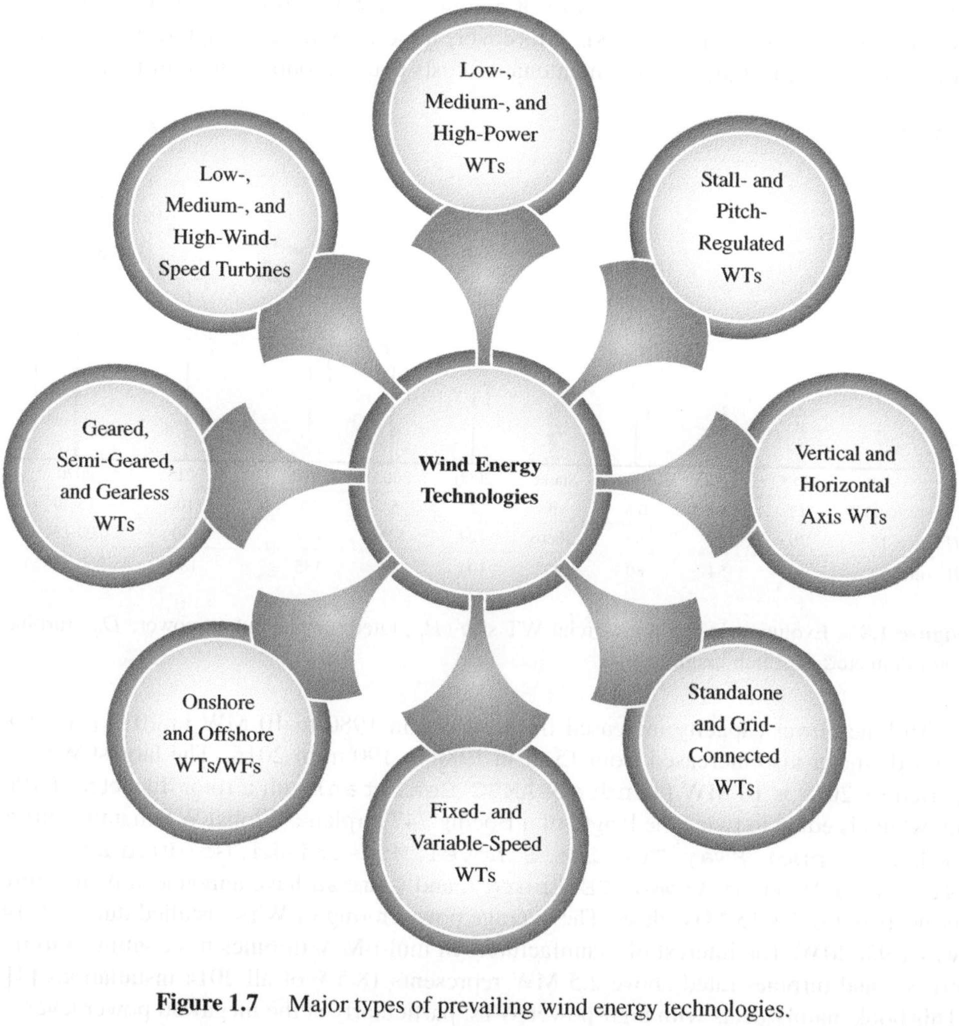

Figure 1.7 Major types of prevailing wind energy technologies.

(1) Low-, Medium-, and High-Power Turbines: According to the nominal electric output power capacity, WTs are classified into three major categories:

- **Low-Power (<30 kW) Turbines:** These turbines are primarily used for battery charging in remote areas, and household electricity generation.
- **Medium-Power (30 to 300 kW) Turbines:** These turbines are used in distributed generation (DG) in conjunction with other renewable energy sources and energy storage systems to supply the electricity needs of small communities.
- **High-Power (>300 kW) Turbines:** These turbines are primarily deployed in wind farms for bulk electricity generation.

By referring to Equation (1.2), $P_T \propto \rho r_T^2 v_w^3 C_p$. Turbine output power can be increased as follows: (1) selecting a turbine erection site with high ρ value, (2) increasing rotor radius size r_T (P_T increases fourfold when r_T is doubled), (3) selecting a site with strong wind speeds or by increasing the tower height to access stable wind speeds (P_T increases eightfold when v_w is doubled), and (4) designing turbine rotor blades with high C_p value. Considering these facts, the size of commercial WTs has exponentially increased over the past 35 years (Figure 1.8). Moreover, large turbines can capture higher wind power with lower installation and maintenance costs than a group of small turbines.

	1980	1985	1990	1995	Statue of Liberty	2000	2005	2010	2015	2020
P_T (MW)	0.05	0.1	0.5	0.8		2	5	7.5	10	15-20
D_T (m)	15	20	40	50		124	126	127	190	200-250
H_T (m)	24	43	54	80	93	104	114	135	180	200-250

Figure 1.8 Evolution in the commercial WT size (P_T, rated turbine output power; D_T, turbine rotor diameter; H_T, hub height).

Turbine power capacity increased from 50 kW in 1980 to 10 MW in 2015 [1]. WT rotor diameter also increased from 15 m in 1980 to 190 m in 2015. The largest WT reported by 2015 is 10 MW (`Windtec-AMSC SeaTitan`) with a rotor diameter of 190 m, which is equal to twice the length of a Boeing 747 airplane. Global WT manufacturers such as `Clipper`, `Sway Turbine`, `Sinovel`, `Mitsubishi`, `Goldwind`, `Mecal`, `MingYang`, `United Power`, `GE Energy`, and `Gamesa` have announced their future projects in the 10–15 MW class. The average power rating of WTs installed during 2014 was 1.958 MW. The interest of manufacturers in multi-MW turbines has continued to increase, and turbines rated above 2.5 MW represents 18.5% of all 2014 installations [4]. This book mainly deals with high-power WTs, particularly at the megawatt power level.

(2) Passive Stall-, Active Stall-, and Pitch-Regulated Turbines: By theory, wind turbines can produce more power than the rated (nameplate) value when the wind speed v_w is above its rated value. However, to guarantee secure and reliable function and avoid the overloading of turbine blades, the turbine mechanical output power P_T must be limited to its rated value during strong wind-speed conditions. This process is usually achieved by aerodynamic power control, which manipulates the flow of air stream over rotor blades. Three aerodynamic power regulation methods are available: passive stall, active stall, and pitch control [11]. The characteristics of mechanical output power limitation by these three control methods are depicted in Figure 1.9.

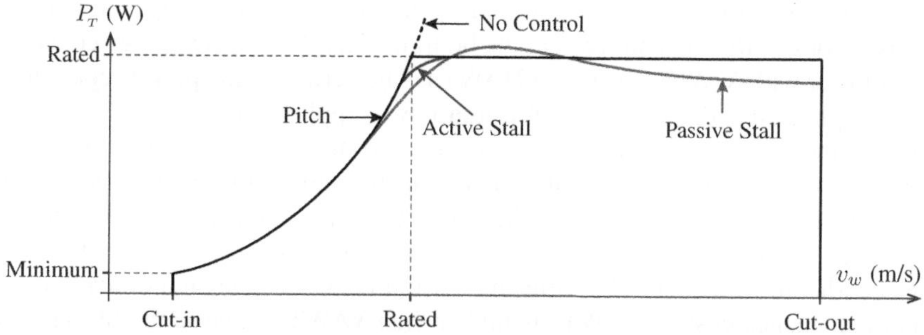

Figure 1.9 Aerodynamic power regulation of WTs by different control methods [11].

The passive stall or simply stall control is the first-generation power regulation method employed for WTs and is the simplest method among the three classes. This approach does not use any sensors, electronic controller, or actuator, thus making it less expensive and robust. In stall-regulated turbines, the rotor blades are firmly fixed (bolted) to the rotor hub at a fixed angle. When wind speed increases beyond a certain limit (for example, 15 m/s), the turbulence created on the rotor surface causes the airfoils to lose lift force, thus decreasing the power captured from the wind. P_T exceeds the rated value with an overshoot at a certain wind speed. The overshoot in P_T is an undesirable function. As v_w increases above the rated value, P_T gradually decreases, thus leading to low energy conversion efficiency. Passive stall control is used in low- to medium-power WTs.

Active stall control is an advanced version of passive stall control with adjustable rotor blades. When v_w increases above the rated value, P_T is reduced by moving (pitching) the blades into the wind, thus causing turbulence (stall mechanism) over the blades. This approach improves wind energy conversion efficiency at low wind speeds and ensures that P_T does not exceed the rated value during high-wind-speed conditions (refer to Figure 1.9). This method is employed in medium- to high-power WTs.

In pitch-regulated turbines, the rotor blades are adjustable similar to active stall turbines. The pitch control mechanism is assisted by an electronic controller and motor (or hydraulic) drives. During high wind speeds, the electronic controller sends a control signal to the motor to turn the rotor blades along the longitudinal axis (pitching) such that angle of attack of the blades is reduced. The active stall method turns the blades "into wind" to create a stall mechanism, whereas pitch control turns the blades "out of wind." The mechanical output power is tightly regulated by the pitch control as highlighted in Figure 1.9. This control provides faster responses/actions than the passive stall and active stall controls and is extensively used in modern high-power WTs [12]. Unless otherwise stated, the default aerodynamic power regulation employed in this book is pitch control.

(3) Vertical and Horizontal Axis Turbines: In the present wind industry, two basic types of WTs are available depending on the orientation of the wind generator and drivetrain shaft with respect to the ground: vertical-axis WTs (VAWTs) and horizontal-axis WTs (HAWTs) [13]. Both WT types convert wind kinetic energy into electric energy but have different spinning axes with respect to ground.

In VAWTs, the shaft of wind generator is placed perpendicular to the ground. The turbine rotor is implemented by vertically mounted curved airfoils with different designs for rotor shapes. The installation and maintenance of VAWTs is easier than HAWTs because the generator and gearbox are placed close to the ground. VAWTs accept wind from any direction; thus, the orientation of the WT blades is not necessary. The rotor shaft is usually long and is more prone to mechanical vibrations. The rotor blades are subjected to uneven wind speeds: strong at the top and weak at the bottom. Therefore, wind energy conversion efficiency is lower with VAWTs than with HAWTs. The aerodynamic power regulation of VAWTs is complicated and is unsuitable for high-power applications.

In HAWTs, the wind generator shaft is placed horizontally to the ground. Major mechanical and electrical components are placed in a nacelle, and the tower elevates the height of the nacelle to allow sufficient space for the blades to rotate. Three rotor blades are commonly used in commercial HAWTs. The rotor blades are subjected to strong winds because of the elevated height; thus, HAWTs offer high-wind-energy conversion efficiency. The initial and maintenance cost of HAWTs is higher than VAWTs because the blades, gearbox, and generator of the former are placed away from the ground. Furthermore, HAWTs have easier aerodynamic power regulation and lower mechanical vibrations than VAWTs. HAWTs are predominantly used in medium- and high-power applications, and this book considers the HAWT technology only.

(4) Standalone and Grid-Connected Turbines: According to the type of application to feed electric output power, WTs are classified as standalone and grid-connected systems. Standalone DG is an alternative solution to power consumers in locations wherein the expansion of the electric grid is prohibitive and expensive. In standalone DG, load demand varies with respect to time and turbine output power changes with respect to wind speed. To provide stable, secure, and reliable electricity, WTs are used in conjunction with other power generation units such as photovoltaic (PV), mini/micro hydro, biomass, and diesel generators, as well as energy (usually battery energy) storage systems [14]. Low- to medium-power WTs are commonly employed in standalone applications and represent only a fraction of global cumulative installed wind power capacity.

Low- and medium-power WTs can also be used in grid-connected DG to reduce electricity consumption from the utility grid. When the electricity produced by the WT exceeds the household or community requirements, the electricity is sold to the utility without employing any energy storage. By contrast, the majority of commercial high-power WTs are employed in grid-connected applications. The power produced by high-power WTs is fed directly into the grid. The output of WTs is usually less than 1000 V but is boosted to a few kV by using a step-up transformer to send power via transmission lines. Wind farm power is fed into a grid by high-voltage alternating current (HVAC) or high-voltage direct current (HVDC) transmission lines. The power to be delivered and the distance of the wind farm to the nearby utility grid are two crucial factors that play an important role in deciding between HVAC and HVDC transmission. HVAC transmission is favorable for small-scale wind farms, which are located close to the utility grid. For power ratings and distances greater than 400 MW and 60 km, respectively, HVDC transmission is an excellent choice [15]. This book mainly deals with grid-connected high-power WTs.

(5) Fixed-Speed and Variable-Speed Turbines: On the basis of rotational speed, grid-connected turbines are classified as FSWTs and VSWTs. To aid in the discussion, tip speed ratio (TSR), which is a vital WT design parameter, is introduced here. TSR is defined as the ratio between the tangential speed of the blade tip and the wind speed [11]:

$$\lambda_T = \frac{\omega_T\, r_T}{v_w} \tag{1.4}$$

where ω_T is the angular speed of the turbine (rad/s) and v_w is the wind speed (m/s).

The power coefficient C_p discussed in (1.2) is a function of the TSR and blade pitch angle. The C_p versus λ_T curve is shown in Figure 1.10(a) when the pitch angle is held at a rated (zero) value. At a rated wind speed $v_{w,R}$, both FSWT and VSWT operate at an optimal TSR, λ_T^{op}, and maximum (optimal) power coefficient, C_p^{op}. The turbine output power P_T corresponding to the rated wind speed is also maximum (Figure 1.10(b)).

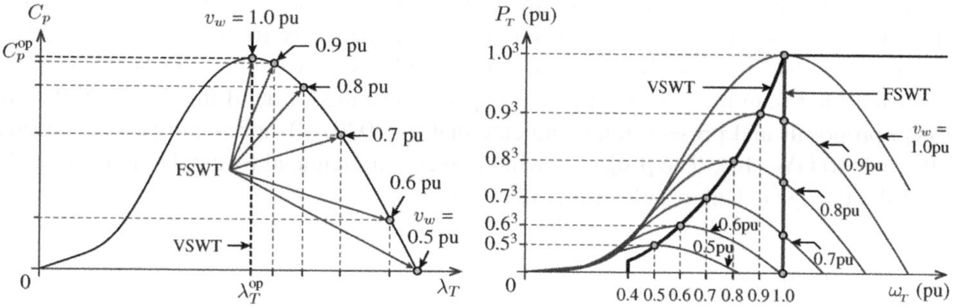

(a) C_p versus λ_T curve (b) P_T versus ω_T curve

Figure 1.10 Characteristics of FSWT and VSWT during different wind-speed conditions.

The first generation of grid-connected WTs developed during the 1980s were based on fixed-speed technology. These WTs are connected to the utility grid directly without any power converter, thus making the FSWT configuration simple and cost effective. In FSWTs, the wind generator operates at an almost constant speed, regardless of the wind speed. In FSWTs, the speed ω_T is decided according to the gear ratio, number of poles in the wind generator, and grid frequency. By referring to (1.4), the TSR value (λ_T) increases when wind speed v_w decreases at a constant ω_T. According to the WT characteristics, power coefficient C_p decreases when λ_T increases (Figure 1.10(a)). As demonstrated in Figure 1.10(b), the turbine output power P_T also decreases rapidly because of the reduction in both C_p and v_w. With the fixed-speed operation, the wind energy conversion efficiency decreases and mechanical stress on the drivetrain increases.

In VSWTs, the rotor speed ω_T is changed according to the wind speed v_w, such that the turbine always operates at an optimal TSR value λ_T^{op}. For example, when the wind speed changes to 0.9 pu, the turbine/generator speed ω_T also changes to 0.9 pu. Given this, C_p value is maintained at an optimal value, C_p^{op}, and P_T always holds a cubic relationship with v_w. Therefore, wind energy conversion efficiency in VSWTs is higher than with FSWTs (see Figure 1.10(b)). The initial cost of VSWTs is high because of the power converter; however, the high-energy yield of the VSWTs compensates the high initial cost and power losses in the converter. In the present wind energy industry, VSWTs are dominantly used. This book mainly deals with variable-speed WTs.

(6) Onshore and Offshore Turbines/Farms: A group of WTs is often placed over an extended area to form a wind farm and is connected to a national electric grid. Wind farms can be located on the land (onshore) or in the sea (offshore). Traditionally, onshore wind farms have been developed for the ease of construction, low initial and maintenance costs, improved proximity to transmission lines, and low-power transmission losses [1].

Nowadays, offshore wind farms are gaining more attention because power production can be increased and stabilized with the help of strong and steady winds. Furthermore, the effect on land use and landscapes can be reduced, audible noise and visual impacts can be mitigated, and opposition by "Not in My Back Yard (NIMBY)" movement will be low. The initial and maintenance costs of offshore wind farms are higher than onshore wind farms for the same power levels because stronger foundations are needed and connections to the onshore grid require submarine AC or DC cables (buried deep under the ocean floor). Offshore technology is another important driving force behind the amazing growth size in WTs (refer to Figure 1.8 on page 10). In 2013, the average power ratings of onshore and offshore WTs were reported to be 1.926 and 3.613 MW, respectively.

The global cumulative installed offshore wind power capacity is displayed in Figure 1.11. The offshore installed capacity increased exponentially from 29 MW in 1997 to 8771 MW in 2014 [2, 16]. By 2014, offshore wind power accounted for approximately 2% of the total (cumulative) wind power capacity installed around the world. Offshore project proposals and present trends indicate that by 2020, offshore wind power capacity will reach 40 GW [16]. The projected wind power is anticipated to supply approximately 4% of the European Union's electricity demand.

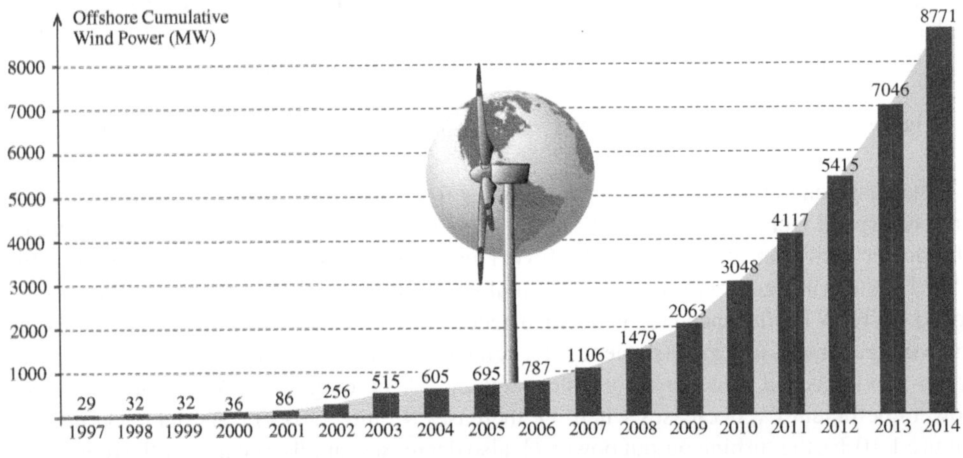

Figure 1.11 Global cumulative installed offshore wind power from 1997 to 2014 [2, 16].

Most offshore projects (nearly three-quarters) are located in European countries. The United Kingdom, Denmark, Belgium, Netherlands, Germany, Sweden, Finland, and Ireland are key players for offshore wind energy in Europe. The largest offshore wind farm to date is the London Array with 630 MW installed capacity. Future offshore wind farms are proposed in the range of 1200–2500 MW. Some of these projects include the following: Blekinge Offshore, Sweden (2500 MW); Korea Offshore, South Korea (2500 MW); and Moray Firth, United Kingdom (1300 MW). This book deals with both onshore and offshore wind energy systems.

(7) Geared, Semi-Geared, and Gearless Turbines: WTs are categorized into three types according to the speed of the wind generator and the type of coupling between the turbine main shaft and wind generator shaft:

- **Geared Turbine:** The low-speed (5–20 rpm) turbine rotor is connected to the high-speed (1500–1800 rpm) wind generator shaft through a three-stage gearbox.
- **Semi-Geared Turbine:** A single- or two-stage gearbox is used to couple the low-speed turbine rotor with the medium-speed (400–600 rpm) wind generator shaft.
- **Gearless Turbine:** The turbine rotor is directly coupled to the generator shaft.

The three-stage gearbox is commonly used in FSWTs and in some VSWTs. This gearbox has several serious issues such as high initial cost, high audible noise, extensive wear and tear, reduced life span, reduced efficiency, and need for regular maintenance [17]. The omission of a gearbox (often referred to as gearless or direct-drive technology) helps overcome the aforementioned problems, particularly in offshore WTs. This concept was first introduced in 1992 by `Enercon`, a German manufacturer, through the E-40/500 kW WT model. In recent years, many turbine manufacturers, such as `Avantis`, `GE Energy`, `Goldwind`, and `Vensys` to name a few, have used direct-drive technology in their commercial products. The direct-drive market in 2014 is 13,740 MW and represents 27% of global WT installations [4].

By contrast, a direct-drive operation leads to some drawbacks in the design of generators, such as large diameter and more weight [18]. To make a compromise between high- and low-speed operations, a medium-speed generator is used in conjunction with a single- or two-stage gearbox. The single-stage gearbox (with a gear ratio of 10) was first introduced by `Multibrid` in their M5000 WT models. Other turbine manufacturers such as `MingYang` and `WinWinD` developed 2-stage gearboxes with a gear ratio of 20–30. This book considers geared, semi-geared, and direct-drive turbines in the analysis.

(8) Low-, Medium-, and High-Wind-Speed Turbines: According to the International Electrotechnical Commission (IEC) 61400-1 standard, WTs are also arranged into different classes on the basis of annual average wind speed, turbulence, and extreme 50-year gust data. The summary of WT classes is shown in Table 1.1. The design and construction phase of wind power plant relies highly on this data to forecast the wind climate that will affect the WTs. All manufacturers provide WT products to meet various IEC classes. For example, `Vestas` offers 3.45 MW WTs in different model numbers to cover various IEC classes: V105-IEC IA, V112-IEC IA, V117-IEC IB/IIA, V126-IEC IIA/IIB, and V136-IEC IIIA.

Table 1.1 Wind turbine classes according to the IEC 61400-1 standard

Wind Class	Annual Average Wind Speed	Turbulence Level	Extreme 50-Year Gust
IA	High Wind Speed – 10 m/s	High Turbulence – 18%	High Gust – 70 m/s
IB	High Wind Speed – 10 m/s	Low Turbulence – 16%	High Gust – 70 m/s
IIA	Medium Wind Speed – 8.5 m/s	High Turbulence – 18%	Medium Gust – 59.5 m/s
IIB	Medium Wind Speed – 8.5 m/s	Low Turbulence – 16%	Medium Gust – 59.5 m/s
IIIA	Low Wind Speed – 7.5 m/s	High Turbulence – 18%	Low Gust – 52.5 m/s
IIIB	Low Wind Speed – 7.5 m/s	Low Turbulence – 16%	Low Gust – 52.5 m/s
IV	6.0 m/s	–	42 m/s

1.3 MAJOR COMPONENTS OF WECS

From the pool of 8 types of WT technologies presented in Section 1.2.3, the following turbine types are considered in order: high-power WT, pitch-controlled WT, horizontal-axis WT, grid-connected WT, variable-speed WT, onshore WT, geared WT, and any IEC class WT. The basic configuration of WT, which integrates the above 8 types of technologies, is depicted in Figure 1.12. A typical MW-WECS is composed of over 8000 components that convert wind kinetic energy into electric energy in a controlled, reliable, and efficient manner. The most visible parts are the tower, nacelle, rotor blades, and step-up transformer. The rest of the components are housed inside the WT. The major components of a WECS are broadly classified into three categories:

- **Mechanical Components:** Includes the rotor blades, rotor hub, rotor bearings, main shaft, mechanical brake, gearbox, pitch drives, yaw drives, wind measurement unit, nacelle, tower, foundation, heat exchange system, and ladder.
- **Electrical Components:** Includes the wind generator, power electronic converter along with generator- and grid-side harmonic filters, step-up transformer, power cables, wind farm collection point, and switch gear.
- **Control Components:** Includes mechanical and electrical-related control systems.

The cost analysis on a `REPower` (now `Senvion`) MM92-5.0 MW WT indicates that the costs (in % of total WT cost) associated with major parts are as follows: tower 26.3%, rotor blades 22.2%, gearbox 12.91%, power converter 5.01%, transformer 3.59%, wind generator 3.44%, main frame 2.80%, and pitch drives 2.66% [19]. The costs associated with the wind generator, power converter, and transformer change slightly according to the type of wind energy system configuration. Further details on various WECS configurations will be discussed in Section 1.5.

1.3.1 Mechanical Components

Mechanical components play an important role in the practical development of multi-megawatt WTs. Mechanical components convert wind kinetic energy into mechanical energy and transmit it to the wind generator through the drivetrain. The "rotor" components include rotor blades, rotor hub, rotor bearings, nose cone, and pitch drive. The "drivetrain" includes mechanical coupling, bearings, low-speed main shaft, high-speed generator shaft, gearbox, and mechanical brakes. A brief description of the major mechanical components is given as follows.

(1) Rotor Blades: The wind kinetic energy is first converted to mechanical energy with the help of airfoil-shaped rotor blades. WT blades are the most distinctive, visible, and important components. WT blades look like airplane wings and share the same aerodynamic characteristics because of their structural and aerodynamic behavior. Over the past three decades, WT blade technology has evolved rapidly in terms of aerodynamic design and materials. Older style WTs were designed with heavier steel blades, thus leading to high inertia. Newer turbines use efficient and composite materials to obtain low rotational inertia such that WTs can accelerate quickly if the wind speed increases. The most commonly used materials in modern WTs are fiber-reinforced polymers, aluminum, polyester resin, balsa wood, glass-reinforced plastics, and carbon-fiber composites. The three-blade design is the standard for current MW WTs because it provides symmetrical loading.

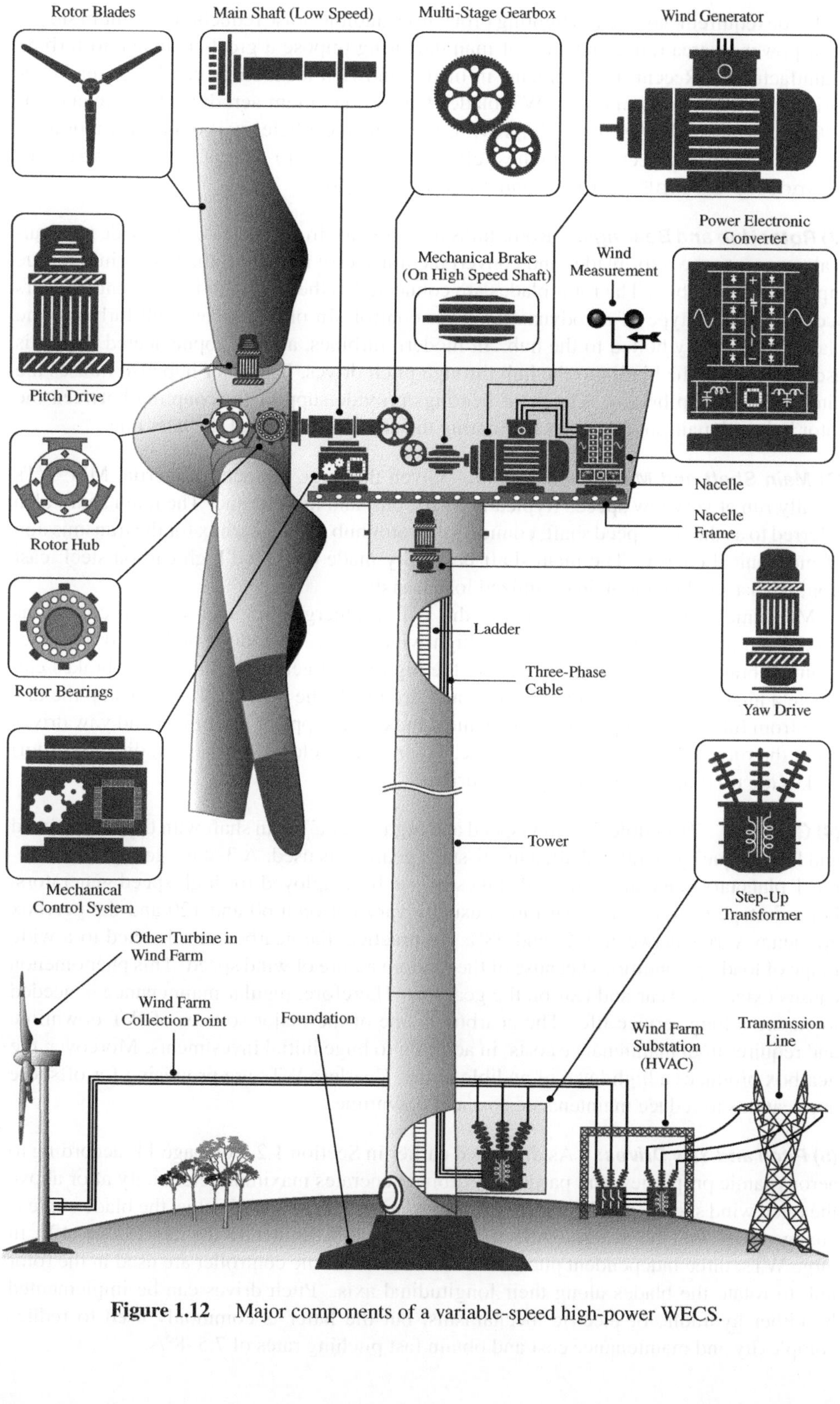

Figure 1.12 Major components of a variable-speed high-power WECS.

Blade requirements such as lighting protection, audible noise reduction, optimum shape, high power-to-area ratio, and ease of manufacturing impose a great challenge to turbine manufacturers. Recent developments in blade sandwich technology can lower manufacturing and transportation costs. WT blades have a maximum aerodynamic efficiency of 59% according to the *Betz* limit. Improving aerodynamic efficiency is of prime importance for mechanical engineers; the highest efficiency achieved to date is 50% in `Enercon`, `Shandong SwissElectric`, and `Wikov` blades [9].

(2) Rotor Hub and Bearings: Rotor hubs made of cast iron or cast steel provide mechanical support to the rotor blades and convert the transverse motion of the blades into torque input for the gearbox. The rotor blades are connected to the rotor hub in different manners according to the type of aerodynamic power control. In older passive stall turbines, the blades are directly bolted to the hub. In modern turbines, a more sophisticated design is used to connect the blades to the hub through pitch drives. The rotor hub also houses the pitch drives or tip brakes. The rotor bearings provide support for coupling between the rotor hub and main shaft besides facilitating the smooth rotation of the later one.

(3) Main Shaft and Mechanical Brake: Given the large moment of inertia, MW WTs usually run at very low speeds (typically 5–20 rpm) and high torque. The main shaft, also referred to as the low-speed shaft, connects the rotor hub to the gearbox for the transmission of mechanical energy. The main shaft is usually made of forged high-carbon steel, cast iron, or cast steel to maintain optimized load transfer.

Mechanical brakes are used to stop the WT in emergencies such as fault conditions (over-heating of gearbox/generator) or high wind gusts. To reduce braking torque, mechanical brakes are mounted directly on the high-speed generator shaft. The brakes can create a huge wear and tear on the shaft, and fire inside the nacelle if used to stop the turbine from full speed. In practice, mechanical brakes are applied after pitch and yaw drives bring the rotor speed to certain low value. Hydraulic or electromechanical disc (or drum) brakes are used in modern high-power turbines.

(4) Gearbox: To couple the "low-speed and high-torque" main shaft with the "high-speed and low-torque" generator shaft, a multi-stage gearbox is used. A 3-stage gearbox containing 1 planetary stage and 2 helical stages is usually employed for high-speed generators. For a 3-stage gearbox, the gear ratios usually vary between 60 and 120 and the gearbox efficiency varies between 95% and 98%. In practice, the gearbox is subjected to a wide range of loading conditions because of the random nature of wind speed. This phenomenon causes extensive wear and tear on the gearbox. Therefore, regular maintenance is needed to ensure a long service life. The gearbox is one of the major sources of WT downtime and requires high maintenance costs, in addition to huge initial investments. Moreover, the gearbox produces a high level of audible noise. Gearless WTs are promising for offshore applications to reduce maintenance cost and downtime.

(5) Pitch and Yaw Drives: As discussed earlier in Section 1.2.3 on page 11, according to aerodynamic properties, any particular turbine generates maximum electricity at or above the rated wind speed only. When wind speed is more than the rated value, the blade angle is changed such that the generator electric output power is limited to the rated value [12]. In large WTs, three independent pitch drives and an electronic controller are used in the rotor hub to rotate the blades along their longitudinal axis. Pitch drives can be implemented by either hydraulic or electric mechanisms, but the latter is commonly used to reduce complexity and maintenance cost and obtain fast pitching rates of 7.5–8°/s.

The yaw drive is used to move the rotor blades and nacelle toward the wind to extract the maximum possible energy. The yaw drive is composed by electric motor drives, yaw gear, gear rim, and bearings and produces high torque to turn the nacelle. When the wind speed is above the cut-out value or when a malfunction occurs, the yaw mechanism helps stop the turbine by moving it out of the wind.

(6) Wind Measurement Unit: WTs are equipped with wind measurement units composed of an anemometer and wind vane to continuously monitor and collect wind data. As mentioned earlier, the pitch and yaw drives require wind speed and direction information, respectively. Electrical control systems also require this information for variable-speed operations. The anemometer consists of a three-cup vertical-axis micro-turbine and a speed transducer to measure wind speed. Ultrasonic anemometers have recently been introduced for WTs to obtain accurate wind data and provide reliable operation compared with mechanical anemometers. A wind vane along with an optoelectronic angle transducer measures the wind direction, and the wind measurement units are usually mounted on the top back part of the nacelle.

(7) Nacelle: The nacelle houses most of the mechanical and electrical components and protects them from extreme weather conditions. The nacelle faces upstream wind during normal operating conditions. The size of the nacelle greatly changes according to the type of gearbox employed. For example, for direct-drive WTs, the nacelle size is small compared with the gearbox-based WT. In recent offshore WTs, a helicopter-hoisting platform is built on top of the nacelle for service personnel to winch down to turbine from a helicopter hovering above it.

(8) Tower and Foundation: The tower, commonly made of steel or concrete, elevates the height of the nacelle and rotor blades to access better wind-speed conditions. The height of the tower increases with the rotor diameter or with turbine power capacity, to allow sufficient clearance from the ground. The requirement for clearance from the ground is higher in onshore WTs than in offshore turbines. In modern turbines, tower height is approximately two to three times the blade length. The power cables (which connect the power converter output to the step-up transformer) and ladder are attached to the inner surface of the tower. In some MW turbines, the step-up transformer and electrical switch gear are also housed inside the tower to reduce space requirements and protect these components from severe weather conditions.

The turbine foundation directly supports the tower, nacelle, and rotor blades. The foundations for onshore turbines are simple and includes slab, monopile and multipile types. Designing the foundations for offshore WTs is a challenging task because of different water depths and soil types, as well as harsh weather conditions. This issue has triggered many research and development activities, and currently many offshore foundation concepts are available such as gravity, monopile, tripod, tripile, jacket, and floating. Gravity and monopile foundations are applicable for shallow waters with depths less than 30 m [20]. Other foundations, particularly floating foundations, which have already been used in offshore oil and gas industries, are promising for future deep water (>40 m) offshore wind farm projects.

(9) Miscellaneous Mechanical Components: Several other miscellaneous components exist that are not directly involved in power conversion. However, they assist the main mechanical components in the operation and maintenance of WTs. These components include the heat dissipation/exchange system, lightning protection system, mechanical vibration dampers, ladder, and service lift.

1.3.2 Electrical Components

(1) Wind Generator: The wind generator converts rotational mechanical-energy input into electric energy output. Over the past 35 years, many generators such as the squirrel-cage induction generator (SCIG), wound rotor induction generator (WRIG), doubly fed induction generator (DFIG), permanent magnet synchronous generator (PMSG), and wound rotor synchronous generator (WRSG) have been developed for WTs [17, 21]. WRIG and DFIG have the same construction process but have different names in literature to distinguish different turbine configurations. The WRSG is also denoted as an electrically excited synchronous generator (EESG) in literature. In an attempt to increase the power density of 10 MW class WTs, high-temperature-superconducting (HTS) synchronous generators (SGs) have been introduced recently in the wind energy industry. HTS-SG eliminates the need for rare earth materials and significantly decreases the size and weight of the generator compared with the standard WRSG/PMSG. AMSC and ABB have announced the use of HTS-SGs in their future 10- and 15-MW WT projects, respectively.

The SCIG is popularly used in first-generation direct grid-connected WTs because of its low cost, simple and rugged construction, and minimum maintenance requirements. The SCIG is also used in present variable-speed MW WECS along with a power converter. An induction generator (IG) with accessible rotor windings (i.e., WRIG and DFIG) enables variable-speed operation by employing a variable resistor or power converter in the rotor circuit. The DFIG is the current workhorse of the wind energy industry with more than 50% market share. The PMSG, WRSG, and HTS-SG are used mainly in direct-drive MW WTs. IGs are used in both fixed speed and variable-speed turbines, whereas SGs are used in variable-speed WECS. IGs operate at high rotational speeds, whereas SGs can operate at low, medium, or high speeds. To achieve low operational speed, the generator needs to be equipped with a large number of poles, which is a feasible solution for SGs. In accommodating the large number of poles, the PMSG stator radius becomes approximately 6 times larger and 4.5 times heavier than three-stage gearbox-based IG [18]. The main features of these generators are summarized in Table 1.2.

Table 1.2 Main features of induction and synchronous wind generators [22, 23]

Generator	Advantages	Disadvantages
SCIG	✔ Simple and rugged construction ✔ Lower initial and maintenance cost ✔ Readily available for MW applications	✖ Consumes reactive power from grid ✖ Limited to high-speed operation only ✖ Requires a soft starter to limit in-rush current
WRIG/ DFIG	✔ Rotor power allows variable speed ✔ High starting torque capability ✔ Flexible reactive power control	✖ Initial and maintenance cost is high ✖ Low-power factor (PF) at light loads ✖ Slip rings need regular care
PMSG	✔ High power density and reliability ✔ No need for excitation and gearbox ✔ Lower rotor losses and high efficiency	✖ High cost for PM material ✖ Possible demagnetization of PMs ✖ Low-speed generators are heavy
WRSG	✔ Independent P and Q control ✔ High-torque operation can be achieved ✔ Eliminates the need for gearbox	✖ High capital and maintenance cost ✖ Requires additional excitation circuit ✖ Frequent attention is needed for slip rings
HTS-SG	✔ Significantly compact and light weight ✔ Lower winding losses ✔ Larger air-gap eliminates tolerance issues	✖ High initial cost ✖ Cooling system design is complicated ✖ Greater tendency for rotor speed instability

(2) Power Converter: Wind generator output voltage and frequency changes with respect to the rotational speed (wind speed). The generator output terminals can be directly coupled to the grid or can be interfaced through a power electronic converter [24]. The power converter changes the generator AC output voltage to DC voltage by a rectifier (AC/DC converter) and then back to AC with a fixed voltage magnitude and frequency by an inverter (DC/AC converter). In most WTs, the configuration of both AC/DC and DC/AC converters is the same and is known as a back-to-back (BTB) connected converter. By arranging the power switching devices in different ways, possibly with DC-link elements such as capacitors or inductors, numerous power converter topologies can be derived. As shall be introduced in Section 1.5, these power electronic converters can be combined with electric generators to form a wide variety of WECS configurations. Switching harmonics are inevitable when using power converters. To solve this issue, harmonic filters are used in wind machine (generator)-side converters (MSCs) and grid-side converters (GSCs). The harmonic filter in the MSC helps reduce the harmonic distortion of the generator currents and voltages. This process leads to a reduction in harmonic losses incurred in the magnetic core and winding of the generator. The harmonic filter in the GSC helps meet the strict harmonic requirements specified by the grid codes [25].

On the basis of operating voltage, power converters are broadly classified as low-voltage (LV) and medium-voltage (MV) converters. LV or MV operation for a power converter is decided according to the wind generator output voltage magnitude. A low voltage power converter is used with a low voltage wind generator; similarly, a medium voltage power converter is employed with a medium voltage wind generator. The definition of WECS LV and MV operating voltages in the North American and European market is summarized in Table 1.3 [26]. The most popular LV and MV ratings for wind generators (and power converters) are <1000 and 3000–4000 V, respectively. A more detailed discussion on LV and MV power converter topologies for WTs is presented in Section 1.6.

Table 1.3 LV and MV classification by region [26]

Region	Standard	Voltage Class
Europe	IEC 60038	LV (<1000 V)
		• 220, 400, 690 V
		MV (1 – 35 kV)
		• 3.3, 6.6, 11, 22, 33 kV
North America	ANSI C84.1	LV (<600 V)
		• 208, 120/240, 480, 575 V
		MV (600 V – 35 kV)
		• 2.4, 4.16, 6.9, 12.47, 13.8, 21, 34.5 kV

(3) Step-Up Transformer: The power converter output voltage (typically 690 V in LV class) is boosted to 34.5 kV (in North America) by using a step-up transformer to connect the WTs to an MV collection point. The step-up transformer is mandatory in a WT; otherwise, the wind generator and power electronic converter should be designed for WF collection-point voltage level. This approach eliminates the cost of a step-up transformer; however, additional costs are incurred with the MV wind generator and power converter. In the present wind energy industry, all WTs use step-up transformers as the default component. Step-Up transformers are available in two forms: dry transformer (cast resin) and liquid (oil) filled transformer.

(4) Power Cables: The power cables connect the output of the power converter, which is usually located in the nacelle to the step-up transformer, which is placed at the bottom of the WT. The tower pendent LV power cables consist of coated copper and a simple insulation system. The cost, size, and power losses associated with LV power cables increase as the power rating of the LV WT increases. For example, let us consider a 6.0-MW power converter operating at 690 V. The *root-mean-square* (*rms*) value of the three-phase line currents becomes 5020 A, which must be carried by high-current capacity power cables. In some WT designs, the transformer is placed in the nacelle to decrease the cable cost and losses. However, this approach requires the size optimization of other electrical and mechanical components.

Another approach to mitigate the above cable issues is to increase the WT voltage to MV level. For a 6.0-MW turbine, the *rms* value of the line currents significantly decreases to 866 A when a 4000-V power converter is employed and leads to a smaller cable size and lower power losses than LV cables. The construction and termination of MV power cables is complicated, and the salary and wages of MV technicians is higher than LV technicians. However, the aforementioned advantages of MV cables outweigh this drawback.

(5) Wind Farm Collection Point: The wind farm collection point, also referred to as the point of common coupling (PCC), connects all WTs in a wind farm. The parallel connection of WTs is the default practice for increasing the power capacity of the wind farm. The most common PCC voltages employed in North American and European wind farms are 34.5 and 33 kV, respectively. Commercial WTs can be connected to the PCC through step-up transformers, regardless of the regional voltage classes. Thus, given the participation of European manufacturers in North America and vice versa, the regional voltage classifications outlined in Table 1.3 are becoming less important.

(6) Miscellaneous Electrical Components: The miscellaneous electrical components include electric switch gear, three-phase contactor, and circuit breaker, which are connected between the WT and PCC or between the PCC and transmission line.

1.3.3 Mechanical and Electrical Control Systems

WTs are fully automated and operate unmanned yearlong under all weather conditions. The control systems have been used with limited functions in the first generation of WTs, and now they perform a large number of functions with respect to turbine, generator and power converter, grid integration, protection standards, and wind farm operation. In WTs, several slave control systems for the mechanical/electrical components and a master control system are used in a cooperative manner to perform the following: (1) extract the maximum possible energy from the wind, (2) operate the turbine in safe mode with control variables such as power, voltage, current, speed, and torque under limits, (3) achieve the desired dynamic and steady-state performance, (4) minimize mechanical stress on the drivetrain, and (5) meet strict grid codes (shall be discussed in the next section). The master control system monitors various variables such as wind speed, wind direction, generator voltages/currents, filter/DC-link voltages, and grid voltages/currents. The master control system then adjusts the system operating states or variables at the reference value or in the set boundaries with the support from slave control systems. The control systems are realized by a personal computer, μC, programmable logic controller, DSP, or FPGA. To avoid confusion with mechanical and electrical control systems, in this book, "digital control" is defined as "electrical control" only. The control of WTs is an important subject matter and is discussed further in Section 1.7.

1.4 GRID CODE REQUIREMENTS FOR HIGH-POWER WECS

The steady growth in the power levels of WTs and wind farms have led to the significant penetration of wind energy systems into the existing electric power system. To ensure grid stability and consumer power quality, many specific technical requirements (documents) often called "grid codes" have been developed and are regularly updated by transmission system operators (TSOs) and/or distribution system operators (DSOs) in many countries [27]. The main elements in grid codes include the following:

- Grid voltage and frequency tolerance
- Active power control
- Reactive power generation (RPG)
- Grid power quality
- Black start and short-circuit capability
- Fault ride-through (FRT)

The above requirements stipulate that a wind power plant is expected to act like an active power generation unit, similar to conventional power plants during normal and abnormal grid conditions. The first requirement indicates that WECS must remain operational when the grid voltage and frequency varies within the set boundaries or tolerance band limits. WECS must control the active power output (by disconnecting some WTs from the grid or by pitch control action) according to the power command given by the TSO/DSO, such that active power curtailment and frequency regulation at the PCC is achieved. Some countries also define active power control with respect to the grid frequency variation. For example, according to Danish grid code, when the frequency increases above 48.7 or 50.15 Hz, active power should be decreased depending on the power reserving strategy [5].

WECS must also be able to control reactive power such that the voltage and PF at the PCC is tightly regulated. The RPG requirement must be fulfilled by WECS independent of the type of WT technology being used. In case the WTs cannot produce the required reactive power, additional static compensation devices such as static VAR compensator (SVC) or static synchronous compensator (STATCOM) should be employed in the wind farm. WECS should also provide reactive power during abnormal conditions to support the grid and assist the fast recovery of grid voltage.

Power quality determines the fitness of the electric power to the utility grid. Power quality involves various items such as transient variations, flickers, and harmonics. According to IEEE 519-1992, for voltages below 69 kV, the maximum total harmonic distortion (THD) of the grid voltage at the PCC must be limited to 5% during normal operations to prevent harm to other electrical equipment connected to the utility grid. For voltages in the range of 69 kV to 161 kV, the maximum voltage THD and individual voltage harmonics should be limited to 2.5% and 1.5%, respectively. For voltages above 161 kV, total and individual voltage distortion should not exceed 1.5% and 1%, respectively. Similarly, harmonic current limits are imposed in terms of total demand distortion (TDD) on the basis of the ratio of the three-phase fault current to maximum demand current, I_{sc}/I_L.

Black-start capability indicates the ability of the power generation unit to restore normal operations from shutdown mode without using an external power network. Short-circuit capability indicates the ability of an electrical apparatus to interrupt without being destroyed or causing an electric arc. FRT indicates the ability of WECS to operate through periods of low or high grid voltage. Among all the grid codes, the FRT and RPG are major concerns for WT and power converter manufacturers.

1.4.1 Fault Ride-Through

Grid disturbances such as grid voltage sags/swells might lead to disconnection of large-scale wind power generation units. The sudden disconnection of generation units stimulates the instability of the utility network. Beginning in early 2003, the German utility operator E.ON imposed FRT requirements to overcome the aforementioned scenario [28]. The TSO and DSO of diverse countries issued similar FRT profiles with different grid voltage dip magnitudes and grid fault durations [29]. The FRT requirement is a broad category covering zero-voltage ride-through (ZVRT), low-voltage ride-through (LVRT), and high-voltage ride-through (HVRT). The ZVRT and LVRT requirements correspond to grid voltage sags: during grid faults, the grid voltage becomes zero in the ZVRT profile; in the LVRT profile, the grid voltage becomes 15–25% of its nominal value. The HVRT corresponds to grid voltage swells.

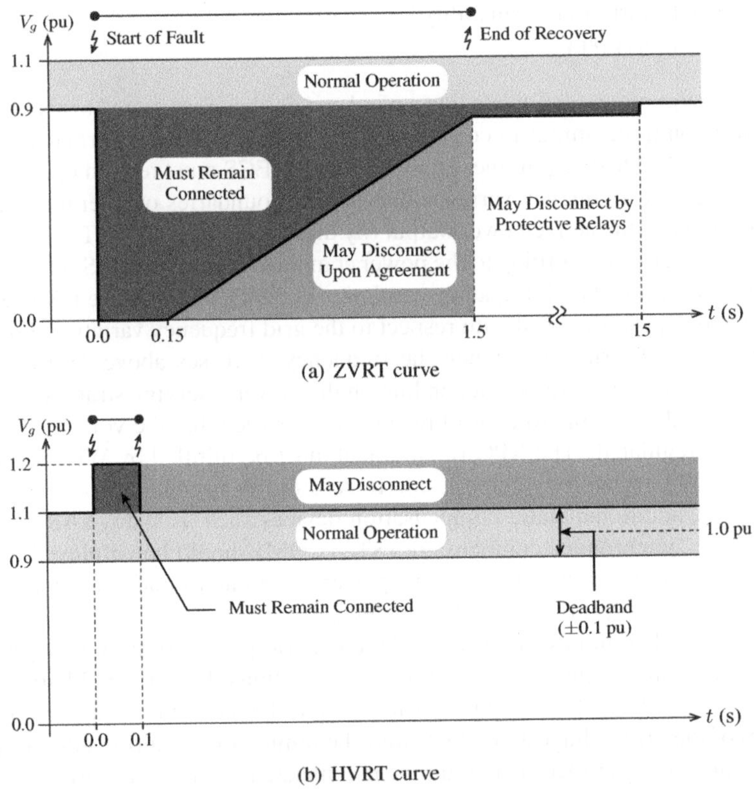

Figure 1.13 FRT requirement according to the E.ON regulation [28].

The ZVRT and HVRT profiles of the E.ON regulation are shown in Figure 1.13. These profiles specify that the WTs must "ride through" instead of "trip off" during transmission faults. During grid voltage sags (refer to Figure 1.13(a)), the FRT function should start within 20 ms (one cycle) when the grid voltage falls below 0.9 pu and should provide 1.0-pu reactive power. The WT must be connected to an electric network if the grid voltage profile is above the ZVRT limit line. The turbine is allowed to disconnect from the grid if the magnitude of the grid voltage falls below the ZVRT limit line. For offshore wind farms, the ±0.1-pu deadband is reduced to ±0.05 pu.

During grid voltage swells (see Figure 1.13(b)), when the grid voltage swells to 1.2 pu, the WECS must ride through for 0.1 s. To ensure grid voltage recovery, WECS should absorb 1.0 pu reactive power (opposite to LVRT requirement). Recently, considerable research has been conducted to address the FRT issue in WECS [1, 27].

1.4.2 Reactive Power Generation

In addition to the FRT operation, another important requirement for WECS is that it should perform "reactive power control" similar to a conventional power plant during both normal and abnormal operations. Reactive power control helps compensate transmission equipment such as cables and transformers, in addition to maintaining voltage stability. Thus, reactive power control is an important grid code regulation to maintain reliable and efficient transmission and distribution grids.

Similar to FRT profiles, many reactive power profiles are defined by diverse TSOs and DSOs with respect to the active power output and grid voltage magnitude [30]. The reactive power Q_g requirement as a function of the active power P_g and grid voltage V_g is shown in Figures 1.14(a) and 1.14(b), respectively. *PQ* dependence according to the German grid code shows the range of Q_g to be delivered by WT GSC with respect to P_g. When WECS delivers rated (1.0 pu) active power, it should be able to supply 0.41 pu or absorb 0.33 pu reactive power. This case implies that WECS should be able to adjust the grid PF anywhere from 0.93 leading to 0.95 lagging when it delivers rated active power. During reduced (0.2 pu) active power output, the range of the PF varies between 0.44 leading and 0.52 lagging (Figure 1.14(a)). The Q_g requirement is variable when P_g is between 0 and 0.2 pu.

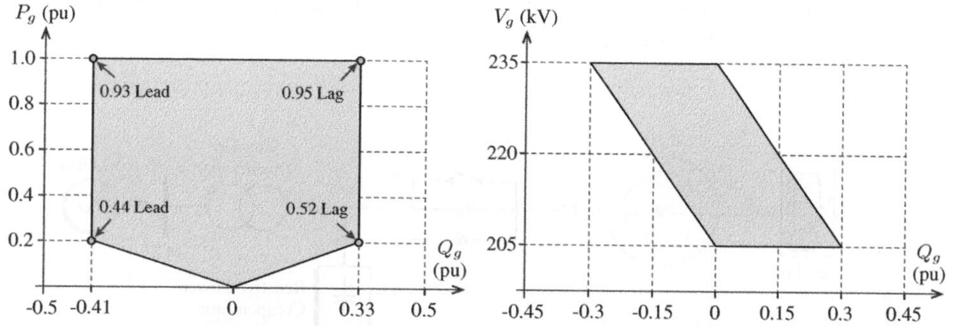

(a) *PQ* dependence in Germany

(b) *VQ* dependence in Spain

Figure 1.14 Reactive power requirement for grid-connected WECS.

As highlighted in Figure 1.14(b), the *VQ* dependence curve according to the Spanish grid code shows the Q_g requirement with respect to the grid voltage V_g. In addition to the reactive power requirement during normal operations, WECS must also provide Q_g under the grid faults. For example, according to German and Danish grid codes, WECS must stop injecting active power during grid voltage dips and should generate 1.0 pu reactive power until the grid voltage recovers to 0.5 pu. For grid voltages in the range of 0.5 pu to 0.9 pu, WECS should provide 0.02 pu reactive current for each 0.01 pu grid voltage dip. Although the RPG requirement is a stringent grid code, this requirement can easily be fulfilled by properly designing the following: (1) the WT GSC with high MVA capacity and (2) high-performance digital control for MSC and GSC.

1.5 WECS COMMERCIAL CONFIGURATIONS

The major mechanical and electrical components in WECS are the wind turbine, gearbox, mechanical speed/torque converter, wind generator, power electronic converter, and step-up transformer. By using different designs and combinations with some or all of these mechanical and electrical components, a wide variety of WECS configurations have been commercialized, such as [1, 17, 23, 24]:

- **Type 1:** Fixed-speed WECS with SCIG
- **Type 2:** Semi-variable-speed WECS with WRIG
- **Type 3:** Semi-variable-speed WECS with DFIG
- **Type 4:** Full-variable-speed WECS with SCIG, PMSG, WRSG, or HTS-SG
- **Type 5:** Full-variable-speed WECS with WRSG and mechanical converter

In this section, the above five configurations are analyzed in detail on the basis of Figures 1.15 to 1.19. The major mechanical and electrical components, operating principle, advantages, and disadvantages of each WECS configuration are discussed.

1.5.1 Type 1 WECS Configuration

A fixed-speed SCIG-based WECS without a power converter interface (Type 1 turbine) is illustrated in Figure 1.15, where the generator is connected to the grid through a soft starter and step-up transformer. This technology is the oldest and very first technology ("Danish" concept) developed for WTs during the 1980s [17]. To reduce initial investment and increase reliability, first-generation WTs were developed by using well-proven mechanical and electrical components already existing in the market.

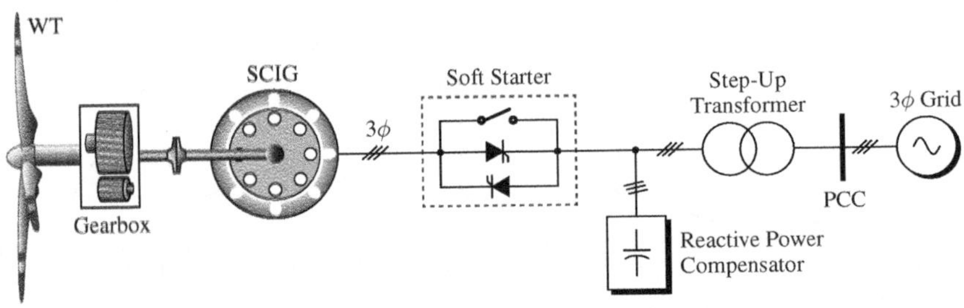

Figure 1.15 Type 1 WECS with fixed-speed (1%) SCIG, three-stage gearbox, soft starter, and reactive power compensator.

In high-power WECS, the SCIG contains four or six poles for 50- or 60-Hz grid frequency operations, respectively. During different wind speeds, the generator speed varies slightly (within 1% above the synchronous speed); hence, this configuration is called fixed-speed WECS. Pole-changeable SCIGs were also practiced in the industry to accomplish two rotational speeds during varying wind speeds. Early WTs employed passive stall aerodynamic power regulation, whereas modern turbines use active stall or pitch control techniques. A three-stage gearbox is normally employed between the turbine rotor and wind generator to match the speed difference between them.

During the startup process, the voltage difference between the wind generator and utility grid causes a high in-rush current. A three-phase soft starter consisting of anti-parallel thyristors and a bypass switch limits the in-rush current to safe limits. The firing angle of the thyristors is gradually adjusted such that the grid voltage applied to the wind generator changes progressively from zero to the rated value. After the startup procedure, the thyristors are bypassed (because of the limited thermal capacity of the thyristors and to limit power losses) by a switch and the system essentially works without any power converter [13]. The SCIG draws reactive power from the grid; to compensate for this mechanism, three-phase capacitor banks are usually employed. To provide optimal reactive power compensation, multiple capacitor banks are used to switch in and out of the three-phase electric system. Major advantages and disadvantages of this configuration are summarized below:

✔ Simple power conversion configuration.

✔ Low initial and maintenance costs because of SCIG, and inexpensive soft starter.

✔ Reliable operation as no power converter is needed.

✖ Lower wind energy conversion efficiency because of narrow (1%) speed range.

✖ Changes in wind speed cause grid frequency stabilization issues.

✖ Grid faults cause severe stress on the mechanical components of the WT [31].

FSWTs are equipped with additional hardware, such as STATCOM, to comply with the grid codes [32]. Despite the drawbacks, this configuration has been accepted by the wind industry and commercial solutions are available in MW range such as: (1) `Vestas` V82 (1.65 MW) and (2) `Siemens` SWT 2.3-101 (2.3 MW). Fixed-speed turbines were famous until a decade back, but this technology is slowly becoming obsolete because of its inherent disadvantages. The turbines in the present study, which have been installed already, are still in operation to generate electricity.

1.5.2 Type 2 WECS Configuration

To overcome the drawbacks associated with Type 1 WTs, the wind energy industry developed semi-variable-speed WTs. The configuration of a semi-variable-speed WECS using WRIG and a partial rated (10%) power converter is shown in Figure 1.16 (Type 2 turbine). The system configuration is similar to a Type 1 turbine; however, the SCIG is replaced by a WRIG with rotor windings connected to the converter-controlled external resistor. The power converter is realized by a three-phase diode-bridge rectifier and an insulated gate bipolar transistor (IGBT)-based chopper circuit .

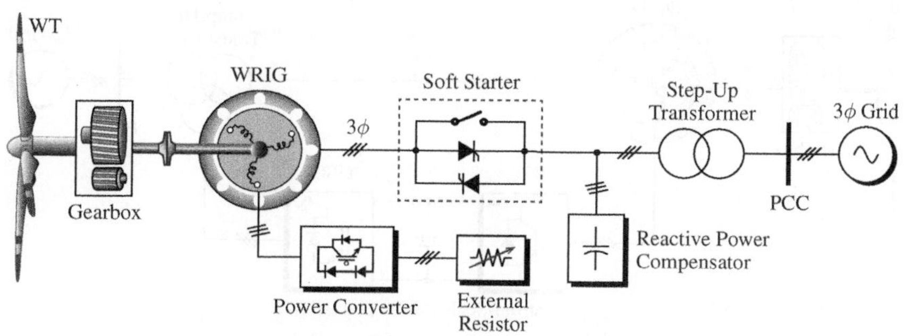

Figure 1.16 Type 2 WECS with semi-variable-speed (10%) WRIG, converter-controlled external resistor, three-stage gearbox, soft starter, and reactive power compensator.

The change in the rotor resistance affects the torque/speed characteristic of the generator, thus enabling the variable-speed operation of the turbine; this configuration is often called Optislip WT [17]. Speed adjustment range is typically limited to approximately 10% above the synchronous speed. With semi-variable-speed operations, WECS can capture slightly higher power from the wind with reduced stress on mechanical components but with energy losses in rotor resistance. This Type 2 configuration also requires a gearbox, soft starter, and reactive power compensation similar to a Type 1 turbine. The main merits and demerits of this configuration are given as follows:

✔ Higher energy conversion efficiency than Type 1 turbines because of extended (10%) speed range.

✔ Lower stress on WT mechanical components compared with Type 1 turbines.

✔ Less effect on grid frequency because of semi-variable-speed operation.

✖ Higher initial cost because of the power converter and higher maintenance cost due to slip rings and brushes in WRIG.

✖ The external resistor causes system losses and lower reliability.

✖ The system still requires soft starter and grid-side reactive power compensation.

The WRIG with variable rotor resistance has been on the market since the mid-1990s with a power rating up to a couple of MWs. A few examples of commercial solutions are as follows: (1) `Vestas` V66-2.0 MW and (2) `Suzlon Energy` S88-2.1 MW. This configuration is also becoming less important among WT manufacturers because of its limited speed range and low energy conversion efficiency.

1.5.3 Type 3 WECS Configuration

In a continued effort to increase the operational speed range of WTs and eliminate soft starter and grid-side reactive power compensator, Type 3 semi-variable-speed turbines were developed. The configuration of Type 3 WECS using a DFIG and power electronic converter is shown in Figure 1.17 [33]. This configuration replaces the converter-controlled external resistor in a Type 2 turbine by a power converter. As the name implies, power from the doubly fed induction generator is fed to the grid through both stator and rotor windings.

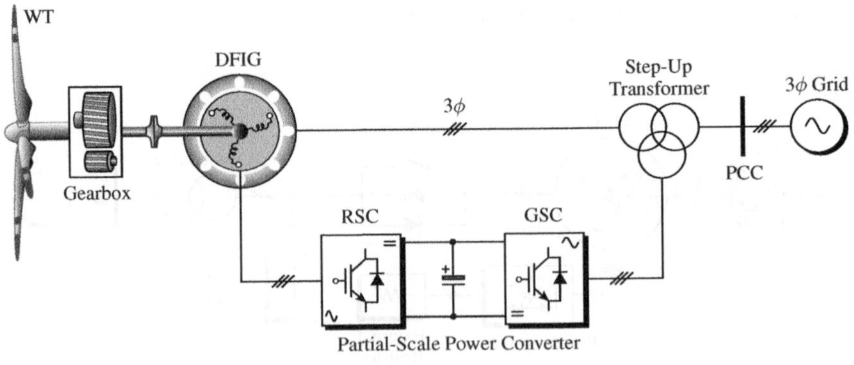

Figure 1.17 Type 3 WECS with semi-variable-speed (±30%) DFIG, three-stage gearbox, and partial-scale (30%) power converter.

The converter in the rotor circuit has to handle only the slip power; thus, the power capacity of this converter should be only 30% of the generator rated power [34]. The use of a partial scale (30%) converter leads to a reduction in cost, weight, and nacelle space requirement. From a cost point of view, a smaller converter capacity makes Type 3 WECS more attractive and popular. The cost of converter in a `REPower` (now `Senvion`) MM92-5.0 MW WT was reported as 5% of the total turbine cost [19]. The power converter is usually made of two-level voltage source converters (VSCs) connected in a BTB configuration. The machine-side VSC, also called a rotor-side converter (RSC), controls the generator torque/speed or active/reactive power, whereas the GSC controls the net DC-bus voltage. The power converter is also equipped with harmonic filters on the rotor-side and grid-side to attenuate switching harmonics. The step-up transformer uses three winding structures with two primary windings on the generator side and one secondary winding on the grid side. The main features of Type 3 WECS are summarized below:

✔ The power converters allows bidirectional power flow in the rotor circuit. The generator speed is adjustable 30% above or below the synchronous speed. Therefore, the energy conversion efficiency is high and stress on the mechanical components is low.

✔ The power converter performs smooth grid connection and grid-side reactive power compensation. Therefore, soft starters and capacitor banks are not needed.

✔ The power converter also provides enhanced dynamic performance and robustness against power system disturbances compared with Type 1 and 2 turbines [35].

✖ The initial cost and system complexity is high because of the power converter.

✖ Unsuitable for offshore wind farms because of the regular maintenance needed by the slip rings and brushes in DFIG and the three-stage gearbox.

✖ FRT compliance is complicated because of the direct grid-connected DFIG stator terminals and partial (30%) rated power converter.

Despite the above drawbacks, Type 3 WECS are one of the dominating technologies in today's wind energy industry with a market share of about 50% [36]. Some examples of high-power Type 3 turbines are: (1) `REpower` (now `Senvion`) 6M (6.0 MW), (2) `Bard` 5.0 (5.0 MW), and (3) `Acconica` AW-100/3000 (3.0 MW).

1.5.4 Type 4 WECS Configuration

To achieve variable-speed operations over the entire wind-speed range, Type 4 turbines were developed during the 1990s. The configuration of Type 4 WECS with a wind generator, and a full-scale power converter composed of machine (generator)-side converter (MSC), DC-link capacitor and GSC is shown in Figure 1.18 [1, 24]. In contrast to Type 3 turbines, where the power converter is connected in the rotor circuit to process slip power, Type 4 turbines employ a power converter between the wind generator stator terminals and utility grid to process all the electric power produced. Therefore, the capacity of the power converter increases from 30% to 100%. The cost of a power converter in Type 3 and 4 WECS is approximately 5% and 7–12% (according to the type of converter topology) of the total WT cost, respectively [19, 37]. A full-scale (100%) power converter leads to full-variable-speed range (0% to 100%) and the energy yield in these turbines is the highest. SCIG, PMSG, WRSG, and HTS-SG have all found applications in this type of configuration with power ratings of up to several megawatts. Among the classes of wind generators, PMSG is the most popular in Type 4 WECS (refer to Table 1.2 on page 20 for advantages of PMSG over other wind generators).

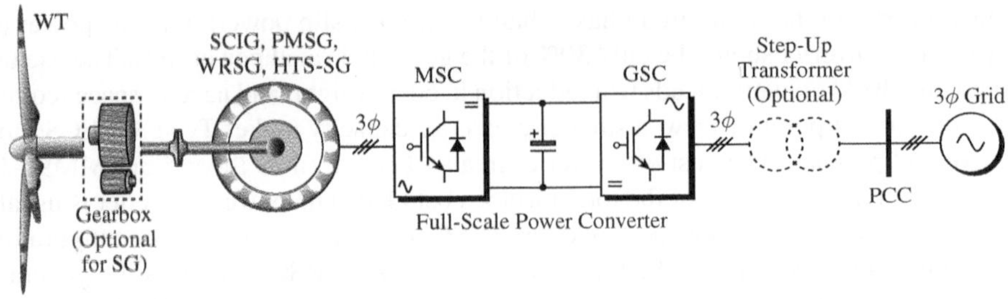

Figure 1.18 Configuration of Type 4 WECS with full-variable speed (0 – 100%) wind generators, optional gearbox, and full-scale power converter.

The BTB connected power converter configuration shown in Figure 1.18 is similar to the one shown in Figure 1.17 in terms of construction, modulation, and digital control. This power converter can provide or consume reactive power; thus, it represents dynamic capacitor/inductor bank and helps maintain power system stability. The step-up transformer can be eliminated by operating the power converters at a MV level. The significant pros and cons of this configuration are as follows:

✔ High-energy yield and no stress on mechanical components because of full-variable-speed (0 – 100%) operation.

✔ Independent active and reactive power control lead to excellent FRT compliance.

✔ The generator is fully decoupled from the grid. The power converters also enable smooth grid connection.

✖ Owing to the full-scale converter, the initial cost and nacelle space requirement increase along with overall system complexity.

✖ Higher power losses in the converter degrade overall efficiency.

✖ The complexity of digital control system design for power converters increases.

Wind energy conversion efficiency in Type 4 WECS is the highest compared with other types of turbines. The high-energy yield compensates the drawbacks of high converter cost and power losses. The need for the gearbox can be eliminated by using a high-pole number SG. This configuration is more robust against power system faults than Type 1, 2, and 3 turbines [38]. The best FRT compliance is achieved by a full-scale converter and an external static reactive power compensator is not needed. Typical commercial turbines include: (1) `Enercon` E126 (7.5 MW), (2) `Multibrid` M5000 (5.0 MW), and (3) `Vestas` V-112 (3.0 MW).

The distributed drivetrain concept is used in recent megawatt Type 4 WTs with a PMSG to obtain a simple design. The gearbox drives multiple generators at high speeds. The distributed drivetrain and multiple generators enable the achievement of high power density [39]. One commercial application is `Clipper Liberty` C99 WT, which uses a quantum drivetrain, four PMSGs, and four converters. The high torque is distributed among the four drivetrains, and the power rating of the converters is one-fourth of the system rating. This configuration also offers an effective fault-tolerant operation. When one converter fails, the other three converters can still deliver the power to the grid. The main disadvantage of using this configuration is the complicated drivetrain.

1.5.5 Type 5 WECS Configuration

The Type 5 WT with a direct grid-connected WRSG (EESG), two-stage helical/planetary gearbox, and mechanical speed/torque converter is shown in Figure 1.19. This configuration, wherein the variable-speed operation is achieved by a mechanical converter rather than an electrical converter, is an old concept for WTs. The torque/speed converter, also known as a variable ratio transmission, converts the variable speed of the WT to a constant speed. The generator operates at a fixed speed and is directly connected to the grid through a synchronizing circuit breaker. The rotor in the WRSG carries a field winding to produce rotor flux. The rotor DC excitation current is provided by a small AC/DC power converter that is directly connected to the three-phase grid (low-voltage side of a step-up transformer, if any). The DC excitation also includes slip rings and brushes (or brushless exciter). The excitation current is adjusted such that the generator output voltage and frequency match the grid code requirements. One of the major drawbacks with this configuration is that the field winding losses in WRSG reduce generator efficiency compared with PMSG.

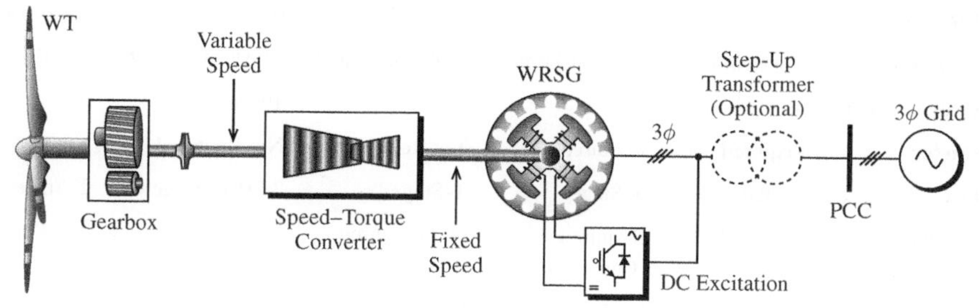

Figure 1.19 Configuration of Type 5 WECS with a fixed-speed WRSG, three-stage gearbox, and steed-torque converter.

The main features of this WECS configuration are summarized as follows:

✔ Similar to the Type 4 WECS, wind energy conversion efficiency is high because of the full-variable-speed range (0 – 100%).

✔ The overall cost and nacelle space requirement is lower than Type 4 WECS because of the absence of a converter.

✔ The generator can be directly connected to the MV collection point without any step-up transformer because there is no restriction imposed by the power electronic converter unlike the Type 4 turbine.

✖ The configuration has high system complexity because of the mechanical speed/torque converter.

✖ Type 5 WECS is unsuitable for offshore applications because of the high maintenance of the gearbox and mechanical speed/torque converter.

✖ The FRT compliance becomes complicated because it must be handled solely by the WRSG excitation system.

Despite the good features of this configuration, it is rarely used in the wind energy industry because of issues related to the mechanical converter. Currently, only three WT manufacturers use this technology: (1) `DeWind` D8.2 (2.2 MW, 4.16/13.8 kV); (2) `Wikov` W2000 (2.0 MW, 6.3/11 kV); (3) `AMSC-Windtec` SuperGear (2.0 MW, 11 kV). These turbines can be directly connected to a medium-voltage distribution line without a step-up transformer because of MV-WRSGs.

1.5.6 Comparison of WECS Configurations

The main features and drawbacks of all five types of wind energy configurations are summarized in Table 1.4. Comparisons and analyses are performed on the generator, power converter configuration, power converter capacity, achievable speed range, soft-starter requirement, gearbox, external reactive power compensator, MPPT ability, aerodynamic power control, compliance with the FRT requirement, technology status, and market penetration. Each configuration has its own merits, demerits, and practical applications.

Table 1.4 Summary of the five types of wind energy system configurations [1]

	Fixed Speed	**Semi-Variable Speed**		**Full-Variable Speed**	
	Type 1 WECS	Type 2 WECS	Type 3 WECS	Type 4 WECS	Type 5 WECS
Generator	SCIG	WRIG	DFIG	SCIG, SG ①	WRSG
Speed Range	1%	10%	±30%	0 – 100%	0 – 100%
Power Converter	Not Required	Diode+Chopper	AC/DC+DC/AC ②	AC/DC+DC/AC ②	Not Required
Converter Capacity	0%	10%	30%	100%	0%
Soft Starter	Required	Required	Not Required	Not Required	Not Required
Gearbox	3–Stage	3–Stage	3–Stage	3/2/1/0–Stage ③	2–Stage
Reactive Power Compensation	Required (Capacitors)	Required (Capacitors)	Not Required	Not Required	Not Required
Aerodynamic Power Control	Active Stall, Stall, Pitch	Pitch	Pitch	Pitch	Pitch
MPPT Operation	Not Possible	Limited	Achievable	Achievable	Unknown
FRT Compliance	By External Hardware	By External Hardware	By Power Converter	By Power Converter	Unknown
Technology Status	Outdated	Outdated	Highly Mature	Mature/Emerging	Old Concept
Current Market Penetration	Few or No Installations	Few or No Installations	Highest Share (>50 %)	2nd Highest Share ④	Few Installations
Example WTs	`Vestas` V82 1.65 MW	`Suzlon` S88 2.1 MW	`REpower` 6M 6.0 MW	`Enercon` E-126 7.5 MW	`DeWind` D82 2.2 MW

① SG includes PMSG, WRSG, and HTS-SG
② Direct AC/AC conversion is also possible without an intermediate DC stage
③ Three-stage gearbox is mandatory with the SCIG
④ PMSG is the most favored industry choice in this category

To demonstrate the popularity of each WECS configuration in the wind energy industry, the top 10 WT manufacturers in 2014 and their priority generator and gearbox technologies are illustrated in Figure 1.20. These manufacturers account for approximately 72% of the 51.5 GWs installed wind power capacity in 2014 [4]. The Danish company `Vestas` remained on the top of the list with a 12.3% share from both onshore and offshore installations. WT manufacturers `Siemens`, `GE Energy`, `Goldwind`, `Enercon`, `Suzlon Group`, `United Power`, `Gamesa`, `Ming Yang`, and `Envision` hold the top 2 to 10 positions in order. Four Chinese companies belong to the top 10 manufacturers list with a cumulative market share of 22.3%.

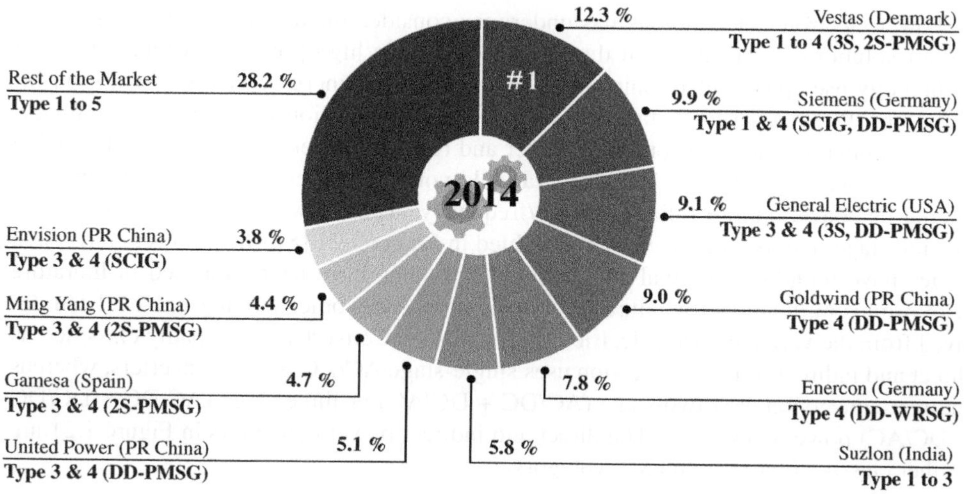

Figure 1.20 Top 10 WT manufacturers in 2014, along with their priority WECS configurations.

The WECS configuration, wind generator, and gearbox employed by the top 10 WT manufacturers are also shown in Figure 1.20. These details are obtained from the product brochures of the companies. Type 3 turbines (DFIG), which have the highest market share, are used by 7 of the top 10 manufacturers. Approximately 100 different DFIG turbine models are offered by the WT manufacturers. Type 4 turbines are produced by 6 manufacturers, and 4 of these manufacturers offer direct-drive solutions. The best-selling WTs in the market use Type 3 and 4 technologies. Future project announcements from WT manufacturers indicate that Type 4 technology will take over the market in the coming years. This book deals with Type 3 and 4 turbines only.

1.6 POWER ELECTRONICS IN WIND ENERGY SYSTEMS

The analysis of all five types of WECS configurations in the previous section indicates that since the production of the first-generation fixed-speed WTs in the 1980s, power electronics technology has an important collaboration with the grid-connected WTs. The power electronics technology contributes to low cost of energy, extraction of maximum possible energy from wind, enhancement of reliability and power density, fault-tolerant operation, reduction of weight and footprint, superior grid power quality, compliance with the strict grid codes, etc. Examples of the applications of power electronics in WECS are as follows:

- As a soft starter for a smooth grid connection (limiting startup in-rush currents) of Types 1 and 2 turbines.
- As a power converter to adjust rotor circuit resistance value and the slip characteristics of WRIG in Type 2 turbines.
- As a partial-scale converter to handle the slip power of DFIG and to increase speed range in Type 3 turbines.
- As a full-scale converter to decouple the generator from the grid and provide a full-speed range in Type 4 turbines.
- As an AC/DC converter to adjust the excitation current for WRSG in Type 5 turbines such that the generator output voltage and frequency matches the grid conditions

Power electronics technology has undergone considerable advancement, and state-of-the-art solutions are available in the form of full-scale high-power converters. Current technology uses power electronics in WTs and WFs for energy conversion and grid integration. Power converters facilitate variable-speed operation in Type 3 and 4 WECS while eliminating the need for a soft starter and reactive power compensation. To enable the grid connection of these WTs, the variable voltage/frequency of the wind generator should be converted into a fixed voltage/frequency. Thus, a wide variety of power conversion stages can be employed as highlighted in Figure 1.21 [8]. Most power conversion stages have found commercial applications, and some have been proposed in literature with promising features for future development, whereas other topologies have been derived from the variable-speed electric drives industry. Converters are broadly classified as direct and indirect: direct conversion uses single-stage AC/AC power converters, whereas indirect conversion uses two-stage (AC/DC + DC/AC) or three-stage (AC/DC + DC/DC + DC/AC) power converters. The direct and indirect power converters in Figure 1.21 are further classified into LV and MV category.

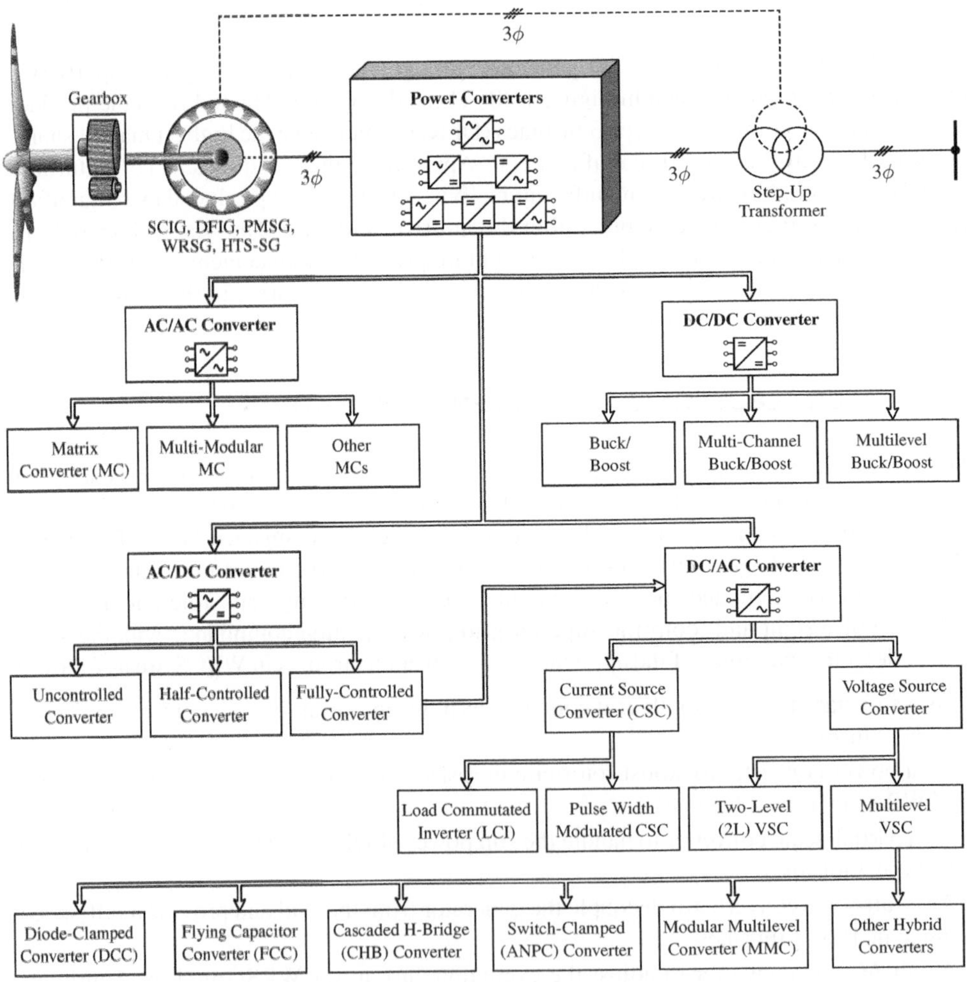

Figure 1.21 Taxonomy of power converters for Type 3 and 4 variable-speed WECS.

Direct AC/AC (matrix) converters compete with two-stage (AC/DC + DC/AC) VSCs in the electric drives industry because of the elimination of DC-link components and high reliability. The AC/AC converters are available in the form of direct and indirect MCs in the LV category or as multi-modular MC in the MV category. Although MCs have not been used in the WECS industry, Yaskawa, a leading matrix converter manufacturer, has developed an MV matrix converter product (Enewin-MX1) for high-power WTs [40].

Indirect two-stage (AC/DC + DC/AC) power converters are widely available in a BTB connected configuration. A fully controlled AC/DC converter, DC-link components such as capacitors and inductors, and a DC/AC converter form the BTB configuration. BTB converters are used as VSCs and CSCs. In the WECS industry, BTB VSCs are well proven, low cost, and reliable. These converters facilitate (inherently) a four-quadrant operation with a relatively simple configuration. Furthermore, BTB VSCs are suitable for both Types 3 and 4 WECS with IGs and SGs. The BTB 2L-VSC and parallel 2L-VSCs are used in the LV category. For high-power WTs, MV converters are preferred. The three-level DCC or neutral-point-clamped (NPC) converter topology is the most attractive choice in the wind energy industry. This topology increases the output voltage without connecting switching devices in series (i.e., no derating), reduces *dv/dt* stress compared with 2L-VSC, minimizes the harmonic filter requirement, and improves the grid power quality. The neutral-point voltage control is a challenging issue and has been extensively researched in literature. MV converters such as FCC, CHB converter, ANPC converter, and PWM CSC are popular in the electric drives industry [41] and are promising for future developments in the WECS industry. MMC is a recent development for HVDC applications, and many current research works study its suitability for the WECS industry.

Indirect three-stage (AC/DC + DC/DC + DC/AC) power converters are also employed by a few WT manufacturers. The generator-side AC/DC conversion is performed by an uncontrolled converter (diode-bridge rectifier). Diode rectifiers are cheap and reliable compared with IGBTs. Intermediate DC/DC converters perform variable-speed operations for WECS. The DC/AC converter is similar to the one employed in two-stage converters. Three-stage power converters are only suitable for Type 4 WECS with SGs because the power flow is unidirectional (from the generator to the grid).

1.7 CONTROL OF WIND ENERGY SYSTEMS

Control theory has evolved as an important discipline in modern WTs and WFs. Control schemes enforce WECS to achieve the desired operation, increase wind energy conversion efficiency, reduce energy cost, increase the lifespan of WT components, decrease structural loading, decrease turbine down times, and provide a superior dynamic and steady-state performance. In this section, the control of mechanical and electrical power conversion units are discussed in detail.

The block diagram of the overall control scheme for a modern variable-speed WECS is shown in Figure 1.22. The analysis given in this section is applicable for Type 3 and 4 WECS. The stator and rotor connections of Type 3 DFIG WECS are shown by the dotted lines. The power conversion system in Type 3 and 4 WECS is realized by RSC + GSC and MSC + GSC, respectively. The WECS mainly consists of six control levels, wherein the Level I control loop involves fast varying variables and the Level VI control loop comprises by slow-varying variables. The tight control of variables in Level I loop is important to fulfill the active and reactive power commands imposed by the TSO/DSO in the Level VI control loop. The control loops also consider normal and abnormal operation for WECS.

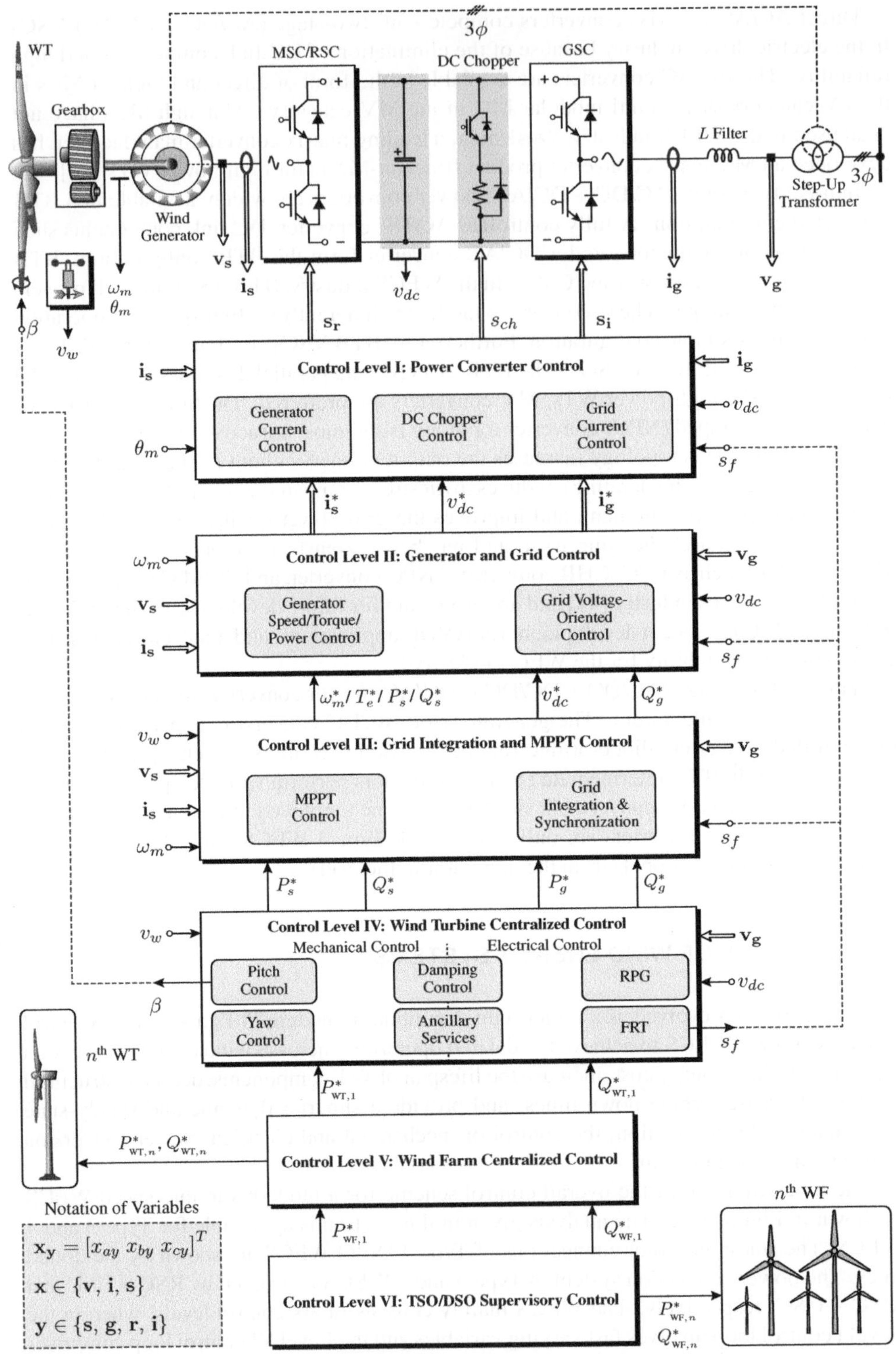

Figure 1.22 Block diagram of the overall control scheme for variable-speed WECS [5, 30, 42].

During grid faults, the FRT control in the Level IV loop issues a fault enable signal s_f. The mechanical and electrical control systems in the Level I to IV loops coordinate for better control performance during grid faults. For example, during grid faults, the GSC stops injecting active power and produces reactive power to the grid, the pitch control system starts working to decrease power extraction, and the DC chopper starts functioning to prevent the DC-bus voltage from exceeding the upper threshold limit.

The feedback signals from the WECS such as grid voltages $\mathbf{v_g}$, grid currents $\mathbf{i_g}$, generator voltages $\mathbf{v_s}$, generator currents $\mathbf{i_s}$, DC-link voltage v_{dc}, generator angular speed ω_m, rotor position angle θ_m, and wind speed v_w are used by various control loops. For DFIG WECS, the rotor currents are measured additionally. The control requirements are met by generating optimal gating signals $\mathbf{s_r}$, $\mathbf{s_i}$, and s_{ch} for the MSC/RSC, GSC, and DC chopper, respectively. The three-phase variables are represented by an equivalent vector, for example, $\mathbf{x_y} = [x_{ay}\ x_{by}\ x_{cy}]^T$. The main variable $\mathbf{x}$ denotes the voltage, current, or switching signal ($\mathbf{x} \in \{\mathbf{v},\ \mathbf{i},\ \mathbf{s}\}$). The subscript $\mathbf{y}$ represents the source (generator), grid, rectifier, or inverter ($\mathbf{y} \in \{\mathbf{s},\ \mathbf{g},\ \mathbf{r},\ \mathbf{i}\}$). The vector variables are denoted by a double line arrow ($\Rightarrow$), and the scalar variables are represented by a single line arrow ($\rightarrow$).

1.7.1 TSO/DSO Supervisory Control (Level VI)

The power output from the WT or WF is highly variable depending on wind-speed conditions. The variable energy in the existing energy mix brought serious reliability and security issues because of the intensive penetration of large-scale WECS into the existing power system. With the efforts made by the TSOs and DSOs through strict grid codes, large-scale wind farms changed their operations over the years from being passive power generation sources to active generation units with grid support characteristics.

Large-scale wind farms are connected to the TSO/DSO dispatch centers through communication networks to continuously share information about active and reactive power generation statuses. The top-level TSO/DSO supervisory control (Level VI) sends active and reactive power commands to each wind farm (similar to conventional power plants) connected to the power system. The first wind farm receives $P^*_{\text{WF},1}$ and $Q^*_{\text{WF},1}$, and the nth wind farm receives $P^*_{\text{WF},n}$ and $P^*_{\text{WF},n}$ commands from TSO/DSO [42]. To simplify the diagram, the communication link and measurements from the wind farm to the TSO/DSO supervisory control are not shown.

1.7.2 Wind Farm Centralized Control (Level V)

The power commands from the Level VI control is received from the WF centralized control. The WTs are connected to the WF centralized control by communication links to share the active and reactive power generation statuses. The supervisory control and data acquisition system is used for WF monitoring. The WF centralized control defines the active and reactive power requirements for each WT. The first WT is commanded with $P^*_{\text{WT},1}$ and $Q^*_{\text{WT},1}$, and the nth WT is supplied with the $P^*_{\text{WT},n}$ and $Q^*_{\text{WT},n}$ commands. The active and reactive power references are calculated such that the frequency and voltage at the PCC are maintained at the desired values. If the WF centralized control learns that the WTs cannot meet the RPG requirement, the wind farm static compensators such as STATCOM or SVC are initiated to support the WTs. The WF centralized control takes every possible effort to command the WTs such that the P and Q references imposed by the superior Level VI control loop are met all the time. The aerodynamic interaction of the WTs is also surpassed by the WF centralized control [43].

1.7.3 WT Centralized Control (Level IV)

As shown in Figure 1.22, the WT centralized control includes both mechanical and electrical controls. The pitch control and yaw control is solely involved in mechanical control, whereas RPG and FRT correspond to electrical control. The damping control includes both mechanical and electrical aspects: mechanical damping control lessens the mechanical resonances in the tower and torsional vibrations in the drivetrain, whereas electrical damping control provides damping for electrical subsynchronous resonances in the grid. The electrical ancillary services include electrical energy storage, uninterruptible power supply (UPS), and power quality. The UPS unit with battery energy storage forms an emergency power supply backup for pitch and yaw drives. Inertia emulation, spinning reserve, and kinetic energy storage are included in mechanical ancillary services [30].

The grid voltage magnitude is continuously monitored by the FRT subsystem. When the grid voltage falls below or above the preset magnitude, it sends a fault enable signal, s_f. This fault signal is sent to the other control loops and other functions inside the Level IV control loop. By coordinating various mechanical and electrical control systems, the WT centralized control provides active power reference P_s^* for WT MSC (P_s^* and reactive power reference Q_s^* for RSC), along with P_g^* and Q_g^* to the GSC. During normal grid conditions, Q_g^* is set to zero to maintain unity grid PF in Type 4 WECS. In DFIG WECS, grid PF is adjusted through Q_s^* command while setting Q_g^* to zero.

As mentioned in Section 1.2.3, modern variable-speed WTs use a pitch mechanism to change the rotation of blades in their longitudinal axis. Three individual electric pitch drives are commonly used in current WTs. The block diagram of the pitch control system is illustrated in Figure 1.23(a). When the wind speed is lower than the rated value, the pitch angle β is kept constant at zero degrees to extract the maximum possible energy from the wind. When the wind speed is higher than the rated value, the proportional-integral (PI) controller produces β such that the generator output power is limited to its rated value. As shown in Figure 1.23(b), as pitch angle increases, C_p decreases along with the extracted wind power (refer to Equation (1.2)), and the generator power comes back to the nominal value. In practice, the pitch control system is realized by a high-performance μC, generator output voltage and current sensors, three DC/AC power converters, three electric motors, and a backup power supply. The μC generates gating signals for power converters on the basis of wind-speed data, measured generator output power, preset power rating, and so on. Power converters change the speed of the three independent electric motors according to the instructions given by the μC.

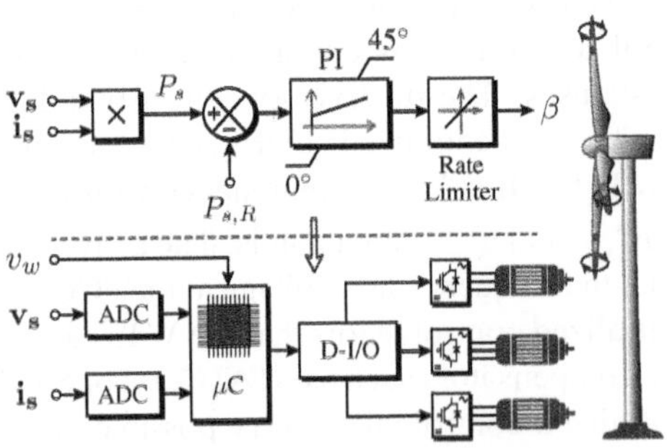

(a) Block diagram and implementation

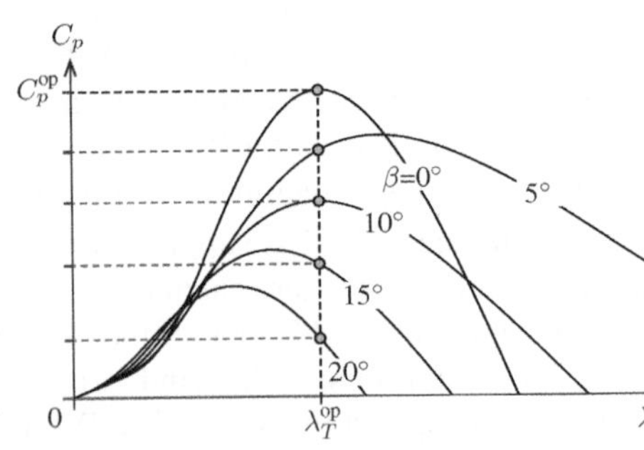

(b) C_p versus λ_T curve with different β values

Figure 1.23 Pitch control system for high-power WTs (D-I/O: digital input/output).

To show the importance of pitch control in WECS, the turbine characteristics during high-wind-speed conditions is plotted without and with pitch control in Figures 1.24(a) and 1.24(b), respectively. In both cases, the wind speed v_w is assumed to change in a trapezoidal manner with minimum and maximum speed velocities of 1.0 and 1.25 pu, respectively. In the first illustration in Figure 1.24(a), the pitch control is deactivated ($\beta = 0°$). As a result, the generator P_s changes in a trapezoidal manner with 1.0 and 1.95 pu as minimum and maximum values, respectively. By neglecting losses, P_s maintains a cubic relationship with respect to v_w. The corresponding generator output current magnitude increases significantly. This result leads to generator and power cable insulation damage and the cascaded failure of the whole WT with possible fire in the nacelle. In the second case, as illustrated in Figure 1.24(b), the pitch angle is adjusted dynamically by the PI controller as wind speed changes; thus, the generator output power and currents are maintained at a rated value. This leads to a safe WT operation with less stress on the electrical and mechanical structural components.

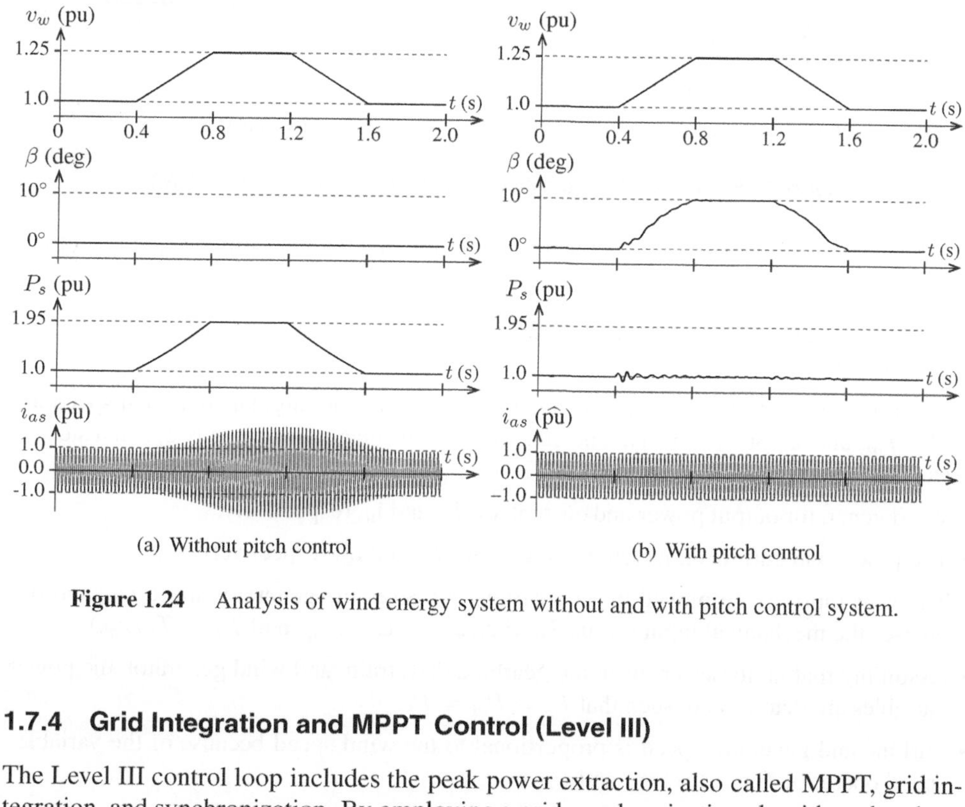

Figure 1.24 Analysis of wind energy system without and with pitch control system.

1.7.4 Grid Integration and MPPT Control (Level III)

The Level III control loop includes the peak power extraction, also called MPPT, grid integration, and synchronization. By employing a grid synchronization algorithm, the phase angle between grid currents and grid voltages can be adjusted accurately with the grid PF requirement. Grid synchronization is performed by zero-crossing detection, grid voltage filtering, or phase-locked loop (PLL). The latter is preferred because of its robustness against grid voltage harmonics and digital control platform friendliness [30]. The control system for a GSC helps in grid synchronization and integration by employing a PLL. The output of the grid integration subsystem is the reference DC-bus voltage v_{dc}^* and reference grid reactive power Q_g^*. For a given grid voltage magnitude v_{dc}^* is usually set constant according to the required modulation index of a GSC.

With the random nature of wind speed v_w, the peak power extraction is important in variable-speed WECS because it increases energy conversion efficiency. For a given v_w value, the MPPT control attempts to obtain the maximum possible power from the wind. As shown in Figure 1.25(a), the maximum power point (MPP) trajectory changes with respect to the wind speed. According to this curve, the operating region for the MPPT control is the cut-in wind speed to the rated wind speed. To facilitate the discussion of various MPPT control techniques, the notation and correlation of variables for a variable-speed WT is shown in Figure 1.25(b).

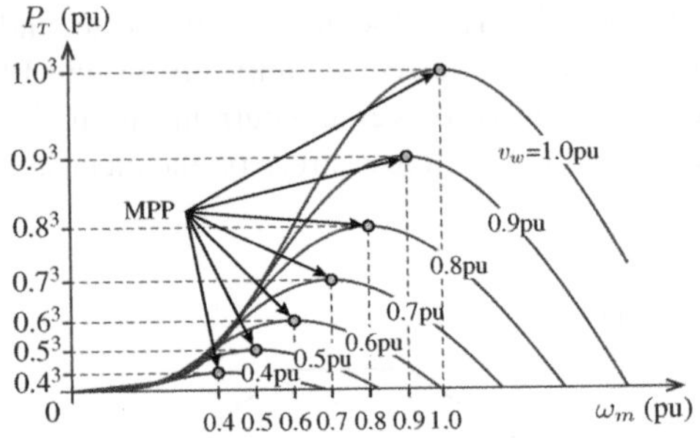

(a) P_T versus ω_m curve

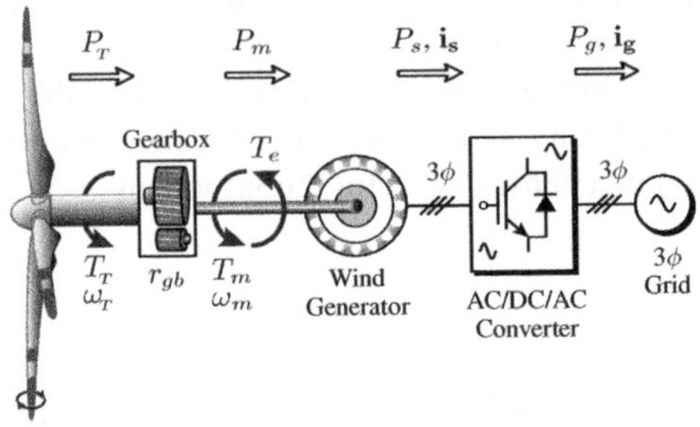

(b) Notation of variables in VSWT

Figure 1.25 Power characteristics and notation of variables for VSWT.

The major observations are listed below, assuming that the blade pitch angle is set to its rated (zero) value:

- Turbine output power, mechanical torque, and angular rotational speed are P_T, T_T, and ω_T, respectively.
- Wind generator input power, mechanical input torque, and angular rotational speed are P_m, T_m, and ω_m, respectively. The generator electromagnetic torque is T_e. In a steady-state condition, T_m becomes equal to T_e.
- Wind generator output power and current are P_s and $\mathbf{i_s}$, respectively.
- The power and currents injected to the grid are P_g and $\mathbf{i_g}$, respectively.
- The gear ratio r_{gb} employed in the gearbox increases the generator speed ω_m and decreases the mechanical input torque T_m (i.e., $\omega_m = \omega_T \times r_{gb}$ and $T_m = T_T / r_{gb}$).
- Assuming that no losses occur in the gearbox, drivetrain, and wind generator, the power variables are determined, such that $P_T = P_m = P_s$.
- Turbine and generator speed is proportional to the wind speed because of the variable-speed operation (i.e., $\omega_T \propto v_w$ and $\omega_m \propto v_w$).
- Based on the WT characteristics in Equation (1.2), turbine output power is proportional to cubic of the wind speed (i.e., $P_T \propto v_w^3$). Rotational speed ω_m is proportional to v_w, hence the above notation can be rewritten as $P_T \propto \omega_m^3$.
- Turbine output power is a product of torque and speed (i.e., $P_T = T_T \times \omega_T$). Similarly, for a wind generator, $P_m = T_m \times \omega_m$.
- According to the above representation, the turbine output torque is proportional to the square of the wind or rotor speed (i.e., $T_T \propto v_w^2$ or $T_T \propto \omega_m^2$). A similar expression can be applied to wind generator: $T_m \propto v_w^2$ or $T_m \propto \omega_m^2$.

By utilizing the above notation, various MPPT control techniques, as shown in Figure 1.26, have been developed in literature and applied in the wind energy industry [13, 44, 45]. These techniques determine the reference speed ω_m^*, reference electromagnetic torque T_e^* or reference power P_s^*, such that control Levels I and II force the WECS to follow the MPP trajectory. MPPT control can be achieved by MSC/RSC or GSC, although the most common approach uses MSC/RSC to reduce complexity. MPPT techniques are broadly classified as wind/rotational speed sensor-based methods or speed sensorless (SSL) methods. These methods also vary according to the required input variable (sensor), prior WT knowledge, control complexity, and memory. A brief description of various MPPT control techniques is given below.

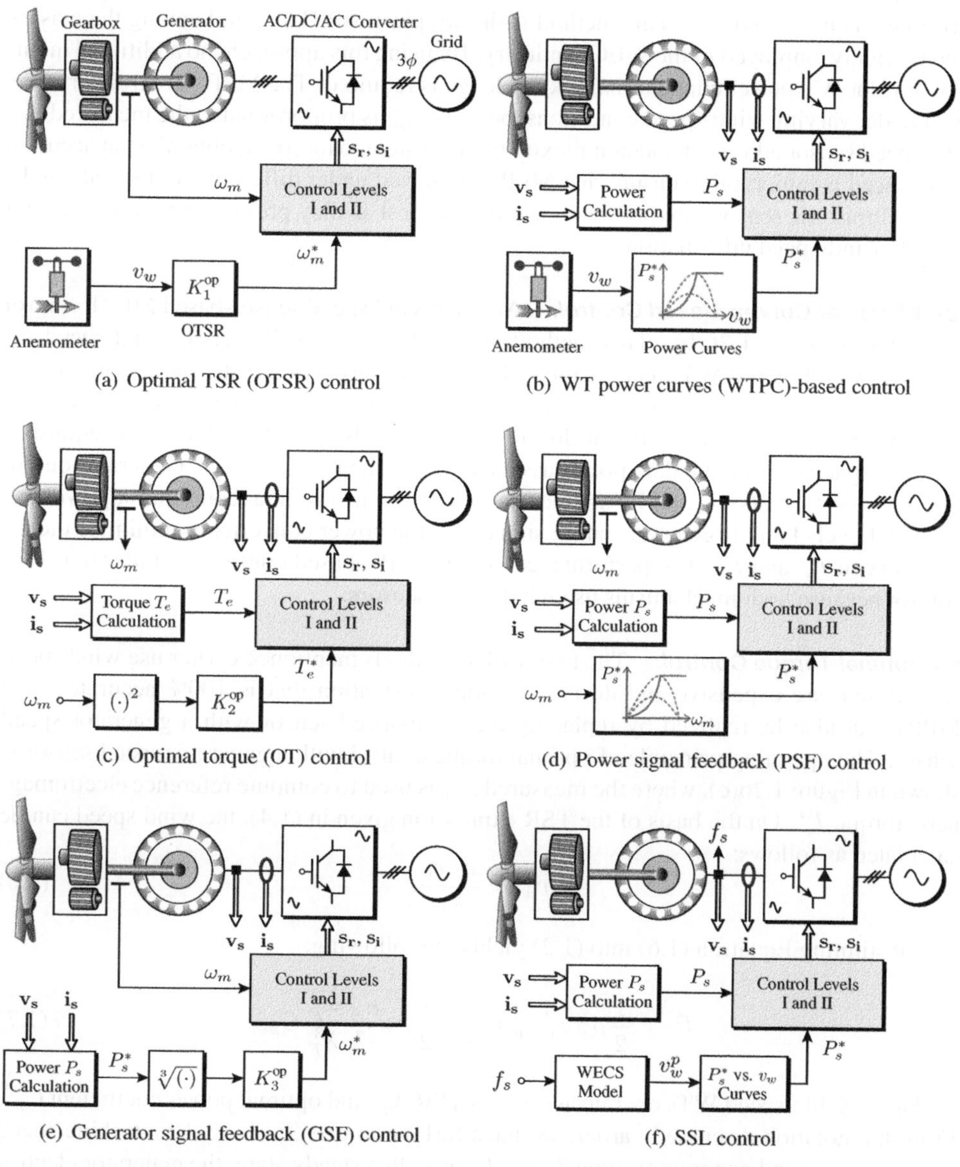

Figure 1.26 MPPT control of variable-speed wind energy systems.

(1) Optimal Tip Speed Ratio Control: The MPPT with optimal tip speed ratio is shown in Figure 1.26(a). This algorithm provides reference speed ω_m^* on the basis of the measured wind speed v_w and other WT parameters [46]. The generator reference speed ω_m^* is adjusted in proportion to v_w, such that the WT always operates at the λ_T^{op} value to reach the MPP. This adjustment is demonstrated as follows on the basis of the TSR expression given in (1.4):

$$\omega_m^* = \omega_T^* \, r_{gb} = \left(\frac{\lambda_T^{op} \, r_{gb}}{r_T} \right) v_w = K_1^{op} \, v_w. \tag{1.5}$$

In the above expression, the gain K_1^{op} can be calculated offline on the basis of the rated turbine parameters. Control Levels I and II use generator feedback speed to minimize the error between ω_m^* and ω_m. This method is the simplest MPPT method among the classes and is widely employed in the WECS industry. By using this approach, no additional memory or prior WT power characteristics knowledge is required. The MPPT control performs well under varying wind speed conditions, because ω_m^* is proportional to the measured v_w. The speed sensor adds cost and complexity to the system. Moreover, obtaining an accurate wind speed is important to ensure that MPP is achieved under different wind speed conditions. Ultrasonic sensors are promising for this control as they provide more accurate and reliable wind-speed information.

(2) WT Power Curves-Based Control: Another wind speed sensor-based MPPT control is shown in Figure 1.26(b). This method uses turbine power P_T versus wind speed v_w curve provided by the WT manufacturer. In initial experimental tests, the generator output power P_s versus wind speed v_w curve can be deducted. By using these WTPCs, the reference power P_s^* is measured at different wind speeds. The MPPT control requires a memory space to store the data points associated with the power curves. Generator output voltages and currents ($\mathbf{v_s}$ and $\mathbf{i_s}$) are measured to compute instantaneous output power P_s. Control Levels I and II generate gating signals for the power converter to minimize the error between P_s^* and P_s. The performance of the WTPC-based control is similar to OTSR control because both mechanisms use wind speed sensors.

(3) Optimal Torque Control: The two MPPT methods mentioned earlier use wind speed sensors that are expensive and do not provide information that is 100% accurate. The MPPT can also be realized by replacing the wind-speed sensor with a generator speed sensor. The operating principle of optimal torque control with a generator speed sensor is shown in Figure 1.26(c), where the measured ω_m is used to compute reference electromagnetic torque T_e^*. On the basis of the TSR expression given in (1.4), the wind speed can be calculated as follows:

$$v_w = \frac{\omega_T \, r_T}{\lambda_T}. \tag{1.6}$$

Substituting Equation (1.6) into (1.2) yields the following:

$$P_T = \frac{1}{2} \rho \pi r_T^2 \, v_w^3 \, C_p = \frac{1}{2} \rho \pi r_T^5 \, \frac{\omega_T^3}{\lambda_T^3} \, C_p. \tag{1.7}$$

The variable-speed WTs operate at optimal TSR λ_T^{op} and optimal power coefficient C_P^{op}. From the notation discussed earlier, we have turbine torque $T_T = P_T/\omega_T$, turbine speed $\omega_T = \omega_m/r_{gb}$, and generator torque $T_m = T_T/r_{gb}$. In a steady state, the generator electromagnetic torque T_e is equal to the mechanical input torque T_m.

Therefore the expression in (1.7) is reorganized as follows:

$$T_e^* = T_m = \frac{T_T}{r_{gb}} = \frac{P_T}{r_{gb}\,\omega_T} = \left(\frac{1}{2}\rho\,\pi\,r_T^5\,\frac{C_p^{\rm op}}{\lambda_T^{\rm op3}\,r_{gb}^3}\right)\omega_m^2 = K_2^{\rm op}\,\omega_m^2. \tag{1.8}$$

The coefficient for OT control, $K_2^{\rm op}$ can be calculated offline according to the rated WT parameters. Thus, its control complexity is similar to the OTSR method. In the above expression, the generator mechanical and electromagnetic torque maintains a quadratic relationship with the rotor speed ω_m. Owing to the Levels I and II control, generator electromagnetic torque T_e (and also mechanical input torque T_m) becomes equal to its reference electromagnetic torque T_e^* in a steady state. Similar to the OTSR, the OT control method is also widely used in commercial WTs.

(4) Power Signal Feedback Control: The block diagram of MPPT with the PSF control is shown in Figure 1.26(d) [11]. This approach combines the merits of the WTPC and OT control methods: the wind-speed sensor is replaced by a generator speed sensor and power curves are used to compute reference power P_s^*. As mentioned earlier, in variable-speed turbines, the generator speed varies in proportion to the wind speed (i.e., $\omega_m \propto v_w$); thus, P_s^* versus v_w curves can also be used as P_s^* versus ω_m curves. The PSF control calculates reference power P_s^* based on the measured generator speed ω_m.

(5) Generator Signal Feedback Control: As shown in Figure 1.26(e), the GSF control uses a slightly different concept from PSF control. The computed generator output power P_s is used to calculate reference generator speed ω_m^*. The relationship between P_s and ω_m^* is shown below:

$$\omega_m^* = K_3^{\rm op}\,\sqrt[3]{P_s}. \tag{1.9}$$

Similar to the OTSR and OT control, the coefficient $K_3^{\rm op}$ for GSF control can be computed offline according to the nameplate parameters.

(6) Speed Sensorless Control: The SSL MPPT control eliminates the need for wind and generator speed sensors. The block diagram of the SSL MPPT control is illustrated in Figure 1.26(f). In this method, autoregressive statistical models of WECS are used to predict wind speed (v_w^p) based on measured generator frequency f_s [47]. Once the wind speed is predicted, reference power P_s^* is calculated similar to WTPC control based on P_s^* versus v_w curves. It is also possible to realize sensorless OTSR control based on the predicted wind speed. This approach eliminates the need for speed sensors; however, it increases the overall complexity of the MPPT control.

(7) Other MPPT Algorithms: MPPT control is a highly researched area in wind energy systems. Many advanced MPPT algorithms based on adaptive control, sliding-mode control (SMC), fuzzy logic control (FLC), artificial neutral networks (ANN), and hybrid combinations have been researched to further increase wind energy conversion efficiency [44]. The perturb and observe (P&O), also called hill-climb searching MPPT control, which is popularly used in PV energy systems, is also studied in wind energy systems. P&O is based on a concept of perturbing the control variable in a small step-size and observing the results until the slope becomes zero. This method fails to follow the MPP trajectory under rapid wind variations because of the large moment of inertia in large WTs. A large step-size can be used to solve this issue but leads to more oscillations around the MPP [45].

A comprehensive comparison of various MPPT control techniques is summarized in Table 1.5 in terms of the input measurement variable(s) needed, prior knowledge of WT power characteristics needed, memory for storing the power curves, complexity of the overall control scheme, and performance of the algorithm with respect to varying wind-speed conditions. Overall, the OTSR and OT control techniques provide the best compromise between the complexity and performance.

Table 1.5 Summary of various MPPT control techniques [44, 45]

	Input Variable	**Prior WT Knowledge**	**Memory Required**	**Complexity of Control**	**Control Performance**
OTSR	v_w	Not Needed	No	Low	Very Good
WTPC	v_w	Needed	Yes	Medium	Very Good
OT	ω_m	Needed	No	Low	Good
PSF	ω_m	Needed	Yes	Medium	Good
GSF	$\mathbf{v_s}$, $\mathbf{i_s}$	Needed	Yes	Medium	Good
SSL	f_s	Needed	Yes	High	Good

1.7.5 Power Converter, Wind Generator, and Grid Control (Level I and II)

The block diagram of the overall control scheme depicted in Figure 1.22 indicates that the Level II control loop is related to the wind generator and grid control, and the Level I control loop corresponds to the power converter control. The Level I and II control loops are shown as two different blocks to clearly indicate the flow of the control variables. However, in reality, the Level I and II control loops are closely intertwined; thus, distinguishing them is difficult. To simplify the discussion, Level I is introduced first followed by Level II. To achieve high-energy conversion and feed the power to the grid, the following control objectives must be fulfilled in high-power WECS:

- MPPT under all wind-speed conditions.
- Net DC-bus voltage control to ensure proper operation for the GSC.
- RPG to meet the grid codes.

An accurate control of wind generators and power converters is necessary to fulfill the above control objectives. The first objective is achieved by the MSC/RSC, whereas the GSC handles last two objectives. Level II control produces the reference generator and grid currents ($\mathbf{i}_s^*$ and $\mathbf{i}_g^*$), and the Level I control produces switching signals ($\mathbf{s}_r$ and $\mathbf{s}_i$), such that the measured generator and grid currents ($\mathbf{i}_s$ and $\mathbf{i}_g$) follow their references ($\mathbf{i}_s^*$ and $\mathbf{i}_g^*$) closely.

The power flow between the wind generator and utility grid is also tightly regulated by the Level I control during both normal and grid fault conditions. In grid faults, the surplus energy between the generator and utility grid is dumped to the resistive load through a DC chopper, thus converting the kinetic energy of the turbine rotation into heat. The control system of the DC chopper dynamically adjusts the amount of energy to be dumped to the resistor. The DC chopper control subsystem reads the fault signal s_f value and generates the switching signal s_{ch} to the DC chopper such that the DC-link voltage v_{dc} never exceeds the upper threshold limit $v_{dc}^{\max}$.

(1) Power Converter Control (Level I): The development of digital control techniques for power converters is an ongoing research topic. A brief summary of classical and advanced control techniques for power converters is shown in Figure 1.27. Classical control techniques are well established in the literature and are widely employed to control converters, variable-speed drives, and energy conversion systems. The recent developments in digital control platforms such as μC, DSP, and FPGA enabled a designer/industry to develop more complex control algorithms to obtain optimal system performance. With the exponential increase in the computational capacity of digital control platforms, often expressed as million instructions per second (MIPS), the designer need not worry about the high number of calculations involved in the control algorithm. For example, the computational capacity offered by the TMS320C14 DSP in 1983 was limited to 10 MIPS, whereas the modern dSPACE DS1103 control platform can perform 2500 MIPS. The advanced control techniques improve the overall system performance by just software reconfiguration.

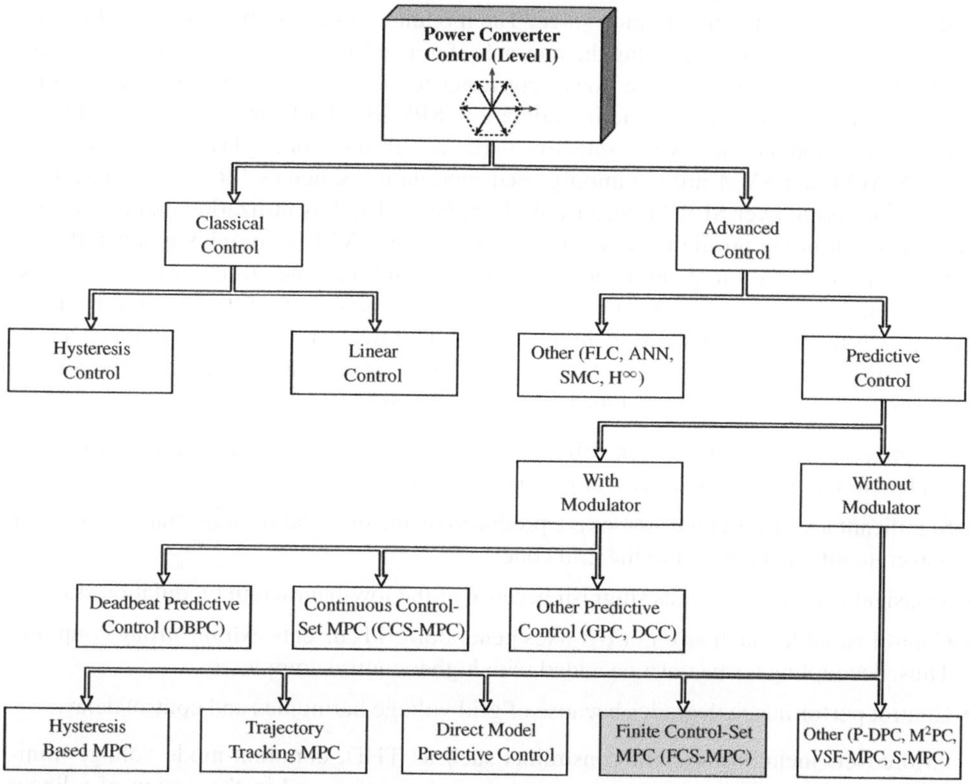

Figure 1.27 Taxonomy of power converter control techniques [48, 49]. (GPC, generalized predictive control; DCC, direct current control; P-DPC, predictive direct power control; M^2PC, modulated MPC; VSF-MPC, variable sampling frequency MPC; and S-MPC, switched MPC)

The classical control schemes include hysteresis (also known as bang–bang or on–off) control and linear control with modulation stage. The hysteresis control properly treats the nonlinear nature of the power converter. The bang–bang controller regulates the control variable x (current, torque, flux, active power, or reactive power) within the hysteresis band limits $(x^* \pm \delta/2)$ and produces switching signals directly (or with the help of a lookup table in multivariable case) for the power converter with no intermediate modulation stage

[50]. On the basis of the variable to be controlled (x), the bang–bang control is used in different forms such as: (1) hysteresis current control to control the generator/grid currents, (2) direct torque control (DTC) to regulate the generator electromagnetic torque and stator flux , and (3) direct power control (DPC) to control the generator/grid active and reactive powers. These control strategies are simple to implement; however, the resultant switching frequency changes with respect to the operating conditions. The variable-switching frequency causes the harmonic spectrum to spread and might ignite resonance in some power converters. Over the years, several articles have been published to obtain constant switching frequency by combining hysteresis control with modulation stage. A very high sampling frequency is usually needed by the digital control platforms to implement hysteresis control.

The linear control with modulation stage treats the power converter as a linear model. In this approach, a PI controller (in most of the cases) regulates the control variable x (usually generator/grid current) at its reference value (x^*) by generating a reference voltage (modulation) signal v^* to the modulation stage. The modulation stage defines the switching actions for the power converter using the internal reference frame transformation and switching strategy. On the basis of the power converter topology, several modulation schemes are available: sinusoidal pulse-width modulation (SPWM), third harmonic injected PWM, space vector modulation (SVM), selective harmonic elimination, hybrid modulation, etc. [41]. SPWM and SVM are commonly used modulation schemes for power converters. SVM is preferred over SPWM because it offers better DC-bus utilization and produces a good load harmonic profile (power quality). However, SVM is complex compared with SPWM and involves a large number of calculations and reference frame transformations. A low switching frequency operation in power converters is an important requirement at a high-power level to minimize switching losses and allow proper heat dissipation. During such operations, a linear control with modulation imposes several technical and operational challenges. A few significant technical challenges are outlined below [51]:

- Unsymmetrical performance in different operating conditions as a linear controller is applied to the power converter that is nonlinear in nature.
- Significant lower-order harmonics are produced by the modulation stage that lead to poor power quality and conflict to the grid codes.
- Transient response becomes sluggish because of the low-bandwidth modulation stage.
- Control variables such as dq or $\alpha\beta$-axes generator/grid currents exhibit strong coupling. Thus, decoupling terms must be added with higher control complexity.
- Control performance degrades because of grid voltage harmonics and control delay.

Moreover, including system constraints such as THD, common-mode voltage minimization, and switching losses reduction is not straightforward in the design of a linear controller. For digital implementation, a designer needs to spend extra time to map the continuous-time (CT) linear controller to the discrete-time (DT) domain using sampled-data models. A favorable approach in using linear control with modulation is the fixed switching frequency (imposed by the carrier frequency in the modulation stage). The field-oriented control (FOC) for wind generators and voltage-oriented control (VOC) for grid-connected converters are high-end versions of the linear control.

Advanced and complex control schemes for power converters include fuzzy logic control, artificial neutral network-based control, neuro-fuzzy control, sliding-mode control, H^∞ control, and predictive control. The classical and advanced control schemes are reviewed further in Chapter 3. In the field of control engineering, predictive control has a long history of development and applications since the 1950s. Initial applications include computer-based supervisory control projects by various oil and petrochemical industries [52]. In modern industrial controls, predictive control is a major success story, and this methodology has been used in hundreds of real-time applications. The predictive control concept for power converters was introduced during the 1980s; however, rigorous research has been conducted in recent years only because of the best responses (high computational capacity) provided by digital control platforms [53]. In the context of energy processing applications, predictive control offers a conceptually different and intuitive approach to control a power converter by treating it as a discontinuous and nonlinear actuator [6].

As outlined in Figure 1.27, predictive control covers a broad range of controllers with different control concepts. However, they all share a common philosophy: predict the future behavior of a control variable in each sampling instant by using the system model and select the optimal actuation based on the predefined optimization criteria [48]. Predictive control schemes can also be recognized as receding horizon controllers with one-sample-ahead prediction horizon. In power converters, predictive control is broadly classified into two categories based on whether a modulation stage is needed or not. DBPC, CCS-MPC, GPC, and DCC belong to the family of modulator-based predictive control schemes [49]. To reduce computational burden, optimizations are solved offline, producing a continuous control-set input variable (duty cycle or modulation index) to the modulation stage. The main advantages of modulator-based predictive control schemes are fast dynamic response and fixed switching frequency operation. However, these schemes are very sensitive to system parameter variations that could possibly lead to system instability.

The second category of predictive control techniques, including hysteresis-based MPC, trajectory-tracking MPC, direct MPC, and finite control-set MPC operate without any modulator; therefore, the switching frequency is not constant. The actuation from these control schemes is optimal switching signals which have a finite (discrete) control-set nature. Hysteresis-based MPC forces the control variable to remain in the hysteresis boundaries (similar to the hysteresis current control), whereas in trajectory-tracking MPC, the control variable follows the predefined control trajectory (similar to the sliding-mode current control). Direct MPC allows long prediction horizons to improve overall control performance. In recent years, FCS-MPC has emerged as a simple, intuitive, powerful, and promising alternative to control power converters that have a finite number of switching states [54]. FCS-MPC replaces the cascaded control structure and modulation stage and provides a fast and dynamic response in addition to good steady-state reference tracking. This technique offers simplicity and great flexibility for a designer to incorporate several control objectives by simply changing the optimization criteria (cost function). One of the best features of the FCS-MPC strategy is that system nonlinearities and limitations can be incorporated directly into the system model. FCS-MPC operates with variable-switching frequency similar to hysteresis control; however, a designer can control the number of switch changes (switching frequency) by imposing a restriction in the cost function. The finite control-set model predictive control is the subject of this book.

Some research has combined the operating principles of MPC and SVM to obtain the best qualities of each class: fast dynamic response as in MPC and constant switching frequency as in SVM. A few of these novel control strategies were introduced very recently in 2015: P-DPC [55], M^{2}PC [56, 57], VSF-MPC [58], and S-MPC [59]. These strategies employ predefined power converter voltage space vector sequences in the modeling of the system. These control schemes are relatively new and premature, hence they are not discussed in the reminder of this book.

(2) Wind Generator and Grid Control (Level II): The design, construction, operation, and control of a wind generator and electric motor are similar. Based on the sign of mechanical input torque T_m, an electric machine can be used as a generator or motor (negative T_m for generator and positive T_m for motor). All theories and practical developments in electric motor control can therefore be attributed to wind generator control. In comparison to electric motors control, the wind generators employ additional MPPT control loop to generate reference speed ω_m^*, reference torque T_e^*, or reference power P_s^*. The wind generator control schemes mentioned in this book are derived from the electric motor control theory that is well documented in literature. Wind generator control schemes are broadly classified into two classes: classical control and predictive control. The taxonomy of wind generator control schemes is presented in Figure 1.28 considering both IGs and SGs.

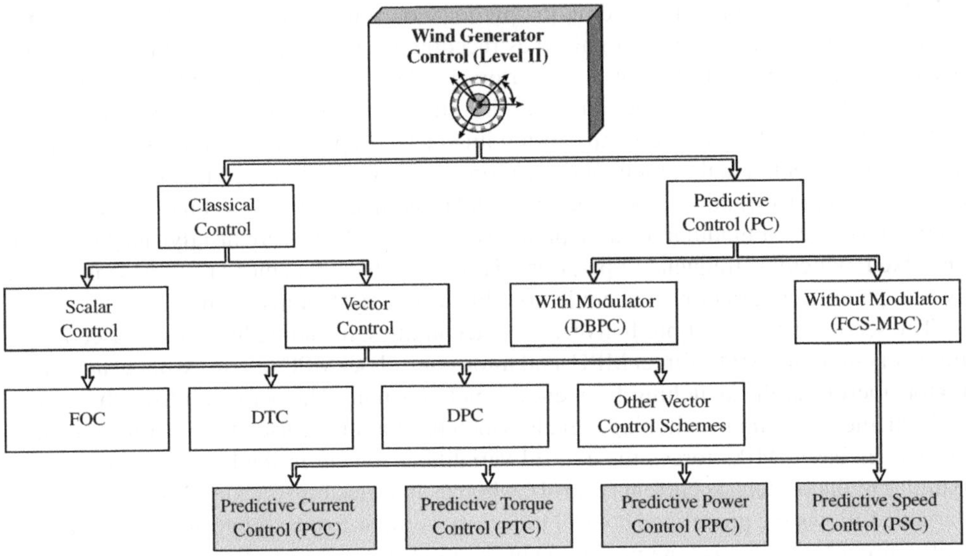

Figure 1.28 Taxonomy of wind generator control techniques [60].

Scalar and vector control techniques form classical control schemes, and the latter is extensively used in electric drives and the wind energy industry. In scalar control, only the magnitude and frequency of space vector (voltage, current, or flux linkage) is adjustable; therefore, an acceptable performance is guaranteed during steady-state operation. In vector control, in addition to magnitude and frequency, the instantaneous position (phase angle) of the space vector is controlled; as a result, a high-performance operation is obtained in both steady-state and transient conditions. Vector control of a wind generator is a general and decoupled control philosophy that can be further classified as FOC, DTC, DPC, and other forms, such as feedback linearization control and passivity-based control [60]. Vector control schemes are implemented by transforming the generator dynamic equations into field coordinates that rotate in synchronism with the rotor or stator flux or voltage vector. The FOC for SCIG and SG uses rotor flux orientation; FOC for DFIG uses stator flux or voltage orientation; and DTC uses stator flux orientation. It has been well established in the literature that field coordinates provide decoupled control in AC machines similar to the separately excited DC machine. As mentioned earlier, FOC of wind generator and VOC of GSC belong to the class of linear control schemes, and the DTC and DPC belong to the nonlinear hysteresis control.

FOC and DTC are two frequently-used methods in the present industry to obtain high control performance [61]. In the FOC scheme, wind generator three-phase currents are transformed into two orthogonal components that define the magnetic flux and electromagnetic torque. To force the measured currents to follow their reference currents, two PI controllers are used to generate a reference voltage vector for the modulation stage. The modulation stage compares the reference voltages with the carrier waveform and generates switching signals for the power converter. FOC is a linear control method, and its switching frequency is constant (set by modulator). FOC schemes are available in two forms: direct or feedback FOC (DFOC or *Blaschke* control) and indirect or feed-forward FOC (IFOC or *Hasse* control). The latter is widely used because of its simplicity throughout the speed range. The rotor field-oriented FOC is used to control SCIG, stator field-oriented FOC is employed in DFIG, and rotor flux-oriented maximum torque per ampere control (essentially FOC scheme) is used in PMSG [13].

DTC represents a viable alternative to the FOC scheme. Instead of a decoupled generator current control, the DTC scheme directly controls the generator stator flux and electromagnetic torque by employing hysteresis comparators and a lookup table. The required flux and torque are computed on the basis of measured generator voltages and currents. In comparison with FOC, the DTC scheme is relatively simple and eliminates the need for coordinate transformation, PI controllers, and pulse-width modulation. However, the number of online calculations with DTC is higher than FOC, thus making DTC a highly computational intensive control. Classical DTC is a nonlinear control and operates with variable-switching frequency. DTC control is available in two forms based on the selection of switching sectors: circular stator flux DTC (*Takahashi* and *Noguchi* control) and hexagonal stator flux (*Depenbrock* control). The DTC technique has been applied to SCIG, DFIG, and interior PMSG (IPMSG). In the DPC method, the generator active and reactive powers are controlled by hysteresis comparators and a lookup table, similar to DTC.

The operating principles of FOC, DTC, and DPC vector control schemes are used by FCS-MPC to design predictive current control (PCC), predictive torque control (PTC), and predictive power control (PPC), respectively. All control variables of FOC, DTC, and DPC are regulated by PCC, PTC, and PPC without employing any internal PI or hysteresis controllers. All model predictive control schemes are modulation-stage free; thus switching frequency is variable. In both classical and predictive control schemes, the outer speed control loop employs a PI controller to regulate the generator speed at its reference value. The PCC, PTC, and PPC schemes only replace the internal control loops of FOC, DTC, and DPC schemes respectively that need faster control actions than the sluggish outer speed control loop. The predictive speed control (PSC) scheme represents a recent development in this area. This scheme eliminates the speed PI controller and internal PI controllers, thus making it a completely cascade-free control structure [62]. This approach uses PCC, PTC, or PPC internally to simultaneously control the generator speed and control variables. Compared with PCC, PTC, and PPC, the control complexity of PSC increases slightly because of higher-order modeling and *ad hoc* cost function definition.

GSCs in variable-speed WECS are controlled by decoupled VOC or DPC [50]. In VOC, through the regulation of orthogonal-axis grid currents, the net DC-bus voltage (active power) and reactive power are controlled, respectively. In the DPC scheme, similar to the generator-side control scheme, the grid active and reactive powers are controlled independently. In summary, the main objectives for both classical vector control and predictive control schemes is to regulate independently: (1) the orthogonal-axis currents in FOC/VOC/PCC schemes, (2) generator torque and flux in DTC/PTC schemes, and (3) active and reactive powers in DPC/PPC of wind generator or GSC.

1.8 FINITE CONTROL-SET MODEL PREDICTIVE CONTROL

FCS-MPC offers several advantages that make it suitable for the optimal control of power converter-based energy processing applications [6]. FCS-MPC is an attractive solution for researchers in the industry and academe. Recent scholarly works demonstrated that FCS-MPC can easily be applied to a wide range of power converters with different output harmonic filter configurations, adjustable-speed motor drives, power quality, HVDC, and wind and PV energy conversion applications [63]. These features come with few technical challenges (research opportunities) that need to be addressed. In this section, the main features and challenges of FCS-MPC are analyzed in detail.

1.8.1 Main Features of FCS-MPC

As illustrated in Figure 1.29, the main features of the FCS-MPC are classified into eight categories. A brief description of these main features is presented below.

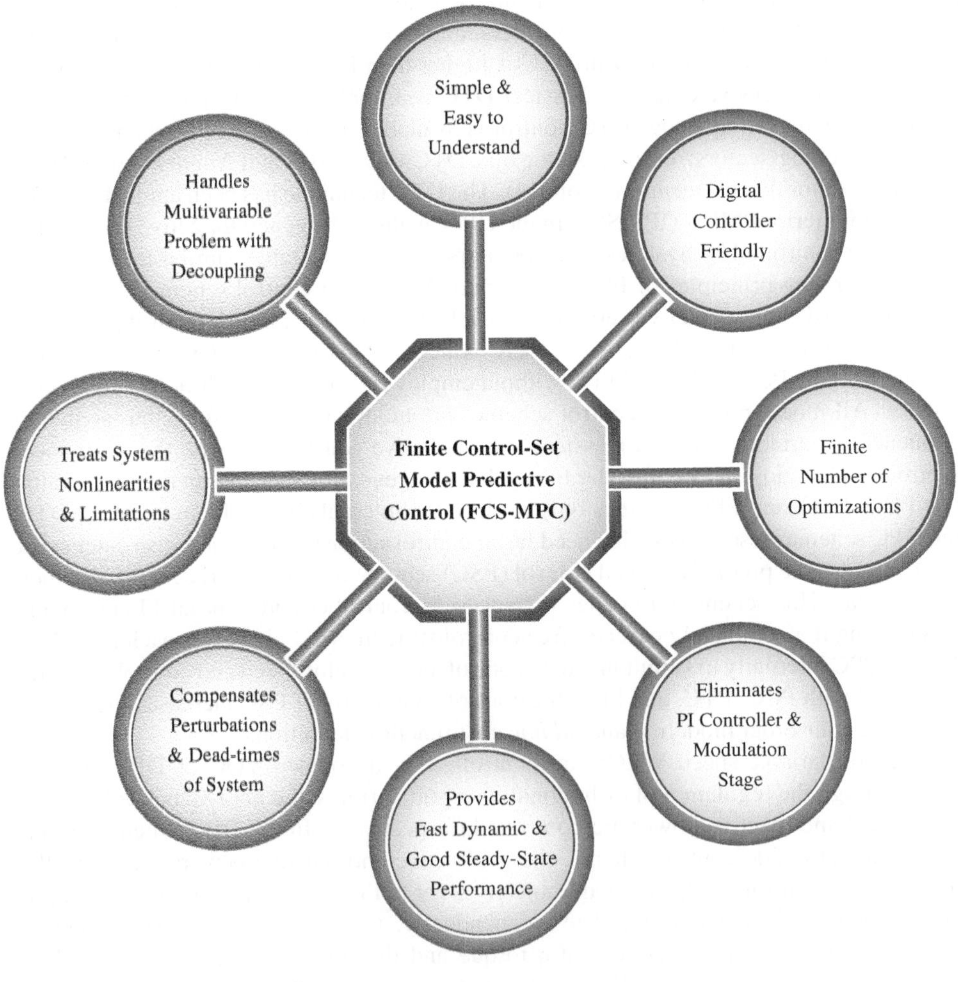

Figure 1.29 Main features of finite control-set model predictive control.

(1) Simple and Easy to Understand: FCS-MPC uses a simple concept, and it is easy to understand. Regardless of control application, FCS-MPC always uses four main subsystems, as presented below:

- **References Calculation:** This subsystem calculates the reference control variable $\mathbf{x}^*(k)$ ($\mathbf{x} \in$ voltage, current, power, torque, flux, etc.) according to the type of application. The references calculation is a common design step for classical and MPC schemes.
- **Extrapolation:** In this subsystem, the future value of the reference control variable $\widehat{\mathbf{x}}^*(k+1)$ is estimated based on the present and past sample values ($\mathbf{x}^*(k)$, $\mathbf{x}^*(k-1)$, etc.), or based on present sample value and angle of reference frame ($\mathbf{x}^*(k)$ and $\theta^*(k)$).
- **Predictive Model:** Possible future values of control variables $\mathbf{x}^{\mathrm{p}}(k+1)$ are predicted by this subsystem based on DT model and parameters of system, feedback signal values, and converter switching state combinations, $\mathbf{s}(k)$.
- **Cost Function Minimization:** The error between the predicted and extrapolated control variable $g = \widehat{\mathbf{x}}^*(k+1) - \mathbf{x}^{\mathrm{p}}(k+1)$ is calculated by this subsystem. The switching state $\mathbf{s}(k)$ that produces the "minimum" cost function error is selected as an optimal actuation (output) of FCS-MPC, and the output is applied to the power converter directly.

According to the application, additional blocks in FCS-MPC can appear; however, these blocks assist main subsystems. For example, a reference frame transformation block converts three-phase generator/grid currents into orthogonal components.

(2) Digital Controller Friendly: The inherent and discrete nature of power converters is considered in the design of a "predictive model" subsystem of FCS-MPC. Simulation models of FCS-MPC execute the algorithm with a discrete step size; consequent analysis of the control scheme is conducted in a DT domain. Real-time implementation of FCS-MPC by digital control platforms is, therefore, intuitive and natural. Hence, the time required for a designer to switch from a simulation stage to a real-time implementation stage is negligible; thus the overall FCS-MPC design framework is simple.

(3) Finite Number of Optimizations: FCS-MPC is a model-based optimization algorithm that performs a set of calculations during each sampling period. The number of iterations in the optimization algorithm are determined according to the possible number of converter switching states $\mathbf{s}(k)$. The power semiconductor switch in any converter possesses two discrete states: *turn-on* ('1') or *turn-off* ('0'). Thus the number of switching combinations (switching states) in any converter are limited to a finite set. For example, 8 and 27 switching states are available for 2L-VSC and NPC converter, respectively. Optimizations are greatly simplified because of the finite number of switching states, making it possible to implement FCS-MPC by digital control platforms available in the market.

(4) Eliminates PI Controller and Modulation Stage: FCS-MPC uses the DT model of system to predict the future behavior of control variables for every possible actuation sequence. Cost function minimization is used as an optimization criteria to evaluate system performance for all possible switching states. An optimal switching state that minimizes the cost function is directly applied to the converter. This approach eliminates the need for linear PI controllers, hysteresis regulators, intermediate modulation stage, and the lookup table in the control loop. FCS-MPC is a nonlinear control method, and it provides a better approach to control the power converters that are also nonlinear in nature.

(5) Provides Fast Dynamic and Good Steady-State Performance: Linear control techniques, such as FOC and VOC, treat the power converter as a linear actuator. The nonlinear nature of the power converter becomes more predominant at a lower switching frequency operation. Under such conditions, FCS-MPC provides better control in dynamic response and steady-state reference tracking. The resultant dynamic performance of FCS-MPC is superior to linear control because of the elimination of the low-bandwidth modulation stage.

(6) Compensates Perturbations and Dead-Times of System: Power conversion system perturbations, power converter dead-times, and on-state voltage drop of semiconductor switches are easily compensated by FCS-MPC. Although the output harmonic filter and internal DC-link filter parameters change, the controller can mitigate the perturbations by choosing a switching state that produces the minimal cost function error. Reference tracking is slightly affected by system perturbations; however, the transient response remains fast compared with linear control.

(7) Treats System Nonlinearities and Limitations: FCS-MPC treats various power converter topologies as discontinuous and nonlinear actuators that are the closest approximations to a real-time scenario. One of the best features of FCS-MPC is that the nonlinearities and limitations of the power converter can be incorporated directly into the system model. The cost function definition is flexible; several constraints and technical requirements such as maximum current limitation, switching frequency reduction, spectrum shaping, common-mode voltage minimization, power losses reduction, THD, boundary limits for electrical and mechanical variables, etc., can be incorporated in the design and operation of the controller to achieve a safe and reliable operation.

(8) Handles Multivariable Problem with Decoupling: For a wide variety of systems, FCS-MPC handles multivariable control problems in a decoupled manner without employing compensation terms outside the control loop. The "soft constraint" and "hard constraint" handled in a multivariable control framework is naturally accommodated by FCS-MPC through proper selection of weighting factors. This feature is particularly preferred in electromechanical energy conversion applications where both electrical and mechanical control variables need to be handled by FCS-MPC.

1.8.2 Challenges of FCS-MPC

Despite the simplicity and benefits of FCS-MPC, several challenges exist in state-of-the-art research that include, but not limited to:

(1) Large Number of Calculations: FCS-MPC performs a large number of online calculations during each sampling interval, thus leading to higher computational burden than the linear control scheme. This finding was a major obstacle for digital control designers until the last decade. However, modern DSPs can perform extensive calculations at a reduced cost. For example, the FCS-MPC for 2L-VSC (8 switching states), 3L-DCC (27 states), 4L-DCC (64 states), and 5L-DCC (125 states) can be realized with a minimum execution time (T_s) of 7, 18, 44, and 93 μs, respectively [8]. With the industry standard microprocessors, more freedom is obtained in the simultaneous control of BTB connected power converters. For some power converter topologies such as CHB and MMC, the number of switching states increases exponentially with the number of modules in each phase. To develop FCS-MPC with less computational burden, the closest vectors to the reference voltage vector should be selected, which is similar to the methodology employed in SVM.

(2) Variable-Switching Frequency: The main drawback of FCS-MPC compared with the linear control is that the converter switching frequency varies with respect to the operating conditions. This finding leads to spread spectrum in the control variable harmonic profile. The cost function can be penalized to control the switching frequency to some extent. For example, the average switching frequency of the converter can be regulated between two close boundary limits by the online adjustment of weighting factors [64].

(3) Heuristic Selection of Weighting Factors: The cost function in FCS-MPC includes several control objectives (variables) simultaneously. The relative importance of one objective over the other can be set through the weighting factors. Control variables possess different physical natures (current, voltage, power, etc.), and these variances lead to coupling effects; thus the selection of suitable weighting factors becomes tedious. The numerical procedure for the weighting factor selection is still an open research topic; however, some guidelines can be found in the literature [65]. In this book, a few case studies are designed to establish the theoretical framework for the calculation of weighting factors.

(4) Need for Accurate Model of System: The control performance obtained by the FCS-MPC depends greatly on the DT system model and prediction horizon. In the field of power electronics, CT mathematical models of various power converters and wind generators that are highly accurate and precise are readily available. Mapping CT models to the DT models is a mature subject matter in the field of control theory. In the current literature, DT models for various power conversion-related applications are well documented. For linear time-varying CT (for example, induction, and synchronous machine) models, calculation of corresponding exact DT models is impossible. In such cases, approximate models with high precision improve the system performance [66].

The challenges of using FCS-MPC that are related to the nature of the control of variable-switching frequency, weighting factors selection, control delay compensation, and robustness analysis are properly addressed in this book.

1.9 CLASSICAL AND MODEL PREDICTIVE CONTROL OF WECS

This section demonstrates the conceptual difference between classical FOC and PCC of variable-speed WECS. A Type 4 WECS with PMSG and BTB 2L-VSC is considered. To simplify the analysis, only the generator-side control scheme is presented considering Levels I to III control loops.

Stator voltage and electromagnetic torque dynamics of a permanent magnet synchronous machine (motor case) are described in the synchronous (dq) reference frame as

$$v_{ds} = R_s\, i_{ds} - \omega_r\, L_{qs}\, i_{qs} + L_{ds}\frac{d\, i_{ds}}{dt} \tag{1.10}$$

$$v_{qs} = R_s\, i_{qs} + \omega_r\, L_{ds}\, i_{ds} + L_{qs}\frac{d\, i_{qs}}{dt} + \omega_r\, \psi_r \tag{1.11}$$

$$T_e = 1.5\, P_p\, [\psi_r\, i_{qs} + (L_{ds} - L_{qs})\, i_{ds}\, i_{qs}] \tag{1.12}$$

where v_{ds} and v_{qs} are machine stator voltages in dq frame, i_{ds} and i_{qs} are machine dq frame stator currents, ψ_r is the peak value of rotor flux linkage produced by permanent magnets, R_s is the machine stator resistance, L_{ds} and L_{qs} are machine dq-axes inductances, ω_r is the rotor electrical speed, and P_p is the machine number of pole pairs.

As mentioned earlier, a motor model can be used for generator control by simply measuring the currents in opposite direction. To simplify the analysis, PMSG is assumed to have surface permanent magnet (non-salient-pole) configuration with $L_{ds} = L_{qs} = L_s$. FOC and PCC of PMSG WECS with a generator-side two-level voltage source rectifier (2L-VSR) is presented below, assuming that the GSC regulates the net DC-bus voltage and grid reactive power. Grid-side control scheme has an identical concept for both the classical and predictive control.

1.9.1 Classical Control of WECS

The block diagram of the classical FOC scheme is shown in Figure 1.30. The CT feedback measurements, such as three-phase generator currents $\mathbf{i_s}$, DC-link voltage v_{dc}, generator speed and position (ω_m and θ_m), and wind speed v_w are interfaced to the digital control through analog-to-digital converter (ADC) channels. Switching signals $\mathbf{s_r}$ produced by the digital control are sent to the 2L-VSR through the digital input/output channel and interface board containing a signal conditioner and gate drivers.

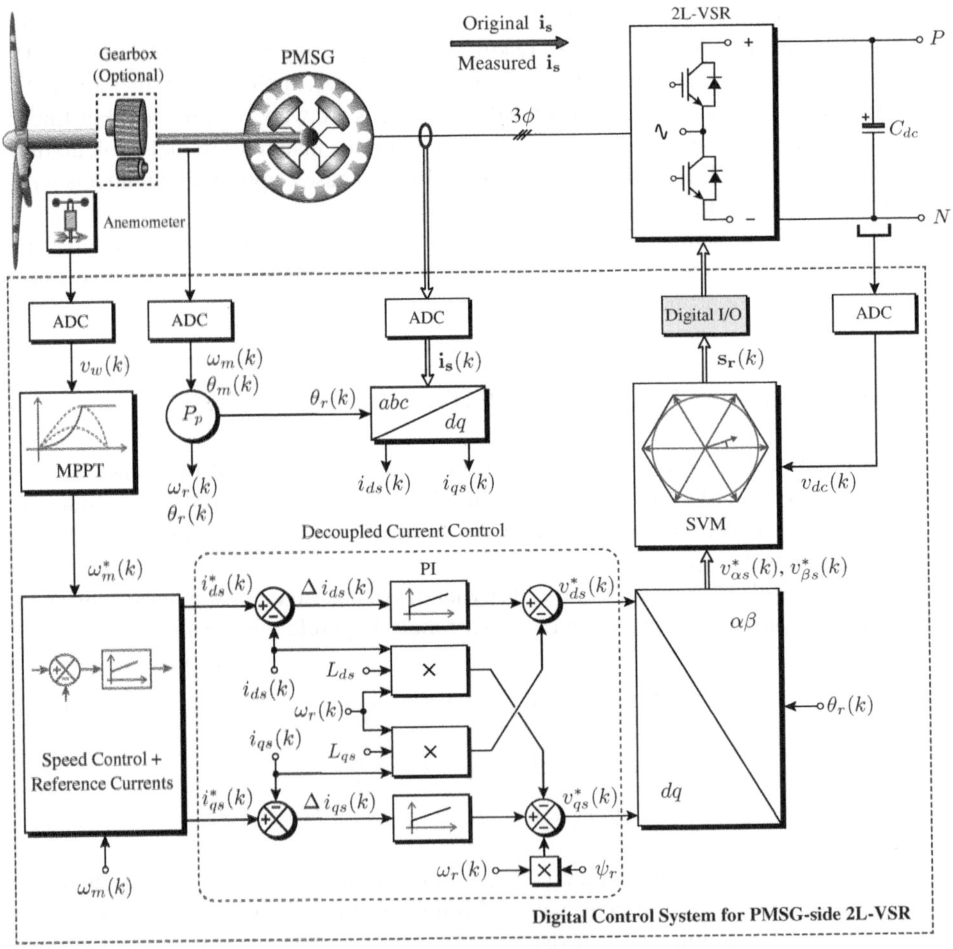

Figure 1.30 Classical zero d-axis current control of 2L-VSR-based PMSG WECS.

OTSR MPPT control produces reference speed ω_m^* based on varying v_w input values, such that the turbine operates at the optimal tip speed ratio λ_T^{op}. Mechanical speed and position values (ω_m and θ_m) are converted to corresponding electrical values (ω_r and θ_r) by multiplying these with the generator number of pole pairs. The three-phase currents are converted to dq-axes to facilitate control algorithm implementation in synchronous frame. The digital control uses cascaded control structure with an outer speed control loop and inner current control loop. The PI controller in the speed control loop regulates ω_m at its reference value ω_m^*, and the output of this loop is reference electromagnetic torque, T_e^*. Based on (1.12), electromagnetic torque production in PMSG is proportional to i_{qs} when $L_{ds} = L_{qs}$. With this criteria, d-axis reference current i_{ds}^* is set to zero, and the q-axis reference current i_{qs}^* is calculated from T_e^*, that is,

$$i_{ds}^* = 0, \qquad i_{qs}^* = \frac{T_e^*}{1.5\, P_p\, \psi_r}. \tag{1.13}$$

For this reason, the rotor flux-oriented FOC of PMSG is called zero d-axis current (ZDC) control. In this control, a linear relationship between torque and stator current is obtained when rotor flux linkage ψ_r is kept constant. This operation is identical to the separately excited DC machine, where field flux is maintained constant and torque is controlled by varying armature current [13].

The feedback dq-axis currents i_{ds} and i_{qs} are compared with their reference currents i_{ds}^* and i_{qs}^*, respectively. Error currents Δi_{ds} and Δi_{qs} are sent to two PI controllers. The output of the PI controller in the d-axis loop along with compensation term $\omega_r L_{qs} i_{qs}$ is used to calculate d-axis reference voltage v_{ds}^*. Similarly, q-axis reference voltage v_{qs}^* is calculated as a summation of the q-axis loop PI controller output and compensation terms $\omega_r L_{ds} i_{ds}$ and $\omega_r \psi_r$. Stator winding resistance R_s is negligible in high-power generators; hence the terms $R_s i_{ds}$ and $R_s i_{qs}$ are not included in the compensation terms. The compensation terms in dq-axes loops provide decoupled control for i_{ds} and i_{qs} while improving the performance during transient conditions [67]. The 2L-VSR synchronous frame reference voltages v_{ds}^* and v_{qs}^* are transformed into stationary frame reference voltages via $dq/\alpha\beta$ transformation using electrical position angle θ_r. More details about the reference frame transformation are presented in Chapter 3.

The two-phase sinusoidal reference voltages $v_{\alpha s}^*$ and $v_{\beta s}^*$ are used by the SVM stage. The SVM stage involves a large number of internal calculations to produce three-phase gating signals $\mathbf{s_r}$, such that the MPPT is achieved during each sampling interval. With the successful MPPT, the generator current, electromagnetic torque, and rotational speed are strictly regulated.

1.9.2 Model Predictive Control of WECS

The block diagram of the predictive current control scheme for a 2L-VSR-based PMSG WECS is shown in Figure 1.31. The control scheme is very easy to understand. It consists of four main subsystems: reference currents calculation, extrapolation, predictive model, and cost function minimization. Notation of variables, calculation of reference speed ω_m^*, and reference currents i_{ds}^* and i_{qs}^* is similar to the ZDC control discussed earlier. Once the reference currents are obtained in $(k)^{\text{th}}$ instant, these are extrapolated to $(k+1)$ sampling instant for use with the cost function. The PCC scheme replaces the decoupled PI controller, $dq/\alpha\beta$ transformation, and SVM stage. PCC considers one-sample-ahead prediction horizon to simplify the analysis.

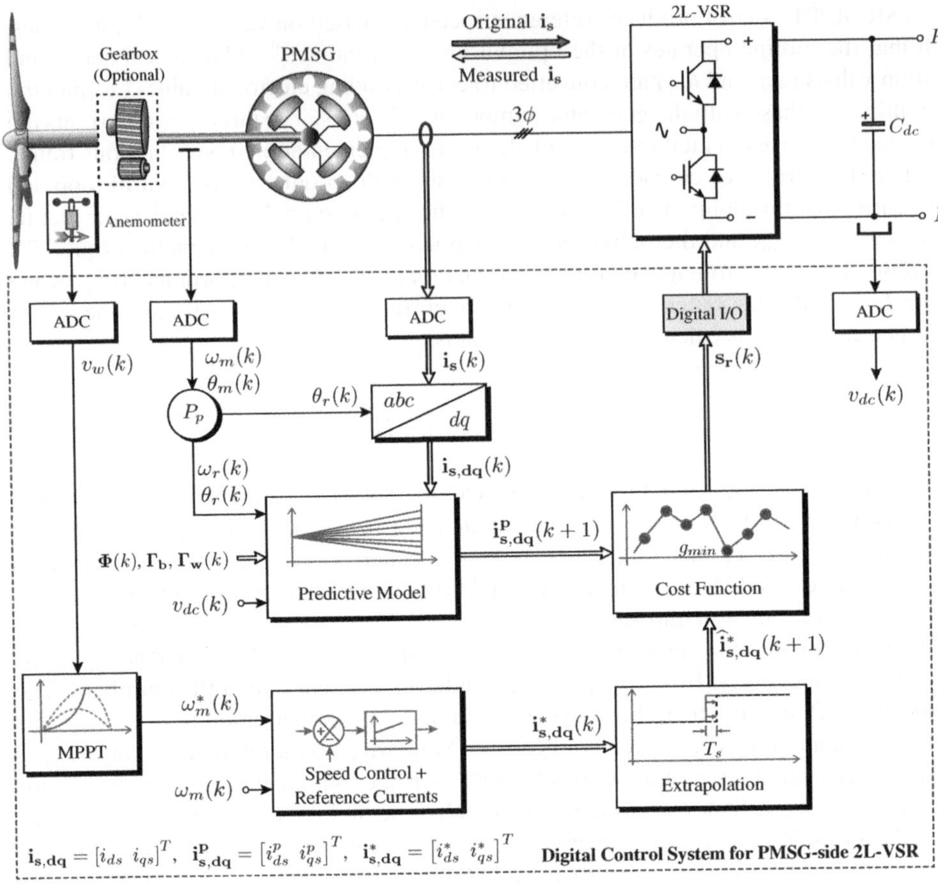

Figure 1.31 Predictive current control of 2L-VSR-based PMSG WECS.

PCC uses the DT model of PMSG and 2L-VSR to predict the future behavior of generator currents. Using the PMSG stator voltage models given in (1.10) and (1.11), the DT model of generator dq-axes currents is obtained by forward Euler approximation as follows:

$$\begin{bmatrix} i^p_{ds}(k+1) \\ i^p_{qs}(k+1) \end{bmatrix} = \underbrace{\begin{bmatrix} 1-\frac{R_s T_s}{L_s} & \omega_r(k)\,T_s \\ -\omega_r(k)\,T_s & 1-\frac{R_s T_s}{L_s} \end{bmatrix}}_{\mathbf{\Phi}(k)} \begin{bmatrix} i_{ds}(k) \\ i_{qs}(k) \end{bmatrix} + \underbrace{\begin{bmatrix} \frac{T_s}{L_s} & 0 \\ 0 & \frac{T_s}{L_s} \end{bmatrix}}_{\mathbf{\Gamma_b}} \underbrace{\begin{bmatrix} v^p_{ds}(k) \\ v^p_{qs}(k) \end{bmatrix}}_{\text{2L-VSR Model}}$$
$$+ \underbrace{\begin{bmatrix} 0 \\ -\frac{\omega_r(k)\,\psi_r\,T_s}{L_s} \end{bmatrix}}_{\mathbf{\Gamma_w}(k)}, \qquad \underbrace{\begin{bmatrix} v^p_{ds}(k) \\ v^p_{qs}(k) \end{bmatrix}}_{\text{2L-VSR Model}} = v_{dc}(k) \begin{bmatrix} s^p_{dr}(k) \\ s^p_{qr}(k) \end{bmatrix} \tag{1.14}$$

where T_s is sampling time. The subscript p denotes the predicted variable. As shown in the expression above, future generator currents are predicted based on the measured i_{ds}, i_{qs}, and v_{dc}, generator parameters R_s and L_s, and predicted 2L-VSR voltages v^p_{ds} and v^p_{qs}. The DT parameter matrix $\mathbf{\Gamma_b}$ can be calculated offline; however, $\mathbf{\Phi}(k)$ and $\mathbf{\Gamma_w}(k)$ must be computed online based on $\omega_r(k)$ measurement. The prediction of 2L-VSR voltages uses eight possible switching states (represented in a synchronous reference frame) and the measured DC-link voltage, v_{dc}. Based on this state-space model, the future behavior of the generator currents is controlled by properly choosing an optimal one among the eight possible switching states.

The control objective is to regulate i_{ds} and i_{qs} at their reference values i^*_{ds} and i^*_{qs}, respectively. As the final stage, a cost function is defined to fulfill the control objective:

$$g(k) = \left[\widehat{i}^*_{ds}(k+1) - i^p_{ds}(k+1)\right]^2 + \left[\widehat{i}^*_{qs}(k+1) - i^p_{qs}(k+1)\right]^2. \tag{1.15}$$

The ideal minimum of the cost function is zero that represents the perfect regulation of the generator currents. During each sampling instant, switching signals that minimize the cost function in (1.15) are chosen and applied to the 2L-VSR directly. Additional constraints such as generator current limitation during transient condition, switching frequency minimization, common-mode voltage mitigation, etc., can also be included in the cost function with suitable weighting factors. A fast dynamic response is guaranteed by PCC because it eliminates the linear PI controllers and SVM stage.

1.9.3 Comparison of Classical and Model Predictive Control

Classical and model predictive control schemes are compared based on abstract details given in the previous two sections and summarized, as shown in Table 1.6 [68]. The classical DTC and PTC are also included for the sake of comparison although these have not been analyzed in detail in this chapter. The digital control schemes are compared based on the nature of controller, critical design stage, requirement for modulation stage, control complexity, computational burden (number of online calculations), nature of converter switching frequency, and performance during transient conditions.

Switching frequency is constant with classical FOC, and all other control schemes operate with variable-switching frequency. The steady-state response obtained in all the methods is more or less same. The dynamic response obtained by predictive control schemes is better compared to classical FOC and DTC. In general, the model predictive control provides opportunities for a designer or industry to control wind generators and power converters in a simple manner.

Table 1.6 Summary of comparison between classical and predictive control schemes [68]

	Classical FOC	Classical DTC	Predictive FOC (PCC)	Predictive DTC (PTC)
Control Scheme	Figure 1.30	–	Figure 1.31	–
Nature of Controller	Linear	Nonlinear	Nonlinear	Nonlinear
Critical Design Stage	PI + Modulation	Lookup Table	Cost Function	Cost Function
Modulation Stage	PWM/SVM	Not Required	Not Required	Not Required
Control Complexity	Very High	High	Low	Low
Computational Burden	Moderate	Very High	High	High
Switching Frequency	Fixed	Variable	Variable	Variable
Transient Response	Moderate	Good	Excellent	Excellent

1.10 CONCLUDING REMARKS

In this chapter, a comprehensive review of high-power wind energy systems is presented with respect to cumulative and annual installed capacity, wind energy popularity around the globe, principle of wind kinetic energy to electric energy conversion, various classes of WT technologies, major mechanical and electrical components of MW-WECS, fixed- and variable-speed operation of WECS, overall control of WECS, main features and challenges of FCS-MPC, and classical and model predictive control of WECS. A detailed technical background on WT power curves, aerodynamic power regulation including pitch control, TSR concept, grid code requirements such as FRT and RPG, MPPT control schemes, grid integration, wind generator and power converter control is also presented.

Link to Next Chapters:

The material in this chapter will complement the in-depth technical analysis to be conducted in other chapters of this book. The generator–converter configurations presented in Section 1.5 are discussed further in Chapter 2 with respect to power converter topologies for Types 3 and 4 variable-speed WECS. The WECS control discussed in Section 1.7 is studied again in Chapters 3 and 4 with more focus on electrical control systems and the fundamentals of model predictive control. The IGs and SGs presented in Section 1.3.2 are further studied in Chapter 6 for modeling and model predictive control.

REFERENCES

1. V. Yaramasu, B. Wu, P. C. Sen, S. Kouro, and M. Narimani, "High-power wind energy conversion systems: State-of-the-art and emerging technologies," *Proceedings of the IEEE*, vol. 103, no. 5, pp. 740–788, May 2015.
2. Global Wind Energy Council (GWEC), "Global wind report: Annual market update," 2014, available at: http://www.gwec.net, accessed on April 2015.
3. Renewable Energy Policy Network for the 21st Centuray (REN21), "Renewables 2014: Global status report," 2014, available at: http://www.ren21.net, accessed on April 2015.
4. Navigant Research, "A BTM wind report: World wind energy market update," 2015, available at: http://www.navigantresearch.com, accessed on April 2015.
5. F. Blaabjerg and K. Ma, "Future on power electronics for wind turbine systems," *IEEE Journal of Emerging and Selected Topics in Power Electronics*, vol. 1, no. 3, pp. 139–152, September 2013.
6. J. Rodriguez and P. Coftes, *Predictive Control of Power Converters and Electrical Drives*, 1st ed. Chichester, UK: IEEE Wiley Press, March 2012.
7. L. Wang, S. Chai, D. Yoo, L. Gan, and K. Ng, *PID and Predictive Control of Electrical Drives and Power Converters using MATLAB/Simulink*. Singapore: Wiley-IEEE Press, March 2015.
8. V. Yaramasu, "Predictive control of multilevel converters for megawatt wind energy conversion systems," Ph.D. Dissertation, Ryerson University, Toronto, ON, Canada, 2014, available at: http://digital.library.ryerson.ca/islandora/object/RULA%3A3459.
9. C. Carrillo, A. Obando Montaño, J. Cidrás, and E. Díaz-Dorado, "Review of power curve modelling for wind turbines," *Renewable and Sustainable Energy Reviews*, vol. 21, pp. 572–581, 2013.
10. J. Earnest and T. Wizelius, *Wind Power Plants and Project Development*. New Delhi, India: PHI Learning, May 2011.
11. Z. Chen, J. Guerrero, and F. Blaabjerg, "A review of the state of the art of power electronics for wind turbines," *IEEE Transactions on Power Electronics*, vol. 24, no. 8, pp. 1859–1875, August 2009.
12. E. Muljadi and C. Butterfield, "Pitch-controlled variable-speed wind turbine generation," *IEEE Transactions on Industry Applications*, vol. 37, no. 1, pp. 240–246, January 2001.
13. B. Wu, Y. Lang, N. Zargari, and S. Kouro, *Power Conversion and Control of Wind Energy Systems*, 1st ed., ser. IEEE Press Series on Power Engineering. Hoboken, NJ: Wiley-IEEE Press, July 2011.
14. J. Guerrero, F. Blaabjerg, T. Zhelev, K. Hemmes, E. Monmasson, S. Jemei, M. Comech, R. Granadino, and J. Frau, "Distributed generation: Toward a new energy paradigm," *IEEE Industrial Electronics Magazine*, vol. 4, no. 1, pp. 52–64, March 2010.

15. P. Bresesti, W. Kling, R. Hendriks, and R. Vailati, "HVDC connection of offshore wind farms to the transmission system," *IEEE Transactions on Energy Conversion*, vol. 22, no. 1, pp. 37–43, March 2007.

16. The European Wind Energy Association (EWEA), "Offshore wind," 2014, available at: http://www.ewea.org, accessed on April 2015.

17. H. Li and Z. Chen, "Overview of different wind generator systems and their comparisons," *IET Renewable Power Generation*, vol. 2, no. 2, pp. 123–138, June 2008.

18. M. Liserre, R. Cardenas, M. Molinas, and J. Rodríguez, "Overview of multi-MW wind turbines and wind parks," *IEEE Transactions on Industrial Electronics*, vol. 58, no. 4, pp. 1081–1095, April 2011.

19. International Renewable Energy Agency (IRENA), "Renewable energy technologies: Cost analysis series – wind power," vol. 1: Power Sector, no. 5/5, June 2012, available at: http://www.irena.org/, accessed on April 2015.

20. European Commision, "2013 JRC wind status report: Technology, market and economic aspects of wind energy in Europe," available at: http://setis.ec.europa.eu/, accessed on April 2015.

21. H. Polinder, J. Ferreira, B. Jensen, A. Abrahamsen, K. Atallah, and R. McMahon, "Trends in wind turbine generator systems," *IEEE Journal of Emerging and Selected Topics in Power Electronics*, vol. 1, no. 3, pp. 174–185, September 2013.

22. J. A. Baroudi, V. Dinavahi, and A. M. Knight, "A review of power converter topologies for wind generators," *International Journal of Renewable Energy*, vol. 32, no. 14, pp. 2369–2385, 2007.

23. Z. Zhu and J. Hu, "Electrical machines and power-electronic systems for high-power wind energy generation applications: Part I – market penetration, current technology and advanced machine systems," *COMPEL: The International Journal for Computation and Mathematics in Electrical and Electronic Engineering*, vol. 32, no. 1, pp. 7–33, 2013.

24. F. Blaabjerg, M. Liserre, and K. Ma, "Power electronics converters for wind turbine systems," *IEEE Transactions on Industry Applications*, vol. 48, no. 2, pp. 708–719, March 2012.

25. F. Blaabjerg, R. Teodorescu, M. Liserre, and A. Timbus, "Overview of control and grid synchronization for distributed power generation systems," *IEEE Transactions on Industrial Electronics*, vol. 53, no. 5, pp. 1398–1409, October 2006.

26. W. Erdman and M. Behnke, *Low Wind Speed Turbine Project Phase II: The Application of Medium-Voltage Electrical Apparatus to the Class of Variable Speed Multi-Megawatt Low Wind Speed Turbines.* National Renewable Energy Laboratory (N.R.E.L.), Golden, CO, USA, 2012.

27. M. Tsili and S. Papathanassiou, "A review of grid code technical requirements for wind farms," *IET Renewable Power Generation*, vol. 3, no. 3, pp. 308–332, September 2009.

28. E.ON Netz Gmbh, "Grid code – high and extra high voltage," April 2006.

29. F. Iov, A. D. Hansen, P. Sorensen, and N. A. Cutululis, "Mapping of grid faults and grid codes," Riso National Laboratory, Technical University of Denmark, Roskilde, Denmark, Technical Report Riso-R-1617(EN), July 2007.

30. R. Teodorescu, M. Liserre, and P. Rodriguez, *Grid Converters for Photovoltaic and Wind Power Systems.* Chichester, UK: Wiley-IEEE Press, January 2011.

31. S. Papathanassiou and M. Papadopoulos, "Mechanical stresses in fixed-speed wind turbines due to network disturbances," *IEEE Transactions on Energy Conversion*, vol. 16, no. 4, pp. 361–367, December 2001.

32. M. Hossain, H. Pota, V. Ugrinovskii, and R. Ramos, "Simultaneous STATCOM and pitch angle control for improved LVRT capability of fixed-speed wind turbines," *IEEE Transactions on Sustainable Energy*, vol. 1, no. 3, pp. 142–151, October 2010.

33. R. Pena, J. Clare, and G. Asher, "Doubly fed induction generator using back-to-back PWM converters and its application to variable-speed wind-energy generation," *IET Electric Power Applications*, vol. 143, no. 3, pp. 231–241, May 1996.

34. S. Muller, M. Deicke, and R. De Doncker, "Doubly fed induction generator systems for wind turbines," *IEEE Industry Applications Magazine*, vol. 8, no. 3, pp. 26–33, May 2002.

35. J. Ekanayake and N. Jenkins, "Comparison of the response of doubly fed and fixed-speed induction generator wind turbines to changes in network frequency," *IEEE Transactions on Energy Conversion*, vol. 19, no. 4, pp. 800–802, December 2004.

36. R. Cardenas, R. Pena, S. Alepuz, and G. Asher, "Overview of control systems for the operation of DFIGs in wind energy applications," *IEEE Transactions on Industrial Electronics*, vol. 60, no. 7, pp. 2776–2798, July 2013.

37. J. Carrasco, L. Franquelo, J. Bialasiewicz, E. Galvan, R. Guisado, M. Prats, J. Leon, and N. Moreno-Alfonso, "Power-electronic systems for the grid integration of renewable energy sources: A survey," *IEEE Transactions on Industrial Electronics*, vol. 53, no. 4, pp. 1002–1016, June 2006.

38. J. Conroy and R. Watson, "Frequency response capability of full converter wind turbine generators in comparison to conventional generation," *IEEE Transactions on Power Systems*, vol. 23, no. 2, pp. 649–656, May 2008.

39. A. S. Mikhail, K. L. Cousineau, L. H. Howes, E. William, and H. William, "Variable speed distributed drive train wind turbine system," May 2006, United States Patent, US 7,042,110 B2.

40. J. Kang, N. Takada, E. Yamamoto, and E. Watanabe, "High power matrix converter for wind power generation applications," in *International Conference on Power Electronics and Energy Conversion Congress and Exposition Asia*, Jeju, South Korea, June 2011, pp. 1331–1336.

41. S. Kouro, M. Malinowski, K. Gopakumar, J. Pou, L. Franquelo, B. Wu, J. Rodríguez, M. Perez, and J. Leon, "Recent advances and industrial applications of multilevel converters," *IEEE Transactions on Industrial Electronics*, vol. 57, no. 8, pp. 2553–2580, August 2010.

42. G. Abad, J. Lopez, M. Rodriguez, L. Marroyo, and G. Iwanski, *Doubly Fed Induction Machine: Modeling and Control for Wind Energy Generation Applications*, ser. IEEE Press Series on Power Engineering. Wiley-IEEE Press, 2011.

43. L. Y. Pao and K. Johnson, "Control of wind turbines," *IEEE Control Systems Magazine*, vol. 31, no. 2, pp. 44–62, April 2011.

44. S. Musunuri and H. Ginn, "Comprehensive review of wind energy maximum power extraction algorithms," in *IEEE Power and Energy Society (PES) General Meeting*, San Diego, CA, USA, July 2011, pp. 1–8.

45. M. Abdullah, A. Yatim, C. Tan, and R. Saidur, "A review of maximum power point tracking algorithms for wind energy systems," *Renewable and Sustainable Energy Reviews*, vol. 16, no. 5, pp. 3220–3227, 2012.

46. E. Koutroulis and K. Kalaitzakis, "Design of a maximum power tracking system for wind-energy-conversion applications," *IEEE Transactions on Industrial Electronics*, vol. 53, no. 2, pp. 486–494, April 2006.

47. K. Tan and S. Islam, "Optimum control strategies in energy conversion of PMSG wind turbine system without mechanical sensors," *IEEE Transactions on Energy Conversion*, vol. 19, no. 2, pp. 392–399, June 2004.

48. P. Cortes, M. Kazmierkowski, R. Kennel, D. Quevedo, and J. Rodríguez, "Predictive control in power electronics and drives," *IEEE Transactions on Industrial Electronics*, vol. 55, no. 12, pp. 4312–4324, December 2008.

49. A. Linder, R. Kanchan, R. Kennel, and P. Stolze, *Model-Based Predictive Control of Electric Drives*. Germany: Cuvillier Verlag Göttingen, 2010.

50. M. Kazmierkowski and L. Malesani, "Current control techniques for three-phase voltage-source PWM converters: a survey," *IEEE Transactions on Industrial Electronics*, vol. 45, no. 5, pp. 691–703, October 1998.

51. V. Yaramasu, B. Wu, S. Alepuz, and S. Kouro, "Predictive control for low-voltage ride-through enhancement of three-level-boost and NPC-converter-based PMSG wind turbine," *IEEE Transactions on Industrial Electronics*, vol. 61, no. 12, pp. 6832–6843, December 2014.

52. J. H. Lee, "Model predictive control: Review of the three decades of development," *International Journal of Control, Automation and Systems*, vol. 9, no. 3, pp. 415–424, 2011.

53. J. Rodríguez, J. Pontt, C. A. Silva, P. Correa, P. Lezana, P. Cortes, and U. Ammann, "Predictive current control of a voltage source inverter," *IEEE Transactions on Industrial Electronics*, vol. 54, no. 1, pp. 495–503, February 2007.

54. S. Kouro, P. Cortes, R. Vargas, U. Ammann, and J. Rodríguez, "Model predictive control-A simple and powerful method to control power converters," *IEEE Transactions on Industrial Electronics*, vol. 56, no. 6, pp. 1826–1838, June 2009.

55. S. Vazquez, A. Marquez, R. Aguilera, D. Quevedo, J. Leon, and L. Franquelo, "Predictive optimal switching sequence direct power control for grid-connected power converters," *IEEE Transactions on Industrial Electronics*, vol. 62, no. 4, pp. 2010–2020, April 2015.

56. L. Tarisciotti, P. Zanchetta, A. Watson, J. Clare, M. Degano, and S. Bifaretti, "Modulated model predictive control for a three-phase active rectifier," *IEEE Transactions on Industry Applications*, vol. 51, no. 2, pp. 1610–1620, March 2015.

57. M. Rivera, M. Perez, V. Yaramasu, B. Wu, L. Tarisciotti, P. Zanchetta, and P. Wheeler, "Modulated model predictive control (M^2PC) with fixed switching frequency for an NPC converter," in *International Conference on Power Engineering, Energy and Electtical Drives (POWERENG)*, Riga, Latvia, May 2015, pp. 623–628.

58. J. Rubinic, V. Yaramasu, B. Wu, and N. Zargari, "Model predictive control of neutral-point clamped inverter with harmonic spectrum shaping," in *IEEE Energy Conversion Congress and Exposition (ECCE)*, Montreal, Canada, September 2015, pp. 717–722.

59. R. Aguilera, P. Lezana, and D. Quevedo, "Switched model predictive control for improved transient and steady-state performance," *IEEE Transactions on Industrial Informatics*, vol. 11, no. 4, pp. 968–977, August 2015.

60. G. Buja and M. Kazmierkowski, "Direct torque control of PWM inverter-fed AC motors - a survey," *IEEE Transactions on Industrial Electronics*, vol. 51, no. 4, pp. 744–757, August 2004.

61. D. Casadei, F. Profumo, G. Serra, and A. Tani, "FOC and DTC: two viable schemes for induction motors torque control," *IEEE Transactions on Power Electronics*, vol. 17, no. 5, pp. 779–787, September 2002.

62. E. Fuentes, D. Kalise, J. Rodriguez, and R. Kennel, "Cascade-free predictive speed control for electrical drives," *IEEE Transactions on Industrial Electronics*, vol. 61, no. 5, pp. 2176–2184, May 2014.

63. J. Rodríguez, M. P. Kazmierkowski, J. R. Espinoza, P. Zanchetta, H. Abu-Rub, H. A. Young, and C. A. Rojas, "State of the art of finite control set model predictive control in power electronics," *IEEE Transactions on Industrial Informatics*, vol. 9, no. 2, pp. 1003–1016, May 2013.

64. V. Yaramasu, B. Wu, and J. Chen, "Model-predictive control of grid-tied four-level diode-clamped inverters for high-power wind energy conversion systems," *IEEE Transactions on Power Electronics*, vol. 29, no. 6, pp. 2861–2873, June 2014.

65. P. Cortes, S. Kouro, B. La Rocca, R. Vargas, J. Rodríguez, J. Leon, S. Vazquez, and L. Franquelo, "Guidelines for weighting factors design in model predictive control of power converters and drives," in *IEEE International Conference on Industrial Technology (ICIT)*, Gippsland, VIC, Australia, February 2009, pp. 1–7.

66. H. Miranda, P. Cortes, J. Yuz, and J. Rodríguez, "Predictive torque control of induction machines based on state-space models," *IEEE Transactions on Industrial Electronics*, vol. 56, no. 6, pp. 1916–1924, June 2009.

67. M. Chinchilla, S. Arnaltes, and J. Burgos, "Control of permanent-magnet generators applied to variable-speed wind-energy systems connected to the grid," *IEEE Transactions on Energy Conversion*, vol. 21, no. 1, pp. 130–135, March 2006.

68. J. Rodríguez, M. Rivera, J. Kolar, and P. Wheeler, "A review of control and modulation methods for matrix converters," *IEEE Transactions on Industrial Electronics*, vol. 59, no. 1, pp. 58–70, January 2012.

CHAPTER 2

REVIEW OF GENERATOR–CONVERTER CONFIGURATIONS FOR WECS

2.1 INTRODUCTION

Five types of wind energy configurations have been analyzed in the previous chapter. Type 1 and 5 wind energy conversion systems (WECS) are directly connected to the grid without any power electronic converter interfaces (except for soft start and rotor field excitation). A Type 2 WECS uses a converter-controlled external resistor to achieve variable-speed operation, but in a limited speed range (10%). Type 1, 2, and 5 technologies are either outdated or not popularly used in the wind energy market. Semi-variable-speed Type 3 wind turbines (WTs) dominate the current wind energy market with more than 50% market share. Full-variable-speed Type 4 turbines hold the second highest market share and are expected to dominate the market in the coming years [1].

Power electronic converters are key components in Type 3 and 4 WECS. A Type 3 WECS uses a partial-scale (30%) power converter, whereas a Type 4 WECS employs a full-scale (100%) power converter. For example, a 2.5 MW Type 3 WECS requires only a 0.75-MW power converter, whereas a 2.5-MW Type 4 WECS needs a 2.5-MW power converter. Power electronics technology has been developed rigorously to decrease the cost of energy (COE), increase wind energy conversion efficiency, reliability, and power density, and comply with stringent grid codes. Several commercial and promising power converter solutions are available at different power levels for Type 3 and 4 WECS. Technical, operational, and legal requirements, such as maximum power point tracking (MPPT), generator speed/torque control, smooth grid connection, grid active/reactive power control, and fault-ride through, are fulfilled by the proper control of power converters.

Model Predictive Control of Wind Energy Conversion Systems, First Edition. Venkata Yaramasu and Bin Wu.

Companion Website: www.wiley.com/go/yaramasu_wu_mpc_wecs

The doubly fed induction generator (DFIG) is a standard solution for Type 3 WECS. The Type 4 WECS uses either a squirrel-cage induction generator (SCIG), permanent magnet synchronous generator (PMSG), or wound rotor synchronous generator (WRSG). The high-temperature super conducting synchronous generator (HTS-SG) is a new class of wind generator that is favorable for WTs rated above 6.0 MW. This chapter considers DFIG, SCIG, PMSG, and WRSG to analyze the power converter configurations. The HTS-SG is omitted from the discussion; however, the high-power converter configurations analyzed with PMSG and WRSG are also applicable to HTS-SG. Induction generators, such as SCIG and DFIG, operate at high rotational speeds; thus, they need a three-stage gearbox to couple to the WT main shaft. Synchronous generators can operate at low, medium, or high speed; thus, their gear ratio and number of gear stages vary according to the speed of operation. No gearbox is needed for low-speed applications. This is referred to as a direct-drive or gearless solution.

Various power conversion configurations can be obtained for WECS by combining wind generators with power converters. The current technology uses power electronics at WT and wind farm levels for energy conversion and grid integration. This chapter reflects the research and development activities by industry and academia and provides a detailed review on wind energy generator–converter configurations at the WT level. All commercial and promising configurations, including more complex topologies, are classified into four categories: back-to-back (BTB) connected power converters, passive generator-side (PGS) power converters, converters for multiphase wind generators, and converters without an intermediate DC link. These configurations are investigated for high-power WECS and are presented in low-voltage (LV) and medium-voltage (MV) categories. The power converters are analyzed on the basis of component count, modularity, reliability, converter/control complexity, device voltage stress, voltage of operation, and achievable power levels.

The model predictive control of wind energy systems requires the model of both the wind generator and power converter to develop the control scheme. This chapter not only provides the basics for wind energy, but also provides enough background to select wind generator-power converter configurations for modeling and predictive control. The most standard generator–converter configurations are analyzed in the subsequent chapters for model predictive control.

Chapter Overview

- The technical and operational requirements for MW-class wind energy power converters are presented in Section 2.2. This section provides the motivation behind the development of different power converter configurations.
- The classification of power converters for variable-speed WECS is discussed in Section 2.3. The four classes of power converter configurations are briefly introduced.
- BTB connected and PGS power converters are presented in Sections 2.4 and 2.5, respectively. These power converters are further classified into LV and MV categories.
- Section 2.6 depicts the power converters for multiphase wind generators including six phase and open-winding generators. These power converters are further classified as isolated/parallel/distributed and hybrid/cascaded. Single-stage power converters without intermediate DC link are discussed in Section 2.7.
- Section 2.8 presents the concluding remarks. The technical merits and demerits of the four categories of wind energy power converters are also tabulated.

2.2 REQUIREMENTS FOR POWER CONVERTERS IN MW-WECS

Power converters in first generation of WTs (Type 1) are used only for smooth grid connections. These power converters are disconnected from the circuit once the generator is tied to the grid. However, modern Type 3 and 4 WECS require power converters to meet the technical, operational, and legal requirements listed below [1–3]:

- **Initial Cost:** This factor has prime importance in achieving low COE and in competing with other energy sources. The initial cost of a power converter is only a fraction (approximately 7–12%) of the overall WT cost. Although the cost is a small part, huge cost savings can be accomplished for a wind farm consisting of hundreds of WTs.
- **Reliability, Modularity, and Maintenance Cost:** In addition to the initial cost of a power converter, the maintenance cost (i.e., replacement cost for components and salary for technicians) must be minimal to achieve low COE. According to the latest reports on turbine faults, electric generators and power converters are on the top of the list with an average failure rate of 13% to 20%. Power converter faults increase the downtime of WT operation and the overall COE. Thus, power converters for WTs, particularly offshore turbines, must have high reliability. Power converters with a modular structure are preferable because even if one power converter fails, the WT can still work with reduced capacity and mitigate downtime.
- **Efficiency:** At the MW power level, efficiency is an important factor in reducing COE. Even a 1% improvement of power converter efficiency saves millions of dollars in a wind farm, which consists of hundreds of power-converter-based WTs. The losses in power converter are categorized as conduction and switching losses. Conduction losses are caused by the switching device type, whereas switching losses are caused by the modulation/control scheme. Power losses, which directly affect efficiency, should be minimized by using highly efficient switching devices, an optimal arrangement of switching devices (converter topology), a cooling system, and modulation/control schemes. Low switching frequency operation is an important requirement at the MW power level to minimize switching losses and allow proper heat dissipation.
- **Power Quality:** Power quality is attributed to the various parameters of a power converter. The output voltage waveform must be similar to the sinusoidal waveform, and this parameter is also described as the number of steps in the output voltage waveform (dv/dt). The increase in the number of steps in the waveform corresponds to the decrease of dv/dt and output harmonic filter requirement. Moreover, the electromagnetic interference decreases with decreasing dv/dt. The total harmonic distortion (THD) of generator and grid currents must be low to decrease the generator shaft oscillations and feed quality currents to the utility grid, respectively.
- **Common Mode Voltage:** The fast switching actions of semiconductor devices in a power converter also generate common-mode voltage (CMV), in addition to high dv/dt. CMV is undesirable because it leads to the premature failure of wind generator winding insulation. CMV in odd-level (e.g., three-level and five-level) power converters can be completely mitigated by the proper selection of the modulation and control schemes. However, CMV in even-level (e.g., two-level and four-level) power converters cannot be made zero unless certain external hardware is used. The power converter configuration and its associated control scheme must suppress or minimize CMV to increase the life expectancy of the wind generator.

- **Grid Code Compliance:** This factor is one of the prime requirements for grid-connected MW WTs. Utility standards, such as IEEE 519-1992, dictate strict guidelines for gird current harmonic regulation. Power converters must feed currents to the grid with low THD (i.e., less than 5%), provide reactive power whenever requested by the grid operator, provide ride-through during grid faults, and provide voltage/frequency support, among other requirements. The power converter must accomplish these requirements without requesting support from external hardware/components, such as static VAR compensator (SVC), static synchronous compensator (STATCOM), or FACTS.
- **Footprint and Weight:** A limited space is available in the nacelle of WTs unlike in electric drives. A power converter (also an electric generator) must have high power density to achieve a small footprint and weight. This is a major requirement, particularly for offshore WTs.
- **Cable Size and Losses:** The generator/converter AC output is connected to the step-up transformer/substation through AC cables. The typical hub heights for modern WTs are in the range of 60 m to 150 m; thus, the cost of cables and associated losses are high. A power conversion system must take this factor into account and decrease cable cost and losses to the maximum possible extent.

An ideal power converter must possess all the above features; however, designing such a power converter is impossible. An engineering approach would be to satisfy the most important requirements while sacrificing the least important conditions. A best-selling power converter obviously incorporates most of the above features for WTs.

2.3 OVERVIEW OF POWER CONVERTERS FOR WECS

In pursuit of achieving the aforementioned requirements, various generator–converter configurations have been developed by WT manufacturers and their supporting power converter companies. On the other hand, some new WT manufacturers have introduced novel power conversion topologies to differentiate themselves from existing turbine/converter manufacturers. The classification of state-of-the-art WECS power converters is a complex subject matter. Classifying all converters on the basis of one parameter/operation is impossible. In this book, the generator–converter configurations are classified into four different groups (see Figure 2.1) to facilitate easier discussion. A brief introduction to each converter category is given below:

- **Back-to-Back Connected Power Converters:** Power converters that are identical on the generator-side and grid-side and are linked through a DC link are classified as BTB connected converters. BTB converters include both voltage source converters (VSCs) and current source converters (CSCs), with a DC link composed of capacitors or inductors. BTB converters convert the variable voltage/frequency output of the generator to DC and then DC to AC with a fixed voltage/frequency for the grid connection. Considering that power conversion has two stages (i.e., AC/DC and DC/AC), BTB converters are often referred to as two-stage power converters. The power flow is also bidirectional: generator to the grid and vice versa. Thus, BTB converters can be used with SCIG, DFIG, PMSG, or WRSG. The topology, construction, modulation, and control of a machine (generator)-side converter (MSC) (alternatively called as rotor-side converter (RSC) in Type 3 WECS) and a grid-side converter (GSC) are the same. Hence, the production and operation of these converters are simple and convenient. BTB power converters are popularly used in current Type 3 and 4 WTs and hold the highest market share.

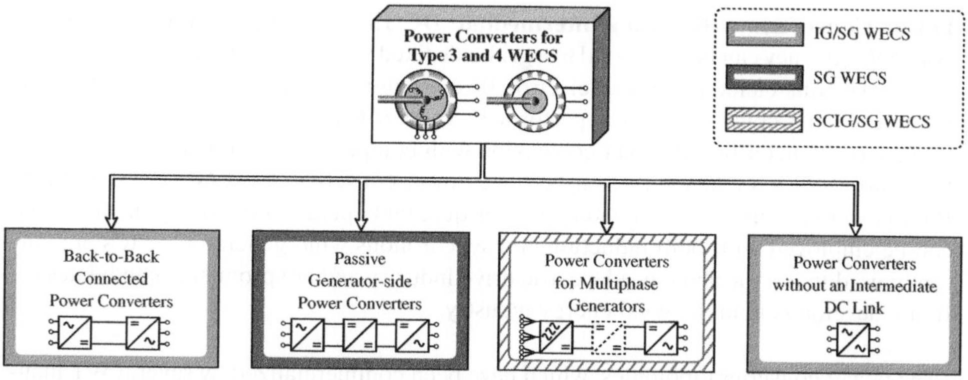

Figure 2.1 Classification of power converters for variable-speed wind energy systems.

- **Passive Generator-side Power Converters:** BTB converters enable a four-quadrant operation for wind generators. However, the power flow in WECS is unidirectional, that is, from the generator to the grid. Therefore, the generator-side AC to DC conversion can be obtained by diode-bridge rectifiers (often referred to as PGS converters) instead of pulse-width-modulated (PWM) active converters. Diode-bridge rectifiers are less expensive and inherently more reliable than PWM converters. The rotor flux in PMSG and WRSG is generated by permanent magnets and rotor field excitation, respectively. Thus, the generator-side power conversion system in PMSG/WRSG WTs can be realized by using diode-bridge rectifiers [4, 5]. Induction generators, such as SCIG and DFIG, require magnetizing current during its operation; thus, this type of generators cannot allow diode-bridge rectifiers on the generator side. A PGS power conversion system employing a diode-bridge rectifier also uses a DC/DC converter to enable MPPT operation; thus, they are noted as three-stage power converters (i.e., AC/DC + DC/DC + DC/AC). PGS converters have been used in practical WECS ranging from a few kWs to MWs but with a limited market share at the MW power level.

- **Power Converters for Multiphase Wind Generators:** Parallel power converters increase power handling capacity but with the drawbacks of circulating currents and the subsequent power converter derating. The wind energy industry also promotes multiphase wind generators and distributed converters as an alternative approach to increase power handling capacity [6]. Issues with circulating currents are surpassed by multiphase wind generators because the windings provide inherent electrical isolation. A generator that has two sets of three-phase windings is denoted as a six-phase generator. Generators with more than six phases are called open-winding generators. The insulation requirement also decreases with multiphase generators because the power handling capacity of each set of windings is only a fraction of the total power to be handled. Moreover, multiphase wind generators can be constructed with a phase shift between the windings, such that low-order harmonics in the generator currents are cancelled. For example, Commercial WTs, `Vensys` V70/77, and `Goldwind` GW70/77 employ a six-phase PMSG with 30° phase shift to cancel 5th and 7th harmonics in the generator currents. The low THD for generator currents leads to a small or no harmonic filter in MSC. In practical WTs, both the BTB and PGS power converters are employed with multiphase wind generators. The market share of multiphase wind generators is limited because they require special construction processes and involve high initial costs.

- **Power Converters without an Intermediate DC Link:** The generator-side variable voltage/frequency can be converted to a grid-side fixed voltage/frequency without involving any DC link elements such as electrolytic capacitors which are bulky and have short lifetime. Matrix converters (MCs) perform direct AC/AC power conversion, thus leading to increased silicon-based power conversion with compact size and high efficiency. For this reason, MCs are labeled as single-stage power converters (i.e., AC/AC). Similar to BTB power converters, MCs enable the four-quadrant operation of wind generators and can be employed with both induction and synchronous wind generators. MCs are currently used in the variable-speed electric drive industry and are promising candidates for future development in the wind energy industry.

The power converter topologies, which have been commercialized by several WT manufacturers and proposed in literature with promising features, have the following four distinct categories (Figure 2.2): LV power converters, MV power converters, bidirectional power flow converters, and unidirectional power flow converters. LV class power converters are rated for voltages below 1 kV, whereas MV class power converters are rated for voltages within 3000–4000 V. The LV and MV power converters are usually employed with LV and MV wind generators, respectively. For example, a 690-V power converter is used with a 690-V Type 4 wind generator. Bidirectional power flow converters include the BTB, matrix, and some multiphase converters. Unidirectional power flow converters consist mainly of PGS converters, some of which are employed with multiphase generators.

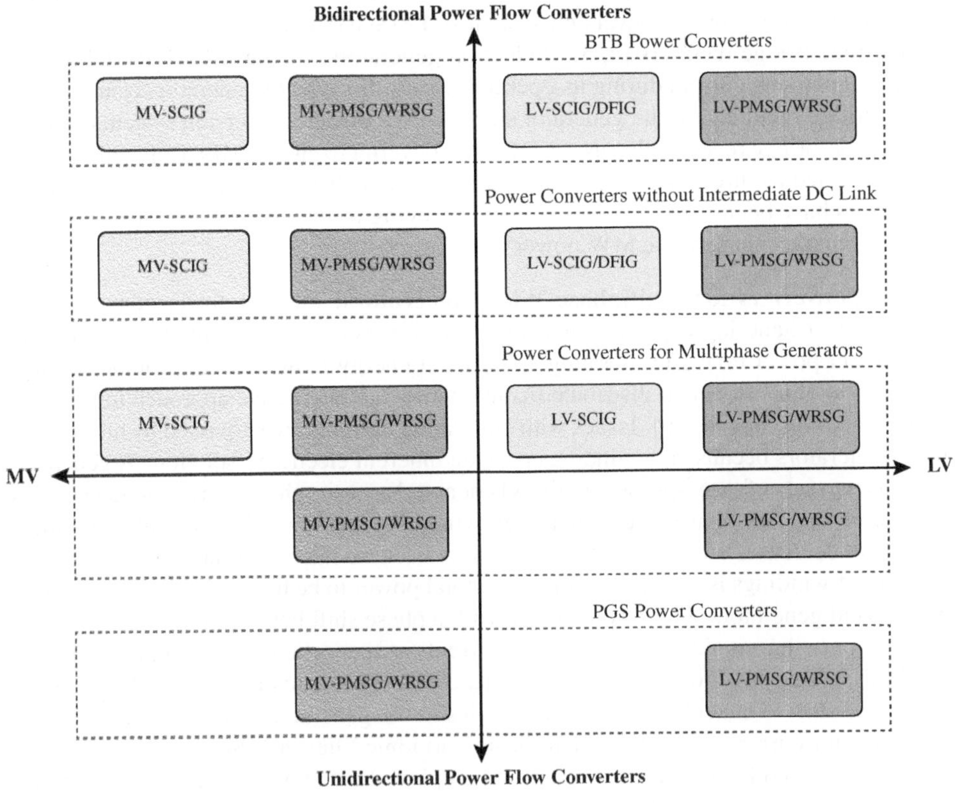

Figure 2.2 Classification of power converters with respect to voltage of operation and power flow.

LV power converters are efficient and cost effective at power levels lower than 3.0 MW. The MV operation of WECS is the most suitable, reliable, and economical approach for turbine power ratings greater than 3.0 MW [7]. The MV operation is a mature technology in the electric drive industry [8]. However, WT manufacturers are reluctant to move from LV to MV technology because of the limited availability of MV generators and the lack of available information on the MV operation of turbines. In [9], a detailed cost analysis has been conducted between the LV and MV operations of WTs. The results show that MV operation decreases COE production by 2% to 4%. Only four MV WTs are currently operational. However, seven manufacturers have announced that their future projects are based on MV technology. The four categories of power converters are available in both LV and MV classes. The wind generators applicable for each category of power converters are also shown in Figure 2.2. The key notes regarding the application of wind generators in high-power WECS are given below:

- SCIGs, PMSGs, and WRSGs are available in the MV category. However, DFIGs are currently limited to LV rotor voltages only. In this regard, DFIGs are omitted in the analysis of MV power converters.
- The realization of multiphase wind generators is possible with SCIGs, PMSGs, and WRSGs. The construction of DFIGs with multiphase rotor windings is complicated. Thus, DFIGs are unfavorable candidates as multiphase wind generators.
- SCIG and DFIG are only employed with bidirectional power flow converters because such generators draw magnetizing current from the grid during startup.

Semiconductor switching devices are key elements in any power converter. The evolution and technological improvements in semiconductor devices directly affect the power converter technology and their key operating characteristics (i.e., such as cost, reliability, efficiency, power density, operating voltage levels, and power limits) [10]. In the wind energy industry, the insulated gate bipolar transistor (IGBT) is dominantly used for LV and MV power converters. The asymmetrical integrated gate-commutated thyristor (IGCT) is used for MV VSC, whereas the symmetrical gate-commutated thyristor (SGCT) is used in CSC. Notably, gate turn-off thyristors applicable for the CSC are outdated and currently being replaced by SGCTs. The gate driver complexity is lower with IGBT than with IGCT and SGCT. Moreover, IGBTs can operate at higher switching frequencies than GCT switching devices, and increase power quality and power density [11]. IGBTs are available as modules and press packs; however, GCTs are available only in the form of press packs. IGBTs are connected BTB with a common emitter or common collector to form a bidirectional switch. They are often referred to as reverse blocking IGBT (RB-IGBT) and used as a switching device to realize the MC.

Next-generation power semiconductor devices are based on wide bandgap materials. These devices have substantial performance improvements over their silicon-based counterparts. These devices are also able to operate at higher temperatures, power densities, voltages, and switching frequencies than their silicon-based counterparts. The two wide bandgap materials for power applications are gallium nitride (GaN) and silicon carbide (SiC). The SiC devices have exhibited success at higher voltages (1200 V or higher) and are smaller than silicon-based IGBTs. GaN devices are available at low voltages and are expensive compared with Si- and SiC-based switching devices. GaN and SiC devices can have a positive impact on the next-generation high-power and high-voltage wind energy power converters.

The main features and drawbacks of each power converter configuration are discussed in the following sections, with the important observations tabulated in Section 2.8.

2.4 BACK-TO-BACK CONNECTED POWER CONVERTERS

The BTB connected four-quadrant power converters, realized as MSC/RSC + DC link + GSC, are the dominant technology not only in wind energy but in many power conversion applications. Different BTB converters that can be used in variable-speed WECS are summarized in Figure 2.3. BTB converters are broadly classified as LV and MV converters. The LV class includes VSC, whereas the MV class includes both VSC and CSC. As depicted in Figure 2.3, switching devices such as LV-IGBT, MV-IGBT, IGCT, and SGCT are used as building blocks for these power converters. LV converters can be employed with all wind generators, but MV converters are only applicable for SCIG, PMSG, and WRSG.

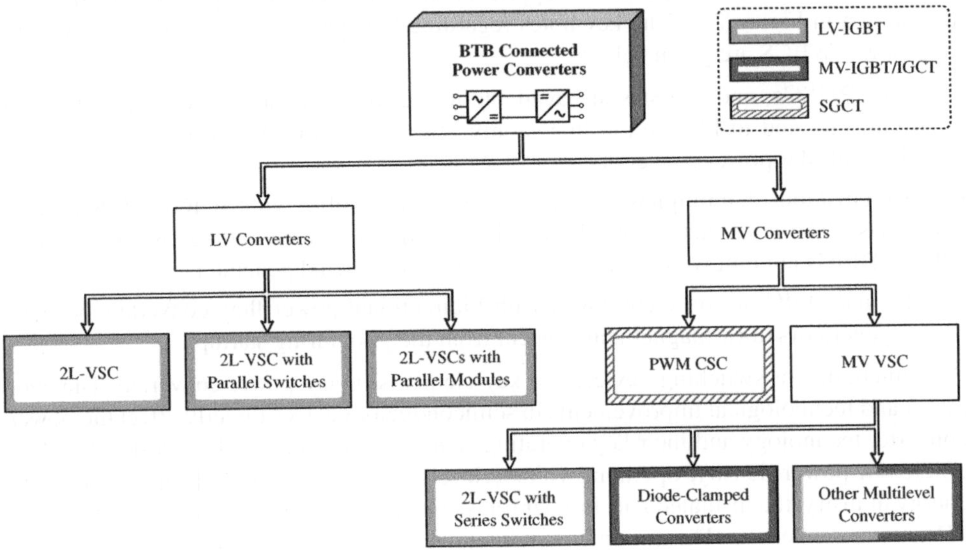

Figure 2.3 Classification of BTB connected power converters for variable-speed WECS.

2.4.1 Low-Voltage BTB Converters

In this subsection, four LV converter configurations are analyzed. The standard voltages used by many commercial WT manufacturers for the LV grid connection are 690 and 575 V. Therefore, the wind generator, power converter, and primary side of step-up transformer are designed for 690 or 575 V. The other voltages being used by very few LV WT manufacturers are 600, 660, 750, 850, and 950 V.

(1) Full-Scale BTB 2L-VSC: A typical Type 4 WECS using a full-scale BTB connected two-level (2L) VSC is shown in Figure 2.4. The MSC and GSC are realized by 2L voltage source rectifier (2L-VSR) and 2L voltage source inverter (2L-VSI), respectively. The MSC and GSC are linked by a DC-link capacitor. The 2L-VSR and 2L-VSI are realized by LV-IGBTs arranged in matrix form. The DC link is implemented by using series/parallel string of electrolytic capacitors to achieve the required voltage and capacitance level. Moreover, the DC link provides decoupling between the generator and grid. Thus, the transients in the generator do not appear on the grid side. Wind generators, such as PMSG [12], WRSG [13], and SCIG [14], are used with this configuration.

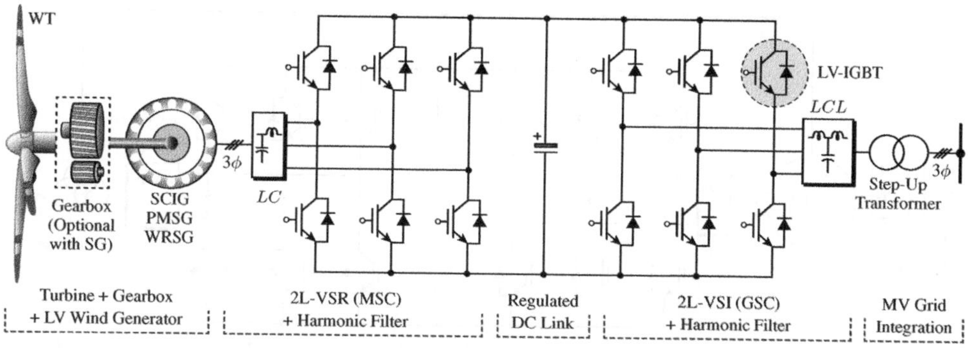

Figure 2.4 Type 4 WECS with SCIG/SG and full-scale (100%) BTB connected 2L-VSC.

The power rating of the BTB converter is usually equal to the generator output power. A 0.75-MW electric generator is connected to the grid through a 0.75-MW power converter. To meet the recent grid codes such as reactive power generation, the GSC is designed with higher MVA capacity than the MSC. The generator torque and speed are controlled by MSC, whereas GSC controls the net DC-bus voltage and grid reactive power. The net DC-bus voltage is maintained higher than the peak of grid line-to-line voltage to ensure the proper operation of the GSC. The switching frequency of VSR and VSI is maintained at 1–3 kHz to achieve low switching losses and high power density [15].

The grid current contains high THD, and LCL filters are used on the grid-side to meet the grid codes. To reduce harmonic distortion in the generator currents, an LC filter is used with L and C respectively connected to the VSR and generator terminals. The entire power converter including the generator-side filter, VSR, DC link, VSI, and grid-side harmonic filter are packed in a cabinet and placed in the nacelle. The output of the grid-side LCL filter is connected through three-phase AC cables to the step-up transformer located at the bottom of the tower. The AC cables present significant cost and losses because they are rated for LV and high current operation. The cost of these converters is low because of the mass production. One of the most widely used commercial VSC modules is SKiiP by `Semikron`, which is equipped with its own heat sink, semiconductor switching devices, and gate drivers and feature a compact design with high power density [16]. This configuration is mature in terms of technology status and market penetration and is used by about 90% of Type 4 WTs rated below 0.75-MW power level.

(2) Partial-Scale BTB 2L-VSC: Type 3 WECS with a partial-scale BTB VSC, often realized as RSC + DC-link + GSC, is shown in Figure 2.5. The stator of the generator is directly connected to the grid via step-up transformer, whereas the rotor of the generator is connected to the utility grid through a power converter and step-up transformer [17, 18]. Both the stator and rotor winding are designed for LV operation; however, the rotor voltages are only a fraction of stator voltages. Therefore, the turns ratio used in step-up transformer primary windings connected to the DFIG stator and rotor terminals are different. The power rating of the converter is usually 30% of the generator-rated power. For example, a 2.5-MW DFIG requires only a 0.75-MW power converter. The BTB converters in the rotor circuit enable bidirectional power flow; thus, the power can flow from the rotor to the grid or vice versa. The RSC in DFIG WECS controls the generator active and reactive power, whereas DC-link voltage is controlled by the GSC [19]. The grid power factor is solely adjusted by the RSC through the control of DFIG reactive power.

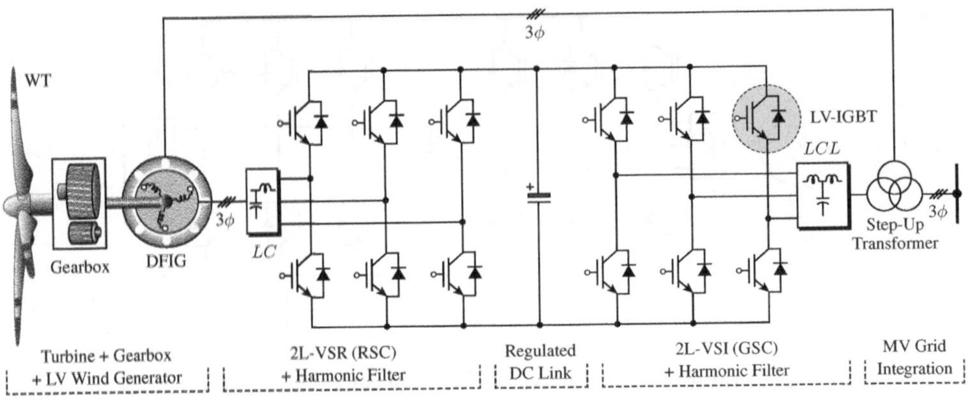

Figure 2.5 Type 3 WECS with DFIG and partial-scale (30%) BTB connected 2L-VSC.

The speed range achievable is decided according to the power converter rating. A power converter with 30% rated capacity can control the DFIG speed in ±30% around synchronous speed, which is sufficient to perform variable-speed operation. The use of partial-scale converter decreases the size and weight of Type 3 turbine in comparison with Type 4 turbine. The reduced size and weight of power converter allow more room in nacelle. Thus, the step-up transformer can be placed in the nacelle to decrease cable cost and losses. This approach has been applied in the ECO110 model of `Alstom` WTs. The FRT operation of Type 3 WTs is complicated and challenging compared with Type 4 WTs and significant research has been conducted in this area [20]. This is a de facto configuration used in hundreds of commercial Type 3 WTs.

(3) Parallel BTB 2L-VSCs with Common DC Link: For power ratings greater than 0.75 MW in Type 4 turbines, the current carrying (power handling) capability can be increased by connecting the three-phase VSCs in parallel along with harmonic filters. For example, two 2L-VSC modules are connected in parallel to achieve a power rating of 1.5 MW in Type 4 turbines (see Figure 2.6). Similarly, a 5.0-MW Type 3 WT needs to process the slip power of 1.5 MW. This can be accomplished by connecting two 2L-VSCs in parallel in the rotor circuit of DFIG.

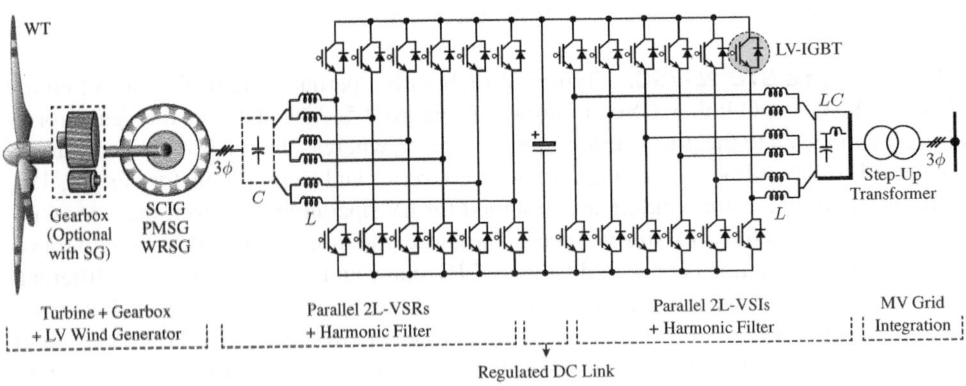

Figure 2.6 Type 4 WECS with two VSCs in parallel for high current/power applications (common DC-link configuration; also applicable for Type 3 WECS).

Figure 2.6 only shows the configuration for Type 4 WECS. However, a similar interpretation can be made to Type 3 WECS by connecting the stator directly to the grid and the power converter in the rotor circuit. More converter modules are connected in parallel to increase the power ratings. The DC link is configured as a common element for all converters to reduce cost and space [21]. This configuration offers energy efficiency and redundancy. For example, one or more converters can be turned-off to achieve high system efficiency when the wind speed is low. When a converter fails, other converters can still deliver the power but with reduced capacity. Moreover, the equivalent converter switching frequency is increased by operating the converters in interleaving (phase-shifting) mode, thus producing less THD in the generator- and grid-side currents.

Circulating currents exist in both the MSCs and GSCs because of a mismatch in the converter and grid-side filter parameters. This issue should be considered in the design of the controller. On the generator side, L filters are connected between each converter to reduce the circulating currents. On the grid side, an LC filter (forming equivalent LCL filter) is used in addition to the L filters to reduce the THD of grid currents. Given that DC link is common, the MSCs along with harmonic filters are placed close to the generator in nacelle. On the other hand, the DC link, GSCs, and a step-up transformer can be placed at the bottom of the tower. The MSC is connected to the DC link through DC cables leading to lower cable cost and losses and reduced nacelle weight. However, this configuration suffers from low reliability because DC-link capacitors are prone to failure and have short lifetime. In the `Enercon` E-126 model, more than 10 power converters are connected in parallel to reach a power rating of 7.5 MW. By contrast, a large number of converter modules lead to complex configuration and control.

(4) Parallel BTB 2L-VSCs with Individual DC Links: To solve the circulating currents and reliability issues, the DC links can be configured as individual elements, as shown in Figure 2.7, without losing the best qualities such as power handling capability, modularity, redundancy, and efficiency. However, the individual DC link in each converter module leads to high system cost and footprint. Moreover, the complete power converter must be placed in the nacelle to decrease the voltage drop in the generator-side cables, thus increasing the overall nacelle space requirement. Nevertheless, this configuration remains as the dominant technology in Type 4 WECS.

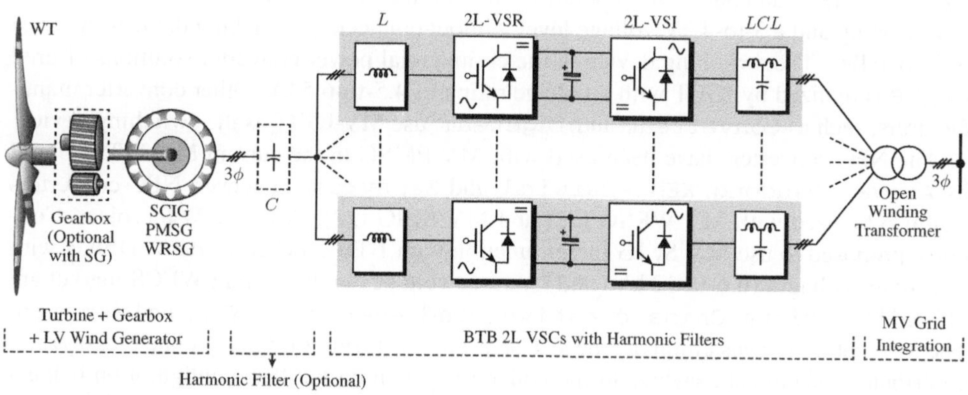

Figure 2.7 Type 4 WECS with multiple VSCs in parallel for high current/power application (individual DC-link configuration; also applicable for Type 3 WECS).

Open-winding transformers can also be used at the grid-side to minimize the circulating currents. An open-winding transformer isolates the converters at a high cost. The configuration of harmonic filters on the grid side is simpler than the previous topology in Figure 2.6. The LCL filters are used separately for each GSC. The circulating currents remain on the generator side, for which open-winding generators should be used. The discussion on open-winding generators is presented in Section 2.6.

2.4.2 Medium-Voltage BTB Converters

The increase in power rating of a WT corresponds to the increase of the number of parallel LV converter modules. Thus, the size, cost, and complexity of the system increase. The increasing power level in WTs is driving the power electronics technology toward MV operation. The MV power converters will dominate the next generation of multi-MW WTs because they offer a cost effective, reliable, and compact design. The power converter configurations for the MV operation of WTs are discussed as follows:

(1) 2L-VSC with Series Connected Switches: The 2L-VSC shown in Figure 2.4 can also be used for MV applications by connecting switching devices in series. This configuration is a simple solution for MV operation. However, the converter capacity decreases (also called as converter derating) because of the mismatch in IGBT and gate driver static and dynamic characteristics. A proper voltage equalization scheme should be implemented to ensure equal voltages among the series switches. The commercial converter model `Converteam` VDM5000 with a maximum power and a voltage rating of 7.2 MW and 4.16 kV, respectively, features this configuration for the MV drive application [8]. However, this approach is not practiced in the wind energy industry.

(2) BTB Neutral-Point Clamped Converter: As an alternative solution, the three-level (3L) diode-clamped converter (DCC), most popularly known as neutral-point-clamped (NPC) converter, has been widely studied in literature for Type 4 turbines [22–24]. In this configuration, two 2L-VSCs are stacked one over the other by using split DC-link capacitors and clamping diodes. The converter output phase voltage contains three levels, thus leading to reduced *dv/dt* and electromagnetic interference compared with 2L-VSCs [25, 26]. Moreover, the NPC converters lead to reduced switching losses, low output current ripple, and small harmonic filter requirement than 2L-VSCs. As shown in Figure 2.8, NPC converters can enable MV operation and commercial WTs have reached a 6.0-MW power rating and a 3-to-4 kV voltage level without connecting switching devices in series or in parallel. The switching device in the commercial power converter solutions offered by `ABB` is realized by IGCT with a voltage rating of 4.5-to-6.5 kV. Other converter manufacturers, such as `Converteam` and `Ingeteam`, use MV-IGBT as the switching device.

The NPC converters have been used with MV PMSG in the current MV WTs offered by `Areva`, `Shandong`, `XEMC-Darwind`, and `Zephyros`. However, NPC converters can also be used with MV WRSG [23] and MV SCIG [22]. A future WT project, Condor6, proposed to use MV SCIG in conjunction with BTB NPC converters. DFIGs with MV stator voltages (6.6-to-12 kV) and LV rotor voltages in the current WECS market are offered by `Acciona`, `China creative wind energy (CCWE)`, and `Senvion`. The MV stator voltages eliminate the WT step-up transformers (and associated losses) and contribute to significant savings in the collector system costs. This configuration is ideal for wind farms which are in close proximity to the substation (less than 5 km). The LV rotor windings are connected to grid through BTB 2L-VSC. Thus, DFIG WECS are omitted in the discussion of MV power conversion systems.

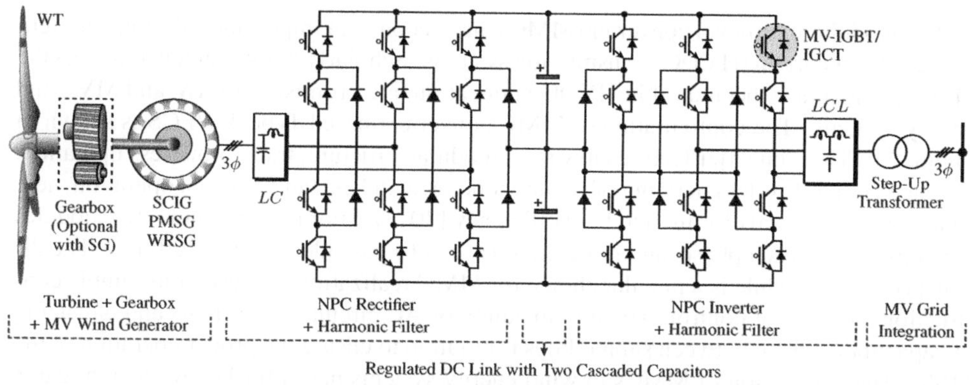

Figure 2.8 Type 4 WECS with BTB connected NPC converter.

The semiconductor active switching devices, clamping diodes, and DC-link capacitors are rated for half of the net DC-bus voltage. The switching actions of the semiconductor devices lead to a drift in the capacitors voltage. If the capacitors voltage are not balanced, the stress on the semiconductor switches will be high and these switches will be damaged. This issue has been extensively studied by both the industry and academia [25]. The use of external hardware can mitigate the problem of unbalanced capacitors voltage. In the classical linear control, carrier-based PWM with zero-sequence voltage injection or space vector modulation with redundant switching states selection is used to balance capacitors voltage [26–28]. In the model predictive control, the objective of capacitors voltage regulation is simply included in the cost function formulation [7]. Given that high-power NPC converters have been marketed by many manufacturers, the problem of capacitors voltage imbalance can be assumed to have been solved.

The semiconductor device switching frequency is limited to few hundred hertz to minimize the switching losses and allow proper heat dissipation. The outer switching devices, which are connected to the positive and negative DC bus, operate at higher switching frequency. On the other hand, the inner switches, which are connected to the split DC bus, work with lower switching frequency. This phenomenon leads to uneven power losses, heat dissipation, and the possible derating of converter. As a result, designing the mechanical layout of the semiconductor switches can be difficult.

(3) Other Emerging Multilevel Converters: High-power density converters are needed for the WT manufacturers because of the limited space in nacelle. A customized and efficient design along with proper cooling system decreases the overall size of power converters and nacelle space requirement. Many MV power converter topologies have been recently studied for wind energy application.

The uneven power loss challenge associated with the NPC converter can be solved by using active neutral-point clamped (ANPC) converter. The clamping diodes (passive switching devices) in Figure 2.8 are replaced by the MV-IGBTs (active switching devices), giving more redundancy to maintain equal switching frequency and losses among all IGBTs. Similarly, the BTB 3L-ANPC converters are capable of handling of 32% higher power (up to 7.12 MW) and 57% higher switching frequency (1650 Hz) compared with the BTB NPC converters [29]. This configuration is applied more recently in the MV drives industry and can also be used in the WECS industry. One of the top WT manufacturers, `Vestas` is currently investigating this power converter topology for Type 4 WECS.

The modular multilevel converter (MMC) has become an important milestone in high-voltage direct current (HVDC) transmission systems. Scholarly works demonstrate that the MMCs, which have been used in HVDC application, are promising for LV and MV wind energy systems. The configuration of MMC converters-based Type 4 WECS is shown in Figure 2.9. This converter features low size, modular structure, transformerless operation, enhanced reliability, less derating of semiconductor switches, fault-tolerant operation, near sinusoidal output waveforms, and high efficiency [30, 31]. For LV operation, two modules are employed in each phase, and more modules can be cascaded for MV turbines. The 2L half-bridge submodule is a popular choice for MMC realization because of its simple construction and ease of control. The minimization of AC circulating currents and balancing of capacitors voltage between submodules is required to ensure the proper operation of the MMC. The application of MMCs to wind energy systems needs further evaluation and is an on-going research topic.

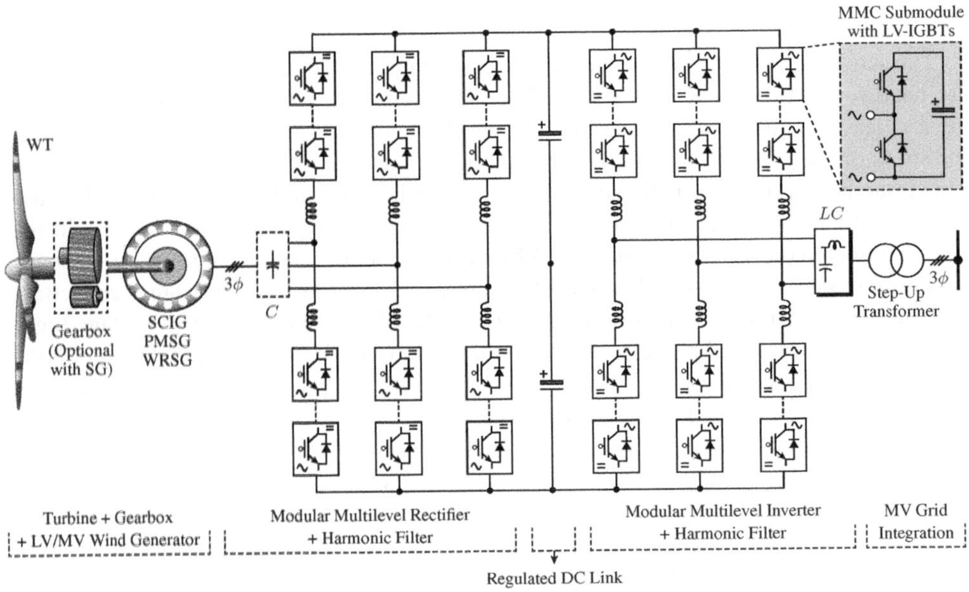

Figure 2.9 Type 4 WECS with BTB connected MMC.

Novel multilevel converters may be used to connect WTs to MV collection point of 6.9 kV (North America) or 6.6 kV (Europe) without using a step-up transformer and decrease grid-side filter requirement. By operating power converters at higher MV operation levels, the cable cost and losses are minimized further. The classical multilevel converter topologies (such as flying capacitor converter, cascaded H-bridge (CHB) converter, and multilevel DCC) and hybrid converter topologies (such as multilevel ANPC converter and nested neutral-point clamped converter) are also promising candidates for MV operation of WTs. Thus, research and development activities in the MV power converters area should become more rigorous in the coming years.

(4) BTB Current Source Converter: The previously discussed topologies belong to VSC class. As shown in Figure 2.10, the CSCs can also be used for MV WTs. The CSCs in power conversion applications are the duality of VSCs. The CSC configuration is most favorable for WT power ratings greater than 5.0 MW.

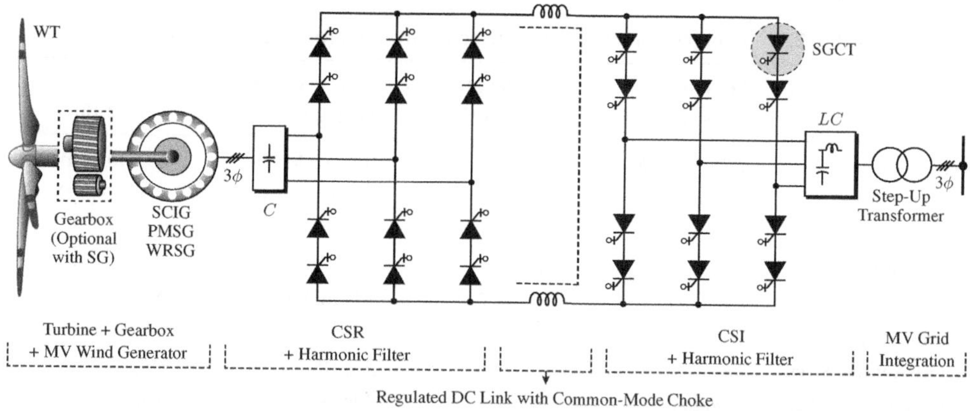

Figure 2.10 Type 4 WECS with BTB connected CSC.

The topology consists of SGCT-based current source rectifier (CSR) and current source inverter (CSI). The SGCTs in CSR and CSI are connected in series to reach required MV operation level. The CSR and CSI are linked by a DC choke. Similar to VSCs, decoupling between the generator and grid can be achieved with CSCs [32]. The CSCs can be employed with both MV SCIG and MV synchronous generator. Three-phase capacitor banks are used on the AC sides of CSR and CSI to assist the commutation of semiconductor devices and mitigate switching harmonics [2].

This topology features a simple structure, low switch count, low switching *dv/dt*, and reliable short-circuit protection compared with VSCs. The DC-link capacitors in VSC are bulky components, whereas the DC choke is a bulky component in CSC. The dynamic response of CSCs is sluggish compared with that of VSCs because of the bulky DC choke and lower switching frequency operation. The net DC-bus voltage is maintained lower than the peak of grid line-to-line voltage to ensure proper operation of CSI. Similar to the VSC-WECS, CSR controls the generator speed/torque, whereas CSI controls the DC-link current and grid reactive power.

CSCs in the MV drives industry are competent to the multilevel VSCs because of transformerless operation. However, a step-up transformer is mandatory in WECS to connect WTs to MV collection point. The cost of CSC production could be 1% to 2% lower compared to the VSCs [9]. CSC technology is successfully applied to the multi-megawatt MV drives [33]; however, it has not been used in the MV-WECS.

2.4.3 Comparison of BTB Power Converters

The summary of comparison between BTB converters is given in Table 2.1 regarding power/voltage rating, semiconductor/passive component count, voltage stress of switches, reliability, power quality, converter and control complexity, grid code compliance, technology status, and market penetration. The CSC is assumed to have two series-connected SGCTs. The commercially practiced BTB power converters in the present wind energy industry are limited to 2L-VSC, parallel 2L-VSCs, and NPC converter. The analysis given summarizes the feasibility of applying various power converters in LV and MV WECS. The multilevel converters are promising for next-generation WTs because of the technical merits they exhibit.

Table 2.1 Comparison of BTB connected converters for megawatt WTs [1]

	2L-VSC (Figs. 2.4/2.5)	Parallel 2L-VSCs (Figs. 2.6/2.7)	NPC Converter (Fig. 2.8)	CSC (Fig. 2.10)
Typical Power	0.75/3.0 MW	3.0 MW	3.0–6.0 MW	3.0–10.0 MW
Typical Voltage	690 V	690 V	3000–4000 V	3000–6600 V
No. of Converters	1	4	1	1
No. of Switches	12	48	24	24
Switching Devices	LV-IGBT	LV-IGBT	MV-IGBT/ IGCT	SGCT
No. of Diodes	0	0	12	0
Device Voltage Stress	v_{dc}	v_{dc}	$v_{dc}/2$	$v_{dc}/2$
Reliability of System	Medium	High	Medium	High
Converter Complexity	Low	Medium	Medium	Low
Control Complexity	Low	High	High	Medium
Grid Code Compliance	Acceptable	Acceptable	Good	Good
Technology Status	Highly Mature	Highly Mature	Emerging	Research Only
Current Market Share	≈ 50%	2nd Highest	Few Models Only	No Models
Example WT Models	`Acciona`-AW125 `DeWind`-D8.0	`Enercon`-E126 `Vestas`-V112	`Multibrid`-M5000 `Marvento`-M3.6	—

2.5 PASSIVE GENERATOR-SIDE POWER CONVERTERS

The PGS converters decrease the overall cost of power converters and improve the reliability of generator-side power conversion. The PGS converters are solely used with the synchronous generators, both PMSG and WRSG. A few examples of MW WTs include the `Enercon` E82, `Clipper` Liberty C89, `Vensys` V70/77, and `Goldwind` GW70/77. The classification of PGS converters for Type 4 WECS is shown in Figure 2.11 along with the switching device used. This section describes the PGS converter topologies applicable for the Type 4 LV and MV WECS.

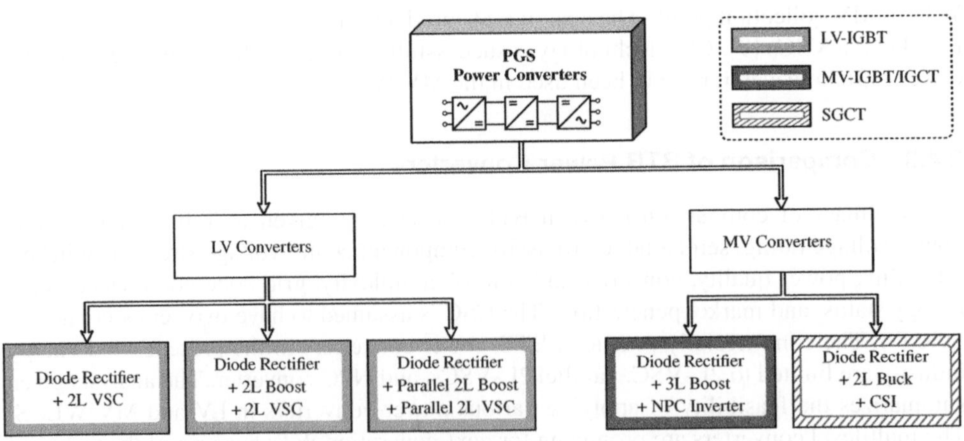

Figure 2.11 Classification of PGS power converters for Type 4 WECS.

2.5.1 Low-Voltage PGS Converters

Three major PGS power converter configurations are available in LV class. The description of these topologies is given below:

(1) Diode Rectifier + 2L-VSI: The power converter configuration for PMSG/WRSG WECS with three-phase diode rectifier and 2L-VSI is shown in Figure 2.12. This topology offers a low cost, lightweight solution compared with the BTB 2L-VSC. The generator output AC voltage is converted to DC by the diode rectifier. The DC voltage is then converted back to AC voltage by 2L-VSI. The inductor in the DC link helps smoothen the diode rectifier output current. The output voltage of wind generator and diode rectifier changes with respect to the varying wind speed conditions. However, the maximum diode rectifier output voltage value is limited by the rated speed of generator. Thus, the first DC link is denoted as an unregulated DC link. The diode rectifier output voltage becomes significantly low during low wind speeds. The DC-link voltage must be higher than the peak value of grid line-to-line voltage to transfer the generated power to grid. To ensure this condition, the generator should be overrated (i.e., the rated generator voltage must be higher than the grid voltage).

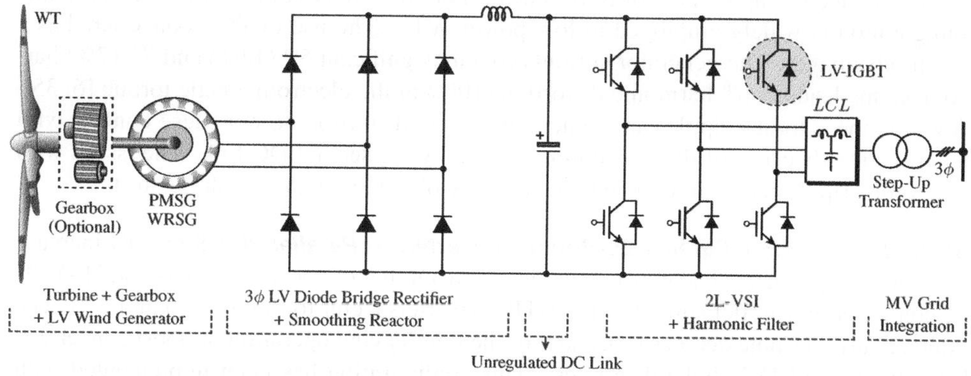

Figure 2.12 Type 4 SG WECS with diode bridge rectifier and 2L-VSI.

This configuration is used in the `Clipper Liberty` 2.5 MW WTs along with the quantum drivetrain [34]. The absence of intermediate DC/DC converter decreases the degree of control freedom by one; that is, the control system cannot incorporate either MPPT operation or regulation of DC-link voltage. In the `Clipper Liberty` WTs, MPPT is achieved by controlling the GSC, whereas DC-link voltage is allowed to vary with respect to the wind speed. The GSC is designed according to the maximum possible DC-link voltage.

(2) Diode Rectifier + 2L-Boost Converter + 2L-VSI: A boost converter can be employed as an intermediate stage to increase the control freedom by one. The WECS with a diode-rectifier, a 2L boost converter, and a 2L-VSI is shown in Figure 2.13. The variable output voltage of the generator is converted to DC by the diode rectifier. The boost converter increases the unregulated DC-link voltage to a higher level that is suitable for the 2L-VSI. The capacitor in the first DC link filters the ripple in diode rectifier output DC voltage and is an optional component [35]. The boost converter also enables variable-speed operation for the PMSG/WRSG WECS by performing MPPT. Therefore, the wind energy conversion efficiency improves, particularly during low wind speeds.

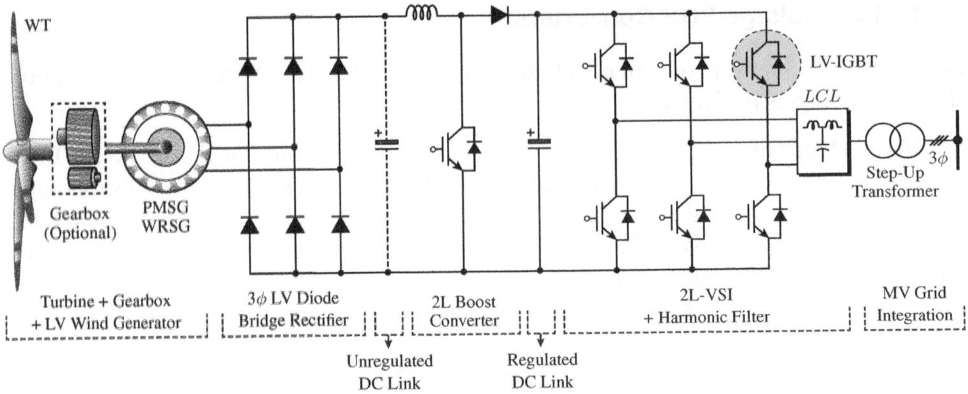

Figure 2.13 Type 4 SG WECS with diode bridge rectifier, 2L boost converter, and 2L-VSI.

The control system development for the MPPT is less complicated with PGS converters. In comparison to 2L-VSR, IGBT gate drivers have a cost and complexity that decrease with decreasing number of IGBTs from six to one. In addition to MW turbines, this configuration is widely employed in low-power WTs. The use of PGS converters has a few disadvantages. The generator currents contain significant 5th (14%) and 7th (7%) harmonics; this leads to 6th harmonic distortion (10%) in the electromagnetic torque [6, 35]. However, these torque ripples and generator current distortion do not cause conflict with the grid codes because of the decoupling offered by the second DC link. The issues with high torque ripple can be mitigated by using a six-phase SG with 30° phase-shift.

(3) Diode Rectifier + Parallel 2L-Boost Converters + Parallel 2L-VSIs: To increase power handling (current carrying) capacity, a high number of boost converters and 2L-VSIs are connected in parallel similar to the BTB converters (refer to Figures 2.6 and 2.7). The inductor current ripple decreases because of the interleaving operation of boost converters. Thus, the size of DC choke decreases. This configuration has been implemented with WRSG in `Enercon` E82 turbines. The market share of LV PGS converters is only a fraction of BTB LV power converters.

2.5.2 Medium-Voltage PGS Converters

The diode rectifiers can also be used in MV WECS. However, the PGS converters in the present wind energy industry are not used at the MV operation level. This section describes two promising MV PGS power converters for Type 4 WECS.

(1) Diode Rectifier + 3L-Boost Converter + NPC Inverter: The 3L PGS power converter-based MV WECS is shown in Figure 2.14. The diode-bridge rectifier features series-connected diodes owing to the MV generator. The intermediate 3L boost converter enables MV operation for the DC/DC stage. The output of the 3L boost converter directly fits the two DC-link capacitors of the NPC inverter. Thus, the advantages of generator-side PGS converters and grid-side multilevel converters are combined [7]. The voltage rating for the IGBT/IGCT switches and diodes in a 3L boost converter is half of the net DC-bus voltage. The 3L boost converter offers many benefits compared with the standard 2L boost converter in Figure 2.13: reduced switching and reverse recovery losses, and ability to control neutral-point voltage (balancing of capacitors voltage).

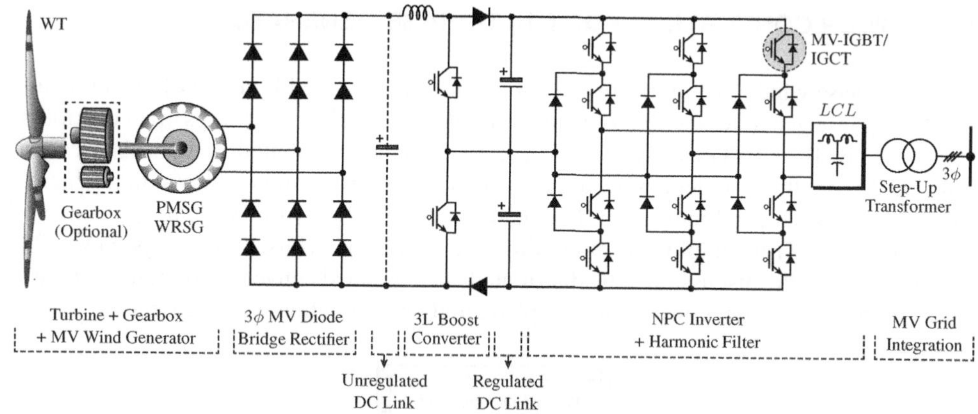

Figure 2.14 Type 4 SG WECS with diode bridge rectifier, 3L boost converter, and NPC inverter.

The number of active switches in MSC decreases from 12 to 2 compared with the BTB NPC converters. This phenomenon leads to a cost-effective solution for MV turbines. This configuration leads to enhanced power quality and better grid code compliance compared with the BTB-NPC converters. The control complexity for the NPC inverter decreases because the neutral-point voltage is controlled by the 3L boost converter during all operating conditions. Similar to the 3L boost converter, a 4L boost converter can be used in conjunction with the 4L-DCC [4, 36].

(2) Diode Rectifier + Buck Converter + CSI: Similar to VSCs, the CSCs can also be designed with PGS converters. The configuration of WECS with diode rectifier, 2L buck converter, and PWM CSI is shown in Figure 2.15 [6]. The switching devices in buck converter and CSI are connected in series to withstand MV operation. This configuration is simple and less complicated compared with PGS VSCs. The DC link voltage in VSCs is boosted, whereas DC link current in CSCs is boosted. Similar to other PGS converters, this configuration suffers from high generator torque ripple.

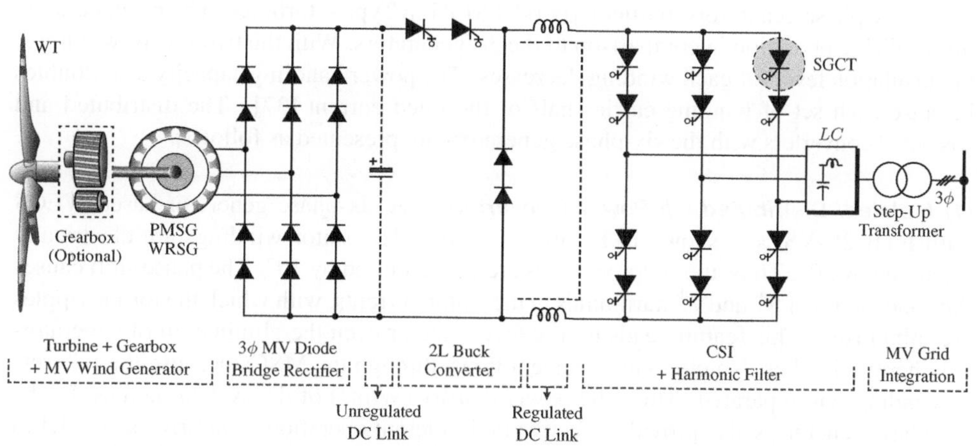

Figure 2.15 Type 4 SG WECS with diode rectifier, 2L buck converter, and CSI.

2.6 POWER CONVERTERS FOR MULTIPHASE GENERATORS

The multiphase wind generators are specialized technology used by few WT manufacturers. Multiphase generators with more than one set of three-phase windings open the doors to connect BTB and PGS converters in parallel/isolated/distributed or hybrid/cascaded manner. The multiphase generators are classified as six-phase and open-winding generators, as demonstrated in Figure 2.16. It should be noted that only full-scale power converters (Type 4 WECS) are employed with multiphase generators. Thus, DFIG (Type 3) WECS are omitted from the discussion. The distributed and cascaded converter configurations are presented in this section, considering six-phase and open-winding wind generators.

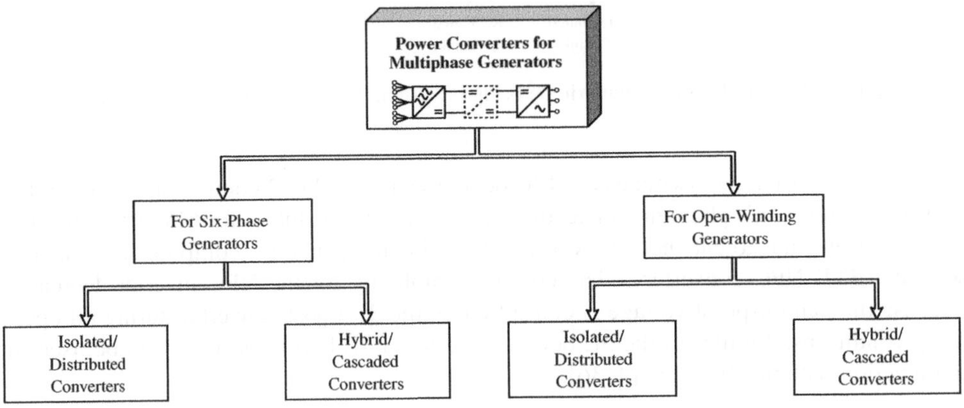

Figure 2.16 Classification of power converters for multiphase generator-based Type 4 WECS.

2.6.1 Power Converters for Six-Phase Generators

The six-phase generators, both induction (excluding DFIG) and synchronous generators, are widely used in the electric drives industry. The wind energy industry also adopted the use of six-phase generators, particularly 6ϕ-PMSG in Type 4 turbines. The 6ϕ generators offer reliable operation in comparison to the 3ϕ generators. With the two sets of windings, the insulation level for each winding decreases. The power handling capacity also doubles because each set of winding carries half of the rated current [37]. The distributed and cascaded converters with the six-phase generators are presented as follows:

(1) Isolated/ Distributed 6ϕ Power Converters: A six-phase generator-based WECS with BTB 2L-VSCs is shown in Figure 2.17 [38]. The stator windings are electrically separated by 30°. Thus, the stator voltages are phase shifted by 30°. The phase shift causes the cancelation of 5th and 7th harmonics in the stator currents, with which the torque ripples are minimized. This feature leads to a reduced size or even the elimination of generator-side harmonic filter. No circulating current flows through the MSCs because the two sets of windings are separated. Thus, the power density (W/m^3) of the system increases. The six-phase generators also provide effective fault tolerant operation. The `Envision` E128 3.6-MW WTs use such a six-phase configuration with direct-drive PMSG. The present wind energy industry also investigates the use of BTB-NPC converters in place of BTB 2L-VSCs for the MV operation of six-phase generators.

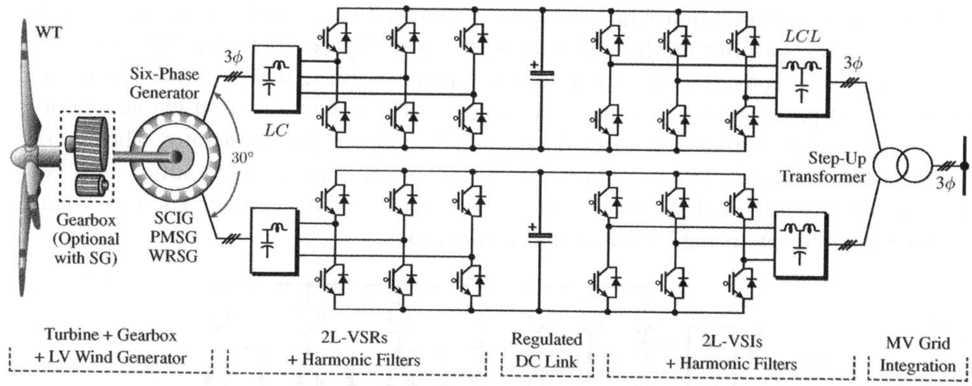

Figure 2.17 Type 4 WECS with a six-phase generator and distributed 2L-VSCs.

As discussed earlier, the PGS converters produce significant generator torque ripples. The disadvantage of high torque ripples can be mitigated by employing a six-phase generator with 30° phase shift [5]. A six-phase configuration employed by Vensys V70/77 and Goldwind GW70/77 WTs is shown in Figure 2.18 [39]. The power converter is realized by a six-phase diode-bridge rectifier, a three-channel boost converter, and a two-channel VSI to handle 1.5-MW power. The DC link is configured as a common element for VSCs. The boost converters and VSCs operate in interleaving mode to decrease the inductor and grid current ripples, respectively.

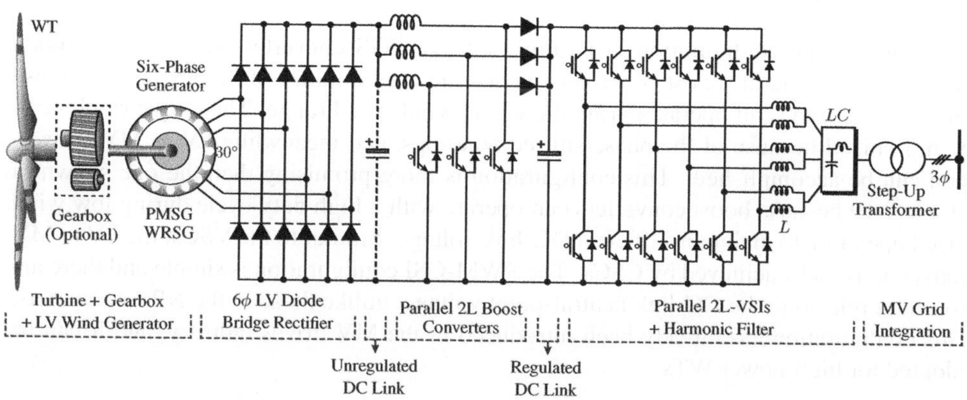

Figure 2.18 Configuration of Type 4 WECS with six-phase PMSG/WRSG, parallel 2L-boost converters, and parallel 2L-VSIs.

(2) Hybrid/ Cascaded 6ϕ Power Converters: The MV operation of WTs is cost effective and a promising solution for MW WTs. In the standard approach, the output of MV generator is connected to the MV grid through an MV power converter. However, one of the shortcomings in the current wind energy industry is the lack of MV generators. The use of LV converters at the generator side and MV converters at the grid side represents a promising approach. Off-the-shelf LV generators and MV converters can be rearranged to increase the efficiency of wind energy conversion and reduce the overall cost.

A six-phase configuration with two series-connected LV converters on the generator side and an NPC inverter on the grid-side is shown in Figure 2.19 [40]. The outputs of generator-side 2L converters are connected in series to achieve high DC-link voltage and provide a mid-point for the grid side NPC inverter. The power output of the generator is distributed among the two converters, decreasing the ampere-per-phase ratio. The neutral-point voltage control is a challenging issue with this particular configuration. This configuration may not work during the whole wind speed range because the DC-link voltage becomes insufficient during low-wind-speed conditions.

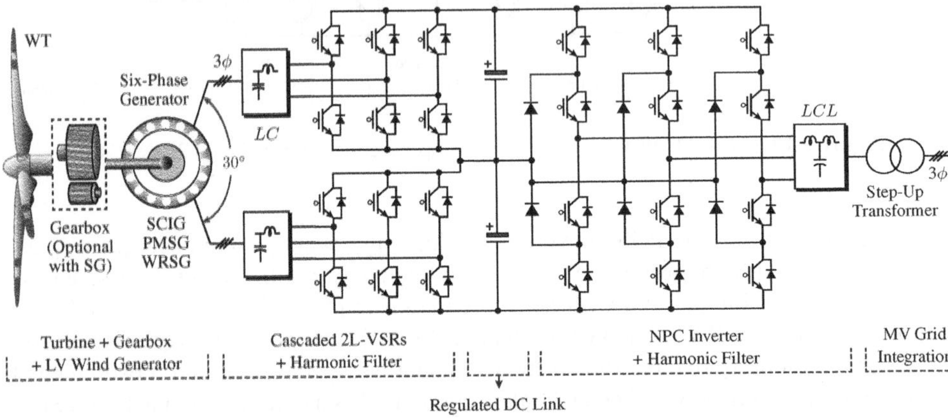

Figure 2.19 Type 4 WECS with a six-phase generator, cascaded 2L-VSRs on the generator side, and an NPC inverter on the grid side.

Similar to the configuration in Figure 2.19, the PGS converters composed of diode bridge rectifier and 2L boost converter can also be used in place of 2L-VSRs to decrease the cost of turbine and operate during the whole wind speed range. The torque ripples can be minimized because of the phase-shifted windings, and meanwhile LV to MV conversion can be accomplished. This configuration is more promising than the one shown in Figure 2.19 because boost converters can operate with a high duty cycle during low wind speed operation to maintain sufficient DC-link voltage. Similar to the VSCs, the LV to MV conversion can be achieved by CSCs. The PWM-CSI configuration is simple and there are no issues related to the DC-link neutral-point voltage, unlike those in the NPC converters. This configuration has already been introduced in the MV drives industry, and it can be adopted for high-power WTs.

2.6.2 Power Converters for Open-Winding Generators

The open-winding generator is equipped with more than two sets of windings and allows power converters to be connected in a distributed or cascaded manner.

(1) Isolated/ Distributed Power Converters: The distributed converters are similar to the one shown in Figure 2.17, except that the number of phases or converter channels are more than two. The configuration of Type 4 WECS with open-winding generator and distributed converters is shown in Figure 2.20 [10]. The Gamesa G10x 4.5 MW WTs employs this concept with 6 BTB 2L-VSC modules in parallel. Each module is rated for 690 V with its own harmonic filters, circuit breakers, measurement units, and control boards. The design also includes the step-up transformer in the nacelle to decrease cable cost and power losses.

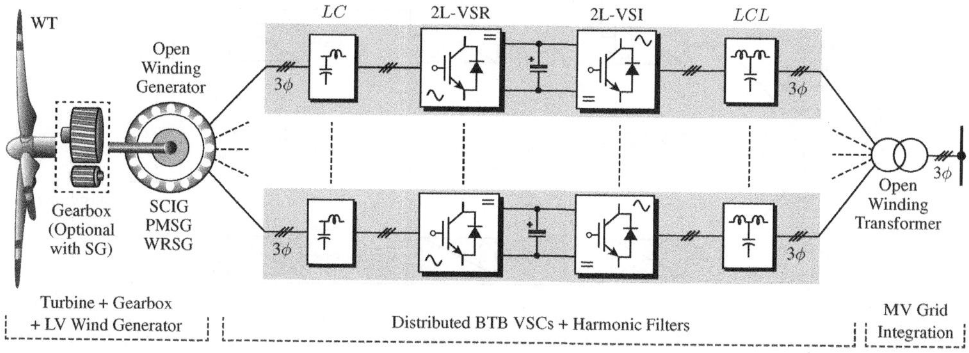

Figure 2.20 Block diagram of Type 4 WECS with open winding generator, distributed BTB VSCs, and open winding transformer.

The studies in [15, 41] also investigated the use of different numbers of converter channels with open-winding generators. Compared with the parallel VSC modules discussed in Section 2.4.1, the circulating currents in the MSCs are eliminated. Thus, the power rating of the overall system increases in proportion to the number of channels employed. The Gamesa G10x 4.5 MW WTs uses liquid cooling and an optimized design, thus leading to a 30% smaller foot print than competitor products and higher power density of 0.58 MW/m^3 [10]. The distributed windings also provide insulation between the converters. These generators are particularly designed at a high cost. Open-winding transformers are used on grid-side to minimize the circulating current in GSCs.

(2) Hybrid/ Cascaded Power Converters: A high-power MV WT using cascaded converters is shown in Figure 2.21. This converter configuration is similar to the CHB converter used in the electric drives industry. The CHB converter requires isolated DC sources, which can be easily generated by the open-winding generator. The generator contains multiple sets of electrically isolated two-phase windings with a phase displacement of 90°.

System voltage and power rating increase when more converter cells are connected in series. Each module can be realized by using the power converters introduced in Figure 2.22 [42–44]. The transformerless grid connection is also possible with this configuration because system voltages of 10–35 kV can be reached by connecting more modules in series. The LV switching devices are used in the power converter to achieve MV operation levels. The modularity, redundancy, and fault-tolerant operation are other added advantages of this configuration. Moreover, the output power quality significantly improves because of the cascaded structure. Thus, the generator- and grid-side harmonic filters can be eliminated. Given the cascaded structure, no issues related to the circulating currents and derating of switching devices is found.

A 3L H-bridge converter is shown in Figure 2.22(a). This converter is a simple structure and its output phase voltage contains three levels. This converter can be realized by a mass-produced 2L VSCs. With three cells in series as shown in Figure 2.21 and with a 3L H-bridge converter, the output voltage contains seven levels. Thus, the *dv/dt* and harmonic filter size decrease significantly. NPC H-bridge modules can also be used to form a 5L H-bridge converter and achieve high system voltage operation with a low number of modules. To accomplish this operation, the electric generator windings should also be designed for MV operation level.

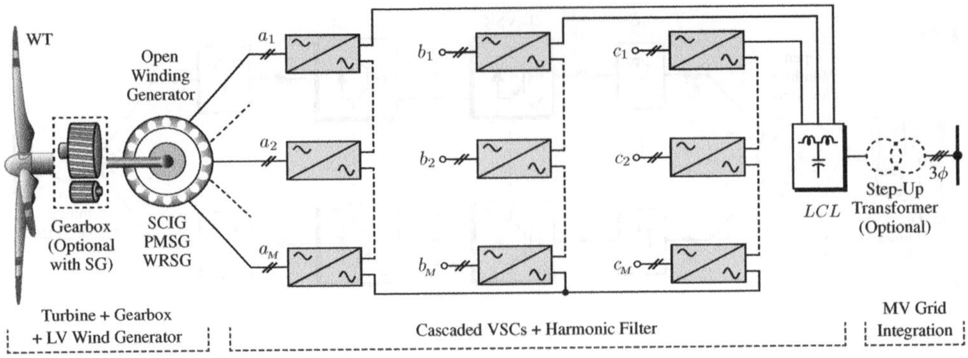

Figure 2.21 Type 4 WECS with open winding generator and cascaded VSCs.

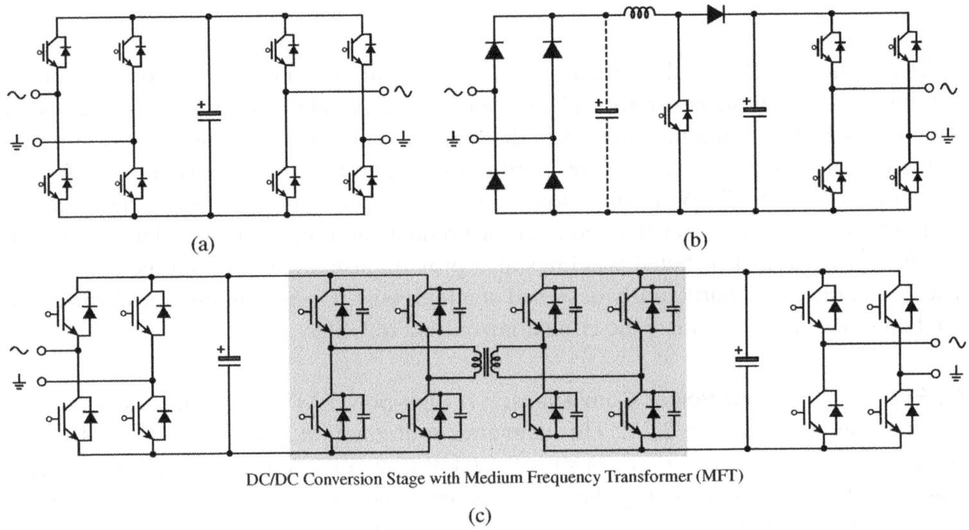

Figure 2.22 Submodule configurations for Figure 2.21: (a) BTB 3L H-bridge, (b) 1ϕ diode bridge rectifier + 2L boost converter + 3L H-bridge, and (c) BTB 3L H-bridge with an intermediate DC/DC conversion stage.

The cost associated with MSCs can be decreased by using a single-phase diode-rectifier and a 2L boost converter, as depicted in Figure 2.22(b) [42]. Notably, the single-phase diode-rectifier produces more harmonic distortion in generator currents than three-phase diode-rectifier. However, the phase shift between the generator windings helps to cancel low-order harmonics in the generator stator currents.

Another possibility is to use MFTs in the BTB H-bridge modules (refer to Figure 2.22(c)). This concept has been proposed in [44] for a UNIFLEX-PM project. The DC/DC converters provide galvanic insulation; thus, the grid-side transformer can be eliminated by connecting more modules in series (refer to Figure 2.21). The MFT operates at several kHz and reduces the overall size of the converter significantly [11]. A large number of components decrease the overall reliability of this configuration. The cascaded power converters are not practiced in the present wind energy industry; however, they are very promising for the future development owing to their best features.

2.7 POWER CONVERTERS WITHOUT AN INTERMEDIATE DC LINK

High-power density converters can be realized by eliminating the bulky DC link components. The MCs provide direct AC/AC conversion without any intermediate DC link, thus leading to more silicon-based power conversion with a low cost and small foot print [45, 46]. Cycloconverters also perform direct AC/AC conversion; however, their output voltage/frequency is only a fraction of the input voltage/frequency [33]. Compared with cycloconverters, MCs can produce a wide range of output voltages/frequencies. MC can convert variable generator output voltage/frequency to fixed voltage/frequency to connect the WT to a three-phase grid. Moreover, MCs offer reliable solution for WTs, particularly when they are employed in offshore.

The classification of MCs for high-power WTs is given in Figure 2.23 with respect to the LV and MV operation. The discussion of LV and MV matrix converters is given below.

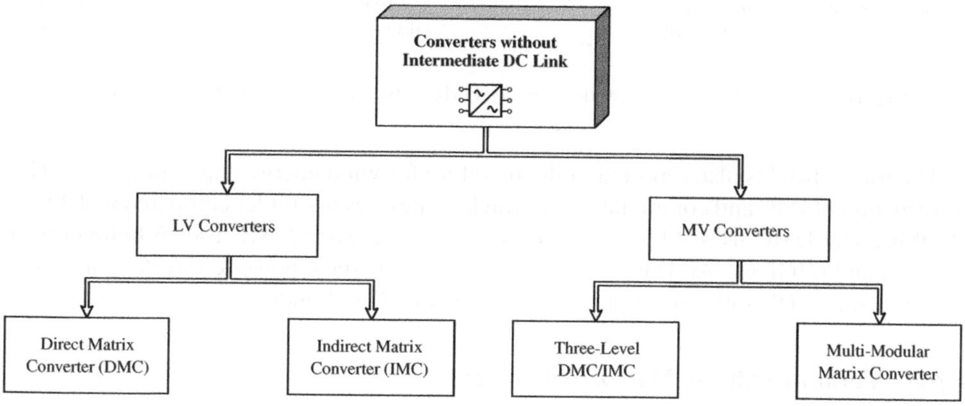

Figure 2.23 Classification of matrix converters for variable-speed WECS.

2.7.1 Low-Voltage Matrix Converters

The DMC-based LV WT is shown in Figure 2.24, in which the generator and grid are rated for LV operation. It employs bidirectional switches (with a common emitter or collector) and an *LC* filter on the grid side that is similar to the CSCs discussed before. The bidirectional switch is realized by a RB-IGBT. The power flow is bidirectional; thus, this converter can be employed with both induction and synchronous generators [47, 48]. Through proper modulation and control, the bidirectional switches directly connect the generator terminals to the grid. Given no intermediate DC link, the variations in the generator-side variables strongly affect the grid-side variables and vice versa. The development of a proper control system is important to ensure the correct operation of MC when employed in WTs. LV gain is a major challenge for DMC compared with the VSC and leads to poor semiconductor device utilization. The large number of semiconductor devices and gate drives is another drawback of the matrix converter. DMCs are currently used in low power and LV electric drive applications; however, they are not practiced in WECS yet. Given that the DMCs are commercially available for low power and LV operation, they are promising candidates for Type 3 WECS because DFIG rotor voltages are only a fraction of stator voltages and the rotor circuit processes slip power which is 30% of rated DFIG capacity.

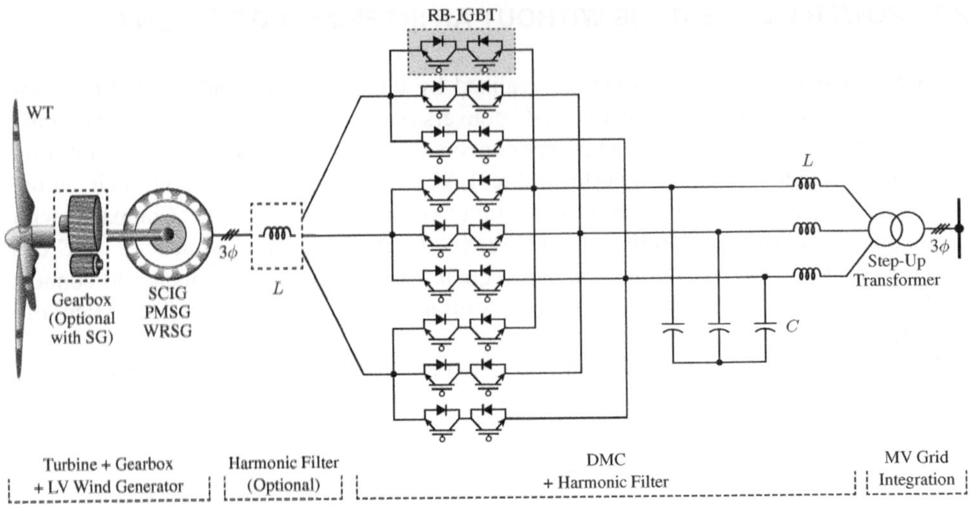

Figure 2.24 Type 4 WECS with 3ϕ to 3ϕ DMC (also applicable for Type 3 WECS).

The use of IMC is also reported in the literature for wind energy application [49]. The construction of IMC and commutation of switching devices is simpler compared with those of DMC. The DMC uses 9 bidirectional switches, whereas an IMC uses 6 bidirectional switches and 6 IGBTs. An IMC can be treated as a two-stage power converter composed of VSC with no DC-link capacitor and CSC with no DC-link inductor.

2.7.2 Medium-Voltage Matrix Converters

By connecting switching devices of DMC and IMC in series, MV matrix converters can be realized but with reduced converter capacity. Arranging MCs in a modular way is another possibility for the MV turbines. A configuration of MV WT with nine MC modules is shown in Figure 2.25. The equivalent representation of a nine-module matrix converter is depicted in Figure 2.26. The basic power cell is implemented with a three phase to one-phase matrix converter (single-phase matrix converter), as shown in Figure 2.27. More modules can be connected in series to further increase MV operation.

This configuration, called a multi-modular matrix converter (MMMC) or cascaded matrix converter, offers sinusoidal input and output currents [50]. In addition to the step-up transformer, this configuration requires a phase-shifting transformer to ensure the series connection of MC modules. The phase-shifting transformer cancels the low-ordered harmonics, thus improving grid current quality. The phase-shifting transformers are more expensive and bulky than standard transformers.

The semiconductor device voltage rating is a fraction of the system operating voltage; however, the component count is significantly high. For example, the 9-module MMMC requires 108 switches in contrast to 72 switches used in the 9-module cascaded VSC shown in Figure 2.21. The output waveform contains 7 levels, similar to the one obtained by Figure 2.21; therefore, the output voltage waveform is close to sinusoidal. The system reliability decreases because of the large number of switches. This configuration also requires a complex modulation scheme and control system. This technology has not been applied in the WTs; however, a 9-module MMMC is available in the market for WT applications (Yaskawa Enewin-MX1) [48].

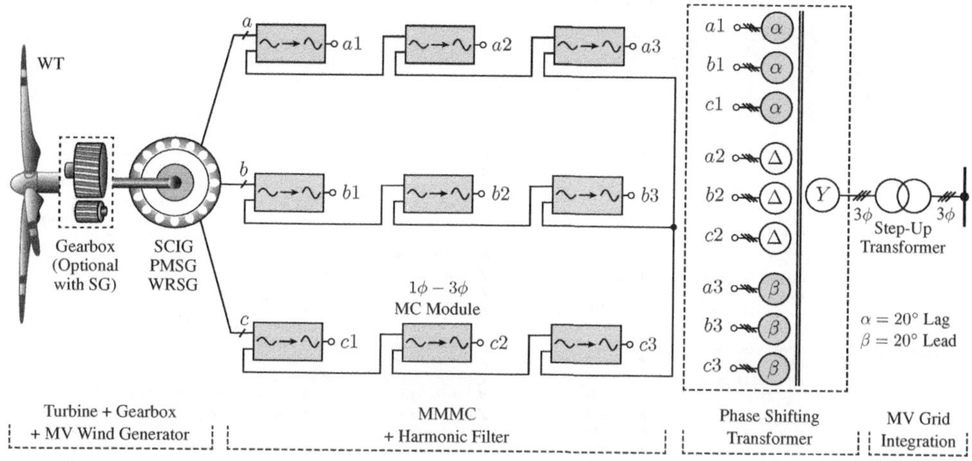

Figure 2.25 Block diagram of Type 4 WECS with MMMC.

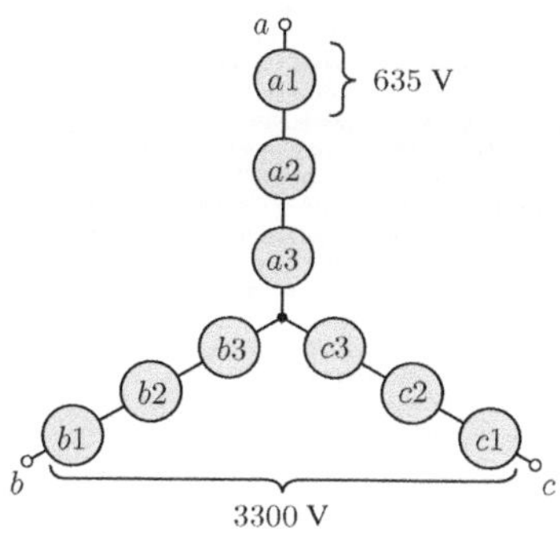

Figure 2.26 Simplified representation of multi-modular matrix converter.

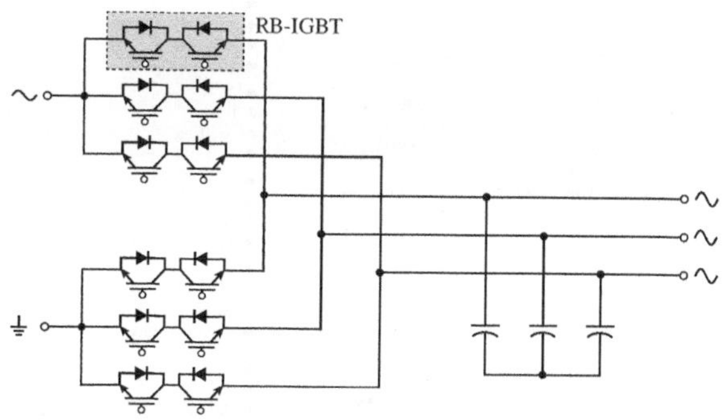

Figure 2.27 Configuration of single-phase to three-phase matrix converter module.

2.8 CONCLUDING REMARKS

The generator–converter configurations applicable for the variable-speed (Type 3 and 4) WECS are presented and analyzed in this chapter. The state-of-the-art and emerging power converter configurations for SCIG, DFIG, PMSG, and WRSG WECS are discussed. Various technical issues related to these configurations are discussed, including generator types, power converter topologies, active power control, energy conversion efficiency, and grid-side reactive power compensation. BTB connected power converters, PGS power converters, converters for six-phase and open-winding generators, and power converters without an intermediate DC link are analyzed in detail. The main features of the power converters discussed in Sections 2.4 to 2.7 are listed in Table 2.2 [1]. The MV power converters and advanced control schemes will play an important role in the development of multi-MW WTs. MV technology is a low-risk choice because it is a well-proven technology in the high-power electric drives industry. The perceived technical risks associated with the MV operation of WTs are therefore mitigated. In this book, both LV and MV power converters are equally treated and analyzed.

Table 2.2 Main features of power converters for megawatt WTs [1]

Power Converter Category	Power Converter Topology (Reference Figure)	• Present No. of WTs • Future No. of WTs • Technology Status • Voltage Class	Main Features
Back-to-Back Connected Power Converters	Partial-Scale (30 %) BTB 2L-VSC (Fig. 2.5)	• 33 (4 with MV stator) • 3 (1 with MV stator) • Highly Mature • LV (up to 1 kV)	✓ Reduced converter cost and complexity; higher efficiency compared with FS converter ✓ Single converter for 3MW WT; no need for soft starter and reactive compensation ✗ 3S-gearbox is mandatory; increases maintenance cost for offshore application ✗ FRT compliance is complicated; grid current THD is high because of 2L waveform
	Full-Scale (100 %) BTB 2L-VSC (Fig. 2.4)	• 37 (28-PMSG, 6-WRSG, 3-SCIG) • 9 (20 more unknown) • Highly Mature • LV (up to 1 kV)	✓ Generator is fully decoupled from the grid; excellent FRT compliance ✓ Well-established technology; mass produced by many; power levels up to 0.75 MW ✗ Higher initial converter cost; high cable costs; requires bulky transformer ✗ Lower efficiency; higher grid current THD
	Parallel Connected BTB 2L-VSCs with Common DC Link (Fig. 2.6)		✓ Highly redundant and modular approach; decreases nacelle space requirement ✓ Grid current quality can be improved by applying interleaving technique ✗ High cost; more losses; high circulating currents due to common DC link ✗ Less reliable because DC link capacitors have shorter life
	Parallel Connected BTB 2L-VSCs with Individual DC Links (Fig. 2.7)		✓ Most commonly used for high-power LV turbines; flexible/modular/redundant ✓ Highly reliable operation; easy maintenance and replacement ✗ High cost and weight; less efficient due to losses; increases nacelle space requirement ✗ Increased cable costs and losses; complicated converter/controller structure
	BTB NPC Converter (Fig. 2.8)	• 4 (with PMSG) • 7 (6-PMSG, 1-SCIG) • Established • MV (up to 4 kV)	✓ Popularly used for high-power MV WECS; less dv/dt and losses; high efficiency ✓ Excellent compliance with grid codes; decreases the size of cables and transformer ✗ Balancing of capacitors voltage is challenging; sluggish response due to low f_{sw} ✗ Possible derating of converter due to uneven power losses and heat distribution
	BTB CSC (Fig. 2.10)	• 0 • 0 • Available for VSD • MV (up to 4 kV)	✓ Simple converter configuration; grid friendly waveforms ✓ Reliable short-circuit protection; well proven technology in drives industry ✗ Sluggish dynamic response due to bulky inductor; possible *LC* resonances ✗ Possible derating due to series connected switches
Passive Generator-Side Power Converters	Diode Rectifier + 2L-VSI (Fig. 2.12)	• 1 • 1 (for MV application) • Available • LV (690 V)	✓ Low cost and footprint; simple converter configuration ✓ Diodes are more reliable; generator-side control scheme is simple ✗ Generator must be over rated; generator current THD is high ✗ Significant torque oscillations; no MPPT or DC link voltage control
	Diode Rectifier + 2L-Boost Converter + 2L-VSI (Fig. 2.13)	• 2 • 0 • Available • LV (up to 1 kV)	✓ Low cost; high reliability; MPPT is performed by dc/dc converter stage ✓ Improves WECS efficiency during low wind speeds; simple control scheme ✗ Non-sinusoidal generator currents; grid current THD is high due to 2L waveform ✗ Increases load on drive-train due to torque oscillations
	Diode Rectifier + 3L-Boost + NPC Inverter (Fig. 2.14)	• 0 • 0 • Research Only • MV (up to 4 kV)	✓ Promising topology; cost effective compared to the BTB NPC; better power quality ✓ Capacitor voltages are controlled by the boost converter; simple control for NPC ✗ Torque oscillations and non-sinusoidal generator currents are still present ✗ Less documented (unknown issues may arise)
	Diode Rectifier + 2L-Buck + CSI (Fig. 2.15)	• 0 • 0 • Available for VSD • MV (up to 4 kV)	✓ Less component count; decreases the cost and footprint compared with BTB CSC ✓ Improves the wind energy conversion efficiency ✗ Dynamic response is still sluggish ✗ Possible derating of buck converter due to series switches
Power Converters for Multiphase Generators	6ϕ - Generator with Distributed Modules (Figs. 2.17 and 2.18)	• 1 • 4 • Available • LV (up to 1 kV)	✓ Modular; low torque ripple; most favorable with passive converters ✓ Lower current carrying capacity for each converter ✗ Generator requires special design ✗ Power handling capacity is limited to 1.5 MW
	6ϕ - Generator with Cascaded Modules (Fig. 2.19)	• 0 • 0 • Research Only • LV (up to 3 kV)	✓ Hybrid configuration; LV generator can be used for MV grid integration ✓ Low cost compared to BTB NPC; less torque ripple ✗ Limited speed range; hybrid structure is less favored by industry ✗ Less documented (unknown issues may arise)
	Open-Winding Generator with Distributed Modules (Fig. 2.20)	• 1 • 2 • Mature • LV (up to 1 kV)	✓ High power density and reliability; modular; fault-tolerant ✓ No circulating currents on generator-side; no derating of converters ✗ Complicated generator and open-winding transformer design ✗ High component count due to the large number of converters
	Open-Winding Generator with Cascaded Modules (Fig. 2.21)	• 0 (but have a future) • 0 • Available for VSD • MV (up to 13 kV)	✓ Modular structure; eliminates the step-up transformer; excellent grid current quality ✓ Increases redundancy; LV devices can be used to achieve MV operation ✗ The generator design is complex as the number of phases increase ✗ Fault tolerant operation should be improved
Power Converters without Intermediate DC Link	DMC (Fig. 2.24)	• 0 • 0 • Available for VSD • LV (up to 1 kV)	✓ Low cost; highly reliable as no DC link capacitors are needed ✓ Compact design and less weight ✗ Low voltage gain; requires bidirectional switches ✗ Complicated FRT compliance
	MMMC (Fig. 2.25)	• 0 • 1 (by Yaskawa) • Available for VSD • MV (up to 4 kV)	✓ Compact design and less footprint; no DC capacitors and no balancing issues ✓ Excellent grid current quality due to less dv/dt and phase-shifting transformer ✗ Component count is high compared to BTB NPC; less reliable due to more devices ✗ Possible *LC* resonances; requires bulky phase-shifting transformer

Link to Next Chapters:

- Two-level and multilevel VSCs, as well as CSCs, are reviewed in Chapter 5 to establish the correlation (modeling) between the input and output variables in terms of converter switching states. The induction and synchronous generators are studied again in Chapter 6 for dynamic modeling and model predictive control.
- The wind generator and power converter configurations are analyzed in Chapters 9 to 12 to study WECS performance during normal grid operating conditions.

REFERENCES

1. V. Yaramasu, B. Wu, P. C. Sen, S. Kouro, and M. Narimani, "High-power wind energy conversion systems: State-of-the-art and emerging technologies," *Proceedings of the IEEE*, vol. 103, no. 5, pp. 740–788, May 2015.
2. B. Wu, *High-Power Converters and AC Drives*, 1st ed. Hoboken, NJ: Wiley-IEEE Press, March 2006.
3. S. Kouro, J. Leon, D. Vinnikov, and L. Franquelo, "Grid-connected photovoltaic systems: An overview of recent research and emerging PV converter technology," *IEEE Industrial Electronics Magazine*, vol. 9, no. 1, pp. 47–61, March 2015.
4. V. Yaramasu, "Predictive control of multilevel converters for megawatt wind energy conversion systems," Ph.D. Dissertation, Ryerson University, Toronto, ON, Canada, 2014, available at: http://digital.library.ryerson.ca/islandora/object/RULA%3A3459.
5. A. Di Gerlando, G. Foglia, M. Iacchetti, and R. Perini, "Analysis and test of diode rectifier solutions in grid-connected wind energy conversion systems employing modular permanent-magnet synchronous generators," *IEEE Transactions on Industrial Electronics*, vol. 59, no. 5, pp. 2135–2146, May 2012.
6. B. Wu, Y. Lang, N. Zargari, and S. Kouro, *Power Conversion and Control of Wind Energy Systems*, 1st ed., ser. IEEE Press Series on Power Engineering. Hoboken, NJ: Wiley-IEEE Press, July 2011.
7. V. Yaramasu and B. Wu, "Predictive control of a three-level boost converter and an NPC inverter for high power PMSG-based medium voltage wind energy conversion systems," *IEEE Transactions on Power Electronics*, vol. 29, no. 10, pp. 5308–5322, October 2014.
8. S. Kouro, J. Rodríguez, B. Wu, S. Bernet, and M. Perez, "Powering the future of industry: High-power adjustable speed drive topologies," *IEEE Industry Applications Magazine*, vol. 18, no. 4, pp. 26–39, July/August 2012.
9. W. Erdman and M. Behnke, *Low Wind Speed Turbine Project Phase II: The Application of Medium-Voltage Electrical Apparatus to the Class of Variable Speed Multi-Megawatt Low Wind Speed Turbines*. National Renewable Energy Laboratory (N.R.E.L.), Golden, CO, USA, 2012.
10. B. Andresen and J. Birk, "A high power density converter system for the Gamesa G10x 4.5 MW wind turbine," in *European Conference on Power Electronics and Applications (EPE)*, Aalborg, Denmark, September 2007, pp. 1–8.
11. F. Blaabjerg and K. Ma, "Future on power electronics for wind turbine systems," *IEEE Journal of Emerging and Selected Topics in Power Electronics*, vol. 1, no. 3, pp. 139–152, September 2013.
12. M. Chinchilla, S. Arnaltes, and J. Burgos, "Control of permanent-magnet generators applied to variable-speed wind-energy systems connected to the grid," *IEEE Transactions on Energy Conversion*, vol. 21, no. 1, pp. 130–135, March 2006.
13. M. Fischer, A. Mendonca, and P. Godin, "Voltage control with wind farms: Current practice with type 4 WTG in Canada," in *IEEE Electrical Power and Energy Conference (EPEC)*, Calgary, AB, Canada, November 2014, pp. 165–169.
14. R. Cardenas and R. Pena, "Sensorless vector control of induction machines for variable-speed wind energy applications," *IEEE Transactions on Energy Conversion*, vol. 19, no. 1, pp. 196–205, March 2004.
15. J. Chivite-Zabalza, C. Girones, A. Carcar, I. Larrazabal, E. Olea, and M. Zabaleta, "Comparison of power conversion topologies for a multi-megawatt off-shore wind turbine, based on commercial power electronic building blocks," in *IEEE Industrial Electronics Conference (IECON)*, Vienna, Austria, November 2013, pp. 5242–5247.
16. Semikron, "Power modules," December 2012, available at: http://www.semikron.com, accessed on August 2013.
17. R. Pena, J. Clare, and G. Asher, "Doubly fed induction generator using back-to-back PWM converters and its application to variable-speed wind-energy generation," *IET Electric Power Applications*, vol. 143, no. 3, pp. 231–241, May 1996.
18. S. Muller, M. Deicke, and R. De Doncker, "Doubly fed induction generator systems for wind turbines," *IEEE Industry Applications Magazine*, vol. 8, no. 3, pp. 26–33, May 2002.
19. R. Cardenas, R. Pena, S. Alepuz, and G. Asher, "Overview of control systems for the operation of DFIGs in wind energy applications," *IEEE Transactions on Industrial Electronics*, vol. 60, no. 7, pp. 2776–2798, July 2013.
20. J. Morren and S. W. H. De Haan, "Ridethrough of wind turbines with doubly-fed induction generator during a voltage dip," *IEEE Transactions on Energy Conversion*, vol. 20, no. 2, pp. 435–441, June 2005.
21. Z. Xu, R. Li, H. Zhu, D. Xu, and C. Zhang, "Control of parallel multiple converters for direct-drive permanent-magnet wind power generation systems," *IEEE Transactions on Power Electronics*, vol. 27, no. 3, pp. 1259–1270, March 2012.
22. E. Bueno, S. Cobreces, F. Rodriguez, A. Hernandez, and F. Espinosa, "Design of a back-to-back NPC converter interface for wind turbines with squirrel-cage induction generator," *IEEE Transactions on Energy Conversion*, vol. 23, no. 3, pp. 932–945, September 2008.
23. A. Yazdani and R. Iravani, "A neutral-point clamped converter system for direct-drive variable-speed wind power unit," *IEEE Transactions on Energy Conversion*, vol. 21, no. 2, pp. 596–607, June 2006.
24. S. Alepuz, A. Calle, S. Busquets-Monge, S. Kouro, and B. Wu, "Use of stored energy in PMSG rotor inertia for low-voltage ride-through in back-to-back NPC converter-based wind power systems," *IEEE Transactions on Industrial Electronics*, vol. 60, no. 5, pp. 1787–1796, May 2013.
25. J. Rodríguez, S. Bernet, P. Steimer, and I. Lizama, "A survey on neutral-point-clamped inverters," *IEEE Transactions on Industrial Electronics*, vol. 57, no. 7, pp. 2219–2230, July 2010.
26. S. Kouro, M. Malinowski, K. Gopakumar, J. Pou, L. Franquelo, B. Wu, J. Rodríguez, M. Perez, and J. Leon, "Recent advances and industrial applications of multilevel converters," *IEEE Transactions on Industrial Electronics*, vol. 57, no. 8, pp. 2553–2580, August 2010.

27. J. Pou, J. Zaragoza, S. Ceballos, M. Saeedifard, and D. Boroyevich, "A carrier-based PWM strategy with zero-sequence voltage injection for a three-level neutral-point-clamped converter," *IEEE Transactions on Power Electronics*, vol. 27, no. 2, pp. 642–651, February 2012.

28. N. Celanovic and D. Boroyevich, "A fast space-vector modulation algorithm for multilevel three-phase converters," *IEEE Transactions on Industry Applications*, vol. 37, no. 2, pp. 637–641, March 2001.

29. O. Senturk, L. Helle, S. Munk-Nielsen, P. Rodriguez, and R. Teodorescu, "Power capability investigation based on electrothermal models of press-pack IGBT three-level NPC and ANPC VSCs for multimegawatt wind turbines," *IEEE Transactions on Power Electronics*, vol. 27, no. 7, pp. 3195–3206, July 2012.

30. M. Perez, S. Bernet, J. Rodriguez, S. Kouro, and R. Lizana, "Circuit topologies, modeling, control schemes, and applications of modular multilevel converters," *IEEE Transactions on Power Electronics*, vol. 30, no. 1, pp. 4–17, January 2015.

31. S. Debnath, J. Qin, B. Bahrani, M. Saeedifard, and P. Barbosa, "Operation, control, and applications of the modular multilevel converter: A review," *IEEE Transactions on Power Electronics*, vol. 30, no. 1, pp. 37–53, January 2015.

32. J. Dai, D. Xu, and B. Wu, "A novel control scheme for current-source-converter-based PMSG wind energy conversion systems," *IEEE Transactions on Power Electronics*, vol. 24, no. 4, pp. 963–972, April 2009.

33. B. Wu, J. Pontt, J. Rodríguez, S. Bernet, and S. Kouro, "Current-source converter and cycloconverter topologies for industrial medium-voltage drives," *IEEE Transactions on Industrial Electronics*, vol. 55, no. 7, pp. 2786–2797, July 2008.

34. A. S. Mikhail, K. L. Cousineau, L. H. Howes, E. William, and H. William, "Variable speed distributed drive train wind turbine system," May 2006, United States Patent, US 7,042,110 B2.

35. Y. Xia, J. Fletcher, S. Finney, K. Ahmed, and B. Williams, "Torque ripple analysis and reduction for wind energy conversion systems using uncontrolled rectifier and boost converter," *IET Renewable Power Generation*, vol. 5, no. 5, pp. 377–386, September 2011.

36. V. Yaramasu, B. Wu, M. Rivera, and J. Rodriguez, "A new power conversion system for megawatt PMSG wind turbines using four-level converters and a simple control scheme based on two-step model predictive strategy – Part I: Modeling and theoretical analysis," *IEEE Journal of Emerging and Selected Topics in Power Electronics*, vol. 2, no. 1, pp. 3–13, March 2014.

37. E. Levi, "Multiphase electric machines for variable-speed applications," *IEEE Transactions on Industrial Electronics*, vol. 55, no. 5, pp. 1893–1909, May 2008.

38. Z. Zhu and J. Hu, "Electrical machines and power-electronic systems for high-power wind energy generation applications: Part I – market penetration, current technology and advanced machine systems," *COMPEL: The International Journal for Computation and Mathematics in Electrical and Electronic Engineering*, vol. 32, no. 1, pp. 7–33, 2013.

39. X. Xin and L. Hui, "Research on multiple boost converter based on MW-level wind energy conversion system," in *IEEE International Conference on Electrical Machines and Systems (ICEMS)*, Nanjing, China, vol. 2, September 2005, pp. 1046–1049.

40. H. S. Che, E. Levi, M. Jones, M. Duran, W.-P. Hew, and N. Abd Rahim, "Operation of a six-phase induction machine using series-connected machine-side converters," *IEEE Transactions on Industrial Electronics*, vol. 61, no. 1, pp. 164–176, January 2014.

41. S. Brisset, D. Vizireanu, and P. Brochet, "Design and optimization of a nine-phase axial-flux PM synchronous generator with concentrated winding for direct-drive wind turbine," *IEEE Transactions on Industry Applications*, vol. 44, no. 3, pp. 707–715, May 2008.

42. X. Yuan, J. Chai, and Y. Li, "A transformer-less high-power converter for large permanent magnet wind generator systems," *IEEE Transactions on Sustainable Energy*, vol. 3, no. 3, pp. 318–329, July 2012.

43. M. Parker, C. Ng, and L. Ran, "Fault-tolerant control for a modular generator-converter scheme for direct-drive wind turbines," *IEEE Transactions on Industrial Electronics*, vol. 58, no. 1, pp. 305–315, January 2011.

44. F. Iov, F. Blaabjerg, J. Clare, P. Wheeler, A. Rufer, and A. Hyde, "UNIFLEX-PM - A key-enabling technology for future European electricity networks," *European Power Electronics and Drives Association Journal*, vol. 19, no. 4, pp. 6–16, 2009.

45. E. Yamamoto, H. Hara, T. Uchino, M. Kawaji, T. Kume, J.-K. Kang, and H.-P. Krug, "Development of MCs and its applications in industry [Industry Forum]," *IEEE Industrial Electronics Magazine*, vol. 5, no. 1, pp. 4–12, March 2011.

46. P. Wheeler, J. Rodríguez, J. Clare, L. Empringham, and A. Weinstein, "Matrix converters: a technology review," *IEEE Transactions on Industrial Electronics*, vol. 49, no. 2, pp. 276–288, April 2002.

47. R. Cardenas, R. Pena, P. Wheeler, J. Clare, and G. Asher, "Control of the reactive power supplied by a WECS based on an induction generator fed by a matrix converter," *IEEE Transactions on Industrial Electronics*, vol. 56, no. 2, pp. 429–438, February 2009.

48. J. Kang, N. Takada, E. Yamamoto, and E. Watanabe, "High power matrix converter for wind power generation applications," in *International Conference on Power Electronics and Energy Conversion Congress and Exposition Asia*, Jeju, South Korea, June 2011, pp. 1331–1336.

49. R. Pena, R. Cardenas, E. Reyes, J. Clare, and P. Wheeler, "Control of a doubly fed induction generator via an indirect matrix converter with changing DC voltage," *IEEE Transactions on Industrial Electronics*, vol. 58, no. 10, pp. 4664–4674, October 2011.

50. J. Wang, B. Wu, D. Xu, and N. Zargari, "Multimodular matrix converters with sinusoidal input and output waveforms," *IEEE Transactions on Industrial Electronics*, vol. 59, no. 1, pp. 17–26, January 2012.

CHAPTER 3

OVERVIEW OF DIGITAL CONTROL TECHNIQUES

3.1 INTRODUCTION

The previous chapter clarifies that power electronic converters are important elements for Type 3 and 4 variable-speed wind energy conversion systems (WECS). The general, technical, operational, and legal requirements for these systems (e.g., low cost of energy, extraction of maximum possible energy from wind, superior grid power quality, and compliance with the strict grid codes (refer to Chapter 2, Section 2.2 for more details)) are often fulfilled by the following major items/research areas:

- **Semiconductor Devices:** Narrow and wide bandgap materials (e.g., silicon (Si), silicon carbide (SiC), and gallium nitride (GaN)) are used to realize highly energy efficient "semiconductor devices," including insulated-gate bipolar transistors (IGBTs) and integrated gate-commutated thyristors (IGCTs).
- **Power Converter:** An optimal arrangement of the aforementioned semiconductor devices with or without DC-link elements (e.g., inductors or capacitors), as well as generator- and grid-side harmonic filters, leads to a configuration called "power converter." This concept applies to both classical and advanced power converter topologies.
- **Digital Control:** The proper switch mode operation (i.e., turn *on/off*) of semiconductor devices in a power converter is often accomplished by using a "digital controller." In particular, data on wind speed, generator/grid/harmonic-filter voltages/currents, and DC-link voltages are used as feedback signals to the digital control system to generate optimal gating signals for semiconductor devices in a power converter.

Model Predictive Control of Wind Energy Conversion Systems, First Edition. Venkata Yaramasu and Bin Wu.

Companion Website: www.wiley.com/go/yaramasu_wu_mpc_wecs

Figure 3.1 shows a variable-speed WECS that possesses the distinct categories explained in the preceding paragraph. A power converter consisting of semiconductor devices, DC-link elements, generator- and grid-side harmonic filters, and a digital control system are enclosed together in a cabinet and placed in a nacelle or in the top portion of a turbine tower. The operational requirements for WECS are often fulfilled by properly controlling the power converters. In this chapter, "digital control" is only applicable for electrical energy conversion systems. To simplify the analysis, wind turbine pitch control, yaw control, and so on, are excluded from the scope of digital control.

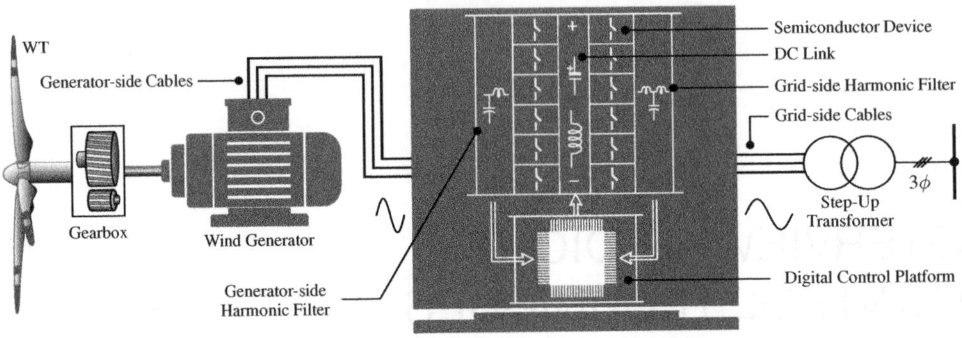

Figure 3.1 Type 3/4 WECS with a power converter and digital control system.

The previous chapter is focused on the discussion of semiconductor devices and power converters. The digital control of power converters is an active research topic and it is constantly evolving according to the technological developments in semiconductor devices, control platforms, control requirements, power quality standards, and grid code requirements [1–3]. Chapter 1, Section 1.7.5 provided an overview of the classical and advanced digital control techniques. This chapter extends the discussion regarding these techniques, including predictive control. In particular, the operating principle of these control techniques is analyzed with an intuitive example of generator/grid current control.

Chapter Overview

- Section 3.2 discusses the past, present, and future of digital control platforms. Section 3.3 presents reference frame theory, which is the backbone for designing and analyzing digital control techniques.
- Section 3.4 introduces the digital control of power conversion systems and explains the block diagram of digital control system for the current control of two-level (2L) voltage source converter (VSC).
- Section 3.5 analyzes the classical control techniques, including hysteresis and linear controls. The procedures for designing the modulation stage and tuning of linear controllers are also presented. Section 3.6 briefly explains advanced control techniques, such as sliding mode, intelligent, and predictive controls.
- Section 3.7 explores predictive control techniques with and without modulation stage. The operating principles of deadbeat predictive control and finite control-set model predictive control (FCS-MPC) are also examined.
- Section 3.8 compares the digital control techniques and enumerates the important observations. Finally, Section 3.9 presents the summary of the chapter.

3.2 THE PAST, PRESENT, AND FUTURE OF CONTROL PLATFORMS

As mentioned earlier, the control theory for power electronics emerged in a rigorous manner, in parallel with the development of semiconductor devices and ahead of the evolution of control platforms. This section discusses the evolution in control platforms and how these platforms help power electronics implement control schemes in real time. Figure 3.2 graphically presents the evolution in semiconductor devices and control platforms.

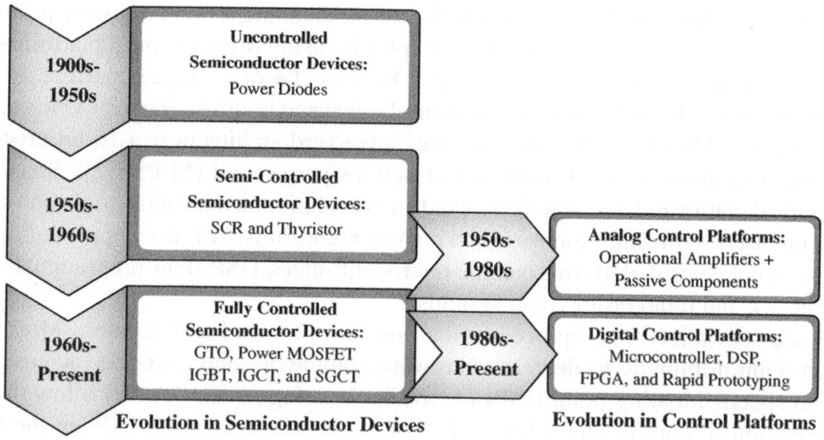

Figure 3.2 Evolution in semiconductor devices and associated control platforms.

An ideal switch is either fully *opened* or *closed* without absorbing any power. When closed, a switch allows the passage of any amount of current without any on-state voltage drop. By contrast, an open switch can withstand the voltage applied across terminals and does not allow the passage of any current. Since the 1900s, many semiconductor devices have been developed for a switch to obtain such ideal characteristics. These developed devices include diodes, power diodes, silicon-controlled rectifier (SCR), gate turn-off thyristor (GTO), power MOSFET, IGBT, reverse blocking (RB)-IGBT, IGCT, and symmetrical GCT (SGCT) [4]. These devices are broadly classified as uncontrolled, semi-controlled, and fully controlled.

The power diode belongs to the family of uncontrolled semiconductor devices because its switching *on* and *off* instances depend on natural commutation and power circuits. This device has been widely used in industrial applications from the 1950s and is still widely used. The operation of power diodes does not require any additional control platform.

Gating terminals have been incorporated into second-generation semiconductor devices to turn *on/off* switches according to the desired performance. Early developments in these devices include SCR and thyristor, which paved the way to the power electronics era. These switching devices are turned-on by the control scheme, but their turn-off instance cannot be controlled similarly to power diodes. Therefore, SCR and thyristor belong to the family of semi-controlled devices. The switching signals (or firing angles) for SCR are generated by analog control platforms to regulate the output voltage. Analog control platforms are usually realized by using operational amplifiers and passive components, and feature simplicity and low implementation cost. However, these platforms present several drawbacks, such as large number of components, reduced system reliability, sensitivity to environmental disturbances (e.g., noise and temperature), and aging.

Since 1960s, fully controlled semiconductor devices (e.g., power MOSFET, IGBT, RB-IGBT, IGCT, and SGCT) have been developed to access turn-on and turn-off instants and obtain characteristics close to an ideal switch. In the early days, analog control platforms were used to generate gating signals based on pulse width modulation (PWM) or hysteresis control. To achieve high-performance operation, advanced control schemes were developed in the 1970s that involved powerful calculations, complex strategies, and math-intensive algorithms. However, the poor computational capability of analog circuits did not support the real-time implementation of these advanced control schemes.

Control algorithms involving large number of mathematical operations are performed quickly and repeatedly on a series of data samples by using digital control platforms. The introduction of the digital signal processor (DSP) by `Texas Instruments` in 1983 brought breakthroughs in the implementation of advanced control schemes. One example of early DSPs is TMS320C14, which features a Harvard architecture, a 16-bit processor, and an operational power of 10 million instructions per second (MIPS) [2]. Since then, digital control platforms have rapidly evolved and revolutionized the industrial control field by allowing users to develop sophisticated control algorithms. A few examples of modern real-time digital control platforms include microcontrollers, DSP, field-programmable gate array (FPGA), and rapid prototyping systems.

DSP is a specialized microprocessor that has an optimized architecture. Most DSPs use fixed-point arithmetic to decrease hardware complexity and cost and increase computational speed. Floating-point DSPs are available at high cost, but they allow the easy implementation of control algorithms [3]. The most popularly used DSPs in the power electronics control area are fixed-point TMS320F2812 and floating-point TMS320F28335, which offer 150 MIPS computational power with a clock frequency of 150 MHz. `Texas Instruments`, the market leader in general-purpose DSPs, produces the C6000 series fixed- and floating-point DSPs. The top models have a clock speed of 1.2 GHz and can achieve a computational power of 8000 MIPS. FPGAs were extensively developed in the 1990s. By the end of the decade, these digital control platforms were applied in the power electronics field to implement complex digital computations [5]. The advantages of FPGAs include flexible hardware, parallel computations, high speeds, reliability, and fast I/Os. DSP and FPGA platforms can be collectively used to share computations and execute complex algorithms but at the expense of high cost [6]. `Xilinx` and `Altera` are the current FPGA market leaders with over 80% of market share.

Rapid prototyping platforms, such as `dSPACE` DS1103/1104, use floating-point DSPs internally and link them to simulation softwares (e.g., `MATLAB/Simulink`). DS1103 can attain a computational power of 2500 MIPS with a clock frequency of 1 GHz. Therefore, this platform is more powerful than the earlier versions of DSP (TMS320C14). Rapid prototyping platforms are widely used in the academia to quickly move from the "concept/idea" phase to the "experimental prototyping" phase. Given that rapid prototyping platforms use the same microprocessor of industrial products, software programming can be transferred to industry (or product development) "as is" or with slight modifications.

FCS-MPC requires digital control platforms with high computational power to perform several calculations and online optimizations. Many scholarly works have proven that state-of-the-art digital control platforms, such as DSPs and FPGAs, can easily handle the high computational capacity needed by FCS-MPC [7]. More powerful digital control platforms are expected to be developed in the near future, thus resulting in lower costs and higher computational capacity than existing control platforms. As such, users do not need to worry about the computational power needed to develop advanced control schemes and can incorporate many constraints to improve system response.

3.3 REFERENCE FRAME THEORY

Reference frame theory uses mathematical transformation to simplify the modeling, analysis, and simulation of balanced three-phase circuits, power converters, and electric machines. Advanced digital control schemes greatly rely on this theory to simplify the design and complexity of algorithms. Three major types of reference frames are widely used in electrical engineering [8]. These reference frames are classified on the basis of the speed of reference frame and nature of the variables involved.

- **Natural** (abc) **Reference Frame:** Reference frame has a speed (ω) of zero and is referred to as a three-phase stationary frame. Three-phase variables exhibit a time-varying (AC) nature with a 120° phase displacement. This frame represents the true correlation between the machine or power converter construction and the corresponding mathematical model. For example, Chapter 5 illustrates that the relationship between the power converter output voltages and switching states is formulated in the natural frame.
- **Stationary** ($\alpha\beta$) **Reference Frame:** The speed of the reference frame (ω) is still zero, but the variables are represented as two-phase variables, which exhibit a time-varying (AC) nature with a 90° phase displacement. Hence, these variables are also denoted in the literature as a two-phase stationary frame. The transformation of three-phase time-varying quantities into two-phase time-varying quantities simplifies the modeling and analysis. $\alpha\beta$-frame is widely used in the direct torque control (DTC) and space vector modulation (SVM) of wind generators and power converters, respectively. In FCS-MPC, such a frame is extensively used to reduce the computational burden.
- **Synchronous** (dq) **Reference Frame:** This reference frame rotates at a synchronous angular speed ω in space and is often called a two-phase rotating frame. Two-phase variables exhibit a time-invariant (DC) nature. A dq-frame is commonly used in the field-oriented control (FOC) and voltage-oriented control (VOC) of wind generators and grid-connected power converters, respectively. The reason behind this condition is that the proportional-integral (PI) regulator is tuned in a simplified manner.

This section discusses the transformation of variables between the above reference frames. Electrical variables are expressed in generic form f, which represents voltages, currents, flux linkages, and switching signals. To simplify the analysis, the space phasor concept is introduced with three-phase balanced variables. This concept is then extended to two-phase stationary and synchronous variables.

3.3.1 Definition of Natural Frame Space Phasor

Space phasor carries information on the amplitude, phase angle, and frequency of the corresponding three-phase function [9]. The generic three-phase balanced electrical variables are considered as follows:

$$
\begin{aligned}
f_a(t) &= F\cos\left(\omega t + \theta_0\right) \\
f_b(t) &= F\cos\left(\omega t + \theta_0 - \frac{2\pi}{3}\right) \\
f_c(t) &= F\cos\left(\omega t + \theta_0 - \frac{4\pi}{3}\right)
\end{aligned}
\tag{3.1}
$$

where F, ω, θ_0, and t are the peak amplitude (V, A, or Wb), angular (radian) frequency (rad/s), initial phase angle (rad), and time value (s), respectively.

The three-phase function in (3.1) is depicted by an equivalent natural frame space phasor as shown below:

$$\mathbf{f_{abc}}(t) = \frac{2}{3}\left[f_a(t) + \mathfrak{a}\, f_b(t) + \mathfrak{a}^2\, f_c(t)\right] \tag{3.2}$$

where $\mathfrak{a} = e^{j\frac{2\pi}{3}}$ is a Fortescue operator. This operator is a unitary vector that represents 120° displacement among phases and yields the expression $\mathfrak{a}^0 + \mathfrak{a}^1 + \mathfrak{a}^2 = 0$.

The relationship between the space phasor and corresponding three-phase variables is illustrated in Figure 3.3, where phase-a axis is the reference point. The three-phase a-, b-, and c-axes are $2\pi/3$ apart in space and are stationary.

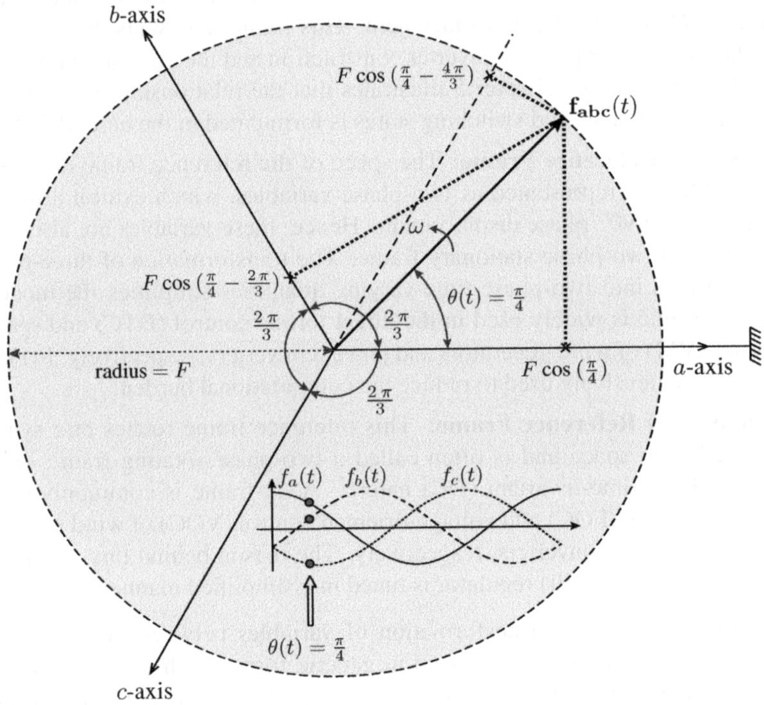

Figure 3.3 Representation of three-phase variables and corresponding space phasor at $\theta(t) = \frac{\pi}{4}$.

The space phasor $\mathbf{f_{abc}}(t)$ rotates counterclockwise at an angular speed ω with respect to the reference point. In a fundamental cycle, the space phasor makes one complete revolution in space. The equivalent phase angle of the space phasor is given by the following:

$$\theta(t) = \theta_0 + \int_0^t \omega(\tau)\, d\tau. \tag{3.3}$$

The three-phase variables are also shown at the bottom of Figure 3.3 for the phase angle $\theta(t) = \pi/4$ and initial phase angle $\theta_0 = 0°$. The magnitude of $f_a(t)$ is greater than that of $f_b(t)$, and the magnitude of $f_c(t)$ is negative. These magnitude values change over time as the space phasor rotates in space. Nonetheless, the tip of $\mathbf{f_{abc}}(t)$ always moves along the circumference of a circle centered at the complex plane. In other words, the tip of $\mathbf{f_{abc}}(t)$ is always equal to the radius value, F.

3.3.2 Transformation Between Natural and Stationary Frames

The natural frame to stationary frame transformation projects the three-phase time-varying variables along the a-, b-, and c-phase axes onto two-phase time-varying variables along a pair of orthogonal axes (i.e., α and β). The graphical representation of natural to stationary frame transformation is shown in Figure 3.4. The c-phase axis is not shown to simplify the illustration. The α and β axes are stationary and do not rotate in space.

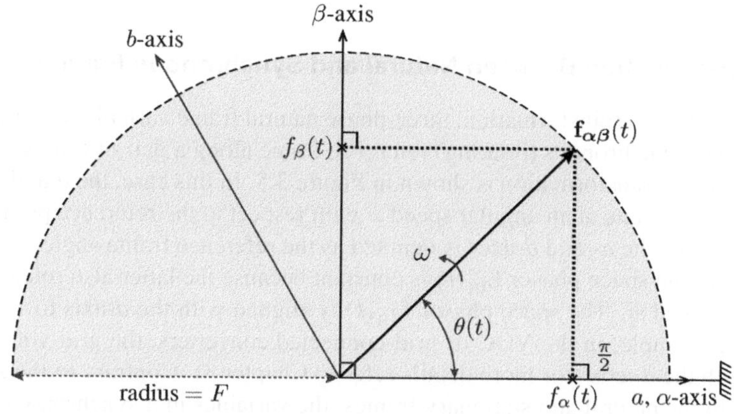

Figure 3.4 Representation of stationary frame variables and corresponding space phasor.

The natural frame space phasor $\mathbf{f_{abc}}(t)$ is decomposed into real and imaginary components to form a stationary frame space phasor:

$$\mathbf{f}_{\alpha\beta}(t) = [f_\alpha(t) + jf_\beta(t)] \tag{3.4}$$

where sinusoidal functions $f_\alpha(t)$ and $f_\beta(t)$ are α- and β-axis components of $\mathbf{f}_{\alpha\beta}(t)$, respectively. Space phasor $\mathbf{f}_{\alpha\beta}(t)$ moves at an angular speed ω with respect to the reference point. When the space phasor moves in space, the magnitude values of $f_\alpha(t)$ and $f_\beta(t)$ change over time similar to the case of f_a, f_b, and f_c.

The phase angle of the space phasor is deducted from the real and imaginary components of $\mathbf{f}_{\alpha\beta}(t)$, as shown below:

$$\theta(t) = \tan^{-1}\left[\frac{f_\beta(t)}{f_\alpha(t)}\right]. \tag{3.5}$$

The transformation of natural frame variables into stationary frame variables is also referred to as the $abc/\alpha\beta$ transformation:

$$\begin{bmatrix} f_\alpha(t) \\ f_\beta(t) \end{bmatrix} = \underbrace{\frac{2}{3}\begin{bmatrix} 1 & -\frac{1}{2} & -\frac{1}{2} \\ 0 & \frac{\sqrt{3}}{2} & -\frac{\sqrt{3}}{2} \end{bmatrix}}_{\mathbf{T}_{abc/\alpha\beta}} \begin{bmatrix} f_a(t) \\ f_b(t) \\ f_c(t) \end{bmatrix} \tag{3.6}$$

where $\mathbf{T}$ denotes the transformation matrix. For a three-phase balanced system, $f_a(t) + f_b(t) + f_c(t) = 0$; thus, $f_\alpha(t)$ becomes equal to $f_a(t)$. The coefficient 2/3 is arbitrarily added to the equation to preserve equal magnitudes for three- and two-phase variables after transformation [10].

The inverse transformation (i.e., $\alpha\beta/abc$) from the stationary frame variables into the natural frame variables is obtained as follows:

$$\begin{bmatrix} f_a(t) \\ f_b(t) \\ f_c(t) \end{bmatrix} = \underbrace{\begin{bmatrix} 1 & 0 \\ -\frac{1}{2} & \frac{\sqrt{3}}{2} \\ -\frac{1}{2} & -\frac{\sqrt{3}}{2} \end{bmatrix}}_{\mathbf{T}_{\alpha\beta/abc}} \begin{bmatrix} f_\alpha(t) \\ f_\beta(t) \end{bmatrix}. \tag{3.7}$$

3.3.3 Transformation Between Natural and Synchronous Frames

Similar to the $abc/\alpha\beta$ transformation, three-phase natural frame variables are transformed into a two-phase synchronous (rotating) reference frame along a pair of orthogonal axes d and q. The abc/dq transformation is shown in Figure 3.5. In this case, the d and q axes are not stationary and rotate at an angular speed ω with respect to the reference point (a-axis). The angle between the a- and d-axes is denoted as the reference frame angle. The angle ϕ between d-axis and space phasor $\mathbf{f_{dq}}(t)$ is constant because the latter also rotates in space at an angular speed ω. The space phasor $\mathbf{f_{dq}}(t)$ is aligned with the d-axis to simplify the analysis. For example, in the VOC of grid-connected converters, the grid voltage vector is aligned with the d-axis (for more details refer to Chapter 8). Contrary to the AC nature of variables in the natural and stationary frames, the variables in a synchronous reference frame exhibit a time-invariant (DC) nature because the dq axes rotate with respect to the reference point.

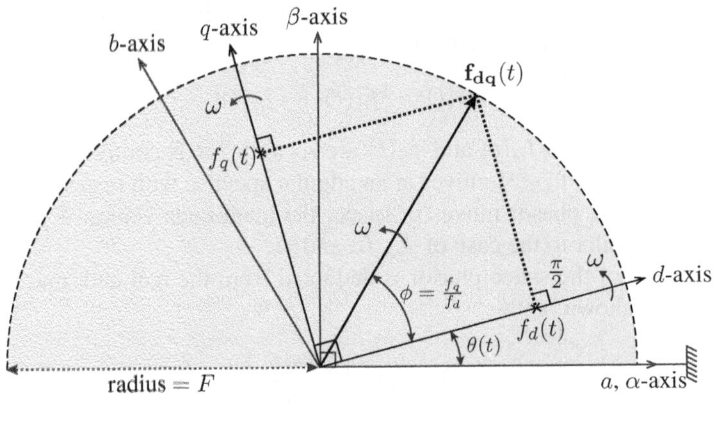

Figure 3.5 Natural (abc) frame voltages and corresponding synchronous (dq) frame voltages.

Similar to (3.4), the synchronous frame space phasor is defined in terms of orthogonal dq axes variables:

$$\mathbf{f_{dq}}(t) = [f_d(t) + j f_q(t)]. \tag{3.8}$$

By simple trigonometrical calculations, the transformation of the abc-frame to the dq-frame is obtained as follows:

$$\begin{bmatrix} f_d(t) \\ f_q(t) \end{bmatrix} = \underbrace{\frac{2}{3}\begin{bmatrix} \cos\theta & \cos\left(\theta - \frac{2\pi}{3}\right) & \cos\left(\theta - \frac{4\pi}{3}\right) \\ -\sin\theta & -\sin\left(\theta - \frac{2\pi}{3}\right) & -\sin\left(\theta - \frac{4\pi}{3}\right) \end{bmatrix}}_{\mathbf{T}_{abc/dq}} \begin{bmatrix} f_a(t) \\ f_b(t) \\ f_c(t) \end{bmatrix}. \tag{3.9}$$

By contrast, the transformation of the dq-frame to the abc-frame is determined by the following equation:

$$\begin{bmatrix} f_a(t) \\ f_b(t) \\ f_c(t) \end{bmatrix} = \underbrace{\begin{bmatrix} \cos\theta & -\sin\theta \\ \cos\left(\theta - \frac{2\pi}{3}\right) & -\sin\left(\theta - \frac{2\pi}{3}\right) \\ \cos\left(\theta - \frac{4\pi}{3}\right) & -\sin\left(\theta - \frac{4\pi}{3}\right) \end{bmatrix}}_{\mathbf{T}_{dq/abc}} \begin{bmatrix} f_d(t) \\ f_q(t) \end{bmatrix}. \tag{3.10}$$

3.3.4 Transformation Between Stationary and Synchronous Frames

Stationary frames variables are transformed into synchronous frame (i.e., $\alpha\beta/dq$ transformation) to convert two-phase stationary variables into rotating variables. Figure 3.5 describes the relationship between the stationary and synchronous frame variables. This transformation matrix uses the angle of the space phasor as defined below:

$$\begin{bmatrix} f_d(t) \\ f_q(t) \end{bmatrix} = \underbrace{\begin{bmatrix} \cos\theta & \sin\theta \\ -\sin\theta & \cos\theta \end{bmatrix}}_{\mathbf{T}_{\alpha\beta/dq}} \begin{bmatrix} f_\alpha(t) \\ f_\beta(t) \end{bmatrix}. \tag{3.11}$$

The inverse transformation (i.e., $dq/\alpha\beta$ transformation) from the synchronous reference frame into stationary reference frame is given by the following:

$$\begin{bmatrix} f_\alpha(t) \\ f_\beta(t) \end{bmatrix} = \underbrace{\begin{bmatrix} \cos\theta & -\sin\theta \\ \sin\theta & \cos\theta \end{bmatrix}}_{\mathbf{T}_{dq/\alpha\beta}} \begin{bmatrix} f_d(t) \\ f_q(t) \end{bmatrix}. \tag{3.12}$$

3.4 DIGITAL CONTROL OF POWER CONVERSION SYSTEMS

As mentioned in Chapter 1 and Section 1.7, the Level I inner current control loops adjust the generator currents ($\mathbf{i_s}$) and grid currents ($\mathbf{i_g}$) at the required values (i.e., $\mathbf{i_s^*}$ and $\mathbf{i_g^*}$) by controlling the "on" and "off" time duration of the power converter switching devices. The overall control performance of WECS greatly depends on the effectiveness of the Level I current control strategy. In fact, the feedback current control of converters is one of the most extensively researched classical subjects in the power electronics field [11].

3.4.1 Block Diagram of Digital Current Control

Figure 3.6 presents the generic block diagram of digital current control for a three-phase power converter to discuss the classical and advanced current control schemes. The power for the grid-connected case flows from the DC link to the AC side of the converter (similar to passive load), whereas the power in the generator case flows from the converter AC side to the DC link (opposite the passive load). The design of current control schemes (except the measurement of current direction) and the performance indices are the same for both grid and generator current controls. Grid voltages and wind generator back EMF do not affect the quality of the current control schemes. Therefore, a simple resistive-inductive (RL) load is considered in this investigation to simplify the analysis.

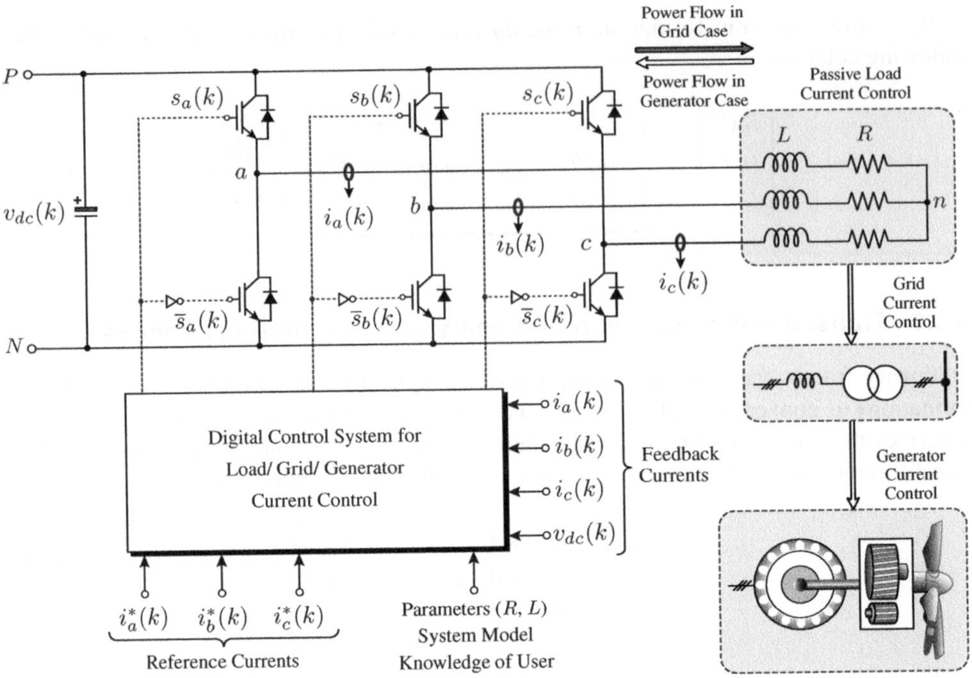

Figure 3.6 Simplified block diagram of digital control system for a current control of 3ϕ-VSC.

The above figure demonstrates that the generic digital control system requires feedback measurements (i.e., v_{dc}, i_a, i_b, and i_c), load parameters (i.e., R and L), power converter model, and prior knowledge of the user about power converter operations. The *abc*-frame reference currents (i.e., i_a^*, i_b^*, and i_c^*) are shown to compare the operating principle of various control schemes. However, for some control schemes, $\alpha\beta$- or dq-frames provide optimal performance with simplicity. For example, the dq-frame is preferred for linear control to achieve good reference tracking and the $\alpha\beta$-frame is favored for predictive control to reduce the number of online calculations. The basic control structure of both classical and advanced control schemes is considered in the discussion. Nevertheless, many other classes or variations are available in each category of control schemes.

3.4.2 Model of Two-Level VSC for Digital Current Control

The 2L-VSC is composed of two IGBTs in each phase, thus leading to a matrix format for a three-phase configuration. In this study, positive and negative DC buses are denoted as P and N, respectively. The DC-link voltage v_{dc} is assumed constant and ripple free. The switching signal for the top and bottom switches operates in a complementary manner. In other words, only one switch is conducted in each phase at any time instant. This condition is expressed as $s_x + \overline{s}_x =$ '1' for all $x \in \{a, b, c\}$. With two switching states in each phase (i.e., $s_x =$ '1' or '0'), the total number of possible switching states for a three-phase 2L-VSC becomes 8 (2^3). Table 3.1 provides a summary of the 2L-VSC switching vector, upper-leg switching signals, and voltage vector. The switching signals dynamically connect load terminals a, b, and c to P or N, thus allowing the converter output phase voltage waveform to switch between two distinct values (i.e., v_{dc} and zero).

Table 3.1 Summary of switching signals and voltage vectors for a 2L-VSC

Switching Combination	Switching Vector	Upper-Leg Switching Signals			Voltage Vector
$\mathbf{s}$	$\mathbf{S}_{\mathbf{abc}}$	s_a	s_b	s_c	$\mathbf{V}$
$\mathbf{s}_0$	[000]	'0'	'0'	'0'	$\mathbf{V}_0 = 0 + j\,0$
$\mathbf{s}_1$	[100]	'1'	'0'	'0'	$\mathbf{V}_1 = +\frac{2}{3}\,v_{dc} + j\,0$
$\mathbf{s}_2$	[110]	'1'	'1'	'0'	$\mathbf{V}_2 = +\frac{1}{3}\,v_{dc} + j\,\frac{\sqrt{3}}{3}\,v_{dc}$
$\mathbf{s}_3$	[010]	'0'	'1'	'0'	$\mathbf{V}_3 = -\frac{1}{3}\,v_{dc} + j\,\frac{\sqrt{3}}{3}\,v_{dc}$
$\mathbf{s}_4$	[011]	'0'	'1'	'1'	$\mathbf{V}_4 = -\frac{2}{3}\,v_{dc} + j\,0$
$\mathbf{s}_5$	[001]	'0'	'0'	'1'	$\mathbf{V}_5 = -\frac{1}{3}\,v_{dc} - j\,\frac{\sqrt{3}}{3}\,v_{dc}$
$\mathbf{s}_6$	[101]	'1'	'0'	'1'	$\mathbf{V}_6 = +\frac{1}{3}\,v_{dc} - j\,\frac{\sqrt{3}}{3}\,v_{dc}$
$\mathbf{s}_7$	[111]	'1'	'1'	'1'	$\mathbf{V}_7 = 0 + j\,0$

With respect to the negative DC bus, the 2L-VSC output voltage is expressed in terms of upper-leg switching signals and DC-link voltage as follows:

$$v_{xN} = v_{dc}\, s_x, \quad \forall x \in \{a, b, c\}. \tag{3.13}$$

The above expression is valid for analyzing the converter performance in continuous-time (CT) or for designing a controller in a discrete-time (DT) frame. The isolated load neutral voltage v_{nN} is measured between the load neutral point n and negative DC-bus N. This voltage is related to three-phase converter phase voltages as demonstrated below:

$$v_{nN} = \frac{v_{aN} + v_{bN} + v_{cN}}{3}. \tag{3.14}$$

The converter output phase-to-neutral voltage is deduced from the phase voltages and v_{nN} as follows:

$$\begin{bmatrix} v_{an} \\ v_{bn} \\ v_{cn} \end{bmatrix} = \begin{bmatrix} v_{aN} \\ v_{bN} \\ v_{cN} \end{bmatrix} - \begin{bmatrix} v_{nN} \\ v_{nN} \\ v_{nN} \end{bmatrix}, \Rightarrow \begin{bmatrix} v_{an} \\ v_{bn} \\ v_{cn} \end{bmatrix} = \begin{bmatrix} \frac{2}{3} & -\frac{1}{3} & -\frac{1}{3} \\ -\frac{1}{3} & \frac{2}{3} & -\frac{1}{3} \\ -\frac{1}{3} & -\frac{1}{3} & \frac{2}{3} \end{bmatrix} \begin{bmatrix} v_{aN} \\ v_{bN} \\ v_{cN} \end{bmatrix}. \tag{3.15}$$

The *abc*-frame CT dynamic model of load current with a passive RL load is formulated in matrix notation as follows:

$$\begin{bmatrix} v_{an}(t) \\ v_{bn}(t) \\ v_{cn}(t) \end{bmatrix} = \begin{bmatrix} R & 0 & 0 \\ 0 & R & 0 \\ 0 & 0 & R \end{bmatrix} \begin{bmatrix} i_a(t) \\ i_b(t) \\ i_c(t) \end{bmatrix} + \begin{bmatrix} L & 0 & 0 \\ 0 & L & 0 \\ 0 & 0 & L \end{bmatrix} \frac{d}{dt} \begin{bmatrix} i_a(t) \\ i_b(t) \\ i_c(t) \end{bmatrix} \tag{3.16}$$

which can be simplified as

$$\mathbf{v}(t) = R\,\mathbf{i}(t) + L\,\frac{d\,\mathbf{i}(t)}{dt}. \tag{3.17}$$

The *abc*-frame model illustrated in (3.16) can be converted into $\alpha\beta$- and dq-frames by applying the corresponding transformation matrices.

3.5 CLASSICAL CONTROL TECHNIQUES

This section discusses two industry-standard classical current control techniques, namely, hysteresis and linear controls.

3.5.1 Hysteresis Control

A hysteresis control scheme for regulating three-phase load currents is displayed in Figure 3.7. Hysteresis control is recognized as a nonlinear control technique [12]. The measured load currents (i.e., i_a, i_b, and i_c) are compared with their respective reference currents (i.e., i_a^*, i_b^*, and i_c^*), and the error in the currents (i.e., Δi_a, Δi_b, and Δi_c) are passed to the relays (also called hysteresis comparators or bang–bang controllers). The relay in each phase determines the switching signals such that the measured loads are confined within the upper and lower band limits set by the user-defined hysteresis bandwidth δ.

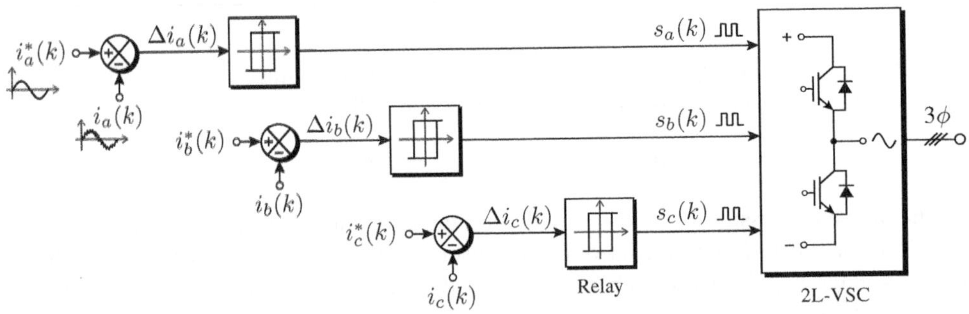

Figure 3.7 Hysteresis current control scheme for regulating three-phase load currents.

The operating principle of single-phase hysteresis relay is exhibited in Figure 3.8 assuming that total bandwidth is symmetrically distributed around the sinusoidal reference current. When the load current hits the upper band limit ($i_a^* + \frac{\delta}{2}$), the switching signal s_a is set to a logic low '0'. On the basis of the notation in (3.13), the converter output voltage v_{aN} becomes zero, thus causing i_a to decay. Similarly, the s_a is set to a logic high '1' when the load current crosses the lower band limit ($i_a^* - \frac{\delta}{2}$). v_{aN} then becomes equal to v_{dc}, thus forcing i_a to increase. This operating principle is summarized as follows:

$$s_x = \begin{cases} \text{`0'} & \text{if} \quad i_x > i_x^* + \frac{\delta}{2} \qquad \text{(forces current to decay)} \\ \text{`1'} & \text{if} \quad i_x < i_x^* - \frac{\delta}{2} \qquad \text{(forces current to rise)} \end{cases} \quad \forall\ x \in \{a, b, c\}. \quad (3.18)$$

The current waveform exhibits a saw-tooth nature, with tips periodically touching the hysteresis band limits. However, for three-phase applications, the shape of current waveform does not periodically exhibit a saw-tooth nature because of the interaction among switching device commutations in three phases. The error in reference tracking can be minimized by reducing the hysteresis bandwidth δ. However, the converter switching frequency increases, thus leading to high switching losses and heat output from the converter.

Hysteresis current control does not require the system parameters, converter model, or prior knowledge of the user, thus making it one of the simplest and robust current control techniques. This approach has been widely used in the power electronics industry since the era of analog control platforms [13]. This approach also does not involve any complex mathematical expressions or reference frame theory and does not need any compli-

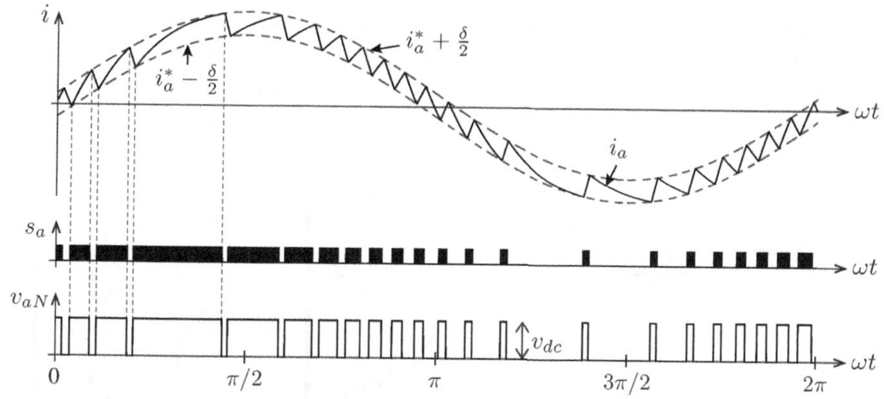

Figure 3.8 Operating principle of single-phase hysteresis current control scheme.

cated analog circuits or high precision digital controllers. During step changes in reference currents, hysteresis control exhibits fast dynamic response because of its modulator-free control structure. The dynamic response time in this method is mainly limited by the switching frequency of converter, load time constant, magnitude of DC-link voltage, and back electromagnetic force of the machine, if any.

The switching frequency of 2L-VSC varies according to filter/load parameters, operating conditions, and hysteresis bandwidth. For some applications, the variable switching frequency is undesirable because of issues on filter resonance and electromagnetic interference. The uncontrollable switching frequency is one of the major drawbacks of this controller but has since been significantly resolved through several attempts. One of the solutions proposed to solve the above-mentioned issues is delta modulation, which compares the current error values at fixed sample times. Another drawback in using hysteresis control is that constraint inclusion (e.g., common-mode voltage reduction and spectrum shaping) cannot be achieved.

As mentioned in Chapter 1, the advanced versions of hysteresis control include DTC and direct power control (DPC). The torque and flux of the drive system are controlled in DTC, whereas the active and reactive powers of the grid are directly controlled in DPC. Hysteresis controllers produce error signals. The relative magnitudes of these error signals are then used by a lookup table to generate the switching signals to the converter. To implement these methods on digital platforms, a high-sampling frequency is needed to force the variables within the hysteresis band limits.

3.5.2 Linear Control

Figure 3.9 demonstrates a linear current control scheme with a PI regulator and modulation stage. The scheme linearizes the nonlinear power converter by employing a cascaded control structure. This technique is well established in literature and widely deployed in practical applications. The measured load currents $\mathbf{i}(k)$ are compared with the reference currents $\mathbf{i}^*(k)$, and the current error $\Delta\mathbf{i}(k)$ is sent as an input to the PI controller. The PI controller then processes the current error and produces reference voltage (also called modulating waveform) $\mathbf{v}^*(k)$ for use with the modulation stage. The modulation stage requires a carrier (triangular) waveform v_{cr} and DC-link voltage v_{dc}, reference frame transformation, comparators, and switching logic internally to produce gating signals for

2L-VSC. PWM, SVM, and selective harmonic elimination (SHE) are the most commonly used modulation schemes. The linear control scheme using PWM/SVM/SHE is operated with a fixed switching frequency dictated by the carrier frequency. The performance of this control scheme greatly depends on the PI controller parameters and modulation type, which are thoroughly discussed in this part.

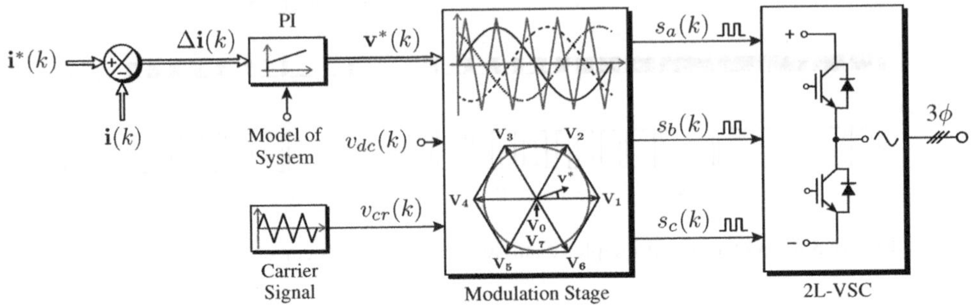

Figure 3.9 Linear current control scheme using PI regulator and modulation stage.

(1) Selection of PI Controller Parameters: The closed-loop response of linear control significantly depends on the nature of reference currents (i.e., AC or DC). As mentioned earlier, the variables in the abc and $\alpha\beta$ reference frames exhibit a sinusoidal nature (AC), whereas those in the dq-frame are time invariant (DC). In the abc and $\alpha\beta$ frames, the load currents follow the reference currents but with a definite steady-state error in both the amplitude and phase angle. In such a condition, the control system must be designed with additional compensation terms, thus resulting in high control complexity. By contrast, the load currents in the dq-frame precisely follow the reference currents without any steady-state errors, during which no phase-shift issues for DC variables are encountered. In practice, current control schemes are implemented in the synchronous frame with additional decoupling terms to compensate for the induced voltages. The measured load/generator/grid currents are usually converted from the abc-frame to the dq-frame. The PI controller acts independently on dq-axes currents and generates the converter reference voltages in dq-frame. These reference voltages are converted back to the abc-frame for use with the modulation stage.

The PI controller parameters should be properly tuned to ensure fast transient response and less steady-state errors. The CT mathematical model of the RL load (first-order plant) and PI controller is used to calculate the proportional and integral gain values, k_p and k_i, respectively. Figure 3.10(a) illustrates a simplified block diagram of closed-loop control scheme for a 2L-VSC, and Figure 3.10(b) depicts the step response of closed-loop control with optimal k_p and k_i values. The design procedure for calculating the k_p and k_i values is presented in this section according to the pole placement methodology given in [9]. The pole placement method helps in exploiting the PI controller parameters such that the control system can provide the desired performance. By using a simple design procedure, this method places the closed-loop poles of the plant in predetermined locations.

As a first step in the design procedure, the open-loop gain of the control system in Figure 3.10(a) is calculated in the Laplace domain as follows:

$$\ell(s) = K(s)\,G(s) = \frac{k_p\,s + k_i}{s} \times \frac{1}{L\,s + R} = \frac{k_p}{s}\left(\frac{s + \frac{k_i}{k_p}}{s + \frac{R}{L}}\right). \tag{3.19}$$

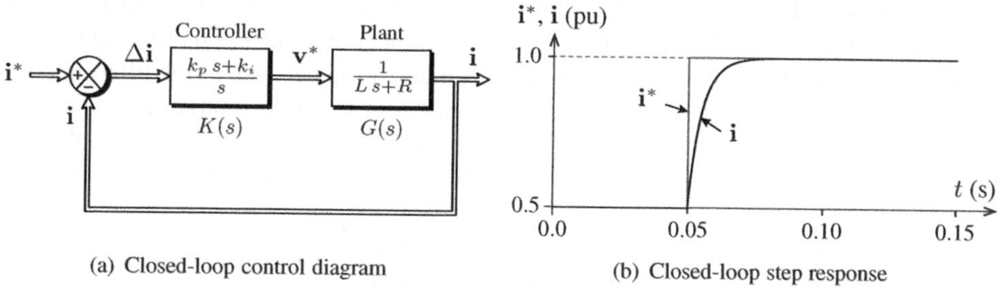

(a) Closed-loop control diagram (b) Closed-loop step response

Figure 3.10 Simplified representation of closed-loop control scheme for a 2L-VSC.

The above expression indicates that the open-loop control system has a stable pole at $p = -R/L$. However, the pole is located close to the origin; hence, the system response is sluggish. To improve the open-loop system response, the pole ($p = -R/L$) and zero ($z = -k_i/k_p$) of the PI controller can be designed in a manner that they cancel each other. Thus, the k_p and k_i values yield the following formulas:

$$p = z \quad \Rightarrow \quad \frac{k_i}{k_p} = \frac{R}{L}, \text{ and } \frac{k_p}{L} = \frac{1}{\tau_i} \quad \Rightarrow \quad k_p = \frac{L}{\tau_i}, \text{ and } k_i = \frac{R}{\tau_i} \tag{3.20}$$

where τ_i represents the desired time constant of the closed-loop system, and $1/\tau_i$ is the closed-loop bandwidth of the system. For high-power applications, setting τ_i value as 10 times smaller than the converter switching frequency f_{sw} is fair enough (i.e., $\tau_i = 1/(f_{sw}/10)$ [9]). From the notations in (3.19) and (3.20), the closed-loop transfer function is obtained as follows:

$$G_i(s) = \frac{\mathbf{i}}{\mathbf{i}^*} = \frac{\ell(s)}{1+\ell(s)} = \frac{\frac{k_p}{L\,s}}{1+\frac{k_p}{L\,s}} = \frac{1}{\tau_i\, s + 1} \tag{3.21}$$

this transfer function is also a first-order transfer function similar to the plant model.

A simple example with 2L-VSC is considered: $R = 10\ \Omega$, $L = 10$ mH, and $f_{sw} = 2000$ Hz. Therefore,

$$\begin{aligned} \tau_i &= 1/(f_{sw}/10) = 1/(2000/10) = 5 \text{ ms} \\ k_p &= L/\tau_i = (10 \times 10^{-3})/(5 \times 10^{-3}) = 2 \\ k_i &= R/\tau_i = 10/(5 \times 10^{-3}) = 2000. \end{aligned} \tag{3.22}$$

Figure 3.10(b) exhibits the closed-loop system response with k_p and k_i values during a step change in reference signal from 0.5 pu to 1.0 pu. The feedback signal follows the reference command with an acceptable rise time. As outlined above, the PI controller parameters are designed on the basis of the plant model and rated parameters. During the other operating conditions and plant parameter variations, the control performance deteriorates, and stability-related issues arise.

(2) Sinusoidal PWM: Sinusoidal PWM (SPWM; or carrier-based PWM) is a simple and popular modulation technique for power converters. Figure 3.11 shows the operating principle of this technique for a 2L-VSC considering phase-a modulating signal v_a^* and triangular carrier signal v_{cr}. The conduction-time duration of the upper-leg switching devices is determined by comparing the carrier signal with the three-phase modulating signals. When

v_{cr} is greater than the magnitude of v_a^*, the gating signal s_a is set to a logic low ('0'). Correspondingly, the 2L-VSC output phase voltage v_{aN} becomes zero, thus forcing the load current i_a to decay. Moreover, i_a exhibits a sinusoidal nature because of the filtering effect of load inductance. Contrarily, when v_{cr} is less than or equal the magnitude of v_a^*, the gating signal s_a is set to a logic high ('1'), thus resulting in $v_{aN} = v_{dc}$. As such, the load current i_a begins to increase. The SPWM operating principle is summarized below:

$$s_x = \begin{cases} \text{'0'} & \text{if} \quad v_{cr} > v_x^* \quad \text{(forces current to decay)} \\ \text{'1'} & \text{if} \quad v_{cr} \le v_x^* \quad \text{(forces current to rise)} \end{cases} \qquad \forall \ x \in \{a, b, c\}. \tag{3.23}$$

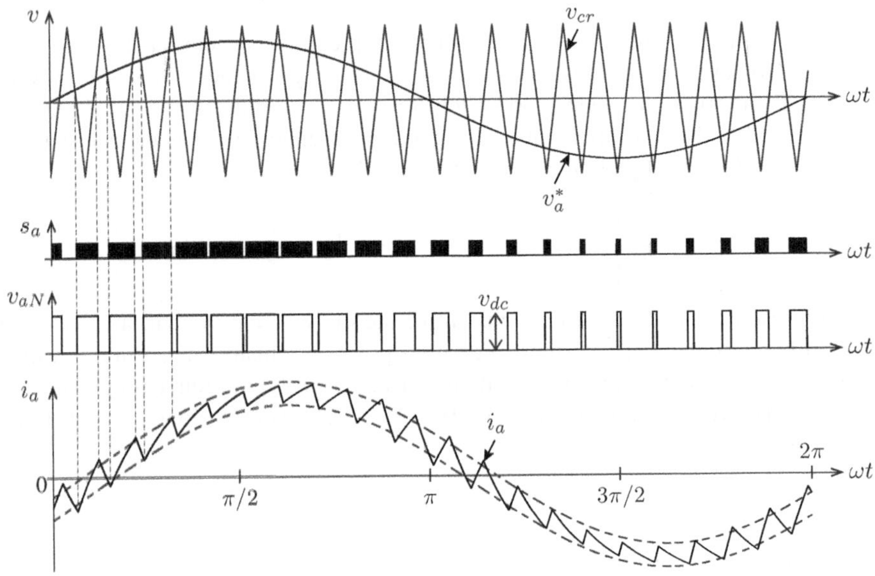

Figure 3.11 Operating principle of SPWM.

The amplitude and frequency modulation indices (m_a and m_f) are important design factors for adjusting the magnitude and frequency of the fundamental frequency component of the 2L-VSC output voltage $v_{ab,1}$, respectively. The m_a is adjusted by varying the amplitude of the modulating signals while maintaining the fixed value of the carrier signals. The m_f is modified by changing the carrier signal frequency while retaining the modulating signal frequency. These mechanisms are expressed as follows:

$$m_a = \frac{\widehat{V}_m^*}{\widehat{V}_{cr}^*}, \qquad m_f = \frac{f_{cr}}{f_m} \tag{3.24}$$

where $\widehat{V}_m^*$ and $\widehat{V}_{cr}^*$ are the peak values of the modulating and carrier signals, respectively, and f_m and f_{cr} are the modulating and carrier signal frequencies, respectively. For a 2L-VSC, the device switching frequency f_{sw} is equal to the carrier frequency f_{cr} ($m_f \times f_m$). An integer m_f value (multiple of the fundamental frequency) leads to a digital synchronous PWM, which is easily implemented in digital control platforms. Characteristic voltage harmonics are concentrated around the switching frequency f_{sw} and its multiples (e.g., $2f_{sw}$ and $3f_{sw}$) [4]. The m_a is a variable ranging from zero to one, and the magnitude of

$v_{ab,1}$ increases linearly with m_a. For $m_a = 1$, the maximum DC-bus voltage utilization is 61.2% (i.e., $v_{ab,1}^{\max} = 0.612\, v_{dc}$), which represents the narrow range of linearity for SPWM. The DC-bus voltage utilization can be increased by adding zero-sequence voltage to the modulating signals. This process is also called the third harmonic injection PWM.

The assumption of the linear model for power converter provides a favorable performance only if a high bandwidth modulation is used, thus leading to a high switching frequency operation. However, low switching frequency operation is important in high-power applications. During such conditions, the magnitude of low-order harmonics significantly increases and causes poor power quality. Therefore, simultaneous low switching frequency operation and high bandwidth closed-loop control are important requirements in power electronics for high-power applications.

(3) Space Vector Modulation: SVM is another widely used modulation technique for the digital control of 2L-VSCs. This method is considered efficient because it offers better harmonic performance and enhanced DC-bus voltage utilization compared with SPWM. With SVM, the vectors closest to the reference voltage vector are selected based on dwell time calculation, and optimal gating signals are produced by switching scheme designs.

Figure 3.12 depicts a simulation model of SVM with six subsystems. The input variables are the three-phase modulating signals (i.e., v_a^*, v_b^*, and v_c^*) produced by a PI controller, and the output variables are upper-leg switching signals. SVM involves several design steps and complex modeling, thus incurring a high computational burden than SPWM. The step-by-step procedure for designing SVM for a 2L-VSC is presented below.

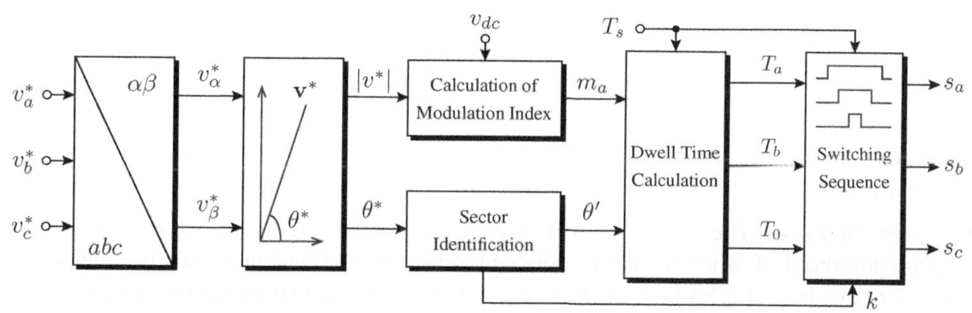

Figure 3.12 Simulation model of SVM for a 2L-VSC [10].

- ***Design Step – 1:*** The three-phase instantaneous modulating signals (i.e., v_a^*, v_b^*, v_c^*) are converted into stationary frame modulating signals (i.e., v_α^* and v_β^*) by using the $abc/\alpha\beta$ transformation given in (3.6). The reference voltage vector in $\alpha\beta$-frame is represented as $\mathbf{v}^* = v_\alpha^* + j\, v_\beta^*$.
- ***Design Step – 2:*** The stationary frame modulating signals are then transformed from the Cartesian coordinates to the polar coordinates:

$$\mathbf{v}^* = |v^*|\, e^{j\theta^*}, \quad |v^*| = \sqrt{{v_\alpha^*}^2 + {v_\beta^*}^2}, \quad \theta^* = \tan^{-1}\left(\frac{v_\beta^*}{v_\alpha^*}\right) \tag{3.25}$$

where $|v^*|$ is the length of the reference vector, and θ^* is the angle between $\mathbf{v}^*$ and α-axis shown in Figure 3.13(a). $|v^*|$ and θ^* are the peak value and frequency of the converter output voltage, respectively. By controlling the reference vector $\mathbf{v}^*$, the magnitude and frequency of the output voltage waveform are precisely regulated.

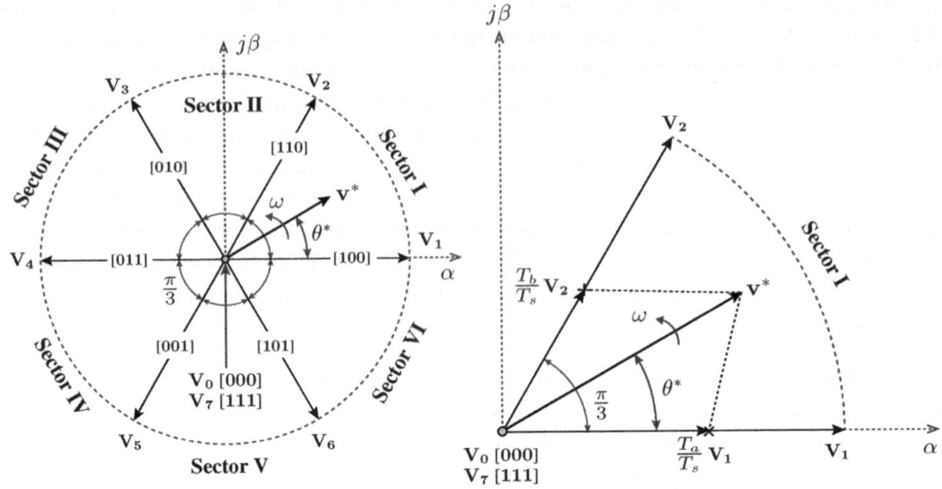

(a) Voltage vectors and sectors

(b) Calculation of reference vector

Figure 3.13 Operating principle of SVM for a 2L-VSC.

- ***Design Step – 3:*** The modulation index m_a is calculated based on the magnitude of the reference voltage and measured DC-link voltage:

$$m_a = \frac{\sqrt{3}\,|v^*|}{v_{dc}}. \tag{3.26}$$

The DC-bus voltage utilization with SVM is 70.7% (i.e., $v_{ab,1}^{\max} = 0.707\, v_{dc}$), which is 15.5% higher than that with SPWM.

- ***Design Step – 4:*** The sector number is computed according to the value of θ^*. The sector number is I when θ^* varies between zero and $\pi/3$ and increases by one when θ^* increases in multiples of $\pi/3$. Figure 3.13(a) illustrates a space vector diagram with voltage vectors and sector numbers for a 2L converter. The six active voltage vectors (i.e., $\mathbf{V_1}$ to $\mathbf{V_6}$) form a regular hexagon with six equal sectors (i.e., I to VI). The zero voltage vectors $\mathbf{V_0}$ and $\mathbf{V_7}$ lie on the center of the hexagon and do not move in space. The redundant zero voltage vectors are used to minimize the VSC switching frequency or to accomplish other control objectives. The reference vector $\mathbf{v}^*$ rotates in space at an angular speed ω. A modified reference voltage vector angle is also computed in a manner that its value always resides within zero and $\pi/3$, as demonstrated below.

$$\theta' = \theta^* - \frac{\pi}{3}(k-1) \tag{3.27}$$

where $k = 1, \cdots, 6$ for sectors I to VI, respectively.

- ***Design Step – 5:*** As shown in Figure 3.13(b), $\mathbf{v}^*$ is synthesized by three nearby stationary vectors (i.e., two active vectors and one zero vector), based on which the gating signals are generated for the converter. Dwell times define the time duration wherein the reference vector $\mathbf{v}^*$ is switched onto the adjacent vectors during a sampling period T_s. The dwell times for two active vectors (i.e., T_a and T_b) and one zero vector (i.e., T_0) are determined with the following formulas:

$$\begin{aligned} T_a &= T_s\, m_a \sin\left(\frac{\pi}{3} - \theta'\right) \\ T_b &= T_s\, m_a \sin\theta' \\ T_0 &= T_s - T_a - T_b. \end{aligned} \tag{3.28}$$

- ***Design Step – 6:*** According to the information about the dwell times, sampling time, and sector number, a switching logic is designed to produce the gating signals for the converter. Figure 3.14 demonstrates a seven-segment switching sequence when $\mathbf{v}^*$ resides in all six sectors. The minimum number of switchings is involved in moving from one switching state to the next. The sampling period T_s is divided into seven segments for the selected vectors.

Sector I	$\mathbf{V}_0$ [000]	$\mathbf{V}_1$ [100]	$\mathbf{V}_2$ [110]	$\mathbf{V}_7$ [111]	$\mathbf{V}_2$ [110]	$\mathbf{V}_1$ [100]	$\mathbf{V}_0$ [000]
Sector II	$\mathbf{V}_0$ [000]	$\mathbf{V}_3$ [010]	$\mathbf{V}_2$ [110]	$\mathbf{V}_7$ [111]	$\mathbf{V}_2$ [110]	$\mathbf{V}_3$ [010]	$\mathbf{V}_0$ [000]
Sector III	$\mathbf{V}_0$ [000]	$\mathbf{V}_3$ [010]	$\mathbf{V}_4$ [011]	$\mathbf{V}_7$ [111]	$\mathbf{V}_4$ [011]	$\mathbf{V}_3$ [010]	$\mathbf{V}_0$ [000]
Sector IV	$\mathbf{V}_0$ [000]	$\mathbf{V}_5$ [001]	$\mathbf{V}_4$ [011]	$\mathbf{V}_7$ [111]	$\mathbf{V}_4$ [011]	$\mathbf{V}_5$ [001]	$\mathbf{V}_0$ [000]
Sector V	$\mathbf{V}_0$ [000]	$\mathbf{V}_5$ [001]	$\mathbf{V}_6$ [101]	$\mathbf{V}_7$ [111]	$\mathbf{V}_6$ [101]	$\mathbf{V}_5$ [001]	$\mathbf{V}_0$ [000]
Sector VI	$\mathbf{V}_0$ [000]	$\mathbf{V}_1$ [100]	$\mathbf{V}_6$ [101]	$\mathbf{V}_7$ [111]	$\mathbf{V}_6$ [101]	$\mathbf{V}_1$ [100]	$\mathbf{V}_0$ [000]
	$\frac{T_0}{4}$	$\frac{T_b}{2}$	$\frac{T_a}{2}$	$\frac{T_0}{2}$	$\frac{T_a}{2}$	$\frac{T_b}{2}$	$\frac{T_0}{4}$
				T_s			

Figure 3.14 SVM seven-segment switching scheme for all six sectors.

SVM provides better control and harmonic performance than SPWM, but at the expense of complex modeling and design procedures. The complexity of SVM increases when it is applied to multilevel converters, in which additional constraints (e.g., load THD improvement, low switching frequency operation, common-mode voltage minimization, and DC-link neutral-point voltage control) should be satisfied by the modulation stage. SHE is an offline modulation scheme that is mainly based on switching angle calculation, wherein specific low-order harmonics in the output waveform are eliminated.

Linear control techniques have been widely used in the drives industry and energy systems. To control the motor torque and flux in a decoupled manner, FOC is used. This process involves the reference frame transformations, two dq-axes PI controllers, and a modulation stage. Similar to FOC, VOC is used to control the grid-connected converter. The net DC-bus voltage and grid reactive power are controlled in a decoupled manner by using VOC. The grid voltage harmonics in this scheme deteriorate the control performance. With hysteresis and linear control techniques, the system constraints and technical requirements (e.g., maximum current, load voltage spectrum shaping, switching losses reduction, and THD minimization) cannot be easily incorporated into the design of the control scheme.

3.6 ADVANCED CONTROL TECHNIQUES

3.6.1 Sliding Mode Control

Figure 3.15 exhibits a sliding mode control (SMC) with a modulation stage for the load current of 2L-VSC. SMC replaces the PI controller in linear control, and the rest of the control block diagram is similar to that displayed in Figure 3.9. Similar to the PI regulator, SMC also requires the converter model and system parameters. SMC is an advanced digital control technique and belongs to the family of variable structure control and adaptive control [14]. This control technique is discrete and nonlinear and properly treats the nonlinear nature of the power converter. The current errors are processed by SMC, which accordingly generates the load voltage references $\mathbf{v}^*$. The control variable (load current) is forced to track or slide along the predefined trajectory [15]. By using SMC, the structure of the controller is intentionally changed according to a preset structure control law to achieve a robust and stable response during system parameter variations and load disturbances.

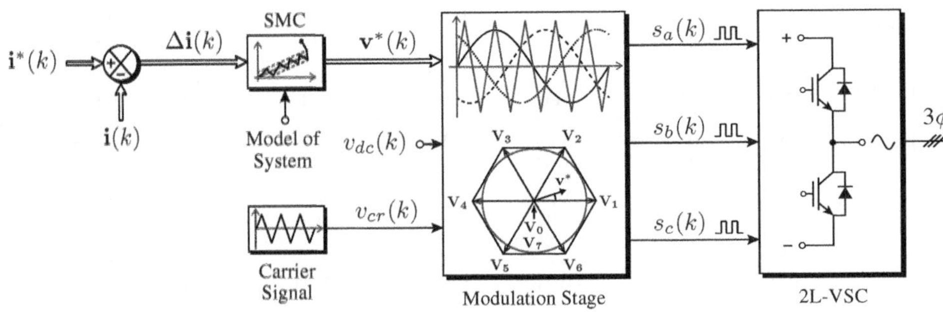

Figure 3.15 Block diagram of SMC with a modulation stage.

Compared with the classical controllers discussed in the preceding sections, SMC provides good dynamic response and optimal performance during all operating conditions. However, this scheme also has drawbacks such as complex mathematical modeling and chattering phenomena. Thus, SMC is not extensively considered for practical applications compared with the classical control techniques.

3.6.2 Intelligent Control

Intelligent control includes various control techniques, such as expert systems, fuzzy logic, artificial neutral network (ANN), and genetic algorithms [16]. In this section, the load current regulation by fuzzy and ANN control are briefly discussed.

(1) Fuzzy Logic Control: As demonstrated in Figure 3.16, the PI controller in linear control is replaced by the fuzzy logic controller (FLC), which is a class of nonlinear control techniques and the best among the adaptive controllers. The load current reference tracking error and its derivative are used as the input to the fuzzy controller, which in turn embeds the experience, knowledge, and intuition of the designer in the form of membership functions and "if-then" control rules.

Given that power converters are nonlinear in nature, the robustness of the system during parameter variations can be improved by using the fuzzy controller even without knowing the exact converter model and its parameters. Compared with the linear PI regulator, FLC

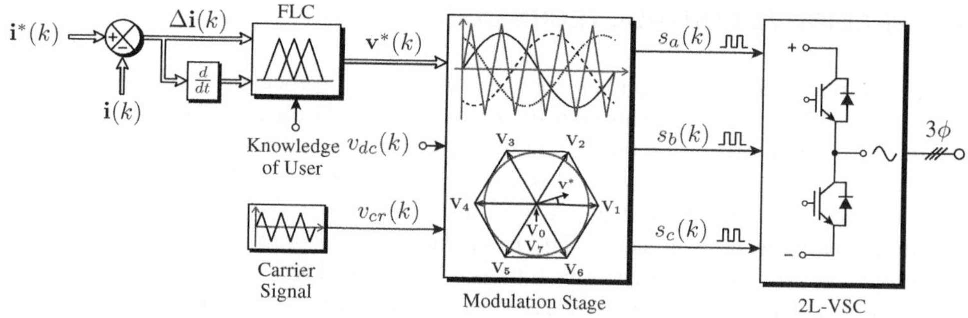

Figure 3.16 Block diagram of FLC with a modulation stage.

offers superior dynamic response during all operating conditions. Nonetheless, this scheme can only achieve a good control performance with the heuristic-reasoning-based expert knowledge of designers and precise control rules.

(2) ANN-Based Control: ANN-based load current regulation is shown in Figure 3.17 [17]. Compared with other intelligent controllers, ANN represents the most generic form of the process of human thinking (biological nervous system). In this control scheme, the current tracking error signals are entered through a suitable gain or scaling factor (K). The ANN modulator produces switching signals to the power converter. The modulation stage is realized by a network of neurons with corresponding weights. Each neuron then adjusts its weight with a proper training process to achieve the required control performance. A constant switching frequency operation is achieved with this approach. The major disadvantage of ANN-based control is that it requires offline training before it can be used in practical applications.

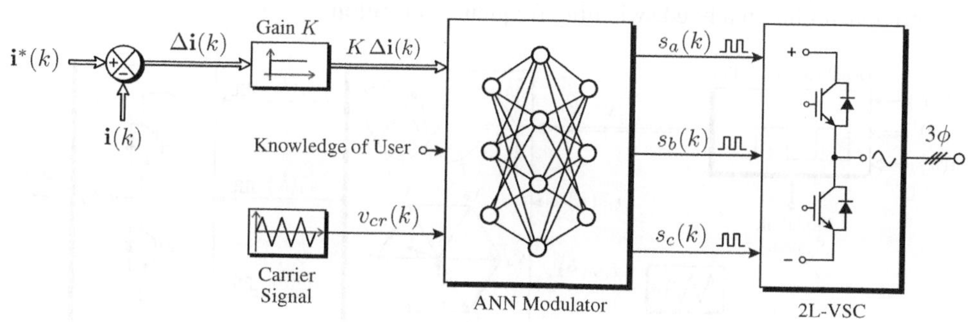

Figure 3.17 Block diagram of ANN-based load current control scheme.

The merits of fuzzy and ANN can also be combined (called neuro-fuzzy control) to achieve the best control performance. Although the intelligent controllers do not need a converter model, they require precise knowledge about the converter operation/behavior. Compared with the classical control techniques, intelligent control techniques are not widely utilized for power conversion applications.

3.7 PREDICTIVE CONTROL TECHNIQUES

The predictive control philosophy emerged in the 1950s for process control, which is extremely slow in nature [18]. This control approach has been used to obtain a systematic solution for solving the multivariable-constrained control problems of the oil and chemical industries and is widely accepted because of its simplicity [19]. The revolution in DSPs enabled the application of predictive control schemes to electrical engineering applications involving fast variables. The research works on predictive control of power electronic converters can be traced back to as early as the 1980s [20, 21]. As explained in Chapter 1, a wide range of predictive control techniques with different control concepts have been developed over the years. In the context of power converters, predictive control techniques are broadly classified into two groups, namely, (1) modulator-based predictive control techniques (e.g., deadbeat predictive control, continuous control-set MPC, generalized predictive control, and direct current control) and (2) modulator-free predictive control schemes (e.g., hysteresis-based MPC, trajectory-tracking MPC, direct MPC, and FCS-MPC) [22, 23]. A common control philosophy behind all predictive control techniques is that they use the system DT model in predicting the future behavior of the plant and implement the optimal control actions on the basis of predefined control objectives. In this section, the two most popular predictive control techniques are explained.

3.7.1 Predictive Control with Modulation

Figure 3.18 exhibits a deadbeat predictive control scheme (also known as the single-step predictive control) with a modulation stage. In this scheme, the PI regulator in linear control is replaced by the deadbeat controller [24]. Deadbeat control is one of the most well-known control techniques; in fact, it is the very first type of predictive controller developed for power converters [21]. Deadbeat controller uses the load and converter models to predict future current behaviors during every sampling period. Correspondingly, the controller selects the most appropriate reference voltage vector $\mathbf{v}^*$ and applies it to the modulation stage. A PWM/SVM modulator then produces gating signals to the power converter, resulting in a fixed switching frequency operation.

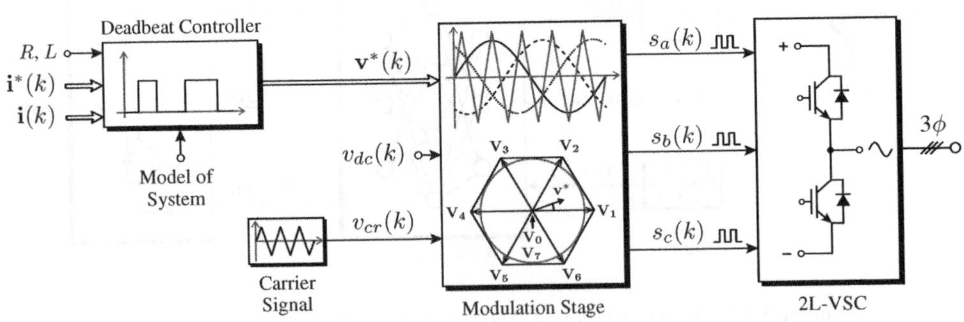

Figure 3.18 Deadbeat predictive current control with a modulation stage.

The load current dynamics from the converter model given in (3.17) on page 101 are considered as the state variables, yielding the following modified expression:

$$\frac{d\mathbf{i}(t)}{dt} = -\frac{R}{L}\mathbf{i}(t) + \frac{1}{L}\mathbf{v}(t). \tag{3.29}$$

The above CT dynamic model of load current is converted into a DT model, as depicted below, for a sampling time T_s:

$$\mathbf{i}(k+1) = \mathbf{\Phi}\,\mathbf{i}(k) + \mathbf{\Gamma}\,\mathbf{v}(k) \tag{3.30}$$

where $\mathbf{\Phi}$ and $\mathbf{\Gamma}$ are the DT equivalents of the CT parameters $-R/L$ and $1/L$, respectively, and $\mathbf{i}(k+1)$ is the future load current. The variables between discrete sampling instants are assumed constant. Discretization methods are thoroughly explained in Chapter 7.

Based on the DT model in (3.30), $\mathbf{v}^*(k)$ is calculated as

$$\mathbf{v}^*(k) = \mathbf{\Gamma}^{-1}\left[\widehat{\mathbf{i}}^*(k+1) - \mathbf{\Phi}\,\mathbf{i}(k)\right] \tag{3.31}$$

where $\widehat{\mathbf{i}}^*(k+1)$ is the extrapolated (or estimated) reference current. Extrapolation techniques are discussed in the subsequent chapter.

An optimal reference voltage vector $\mathbf{v}^*(k)$ ideally produces zero current error at the next sampling instant $(k+1)$. Deadbeat control uses a simple concept and does not involve any tuning or optimization procedures. As such, this control technique provides fast dynamic response compared with the classical controllers. However, system parameter variations and perturbations, measurement noise, and control delay deteriorate the performance of basic deadbeat control and trigger instability issues. Correspondingly, many solutions have been proposed in the literature to make deadbeat control robust and unsusceptible to system perturbations [22].

3.7.2 Predictive Control without Modulation

Figure 3.19 [25] presents an FCS-MPC strategy for regulating load currents. Compared with linear control, this scheme eliminates the need for linear PI regulators and modulation stage, and it offers a conceptually different approach for controlling power converters. All control techniques discussed earlier take necessary control actions after the error has occurred. Contrarily, FCS-MPC implements preventive control actions even before a huge current error arises. This control scheme mainly consists of the following subsystems:

- Extrapolation of reference currents from $(k)^{\text{th}}$ to $(k+1)$ sampling instant.
- Predictive model for calculating future load currents.
- Cost function for evaluating predictions and generating gating signals.

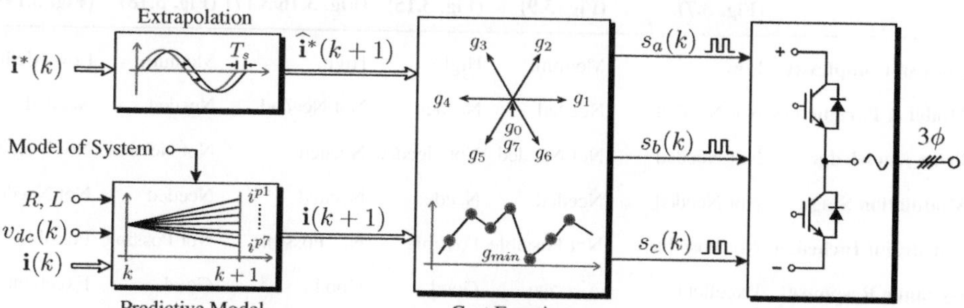

Figure 3.19 Block diagram of FCS-MPC for regulating load currents.

The predictive model in (3.30) is used to calculate the future load current values. The finite number of switching states allows these values to be enumerated for all feasible switching states. The predicted load currents are then evaluated using a cost function defined as an absolute error between the reference and predicted currents:

$$g(k) = \left| \hat{\mathbf{i}}^*(k+1) - \mathbf{i}(k+1) \right|. \tag{3.32}$$

The switching state, which minimizes the current error in the next sampling instant is selected as an optimal actuation and is directly applied to the converter. FCS-MPC offers a dynamic response that is faster than that of the classical control techniques, along with good steady-state reference tracking that is comparable to that of the classical control techniques. Compared with deadbeat predictive control, the cost function definition in FCS-MPC gives more flexible criterion for selecting optimal switching signals and a possibility to incorporate multiobjective control requirements (e.g., system nonlinearities and restrictions). In deadbeat control, reference vector $\mathbf{v}^*(k)$ is calculated directly (without any online optimizations or iterations) by measurements and system parameters (refer to (3.31) for details). The online optimizations in FCS-MPC increase the computational burden. However, this issue is trivial because the current industry-standard digital control platforms can address the computational capacity needed by FCS-MPC.

3.8 COMPARISON OF DIGITAL CONTROL TECHNIQUES

Table 3.2 summarizes the main features of classical and advanced control techniques with respect to control complexity, requirement for converter model and parameters, need for prior knowledge of converter operation, requirement for modulation stage, flexibility to include constraints and restrictions, dynamic response, and nature of converter switching frequency. The classical control schemes with PI regulator and modulation stage are well documented in the literature and widely employed by the power electronics industry. Alternatively, the advanced control schemes involve new concepts, designer's intuition, complex calculations, reference frame transformations, and often demand powerful microprocessors.

Table 3.2 Summary of main features of digital current control techniques

	Hysteresis (Fig. 3.7)	**Linear (Fig. 3.9)**	**SMC (Fig. 3.15)**	**Intelligent (Fig. 3.16/3.17)**	**DBPC (Fig. 3.18)**	**FCS-MPC (Fig. 3.19)**
Control Complexity	Low	Medium	High	High	Medium	Low-Medium
Model & Parameters	Not Needed	Needed	Needed	Not Needed	Needed	Needed
Prior Knowledge	Not Needed	Not Needed	Not Needed	Needed	Not Needed	Not Needed
Modulation Stage	Not Needed	Needed	Needed	Needed	Needed	Not Needed
Constraint Inclusion	Not Possible	Not Possible	Possible	Not Possible	Not Possible	Possible
Dynamic Response	Excellent	Average	Good	Good	Good	Excellent
Switching Frequency	Variable (Uncontrollable)	Fixed	Fixed	Fixed	Fixed	Variable (Controllable)

Linear control is assumed to be implemented in dq-frame with compensation terms and an SVM. Among all control techniques, hysteresis control is the simplest method, followed by FCS-MPC. Hysteresis and intelligent control techniques do not need system model and parameters, whereas the other schemes do. Intelligent control techniques particularly need the prior knowledge and intuition of designers. Both SMC and FCS-MPC allow constraints to be included in the design of controller, but the latter is the more flexible and easily implementable approach. The dynamic response obtained by hysteresis and FCS-MPC is excellent because of their modulator-free operation. The switching frequency of hysteresis and FCS-MPC techniques is variable. However, the cost function in FCS-MPC can be penalized to control switching frequency. The future development or application of any advanced control scheme is greatly influenced by two critical factors, namely, the (1) simplicity in digital implementation and (2) performance improvement compared with the classical control. FCS-MPC satisfies these requirements by being a simple and intuitive approach for digital implementation, while providing optimal performance during all operating conditions.

3.9 CONCLUDING REMARKS

This chapter exhaustively discussed the fundamentals of digital and predictive controls. Likewise, the evolution in digital control platforms is reviewed with respect to the advancements in semiconductor switching devices and advanced control techniques. The basics of reference frame theory, including the transformations among natural, stationary, and synchronous reference frames, are also presented.

Over the past few decades, numerous digital current schemes for controlling power converters have been developed by the power electronics industry and academia. The research on the digital control schemes is ongoing. This chapter analyzed the state-of-the-art digital control techniques, including hysteresis control, linear control with PI regulator and modulation stage, SMC, FLC, ANN-based control, deadbeat predictive control, and FCS-MPC along with their main features and challenges considering a case study of load (grid/generator) current control. The guidelines for tuning PI controller parameters are presented, and the operating principles of SPWM and SVM are extensively examined. Likewise, the main features of different digital control strategies are compared. The technical details given in this chapter are also applicable for other power conversion applications, such as active front-end rectifiers, active power filters, switched mode power supplies, variable-speed electric drives, electric vehicles, electric traction, photovoltaic energy, distributed generation, and high-voltage direct current transmission, which employ a digital control structure with an inner current control loop.

Link to Next Chapters:

- Reference frame theory, which is introduced in Section 3.3, is used in Chapters 5 and 6 for modeling power converters and wind generators, respectively. Chapters 8 to 12 also employ this theory to design and analyze control schemes.

- The FCS-MPC principle introduced in Section 3.7.2 is further analyzed in Chapter 4 with respect to its design procedure, implementation, stability issues, cost function flexibility, weighting factor selection, delay compensation, and extrapolation of reference variables.

REFERENCES

1. V. Yaramasu, B. Wu, P. C. Sen, S. Kouro, and M. Narimani, "High-power wind energy conversion systems: State-of-the-art and emerging technologies," *Proceedings of the IEEE*, vol. 103, no. 5, pp. 740–788, May 2015.
2. C. Buccella, C. Cecati, and H. Latafat, "Digital control of power converters - A survey," *IEEE Transactions on Industrial Informatics*, vol. 8, no. 3, pp. 437–447, August 2012.
3. J. Rodriguez and P. Coftes, *Predictive Control of Power Converters and Electrical Drives*, 1st ed. Chichester, UK: IEEE Wiley press, March 2012.
4. B. Wu, *High-Power Converters and AC Drives*, 1st ed. Hoboken, NJ: Wiley-IEEE Press, March 2006.
5. E. Monmasson, L. Idkhajine, M. Cirstea, I. Bahri, A. Tisan, and M.-w. Naouar, "FPGAs in industrial control applications," *IEEE Transactions on Industrial Informatics*, vol. 7, no. 2, pp. 224–243, May 2011.
6. T. Atalik, M. Deniz, E. Koc, C. Gercek, B. Gultekin, M. Ermis, and I. Cadirci, "Multi-DSP and -FPGA-based fully digital control system for cascaded multilevel converters used in FACTS applications," *IEEE Transactions on Industrial Informatics*, vol. 8, no. 3, pp. 511–527, August 2012.
7. J. Rodríguez, M. P. Kazmierkowski, J. R. Espinoza, P. Zanchetta, H. Abu-Rub, H. A. Young, and C. A. Rojas, "State of the art of finite control set model predictive control in power electronics," *IEEE Transactions on Industrial Informatics*, vol. 9, no. 2, pp. 1003–1016, May 2013.
8. P. Krause, O. Wasynczuk, S. Sudhoff, and S. Pekarek, *Analysis of Electric Machinery and Drive Systems*, 3rd ed., ser. IEEE Press series on power engineering. Hoboken, NJ: Wiley-IEEE Press, June 2013.
9. A. Yazdani and R. Iravani, *Voltage-Sourced Converters in Power Systems*. Hoboken, NJ: Wiley-IEEE Press, January 2010.
10. B. Wu, Y. Lang, N. Zargari, and S. Kouro, *Power Conversion and Control of Wind Energy Systems*, 1st ed., ser. IEEE Press Series on Power Engineering. Hoboken, NJ: Wiley-IEEE Press, July 2011.
11. M. Kazmierkowski and L. Malesani, "Current control techniques for three-phase voltage-source PWM converters: A survey," *IEEE Transactions on Industrial Electronics*, vol. 45, no. 5, pp. 691–703, October 1998.
12. D. M. Brod and D. Novotny, "Current control of VSI-PWM inverters," *IEEE Transactions on Industry Applications*, vol. IA-21, no. 3, pp. 562–570, May 1985.
13. D. Maksimovic, R. Zane, and R. Erickson, "Impact of digital control in power electronics," in *IEEE International Symposium on Power Semiconductor Devices and ICs (ISPSD)*, Fukuoka, Japan, May 2004, pp. 13–22.
14. J. Hung, W. Gao, and J. Hung, "Variable structure control: A survey," *IEEE Transactions on Industrial Electronics*, vol. 40, no. 1, pp. 2–22, February 1993.
15. J. Silva, "Sliding-mode control of boost-type unity-power-factor PWM rectifiers," *IEEE Transactions on Industrial Electronics*, vol. 46, no. 3, pp. 594–603, June 1999.
16. P. Vas, A. Stronach, and M. Neuroth, "DSP-controlled intelligent high-performance AC drives present and future," in *IEE Colloquium on Vector Control and Direct Torque Control of Induction Motors*, London, England, October 1995, pp. 7/1–7/8.
17. B. Bose, *Power Electronics and Motor Drives: Recent Advances and Trends*. Academic Press, 2006.
18. M. Morari and J. H. Lee, "Model predictive control: past, present and future," *International Journal of Computers and Chemical Engineering*, vol. 23, no. 4–5, pp. 667–682, 1999.
19. J. H. Lee, "Model predictive control: Review of the three decades of development," *International Journal of Control, Automation and Systems*, vol. 9, no. 3, pp. 415–424, 2011.
20. Y. Dote, T. Sriram, and R. G. Hoft, "Parameter and state estimation for dead-beat DC drive speed control," in *IEEE Power Electronics Specalists Conference (PESC)*, Atlanta, GA, USA, June 1980, pp. 335–340.
21. K. P. Gokhale, A. Kawamura, and R. G. Hoft, "Dead beat microprocessor control of PWM inverter for sinusoidal output waveform synthesis," in *IEEE Power Electronics Specalists Conference (PESC)*, Toulouse, France, June 1985, pp. 28–36.
22. P. Coftes, M. Kazmierkowski, R. Kennel, D. Quevedo, and J. Rodríguez, "Predictive control in power electronics and drives," *IEEE Transactions on Industrial Electronics*, vol. 55, no. 12, pp. 4312–4324, December 2008.
23. A. Linder, R. Kanchan, R. Kennel, and P. Stolze, *Model-Based Predictive Control of Electric Drives*. Germany: Cuvillier Verlag Göttingen, 2010.
24. L. Malesani, P. Mattavelli, and S. Buso, "Robust dead-beat current control for PWM rectifiers and active filters," *IEEE Transactions on Industry Applications*, vol. 35, no. 3, pp. 613–620, May/June 1999.
25. J. Rodríguez, J. Pontt, C. A. Silva, P. Correa, P. Lezana, P. Coftes, and U. Ammann, "Predictive current control of a voltage source inverter," *IEEE Transactions on Industrial Electronics*, vol. 54, no. 1, pp. 495–503, February 2007.

CHAPTER 4

FUNDAMENTALS OF MODEL PREDICTIVE CONTROL

4.1 INTRODUCTION

Model predictive control (MPC) is available under different names, such as receding (or moving) horizon control, rolling horizon planning, dynamic matrix control, model algorithmic control, generalized predictive control, and dynamic linear programming [1]. However, the underlying idea or operating principle of all these control techniques is similar: they all have the ability to predict future events and take control actions on the basis of objective function minimization. As mentioned in Chapter 3, in the field of power electronics, digital control techniques, including predictive control, have been rigorously researched for almost three decades. Finite control-set MPC (FCS-MPC) was proposed almost a decade ago for a two-level voltage source converter (2L-VSC) [2]. Since then, the research and development activities in this area have become active and significant. FCS-MPC has been theoretically analyzed and experimentally tested for a wide range of low-, medium-, and high-power converter topologies and power conversion applications. FCS-MPC has now reached a sufficient maturity level to find practical applications in the power electronics and energy conversion industry.

The performance of FCS-MPC with respect to the power electronics industry-standard hysteresis and linear control has been investigated in numerous research articles. A comparison between the classical linear control and FCS-MPC strategy for a 2L-VSC is shown in Table 4.1 on the basis of the findings in [3–6]. Unlike in linear control, linearizing the plant model with FCS-MPC is unnecessary; therefore, system nonlinearities are treated very well. The discrete nature of the power converter and modulator-free control structure

Model Predictive Control of Wind Energy Conversion Systems, First Edition. Venkata Yaramasu and Bin Wu.

Companion Website: www.wiley.com/go/yaramasu_wu_mpc_wecs

greatly simplifies the FCS-MPC optimization. The dynamic response obtained by FCS-MPC outperforms the linear control even during system parameter variations and perturbations. The simplicity of the concept and the high-performance operation are important factors that decide the success of any new control technique. The analysis suggests that the FCS-MPC strategy is an intuitive and powerful tool for controlling power converters compared with the classical linear control with proportional-integral (PI) regulator and space vector modulation (SVM). The merits of FCS-MPC are significantly highlighted when multilevel topologies, such as the neutral-point-clamped (NPC) converter, are considered.

Table 4.1 Comparison of linear control with PI regulator and SVM, and FCS-MPC

Description	Linear Control	FCS-MPC
Model	Linear Load Model for PI & Converter Model for SVM	Discrete-Time (DT) Model of Complete System
Controller Design	PI Adjustment + Modulator Design	Cost Function Definition
Nature of Controller	Linear	Nonlinear
Implementation Platform	Analog or Digital	Digital
Modulation	PWM/SVM	Not Required
Switching Frequency	Fixed	Variable (but controllable)
Multivariable	Coupled	Decoupled
Constraints Inclusion	Not Possible	Easy to Include
Complexity of Concept	Medium with SVM	Simple and Intuitive
Steady-State Performance	Good in dq frame	Good in abc, $\alpha\beta$, and dq frames
Transient Performance	Moderate	Excellent
Computational Burden	Medium with SVM	High

The concerns related to the predictive control of power electronics have been addressed in various research works, and several standard solutions are available for the selection of the cost function, design of the weighting factors, compensation of control delay, extrapolation of reference control variables, and selection of the sampling time for the real-time implementation. The fundamentals of FCS-MPC are presented in this chapter to supplement the basic FCS-MPC concept introduced in Chapter 3.

Chapter Overview

- The theory of the sampled-data model for the DT design and analysis of FCS-MPC is introduced in Section 4.2.
- The operating principle, design procedure, implementation, and stability issues of the FCS- MPC are discussed in Section 4.3.
- The flexibility of the cost function in incorporating several primary and secondary control objectives is discussed in Section 4.4.
- The weighting factor selection approaches, delay compensation methods, and extrapolation techniques are presented along with their advantages and disadvantages in Sections 4.5, 4.6, and 4.7, respectively.
- The sampling time selection for the real-time implementation of FCS-MPC is discussed in Section 4.8. The concluding remarks of this chapter are presented in Section 4.9.

4.2 SAMPLED-DATA MODEL

A sampled-data system exemplifies a continuous-time (CT) plant controlled by a digital device [7]. The process of transforming (or mapping) continuous functions, models, and equations into discrete counterparts is related to discrete mathematics and is often referred to as discretization. FCS-MPC employs an online optimization algorithm that is usually formulated in the DT domain with fixed sampling interval T_s. To achieve optimal control performance with FCS-MPC, sophisticated sampled-data models are often needed from the corresponding CT models.

Suppose that the CT model of a generic power conversion system has a first-order nature as shown below:

$$\frac{d\mathbf{x}(t)}{dt} = \dot{\mathbf{x}}(t) = \mathbf{A}\,\mathbf{x}(t) + \mathbf{B}\,\mathbf{u}(t) \tag{4.1}$$

where $\mathbf{A}$ and $\mathbf{B}$ denote the CT parameters of the converter (filter inductance, load resistance, DC-link capacitance, etc.). $\mathbf{x}(t)$ represents the state vector to be controlled (voltage, current, power, torque, flux, etc.), and $\mathbf{u}(t)$ is the input vector (DC voltage, grid voltage, etc.) in continuous time.

The discretization methods are broadly classified as exact and approximate methods. The former method has the highest accuracy in transforming the models. Exact discretization is performed by the matrix exponential method (also called zero-order hold (ZOH) method). The approximation methods include forward Euler, backward Euler, bilinear transformation, and truncated Taylor series.

The first-order nature of state equations describes the model in (4.1); hence, a first-order approximation for the derivative can be obtained by the forward Euler method, considering the present sample (k) and future sample ($k+1$) as demonstrated below:

$$\left\{\frac{dx(t)}{dt}\right\}_{t=k} \approx \frac{x(k+1) - x(k)}{T_s} \tag{4.2}$$

where T_s is the discretization step-size.

By substituting (4.2) into (4.1), an approximate DT model is obtained as follows:

$$\frac{\mathbf{x}(k+1) - \mathbf{x}(k)}{T_s} = \mathbf{A}\,\mathbf{x}(k) + \mathbf{B}\,\mathbf{u}(k) \tag{4.3}$$

which can be simplified to

$$\mathbf{x}(k+1) = \mathbf{\Phi}\,\mathbf{x}(k) + \mathbf{\Gamma}\,\mathbf{u}(k) \tag{4.4}$$

with

$$\mathbf{\Phi} \approx \mathbf{I} + \mathbf{A}\,T_s, \quad \mathbf{\Gamma} \approx \mathbf{B}\,T_s \tag{4.5}$$

where $\mathbf{I}$ is a unitary matrix with a dimension equal to the state matrix $\mathbf{A}$ dimension.

The DT model obtained by the exact discretization is similar to the one shown in (4.4). However, the computation of $\mathbf{\Phi}$ and $\mathbf{\Gamma}$ involves a matrix exponential to obtain an accurate model, as demonstrated below:

$$\mathbf{\Phi} = e^{\mathbf{A}T_s}, \quad \mathbf{\Gamma} = \int_0^{T_s} e^{\mathbf{A}\tau}\,d\tau\,\mathbf{B} = \mathbf{A}^{-1}\,(\mathbf{\Phi} - \mathbf{I})\,\mathbf{B}. \tag{4.6}$$

The system in Equation (4.1) is called linear time invariant (LTI) if $\mathbf{A}$ and $\mathbf{B}$ are constant and remains unchanged with time. When $\mathbf{A}$ and $\mathbf{B}$ change over time (for example, as in the induction machine), the system is called linear time variant (LTV). For the LTI systems, the $\mathbf{\Phi}$ and $\mathbf{\Gamma}$ can be computed offline; thus, either the forward Euler or ZOH method can be employed without any complexity. For the LTV systems, exact discretization is highly complex and is impossible in several cases. The forward Euler method presents a simple approach in such conditions. However, the accuracy of this method is slightly lower than that of ZOH method. To improve the accuracy in discretization, approximation methods with high accuracy, such as bilinear transformation and truncated Taylor series, are employed [8]. A detailed discussion on the discretization methods for LTI and LTV systems has not been reported in literature, particularly for the MPC of power conversion systems. Chapter 7 deals with the exact, approximate, and quasi-exact discretization methods for MPC, along with example problems.

4.3 BASICS OF MODEL PREDICTIVE CONTROL

The generic operating principle, design procedure, control scheme implementation, and stability-related issues are presented in this section.

4.3.1 Operating Principle

The generic working principle of an MPC is shown in Figure 4.1. The reference control trajectory $\mathbf{x}^*$ is set constant to simplify the analysis. The MPC is formulated in DT by allowing the variables to change their values at discrete sampling instants only. Compared with the classical hysteresis and linear control techniques, which observe the past and present values of the variable, the MPC strategy deals with the past, present, and future values of the variable. In other words, the MPC predicts future errors and takes preventive control actions such that the system will not be subjected to huge errors, thereby making the overall system robust. The principle of the MPC mainly consists of three parts/components as demonstrated below [9]:

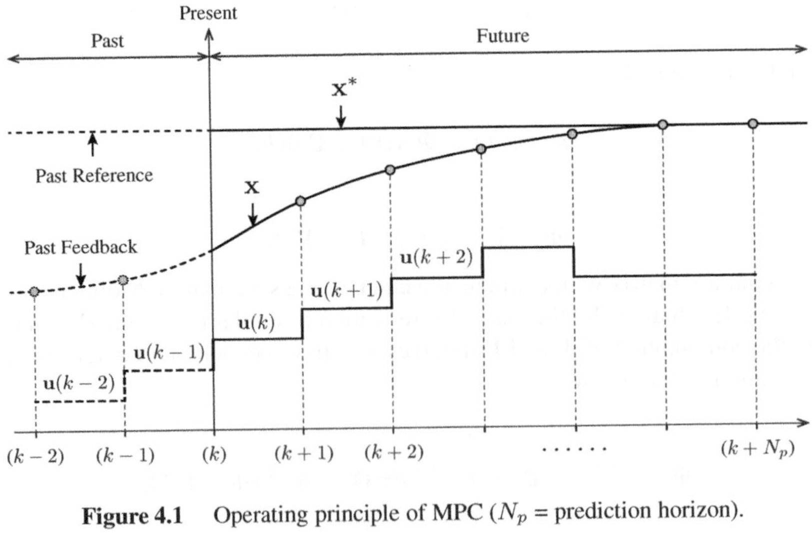

Figure 4.1 Operating principle of MPC (N_p = prediction horizon).

- **Prediction:** The MPC is based on numerical optimizations or online iterations. This step mainly involves the DT model, similar to the one shown in (4.4). The future values of state variable $\mathbf{x}$ and control input sequence $\mathbf{u}$ are predicted for a prediction horizon N_p by using the system model and feedback measurements at the $(k)^{\text{th}}$ instant.
- **Optimization:** The predictions are evaluated by a cost function that defines the system control objectives or desired behavior. A generic form of cost function is defined as a nonlinear function of reference, predicted, and input sequences:

$$J(k) = \mathbf{f}\left[\mathbf{x}^*(k), \cdots, \mathbf{x}^*(k+N_p),\, \mathbf{x}(k),\, \mathbf{u}(k), \cdots, \mathbf{u}(k+N_p)\right]. \tag{4.7}$$

The input control sequence that minimizes $J(k)$ is selected as an optimal actuation:

$$\mathbf{u}^{\mathbf{op}}(k) = \arg\min_{\mathbf{u}} J(k). \tag{4.8}$$

- **Receding Horizon Strategy:** The first element of the optimal input sequence $\mathbf{u}(k) = \mathbf{u}^{\mathbf{op}}(k)$ is applied to the plant. As a result, the state variable $\mathbf{x}$ moves toward the reference trajectory $\mathbf{x}^*$. The process of measuring new feedback variables, predicting new system behavior, and optimizing performance cost is repeated during each sampling interval. This procedure is called the receding horizon strategy.

Consider a plant with the DT model, as indicated in (4.4). The control goal is to maintain $\mathbf{x}$ at its set point $\mathbf{x}^*$. The state variable $\mathbf{x}$ is forced to match the value of $\mathbf{x}^*$, by choosing $\mathbf{u}$ as a steering variable, as demonstrated below:

$$\widehat{\mathbf{x}}^*(k+1) - \mathbf{x}(k+1) = 0 \quad \Rightarrow \quad \underbrace{\widehat{\mathbf{x}}^*(k+1)}_{\text{Set Point}} - \mathbf{\Phi}\ \underbrace{\mathbf{x}(k)}_{\text{Feedback}}\ - \mathbf{\Gamma}\ \underbrace{\mathbf{u}(k)}_{\text{Steering}} = 0. \tag{4.9}$$

The state variable is maintained at its set-point value by *steering* the $\mathbf{u}(k)$ value optimally during each sampling interval, despite slight changes in $\mathbf{\Phi}$ and $\mathbf{\Gamma}$. Therefore, the MPC provides a degree of robustness to modeling errors and uncertainty.

4.3.2 Design Procedure

The FCS-MPC for load current regulation has been discussed earlier in Chapter 3. In the current section, this concept is presented in more detail, with consideration to novice readers. A step-by-step procedure for the design of the FCS-MPC implementation is provided. The block diagram of the digital current control is illustrated in Figure 4.2 by combining Figures 3.6 and 3.19 in Chapter 3. The control scheme is presented in the natural (*abc*) reference frame to simplify the analysis. The FCS-MPC scheme is realized by five subsystems: (1) calculation of reference currents, (2) extrapolation of reference currents, (3) list of possible switching state combinations, (4) DT predictive model, and (5) cost function minimization. The main control objective is to force load currents ($\mathbf{i}$) to track the reference currents ($\mathbf{i}^*$) during all operating conditions. The FCS-MPC shown in Figure 4.2 is intuitive to understand and easy to implement in real time. This approach mainly consists of five design steps, as described below [10]:

Design Step 1: Measurement of Feedback Variables: The first step is to measure the feedback DC-link voltage v_{dc} and three-phase load currents (i_a, i_b, and i_c) and interface them to the digital control system through the ADC channels. For balanced loads, the sum of instantaneous three-phase currents is zero; thus, the i_c can be easily deducted by measuring i_a and i_b. Therefore, one voltage sensor and two current sensors are sufficient to design the digital current control system by FCS-MPC.

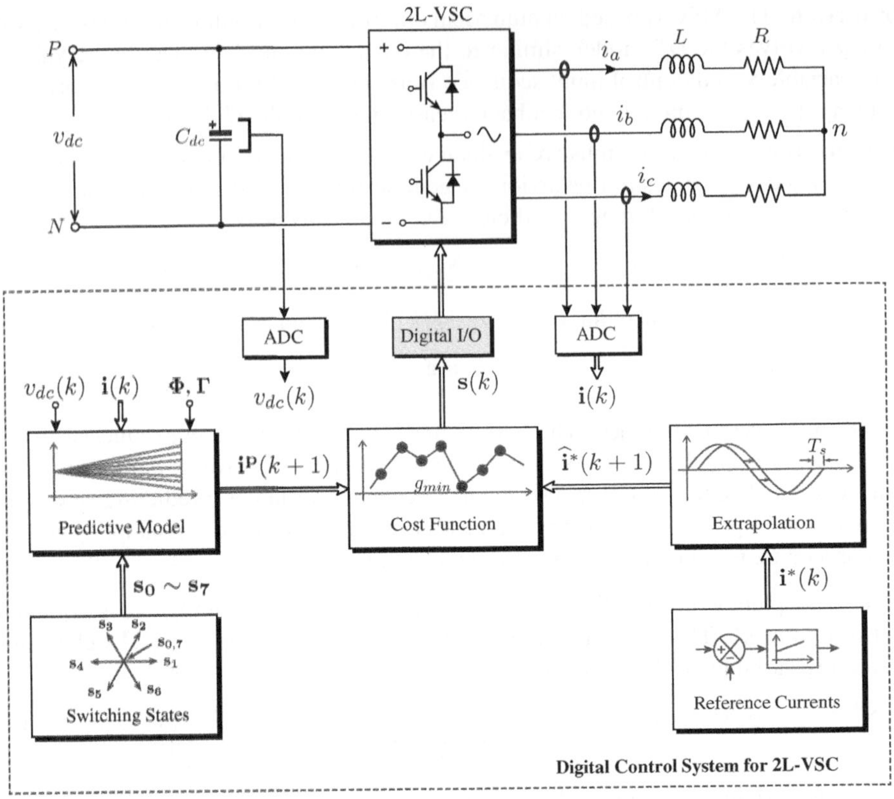

Figure 4.2 Predictive current control (PCC) of three-phase 2L-VSC with passive load.

Design Step 2: Calculation of Reference Currents: The next step is to calculate the three-phase reference currents $\mathbf{i}^*(k)$ based on the specific application. As mentioned in Chapter 1, the field-oriented control (FOC) and voltage-oriented control (VOC) schemes are used to calculate the reference currents for generator- and grid-side converters (GSCs), respectively. In this particular example, the $\mathbf{i}^*(k)$ is generated simply by the user-defined three-phase sinusoidal sources. This control scheme can easily be adapted to other power conversion applications by just changing the reference currents.

Design Step 3: Extrapolation of Reference Currents: The cost function deals with current error calculation in the $(k+1)$ sampling instant. To support this function, the reference currents (not load currents) are extrapolated from the $(k)^{\text{th}}$ instant to the $(k+1)$ sampling instant. A third-order Lagrange extrapolation uses present and three past samples to calculate the future reference currents as demonstrated below [11]:

$$\hat{\mathbf{i}}^*(k+1) = 4\,\mathbf{i}^*(k) - 6\,\mathbf{i}^*(k-1) + 4\,\mathbf{i}^*(k-2) - \mathbf{i}^*(k-3). \tag{4.10}$$

Design Step 4: Prediction of State Variables: The sub-steps for the prediction of load currents are given as follows:

- From (3.17) in Chapter 3, the CT load current model is obtained as follows:

$$\frac{d\,\mathbf{i}(t)}{dt} = \frac{1}{L}\left[\mathbf{v}(t) - R\,\mathbf{i}(t)\right]. \tag{4.11}$$

This model changes according to the type of power conversion application.

- By using the forward Euler approximation in (4.2), the CT model in (4.11) is mapped to the DT as follows:

$$\mathbf{i}^{\mathrm{p}}(k+1) = \underbrace{\left(1 - \frac{RT_s}{L}\right)}_{\mathbf{\Phi}} \mathbf{i}(k) + \underbrace{\left(\frac{T_s}{L}\right)}_{\mathbf{\Gamma}} \mathbf{v}^{\mathrm{p}}(k) \tag{4.12}$$

where superscript $\mathbf{p}$ denotes the predicted variable.

- The DT parameters $\mathbf{\Phi}$ and $\mathbf{\Gamma}$ are computed offline by using R, L, and T_s, and they are provided as inputs to the predictive model subsystem.
- The model in (4.12) indicates that the future load currents are calculated on the basis of the prediction of converter voltages $\mathbf{v}^{\mathrm{p}}(k)$. Through the modeling of 2L-VSC in Section 3.4.2 on page 100, the converter voltages $\mathbf{v}(k) = [v_a\, v_b\, v_c]^T$ are found to be related to the switching states $\mathbf{s}(k) = [s_a\, s_b\, s_c]^T$. Therefore, identifying the possible switching state combinations for the converter under study is important. Table 3.1 on page 101 shows that the switching signals s_a, s_b, and s_c lead to eight possible combinations: $\mathbf{s}_0$ = ['0','0','0'], $\mathbf{s}_1$ = ['1','0','0'], $\mathbf{s}_2$ = ['1','1','0'], $\mathbf{s}_3$ = ['0','1','0'], $\mathbf{s}_4$ = ['0','1','1'], $\mathbf{s}_5$ = ['0','0','1'], $\mathbf{s}_6$ = ['1','0','1'], and $\mathbf{s}_7$ = ['1','1','1'].
- The converter voltages with respect to the load neutral are predicted by using the measured DC-link voltage and eight possible switching states. A combination of (3.13) and (3.15) result in the following:

$$\mathbf{v}^{\mathrm{p}}(k) = \begin{bmatrix} v_a^p(k) \\ v_b^p(k) \\ v_c^p(k) \end{bmatrix} = v_{dc}(k) \begin{bmatrix} \frac{2}{3} & -\frac{1}{3} & -\frac{1}{3} \\ -\frac{1}{3} & \frac{2}{3} & -\frac{1}{3} \\ -\frac{1}{3} & -\frac{1}{3} & \frac{2}{3} \end{bmatrix} \begin{bmatrix} s_a(k) \\ s_b(k) \\ s_c(k) \end{bmatrix}. \tag{4.13}$$

- Owing to the finite number of switching state combinations, the number of online iterations in the FCS-MPC scheme is significantly simplified. The eight possible switching state combinations lead to eight different predictions for converter output voltages, which in turn produces eight different predictions for the load currents.

Design Step 5: Cost Function Minimization: As a final design step, the absolute error between the eight predicted currents $\mathbf{i}^{\mathrm{p}}(k+1)$ and the extrapolated reference currents $\widehat{\mathbf{i}}^*(k+1)$ are evaluated by a cost function $g(k)$:

$$g(k) = \left| \widehat{\mathbf{i}}^*(k+1) - \mathbf{i}^{\mathrm{p}}(k+1) \right|. \tag{4.14}$$

An ideal cost function g value is zero and represents the perfect regulation of load currents. However, in practice, a zero g value is impossible. The cost evaluation in (4.14) leads to eight different cost function values ($g_0 \sim g_7$). The objective of the cost function minimization subsystem is to identify the minimum cost function value and corresponding switching state combination. For example, assuming that the cost function g_5 has a minimum value among the pool, the corresponding switching state $\mathbf{s}_5$ with s_a = '0', s_b = '1', and s_c = '1' is chosen as an optimal actuation and is directly applied to the converter. The whole design procedure is performed during the $(k)^{\text{th}}$ sampling interval, and the optimal switching state is applied to the converter at the next sampling instant. As a final note, the FCS-MPC involves neither PI controllers nor a modulation stage.

✍ EXAMPLE 4.1 Sample-Based Analysis for PCC of 2L-VSC

Consider a predictive current controlled 2L-VSC as shown in Figure 4.2. The system parameters are as follows: $R = 10\ \Omega$ and $L = 10$ mH. For a given sampling instant, the instantaneous feedback measurements are as follows: $v_{dc}(k) = 400$ V, $i_a(k) = 9.61$ A, $i_b(k) = 5.75$ A, and $i_c(k) = -12.86$ A. The extrapolated reference currents are $\widehat{i_a^*}(k+1) = 9.42$ A, $\widehat{i_b^*}(k+1) = 4.43$ A, and $\widehat{i_c^*}(k+1) = -13.85$ A. The following is calculated with a sampling time $T_s = 100\ \mu$s: (a) DT system parameters, (b) predicted converter voltages, (c) predicted load currents, (d) cost function values, and (e) optimal switching signals for the next sampling interval.

Solution:

(a) The DT system parameters that correspond to the CT counterparts are as follows:

$$\mathbf{\Phi} = 1 - \frac{RT_s}{L} = 1 - \frac{10 \times 0.0001}{0.01} = 0.9$$

$$\mathbf{\Gamma} = \frac{T_s}{L} = \frac{0.0001}{0.01} = 0.01.$$

(b) By using the model in (4.13), the prediction of converter output voltages for a switching state combination $\mathbf{s_2} \in$ ['1','1','0'] is obtained as follows:

$$\mathbf{v_2^p}(k) = v_{dc}(k) \begin{bmatrix} \frac{2}{3} & -\frac{1}{3} & -\frac{1}{3} \\ -\frac{1}{3} & \frac{2}{3} & -\frac{1}{3} \\ -\frac{1}{3} & -\frac{1}{3} & \frac{2}{3} \end{bmatrix} \begin{bmatrix} s_a(k) \\ s_b(k) \\ s_c(k) \end{bmatrix} = \begin{bmatrix} 133.33 \\ 133.33 \\ -266.67 \end{bmatrix}.$$

(c) The prediction for load currents is obtained for $\mathbf{v_2^p}(k)$ from the model in (4.12):

$$\mathbf{i_2^p}(k+1) = \mathbf{\Phi}\,\mathbf{i}(k) + \mathbf{\Gamma}\,\mathbf{v_2^p}(k) = 0.9 \begin{bmatrix} 9.61 \\ 5.75 \\ -12.86 \end{bmatrix} + 0.01 \begin{bmatrix} 133.33 \\ 133.33 \\ -266.67 \end{bmatrix} = \begin{bmatrix} 9.98 \\ 6.51 \\ -14.24 \end{bmatrix}.$$

(d) The cost function value for switching state combination $\mathbf{s_2}$ is

$$\begin{aligned} g(k) &= \left|\widehat{i_a^*}(k+1) - i_a^p(k+1)\right| + \left|\widehat{i_b^*}(k+1) - i_b^p(k+1)\right| + \left|\widehat{i_c^*}(k+1) - i_c^p(k+1)\right| \\ &= |9.42 - 9.98| + |4.43 - 6.51| + |-13.85 + 14.24| = 3.031. \end{aligned}$$

The predictions for converter output voltages and load currents, and cost function values are evaluated by repeating the calculations in (b)-(d) for all other possible switching state combinations. A summary of these values is listed in Table 4.2.

Table 4.2 Calculation of $\mathbf{v^p}(k)$, $\mathbf{i^p}(k+1)$, and $g(k)$ for all possible switching states of a 2L-VSC

$\mathbf{s}$	$s_a(k)$	$s_b(k)$	$s_c(k)$	$v_a^p(k)$	$v_b^p(k)$	$v_c^p(k)$	$i_a^p(k+1)$	$i_b^p(k+1)$	$i_c^p(k+1)$	$g(k)$
$\mathbf{s_0}$	'0'	'0'	'0'	0	0	0	8.65	5.18	–11.57	3.792
$\mathbf{s_1}$	'1'	'0'	'0'	266.67	–133.33	–133.33	11.32	3.84	–12.91	3.427
$\mathbf{s_2}$	'1'	'1'	'0'	133.33	133.33	–266.67	9.98	6.51	–14.24	**3.031**
$\mathbf{s_3}$	'0'	'1'	'0'	–133.33	266.67	–133.33	7.32	7.84	–12.91	6.459
$\mathbf{s_4}$	'0'	'1'	'1'	–266.67	133.33	133.33	5.98	6.51	–10.24	9.125
$\mathbf{s_5}$	'0'	'0'	'1'	–133.33	–133.33	266.67	7.32	3.84	–8.91	7.635
$\mathbf{s_6}$	'1'	'0'	'1'	133.33	–266.67	133.33	9.98	2.51	–10.24	6.093
$\mathbf{s_7}$	'1'	'1'	'1'	0	0	0	8.65	5.18	–11.57	3.792

(e) Table 4.2 shows that the switching state combination $\mathbf{s}_2$ gives a minimum cost function value of 3.031. Therefore, the corresponding optimal switching signals to be applied to the converter at the next sampling interval are as follows: $s_a(k) =$ '1', $s_b(k) =$ '1', and $s_c(k) =$ '0'. ♦

4.3.3 Implementation of Control Scheme

The predictive control scheme is basically an optimization algorithm; thus it is digitally implemented in the microprocessor-based hardware. The flowchart for the digital implementation of PCC algorithm is shown in Figure 4.3.

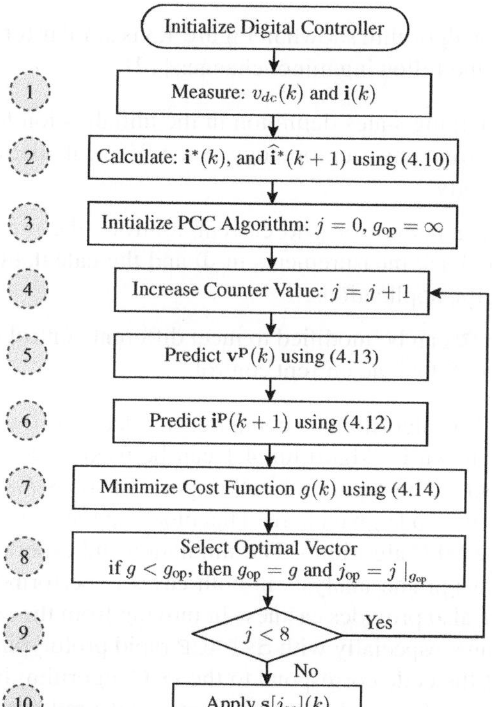

Figure 4.3 Flowchart for 2L-VSC predictive current control algorithm.

The PCC algorithm consists mainly of ten blocks/steps as discussed below:

① The measured DC-link voltage and load currents are used by the predictive controller.

② The reference currents are defined by the user. The extrapolated reference currents are calculated according to (4.10).

③ The algorithm is initialized by setting the switching state combination j to 0 and the optimal cost function (g_{op}) value to ∞, thus erasing g_{op} in the previous iteration.

④ The algorithm then enters the loop, where a counter increases j value in steps.

⑤ The converter voltages are predicted based on the measured DC-link voltage and all possible switching states defined by the user (refer to (4.13) for details).

⑥ On the basis of the predicted converter voltages, the load currents are predicted by using DT load current model in (4.12).

⑦ The predicted load currents are evaluated by a cost function g. The absolute error between the extrapolated reference currents $\widehat{\mathbf{i}}^*(k+1)$ and predicted currents $\mathbf{i}^p(k+1)$ is calculated according to (4.14).

⑧ During any iteration, if $g < g_{op}$, the minimum g value is stored as an optimal value g_{op} and the corresponding switching state combination is stored as j_{op}.

⑨ The switching state combination number j is counted by the algorithm. When the loop runs for eight times ($0 \sim 7$), the optimal j_{op} number is released by the algorithm.

⑩ On the basis of the j_{op} value, the switching signals for the 2L-VSC are deduced from the predefined list of switching states.

The predictive control algorithm shown in Figure 4.3 is similar for all power conversion applications, except for the following minor changes [12]:

- By changing the switching states definition in the initialization file, the algorithm can be applied for any power converter topology. The $\mathbf{v}^p(k)$ calculation in ⑤ and j counter in ⑨ change accordingly.
- When employed for generator or grid current control, the prediction of currents in ⑥ changes. The feedback measurements in ① and the calculation of $\mathbf{i}^*(k)$ in ② also change to meet specific applications.
- The cost function in ⑦ can be modified to meet different control objectives in addition to the basic function of the load current control.

In the simulation of PCC algorithm by computer simulation software such as `MATLAB/Simulink`, the code shown in Algorithm 4.1 can be used in conjunction with the "***S-Function Builder***" block. The input and output ports of the *S-Function Builder* should be created according to the code given here. This block operates in DT with the sampling time (T_s) dictated by the MPC algorithm. The simulation and experimental results are well matched because the design and analysis in both cases is performed in the DT domain. The *S-Function Builder* also provides easiness in moving from the simulation study to the experimental testing stage, especially with `dSPACE` rapid prototyping.

The lines 1 and 2 of the code correspond to the PCC algorithm initialization. The line 3 defines the counter operation and the maximum count limit for the loop. The line 3 essentially combines ③ and ⑨ in Figure 4.3. The PCC algorithm enters `for` loop at line 4 and exits at line `20`. The converter voltages are predicted in lines `5-7` by using the DC-link voltage and transformed switching state combinations `sn_abc` with respect to the load neutral n. The natural switching signals `s_abc` are multiplied offline with the transformation matrix in (4.13) to obtain the transformed switching states `sn_abc`. The number of online calculations in the PCC algorithm are minimized by computing `sn_abc` offline. In lines `8-10`, the load currents are predicted on the basis of measured currents, DT parameters (`Phi` and `Gamma`), and predicted converter voltages in lines `5-7`. The absolute error in the three-phase currents is evaluated by three sub-cost functions in lines `11-13`. The main cost function is evaluated in line `14` by adding the sub-cost functions together. Lines `15-19` implement the block ⑧ in Figure 4.3 to select the optimal cost function value and switching state combination number. Finally, in lines `21-23`, the optimal switching signals that minimize the cost function are generated on the basis of the j_{op} value. Even though the other types of switching states are used in the prediction, the output of the *S-Function Builder* is always natural switching signals `s_abc` whose value correspond to the Boolean logic high ('1') or low ('0').

ALGORITHM 4.1 PCC Algorithm for 2L-VSC

```
j_op = 1000;
g_op = 1000000000;
for(j = 0; j < 8; j++)
{
    v_abc[0] = vdc[0]*vsc[j].sn_abc[0];
    v_abc[1] = vdc[0]*vsc[j].sn_abc[1];
    v_abc[2] = vdc[0]*vsc[j].sn_abc[2];
    i_abc_k1[0] = Phi*i_abc[0] + Gamma*v_abc[0];
    i_abc_k1[1] = Phi*i_abc[1] + Gamma*v_abc[1];
    i_abc_k1[2] = Phi*i_abc[2] + Gamma*v_abc[2];
    g_a = fabs(i_ref_k1[0]-i_abc_k1[0]);
    g_b = fabs(i_ref_k1[1]-i_abc_k1[1]);
    g_c = fabs(i_ref_k1[2]-i_abc_k1[2]);
    g_k = g_a + g_b + g_c;
    if(g_k < g_op)
        {
            j_op = j;
            g_op = g_k;
        }
}
s_a[0] = vsc[j_op].s_abc[0];
s_b[0] = vsc[j_op].s_abc[1];
s_c[0] = vsc[j_op].s_abc[2];
```

4.3.4 Stability-Related Issues

The closed-loop stability is a classical subject matter in control engineering. In the closed control of power converters, state variables such as voltages, currents, and power have a tendency to diverge from the set-point values. Therefore, the small perturbations in the control (or disturbance) input should be prevented from causing large variations in the state variable behavior; the system should also be ensured to converge to a desired state. The purpose of digital control is to maintain closed-loop stability and reject disturbances that occur in the plant.

The closed-loop stability of the MPC is usually analyzed by Lyapunov stability theory. The infinite-horizon MPC guarantees stability. Increasing the horizon length N_p results in the complexity of the overall MPC. Quite a large number of switching state combinations resulting from higher N_p also increase the number of calculations, thereby hindering the real-time implementation of the control algorithm. For this reason, most FCS-MPC techniques use one-sample-ahead prediction horizon ($N_p = 1$). The closed-loop MPC is not necessarily stable with short prediction horizons [13]. Numerous studies have proven that FCS-MPC is robust against system parameter variations and perturbations [10, 14]. However, an optimal and robust controller is not necessarily a closed-loop stable controller. The MPC becomes a nonlinear control problem when constraints are incorporated; therefore, the system stability properties of LTI cannot be applied. Thus, analyzing the closed-loop stability of FCS-MPC is difficult unless complicated mathematics is employed. Several recent works suggested that practical stability can be guaranteed by carefully choosing the cost function type and adjusting the weight matrices of horizon-one FCS-MPC [15]. In this book, attention is paid more to the robustness analysis of the FCS-MPC with system parameter variations and perturbations. Closed-loop stability is not thoroughly researched in literature; hence, this topic is omitted from the discussion for the time being. A more rigorous research in this direction is expected in the coming years.

4.4 COST FUNCTION FLEXIBILITY

The definition of the cost function is an important design stage in FCS-MPC for fulfilling the operational requirements in various power conversion applications. In general, the cost function design is simple and flexible and can handle multivariable systems with different physical natures, magnitudes, frequencies, and phase angles for control variables. The cost function flexibility and design was highlighted in many research articles, survey papers, books, and dissertations [12, 16–21]. In this section, the cost function flexibility is presented (with respect to the previously published works) by broadly classifying the control objectives as primary and secondary goals. The primary goals are related to the basic operation (such as reference tracking), whereas the secondary goals correspond to the technical requirements, safety constraints, and nonlinearities. In FCS-MPC, the predictions for control variables are enumerated with respect to all the possible switching states of the converter. As demonstrated in Figure 4.4, several primary and secondary objectives (through weighting factors) can be incorporated in the cost function as long as the objectives are a mathematical function ($\mathbf{f}$) of the converter switching states, $\mathbf{s}(k)$ (or simply $\mathbf{s}_k$).

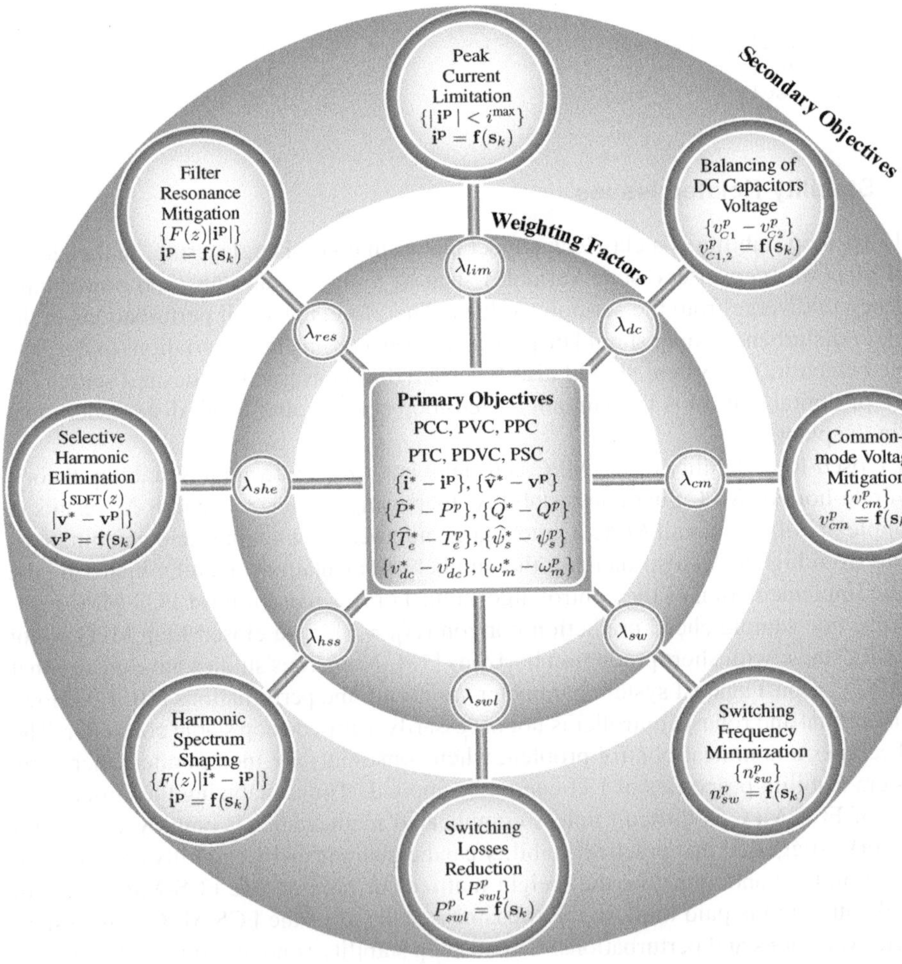

Figure 4.4 Cost function in FCS-MPC with several primary and secondary control objectives.

The main and sub-cost functions shown in Figure 4.4 are represented in generic form $\{\widehat{x}^* - x^p\}$, which can be realized as follows:

$$g(k) = \{\widehat{x}^*(k+1) - x^p(k+1)\} \in \begin{cases} |\widehat{x}^*(k+1) - x^p(k+1)| & \text{(Absolute Cost Function)} \\ [\widehat{x}^*(k+1) - x^p(k+1)]^2 & \text{(Quadratic Cost Function)} \\ \exp|\widehat{x}^*(k+1) - x^p(k+1)| & \text{(Exponential Cost Function)} \\ \frac{|x^*(k)-x(k)|+|\widehat{x}^*(k+1)-x^p(k+1)|}{2} & \text{(Integral Cost Function)} \end{cases} \tag{4.15}$$

where $\widehat{x}^*$ is the extrapolated reference command, and x^p is the predicted value of the control variable; both of them correspond to the $(k+1)$ sampling instant.

The absolute cost function measures the error between the $\widehat{x}^*$ and x^p, and represents a positive value. The quadratic cost function produces an over-proportionate error value (square of the original error value). The exponential cost function produces a similar output as a quadratic cost function. The integral cost function considers the trajectory of the error values between the $(k)^{\text{th}}$ and $(k+1)$ sampling instants, and minimizes the average error between these two sampling instants [22]. This approach leads to better reference tracking with less steady-state errors. However, the approach incurs more number of calculations.

EXAMPLE 4.2 Calculation of Current Error by Different Cost Function Types

The example problem 4.1 is considered again in this example to study the difference between four cost function types in (4.15). The instantaneous feedback currents are $i_a(k)$ = 9.61 A, $i_b(k)$ = 5.75 A, and $i_c(k)$ = –12.86 A. The actual and extrapolated reference currents are $i_a^*(k)$ = 9.81 A, $i_b^*(k)$ = 3.92 A, $i_c^*(k)$ = –13.73 A, $\widehat{i}_a^*(k+1)$ = 9.42 A, $\widehat{i}_b^*(k+1)$ = 4.43 A, and $\widehat{i}_c^*(k+1)$ = –13.85 A. Calculate the current error values with four types of cost functions and tabulate the results.

Solution:

For switching state combination $\mathbf{s}_2$, the current error value with exponential cost function is

$$\begin{aligned} g(k) &= \exp|\widehat{i}_a^*(k+1) - i_a^p(k+1)| + \exp|\widehat{i}_b^*(k+1) - i_b^p(k+1)| \\ &\quad + \exp|\widehat{i}_c^*(k+1) - i_c^p(k+1)| \\ &= \exp|9.42 - 9.98| + \exp|4.43 - 6.51| + \exp|-13.85 + 14.24| = 2.173. \end{aligned}$$

Table 4.3 Calculation of current error values by different cost function types

s	$i_a^p(k+1)$	$i_b^p(k+1)$	$i_c^p(k+1)$	**Abs.** $g(k)$	**Quad.** $g(k)$	**Exp.** $g(k)$	**Int.** $g(k)$
$\mathbf{s}_0$	8.65	5.18	–11.57	3.792	6.329	2.739	3.346
$\mathbf{s}_1$	11.32	3.84	–12.91	3.427	4.828	2.341	3.163
$\mathbf{s}_2$	9.98	6.51	–14.24	**3.031**	**4.788**	**2.173**	**2.966**
$\mathbf{s}_3$	7.32	7.84	–12.91	6.459	16.956	8.624	4.679
$\mathbf{s}_4$	5.98	6.51	–10.24	9.125	29.164	31.267	6.013
$\mathbf{s}_5$	7.32	3.84	–8.91	7.635	29.204	10.010	5.268
$\mathbf{s}_6$	9.98	2.51	–10.24	6.093	17.036	7.429	4.497
$\mathbf{s}_7$	8.65	5.18	–11.57	3.792	6.329	2.739	3.346

A summary of the current error values is listed in Table 4.3 by repeating the calculations for all other switching state combinations and cost function types. The four types of cost functions produce different current error values; however, all the types suggest that $\mathbf{s}_2$ is an optimal switching state combination. These four cost function types might perform differently at other sampling instants and operating conditions. ♦

The first two types of cost functions are commonly used in FCS-MPC and both perform equally when sub-cost functions are excluded. The quadratic cost function performs better than the absolute type when the sub-cost functions are considered simultaneously because the over-proportionate errors help to easily distinguish between the bigger and smaller error values among the control variables [18]. To simplify the analysis, only the absolute and quadratic type cost functions are included in the book.

4.4.1 Primary Control Objectives

To reduce the complexity of notation, the variables $\widehat{x}^*(k+1)$ and $x^p(k+1)$ are simply denoted as $\widehat{x}^*$ and x^p.

(1) Current Control: Numerous power conversion applications employ current control in the inner control loop, irrespective of the power converter topology. By using the PCC scheme, the currents in natural, stationary or synchronous reference frames can be easily controlled in a decoupled manner by defining a cost function as follows [14, 21]:

$$\begin{aligned} g_i(k) &= \{\widehat{\mathbf{i}}^* - \mathbf{i}^p\} \\ &\in \begin{cases} \{\widehat{i}_a^* - i_a^p\} + \{\widehat{i}_b^* - i_b^p\} + \{\widehat{i}_c^* - i_c^p\} & \text{(Natural Frame)} \\ \{\widehat{i}_\alpha^* - i_\alpha^p\} + \{\widehat{i}_\beta^* - i_\beta^p\} & \text{(Stationary Frame)} \\ \{\widehat{i}_d^* - i_d^p\} + \{\widehat{i}_q^* - i_q^p\} & \text{(Synchronous Frame).} \end{cases} \end{aligned} \tag{4.16}$$

(2) Voltage Control: A DC/AC converter with an output inductive-capacitive (LC) filter is used [23, 24] in the uninterruptible power supply system and standalone distributed generation. The load or consumer demands high-quality sinusoidal voltages regardless of the linear/nonlinear, balanced/unbalanced, or single-/three-phase arbitrary load profile. In such applications, ***predictive voltage control (PVC)*** with the following cost function is used in the natural reference frame. Similar to (4.16), the PVC cost function can be defined in other reference frames.

$$g_v(k) = \{\widehat{\mathbf{v}}^* - \mathbf{v}^p\} = \{\widehat{v}_a^* - v_a^p\} + \{\widehat{v}_b^* - v_b^p\} + \{\widehat{v}_c^* - v_c^p\}. \tag{4.17}$$

(3) Active and Reactive Power Control: In grid-connected applications, the grid active and reactive powers are controlled in a decoupled manner by ***predictive power control (PPC)***, similar to the classical direct power control (DPC). The PPC technique is also used to control the wind generators and motor drives. The cost function in the PPC scheme, which minimizes the error in active and reactive powers, is defined as follows [25]:

$$g_{PQ}(k) = \{\widehat{P}^* - P^p\} + \{\widehat{Q}^* - Q^p\}. \tag{4.18}$$

(4) Net DC-Bus Voltage Control: Classical control techniques such as VOC and DPC are used to control GSCs. The PCC scheme replaces the internal current PI controllers and modulation stage in VOC, whereas the PPC scheme eliminates the hysteresis regulators and lookup table in DPC. However, the PCC and PPC schemes still employ a PI controller in the outer loop to regulate the DC-bus voltage, v_{dc} at its reference value v_{dc}^*. The cascaded control structure can be eliminated by simultaneously incorporating the DC-bus voltage control and PCC/PPC objectives in the cost function [26]. This control approach is named as ***predictive DC-bus voltage control (PDVC)***, and its cost function is shown below:

$$g_{vdc}(k) = \{v_{dc}^* - v_{dc}^p\} \cdot \in \begin{cases} \{\widetilde{v}_{dc}^* - v_{dc}^p\} + \lambda_{id}\{\widehat{i}_d^* - i_d^p\} + \lambda_{iq}\{\widehat{i}_q^* - i_q^p\} & \text{(PDVC by PCC)} \\ \{\widetilde{v}_{dc}^* - v_{dc}^p\} + \lambda_P\{\widehat{P}^* - P^p\} + \lambda_Q\{\widehat{Q}^* - Q^p\} & \text{(PDVC by PPC)} \end{cases} \tag{4.19}$$

where $\widetilde{v}_{dc}^*$ is the filtered DC-bus voltage reference. λ_{id}, λ_{iq}, λ_P, and λ_Q are the weighting factors for internal PCC or PPC schemes.

The dynamic response obtained for the DC-bus voltage with this method is excellent because of the elimination of the PI controller in the outer control loop. However, this approach involves complex modeling, and its steady-state performance is highly prone to modeling errors [17]. In state-of-the-art research, the PDVC is verified by the PPC only [26]. However, it can be combined with the PCC scheme, as demonstrated in (4.19).

(5) Torque and Flux Control: The generator/motor torque and flux is controlled in a decoupled manner by ***predictive torque control (PTC)***, which is analogous to the classical direct torque control (DTC) [27]. By using the PTC approach, the nonlinear hysteresis regulators and switching table in the DTC scheme are replaced by predictive model and cost function. The cost function in PTC scheme is defined as follows:

$$g_{T\psi}(k) = \{\widehat{T}_e^* - T_e^p\} + \lambda_\psi\{\widehat{\psi}_s^* - \psi_s^p\} \tag{4.20}$$

where λ_ψ is the weighting factor for stator flux control.

(6) Rotor Speed Control: In the classical FOC and DTC, the outer speed control loops employ PI controllers to generate i_d^*, i_q^*, or T_e^*. The ***predictive speed control (PSC)***, which is analogous to the PDVC of GSCs, eliminates the need for the PI controller in the outer speed control loop. The PSC can be realized through PCC or PTC with the following cost function [28, 29]:

$$g_\omega(k) = \{\omega_m^* - \omega_m^p\} \in \begin{cases} \{\widetilde{\omega}_m^* - \omega_m^p\} + \lambda_{id}\{\widehat{i}_d^* - i_d^p\} + \lambda_{iq}\{\widehat{i}_q^* - i_q^p\} & \text{(PSC by PCC)} \\ \{\widetilde{\omega}_m^* - \omega_m^p\} + \lambda_T\{\widehat{T}_e^* - T_e^p\} + \lambda_\psi\{\widehat{\psi}_s^* - \psi_s^p\} & \text{(PSC by PTC)} \end{cases} \tag{4.21}$$

where $\widetilde{\omega}_m^*$ is the filtered speed reference similar to $\widetilde{v}_{dc}^*$ in PDVC scheme. λ_{id}, λ_{iq}, λ_T, and λ_ψ are the weighting factors for the internal PCC or PTC realization.

Similar to the PDVC scheme, a fast dynamic response for rotor speed control can be obtained; however, this approach involves complex modeling and demands precise rotor speed information. In the literature, the PSC is verified by the PCC scheme only [28, 29]; however, PTC can also be employed inside the PSC scheme.

4.4.2 Secondary Control Objectives

(1) Peak Current Limitation: On the basis of the cost functions defined for the PPC, PDVC, PTC, and PSC schemes, the feedback current dynamics are ignored. During transient conditions, the feedback current can exceed the maximum threshold limit ($i^{\max}$), thus leading to the failure of some components. The classical VOC, FOC, DPC, and DTC schemes employ saturation blocks to mitigate this issue. In FCS-MPC, the saturation effect is obtained by defining a nonlinear sub-cost function as follows [30]:

$$g_{lim}(k) = \{ \,|\,\mathbf{i}^p\,| < i^{\max}\, \} = \begin{cases} \infty & \text{if } |\,\mathbf{i}^p\,| > i^{\max} \\ 0 & \text{if } |\,\mathbf{i}^p\,| \leq i^{\max}. \end{cases} \tag{4.22}$$

(2) Balancing of DC Link Capacitors Voltage: The DC-link capacitors voltage in NPC converters should be maintained at equal values to ensure low semiconductor device stress and reliable operations. This approach is accomplished by defining a sub-cost function:

$$g_{dc}(k) = \{ v^p_{C1} - v^p_{C2} \} \tag{4.23}$$

where v^p_{C1} and v^p_{C2} are the predicted DC-link capacitors voltage in an NPC converter [31]. Similar cost functions can be defined for other multilevel diode-clamped converters (DCCs) and floating capacitor converters.

(3) Common-mode Voltage Mitigation: The common-mode voltage (CMV) causes electromagnetic interference and severe voltage stress on the generator/motor winding insulation; thus, the control scheme must mitigate the CMV in the converter. For an even-level voltage source and matrix converter, the CMV minimization is defined as follows [32]:

$$g_{cm}(k) = \{ \, v^p_{cm} \, \} = \left\{ \frac{v^p_{aN} + v^p_{bN} + v^p_{cN}}{3} \right\} \tag{4.24}$$

where v^p_{aN}, v^p_{bN} and v^p_{cN} are predicted converter output voltages.

(4) Switching Frequency Minimization: In FCS-MPC, the number of commutations between two sampling instants is minimized to reduce the overall switching frequency of the converter. For a three-phase power converter with N_s switching devices per phase, the cost function is defined as follows [31]:

$$g_{sw}(k) = \{ \, n^p_{sw} \, \} = \{ \sum_{j=1}^{N_s} \sum_{x=a,\,b,\,c} |\, s^p_{xj}(k) - s_{xj}(k-1) \,| \} \tag{4.25}$$

where $s^p_{xj}(k)$ and $s_{xj}(k-1)$ are the predicted and past sample optimal switching states of jth device, respectively.

(5) Switching Losses Reduction: Another approach to improve the power conversion efficiency and reduce cooling efforts is to incorporate directly the switching losses in the cost function [33]:

$$g_{swl}(k) = \{ \, P^p_{swl} \, \} = \{ \sum_{j=1}^{N_s} \Delta i^p_{c,j} \cdot \Delta v^p_{ce,j} \, \} \tag{4.26}$$

where $\Delta\, i^{p}_{c,j}$ and $\Delta\, v^{p}_{ce,j}$ represent the predicted changes in the collector current and the collector to the emitter voltage of the jth switching device, respectively. This approach is more effective in improving converter efficiency than the approach in (4.25).

(6) Harmonic Spectrum Shaping: FCS-MPC does not undergo the intermediate modulation stage; therefore, the switching frequency of the converter varies with respect to the sampling frequency and operating conditions. Such a mechanism causes the harmonic spectrum of the output voltage and current waveforms of the converter to spread over a wide range of frequencies. For some applications, a spread spectrum is undesirable because it ignites harmonic resonances and electromagnetic interference. Hence, to obtain a harmonic spectrum shaping similar to that when linear control with PWM is employed, a frequency-dependent DT weight function $F(z)$ is used in the cost function [34], i.e.,

$$g_{hss}(k) = \{\, F(z)\, |\, \mathbf{i}^{*} - \mathbf{i}^{\mathrm{p}}\, |\, \}, \quad F(z) = \frac{z^{0} + b_{1}\, z^{-1} + \cdots + b_{n}\, z^{-n}}{a_{0}\, z^{0} + a_{1}\, z^{-1} + \cdots + a_{n}\, z^{-n}} \tag{4.27}$$

where $F(z)$ is an nth-order discrete filter that imposes a desired harmonic content on the control variables. The coefficients a_0 to a_n and b_1 to b_n are selected according to the type of filtering function required, i.e., low-pass filter, high-pass filter, bandpass filter, or band-stop filter [35]. For load current spectrum shaping, a narrow band stop (notch) filter is employed with the current tracking error terms.

(7) Selective Harmonic Elimination: Selective harmonic elimination (SHE) is a popular modulation technique for high-power converters to eliminate certain lower-order harmonics, thereby reducing switching losses. SHE can be realized with FCS-MPC by forcing the predicted voltage $\mathbf{v}^{\mathrm{p}}$ to follow a predefined converter voltage reference $\mathbf{v}^{*}$ such that the low-order harmonics are eliminated and a small number of commutations are involved. The cost function is implemented by the sliding discrete Fourier transform (SDFT) [36]:

$$\begin{aligned} g_{she}(k) &= \{\, \mathrm{SDFT}(z)\, |\, \mathbf{v}^{*} - \mathbf{v}^{\mathrm{p}}\, |\, \} \\ &= \{\, \mathrm{SDFT}_{f1}(z)\, |\, \mathbf{v}^{*} - \mathbf{v}^{\mathrm{p}}\, |\, \} + \lambda_{f} \sum_{i=2}^{n} \{\, \mathrm{SDFT}_{fi}(z)\, |\, \mathbf{v}^{*} - \mathbf{v}^{\mathrm{p}}\, |\, \} \end{aligned} \tag{4.28}$$

where the first term minimizes the voltage error at fundamental frequency f_1, and the second term eliminates the few selected harmonics (up to n^{th} order). The weighting factor for SHE is denoted by λ_f. Selective harmonic mitigation succeeds SHE and aims to reduce the total harmonic distortion (THD) of control variables below certain limits.

(8) Filter Resonance Mitigation: GSCs are often equipped with output inductive-capacitive-inductive (LCL) filters to reduce low-frequency current harmonics and to improve grid power quality. However, LCL filters demonstrate an intrinsic resonant behavior and are prone to harmonic oscillations caused by current harmonics and step changes in the reference current. To provide active damping for resonance frequencies ω_1 and ω_2 caused by filter parameters L_1, C, and L_2, a frequency-dependent cost function is used similar to the cost function introduced in (4.27) [37]:

$$g_{res}(k) = \{\, F(z)\, |\, \mathbf{i}^{\mathrm{p}}\, |\, \} = \{\, F_1(z)\, |\, \mathbf{i}_1^{\mathrm{p}}\, | + F_2(z)\, |\, \mathbf{i}_2^{\mathrm{p}}\, |\, \} \tag{4.29}$$

where $F_1(z)$ and $F_2(z)$ are discrete first-order bandpass filters tuned at ω_1 and ω_2 for the converter- and grid-side predicted currents ($\mathbf{i}_1^{\mathrm{p}}$ and $\mathbf{i}_2^{\mathrm{p}}$, respectively).

4.5 WEIGHTING FACTOR SELECTION

From the previous section, we noted that the final (or global) cost function in FCS-MPC provides greater flexibility, thus enabling the incorporation of several primary (or main) and secondary (or sub-) control objectives through weighting (or scalar) factors as demonstrated in the following:

$$g(k) = \lambda_{\text{main}} * g_{\text{main}}(k) + \lambda_{\text{sub,1}} * g_{\text{sub,1}}(k) + \cdots + \lambda_{\text{sub,n}} * g_{\text{sub,n}}(k) \tag{4.30}$$

where λ_{main} is the weighting factor associated with the main cost function, g_{main}, and $\lambda_{\text{sub,1}}$ and $\lambda_{\text{sub,n}}$ are the weighting factors for sub-cost functions $g_{\text{sub,1}}$ and $g_{\text{sub,n}}$, respectively.

The advantage of cost function flexibility comes with a significant challenge for the design of weighting factors. In linear control, the PI regulator parameters need to be tuned. By contrast, in FCS-MPC, the weighting factors have to be tuned to obtain optimal control performance. The weighting factors determine the relative importance of one control objective over the other. A large value for weighting factor λ implies a greater penalty to minimize errors in the corresponding variable (or control objective). The variables in the main and sub-cost functions possess different physical natures, such as current, voltage, power, torque, flux, frequency, or switching losses. The different amplitudes of the main and sub-cost function error terms under different operating conditions lead to coupling effects. Therefore, the selection of an optimal weighting factor value is an important design stage in FCS-MPC. In this section, several popular approaches for the weighting factors design are presented with respect to the solutions/methodologies available in literature.

4.5.1 Heuristic Selection

For classical control techniques, various methodologies are documented for the tuning of PI parameters or hysteresis bandwidth limits. However, because of limited knowledge, the heuristic (or empirical) procedure is commonly adopted to select the weighting factors in FCS-MPC. To analyze this approach, an NPC converter with the following global cost function [31] is considered:

$$\begin{aligned} g(k) &= \lambda_i \, g_i(k) + \lambda_{dc} \, g_{dc}(k) \\ &= \lambda_i \,|\, \hat{\mathbf{i}}^*(k+1) - \mathbf{i}^p(k+1) \,| + \lambda_{dc} \,|\, v^p_{C1}(k+1) - v^p_{C2}(k+1) \,| \end{aligned} \tag{4.31}$$

where λ_i and λ_{dc} are the weighting factors to regulate the load current and capacitors voltage, respectively. The current error values in Table 4.2 and arbitrary DC-link capacitor error values are considered and presented in Table 4.4.

Table 4.4 Calculation of global cost function with different weighting factor values

s	$g_i(k)$	$g_{dc}(k)$	$g\ (\lambda_{dc} = 0)$	$g\ (\lambda_{dc} = 1)$	$g\ (\lambda_{dc} = 0.1)$
s_0	3.792	17.036	3.792	20.828	5.496
s_1	3.427	4.828	3.427	8.255	**3.909**
s_2	3.031	16.956	**3.031**	19.987	4.727
s_3	6.459	0.788	6.459	**7.247**	6.538
s_4	9.125	29.204	9.125	38.329	12.045
s_5	7.635	6.329	7.635	13.964	8.268
s_6	6.093	29.164	6.093	35.257	9.009
s_7	3.792	17.036	3.792	28.828	5.496

The global cost functions with different λ_{dc} values are also included in Table 4.4. In (4.31), the primary control goal is to force the load currents to follow their reference currents; thus, the weighting factor associated with this term is 1 ($\lambda_i = 1$). When $\lambda_{dc} = 0$, the DC-link capacitors voltage are not controlled. However, the current shows excellent reference tracking. The corresponding optimal switching state is $\mathbf{s_2}$ as indicated in Table 4.4. When $\lambda_{dc} = 1$, switching state $\mathbf{s_3}$ is selected to assign more importance to the secondary goal. The sub-cost function g_{dc} has a negative impact on the primary cost function g_i, thus resulting in a degraded current reference tracking performance. The optimal λ_{dc} value obtained by this empirical approach is 0.1 and provides equal importance to both g_i and g_{dc} by selecting switching state $\mathbf{s_1}$ [32]. To obtain this optimal λ_{dc} value, the initial trial-and-error value is started from $\lambda_{dc} = 1$ and is decreased in small steps until the drift in capacitors voltage reach 2% of its nominal DC-link voltage.

A similar approach can be used for the selection of weighting factors for other control variables. The common question raised by the beginners in this field is whether to start the initial trail with high λ value or low λ value. The answer is simple: for applications wherein the secondary objectives must be fulfilled simultaneously, the λ value should proceed from high to low. For example, in NPC converters, both load currents and neutral-point voltage should be regulated; therefore, the initial trail should start with a high λ_{dc} value. Control variables, such as CMV mitigation, switching frequency minimization, and switching loss reduction, are intended to improve the performance of a converter, but the converter can still work without addressing these objectives. Hence, for this type of variables, the trial-and-error approach starts with a small λ value. For cases wherein the importance of each control variable is unknown, the branch-and-bound algorithm in [38] helps to design the optimal λ value with a minimum number of trials.

4.5.2 Per-Unit Method

The trial-and-error procedure requires prior knowledge on converter operation, which consumes a significant amount of time in verifying the optimal λ value by computer simulations. The work in [39] proposes an analytical (or numerical) approach to calculate weighting factors online on the basis of torque-ripple minimization in the PTC scheme. The complexity of this approach increases when additional control objectives are incorporated. Another analytical procedure for weighting factor selection employs per-unit values for the error terms. By considering the cost function in (4.31), the current error is divided by the rated reference current I_B^* and the error in DC-link capacitors voltage is divided by the rated DC-link voltage v_{dc}^*. The overall cost function is again multiplied by I_B^* such that the weighting factor for the main cost function becomes one [40]:

$$g(k) = I_B^* \times \left[\frac{|\,\hat{\mathbf{i}}^*(k+1) - \mathbf{i}^p(k+1)\,|}{I_B^*} + \frac{|\,v_{C1}^p(k+1) - v_{C2}^p(k+1)\,|}{v_{dc}^*} \right]. \tag{4.32}$$

By using this procedure, the weighting factors for the main and sub-cost functions become the following:

$$\lambda_i = \frac{I_B^*}{I_B^*} = 1, \quad \lambda_{dc} = \frac{I_B^*}{v_{dc}^*}. \tag{4.33}$$

For example, for a 3.0-MW and 3000-V wind energy system, the rated current and DC voltage are 577 A and 5304 V, respectively. Therefore, $\lambda_{dc} = I_B^*/v_{dc}^* = 577/5304 = 0.109$. This value is close to the optimal value obtained through the heuristic approach (refer to Table 4.4). The same approach can be used for PDVC, PTC, and PSC schemes.

4.5.3 Lookup Table-Based Selection

In some specific applications, the magnitude of error terms or control variables changes according to the operating condition. In such cases, optimal weighting factors are designed for each operating condition by the previous two methods and are stored in a lookup table for online use with FCS-MPC. For example, in GSCs, the average switching frequency holds a linear relationship with the grid active and reactive powers. For all operating conditions, the lookup-table-based approach guarantees switching frequency regulation within the set boundary limits (for example, 750-to-850 Hz) [41].

In this approach, the optimal weighting factors are deduced on the basis of the reference control variables; therefore, FCS-MPC exhibits excellent performance during all operating conditions. The lookup-table-based weighting factor selection is also applicable to other control variables, such as torque and flux control, CMV mitigation, etc.

4.5.4 Multiobjective Ranking Algorithm

The multiobjective ranking algorithm replaces the standard cost function and switching state selection (subsystems ⑦ and ⑧ in Figure 4.3) and eliminates the need for weighting factors. To explain the operating principle of this algorithm, the cost-function values in Table 4.4 are considered and are presented in Table 4.5. In the multiobjective ranking approach, the error terms are ranked: smaller errors have lower rankings, whereas larger errors have higher rankings. On the basis of rankings r_i and r_{dc}, the average ranking is computed and the switching state combination corresponding to the minimum overall ranking is selected and applied to the converter. By considering the sampling instant, both the heuristic approach with $\lambda_{dc} = 0.1$ and multiobjective ranking produce $\mathbf{s}_2$ as the optimal switching state combination. This approach has been verified in the PTC scheme and eliminates the need for weighting factor λ_ψ [42].

Table 4.5 Selection of optimal switching states by multiobjective ranking algorithm

s	$g_i(k)$	$g_{dc}(k)$	$r_i(k)$	$r_{dc}(k)$	$r(k) = 0.5 * (r_i + r_{dc})$	$g\ (\lambda_{dc} = 0.1)$
$\mathbf{s}_0$	3.792	17.036	3	5	4.0	5.496
$\mathbf{s}_1$	3.427	4.828	2	2	**2.0**	**3.909**
$\mathbf{s}_2$	3.031	16.956	1	4	2.5	4.727
$\mathbf{s}_3$	6.459	0.788	5	1	3.0	6.538
$\mathbf{s}_4$	9.125	29.204	7	7	7.0	12.045
$\mathbf{s}_5$	7.635	6.329	6	3	4.5	8.268
$\mathbf{s}_6$	6.093	29.164	4	6	5.0	9.009
$\mathbf{s}_7$	3.792	17.036	3	5	4.0	5.496

The rankings are scalar values; thus, this approach can be applied to any multivariable problem without involving trial-and-error calculations. System performance is improved by the multiobjective formulation and optimization, but this approach incurs some additional computational burden. An approach similar to multiobjective ranking algorithm is the use of Fuzzy decision making in weighting factor selection [43].

4.6 DELAY COMPENSATION METHODS

To aid the discussion of delay compensation, a 2L-VSC with eight switching state combinations ($\mathbf{s}_0 \sim \mathbf{s}_7$) is considered, with the load current used as the control variable. The analysis presented in this section is generic in nature and can be applied to other power converter topologies and control variables (e.g., voltage, power, torque, and flux), as well as other predictive control techniques (e.g., deadbeat and generalized predictive control).

With the prediction horizon length N_p set to one, the design of the resulting FCS-MPC turns out to be simple and the analysis of the overall system becomes less complicated. This approach is essentially called one-sample-ahead prediction (or single prediction (P) or "one-step prediction"). The operating principle of the one-sample-ahead prediction algorithm is presented in Figure 4.5. Eight predictions for the load current are obtained by using the eight switching state combinations, the DT converter model, and the present sampling instant (k) measurements. The algorithm selects a switching state (among the eight states) at the $(k)^{\text{th}}$ instant, which will minimize the cost function error at the $(k+1)$ instant. The operating principle of the one-sample-ahead FCS-MPC is summarized as follows:

$$\left.\begin{aligned} g(k) &= \left[\widehat{\mathbf{i}}^*(k+1) - \mathbf{i}^p(k+1)\right]^2 \\ \mathbf{i}^p(k+1) &= \mathbf{\Phi}\,\mathbf{i}(k) + \mathbf{\Gamma}\,\mathbf{v}^p(k) \end{aligned}\right\} \quad \text{"P" Approach.} \tag{4.34}$$

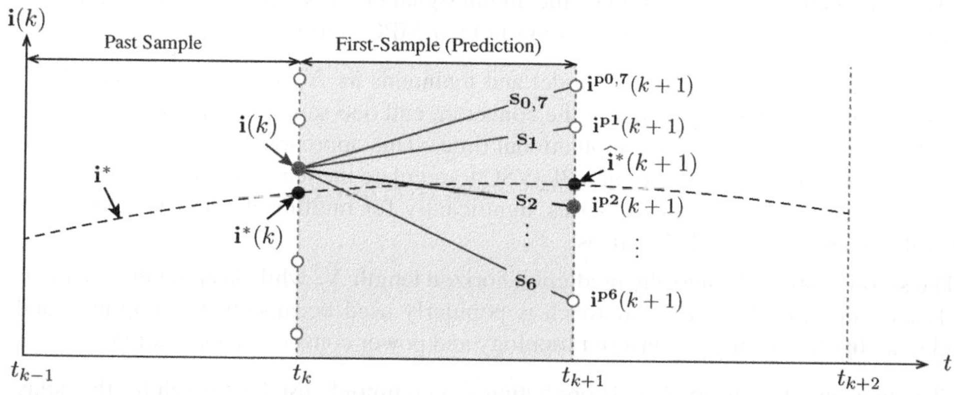

Figure 4.5 One-sample-ahead FCS-MPC based on the single prediction (P) approach (ideal case).

The one-sample-ahead prediction algorithm is used mainly in MATLAB simulations. The computer simulations represent an ideal scenario, wherein the time taken to compute a large number of online optimizations is zero, and the delay in applying optimal actuation is zero. However, in the real-time implementation of FCS-MPC algorithm, the large number of online calculations introduces one sampling time delay between the input and optimal actuation. The delay can be attributed to various factors: (1) digital control platform; (2) feedback voltage, current, and speed sensors; (3) interface board containing gate drivers; (4) switching of the devices. In all sampling rates, the delay associated with Factors 2 to 4 is negligible compared with the delay caused by the digital control platform. The delay caused by the digital control platform deteriorates the overall closed-loop performance if it is not considered in the design of the controller. This problem is common in all types of predictive control techniques implemented with a digital signal processor.

As an example, the performances of a 2L-VSC feeding passive RL load without and with delay compensation methods are illustrated in Figure 4.6. The switching frequency of the converter is kept the same for both cases to ensure a fair comparison. By employing the control delay compensation, the load current average tracking error e_i is decreased from 8.19% to 6.22%. The load current THD is decreased from 7.72% to 5.97%. Therefore, in the real-time implementation of an FCS-MPC algorithm, the computational delay must be treated properly to improve power quality and reduce the current tracking error. The definitions of e_i and THD can be found in Appendix B.

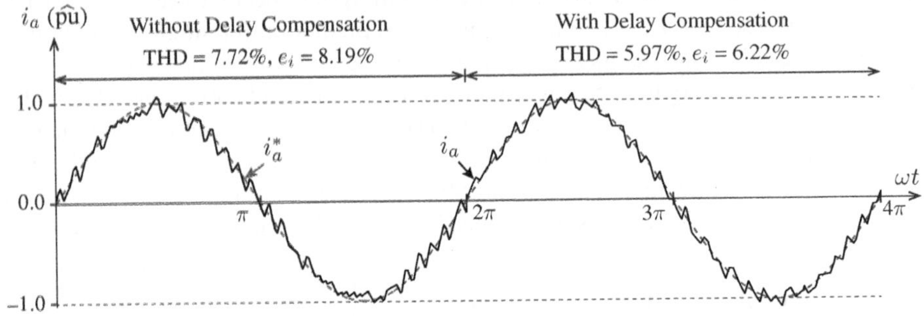

Figure 4.6 Simulation waveforms for FCS-MPC without and with delay compensation.

The computational delay caused by the digital signal processor and other less influential factors can be compensated for in two ways in FCS-MPC algorithms:

- The first approach alters the DT model and maintains an N_p value of one. The delay is incorporated in the DT model of the converter, and one-sample-ahead predictions are performed by using a small computational time. This approach has been employed in the deadbeat predictive control of a 2L-VSC to regulate the load current [44]. However, the complexity of modeling increases significantly for multilevel converter topologies involving various control objectives.
- The second approach alters the prediction horizon length N_p while keeping the DT models the same [45–48]. This approach is popularly used because of its simplicity and adaptability to any power converter topology and power conversion application.

The two-samples-ahead $(k+2)$ prediction is a commonly used approach for the delay compensation of FCS-MPC algorithms. The cost function minimization is performed at the $(k+2)$ sampling instant:

$$g(k) = \left[\widehat{\mathbf{i}}^*(k+2) - \mathbf{i}^p(k+2)\right]^2 \tag{4.35}$$

and the selected optimal switching state is applied at the $(k+1)$ sampling instant. The reference current extrapolated to the $(k+2)$ sampling instant is denoted by $\widehat{\mathbf{i}}^*(k+2)$, and the two-samples-ahead load current prediction is denoted by $\mathbf{i}^p(k+2)$. By this approach, one sampling period (T_s) is used for the control delay compensation and another sampling period is used for online calculations (optimization algorithm).

As shall be discussed in the succeeding sections, the two-samples-ahead load current prediction $\mathbf{i}^p(k+2)$ is performed by three different methods: (1) estimation plus prediction, (2) prediction plus double prediction, and (3) prediction plus prediction, which are respectively illustrated in Figures 4.7, 4.8, and 4.9.

4.6.1 Estimation + Prediction Approach

The control delay compensation based on the *estimation + prediction* (E+P) approach is illustrated in Figure 4.7 [45]. This approach adopts a slight modification of the "P" approach in Figure 4.5; thus, it is often called the "modified one-sample-ahead prediction" approach. The first sample involves the "estimation" of load currents, and the second sample involves the "prediction" of load currents (similar to the "P" approach). The optimal voltage vector $\mathbf{v}^{\mathrm{op}}(k)$ (or switching state combination) from the previous iteration is used to estimate the load currents at $(k+1)$ instant. The estimated current $\widehat{\mathbf{i}}(k+1)$ and all possible switching state combinations are used to predict the load currents at $(k+2)$ instant, i.e.,

$$\left.\begin{aligned} \widehat{\mathbf{i}}(k+1) &= \mathbf{\Phi}\,\mathbf{i}(k) + \mathbf{\Gamma}\,\mathbf{v}^{\mathrm{op}}(k) \\ \mathbf{i}^{\mathrm{p}}(k+2) &= \mathbf{\Phi}\,\widehat{\mathbf{i}}(k+1) + \mathbf{\Gamma}\,\mathbf{v}^{\mathrm{p}}(k+1) \end{aligned}\right\} \text{"E+P" Approach.} \tag{4.36}$$

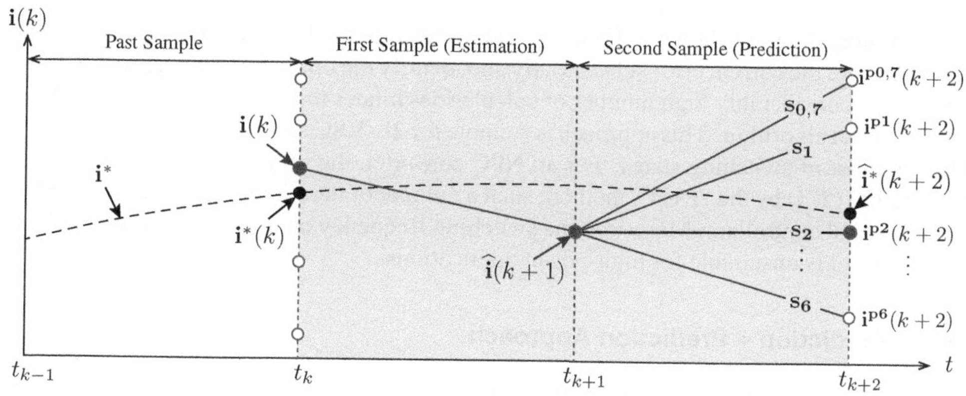

Figure 4.7 Delay compensation based on the E+P approach.

Notably, the calculation of $\widehat{\mathbf{i}}(k+1)$ is performed only once without any online iterations; therefore, no extra computational burden is incurred compared with that in the "P" approach. In the E+P method, the equivalent switching frequency of the converter increases by approximately 170% compared with that in the "P" approach [45]. For high-power applications employing this delay compensation method, a significant penalty should be applied to switching frequency minimization.

4.6.2 Prediction + Double Prediction Approach

The *prediction + double prediction* (P+P^2) approach for control delay compensation is shown in Figure 4.8. This method is often called "standard two-samples-ahead prediction" because the "P" approach is employed in both first and second steps of prediction. The number of predictions in the second sample increases exponentially (8^2) compared with the number of the first sample predictions (8). In addition to delay compensation, this approach improves the steady state and dynamic control performances [46, 47]. At the $(k+2)$ instant, cost function minimization is performed using only the currents predicted over the two samples. The load current predictive model is formulated as follows:

$$\left.\begin{aligned} \mathbf{i}^{\mathrm{p}}(k+1) &= \mathbf{\Phi}\,\mathbf{i}(k) + \mathbf{\Gamma}\,\mathbf{v}^{\mathrm{p}}(k) \\ \mathbf{i}^{\mathrm{p}}(k+2) &= \mathbf{\Phi}\,\mathbf{i}^{\mathrm{p}}(k+1) + \mathbf{\Gamma}\,\mathbf{v}^{\mathrm{p}}(k+1) \end{aligned}\right\} \text{"P+P}^2\text{" Approach.} \tag{4.37}$$

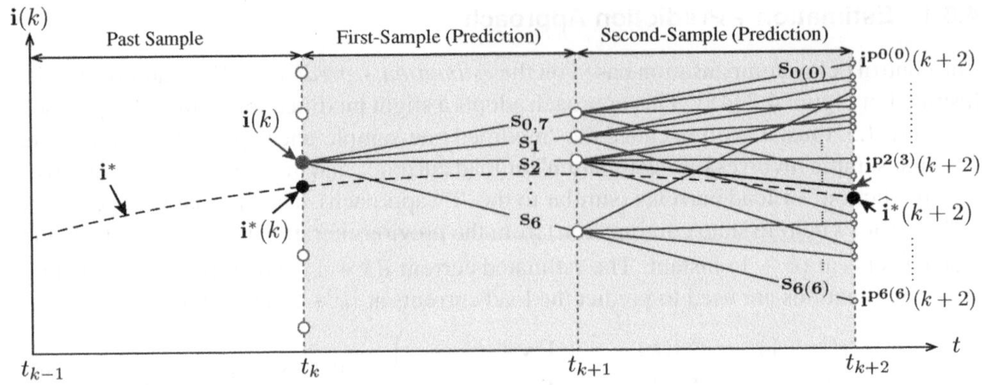

Figure 4.8 Delay compensation based on the P+P^2 approach.

The timing diagram shown in Figure 4.8 indicates that the large number of predictions help minimize the current error substantially and identify the optimal switching state easily. However, a considerably high number of calculations hinder the real-time implementation of this control algorithm. This approach is suitable for 2L-VSC and DC/DC converters with a low number of switching states. For an NPC converter, the number of online iterations become 729 (27^2) by the "P+P^2" method; such a number of iterations cannot be realized by current digital controllers. Moreover, the switching frequency obtained by this approach is very high and is unsuitable for high-power applications.

4.6.3 Prediction + Prediction Approach

The *prediction + prediction* (P+P) approach for simultaneous control delay compensation and performance improvement is shown in Figure 4.9. It uses eight switching state combinations projected onto the $(k+2)$ sampling instant, meaning that the same switching states are used in both $(k+1)$ and $(k+2)$ load current predictions [48]:

$$\left.\begin{aligned} \mathbf{i}^{\mathrm{p}}(k+1) &= \mathbf{\Phi}\,\mathbf{i}(k) &&+ \mathbf{\Gamma}\,\mathbf{v}^{\mathrm{p}}(k) \\ \mathbf{i}^{\mathrm{p}}(k+2) &= \mathbf{\Phi}\,\mathbf{i}^{\mathrm{p}}(k+1) &&+ \mathbf{\Gamma}\,\mathbf{v}^{\mathrm{p}}(k) \end{aligned}\right\} \text{“P+P” Approach} \tag{4.38}$$

and an optimal switching state is chosen on the basis of the cost function minimization in the $(k+2)$ instant.

This approach combines the advantages of lower computation of "P" method and optimal performance of "P+P^2" method. Therefore, this method is often referred to as "simplified two-samples-ahead prediction". The number of online calculations incurred using the "E+P," "P+P^2", and "P+P" methods are 9, 64, and 16, respectively. Compared with "P+P^2" method, the number of calculations in P+P approach is reduced significantly while preserving $N_p = 2$. Therefore, the computational requirement of the "P+P" method is slightly higher than the "E+P" method but is significantly lower than the "P+P^2" method. For an NPC converter, the number of online calculations with the use of "P+P" method becomes 54 (27 + 27), which can be easily implemented by currently available digital control platforms in the market. The switching frequency obtained by this approach is lower than that obtained by using the "E+P" and "P+P^2" methods because calculations are performed over two samples. For example, the switching frequency obtained by using the "P+P" method is approximately 2.8 times lower than when the "E+P" method is employed [48].

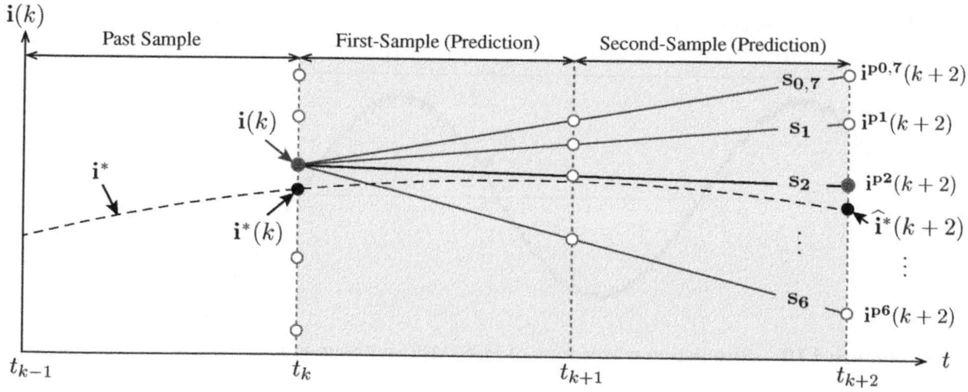

Figure 4.9 Delay compensation based on the P+P approach.

Despite the reduced switching frequency operation, the reference tracking in the "P+P" method is better even during the peaks and valleys of the sinusoidal reference waveforms. For the same switching frequency operation, the "P+P" method compared with the "E+P" method decreases the average tracking error from 6.4% to 3.8% and THD from 5.3% to 3.4%. These features make the "P+P" method a suitable candidate for high-power applications regardless of its drawback of performing double calculations during each sampling interval. In this book, we mainly deal with the "E+P" and "P+P" methods.

4.6.4 Long Prediction Horizons

System behavior can be predicted precisely by increasing N_p further. Long prediction horizons ($N_p > 1$) greatly enhance the closed-loop control performance and stability of a system. Currently, long prediction horizons are verified through direct model predictive control, which is another distinct class of MPC [49]. In the current FCS-MPC literature, only two-samples-ahead predictions ($N_p = 2$) have been presented, thus proving the significant performance improvement from one-sample-ahead prediction to two-samples-ahead prediction. The contribution of longer prediction horizons on the improvement of system performance is negligible compared with that of a two-samples-ahead prediction [18]. Moreover, for real-time FCS-MPC implementation, long prediction horizons significantly increase the number of calculations, particularly for complex converter topologies. Therefore, N_p poses a trade-off between performance and online computational effort. For the time being, the FCS-MPC long prediction horizon is an open research topic.

4.7 EXTRAPOLATION TECHNIQUES

One of the important requirements for an FCS-MPC strategy is reference tracking. As demonstrated in (4.34) and (4.35), cost function minimization is performed in the $(k + 1)$ or $(k + 2)$ sampling instants based on the extrapolated reference current. The performances of FCS-MPC without and with reference currents extrapolation with $T_s = 50\ \mu s$ and $N_p = 1$ are shown in Figure 4.10. Without reference extrapolation, the load current exhibits one-sample delay in reference tracking. By employing reference extrapolation, the average reference tracking error e_i decreases from 4.48% to 3.98%. The load current THD is unaffected by the extrapolation and is 3.95% in both cases.

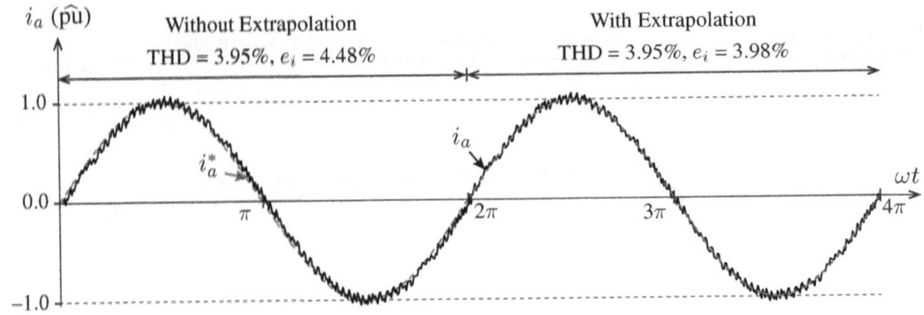

Figure 4.10 Simulation waveforms of FCS-MPC without and with extrapolation.

Similarly, in two-sample-ahead prediction, the load current exhibits two samples delay in reference tracking. When the sampling time T_s is sufficiently small ($< 20\,\mu$s), the e_i obtained without extrapolation is almost close to the one obtained with the use of extrapolation. Therefore, extrapolation is unnecessary when the switching frequency is considerably higher than the fundamental frequency of the variable to be controlled. When the FCS-MPC strategy works with high sampling times (lower switching frequencies), the references should be extrapolated to $(k + 1)$ or $(k + 2)$ sampling instant to effectively minimize the tracking error. In the dq frame, the variables are DC in nature; thus, no extrapolation is required during steady-state operation. However, during transient operation, a delay occurs between the reference and variable to be controlled. To overcome this issue, an extrapolation method should be used. In this section, three most widely used extrapolation methods are discussed.

4.7.1 Discrete Signal Generator

For control applications wherein reference control variables are defined by the user, the future values of references can be obtained by a discrete signal generator. Here, the following generic sinusoidal function is used:

$$x^* = X^* \sin\left(\omega^* t + \theta_0^*\right) \tag{4.39}$$

where X^*, ω^*, and θ_0^* represent the peak amplitude of reference, angular frequency of reference (rad/s), and initial phase angle (rad), respectively.

The sinusoidal function operates in a continuous mode or discrete mode according to the time (t) value. The output of the discrete signal generator may be updated at particular sampling instants only, $t \in \{0 \cdots \infty\}\, T_s$, thus inducing the effect of the "zero-order hold." For a prediction horizon length N_p, the value of θ_0^* is changed such that the "discrete phase shift" in the reference control variable produces the following extrapolation effect:

$$\widehat{x}^*(k + N_p) = X^* \sin\left(\omega^* t + \omega^* N_p T_s\right), \quad t \in \{0 \cdots \infty\}\, T_s \tag{4.40}$$

where $\widehat{x}^*(k + N_p)$ is the estimated reference variable.

The simulation waveforms of $(k + 1)$ and $(k + 2)$ extrapolations with $T_s = 200\,\mu$s and step changes in X^* and ω^* values are shown in Figure 4.11. The CT control variable $x^*(t)$, is also shown for comparison. This approach leads to a simple approach for the extrapolation of references and can be easily extended to multi-phase sinusoidal and non-sinusoidal functions.

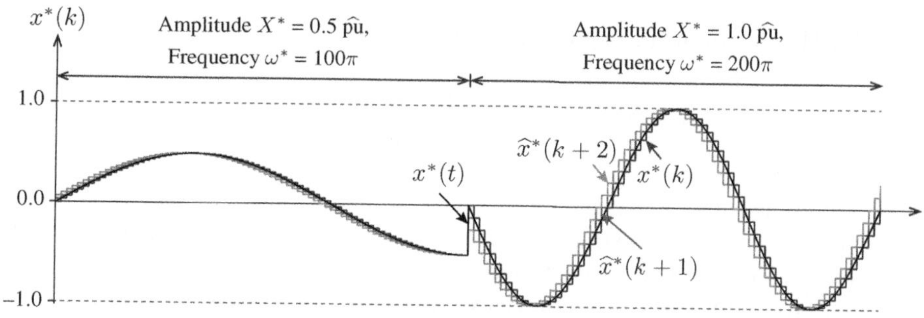

Figure 4.11 Simulation waveforms of present and future references by a discrete signal generator.

4.7.2 Vector Angle Extrapolation

For control applications wherein the reference control variables are generated by an outer control loop (such as in FOC or VOC), vector angle extrapolation is used. In vector angle extrapolation, the abc or dq variables are first converted to the $\alpha\beta$-frame and then converted into polar coordinates ($r\,e^{j\,\theta}$) from Cartesian coordinates ($x + j\,y$). The discrete phase shift $\omega^* N_p T_s$ is then added to the phase angle of the reference variable to enable an extrapolation similar to that by the discrete signal generator. The vector angle extrapolation for a prediction horizon, N_p, is given as follows [20]:

$$\widehat{x}^*(k + N_p) = X^*(k)\, e^{j\,(\theta + \omega^* N_p T s)}. \tag{4.41}$$

The references in polar coordinates are then converted back to Cartesian coordinates for use with the FCS-MPC cost function. The simulation waveforms by vector angle extrapolation with T_s = 200 μs and a step change in X^* from 0.5 $\widehat{\text{pu}}$ to 1.0 $\widehat{\text{pu}}$ are shown in Figure 4.12. Similar to future values obtained by using a discrete signal generator, the future values of the references are calculated precisely by vector angle extrapolation. This approach is unsuitable for single-phase systems and non-sinusoidal functions.

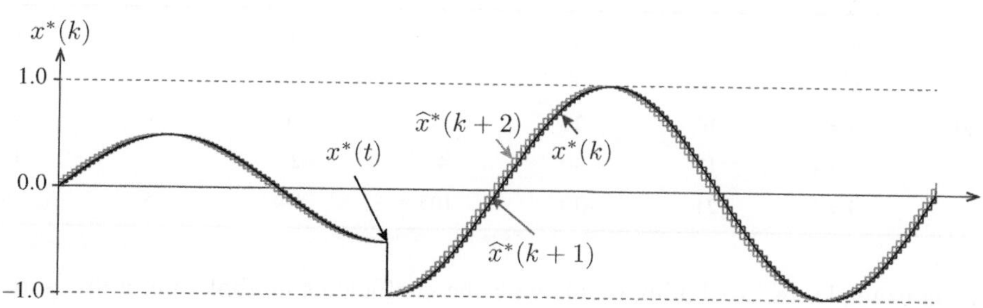

Figure 4.12 Simulation waveforms for present and future references by vector angle extrapolation.

4.7.3 Lagrange Extrapolation

Another simple approach for the extrapolation of multi-phase sinusoidal and non-sinusoidal functions is Lagrange extrapolation, which estimates the future values of control variable on the basis of present and past samples [11]. The one-sample-ahead Lagrange extrapolation is defined by the following expression [50]:

$$\widehat{x}^*(k+1) = \sum_{l=0}^{n} (-1)^{n-l} \frac{(n+1)!}{l!\,(n+1-l)!}\, x^*(k+l-n) \tag{4.42}$$

where x^* represents either voltage, current, power, torque, or flux reference. The variable n corresponds to the order of extrapolation. For sinusoidal reference variables, $n \geq 2$ is used, whereas $n = 1$ is sufficient for DC variables.

Similarly, two-samples-ahead Lagrange extrapolation is expressed as follows:

$$\widehat{x}^*(k+2) = \sum_{l=0}^{n} (-1)^{n-l} \frac{(n+1)!}{l!\,(n+1-l)!}\, x^*(k+1+l-n). \tag{4.43}$$

By solving (4.42) and (4.43), a generic expression for one-sample-ahead and two-samples-ahead Lagrange extrapolation is derived as follows:

$$\widehat{x}^*(k+1) \text{ or } \widehat{x}^*(k+2) = a_x \cdot x^*(k) + a_{x-1} \cdot x^*(k-1) + \cdots + a_{x-n} \cdot x^*(k-n) \tag{4.44}$$

where a_x is the coefficient for reference variable $x^*(k)$. The coefficients for different orders of extrapolation are listed in Table 4.6.

Table 4.6 Coefficients for $(k+1)$ and $(k+2)$ Lagrange extrapolation

Sample	Order	a_x	a_{x-1}	a_{x-2}	a_{x-3}	a_{x-4}	a_{x-5}
	$n = 1$	2	–1	0	0	0	0
	$n = 2$	3	–3	1	0	0	0
$(k+1)$	$n = 3$	4	–6	4	–1	0	0
	$n = 4$	5	–10	10	–5	1	0
	$n = 5$	6	–15	20	–15	6	–1
	$n = 1$	3	–2	0	0	0	0
	$n = 2$	6	–8	3	0	0	0
$(k+2)$	$n = 3$	10	–20	15	–4	0	0
	$n = 4$	15	–40	45	–24	5	0
	$n = 5$	21	–70	105	–84	35	–6

As an example, by combining (4.44) with the coefficients in Table 4.6, a third-order Lagrange method for $(k+1)$ and $(k+2)$ extrapolation is given as follows [11]:

$$\widehat{x}^*(k+1) = 4\,x^*(k) - 6\,x^*(k-1) + 4\,x^*(k-2) - x^*(k-3) \tag{4.45}$$

$$\widehat{x}^*(k+2) = 10\,x^*(k) - 20\,x^*(k-1) + 15\,x^*(k-2) - 4\,x^*(k-3). \tag{4.46}$$

Lagrange extrapolation is simple, does not involve any phase angle calculations, and is applicable to both AC and DC variables. However, this method generates unnecessary spikes during the step change in the reference control variable. These spikes cause high *dv/dt* in the converter output waveform under transient conditions [25].

4.8 SELECTION OF SAMPLING TIME

The selection of sampling time (T_s) in the real-time implementation of a predictive control schemes is an important issue. In a classical linear control with modulation stage, T_s is selected based on carrier (average converter) frequency. Digital signal processors, which realize predictive control, require processing time based on the complexity of the employed algorithm. For predictive control, T_s is selected on the basis of the minimum execution time required to perform all control-related tasks (refer to the PCC flowchart in Figure 4.3). The predictive algorithms (subsystems ③ to ⑩) for 2L-VSC (8 switching states), 3L-DCC (27 switching states), 4L-DCC (64 switching states), and 5L-DCC (125 switching states) require minimum execution times of 6, 10, 30, and 70 μs, respectively. The overall digital control (subsystems ① to ⑩) along with the "P+P" delay compensation of 2L-VSC, 3L-DCC, 4L-DCC, and 5L-DCC feeding RL load require sampling times, T_s, of 13, 19, 44, and 93 μs, respectively [32]. The T_s requirement not only changes with respect to the power converter topology but also with respect to the power conversion application. For example, a grid-connected multilevel converter needs higher T_s than a 2L-VSC feeding RL load [41]. Digital control platforms, such as dSPACE, produce a "task overrun" error when digital control is "built" with a sampling time lower than the minimum execution time. Therefore, to choose the minimum T_s value, a heuristic approach can be used by providing initial guesses with high T_s values.

As demonstrated in Figure 4.13, the average switching frequency f_{sw} of a power converter is related to T_s. By increasing T_s fivefold (from 20 μs to 100 μs), the switching frequency decreases by approximately fivefold and the current tracking error and THD increase by fivefold. The load current waveform with $T_s = 20\ \mu$s and weighting factor $\lambda_{sw} = 0.63$ for switching frequency minimization is also shown in Figure 4.13. Weighting factor-based switching frequency reduction is more efficient than higher T_s selection [51]. Thus, the minimum sampling time is always recommended when implementing predictive control. Additional objectives are incorporated in the cost function to meet specific application requirements. By increasing λ_{sw}, the switching frequency is reduced further. Hence, in predictive control, for a given T_s, the power converter can operate with a wide range of switching frequencies by simply including a constraint in the cost function.

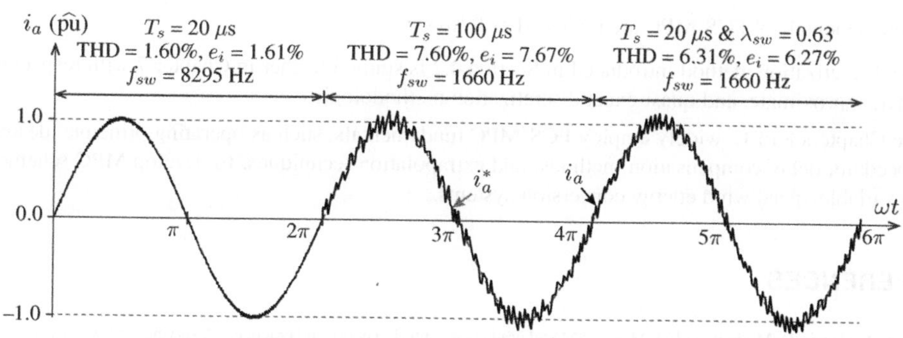

Figure 4.13 Simulation waveforms of FCS-MPC with different sampling times.

As the number of converter switching states increases, the number of calculations during each sampling interval increases accordingly; thus, the minimum execution time also increases. For some multilevel power converters, the switching states increase exponentially. For example, a cascaded H-bridge converter (or half-bridge-based modular multilevel converter) with two cells (or modules) in each phase results in 4096 possible switching states, which is impossible to implement in a standard control platform. A basic predictive control algorithm should be modified such that the total number of online iterations is reduced to a value feasible for current digital control platforms. A feasible solution is to select vectors adjacent to the reference voltage vector, which is a commonly used design criterion for SVM to reduce the output voltage THD. By this approach, irrespective of the number of cells/modules in each phase, the number of online iterations is reduced to only seven predictions, which correspond to the seven adjacent vectors [52]. Therefore, the overall control algorithm can be realized by $T_s = 13\ \mu$s, similar to the computational time needed by a 2L-VSC, but with a slightly sluggish dynamic response. For other emerging multilevel converter topologies, a similar approach can be used to reduce the number of online calculations.

4.9 CONCLUDING REMARKS

The basic concept, sampled-data model, operating principle, step-by-step design procedure, implementation of control scheme, and stability-related issues of FCS-MPC are systematically described. FCS-MPC is a simple concept that can be applied to a wide range of power converters and power conversion applications with only slight modifications in the control algorithm. Cost function flexibility and potentiality in achieving different primary and secondary control objectives are also discussed. By reviewing recent studies in this area, the issues and challenges related to FCS-MPC, such as weighting factor selection, control delay compensation, reference variable extrapolation, and minimum sampling time selection are analyzed. Through a comparison study, FCS-MPC is found to exhibit comparable performance in a steady state but is far superior to classical linear controls with PI regulator and SVM in terms of dynamic response and control flexibility. The analysis presented in this chapter demonstrates that FCS-MPC is conceptually very simple yet powerful tool to control modern high-performance power conversion systems.

Link to Next Chapters:

The fundamentals of FCS-MPC are reviewed as follows:

- The discretization method introduced in Section 4.2 is studied further in Chapter 7 with respect to exact, approximate, and quasi-exact discretization methodologies.
- The Chapters 8 to 12 widely employ FCS-MPC fundamentals, such as operating principle, design procedure, delay compensation methods, and extrapolation techniques, to develop MPC schemes for variable-speed wind energy conversion systems.

REFERENCES

1. C. E. García, D. M. Prett, and M. Morari, "Model predictive control: Theory and practice – A survey," *Automatica*, vol. 25, no. 3, pp. 335–348, 1989.
2. J. Rodríguez, J. Pontt, C. Silva, P. Cortes, U. Ammann, and S. Rees, "Predictive current control of a voltage source inverter," in *IEEE Power Electronics Specalists Conference (PESC)*, Aachen, Germany, vol. 3, June 2004, pp. 2192–2196.
3. H. Young, M. Perez, J. Rodriguez, and H. Abu-Rub, "Assessing finite-control-set model predictive control: A comparison with a linear current controller in two-level voltage source inverters," *IEEE Industrial Electronics Magazine*, vol. 8, no. 1, pp. 44–52, March 2014.
4. J. Rodríguez, R. Kennel, J. Espinoza, M. Trincado, C. Silva, and C. Rojas, "High-performance control strategies for electrical drives: An experimental assessment," *IEEE Transactions on Industrial Electronics*, vol. 59, no. 2, pp. 812–820, February 2012.

5. T. Geyer, "A comparison of control and modulation schemes for medium-voltage drives: Emerging predictive control concepts versus PWM-based schemes," *IEEE Transactions on Industry Applications*, vol. 47, no. 3, pp. 1380–1389, May/June 2011.

6. M. Rivera, A. Wilson, C. Rojas, J. Rodríguez, J. Espinoza, P. Wheeler, and L. Empringham, "A comparative assessment of model predictive current control and space vector modulation in a direct matrix converter," *IEEE Transactions on Industrial Electronics*, vol. 60, no. 2, pp. 578–588, February 2013.

7. G. Goodwin, J. Agüero, M. Cea Garridos, M. Salgado, and J. Yuz, "Sampling and sampled-data models: The interface between the continuous world and digital algorithms," *IEEE Control Systems Magazine*, vol. 33, no. 5, pp. 34–53, October 2013.

8. C. Silva and J. Yuz, "On sampled-data models for model predictive control," in *IEEE Industrial Electronics Conference (IECON)*, Glendale, AZ, USA, November 2010, pp. 2966–2971.

9. J. Maciejowski, *Predictive Control With Constraints*, ser. Pearson Education. Upper Saddle River: Prentice Hall, 2002.

10. V. Yaramasu, M. Rivera, B. Wu, and J. Rodriguez, "Model predictive current control of two-level four-leg inverters – Part I: Concept, algorithm and simulation analysis," *IEEE Transactions on Power Electronics*, vol. 28, no. 7, pp. 3459–3468, July 2013.

11. O. Kukrer, "Discrete-time current control of voltage-fed three-phase PWM inverters," *IEEE Transactions on Power Electronics*, vol. 11, no. 2, pp. 260–269, March 1996.

12. T. Orlowska-Kowalska, F. Blaabjerg, and J. Rodrguez, *Advanced and Intelligent Control in Power Electronics and Drives*, ser. Studies in Computational Intelligence. Springer, 2014.

13. J. Rawlings and D. Mayne, *Model Predictive Control: Theory and Design*. Madison, WI: Nob Hill Publishing, LCC, 2009.

14. J. Rodríguez, J. Pontt, C. A. Silva, P. Correa, P. Lezana, P. Coftes, and U. Ammann, "Predictive current control of a voltage source inverter," *IEEE Transactions on Industrial Electronics*, vol. 54, no. 1, pp. 495–503, February 2007.

15. R. Aguilera and D. Quevedo, "Predictive control of power converters: Designs with guaranteed performance," *IEEE Transactions on Industrial Informatics*, vol. 11, no. 1, pp. 53–63, February 2015.

16. S. Vazquez, J. Leon, L. Franquelo, J. Rodriguez, H. Young, A. Marquez, and P. Zanchetta, "Model predictive control: A review of its applications in power electronics," *IEEE Industrial Electronics Magazine*, vol. 8, no. 1, pp. 16–31, March 2014.

17. J. Rodríguez, M. P. Kazmierkowski, J. R. Espinoza, P. Zanchetta, H. Abu-Rub, H. A. Young, and C. A. Rojas, "State of the art of finite control set model predictive control in power electronics," *IEEE Transactions on Industrial Informatics*, vol. 9, no. 2, pp. 1003–1016, May 2013.

18. S. Kouro, P. Coftes, R. Vargas, U. Ammann, and J. Rodríguez, "Model predictive control-A simple and powerful method to control power converters," *IEEE Transactions on Industrial Electronics*, vol. 56, no. 6, pp. 1826–1838, June 2009.

19. P. Coftes, M. Kazmierkowski, R. Kennel, D. Quevedo, and J. Rodríguez, "Predictive control in power electronics and drives," *IEEE Transactions on Industrial Electronics*, vol. 55, no. 12, pp. 4312–4324, December 2008.

20. J. Rodriguez and P. Coftes, *Predictive Control of Power Converters and Electrical Drives*, 1st ed. Chichester, UK: IEEE Wiley press, March 2012.

21. V. Yaramasu, "Predictive control of multilevel converters for megawatt wind energy conversion systems," Ph.D. dissertation, Ryerson University, Toronto, ON, Canada, 2014, available at: http://digital.library.ryerson.ca/islandora/object/RULA%3A3459.

22. R. Aguilera, P. Lezana, and D. Quevedo, "Finite-control-set model predictive control with improved steady-state performance," *IEEE Transactions on Industrial Informatics*, vol. 9, no. 2, pp. 658–667, May 2013.

23. P. Coftes, G. Ortiz, J. Yuz, J. Rodríguez, S. Vazquez, and L. Franquelo, "Model predictive control of an inverter with output *LC* filter for UPS applications," *IEEE Transactions on Industrial Electronics*, vol. 56, no. 6, pp. 1875–1883, June 2009.

24. V. Yaramasu, M. Rivera, M. Narimani, B. Wu, and J. Rodriguez, "Model predictive approach for a simple and effective load voltage control of four-leg inverter with an output LC filter," *IEEE Transactions on Industrial Electronics*, vol. 61, no. 10, pp. 5259–5270, October 2014.

25. V. Yaramasu and B. Wu, "Model predictive decoupled active and reactive power control for high-power grid-connected four-level diode-clamped inverters," *IEEE Transactions on Industrial Electronics*, vol. 61, no. 7, pp. 3407–3416, July 2014.

26. D. Quevedo, R. Aguilera, M. Perez, P. Coftes, and R. Lizana, "Model predictive control of an AFE rectifier with dynamic references," *IEEE Transactions on Power Electronics*, vol. 27, no. 7, pp. 3128–3136, July 2012.

27. R. Vargas, U. Ammann, B. Hudoffsky, J. Rodríguez, and P. Wheeler, "Predictive torque control of an induction machine fed by a matrix converter with reactive input power control," *IEEE Transactions on Power Electronics*, vol. 25, no. 6, pp. 1426–1438, June 2010.

28. M. Preindl and S. Bolognani, "Model predictive direct speed control with finite control set of PMSM drive systems," *IEEE Transactions on Power Electronics*, vol. 28, no. 2, pp. 1007–1015, February 2013.

29. E. Fuentes, D. Kalise, J. Rodriguez, and R. Kennel, "Cascade-free predictive speed control for electrical drives," *IEEE Transactions on Industrial Electronics*, vol. 61, no. 5, pp. 2176–2184, May 2014.

30. H. Miranda, P. Coftes, J. Yuz, and J. Rodríguez, "Predictive torque control of induction machines based on state-space models," *IEEE Transactions on Industrial Electronics*, vol. 56, no. 6, pp. 1916–1924, June 2009.

31. R. Vargas, P. Cortes, U. Ammann, J. Rodríguez, and J. Pontt, "Predictive control of a three-phase neutral-point-clamped inverter," *IEEE Transactions on Industrial Electronics*, vol. 54, no. 5, pp. 2697–2705, October 2007.

32. V. Yaramasu, B. Wu, M. Rivera, M. Narimani, S. Kouro, and J. Rodriguez, "Generalised approach for predictive control with common-mode voltage mitigation in multilevel diode-clamped converters," *IET Power Electronics*, vol. 8, no. 8, pp. 1440–1450, August 2015.

33. R. Vargas, U. Ammann, and J. Rodríguez, "Predictive approach to increase efficiency and reduce switching losses on matrix converters," *IEEE Transactions on Power Electronics*, vol. 24, no. 4, pp. 894–902, April 2009.

34. P. Cortes, J. Rodríguez, D. Quevedo, and C. Silva, "Predictive current control strategy with imposed load current spectrum," *IEEE Transactions on Power Electronics*, vol. 23, no. 2, pp. 612–618, March 2008.

35. M. Perez and J. Rodriguez, "Predictive frequency spectrum shaping of currents in a three phase inverter," in *IEEE International Symposium on Sensorless Control for Electrical Drives and Predictive Control of Electrical Drives and Power Electronics (SLED/PRECEDE)*, Munich, Germany, October 2013, pp. 1–5.

36. H. Aggrawal, J. Leon, L. Franquelo, S. Kouro, P. Garg, and J. Rodriguez, "Model predictive control based selective harmonic mitigation technique for multilevel cascaded H-bridge converters," in *IEEE Industrial Electronics Conference (IECON)*, Melbourne, VIC, Australia, November 2011, pp. 4427–4432.

37. H. Miranda, R. Teodorescu, P. Rodriguez, and L. Helle, "Model predictive current control for high-power grid-connected converters with output *LCL* filter," in *IEEE Industrial Electronics Conference (IECON)*, Porto, Portugal, November 2009, pp. 633–638.

38. P. Cortes, S. Kouro, B. La Rocca, R. Vargas, J. Rodríguez, J. Leon, S. Vazquez, and L. Franquelo, "Guidelines for weighting factors design in model predictive control of power converters and drives," in *IEEE International Conference on Industrial Technology (ICIT)*, Gippsland, VIC, Australia, February 2009, pp. 1–7.

39. S. Davari, D. Khaburi, and R. Kennel, "An improved FCS-MPC algorithm for an induction motor with an imposed optimized weighting factor," *IEEE Transactions on Power Electronics*, vol. 27, no. 3, pp. 1540–1551, March 2012.

40. M. Narimani, B. Wu, V. Yaramasu, Z. Cheng, and N. Zargari, "Finite control-set model predictive control (FCS-MPC) of nested neutral point clamped (NNPC) converter," *IEEE Transactions on Power Electronics*, vol. 30, no. 12, pp. 7262–7269, December 2015.

41. V. Yaramasu, B. Wu, and J. Chen, "Model-predictive control of grid-tied four-level diode-clamped inverters for high-power wind energy conversion systems," *IEEE Transactions on Power Electronics*, vol. 29, no. 6, pp. 2861–2873, June 2014.

42. C. Rojas, J. Rodríguez, F. Villarroel, J. Espinoza, C. Silva, and M. Trincado, "Predictive torque and flux control without weighting factors," *IEEE Transactions on Industrial Electronics*, vol. 60, no. 2, pp. 681–690, February 2013.

43. F. Villarroel, J. Espinoza, C. Rojas, J. Rodríguez, M. Rivera, and D. Sbarbaro, "Multiobjective switching state selector for finite-states model predictive control based on fuzzy decision making in a matrix converter," *IEEE Transactions on Industrial Electronics*, vol. 60, no. 2, pp. 589–599, February 2013.

44. C. Xia, M. Wang, Z. Song, and T. Liu, "Robust model predictive current control of three-phase voltage source PWM rectifier with online disturbance observation," *IEEE Transactions on Industrial Informatics*, vol. 8, no. 3, pp. 459–471, August 2012.

45. P. Cortes, J. Rodríguez, C. Silva, and A. Flores, "Delay compensation in model predictive current control of a three-phase inverter," *IEEE Transactions on Industrial Electronics*, vol. 59, no. 2, pp. 1323–1325, February 2012.

46. V. Yaramasu, B. Wu, M. Rivera, and J. Rodriguez, "Enhanced model predictive voltage control of four-leg inverters with switching frequency reduction for standalone power systems," in *International Power Electronics and Motion Control Conference (EPE/PEMC)*, Novi Sad, Serbia, September 2012, pp. DS2c.6–1–DS2c.6–5.

47. P. Cortes, J. Rodríguez, S. Vazquez, and L. Franquelo, "Predictive control of a three-phase UPS inverter using two steps prediction horizon," in *IEEE International Conference on Industrial Technology (ICIT)*, Viña del Mar, Chile, March 2010, pp. 1283–1288.

48. V. Yaramasu, M. Rivera, M. Narimani, B. Wu, and J. Rodriguez, "High performance operation for a four-leg NPC inverter with two-sample-ahead predictive control strategy," *International Journal of Electric Power Systems Research*, vol. 123, pp. 31–39, June 2015.

49. P. Karamanakos, T. Geyer, N. Oikonomou, F. Kieferndorf, and S. Manias, "Direct model predictive control: A review of strategies that achieve long prediction intervals for power electronics," *IEEE Industrial Electronics Magazine*, vol. 8, no. 1, pp. 32–43, March 2014.

50. M. Odavic, V. Biagini, P. Zanchetta, M. Sumner, and M. Degano, "One-sample-period-ahead predictive current control for high-performance active shunt power filters," *IET Power Electronics*, vol. 4, no. 4, pp. 414–423, April 2011.

51. P. Cortes, L. Vattuone, and J. Rodríguez, "A comparative study of predictive current control for three-phase voltage source inverters based on switching frequency and current error," in *European Conference on Power Electronics and Applications (EPE)*, Birmingham, UK, September 2011, pp. 1–8.

52. P. Cortes, A. Wilson, S. Kouro, J. Rodríguez, and H. Abu-Rub, "Model predictive control of multilevel cascaded H-bridge inverters," *IEEE Transactions on Industrial Electronics*, vol. 57, no. 8, pp. 2691–2699, August 2010.

PART II

MODELING OF POWER CONVERTERS AND WIND GENERATORS

CHAPTER 5

MODELING OF POWER CONVERTERS FOR MODEL PREDICTIVE CONTROL

5.1 INTRODUCTION

As previously mentioned in Chapters 1 and 2, the power converters in Type 3 and 4 variable-speed wind energy conversion systems (WECS) enable smooth grid connection and reactive power generation, in addition to the basic requirement of power conversion [1]. The power converters employed for WECS can be recognized as single-stage AC/AC, two-stage AC/DC + DC/AC, and three-stage AC/DC + DC/DC + DC/AC converters. A concise list of power converters employed in Type 3 and 4 wind turbines (WTs) is provided in Table 5.1. The summary implies that two-level voltage source converter (2L-VSC) and three-level diode-clamped converter (3L-DCC) (also known as neutral-point clamped (NPC) converter) in back-to-back (BTB) connected configuration are dominantly used in the present variable-speed WECS industry.

We learned in Chapters 3 and 4 that optimal switching signals (also called switching states or gating signals) are generated for the power converters in Type 3 and 4 WECS by the digital control systems. The correlation between power converter input and output variables must be formulated in terms of switching signals to enable the use of model predictive control (MPC) in Type 3 and 4 WECS. This chapter deals with the modeling of power converters employed in Type 3 and 4 WECS, particularly for 2L-VSCs and NPC converters. Other promising converters, such as flying capacitor converter (FCC), current source converter (CSC), direct matrix converter (DMC), and indirect matrix converter (IMC), are briefly analyzed to initiate further studies in this area. A summary of the power converters' component count and switching states is provided in Table 5.2.

Model Predictive Control of Wind Energy Conversion Systems, First Edition. Venkata Yaramasu and Bin Wu.

Companion Website: www.wiley.com/go/yaramasu_wu_mpc_wecs

Table 5.1 Concise list of megawatt power converters employed in Type 3 and 4 turbines

WT Manufacturer	WT Type and Power	WT Model #	Power Converter	Converter Rating
AAER Wind (Canada)	Type 3, 1.65 MW	A1650-77	1x BTB 2L-VSC	500 kW, 690 V
Fuji Ltd. (Japan)	Type 3, 2.0 MW	Subaru 80/2.0	1x BTB 2L-VSC	600 kW, 1400 V
BARD GmbH (Germany)	Type 3, 5.0 MW	BARD5.0	2x BTB 2L-VSC	750 kW, 690 V
Senvion (Germany)	Type 3, 6.2 MW	6.2M126	3x BTB 2L-VSC	620 kW, 690 V
Goldwind (PR China)	Type 4, 1.5 MW	GW70/1500	Diode + 3x Boost + 2x 2L-VSC	Boost: 500 kW VSC: 750 kW
Zephyros (Netherlands)	Type 4, 2.0 MW	Z72	1x BTB 3L-DCC	2.0 MW, 4000 V
Clipper Ltd. (USA)	Type 4, 2.5 MW	Liberty C99	1x (Diode + 2L-VSC)	625 kW, 690 V
Avantis (Germany)	Type 4, 2.5 MW	AV928	4x BTB 2L-VSC	625 kW, 690 V
Areva Wind (Germany)	Type 4, 5.0 MW	M5000-135	1x BTB 3L-DCC	5.0 MW, 3300 V
Siemens (Germany)	Type 4, 7.0 MW	SWT7.0-171	1x BTB 3L-DCC	7 MW, 3300 V
Enercon GmbH (Germany)	Type 4, 7.5 MW	E126/7500	10x BTB 2L-VSC	750 kW, 690 V
Windtec-AMSC (USA)	Type 4, 10.0 MW	SeaTitan	14x BTB 2L-VSC	750 kW, 690 V
Sway Turbine (Norway)	Type 4, 10.0 MW	ST10	1x BTB 3L-DCC	10 MW, 3500 V

Table 5.2 Summary of the component count and switching states for three-phase power converters

Converter Type	Active Switches	Clamping Devices	DC Link Elements	Switching States	Possible Vectors
2L-VSC	6	0	1 [C]	8	7
3L-DCC	12	6 [D]	2 [C]	27	19
mL-DCC	$6(m-1)$	$3(m-1)(m-2)$ [D]	$(m-1)$ [C]	m^3	$m^3-(m-1)^3$
3L-FCC	12	3 [C]	1 [C]	64	19
mL-FCC	$6(m-1)$	$3^{(m-2)}$ [C]	1 [C]	$(m+1)^3$	$m^3-(m-1)^3$
CSC	6	0	1 [L]	9	7
DMC	18	0	0	27	21
IMC	18	0	0	72	14

C = Capacitor, D = Diode, L = Inductor, m = Voltage levels of converter, mL = multilevel.

Chapter Overview

- The objectives for the modeling and their employed notations are discussed in Sections 5.2 and 5.3, respectively.
- 2L-VSC modeling is presented in Section 5.4. The extension of 2L-VSC modeling to multiphase applications and BTB configuration is analyzed in Section 5.5.
- NPC converter modeling is provided in Section 5.6; this modeling concept is extended to multilevel and multiphase applications and BTB topology in Section 5.7.
- The modeling of 3L-FCC, CSC, DMC, and IMC are presented in Section 5.8.
- Section 5.9 concludes the major observations in this chapter.

5.2 OBJECTIVES FOR THE MODELING OF POWER CONVERTERS

As outlined in previous chapters, the regulation of generator currents is a common control objective in Type 3 and 4 WECS to extract the maximum possible energy from wind. The block diagram of Type 4 WECS with a permanent magnet synchronous generator (PMSG), 2L voltage source rectifier (VSR), and predictive current control (PCC) scheme is shown in Figure 5.1. More details about the variables and calculation of reference currents $\widehat{i}_{ds}^*(k+1)$ and $\widehat{i}_{qs}^*(k+1)$ are presented in Chapter 1, Section 1.9.1 on page 55.

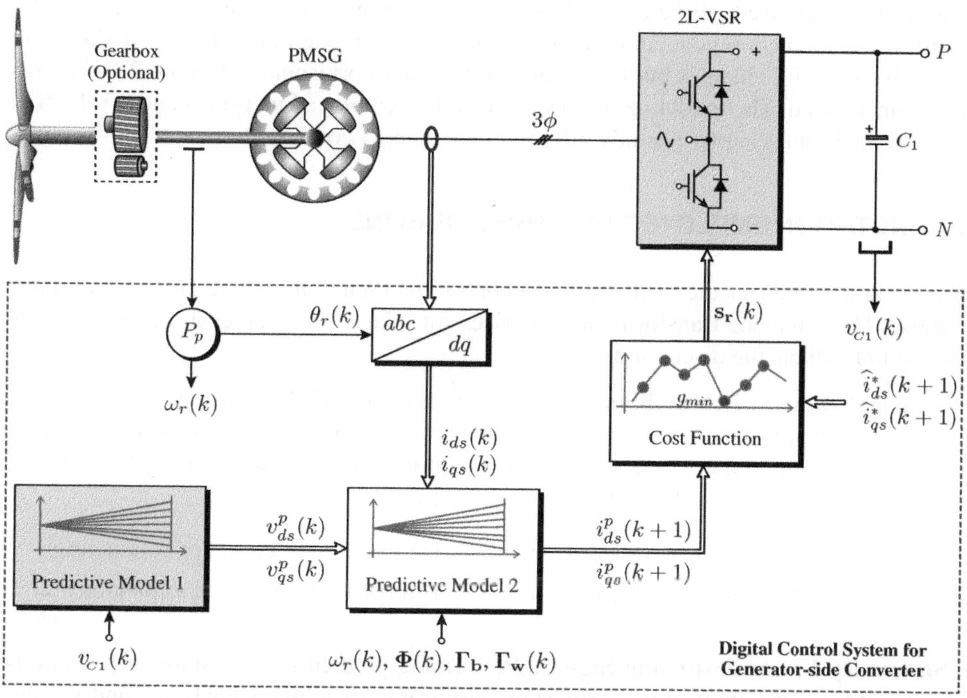

Figure 5.1 Block diagram of the PCC scheme for 2L-VSR-based PMSG WECS.

The MPC shown in Figure 5.1 mainly consists of predictive models and cost function minimization. The overall discrete-time (DT) predictive model for the PMSG with 2L-VSR is provided below:

$$\underbrace{\begin{bmatrix} i_{ds}^p(k+1) \\ i_{qs}^p(k+1) \end{bmatrix} = \mathbf{\Phi}(k) \begin{bmatrix} i_{ds}(k) \\ i_{qs}(k) \end{bmatrix} + \mathbf{\Gamma_b} \underbrace{\begin{bmatrix} v_{ds}^p(k) \\ v_{qs}^p(k) \end{bmatrix}}_{\text{Predictive Model 1}} + \mathbf{\Gamma_w}(k)}_{\text{Predictive Model 2}} \tag{5.1}$$

where superscript p denotes the predicted variable. The matrices $\mathbf{\Phi}(k)$, $\mathbf{\Gamma_b}$, and $\mathbf{\Gamma_w}(k)$ contain the DT equivalents of the machine parameters and rotor speed.

The predictive model 1 in Figure 5.1 corresponds to the particular converter type, whereas the predictive model 2 is related to the generator type. The predictive model 1 uses all possible switching state combinations and measured DC-link voltage to predict the converter output voltages in the synchronous reference frame, dq. The predictive model 2 uses this information along with feedback currents and rotor speed to predict the future value of PMSG dq-axis currents, $i^p_{ds}(k+1)$ and $i^p_{qs}(k+1)$. The eight switching state combinations of 2L-VSR leads to eight different values for $v^p_{ds}(k)$ and $v^p_{qs}(k)$; thus, eight different predictions for $i^p_{ds}(k+1)$ and $i^p_{qs}(k+1)$ are obtained. The predicted currents are compared with the reference currents by the cost function to select the optimal switching states. The MPC performance largely depends on the quality of the system model.

When the 2L converters are replaced by 3L converters in the PMSG WECS, the predictive model 1 changes accordingly but the predictive model 2 remains unchanged. For example, the predictive model 1 generates 27 values for $v^p_{ds}(k)$ and $v^p_{qs}(k)$ when an NPC converter is employed. When the PMSG is replaced by another generator such as the squirrel-cage induction generator (SCIG) or doubly fed induction generator (DFIG), the predictive model 2 changes but the predictive model 1 corresponding to the 2L-VSR remains unchanged. The modeling of power converters employed in Type 3 and 4 WECS is investigated in this chapter to study different wind energy configurations.

5.3 NOTATION EMPLOYED FOR THE MODELING

This section introduces the notation of variables to be used in the modeling of power converters. The reference transformation matrices introduced in Chapter 3 are used in this section to facilitate the discussion.

- **Natural Frame Switching Signal:** These signals (also called gating signals) are binary in nature and are used to control the switching actions of semiconductor switches. These signals are the modulator output in classical control and cost function in MPC. They are expressed for all $x \in \{a, b, c\}$ and $j \in \{1, 2 \cdots\}$ as

$$s_{xj} = \begin{cases} \text{`1'} & \text{if semiconductor switch } T_{xj} \text{ is turned-on or closed or conducting} \\ \text{`0'} & \text{if semiconductor switch } T_{xj} \text{ is turned-off or opened or blocking} \end{cases} \tag{5.2}$$

- **Stationary Frame Switching Signal:** Performing predictions in a stationary frame is required for some applications. The stationary frame switching signals $s_{\alpha j}$ and $s_{\beta j}$ are deducted from the natural-frame switching signal as follows:

$$\begin{bmatrix} s_{\alpha j} \\ s_{\beta j} \end{bmatrix} = \begin{bmatrix} \mathbf{T}_{abc/\alpha\beta} \end{bmatrix} \begin{bmatrix} s_{aj} \\ s_{bj} \\ s_{cj} \end{bmatrix}. \tag{5.3}$$

The stationary frame switching signals are used solely for calculation purposes and are defined offline because $\mathbf{T}_{abc/\alpha\beta}$ is time-invariant.

- **Synchronous Frame Switching Signal:** The switching signals for some applications can also be expressed in dq frame as follows to simplify the calculations:

$$\begin{bmatrix} s_{dj} \\ s_{qj} \end{bmatrix} = \begin{bmatrix} \mathbf{T}_{\alpha\beta/dq} \end{bmatrix} \begin{bmatrix} \mathbf{T}_{abc/\alpha\beta} \end{bmatrix} \begin{bmatrix} s_{aj} \\ s_{bj} \\ s_{cj} \end{bmatrix} = \begin{bmatrix} s_{\alpha j} \cos\theta + s_{\beta j} \sin\theta \\ -s_{\alpha j} \sin\theta + s_{\beta j} \cos\theta \end{bmatrix}. \tag{5.4}$$

Given that θ varies with respect to time, the synchronous-frame switching signals cannot be defined offline; they must be computed online during each sampling interval.

- **Phase to Neutral Switching Signal:** Estimating the converter output voltages with respect to the grid/generator neutral n is usually required for grid/generator-connected VSCs. The "phase to neutral switching signals" are used to directly calculate the converter output voltages with respect to n. The phase-neutral switching signals are calculated offline in the following by using the natural-frame switching signals:

$$\begin{bmatrix} s_{anj} \\ s_{bnj} \\ s_{cnj} \end{bmatrix} = \underbrace{\begin{bmatrix} \frac{2}{3} & -\frac{1}{3} & -\frac{1}{3} \\ -\frac{1}{3} & \frac{2}{3} & -\frac{1}{3} \\ -\frac{1}{3} & -\frac{1}{3} & \frac{2}{3} \end{bmatrix}}_{\mathbf{T}_{abc/abcn}} \begin{bmatrix} s_{aj} \\ s_{bj} \\ s_{cj} \end{bmatrix}. \tag{5.5}$$

- **Switching State Combination:** Expressing a group of switching signals as a variable is always convenient in the analysis of power converters. The switching state combination identifies the different possible combinations for the switching states. For example, switching state combinations in a 2L-VSC are represented as $\mathrm{s_0}$ to $\mathrm{s_7}$.
- **Switching Vector:** The switching vector $\mathbf{S_x}$ is used to represent the group of switching signals in phase-x. Notably, the switching vector is a variable and not the signal. The switching vector value is non-binary in nature, whereas the switching signal is binary. The switching vector is also used in space vector diagrams to represent the switching actions in the three converter phases.
- **Voltage/Current Vectors:** The voltage vector $\mathbf{V}$ and current vector $\mathbf{I}$ are used in space vector diagrams to represent the switching vectors. A voltage or current vector can have more than one switching vector that depends on the converter type. If two or more switching vectors represent a voltage/current vector, they are called redundant vectors.

The notation summary of different variables related to the switching signals is provided in Table 5.3, which considers a 2L-VSC. The switching state combinations $\mathrm{s_0}$ and $\mathrm{s_7}$ are redundant, which both represent the voltage vector $\mathbf{V_0}$.

Table 5.3 Summary of the notation of variables for a 2L-VSC

Switching Combination	Voltage Vector	Switching Vector	Upper-Leg Switching Signals			Phase to Neutral Switching Signals			Stationary Frame Switching Signals	
s	V	$\mathbf{S_{abc}}$	s_{a1}	s_{b1}	s_{c1}	s_{an1}	s_{bn1}	s_{cn1}	$s_{\alpha 1}$	$s_{\beta 1}$
$\mathrm{s_0}$	$\mathbf{V_0}$	[000]	'0'	'0'	'0'	0	0	0	0	0
$\mathrm{s_1}$	$\mathbf{V_1}$	[100]	'1'	'0'	'0'	$\frac{2}{3}$	$-\frac{1}{3}$	$-\frac{1}{3}$	$\frac{2}{3}$	0
$\mathrm{s_2}$	$\mathbf{V_2}$	[110]	'1'	'1'	'0'	$\frac{1}{3}$	$\frac{1}{3}$	$-\frac{2}{3}$	$\frac{1}{3}$	$\frac{\sqrt{3}}{3}$
$\mathrm{s_3}$	$\mathbf{V_3}$	[010]	'0'	'1'	'0'	$-\frac{1}{3}$	$\frac{2}{3}$	$-\frac{1}{3}$	$-\frac{1}{3}$	$\frac{\sqrt{3}}{3}$
$\mathrm{s_4}$	$\mathbf{V_4}$	[011]	'0'	'1'	'1'	$-\frac{2}{3}$	$\frac{1}{3}$	$\frac{1}{3}$	$-\frac{2}{3}$	0
$\mathrm{s_5}$	$\mathbf{V_5}$	[001]	'0'	'0'	'1'	$-\frac{1}{3}$	$-\frac{1}{3}$	$\frac{2}{3}$	$-\frac{1}{3}$	$-\frac{\sqrt{3}}{3}$
$\mathrm{s_6}$	$\mathbf{V_6}$	[101]	'1'	'0'	'1'	$\frac{1}{3}$	$-\frac{2}{3}$	$\frac{1}{3}$	$\frac{1}{3}$	$-\frac{\sqrt{3}}{3}$
$\mathrm{s_7}$	$\mathbf{V_0}$	[111]	'1'	'1'	'1'	0	0	0	0	0

5.4 TWO-LEVEL VOLTAGE SOURCE CONVERTER

BTB 2L-VSCs are popular options among Type 3 and 4 WT manufacturers. Several other practical configurations include diode-rectifier + 2L-VSC and diode-rectifier + boost converter + 2L-VSC. This section presents the model of the grid-connected 2L-VSC, which is a common element in all three configurations.

5.4.1 Power Circuit

Only one phase of 2L-VSC is considered here to simplify the discussion and the analysis can be extended to three-phase and multiphase applications easily. The power circuit of the grid-connected 2L-VSC with low-voltage (LV) insulated gate bipolar transistor (IGBT)-based switching devices is shown in Figure 5.2(a), and its equivalent circuit with a generic model of switching devices is shown in Figure 5.2(b). The DC-link branch and grid currents flow from the converter to the grid because it is a grid-connected case. The modeling of the generator-connected converter is similar, except that the currents flow from the wind generator to the converter.

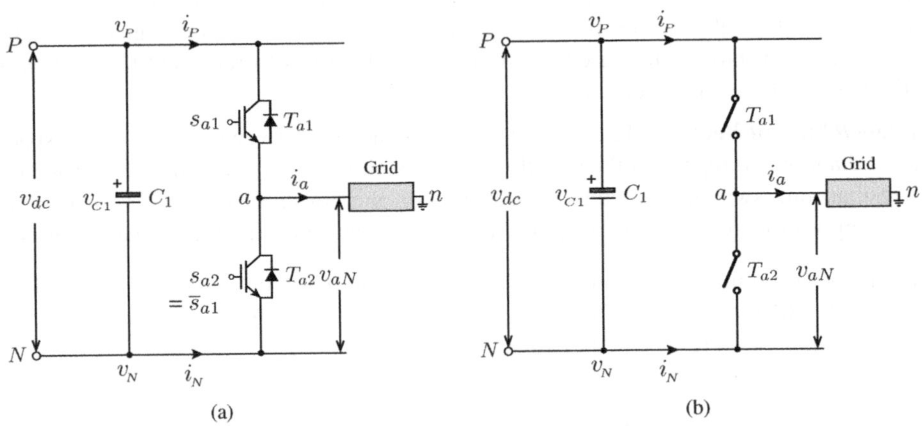

Figure 5.2 Per-phase 2L-VSC power circuit using: (a) IGBT switches, and (b) generic switches.

The properties of the grid-connected 2L-VSC are summarized as follows:

- Each phase leg is composed of two active switches, namely, T_{a1} and T_{a2}, which have corresponding switching (gating) signals of s_{a1} and s_{a2}, respectively.
- Only one switch conducts in any phase at any moment to avoid the short circuit of switches across the DC link. Thus, switches T_{a1} and T_{a2} operate in a complementary manner. The switch signals then yield that $s_{a2} = \overline{s}_{a1}$. This property is extended to all three phases as $s_{x1} + s_{x2} =$ '1' for all $x \in \{a, b, c\}$.
- The semiconductor switches are used to connect the grid-side terminal a to the positive or negative DC bus (P or N).
- The capacitor and net DC-bus voltages are represented as v_{C1} and v_{dc}, respectively. In the case of 2L-VSC, $v_{C1} = v_{dc} = v_P - v_N$.
- The currents i_P and i_N correspond to the positive and negative DC branch currents, respectively. These currents are related to the switching signals of 2L-VSC and three-phase grid currents; thus, they are non-sinusoidal in nature.

5.4.2 Operating Modes

The operating modes for one phase of 2L-VSC are shown in Figure 5.3. The number of operating modes with two switches in a phase is two. A generic model of the semiconductor switch is used to simplify its illustration. The anti-parallel free-wheeling diode across the switching device is ignored to decrease the analysis complexity. The dead time between the upper and lower switching devices is also neglected. Mathematical modeling is more interesting at this point than the accurate analysis of converter commutation actions.

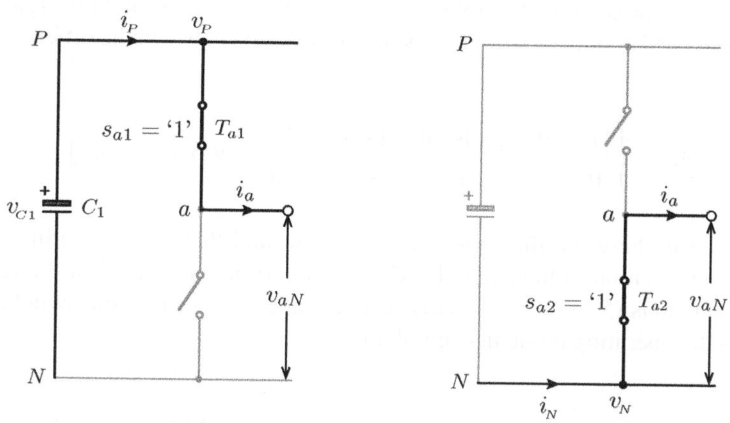

(a) Mode I: $\mathbf{S_a} = [\mathbf{1}]$

(b) Mode II: $\mathbf{S_a} = [\mathbf{0}]$

Figure 5.3 Per-phase representation of 2L-VSC operating modes.

The switching actions and corresponding converter output voltages in the two operating modes are presented as follows:

- **Mode I Operation:** Figure 5.3(a) shows that the upper switch T_{a1} is closed by enabling the switching signal (s_{a1} = '1') in the first mode. The bottom switch T_{a2} is opened by disabling the gating signal (s_{a2} = '0'). The VSC phase-a output voltage with respect to the negative DC-bus N becomes equal to the capacitor voltage, i.e., $v_{aN} = v_{C1}$.
- **Mode II Operation:** The upper switch T_{a1} is turned off with s_{a1} = '0' in this mode, whereas bottom switch T_{a2} is turned on with s_{a2} = '1'. Figure 5.3(b) shows that the terminal a is connected to the negative DC bus; thus, the VSC output voltage with respect to N becomes zero, $v_{aN} = 0$.

The output voltages in the other two phases of VSC yield the same property with respect to the switching signals. The summary of the converter output voltages for a three-phase 2L-VSC is provided as follows for all $x \in \{a, b, c\}$:

$$v_{xN} = \begin{cases} v_{C1} & \text{if } T_{x1} \text{ is on and } T_{x2} \text{ is off} \quad (s_{x1} = \text{`1'},\ s_{x2} = \text{`0'}) \\ 0 & \text{if } T_{x1} \text{ is off and } T_{x2} \text{ is on} \quad (s_{x1} = \text{`0'},\ s_{x2} = \text{`1'}). \end{cases} \tag{5.6}$$

The above representation shows that the corresponding output voltages change between 0 and v_{C1} as the upper switching signal s_{x1} value in any phase changes between '0' and '1'. The operating modes shown in Figure 5.3 also establish the relationship between the DC branch (input) currents, i_P and i_N, and grid (output) current i_a. The positive DC branch current i_P becomes equal to i_a in the first mode. The negative DC branch current i_N and grid current i_a become equal in the second operating mode. Following a similar analysis with the two other phases, the relationship summary between the input and output currents is presented as follows:

$$i_x = \begin{cases} i_P & \text{if } T_{x1} \text{ is on and } T_{x2} \text{ is off} \quad (s_{x1} = \text{`1'},\ s_{x2} = \text{`0'}) \\ i_N & \text{if } T_{x1} \text{ is off and } T_{x2} \text{ is on} \quad (s_{x1} = \text{`0'},\ s_{x2} = \text{`1'}). \end{cases} \tag{5.7}$$

A switching vector is defined as follows in terms of upper-leg switching signals to simplify the analysis and conduct a straightforward comparison between the different variables of 2L-VSC:

$$\mathbf{S_x} = \begin{cases} [\mathbf{1}] & \text{if } T_{x1} \text{ is on} \quad (s_{x1} = \text{`1'}) \\ [\mathbf{0}] & \text{if } T_{x1} \text{ is off} \quad (s_{x1} = \text{`0'}). \end{cases} \quad \forall\ x \in \{a, b, c\} \tag{5.8}$$

The relationship between the switching vector $\mathbf{S_x}$ and their corresponding switching signals, converter output voltages, and DC branch currents are listed in Table 5.4 after combining Equations (5.6) to (5.8). This representation shows that the switching vector $\mathbf{S_x}$ replicates the operating mode in each phase.

Table 5.4 Switching states and output voltages for a 2L-VSC $\forall\ x \in \{a, b, c\}$

Switching Vector	Switching Signals		Output Voltage	Input Currents	
$\mathbf{S_x}$	s_{x1}	s_{x2}	v_{xN}	i_P	i_N
[1]	'1'	'0'	v_{C1}	i_x	0
[0]	'0'	'1'	0	0	i_x

Note: (s_{x1} and s_{x2}) are complementary switching signals.

The possible number of combinations for a three-phase 2L-VSC becomes eight (2^3) with two switching vectors [1] and [0] in each phase. A space vector diagram that contains these eight combinations is shown in Figure 5.4 [2]. The notation of the switching vector $\mathbf{S_x}$ in terms of the switching signals s_{x1} and s_{x2} for a 2L-VSC is graphically represented at the bottom of Figure 5.4. The black circle represents the turn-on status of the switching device, whereas the white circle denotes the turn-off condition.

As previously discussed in Section 5.3, the switching vector combinations are used to represent the voltage vectors. For example, the switching vector combination [**100**] represents the voltage vector $\mathbf{V_1}$. The corresponding switching signal status is also shown in Figure 5.4. The six active voltage vectors $\mathbf{V_1}$ to $\mathbf{V_6}$ form the regular hexagon for a 2L-VSC, which produces nonzero output voltage. The zero voltage vector $\mathbf{V_0}$ lies at the center of the hexagon with two redundant switching vectors [**000**] and [**111**]. As the name implies, the zero voltage vector produces no output voltage; thus, this vector is used to perform additional functions such as switching frequency reduction and common-mode voltage (CMV) mitigation. A detailed summary of the voltage vectors and their corresponding switching vectors, switching states, output AC voltages, and input DC branch currents is presented in the Appendix B.2 and Table B.1.

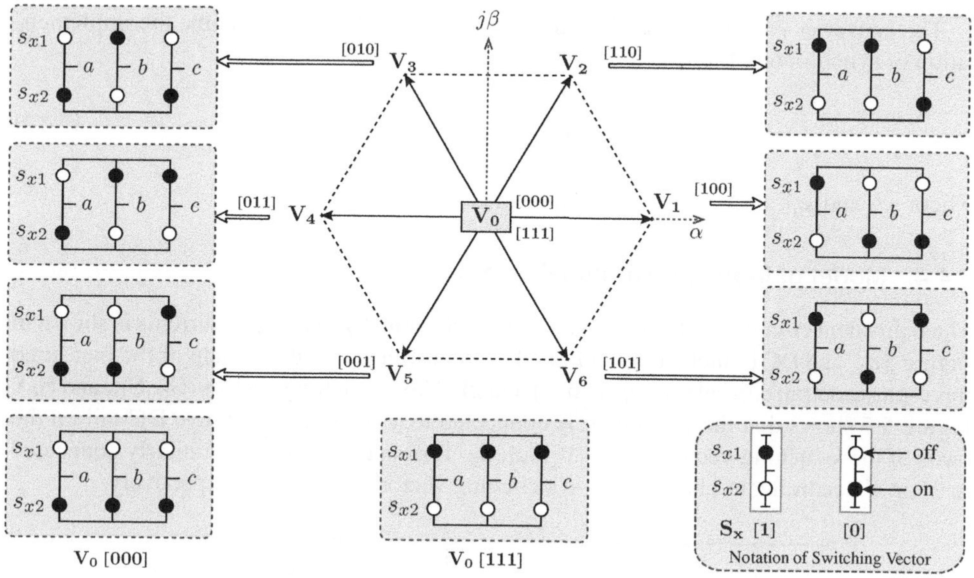

Figure 5.4 Space vector diagram for a 2L-VSC [2] ($x \in \{a, b, c\}$).

5.4.3 Model of Output AC Voltages

Equation (5.6) shows that the three-phase VSC voltages with respect to the negative DC-bus N are expressed in terms of the DC-link capacitor voltage and switching signals as follows [3]:

$$\begin{bmatrix} v_{aN} \\ v_{bN} \\ v_{cN} \end{bmatrix} = v_{C1} \begin{bmatrix} s_{a1} \\ s_{b1} \\ s_{c1} \end{bmatrix}. \tag{5.9}$$

The converter voltages with respect to the grid/generator neutral n are calculated in terms of the measured DC-link capacitor voltage and phase-to-neutral switching signals as demonstrated below:

$$\begin{bmatrix} v_{an} \\ v_{bn} \\ v_{cn} \end{bmatrix} = v_{C1} \left[\mathbf{T}_{abc/abcn}\right] \begin{bmatrix} s_{a1} \\ s_{b1} \\ s_{c1} \end{bmatrix} = v_{C1} \begin{bmatrix} s_{an1} \\ s_{bn1} \\ s_{cn1} \end{bmatrix}. \tag{5.10}$$

The stationary frame control algorithms require the converter $\alpha\beta$ output voltages. The indirect approach applies the transformation matrix (3.6) to the voltages estimated in (5.9). The direct approach combines the DC-link capacitor voltage with the $\alpha\beta$-frame switching signals as formulated below:

$$\begin{bmatrix} v_{\alpha} \\ v_{\beta} \end{bmatrix} = v_{C1} \left[\mathbf{T}_{abc/\alpha\beta}\right] \begin{bmatrix} s_{a1} \\ s_{b1} \\ s_{c1} \end{bmatrix} = v_{C1} \begin{bmatrix} s_{\alpha 1} \\ s_{\beta 1} \end{bmatrix} \tag{5.11}$$

where $s_{\alpha 1}$ and $s_{\beta 1}$ are switching signals in a stationary frame.

The converter voltages in the dq-frame are derived as follows to enable the implementation of synchronous-frame control algorithms:

$$\begin{bmatrix} v_d \\ v_q \end{bmatrix} = v_{C1} \begin{bmatrix} \mathbf{T}_{\alpha\beta/dq} \end{bmatrix} \begin{bmatrix} s_{\alpha 1} \\ s_{\beta 1} \end{bmatrix} = v_{C1} \begin{bmatrix} s_{d1} \\ s_{q1} \end{bmatrix} \tag{5.12}$$

where s_{d1} and s_{q1} are synchronous-frame switching signals.

5.4.4 Model of Input DC Branch Currents

The equivalent circuit of a 2L-VSC with input (DC branch) and output currents is shown in Figure 5.5. The DC branch currents i_P and i_N are estimated with the help of the measured three-phase output currents (i_a, i_b, and i_c) and 2L-VSC switching vectors ($\mathbf{S_a}$, $\mathbf{S_b}$, and $\mathbf{S_c}$). Figure 5.5 shows that the terminal a is connected to a positive or negative DC rail on the basis of the switching vector values $[\mathbf{1}]$ and $[\mathbf{0}]$. Terminals b and c are similarly connected to P or N according to phase-b and -c switching vectors $\mathbf{S_b}$ and $\mathbf{S_c}$, respectively.

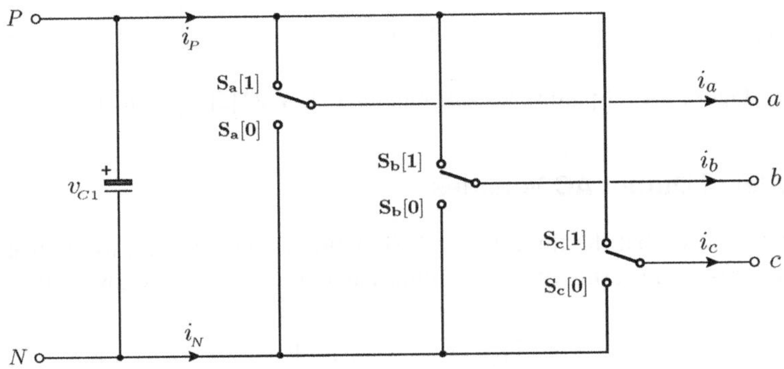

Figure 5.5 Representation of 2L-VSC DC branch (input) and output currents.

The relation between the DC branch and three-phase output currents is obtained in the natural reference frame as follows on the basis of Figure 5.5:

$$\begin{bmatrix} i_P \\ i_N \end{bmatrix} = \begin{bmatrix} \mathbf{S_a}[\mathbf{1}] & \mathbf{S_b}[\mathbf{1}] & \mathbf{S_c}[\mathbf{1}] \\ \mathbf{S_a}[\mathbf{0}] & \mathbf{S_b}[\mathbf{0}] & \mathbf{S_c}[\mathbf{0}] \end{bmatrix} \begin{bmatrix} i_a \\ i_b \\ i_c \end{bmatrix}. \tag{5.13}$$

By substituting Equation (5.8) in the above expression, the relation between the input and output currents is obtained as follows in terms of the switching signals:

$$\begin{bmatrix} i_P \\ i_N \end{bmatrix} = \begin{bmatrix} s_{a1} & s_{b1} & s_{c1} \\ s_{a2} & s_{b2} & s_{c2} \end{bmatrix} \begin{bmatrix} i_a \\ i_b \\ i_c \end{bmatrix}. \tag{5.14}$$

The above expression is transformed to stationary ($\alpha\beta$) and synchronous (dq) reference frames as follows:

$$\begin{bmatrix} i_P \\ i_N \end{bmatrix} = \begin{bmatrix} s_{\alpha 1} & s_{\beta 1} \\ s_{\alpha 2} & s_{\beta 2} \end{bmatrix} \begin{bmatrix} i_\alpha \\ i_\beta \end{bmatrix}, \quad \begin{bmatrix} i_P \\ i_N \end{bmatrix} = \begin{bmatrix} s_{d1} & s_{q1} \\ s_{d2} & s_{q2} \end{bmatrix} \begin{bmatrix} i_d \\ i_q \end{bmatrix}. \tag{5.15}$$

5.5 EXTENSIONS TO 2L-VSC MODELING

The 2L-VSC model is extended to multiphase and BTB VSCs in this section.

5.5.1 Modeling of Multiphase 2L-VSC

As previously mentioned in Chapter 2, multiphase generators are also used in Type 4 WECS. The model of the 2L-VSC output voltages and input DC branch currents can be easily extended to multiphase applications without considering the operating modes and space vector diagram [4–6]. The models for output voltages and input currents in a natural frame are formulated in terms of switching signals as follows:

$$\begin{bmatrix} v_{aN} \\ v_{bN} \\ \vdots \\ v_{yN} \end{bmatrix} = v_{C1} \begin{bmatrix} s_{a1} \\ s_{b1} \\ \vdots \\ s_{y1} \end{bmatrix}, \quad \begin{bmatrix} i_P \\ i_N \end{bmatrix} = \begin{bmatrix} s_{a1} & s_{b1} & \cdots & s_{y1} \\ s_{a2} & s_{b2} & \cdots & s_{y2} \end{bmatrix} \begin{bmatrix} i_a \\ i_b \\ \vdots \\ i_y \end{bmatrix} \tag{5.16}$$

where y represents phase-y of a 2L-VSC.

5.5.2 Modeling of BTB 2L-VSC

The modeling presented earlier for 2L-VSC can be easily extended to the BTB configuration shown in Figure 5.6. The 2L-VSR and 2L voltage source inverter (2L-VSI) are BTB connected through a DC-link capacitor. Subscripts r and i represent the rectifier- and inverter-side variables, respectively. v_{xr} and v_{xi} correspond to the phase-x converter output voltages with respect to the negative DC-bus N. i_{xr} and i_{xi} represent the rectifier and inverter line currents, respectively ($x \in \{a, b, c\}$).

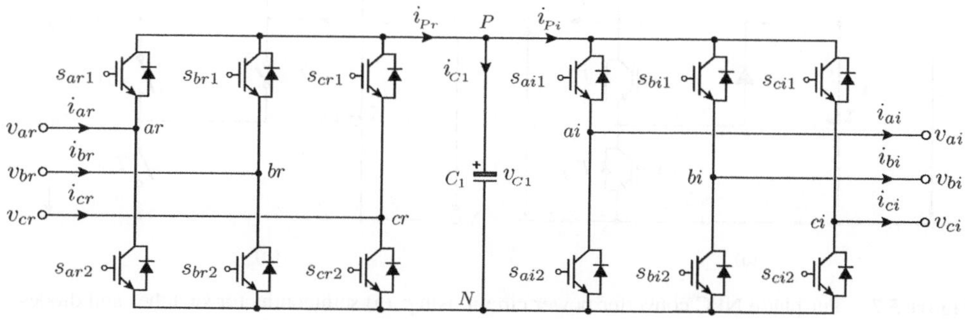

Figure 5.6 Power circuit of BTB 2L-VSC.

Similar to Equation (5.9), the 2L-VSR and 2L-VSI output voltages are calculated as follows:

$$\begin{bmatrix} v_{ar} \\ v_{br} \\ v_{cr} \end{bmatrix} = v_{C1} \begin{bmatrix} s_{ar1} \\ s_{br1} \\ s_{cr1} \end{bmatrix}, \quad \begin{bmatrix} v_{ai} \\ v_{bi} \\ v_{ci} \end{bmatrix} = v_{C1} \begin{bmatrix} s_{ai1} \\ s_{bi1} \\ s_{ci1} \end{bmatrix}. \tag{5.17}$$

The current i_{C1} that flows through the DC-link capacitor C_1 is estimated as follows [7]:

$$i_{C1} = i_{P_r} - i_{P_i} = (s_{ar1}\, i_{ar} + s_{br1}\, i_{br} + s_{cr1}\, i_{cr}) - (s_{ai1}\, i_{ag} + s_{bi1}\, i_{bg} + s_{ci1}\, i_{cg}). \tag{5.18}$$

5.6 NEUTRAL-POINT CLAMPED CONVERTER

Medium-voltage (MV) operation is gaining popularity with the increasing power levels of commercial WTs. Multilevel converters are the most suitable candidates for MV operation, and NPC converters have been applied in commercial WTs [8]. The NPC converter model is presented in this section.

5.6.1 Power Circuit

The per-phase configuration of the grid-connected NPC converter is shown in Figure 5.7. Each converter phase is realized by using four MV IGBT switching devices (T_{a1}, T_{a2}, T_{a3}, and T_{a4}) and two clamping diodes (D_{a1} and D_{a2}). The DC link is composed of two identical capacitors, and their mid-point (neutral-point) is denoted as Z. The mid-point of the clamping diodes in each phase is connected to Z. The net DC-bus voltage is shared equally among the two DC-link capacitors (i.e., $v_{C1} = v_{C2} = v_{dc}/2$). The current that flows through the clamping diodes and neutral point causes an unbalance in the capacitors voltage. The semiconductor switches and clamping diodes are rated for half the net DC-bus voltage, on the assumption that the net DC voltage is shared equally among the DC-link capacitors. The switches and clamping diodes undergo severe voltage stress if the DC-link capacitors voltage are unbalanced [9].

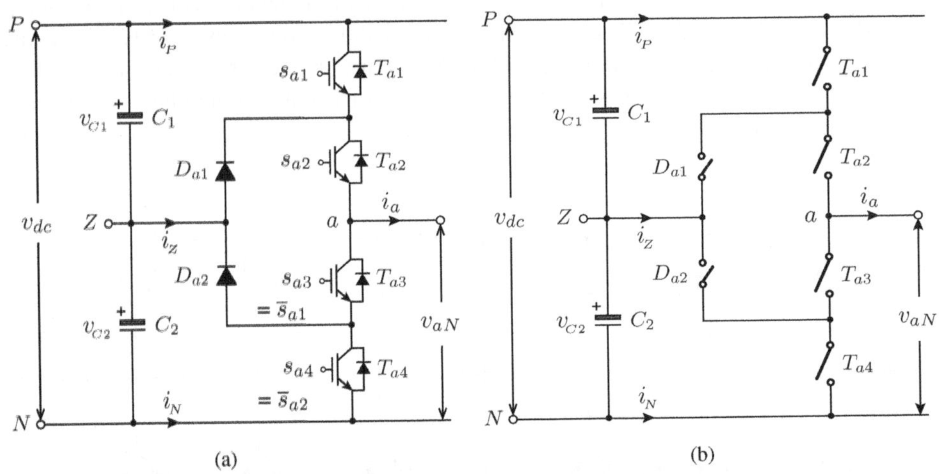

Figure 5.7 Per-phase NPC converter power circuit using: (a) semiconductor switches and diodes, and (b) generic switches and diodes.

The main features of NPC converter are as follows [10, 11]:

- The switching (gating) signals applied to T_{a1}, T_{a2}, T_{a3}, and T_{a4} are s_{a1}, s_{a2}, s_{a3}, and s_{a4}, respectively.
- Semiconductor switches are used to connect the grid-side terminal a to the DC bus terminals P, Z, or N.
- Currents i_P, i_Z, and i_N are related to the switching signals of NPC converter and the three-phase grid currents.

5.6.2 Operating Modes

The modes of operation for an NPC converter are shown in Figure 5.8, where (1) three operating modes are feasible, (2) only two switches conduct at any time, and (3) switch pairs (s_{a1}, s_{a3}) and (s_{a2}, s_{a4}) operate in a complementary manner. For three-phase applications, $s_{x1} + s_{x3} =$ '1' and $s_{x2} + s_{x4} =$ '1' for all $x \in \{a, b, c\}$.

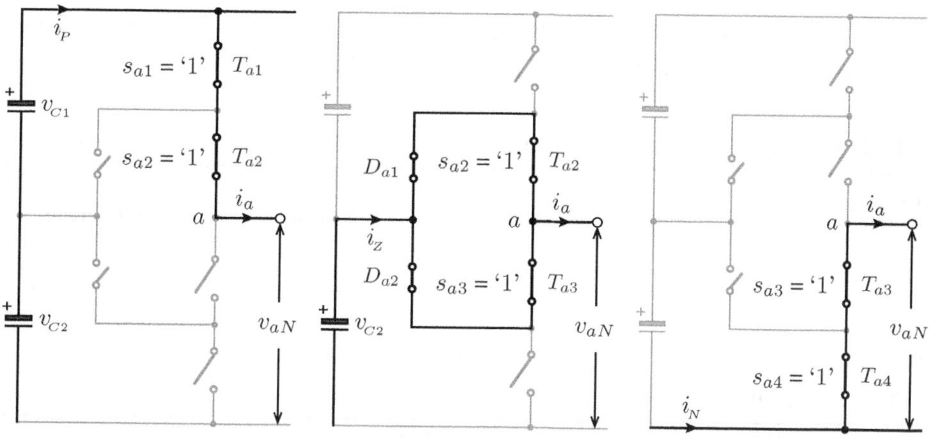

(a) Mode I: $\mathbf{S_a} = [2]$ (b) Mode II: $\mathbf{S_a} = [1]$ (c) Mode III: $\mathbf{S_a} = [0]$

Figure 5.8 Per-phase representation of operating modes for an NPC converter.

The three operating modes in one phase of an NPC converter are summarized as follows:

- **Mode I Operation:** As shown in Figure 5.8(a), the top two switches, T_{a1} and T_{a2}, conduct and terminal a is connected to point P during this mode of operation. The converter voltage becomes a summation of two DC-link capacitors voltage with respect to N, that is, $v_{aN} = v_{C1} + v_{C2}$.
- **Mode II Operation:** Figure 5.8(b) indicates that terminal a is connected to point Z through middle switches T_{a2} and T_{a3} and clamping diodes D_{a1} and D_{a2} during this mode. The voltage of the bottom DC-link capacitor appears at the converter output terminals, $v_{aN} = v_{C2}$.
- **Mode III Operation:** During mode III operation as shown in Figure 5.8(c), no voltage appears at the converter terminal, ($v_{aN} = 0$), as switches T_{a3} and T_{a4} are connected to the negative DC bus, N.

The summary for a three-phase NPC converter output voltages is given as follows:

$$v_{xN} = \begin{cases} v_{C1} + v_{C2} & \text{if } T_{x1}, T_{x2} \text{ are on} \quad (s_{x1} \cdot s_{x2} = \text{`1'}) \\ v_{C2} & \text{if } T_{x2}, T_{x3} \text{ are on} \quad (s_{x2} \cdot s_{x3} = \text{`1'}) \\ 0 & \text{if } T_{x3}, T_{x4} \text{ are on} \quad (s_{x3} \cdot s_{x4} = \text{`1'}). \end{cases} \quad \forall\ x \in \{a, b, c\} \tag{5.19}$$

Similarly, the DC branch currents are related to the output currents as

$$i_x = \begin{cases} i_P & \text{if } T_{x1}, T_{x2} \text{ are on} \quad (s_{x1} \cdot s_{x2} = \text{`1'}) \\ i_Z & \text{if } T_{x2}, T_{x3} \text{ are on} \quad (s_{x2} \cdot s_{x3} = \text{`1'}) \\ i_N & \text{if } T_{x3}, T_{x4} \text{ are on} \quad (s_{x3} \cdot s_{x4} = \text{`1'}). \end{cases} \quad \forall\ x \in \{a, b, c\} \tag{5.20}$$

The switching vectors are defined below in terms of the switching signals of an NPC converter:

$$\mathbf{S_x} = \begin{cases} [2] & \text{if } T_{x1}, T_{x2} \text{ are on} \quad (s_{x1} \cdot s_{x2} = \text{`1'}) \\ [1] & \text{if } T_{x2}, T_{x3} \text{ are on} \quad (s_{x2} \cdot s_{x3} = \text{`1'}) \\ [0] & \text{if } T_{x3}, T_{x4} \text{ are on} \quad (s_{x3} \cdot s_{x4} = \text{`1'}). \end{cases} \quad \forall\ x \in \{a, b, c\} \tag{5.21}$$

By combining the above analyses, the summary of converter output voltages and input DC branch currents for all possible switching states are provided in Table 5.5.

Table 5.5 Switching states and output voltages for an NPC converter $\forall\ x \in \{a, b, c\}$

Switching Vector	**Switching Signals**				**Output Voltage**	**Input Currents**		
$\mathbf{S_x}$	s_{x1}	s_{x2}	s_{x3}	s_{x4}	v_{xN}	i_P	i_Z	i_N
[2]	'1'	'1'	'0'	'0'	$v_{C1} + v_{C2}$	i_x	0	0
[1]	'0'	'1'	'1'	'0'	v_{C2}	0	i_x	0
[0]	'0'	'0'	'1'	'1'	0	0	0	i_x

Note: (s_{x1}, s_{x3}), and (s_{x2}, s_{x4}) are complementary switching signals.

The space vector diagram for a three-phase balanced NPC converter is shown in Figure 5.9. The relation between switching vector $\mathbf{S_x}$ and switching signals s_{x1} to s_{x4} is provided at the bottom of figure. With three modes in each phase, the converter presents 27 different possible ways (switching states) to connect the output terminals to DC-link P, Z, and N points. For an NPC converter, 19 different voltage vectors are available ($\mathbf{V_0}$ to $\mathbf{V_{18}}$) [12].

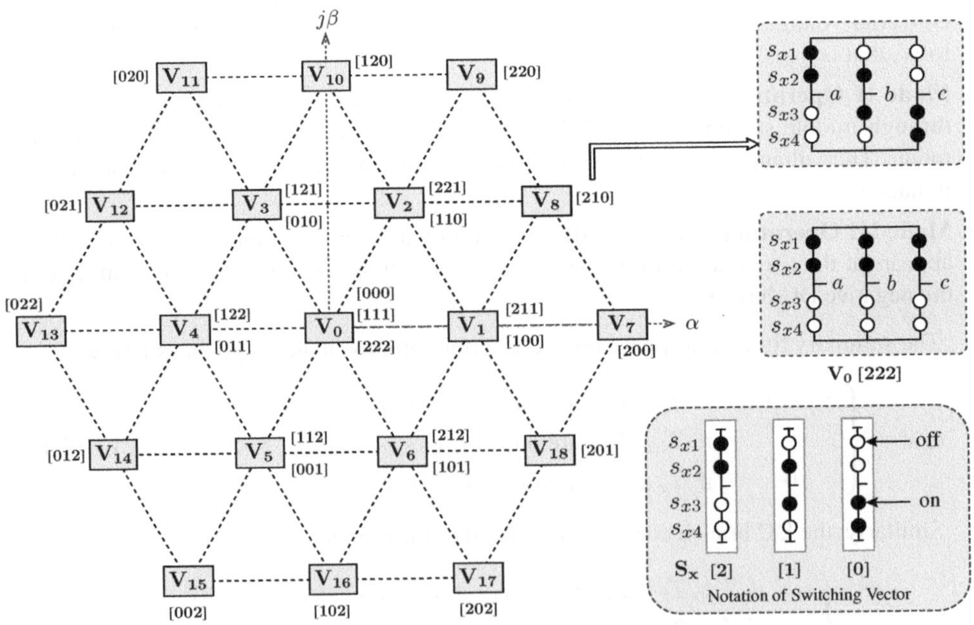

Figure 5.9 Space vector diagram for NPC converter [12, 13].

The redundancy in voltage vectors increases compared to the 2L-VSC. The zero voltage vector $\mathbf{V_0}$ has three redundancies and voltage vectors $\mathbf{V_1}$ to $\mathbf{V_6}$, which form the inner hexagon, have two redundancies. Space vector modulation calculates the nearest voltage vectors to the reference voltage vector and uses the optimal one among them [13]. MPC uses a completely different approach and does not require comprehensive knowledge on the space vector diagram, SV diagram sectors, reference voltage vector, and adjacent vectors. All the voltage vectors are utilized to predict the future behavior of the variable to be controlled, and an optimal voltage vector that fulfils the control objective is selected from them. The summary of voltage vectors, switching states, output AC voltages, and input DC branch currents for an NPC converter is presented in Appendix B.2 and Table B.3 for use with MPC.

5.6.3 Model of Output AC Voltages

In the *abc* frame, the NPC converter AC-side voltages are expressed in terms of switching signals s_{a1}, s_{a2}, s_{b1}, s_{b2}, s_{c1}, and s_{c2} and DC-link capacitors voltage v_{C1} and v_{C2} as follows [14, 15]:

$$\begin{bmatrix} v_{aN} \\ v_{bN} \\ v_{cN} \end{bmatrix} = v_{C1} \begin{bmatrix} s_{a1} \\ s_{b1} \\ s_{c1} \end{bmatrix} + v_{C2} \begin{bmatrix} s_{a2} \\ s_{b2} \\ s_{c2} \end{bmatrix}. \tag{5.22}$$

The above expression can be converted into the required reference frame by using the guidelines presented in Section 5.4.3. The group 1 switching signals (s_{a1}, s_{b1}, and s_{c1}) are used to estimate the corresponding phase-to-neutral, $\alpha\beta$, and dq switching signals s_{an1}, s_{bn1}, s_{cn1}, $s_{\alpha1}$, $s_{\beta1}$, s_{d1}, and s_{q1}. Similarly, group 2 switching signals (s_{a2}, s_{b2} and s_{c2}) are used to estimate s_{an2}, s_{bn2}, s_{cn2}, $s_{\alpha2}$, $s_{\beta2}$, s_{d2}, and s_{q2}.

In even-level (2L and four level (4L)) converters, CMV is calculated as the average value of v_{aN}, v_{bN}, and v_{cN}. In odd-level (3L and five level (5L)) converters, CMV exists between the grid/generator neutral n and mid-point of DC-link capacitors (Z) [16]. Therefore, CMV is expressed as the average value of v_{aZ}, v_{bZ}, and v_{cZ}, that is,

$$v_{cm} = v_{nZ} = \frac{v_{aZ} + v_{bZ} + v_{cZ}}{3}. \tag{5.23}$$

Figure 5.8 illustrates that v_{aZ} in operating modes I, II, and III is v_{C1}, 0, and $-v_{C2}$, respectively. For a three-phase NPC converter, the output voltages with respect to Z are formulated as follows:

$$\begin{bmatrix} v_{aZ} \\ v_{bZ} \\ v_{cZ} \end{bmatrix} = v_{C1} \begin{bmatrix} s_{a1} \\ s_{b1} \\ s_{c1} \end{bmatrix} + v_{C2} \begin{bmatrix} (s_{a2}) - 1 \\ (s_{b2}) - 1 \\ (s_{c2}) - 1 \end{bmatrix}. \tag{5.24}$$

5.6.4 Model of Input DC Branch Currents

Figure 5.10 shows the relation between input DC branch currents and output currents in terms of the three-phase switching vectors. On the basis of the switching vector values [**2**], [**1**], or [**0**], the output terminals are connected to DC bus terminals P, Z, or N, respectively.

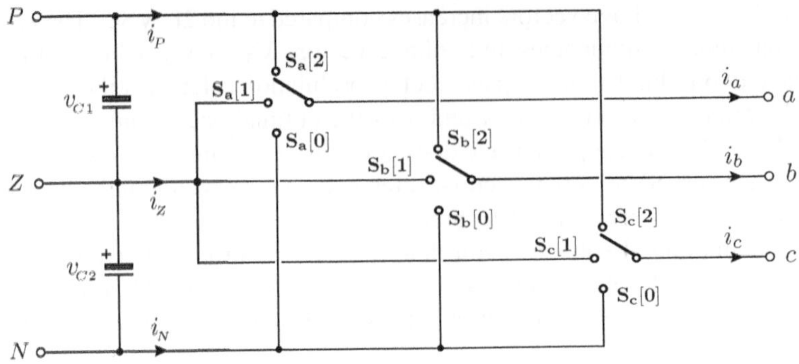

Figure 5.10 Representation of NPC converter DC branch (input) and output currents.

DC branch currents are obtained below with the help of switching vectors:

$$\begin{bmatrix} i_P \\ i_Z \\ i_N \end{bmatrix} = \begin{bmatrix} \mathbf{S_a}[2] & \mathbf{S_b}[2] & \mathbf{S_c}[2] \\ \mathbf{S_a}[1] & \mathbf{S_b}[1] & \mathbf{S_c}[1] \\ \mathbf{S_a}[0] & \mathbf{S_b}[0] & \mathbf{S_c}[0] \end{bmatrix} \begin{bmatrix} i_a \\ i_b \\ i_c \end{bmatrix} \tag{5.25}$$

which is rewritten below in terms of the switching signals by following the notation in Equation (5.21):

$$\begin{bmatrix} i_P \\ i_Z \\ i_N \end{bmatrix} = \begin{bmatrix} s_{a1}s_{a2} & s_{b1}s_{b2} & s_{c1}s_{c2} \\ s_{a2}s_{a3} & s_{b2}s_{b3} & s_{c2}s_{c3} \\ s_{a3}s_{a4} & s_{b3}s_{b4} & s_{c3}s_{c4} \end{bmatrix} \begin{bmatrix} i_a \\ i_b \\ i_c \end{bmatrix}. \tag{5.26}$$

5.7 EXTENSIONS TO NPC CONVERTER MODELING

The NPC converter model presented in the previous section is extended to an m-level DCC and BTB configuration.

5.7.1 Modeling of Multilevel and Multiphase DCC

Similar to (5.22), the terminal voltages for an m-level, y-phase converter are expressed in the *abc* frame as follows [17]:

$$\begin{bmatrix} v_{aN} \\ v_{bN} \\ \vdots \\ v_{yN} \end{bmatrix} = v_{C1} \begin{bmatrix} s_{a1} \\ s_{b1} \\ \vdots \\ s_{y1} \end{bmatrix} + v_{C2} \begin{bmatrix} s_{a2} \\ s_{b2} \\ \vdots \\ s_{y2} \end{bmatrix} + \cdots + v_{C(m-1)} \begin{bmatrix} s_{a(m-1)} \\ s_{b(m-1)} \\ \vdots \\ s_{y(m-1)} \end{bmatrix}. \tag{5.27}$$

Similarly, the input DC branch currents are estimated as follows:

$$\begin{bmatrix} i_P \\ \vdots \\ i_N \end{bmatrix} = \begin{bmatrix} s_{a1} \cdots s_{a(m-1)} & \cdots & s_{y1} \cdots s_{y(m-1)} \\ \vdots & \cdots & \vdots \\ s_{a(m)} \cdots s_{a(2m-2)} & \cdots & s_{y(m)} \cdots s_{y(2m-2)} \end{bmatrix} \begin{bmatrix} i_a \\ \vdots \\ i_y \end{bmatrix}. \tag{5.28}$$

The above expressions are derived on the basis of the assumption that the top DC-link capacitor notation is C_1. MPC is more intuitive than the classical control; without having any deeper knowledge on multilevel converters, closed-loop control can be realized by just adopting the two expressions above. For example, for a five-phase ($abcde$) NPC converter, the output AC voltages and input DC currents are estimated as follows:

$$\begin{bmatrix} v_{aN} \\ v_{bN} \\ v_{cN} \\ v_{dN} \\ v_{eN} \end{bmatrix} = v_{C1} \begin{bmatrix} s_{a1} \\ s_{b1} \\ s_{c1} \\ s_{d1} \\ s_{e1} \end{bmatrix} + v_{C2} \begin{bmatrix} s_{a2} \\ s_{b2} \\ s_{c2} \\ s_{d2} \\ s_{e2} \end{bmatrix}, \quad \begin{bmatrix} i_P \\ i_Z \\ i_N \end{bmatrix} = \begin{bmatrix} s_{a1}s_{a2} & \cdots & s_{e1}s_{e2} \\ s_{a2}s_{a3} & \cdots & s_{e2}s_{e3} \\ s_{a3}s_{a4} & \cdots & s_{e3}s_{e4} \end{bmatrix} \begin{bmatrix} i_a \\ i_b \\ i_c \\ i_d \\ i_e \end{bmatrix} \tag{5.29}$$

which can be transformed to other reference frames using a suitable transformation matrix [18]. As another example, the model of a three-phase 5L-DCC is provided below [19]:

$$\begin{bmatrix} v_{aN} \\ v_{bN} \\ v_{cN} \end{bmatrix} = v_{C1} \begin{bmatrix} s_{a1} \\ s_{b1} \\ s_{c1} \end{bmatrix} + v_{C2} \begin{bmatrix} s_{a2} \\ s_{b2} \\ s_{c2} \end{bmatrix} + v_{C3} \begin{bmatrix} s_{a3} \\ s_{b3} \\ s_{c3} \end{bmatrix} + v_{C4} \begin{bmatrix} s_{a4} \\ s_{b4} \\ s_{c4} \end{bmatrix} \tag{5.30}$$

where v_{C1} to v_{C4} are DC-link capacitors voltage,

$$\begin{bmatrix} i_P \\ i_X \\ i_Y \\ i_Z \\ i_N \end{bmatrix} = \begin{bmatrix} s_{a1}s_{a2}s_{a3}s_{a4} & s_{b1}s_{b2}s_{b3}s_{b4} & s_{c1}s_{c2}s_{c3}s_{c4} \\ s_{a2}s_{a3}s_{a4}s_{a5} & s_{b2}s_{b3}s_{b4}s_{b5} & s_{c2}s_{c3}s_{c4}s_{c5} \\ s_{a3}s_{a4}s_{a5}s_{a6} & s_{b3}s_{b4}s_{b5}s_{b6} & s_{c3}s_{c4}s_{c5}s_{c6} \\ s_{a4}s_{a5}s_{a6}s_{a7} & s_{b4}s_{b5}s_{b6}s_{b7} & s_{c4}s_{c5}s_{c6}s_{c7} \\ s_{a5}s_{a6}s_{a7}s_{a8} & s_{b5}s_{b6}s_{b7}s_{b8} & s_{c5}s_{c6}s_{c7}s_{c8} \end{bmatrix} \begin{bmatrix} i_a \\ i_b \\ i_c \end{bmatrix} \tag{5.31}$$

and Y represents the central mid-point in a 5L-DCC. X and Z are the mid-points close to the positive and negative DC bus, respectively.

5.7.2 Modeling of BTB NPC Converter

Figure 5.11 illustrates the configuration of BTB NPC converter, which can be integrated with SCIG, PMSG, and WRSG [20–23]. The NPC rectifier and inverter are linked by two split DC-link capacitors. The neutral-point Z is connected to the mid-points of the clamping diodes in both the rectifier and inverter. The switching actions of both the rectifier and inverter cause the DC-link capacitors voltage to diverge from their nominal values. Thus, the precise control of both the rectifier and inverter is necessary to ensure equal voltage among the DC-link capacitors. Similar to the BTB 2L-VSC, subscripts r and i denote the variables at the rectifier and inverter side, respectively. The negative DC-bus N is the reference point for converter output voltages v_{xr} and v_{xi} ($x \in \{a, b, c\}$).

The rectifier and inverter output AC voltages are obtained by substituting subscripts r and i in Equation (5.22):

$$\begin{bmatrix} v_{ar} \\ v_{br} \\ v_{cr} \end{bmatrix} = v_{C1} \begin{bmatrix} s_{ar1} \\ s_{br1} \\ s_{cr1} \end{bmatrix} + v_{C2} \begin{bmatrix} s_{ar2} \\ s_{br2} \\ s_{cr2} \end{bmatrix}, \quad \begin{bmatrix} v_{ai} \\ v_{bi} \\ v_{ci} \end{bmatrix} = v_{C1} \begin{bmatrix} s_{ai1} \\ s_{bi1} \\ s_{ci1} \end{bmatrix} + v_{C2} \begin{bmatrix} s_{ai2} \\ s_{bi2} \\ s_{ci2} \end{bmatrix}. \tag{5.32}$$

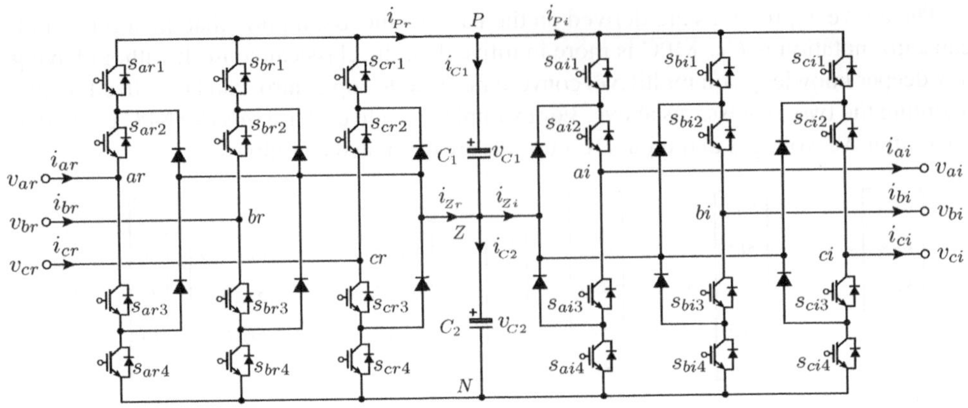

Figure 5.11 Power circuit of BTB connected NPC converter.

Similarly, by substituting subscripts r and i in (5.26), the DC branch currents of the rectifier and inverter are obtained in terms of the measured generator currents, grid currents, and switching states of the converter:

$$\begin{bmatrix} i_{Pr} \\ i_{Zr} \\ i_{Nr} \end{bmatrix} = \begin{bmatrix} s_{ar1}s_{ar2} & s_{br1}s_{br2} & s_{cr1}s_{cr2} \\ s_{ar2}s_{ar3} & s_{br2}s_{br3} & s_{cr2}s_{cr3} \\ s_{ar3}s_{ar4} & s_{br3}s_{br4} & s_{cr3}s_{cr4} \end{bmatrix} \begin{bmatrix} i_{ar} \\ i_{br} \\ i_{cr} \end{bmatrix} \tag{5.33}$$

$$\begin{bmatrix} i_{Pi} \\ i_{Zi} \\ i_{Ni} \end{bmatrix} = \begin{bmatrix} s_{ai1}s_{ai2} & s_{bi1}s_{bi2} & s_{ci1}s_{ci2} \\ s_{ai2}s_{ai3} & s_{bi2}s_{bi3} & s_{ci2}s_{ci3} \\ s_{ai3}s_{ai4} & s_{bi3}s_{bi4} & s_{ci3}s_{ci4} \end{bmatrix} \begin{bmatrix} i_{ai} \\ i_{bi} \\ i_{ci} \end{bmatrix}. \tag{5.34}$$

The relation between the DC-link capacitors voltage (v_{C1}, v_{C2}) and DC-link capacitors current (i_{C1}, i_{C2}) is provided below [14]:

$$\frac{d}{dt}v_{C1} = \frac{1}{C_1}i_{C1}, \quad \frac{d}{dt}v_{C2} = \frac{1}{C_2}i_{C2}. \tag{5.35}$$

By applying KCL to nodes P and Z in Figure 5.11, the relationship between the NPC rectifier and inverter DC branch currents, as well as DC capacitors current, is formulated as follows [21]:

$$\begin{aligned} i_{C1} &= i_{Pr} - i_{Pi} \\ i_{C2} &= i_{C1} + i_{Zr} - i_{Zi} = (i_{Pr} + i_{Zr}) - (i_{Pi} + i_{Zi}). \end{aligned} \tag{5.36}$$

The calculation of i_{C1} is similar to that of the BTB 2L-VSC (refer to (5.18)), and the additional calculation is i_{C2} only. The dynamics in (5.32) to (5.36) suggest that the rectifier and inverter output AC voltages and that the DC-link capacitors voltage are functions of the switching signals. The proper selection of switching signal leads to the accurate control of reference variables, such as generator/grid currents and balancing of DC-link capacitors voltage. By combining this analysis with that in Section 5.7.1, the BTB multilevel DCC model can be easily obtained for use in other power conversion applications.

5.8 MODELING OF OTHER POWER CONVERTERS

Following the detailed analysis in previous sections, the FCC, CSC, DMC, and IMC models are presented here briefly. These converters have not been applied in WECS yet but are promising candidates for future developments.

5.8.1 Three-Level Flying Capacitor Converter

The power circuit of one phase of 3L-FCC using MV IGBTs and generic switches is shown in Figures 5.12(a) and 5.12(b). The clamping diodes in NPC converter are replaced by the FCs and are not connected to the FCs in other phases [24]. The capacitors voltage are floating, thus leading to the name "flying capacitor" converter. For an ideal converter, the flying capacitors share the half DC-bus voltage, that is, $v_{af} = v_{bf} = v_{cf} = v_{dc}/2$. The analysis on operating modes indicates the following: (1) four operating modes are possible, (2) only two switches conduct at any time, and (3) switching pairs (s_{a1}, s_{a4}) and (s_{a2}, s_{a3}) operate in a complementary manner.

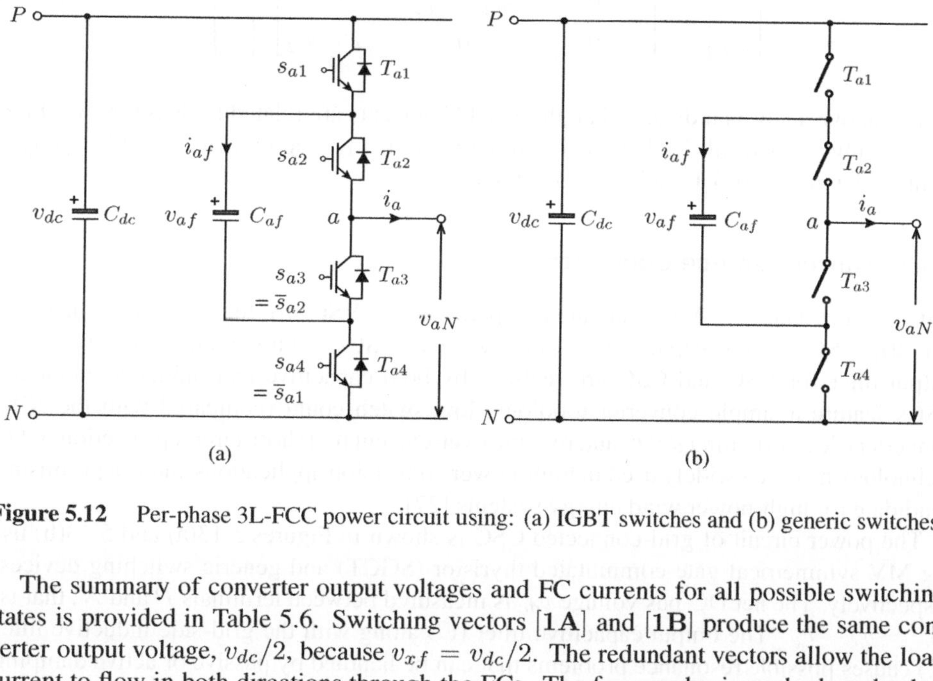

Figure 5.12 Per-phase 3L-FCC power circuit using: (a) IGBT switches and (b) generic switches.

The summary of converter output voltages and FC currents for all possible switching states is provided in Table 5.6. Switching vectors [**1A**] and [**1B**] produce the same converter output voltage, $v_{dc}/2$, because $v_{xf} = v_{dc}/2$. The redundant vectors allow the load current to flow in both directions through the FCs. The four modes in each phase lead to 64 (4^3) switching state combinations. The SV diagram for 3L-FCC is the same as that of the NPC converter, but the switching vector [1] should be substituted by [**1A**] and [**1B**].

In the *abc* frame, the 3L-FCC voltages are provided in terms of the DC-bus voltage, FC voltages, and switching signals as follows [25]:

$$\begin{bmatrix} v_{aN} \\ v_{bN} \\ v_{cN} \end{bmatrix} = v_{dc} \begin{bmatrix} s_{a1} \\ s_{b1} \\ s_{c1} \end{bmatrix} - \begin{bmatrix} v_{af} & 0 & 0 \\ 0 & v_{bf} & 0 \\ 0 & 0 & v_{cf} \end{bmatrix} \begin{bmatrix} s_{a1} - s_{a2} \\ s_{b1} - s_{b2} \\ s_{c1} - s_{c2} \end{bmatrix}. \tag{5.37}$$

Table 5.6 Switching states and output voltages for a 3L-FCC $\forall\ x \in \{a, b, c\}$

Switching Vector	Switching Signals				Output Voltage	FC Currents
$\mathbf{S_x}$	s_{x1}	s_{x2}	s_{x3}	s_{x4}	v_{xN}	i_{xf}
[2]	'1'	'1'	'0'	'0'	v_{dc}	0
[1A]	'1'	'0'	'1'	'0'	$v_{dc} - v_{xf}$	i_x
[1B]	'0'	'1'	'0'	'1'	v_{xf}	$-i_x$
[0]	'0'	'0'	'1'	'1'	0	0

Note: (s_{x1}, s_{x4}) and (s_{x2}, s_{x3}) are complementary switching signals.

The FC currents are estimated on the basis of the output currents and switching signals:

$$\begin{bmatrix} i_{af} \\ i_{bf} \\ i_{cf} \end{bmatrix} = \begin{bmatrix} s_{a1} - s_{a2} & 0 & 0 \\ 0 & s_{b1} - s_{b2} & 0 \\ 0 & 0 & s_{c1} - s_{c2} \end{bmatrix} \begin{bmatrix} i_a \\ i_b \\ i_c \end{bmatrix}. \tag{5.38}$$

The above analysis indicates that phase-x FC currents are related to phase-x switching signals only. This result leads to a simplified approach for the regulation of FC voltages compared with that of DC-link voltages in DCCs.

5.8.2 Current Source Converter

CSCs are the duality of VSCs and are competitive with VSCs for high-power applications [11, 26]. VSCs employ capacitive DC-link, whereas CSCs employ inductive DC-link. The output filter for VSC and CSC are realized by both capacitive and inductive elements. CSCs feature a simple converter topology, low switch count (compared with the NPC converter), low switching *dv/dt*, and reliable over-current (or) short-circuit protection. CSC technology has been widely used in high-power conversion applications and is a promising candidate for high-power wind energy systems [27].

The power circuit of grid-connected CSC is shown in Figures 5.13(a) and 5.13(b) using MV symmetrical gate-commutated thyristor (SGCT) and generic switching devices, respectively. The net DC-bus voltage v_{dc} is measured between terminals P and N, that is, $v_{dc} = v_P - v_N$. The output capacitive filter (C) along with the grid-side inductive filter (L) causes possible resonance problems that can be handled by passive or active damping [28]. The following are observed in a CSC: (1) four operating modes are possible; (2) the short circuit of two switches in any phase is allowed; (3) only two switches operate at any instant; (4) switching pairs s_{a1} and s_{a2} do not need to operate in a complementary manner.

The relationship between the switching vector $\mathbf{S_x}$ and the corresponding switching signals, converter output currents, and DC bus voltage is provided in Table 5.7. With four switching vectors in each phase, the theoretical number of combinations for a three-phase CSC becomes 64 (4^3). However, 55 theoretical combinations become void because of the short-circuit operation of CSC. For example, all DC currents circulate internally regardless of the switching states in the other two phases when T_{a1} and T_{a2} are short-circuited. For this reason, the possible number of switching states available for a CSC are 9 only [29]. A

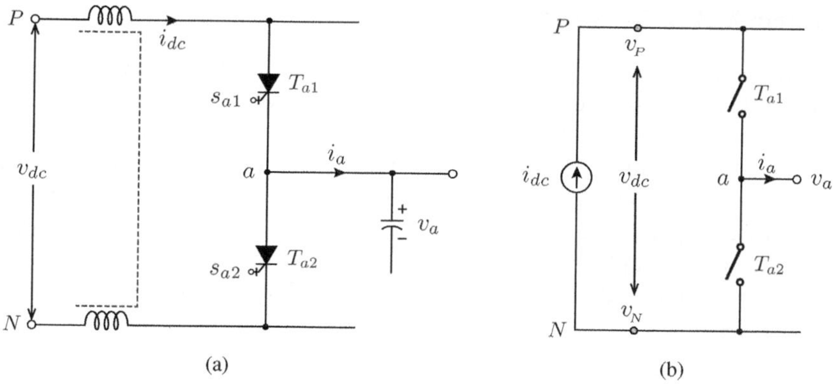

Figure 5.13 Per-phase representation of a CSC power circuit using: (a) SGCT switches, and (b) generic switches, and the DC-link model.

detailed summary of current vectors and their corresponding switching vectors, switching states, output AC currents, and input DC voltage is presented in Appendix B.2, Table B.2.

Table 5.7 Switching states and output voltages for a CSC $\forall\ x \in \{a, b, c\}$

Switching Vector	**Switching Signals**		**Output Current**	**Input Voltage**
$\mathbf{S_x}$	s_{x1}	s_{x2}	i_x	v_{dc}
[3]	'1'	'1'	0	0
[2]	'1'	'0'	i_{dc}	v_x
[1]	'0'	'1'	$-i_{dc}$	$-v_x$
[0]	'0'	'0'	0	0

In the *abc* frame, the CSC AC-side currents are expressed in terms of the switching signals and DC-link current as follows [30]:

$$\begin{bmatrix} i_a \\ i_b \\ i_c \end{bmatrix} = i_{dc} \begin{bmatrix} s_{a1} \\ s_{b1} \\ s_{c1} \end{bmatrix} - i_{dc} \begin{bmatrix} s_{a2} \\ s_{b2} \\ s_{c2} \end{bmatrix}. \tag{5.39}$$

The net DC-bus voltage estimation in terms of the switching signals and measured capacitive filter voltages is provided below:

$$\begin{bmatrix} v_P \\ v_N \end{bmatrix} = \begin{bmatrix} s_{a1} & s_{b1} & s_{c1} \\ s_{a2} & s_{b2} & s_{c2} \end{bmatrix} \begin{bmatrix} v_a \\ v_b \\ v_c \end{bmatrix}, \quad v_{dc} = v_P - v_N. \tag{5.40}$$

5.8.3 Direct Matrix Converter

Matrix converters (MCs) offer single-stage AC/AC power conversion without involving intermediate DC energy store elements, such as capacitors and inductors [31]. The power circuit of three-phase to three-phase DMC using the generic switch model is shown in Figure 5.14. The corresponding IGBT switch models using common emitter and common collector are shown at the bottom of Figure 5.14. The converter uses 9 bidirectional switches with the 3 switches per phase, T_{x1}, T_{x2}, and T_{x3}, for all $x \in \{a, b, c\}$. A three-phase capacitive filter C_e is employed at the grid side to assist the commutation of switching devices. The generator- and grid-side terminals of DMC are denoted as (as, bs, cs) and (ag, bg, cg), respectively.

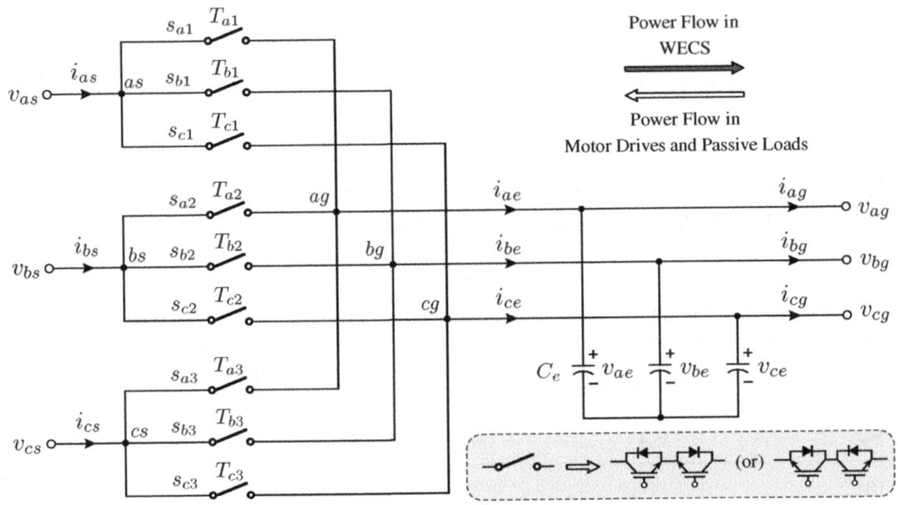

Figure 5.14 Power circuit of DMC using generic switches.

To ensure the safe operation of DMC, grid-side terminals should never be short-circuited and generator-side terminal must never be opened. This approach leads to three operating modes in each phase with 27 possible combinations (switching states) for a three-phase DMC [32]. The relationship between capacitive filter-side measured voltages and generator-side voltages is provided as follows [33]:

$$\begin{bmatrix} v_{as} \\ v_{bs} \\ v_{cs} \end{bmatrix} = \underbrace{\begin{bmatrix} s_{a1} & s_{b1} & s_{c1} \\ s_{a2} & s_{b2} & s_{c2} \\ s_{a3} & s_{b3} & s_{c3} \end{bmatrix}}_{\mathbf{T}_{(d)xe/xs}} \begin{bmatrix} v_{ae} \\ v_{be} \\ v_{ce} \end{bmatrix}. \tag{5.41}$$

The filter- and generator-side currents are expressed in terms of the switching signals:

$$\begin{bmatrix} i_{ae} \\ i_{be} \\ i_{ce} \end{bmatrix} = \underbrace{\begin{bmatrix} s_{a1} & s_{a2} & s_{a3} \\ s_{b1} & s_{b2} & s_{b3} \\ s_{c1} & s_{c2} & s_{c3} \end{bmatrix}}_{\mathbf{T}_{(d)xs/xe}} \begin{bmatrix} i_{as} \\ i_{bs} \\ i_{cs} \end{bmatrix}. \tag{5.42}$$

5.8.4 Indirect Matrix Converter

IMC is another breed of MC technology that uses the fictitious DC-link concept and combines the operation of 2L-VSC and CSC. Therefore, the number of feasible switching states is 72 (8 as in VSC × 9 as in CSC). Compared with DMCs, IMCs employ the same number of switching devices but provide a simple and reliable commutation process [34, 35]. The configuration of three-phase to three-phase IMC using generic switches is shown in Figure 5.15. The generator-side converter can be treated as a 2L-VSC without a DC-link capacitor and the grid-side converter is a CSC without a DC-link inductor. The subscript notation is changed here from r and i to s and g, respectively.

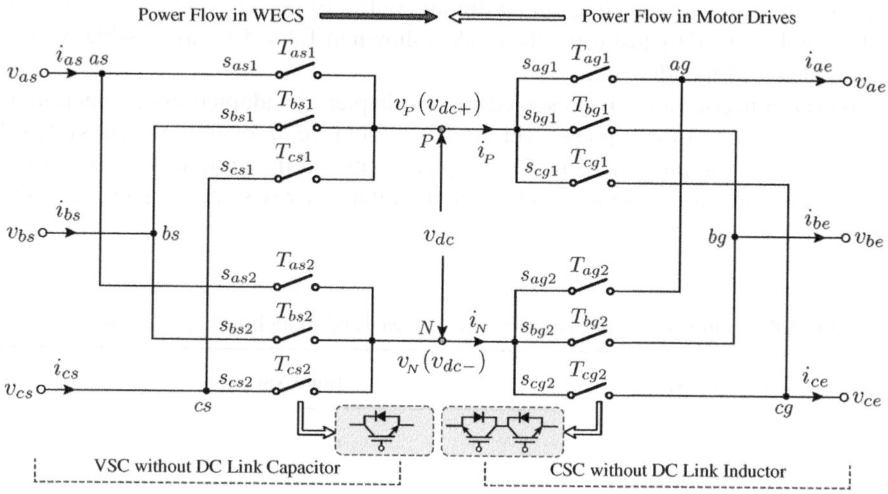

Figure 5.15 Power circuit of IMC using generic switches.

As outlined in (5.40), DC-link voltages (v_P and v_N) are expressed in terms of measured filter capacitors voltage (v_{ae}, v_{be}, and v_{ce}) and switching states (s_{xg1} and s_{xg2}) for the grid-side CSC. As in (5.17), converter terminal voltages (v_{as}, v_{bs}, and v_{cs}) are expressed in terms of DC voltage and switching signals (s_{xs1} and s_{xs2}) for the generator-side VSC. By combining these operations, generator-side voltages are formulated in terms of measured filter capacitors voltage and switching states [36]:

$$\begin{bmatrix} v_{as} \\ v_{bs} \\ v_{cs} \end{bmatrix} = \underbrace{\begin{bmatrix} s_{as1} & s_{as2} \\ s_{bs1} & s_{bs2} \\ s_{cs1} & s_{cs2} \end{bmatrix} \begin{bmatrix} s_{ag1} & s_{bg1} & s_{cg1} \\ s_{ag2} & s_{bg2} & s_{cg2} \end{bmatrix}}_{\mathbf{T}_{(i)xe/xs}} \begin{bmatrix} v_{ae} \\ v_{be} \\ v_{ce} \end{bmatrix}. \tag{5.43}$$

Similarly, the filter-side currents are formulated in terms of the measured generator currents and switching states as follows:

$$\begin{bmatrix} i_{ae} \\ i_{be} \\ i_{ce} \end{bmatrix} = \underbrace{\begin{bmatrix} s_{ag1} & s_{ag2} \\ s_{bg1} & s_{bg2} \\ s_{cg1} & s_{cg2} \end{bmatrix} \begin{bmatrix} s_{as1} & s_{bs1} & s_{cs1} \\ s_{as2} & s_{bs2} & s_{cs2} \end{bmatrix}}_{\mathbf{T}_{(i)xs/xe}} \begin{bmatrix} i_{as} \\ i_{bs} \\ i_{cs} \end{bmatrix}. \tag{5.44}$$

5.9 CONCLUDING REMARKS

The modeling of power converters employed in the Types 3 and 4 WECS is depicted in this chapter. The converter output variables for 2L-VSC, NPC converter, and multilevel DCC, three-level FCC, CSC, DMC, and IMC are formulated in terms of converter input variables and switching states. The summary of modeling for various three-phase power converters is provided in Table 5.8. The generalized approach for the modeling reveals that the models can be extended to multilevel and multiphase applications in an intuitive manner. The operational modes and space vector diagrams of various power converters are presented to aid the discussion and to show proof of concept. However, a sound knowledge of them is unnecessary in practice. The closed-loop digital control of power conversion systems can be readily implemented by just using the models shown in Table 5.8 and possible switching states provided in Appendix B.2.

The power converter models presented in this chapter are adopted in conjunction with the wind generator and three-phase grid models to achieve an overall control system for WECS. The modeling introduced in this chapter is general in nature and can be used for all power conversion applications, such as photovoltaic energy systems, adjustable speed motor drives, power quality, and HVDC transmission.

Table 5.8 Summary of three-phase power converters' modeling $\forall\ x \in \{a, b, c\}$

Converter	Output Variable	Input Variables
2L-VSC	$v_{xN} = v_{C1}\, s_{x1}$	$i_P = \sum_x i_x\, s_{x1}$ $i_N = \sum_x i_x\, s_{x2}$
3L-DCC	$v_{xN} = v_{C1}\, s_{x1} + v_{C2}\, s_{x2}$	$i_P = \sum_x i_x\, s_{x1}\, s_{x2}$ $i_Z = \sum_x i_x\, s_{x2}\, s_{x3}$ $i_N = \sum_x i_x\, s_{x3}\, s_{x4}$
mL-DCC	$v_{xN} = v_{C1}\, s_{x1} + \cdots + v_{C(m-1)}\, s_{x(m-1)}$	$i_P = \sum_x i_x\, s_{x1}\, s_{x2}$ $\vdots$ $i_N = \sum_x i_x\, s_{x(m)}\, s_{x(2m-2)}$
3L-FCC	$v_{xN} = v_{dc}\, s_{x1} - v_{xf}\,(s_{x1} - s_{x2})$	$i_{xf1} = i_x\,(s_{x1} - s_{x2})$
mL-FCC	$v_{xN} = v_{dc}\, s_{x1} - v_{xf}\,(s_{x1} - s_{x2}) - \cdots$ $\cdots - v_{xf(m-2)}\,\big(s_{x(m-2)} - s_{x(m-1)}\big)$	$i_{xf1} = i_x\,(s_{x1} - s_{x2})$ $\vdots$ $i_{xf(m-2)} = i_x\,\big(s_{x(m-2)} - s_{x(m-1)}\big)$
CSC	$i_x = i_{dc}\,(s_{x1} - s_{x2})$	$v_{dc} = \sum_x v_x\,(s_{x1} - s_{x2})$
DMC	$\mathbf{i_e} = \left[\mathbf{T}_{(d)xs/xe}\right] \mathbf{i_s}$	$\mathbf{v_s} = \left[\mathbf{T}_{(d)xs/xe}\right]^T \mathbf{v_e}$
IMC	$\mathbf{i_e} = \left[\mathbf{T}_{(i)xs/xe}\right] \mathbf{i_s}$	$\mathbf{v_s} = \left[\mathbf{T}_{(i)xs/xe}\right]^T \mathbf{v_e}$

Link to Next Chapters:

- The DT conversion of power converter models along with generator/grid models is studied further in Chapter 7.
- The power converters models are employed in Chapters 8 to 12 to design the MPC schemes of variable-speed WECS.

REFERENCES

1. V. Yaramasu, B. Wu, P. C. Sen, S. Kouro, and M. Narimani, "High-power wind energy conversion systems: State-of-the-art and emerging technologies," *Proceedings of the IEEE*, vol. 103, no. 5, pp. 740–788, May 2015.
2. H. Van der Broeck, H.-C. Skudelny, and G. Stanke, "Analysis and realization of a pulsewidth modulator based on voltage space vectors," *IEEE Transactions on Industry Applications*, vol. 24, no. 1, pp. 142–150, January 1988.
3. J. Rodríguez, J. Pontt, C. A. Silva, P. Correa, P. Lezana, P. Coftes, and U. Ammann, "Predictive current control of a voltage source inverter," *IEEE Transactions on Industrial Electronics*, vol. 54, no. 1, pp. 495–503, February 2007.
4. V. Yaramasu, M. Rivera, B. Wu, and J. Rodriguez, "Model predictive current control of two-level four-leg inverters – Part I: Concept, algorithm and simulation analysis," *IEEE Transactions on Power Electronics*, vol. 28, no. 7, pp. 3459–3468, July 2013.
5. C. Lim, E. Levi, M. Jones, N. Rahim, and W. Hew, "FCS-MPC-based current control of a five-phase induction motor and its comparison with PI-PWM control," *IEEE Transactions on Industrial Electronics*, vol. 61, no. 1, pp. 149–163, January 2014.
6. F. Barrero, M. Arahal, R. Gregor, S. Toral, and M. Duran, "A proof of concept study of predictive current control for VSI-driven asymmetrical dual three-phase AC machines," *IEEE Transactions on Industrial Electronics*, vol. 56, no. 6, pp. 1937–1954, June 2009.
7. H. Geng, D. Xu, B. Wu, and G. Yang, "Active damping for PMSG-based WECS with DC-link current estimation," *IEEE Transactions on Industrial Electronics*, vol. 58, no. 4, pp. 1110–1119, April 2011.
8. M. Liserre, R. Cardenas, M. Molinas, and J. Rodríguez, "Overview of multi-MW wind turbines and wind parks," *IEEE Transactions on Industrial Electronics*, vol. 58, no. 4, pp. 1081–1095, April 2011.
9. J. Rodríguez, S. Bernet, P. Steimer, and I. Lizama, "A survey on neutral-point-clamped inverters," *IEEE Transactions on Industrial Electronics*, vol. 57, no. 7, pp. 2219–2230, July 2010.
10. A. Nabae, I. Takahashi, and H. Akagi, "A new neutral-point-clamped PWM inverter," *IEEE Transactions on Industry Applications*, vol. IA-17, no. 5, pp. 518–523, September 1981.
11. B. Wu, *High-Power Converters and AC Drives*, 1st ed. Hoboken, NJ: Wiley-IEEE Press, March 2006.
12. B. McGrath, D. Holmes, and T. Lipo, "Optimized space vector switching sequences for multilevel inverters," *IEEE Transactions on Power Electronics*, vol. 18, no. 6, pp. 1293–1301, November 2003.
13. S. Busquets-Monge, J. Bordonau, D. Boroyevich, and S. Somavilla, "The nearest three virtual space vector PWM - a modulation for the comprehensive neutral-point balancing in the three-level NPC inverter," *IEEE Power Electronics Letters*, vol. 2, no. 1, pp. 11–15, March 2004.
14. R. Vargas, P. Coftes, U. Ammann, J. Rodríguez, and J. Pontt, "Predictive control of a three-phase neutral-point-clamped inverter," *IEEE Transactions on Industrial Electronics*, vol. 54, no. 5, pp. 2697–2705, October 2007.
15. V. Yaramasu and B. Wu, "Predictive control of a three-level boost converter and an NPC inverter for high power PMSG-based medium voltage wind energy conversion systems," *IEEE Transactions on Power Electronics*, vol. 29, no. 10, pp. 5308–5322, October 2014.
16. H.-J. Kim, H.-D. Lee, and S.-K. Sul, "A new PWM strategy for common-mode voltage reduction in neutral-point-clamped inverter-fed AC motor drives," *IEEE Transactions on Industry Applications*, vol. 37, no. 6, pp. 1840–1845, November/December 2001.
17. V. Yaramasu, "Predictive control of multilevel converters for megawatt wind energy conversion systems," Ph.D. dissertation, Ryerson University, Toronto, ON, Canada, 2014, available at: http://digital.library.ryerson.ca/islandora/object/RULA%3A3459.
18. M. Durań, J. Prieto, and F. Barrero, "Space vector PWM with reduced common-mode voltage for five-phase induction motor drives operating in overmodulation zone," *IEEE Transactions on Power Electronics*, vol. 28, no. 8, pp. 4030–4040, August 2013.
19. M. Chaves, E. Margato, J. Silva, S. Pinto, and J. Santana, "Fast optimum-predictive control and capacitor voltage balancing strategy for bipolar back-to-back NPC converters in high-voltage direct current transmission systems," *IET Generation, Transmission and Distribution*, vol. 5, no. 3, pp. 368–375, March 2011.
20. E. Bueno, S. Cobreces, F. Rodriguez, A. Hernandez, and F. Espinosa, "Design of a back-to-back NPC converter interface for wind turbines with squirrel-cage induction generator," *IEEE Transactions on Energy Conversion*, vol. 23, no. 3, pp. 932–945, September 2008.

21. R. Portillo, M. Prats, J. Leon, J. Sanchez, J. Carrasco, E. Galvan, and L. Franquelo, "Modeling strategy for back-to-back three-level converters applied to high-power wind turbines," *IEEE Transactions on Industrial Electronics*, vol. 53, no. 5, pp. 1483–1491, October 2006.

22. S. Alepuz, A. Calle, S. Busquets-Monge, S. Kouro, and B. Wu, "Use of stored energy in PMSG rotor inertia for low-voltage ride-through in back-to-back NPC converter-based wind power systems," *IEEE Transactions on Industrial Electronics*, vol. 60, no. 5, pp. 1787–1796, May 2013.

23. A. Yazdani and R. Iravani, "A neutral-point clamped converter system for direct-drive variable-speed wind power unit," *IEEE Transactions on Energy Conversion*, vol. 21, no. 2, pp. 596–607, June 2006.

24. D. Krug, S. Bernet, S. Fazel, K. Jalili, and M. Malinowski, "Comparison of 2.3-kV medium-voltage multilevel converters for industrial medium-voltage drives," *IEEE Transactions on Industrial Electronics*, vol. 54, no. 6, pp. 2979–2992, December 2007.

25. T. Vyncke, S. Thielemans, and J. Melkebeek, "Finite-set model-based predictive control for flying-capacitor converters: Cost function design and efficient FPGA implementation," *IEEE Transactions on Industrial Informatics*, vol. 9, no. 2, pp. 1113–1121, May 2013.

26. B. Wu, J. Pontt, J. Rodríguez, S. Bernet, and S. Kouro, "Current-source converter and cycloconverter topologies for industrial medium-voltage drives," *IEEE Transactions on Industrial Electronics*, vol. 55, no. 7, pp. 2786–2797, July 2008.

27. J. Dai, D. Xu, and B. Wu, "A novel control scheme for current-source-converter-based PMSG wind energy conversion systems," *IEEE Transactions on Power Electronics*, vol. 24, no. 4, pp. 963–972, April 2009.

28. J. Wiseman and B. Wu, "Active damping control of a high-power PWM current-source rectifier for line-current THD reduction," *IEEE Transactions on Industrial Electronics*, vol. 52, no. 3, pp. 758–764, June 2005.

29. P. Correa, J. Rodríguez, I. Lizama, and D. Andler, "A predictive control scheme for current-source rectifiers," *IEEE Transactions on Industrial Electronics*, vol. 56, no. 5, pp. 1813–1815, May 2009.

30. P. Zavala, M. Rivera, S. Kouro, J. Rodriguez, B. Wu, V. Yaramasu, C. Baier, J. Munoz, J. Espinoza, and P. Melin, "Predictive control of a current source rectifier with imposed sinusoidal input currents," in *IEEE Industrial Electronics Conference (IECON)*, Vienna, Austria, November 2013, pp. 5842–5847.

31. P. Wheeler, J. Rodríguez, J. Clare, L. Empringham, and A. Weinstein, "Matrix converters: a technology review," *IEEE Transactions on Industrial Electronics*, vol. 49, no. 2, pp. 276–288, April 2002.

32. L. Huber and D. Borojevic, "Space vector modulated three-phase to three-phase matrix converter with input power factor correction," *IEEE Transactions on Industry Applications*, vol. 31, no. 6, pp. 1234–1246, November/December 1995.

33. R. Vargas, J. Rodríguez, U. Ammann, and P. Wheeler, "Predictive current control of an induction machine fed by a matrix converter with reactive power control," *IEEE Transactions on Industrial Electronics*, vol. 55, no. 12, pp. 4362–4371, December 2008.

34. R. Pena, R. Cardenas, E. Reyes, J. Clare, and P. Wheeler, "A topology for multiple generation system with doubly fed induction machines and indirect matrix converter," *IEEE Transactions on Industrial Electronics*, vol. 56, no. 10, pp. 4181–4193, October 2009.

35. J. Elizondo, A. Olloqui, M. Rivera, M. Macias, O. Probst, O. Micheloud, and J. Rodriguez, "Model-based predictive rotor current control for grid synchronization of a DFIG driven by an indirect matrix converter," *IEEE Journal of Emerging and Selected Topics in Power Electronics*, vol. 2, no. 4, pp. 715–726, December 2014.

36. P. Correa, J. Rodríguez, M. Rivera, J. Espinoza, and J. Kolar, "Predictive control of an indirect matrix converter," *IEEE Transactions on Industrial Electronics*, vol. 56, no. 6, pp. 1847–1853, June 2009.

CHAPTER 6

MODELING OF WIND GENERATORS FOR MODEL PREDICTIVE CONTROL

6.1 INTRODUCTION

As discussed in Chapter 1, an electric generator is a major component in a wind turbine (WT) because it converts mechanical energy into electric energy. Over the past 35 years, numerous wind generators have been developed for WTs, such as the squirrel-cage induction generator (SCIG), wound rotor induction generator (WRIG), doubly fed induction generator (DFIG), permanent magnet synchronous generator (PMSG), wound rotor synchronous generator (WRSG), and high-temperature superconducting synchronous generator (HTS-SG) [1, 2]. On the basis of the construction approach, wind generators can have different number of poles, stator and rotor construction features, excitation systems, and current carrying capacity. Each wind generator type has its own advantages and disadvantages. Modern Type 3 and 4 wind energy conversion systems (WECS) employ both SGs and IGs along with power converters to inject power to the grid in a controlled manner. IGs usually operate at high rotational speed (through gearbox), whereas SGs operate at low, medium, or high speed [3]. A concise list of wind generators employed in the variable-speed WECS is given in Table 6.1. The manufacturer, generator type, generator power and voltage rating, speed range, and corresponding WT model number are highlighted. Megawatt WTs have also led to the development of medium-voltage generators.

The modeling of power converters has been discussed in the previous chapter. In this chapter, the modeling of wind generators is discussed from the model predictive control (MPC) perspective. Synchronous machines are widely used as generators, whereas induction machines are popular for motor applications. In this chapter, a common notation

Model Predictive Control of Wind Energy Conversion Systems, First Edition. Venkata Yaramasu and Bin Wu.

Companion Website: www.wiley.com/go/yaramasu_wu_mpc_wecs

Table 6.1 List of wind generators employed in the Type 3 and 4 WECS

Manufacturer	Generator	Gen. Rating	Speed (rpm)	Model #
RRB Energy (India)	SCIG	1.8 MW, 690 V	1008 (rated)	PS-1800
Siemens (Germany)	SCIG	3.6 MW, 690 V	500-1300	SWT3.6-120
Condor Ltd. (Germany)	SCIG	6.6 MW, 3300 V	675 (rated)	Condor6
Sinovel (PR China)	DFIG	3.0 MW, 690 V	800-1600	SL3000
Bard GmbH (Germany)	DFIG	5.0 MW, 690 V	679–1358	Bard 5.0 offshore
Senvion (Germany)	DFIG	6.2 MW, 6600/660 V	750–1170	6.2M-152
Unison (South Korea)	PMSG	2.0 MW, 690 V	430–1250	U93
Goldwind (PR China)	PMSG	3.0 MW, 690 V	7–15.3	GW3000
Areva Wind (Germany)	PMSG	5.0 MW, 3300 V	66.1–133.5	M5000-135
Elsewedy (Egypt)	WRSG	1.65 MW, 690 V	6–18	TWT1.68/82
E.N.O Energy (Germany)	WRSG	3.5 MW, 690 V	476–1523.2	E.N.O-126
Enercon GmbH (Germany)	WRSG	7.5 MW, 690 V	5–11.7	E126/7500
Wintec-AMSC (USA)	HTS-SG	10 MW, 690 V	5–11.7	SeaTitan 10MW

is utilized for both synchronous and induction machines, and the modeling is presented accordingly. The models provided in this chapter work for both motor and generator applications on the basis of the sign of mechanical input torque. The machine works as a generator when the mechanical torque is negative, and it acts as a motor when positive. The dynamic modeling of output currents $\mathbf{i_s}$, electrical rotor speed ω_r, electromagnetic torque T_e, stator flux vector $\boldsymbol{\psi}_s$, and active and reactive powers (P_s and Q_s) is presented in natural (abc), stationary ($\alpha\beta$), and synchronous (dq) reference frames for both SG and IG. These models help realize predictive model 2 in Figure 5.1 on page 153, such that various types of MPC schemes can be developed for generator-side power conversion systems. By combining the wind generator models with the power converter models, the MPC scheme for numerous WECS types can be easily achieved.

Chapter Overview

- An overview of electric generators for variable-speed WECS is presented in Section 6.2, along with subclasses and construction features.
- The objectives for dynamic modeling and notation employed for modeling are discussed in Sections 6.3 and 6.4, respectively.
- In Section 6.5, the abc and dq-frame models have been discussed considering both surface and interior PMSGs. The simulation model of PMSG is presented in Section 6.6.
- The dynamic modeling of IG in arbitrary, dq, and $\alpha\beta$ frames is presented in Section 6.7. A generalized model is presented for both SCIG and DFIG. The simulation model of IG is analyzed in Section 6.8.
- Section 6.9 explains how the generator models will be used in various MPC schemes. Section 6.10 draws the conclusions of this chapter.

6.2 OVERVIEW OF WIND GENERATORS FOR VARIABLE-SPEED WECS

6.2.1 Synchronous Generators for WECS

As shown in Figure 6.1, the SGs are classified into two groups according to the rotor construction: WRSG and PMSG. In WRSG and PMSG, the stator carries a three-phase winding similar to that of the IG. In WRSGs, the rotor flux is generated by the external DC excitation. In PMSGs, the rotor flux is created by permanent magnets [4]. WRSGs are classified into two subgroups according to the way the rotor is connected to the external excitation: through slip rings and brushes or brushless exciter. The first type is simple but requires regular maintenance because of the brushes and slip rings. The brushless excitation solves this issue by using a rotating rectifier but at the expense of high initial cost and structure complexity. Moreover, field winding losses are inevitable in WRSGs. Therefore, the efficiency of WRSGs is lower than that of PMSGs.

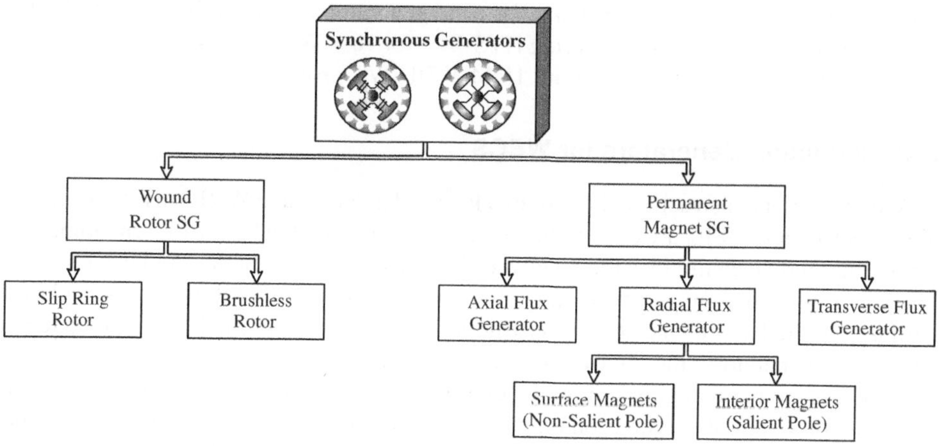

Figure 6.1 Taxonomy of three-phase SGs for variable-speed WECS.

The Type 4 WTs were first introduced with WRSGs (by `Enercon` in 1992), but in recent years PMSGs have become a popular choice because of their advantages such as high efficiency (no field winding losses), large air gap, and high-power density and reliability. However, PMSGs use rare earth materials such as neodymium and neodymium-iron-boron (NdFeB), whose extraction is expensive and not environmentally friendly [5]. PMSGs are popular choice for both low- and high-power applications, particularly at the MW-level. The design of PMSG allows for great flexibility, and numerous generator topologies have been developed for WECS [2].

On the basis of the direction of flux flow, PMSGs are classified as radial flux (RF), axial flux (AF), and transverse flux generators [6]. RF and AF are most commonly used for large WTs, whereas transverse flux generators type is in the research and development stage. In RF-PMSG and AF-PMSG, the magnetic flux crosses the air gap in the radial and axial directions, respectively. RF-PMSGs can be constructed with either inner or outer rotors, and the later designs can easily accommodate the multi-pole structure. The RF-PMSG offers simple and robust construction and is used by many WT manufacturers [7]. The AF-PMSGs have low noise, disc-type structure, short axial length, and high torque density. The sample applications of RF- and AF-PMSGs include the following: (1) `STX Windpower`

STX82 (2.0 MW, 660V) DD WT model with RF-PMSG and (2) Sway Turbine AS ST10 (10 MW, 3500 V) WT model with AF-PMSG.

According to the alignment of permanent magnets in the rotor, RF-PMSGs are further classified into two groups: surface-mount PMSGs (SPMSGs) and interior (inset) PMSGs (IPMSGs) [8]. In SPMSGs, the magnets are placed along the rotor surface. The RF-PMSG with surface mounted magnets is a great choice for MW-class DD WTs [6]. In IPMSGs, the magnets are buried in the rotor, thus leading to low stress. The RF-PMSG with inset magnets is preferable for high-speed applications. SGs are also denoted as salient pole and non-salient pole (cylindrical) according to the shape of the rotor and air-gap flux distribution. Non-salient-pole SGs have symmetrical air-gap flux distributions because of equal d- and q-axis synchronous inductances, namely, L_{ds} and L_{qs} respectively. In salient-pole SGs, the d-axis inductance is lower than the q-axis inductance, that is, $L_{ds} < L_{qs}$, thus leading to uneven flux distribution in the air-gap.

HTS-SGs, which are popular in motor drive applications [9], were introduced in 2010s for variable-speed WTs. Superconductors increase the current carrying capacity, thus leading to high-power density and lower size for wind generators, particularly for 10-MW-class WTs [10, 11]. The AMSC is now offering HTS-SGs for 10 MW DD WTs, and GE Energy plans to employ HTS-SG in their future 15 MW DD WT project [1].

6.2.2 Induction Generators for WECS

Based on the rotor construction, the IGs are classified as SCIG and WRIG, as shown in Figure 6.2. WRIG and DFIG are constructed in the same manner but have different nomenclatures in literature to distinguish them between Type 2 and 3 WTs. SCIG and DFIG have the same stator construction, and both carry three-phase winding over thin silicon steel laminations. On the basis of the stator construction, SCIGs are classified as single-cage and double-cage generators; the later configuration provides high starting torque. In SCIGs, the rotor bars are permanently shorted by end rings, thus leading to zero rotor voltages. In DFIG, the rotor consists of three-phase winding similar to stator winding. The rotor winding is connected to a converter through slip rings and brushes or brushless connection.

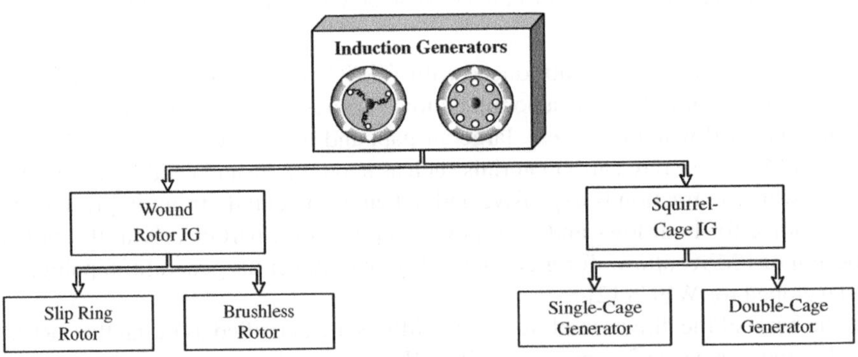

Figure 6.2 Taxonomy of three-phase IGs for variable-speed WECS.

IGs usually operate at high speeds through a gearbox because they cannot accommodate a large number of poles compared with SGs. Practical MW IGs are equipped with four or six poles and are connected to a WT through a three-stage gearbox. Recent developments, such as the Condor6 project, uses an eight-pole SCIG along with two-stage gearbox to achieve medium-speed operation.

6.3 OBJECTIVES FOR THE DYNAMIC MODELING OF WIND GENERATORS

As outlined in Chapter 4, the following generator variables are controlled through a power converter to achieve an optimal operation: (1) output currents, (2) electromagnetic torque and stator flux, (3) output active and reactive power, and (4) rotor speed. To regulate these variables at the set values, the cost function in the MPC is defined as follows:

- **Predictive Current Control in Stationary Frame (PCC-$\alpha\beta$):** The reference $\alpha\beta$ currents are compared with the predicted $\alpha\beta$ generator currents, and to implement this control strategy, the models in the format of $di_{\alpha s}/dt$ and $di_{\beta s}/dt$ should be obtained.
- **Predictive Current Control in Synchronous Frame (PCC-dq):** This method requires di_{ds}/dt and di_{qs}/dt to predict the future values of dq generator currents. The predicted currents are then compared with the reference dq currents.
- **Predictive Torque Control (PTC):** The cost function in this approach includes both electromagnetic torque and stator flux magnitude (T_e, ψ_s) variables. To implement the PTC strategy, T_e and $\boldsymbol{\psi}_s$ should be modeled in terms of generator parameters and feedback variables. Note that the stator flux magnitude ψ_s is derived from the stator flux vector $\boldsymbol{\psi}_s$.
- **Predictive Power Control (PPC):** In this methodology, the active and reactive power (P_s and Q_s) references are compared with the predicted P_s and Q_s. To realize PPC control, the P_s and Q_s needs to be modeled in terms of generator voltages and currents.
- **Predictive Speed Control (PSC):** For this control strategy, the rotor speed dynamics ($d\omega_m/dt$) are needed in addition to the generator current dynamics. The cost function includes both the speed and currents terms.

In summary, to realize various MPC schemes, the generator control variables $\{i_{ys}, T_e, \psi_{ys}, P_s, Q_s\}$ with $y \in \{a, b, c, \alpha, \beta, d, q\}$ needs to be modeled. To predict the future behavior of these variables, the mathematical model of the plant (wind generator) should be obtained in continuous-time (CT) and then converted into discrete-time (DT). For example, let us consider that the generator CT state-space model is in the following form:

$$\begin{aligned} \frac{d\mathbf{x}(t)}{dt} &= \mathbf{A}(t)\,\mathbf{x}(t) + \mathbf{B}(t)\,\mathbf{u}(t) + \mathbf{w}(t) \\ \mathbf{y}(t) &= \mathbf{C}\,\mathbf{x}(t) \end{aligned} \tag{6.1}$$

where $\mathbf{x}$ is the state variable of interest according to the control scheme to be implemented.

The next step in MPC implementation is converting the above CT model to DT using an appropriate discretization method. The DT equivalent of (6.1) is shown below [12]:

$$\begin{aligned} \mathbf{x}(k+1) &= \mathbf{\Phi}(k)\,\mathbf{x}(k) + \mathbf{\Gamma}_{\mathbf{b}}(k)\,\mathbf{u}(k) + \mathbf{\Gamma}_{\mathbf{w}}(k) \\ \mathbf{y}(k) &= \mathbf{H}\,\mathbf{x}(k) \end{aligned} \tag{6.2}$$

where $\mathbf{\Phi}(k)$, $\mathbf{\Gamma}_{\mathbf{b}}(k)$, $\mathbf{\Gamma}_{\mathbf{w}}(k)$, and $\mathbf{H}$ are DT equivalents of CT matrices $\mathbf{A}(t)$, $\mathbf{B}(t)$, $\mathbf{w}(t)$, and $\mathbf{C}$, respectively.

The main objective of this chapter is to obtain the dynamics of various generator types, particularly SPMSG, IMPSG, SCIG, and DFIG, in CT. By using these models, DT conversion is performed in the next chapter via exact or approximate methods.

6.4 NOTATION EMPLOYED FOR THE DYNAMIC MODELING

An electric machine is used as either a generator or motor, depending upon the application. A typical example of using permanent magnet synchronous machine (PMSM) in variable-speed application is shown in Figure 6.3. The first case shown in Figure 6.3(a) uses the machine as a generator in the Type 4 WECS, whereas the second case in Figure 6.3(b) uses the machine as a motor. In variable-speed WECS, the mechanical input torque T_m to the generator is provided by the WT and the machine works as a generator and produces power P_s (or current $\mathbf{i_s}$) to be injected to the three-phase grid through a power converter. In the variable-speed motor drive case, the power P_s is drawn from the grid and supplied to the motor and electric energy is converted to electromagnetic torque T_e by the motor. The T_m has a negative sign in the generator case and has a positive sign in the motor case. On the basis of the sign of T_m, the direction of the current and power changes, whereas the machine dynamics remain same. The motor models are well established in the literature. In this book, all the electric machines employed in the WTs are modeled as motors to be consistent with the literature. These models are utilized for both generator and motor applications by simply changing the sign of mechanical torque T_m.

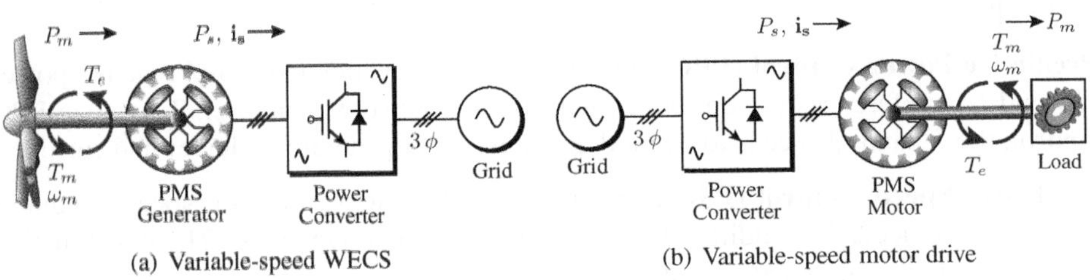

Figure 6.3 Variable-speed applications employing PMS machine.

The mechanical rotor speed dynamics are given as

$$J_m \frac{d}{dt} \omega_m + B_m \omega_m = T_e - T_m \tag{6.3}$$

where

- ω_m is rotor mechanical speed (rad/s).
- T_e and T_m are electromagnetic and shaft mechanical torques (N.m).
- J_m and B_m are shaft moment of inertia (kg.m^2) and viscous friction (N.m.s).

In motor drive applications, T_e and T_m have positive signs. In steady state, the derivative term in (6.3) becomes zero. The value of T_e must be slightly higher than T_m to compensate for the loss due to viscous friction ($T_e - T_m = B_m \omega_m$). In wind energy applications, T_m and T_e have negative signs. In steady state, the T_m is slightly higher than T_e if viscous friction is considered. By substituting negative values for T_m and T_e in (6.3), the overall difference in steady state becomes the same as with the motor drive case, i.e., $T_e - T_m = B_m \omega_m$. The mechanical rotor speed ω_m is related to the electrical rotor speed ω_r as demonstrated below:

$$\omega_m = \frac{1}{P_p} \omega_r \tag{6.4}$$

where P_p represents number of pole pairs in an electric machine.

By combining (6.3) and (6.4), the electrical rotor speed dynamics are given below:

$$\frac{J_m}{P_p}\frac{d}{dt}\omega_r + \frac{B_m}{P_p}\omega_r = T_e - T_m \tag{6.5}$$

from which

$$\frac{d}{dt}\omega_r = \frac{P_p}{J_m}(T_e - T_m) - \frac{B_m}{J_m}\omega_r. \tag{6.6}$$

The mechanical and electrical rotor position angles are given as

$$\theta_m = \frac{d}{dt}\omega_m, \quad \theta_r = \frac{d}{dt}\omega_r, \quad \theta_m = \frac{1}{P_p}\theta_r. \tag{6.7}$$

The stator voltage and current waveforms of an IPMSM with motor and generator convention are shown in shown in Figure 6.4. The waveforms have been obtained by connecting the IPMSM to a resistive load. Note that these waveforms are not to scale.

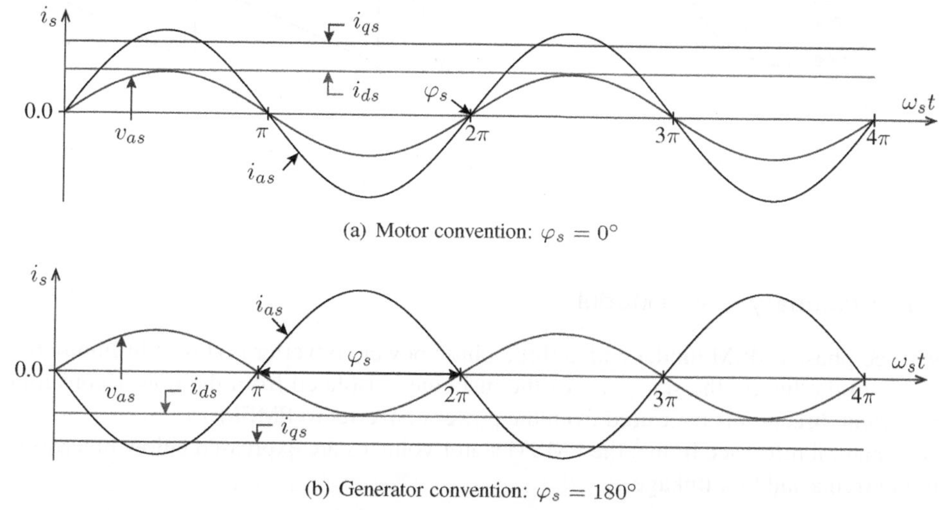

Figure 6.4 PMSM stator voltage and current waveforms using motor and generator convention.

In Figure 6.4(a), with motor mode of operation, the phase angle φ_s between the stator voltage and current waveforms becomes zero. The sign of dq-axis currents is positive, thus enabling the production of positive signs for the electromagnetic torque and output active power waveforms (not shown in Figure 6.4). In motor mode, the range of φ_s is between 0° and 90°, and the stator power factor (PF) has a positive sign. In generator mode, the waveforms for stator voltages and currents are shown in Figure 6.4(b). The phase angle φ_s is noted as 180°. The sign of i_{ds} and i_{qs} is negative, thus leading to a negative sign for T_e and P_m. The phase angle φ_s varies between 90° and 180° in the generator mode of operation, and the PF is negative. These waveforms justify that the well established motor convention for the PMSM can be adapted to WECS, with just a negative sign for T_m.

6.5 MODELING OF PERMANENT MAGNET SYNCHRONOUS GENERATOR

The correlation between the *abc* and *dq* reference frames for PMSM modeling is shown in Figure 6.5. The phase-a axis is considered the reference point. The angle between the a- and d-axis is θ_r, which denotes the PMSM rotor position angle. As shown in Figure 6.6, in one fundamental cycle, the angle θ_r changes from 0 to 2π radians and then becomes zero by the end of cycle. When the PMSM shaft speed changes, the frequency of the phase-a reference rotor flux waveform also changes. The θ_r value changes in accordance with the frequency of ψ_{ar}. In this section, the modeling of three-phase PMSM is introduced in the *abc* frame. Furthermore, the transformation from a time-varying natural frame model to a time-invariant synchronous frame model is presented.

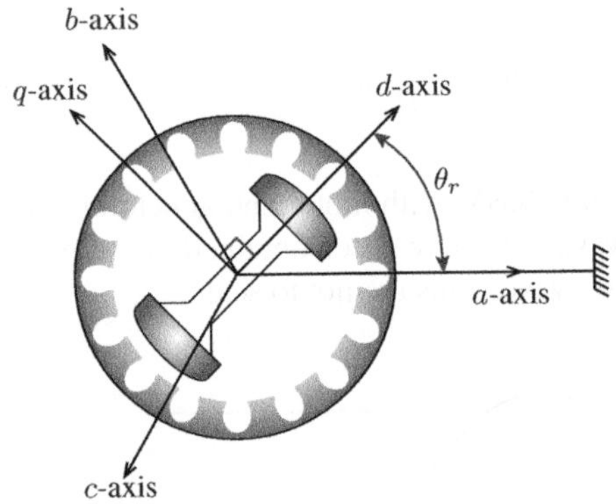

Figure 6.5 Representation of PMSM in natural and synchronous reference frames.

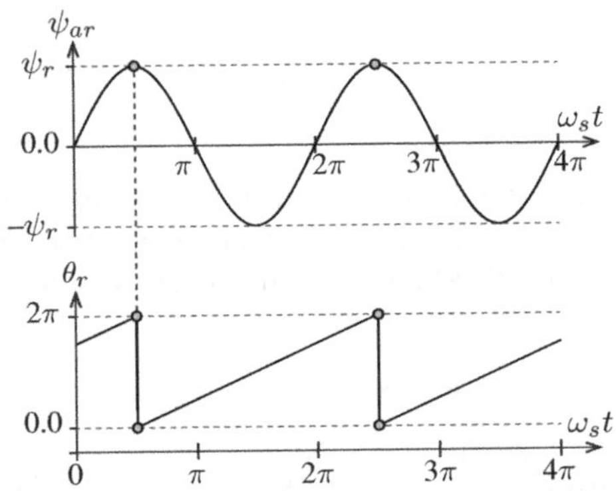

Figure 6.6 Correlation between the PMSM reference phase-a rotor flux ψ_{ar} and θ_r.

6.5.1 Stationary Frame Model

The three-phase PMSM model with a three-phase power converter is shown in Figure 6.7. As seen in the figure, the star point of the machine is isolated from the power converter. The current direction is measured from the power converter to PMSM.

In a natural reference frame, the PMSM stator voltages are expressed below in terms of stator currents and flux linkages:

$$\begin{bmatrix} v_{as} \\ v_{bs} \\ v_{cs} \end{bmatrix} = \begin{bmatrix} R_s & 0 & 0 \\ 0 & R_s & 0 \\ 0 & 0 & R_s \end{bmatrix} \begin{bmatrix} i_{as} \\ i_{bs} \\ i_{cs} \end{bmatrix} + \frac{d}{dt} \begin{bmatrix} \psi_{as} \\ \psi_{bs} \\ \psi_{cs} \end{bmatrix} + \begin{bmatrix} v_{ns} \\ v_{ns} \\ v_{ns} \end{bmatrix} \tag{6.8}$$

where

- v_{as}, v_{bs}, and v_{cs} are the machine phase *abc* stator voltages (V).
- i_{as}, i_{bs}, and i_{cs} are the machine phase *abc* stator currents (A).
- ψ_{as}, ψ_{bs}, and ψ_{cs} are the machine phase *abc* stator flux linkages (Wb).
- v_{ns} is the machine neutral-point voltage (V).
- R_s is the machine stator winding resistance (Ω).

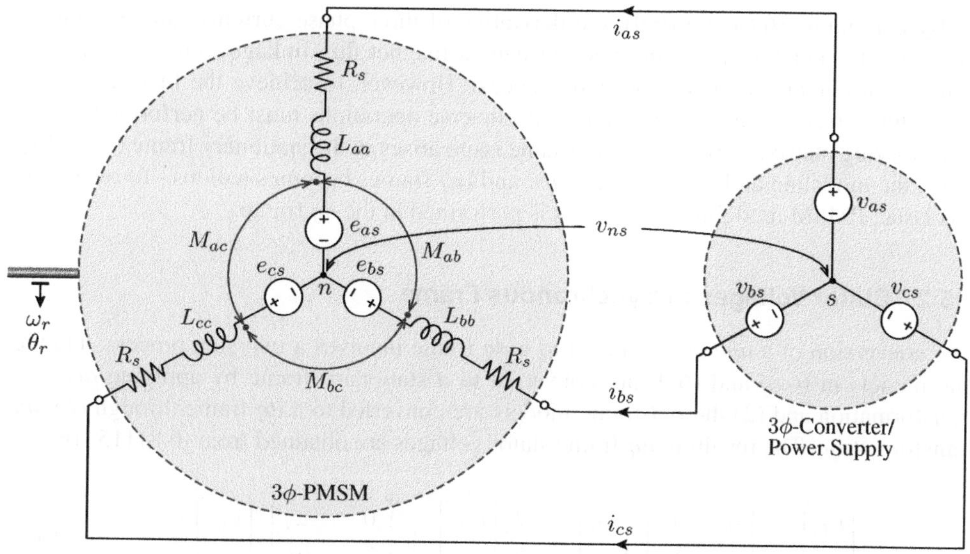

Figure 6.7 Representation of the three-phase PMSM with a power converter.

The *abc* frame stator flux linkages originate from the permanent magnetic flux linkages and the current flowing through the machine self and mutual inductances. This process is demonstrated as follows:

$$\begin{bmatrix}\psi_{as}\\\psi_{bs}\\\psi_{cs}\end{bmatrix} = \begin{bmatrix}L_{aa} & M_{ab} & M_{ac}\\M_{ab} & L_{bb} & M_{bc}\\M_{ac} & M_{bc} & L_{cc}\end{bmatrix}\begin{bmatrix}i_{as}\\i_{bs}\\i_{cs}\end{bmatrix} + \begin{bmatrix}\psi_r\cos\theta_r\\\psi_r\cos\left(\theta_r-\frac{2\pi}{3}\right)\\\psi_r\cos\left(\theta_r-\frac{4\pi}{3}\right)\end{bmatrix} \tag{6.9}$$

where

- L_{aa}, L_{bb} and L_{cc} are the machine phase *abc* self inductances.
- M_{ab}, M_{bc} and M_{ac} are the machine mutual inductances between phase *abc*.
- ψ_r is the peak value of flux linkage created by the permanent magnets.

The last term in (6.9) can be treated as a back EMF of the machine produced by the PM flux linkages. The phase-*a* rotor flux linkage is $\psi_{ar} = \psi_r\cos\theta_r$ and so on; this linkage reaches its maximum and minimum values when the rotor position angles θ_r are 0° and 90°, respectively. The self and mutual inductance values vary with θ_r [13]:

$$\begin{aligned}
L_{aa} &= L_p + L_n\cos[2\,\theta_r], & M_{ab} &= -\frac{L_p}{2} + L_n\cos\left[2\left(\theta_r-\frac{\pi}{3}\right)\right]\\
L_{bb} &= L_p + L_n\cos\left[2\left(\theta_r-\frac{2\,\pi}{3}\right)\right], & M_{bc} &= -\frac{L_p}{2} + L_n\cos\left[2\left(\theta_r-\frac{2\,\pi}{3}\right)\right]\\
L_{cc} &= L_p + L_n\cos\left[2\left(\theta_r-\frac{4\,\pi}{3}\right)\right], & M_{ac} &= -\frac{L_p}{2} + L_n\cos[2\,\theta_r]
\end{aligned} \tag{6.10}$$

where L_p [= 0.5 ($L_{ds} + L_{qs}$)] and L_n [= 0.5 ($L_{ds} - L_{qs}$)] are the machine parameters, which are constant according to the machine dimension and construction; the L_{ds} and L_{qs} are *dq*-axis inductances.

By combining (6.8) and (6.9), the derivative of three-phase currents can be obtained in terms of stator voltages, currents, permanent magnet flux linkages, and a third-order matrix that contains time-varying inductances. However, to achieve the final model for the stator current dynamics, derivative and inverse operations must be performed on the time-varying matrix. Furthermore, the same issue arises in the stationary frame ($\alpha\beta$) [14]. Thus, the modeling and control in the *abc* and $\alpha\beta$ frames becomes tedious. To overcome this issue, PMSM modeling and control is performed in the *dq* frame.

6.5.2 Stator Voltages in Synchronous Frame

The conversion of a *abc* frame model to a *dq* frame involves a two-step process: (1) the *abc* models in (6.8) and (6.9) are converted to a stationary frame by applying $abc/\alpha\beta$ transformation and (2) the $\alpha\beta$-frame models are converted to a *dq* frame through $\alpha\beta/dq$ transformation. The resulting *dq*-frame stator voltages are obtained from (6.8) [15, 16]:

$$\begin{bmatrix} v_{ds} \\ v_{qs} \end{bmatrix} = \begin{bmatrix} R_s & 0 \\ 0 & R_s \end{bmatrix} \begin{bmatrix} i_{ds} \\ i_{qs} \end{bmatrix} + \frac{d}{dt} \begin{bmatrix} \psi_{ds} \\ \psi_{qs} \end{bmatrix} + \begin{bmatrix} 0 & -\omega_r \\ \omega_r & 0 \end{bmatrix} \begin{bmatrix} \psi_{ds} \\ \psi_{qs} \end{bmatrix} \tag{6.11}$$

where

- v_{ds} and v_{qs} are the machine stator voltages in *dq* frame.
- i_{ds} and i_{qs} are the machine stator currents in *dq* frame.
- ψ_{ds} and ψ_{qs} are the machine stator flux linkages in *dq* frame.
- ω_r is the rotor electrical speed.

A comparison of Equation (6.8) with Equation (6.11) shows that the three-phase voltages, currents, and flux linkages are converted to the corresponding *dq* variables. The stator neutral-point voltage v_{ns} is omitted from the machine model during the *abc* to $\alpha\beta$ transformation. The last term in (6.11) corresponds to the speed voltages (induced voltages), and the $\alpha\beta$ to *dq* transformation usually causes speed voltages in all three-phase systems such as electric machines, power converters, and harmonic filters. The *dq*-frame PMSM standard model, which is based on (6.11), is shown in Figure 6.8.

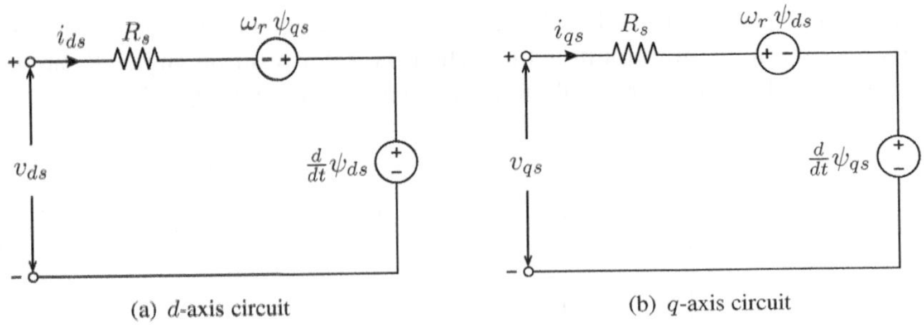

Figure 6.8 Representation of the standard *dq*-axes equivalent circuits for a PMSM through stator flux derivative terms [17].

6.5.3 Stator Flux Linkages in Synchronous Frame

Similarly, the *abc*-frame flux linkages given in (6.9) are converted to the *dq* frame:

$$\begin{bmatrix} \psi_{ds} \\ \psi_{qs} \end{bmatrix} = \begin{bmatrix} L_{ds} & 0 \\ 0 & L_{qs} \end{bmatrix} \begin{bmatrix} i_{ds} \\ i_{qs} \end{bmatrix} + \begin{bmatrix} \psi_r \\ 0 \end{bmatrix}. \tag{6.12}$$

In the *abc* frame, the machine rotor flux linkages and self and mutual inductance values vary according to the rotor electrical position angle θ_r (refer to Equations (6.9) and (6.10)). However, in the *dq* frame, the L_{ds}, L_{qs}, and ψ_r values become time-invariant. These constant values significantly simplify the analysis for PMSM modeling and control.

6.5.4 Stator Current Dynamics in Synchronous Frame

The model of stator voltages in (6.11) includes the derivative of stator flux linkages. In the MPC implementation, the derivative of the stator currents is more interesting than the derivative of the stator flux linkages (see Section 6.3). Thus, after combining Equations (6.11) and (6.12), the stator voltages are expressed in terms of stator currents:

$$\begin{bmatrix} v_{ds} \\ v_{qs} \end{bmatrix} = \begin{bmatrix} R_s & -\omega_r L_{qs} \\ \omega_r L_{ds} & R_s \end{bmatrix} \begin{bmatrix} i_{ds} \\ i_{qs} \end{bmatrix} + \begin{bmatrix} L_{ds} & 0 \\ 0 & L_{qs} \end{bmatrix} \frac{d}{dt} \begin{bmatrix} i_{ds} \\ i_{qs} \end{bmatrix} + \begin{bmatrix} 0 \\ \omega_r \psi_r \end{bmatrix}. \tag{6.13}$$

In Equation (6.13), the stator flux-linkage terms are completely eliminated from the stator voltage dynamic equations. This process leads to a simplified PMSM model in the *dq* frame (refer to Figure 6.9).

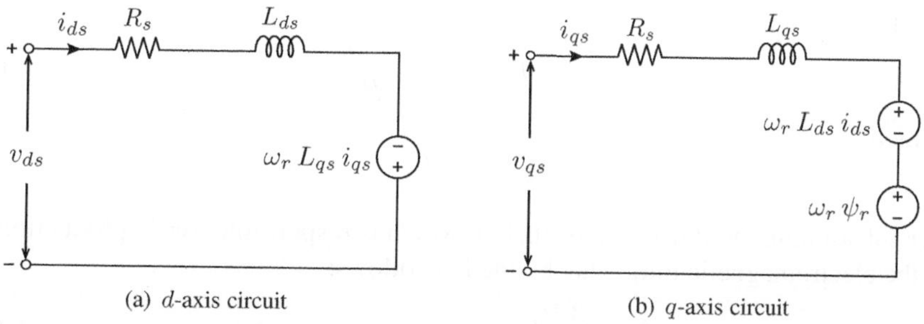

Figure 6.9 Representation of simplified *dq*-axis steady-state circuits for a PMSM (stator current dynamics are ignored).

From the above model, the *dq*-frame stator current dynamics are expressed as follows:

$$\frac{d}{dt} \begin{bmatrix} i_{ds} \\ i_{qs} \end{bmatrix} = \begin{bmatrix} -\frac{R_s}{L_{ds}} & \frac{\omega_r L_{qs}}{L_{ds}} \\ -\frac{\omega_r L_{ds}}{L_{qs}} & -\frac{R_s}{L_{qs}} \end{bmatrix} \begin{bmatrix} i_{ds} \\ i_{qs} \end{bmatrix} + \begin{bmatrix} \frac{1}{L_{ds}} & 0 \\ 0 & \frac{1}{L_{qs}} \end{bmatrix} \begin{bmatrix} v_{ds} \\ v_{qs} \end{bmatrix} + \begin{bmatrix} 0 \\ -\frac{\omega_r \psi_r}{L_{qs}} \end{bmatrix}. \tag{6.14}$$

The above stator current dynamics form the state-space CT model, which can be directly used in the MPC scheme.

6.5.5 Stator Active and Reactive Power

The relationship of the stator apparent power to the dq stator voltages and currents are expressed as follows:

$$S_s = P_s + j\,Q_s = \frac{3}{2}\left(v_{ds} + j\,v_{qs}\right)\left(i_{ds} - j\,i_{qs}\right). \tag{6.15}$$

On the basis of the equation above, the stator active and reactive powers are computed as follows:

$$P_s = \mathfrak{Re}\left(S_s\right) = \frac{3}{2}\left(v_{ds}\,i_{ds} + v_{qs}\,i_{qs}\right) \tag{6.16}$$

$$Q_s = \mathfrak{Im}\left(S_s\right) = \frac{3}{2}\left(v_{qs}\,i_{ds} - v_{ds}\,i_{qs}\right) \tag{6.17}$$

where $\mathfrak{Re}$ and $\mathfrak{Im}$ correspond to the real and imaginary components, respectively.

After substituting the v_{ds} and v_{qs} values from (6.11) into (6.16), the stator active power is expressed as follows:

$$P_s = \frac{3}{2}R_s\left(i_{ds}^2 + i_{qs}^2\right) + \frac{3}{2}\left(i_{ds}\,\frac{d\,\psi_{ds}}{dt} + i_{qs}\,\frac{d\,\psi_{qs}}{dt}\right) + \frac{3}{2}\omega_r\left(\psi_{ds}\,i_{qs} - \psi_{qs}\,i_{ds}\right) \tag{6.18}$$

where the first term represents the Ohmic losses in the stator winding resistance, the second term corresponds to the stored power in the magnetic field, and the third term expresses the air gap power, which is responsible for the production of T_e.

Similarly, the stator reactive power is computed as follows:

$$Q_s = \frac{3}{2}\left(i_{ds}\,\frac{d\,\psi_{qs}}{dt} - i_{qs}\,\frac{d\,\psi_{ds}}{dt}\right) + \frac{3}{2}\omega_r\left(\psi_{ds}\,i_{ds} + \psi_{qs}\,i_{qs}\right). \tag{6.19}$$

6.5.6 Electromagnetic Torque and Rotor Speed

The relationship of the mechanical power of PMSM to mechanical torque and speed is expressed as follows:

$$P_m = T_m\,\omega_m = \frac{T_e\,\omega_r}{P_p} \tag{6.20}$$

from which

$$T_e = \frac{P_m\,P_p}{\omega_r}. \tag{6.21}$$

After substituting the third term of (6.18) (which is responsible for T_e production) into (6.21), the electromagnetic torque is obtained as follows:

$$T_e = \frac{3\,P_p}{2}\left(\psi_{ds}\,i_{qs} - \psi_{qs}\,i_{ds}\right). \tag{6.22}$$

By substituting the ψ_{ds} and ψ_{qs} values from (6.12) into (6.22), we have [18]:

$$T_e = \frac{3\,P_p}{2}\left[\psi_r\,i_{qs} + \left(L_{ds} - L_{qs}\right)i_{ds}\,i_{qs}\right]. \tag{6.23}$$

In SPMSM, $L_{ds} = L_{qs}$. Thus, only i_{qs} is responsible for T_e production. However in IPMSM, $L_{ds} \neq L_{qs}$. Both i_{ds} and i_{qs} contribute to T_e generation. After solving both (6.6) and (6.23), the rotor speed dynamics are expressed as follows [12]:

$$\frac{d\,\omega_r}{dt} = \frac{3\,P_p^2}{2\,J_m}\,\psi_r\,i_{qs} + \frac{3\,P_p^2}{2\,J_m}\left(L_{ds} - L_{qs}\right)i_{ds}\,i_{qs} - \frac{P_p}{J_m}\,T_m - \frac{B_m}{J_m}\omega_r. \tag{6.24}$$

EXAMPLE 6.1 Steady-State Analysis of IPMSM

A 750 kW, 690 V, 9.75 Hz low-speed IPMSM has the following parameters: P_p = 26, R_s = 6.52 mΩ, L_{ds} = 3.74 mH, L_{qs} = 5.88 mH, and ψ_r = 8.295 Wb (*peak*). Note that the IPMSM is popular for three-stage gearbox-based high-speed applications; however this example problem considers low-speed IPMSM for the sake of theoretical analysis and comparison purposes. The block diagram of PMSM is shown in Figure 6.9. When the machine is running at ω_r = 61.26 rad/s, the d- and q-axis currents, i_{ds} and i_{qs}, are noted as –212.7 A and –932.8 A, respectively. Under this operating condition, calculate the following steady-state values by ignoring the machine rotational and core losses: (a) stator flux-linkage values, (b) machine stator voltages, (c) stator active, reactive, and apparent power, (d) machine PF and efficiency, and (e) electromagnetic torque.

Solution:

(a) The stator flux linkages are

$$\psi_{ds} = L_{ds}\, i_{ds} + \psi_r = -3.74 \times 10^{-3} \times 212.7 + 8.295 = 7.5 \text{ Wb } (peak)$$
$$\psi_{qs} = L_{qs}\, i_{qs} = -5.88 \times 10^{-3} \times 932.8 = -5.49 \text{ Wb } (peak)$$
$$\Psi_s = \sqrt{\psi_{ds}^2 + \psi_{qs}^2}/\sqrt{2} = \sqrt{7.5^2 + (-5.49)^2}/\sqrt{2} = 6.57 \text{ Wb } (rms).$$

(b) The steady-state machine stator voltages are

$$v_{ds} = R_s\, i_{ds} - \omega_r\, \psi_{qs} = -6.52 \times 10^{-3} \times 212.7 + 61.26 \times 5.49 = 334.6 \text{ V } (peak)$$
$$v_{qs} = R_s\, i_{qs} + \omega_r\, \psi_{ds} = -6.52 \times 10^{-3} \times 932.8 + 61.26 \times 7.5 = 453.3 \text{ V } (peak)$$
$$V_s = \sqrt{v_{ds}^2 + v_{qs}^2}/\sqrt{2} = \sqrt{334.6^2 + 453.3^2}/\sqrt{2} = 398.4 \text{ V } (rms).$$

(c) The PMSM stator active, reactive and apparent powers are

$$\begin{aligned} P_s &= 1.5\,(v_{ds}\, i_{ds} + v_{qs}\, i_{qs}) \\ &= 1.5 \times (-334.6 \times 212.7 - 453.3 \times 932.8) = -741.1 \times 10^3 \text{ W} \\ Q_s &= 1.5\,(v_{qs}\, i_{ds} - v_{ds}\, i_{qs}) \\ &= 1.5 \times (-453.3 \times 212.7 + 334.6 \times 932.8) = 323.5 \times 10^3 \text{ VAR} \end{aligned}$$
$$S_s = \sqrt{P_s^2 + Q_s^2} = \sqrt{(-741.1 \times 10^3)^2 + (323.5 \times 10^3)^2} = 808.6 \times 10^3 \text{ VA}.$$

(d) The PMSM PF and efficiency are

$$\cos\varphi_s = \frac{P_s}{S_s} = \frac{-741.1}{808.6} = -0.9165, \quad \eta = \left|\frac{P_s}{P_m}\right| \times 100 = \frac{741.1}{750} \times 100 = 98.81\%.$$

(e) The electromagnetic torque is

$$\begin{aligned} T_e &= \frac{3\,P_p}{2}\,(\psi_{ds}\, i_{qs} - \psi_{qs}\, i_{ds}) \\ &= \frac{3 \times 26}{2}\,(-7.5 \times 932.8 + 5.88 \times 212.7) = -318.32 \times 10^3 \text{ N.m.} \end{aligned}$$

Alternatively

$$T_e = \frac{P_m\, P_p}{\omega_r} = \frac{-750 \times 10^3 \times 26}{61.26} = -318.32 \times 10^3 \text{ N.m.}$$

The negative sign for the stator dq-axis currents, stator active power, stator PF, and electromagnetic torque indicate that the PMSM is operating in generator mode. All of the above values match the IPMSG #1 rated values provided in Appendix A and Table A.3. ♦

EXAMPLE 6.2 Steady-State Analysis of SPMSM

A low-speed SPMSM operating in a generator mode with a DD WT is used in this example. The machine parameters are the same as those provided in Example 6.1, except that $L_{ds} = L_{qs} = 3.85$ mH and $\psi_r = 8.532$ Wb (*peak*). The following steady-state values for 0.8-pu rotor speed are calculated: (a) mechanical input power and electromagnetic torque, (b) stator currents, (c) stator output voltage, (d) stator active and apparent power, and (e) generator PF.

Solution:

(a) The mechanical input power and electromagnetic torque are

$$\omega_r = \omega_{r,R} \times \omega_{r,\,\text{pu}} = 61.26 \times 0.8 = 49.008 \text{ rad/s} \quad (0.8 \text{ pu})$$

$$P_m = P_{m,R} \times \omega_{r,\,\text{pu}}^3 = -750 \times 10^3 \times 0.8^3 = -384 \times 10^3 \text{ W} \quad (0.512 \text{ pu})$$

$$T_e = \frac{P_m\,P_p}{\omega_r} = \frac{-384 \times 10^3 \times 26}{49.008} = -203.72 \times 10^3 \text{ N.m} \quad (-0.64 \text{ pu}).$$

(b) The stator currents are expressed as

$$T_e = \frac{3\,P_p}{2}\left[\psi_r\,i_{qs} + (L_{ds} - L_{qs})\,i_{ds}\,i_{qs}\right] \quad \Rightarrow \quad i_{qs} = \frac{2\,T_e}{3\,P_p\,\psi_r}$$

$$i_{qs} = \frac{-2 \times 203.32 \times 10^3}{3 \times 26 \times 8.532} = -612.24 \text{ A } (peak).$$

The above calculation implies that the electromagnetic torque is solely produced by the q-axis current i_{qs}. Thus, the d-axis current in a SPMSM is zero, that is,

$$i_{ds} = 0 \quad \Rightarrow \quad I_s = \sqrt{i_{ds}^2 + i_{qs}^2}/\sqrt{2} = 433 \text{ A } (rms) \quad (0.64 \text{ pu}).$$

(c) By ignoring the derivative terms in (6.13), the steady-state machine stator voltages are calculated as follows:

$$\begin{aligned} v_{ds} &= R_s\,i_{ds} - \omega_r\,L_{qs}\,i_{qs} \\ &= 0 + 49.008 \times 3.85 \times 10^{-3} \times 612.24 = 115.5 \text{ V } (peak) \\ v_{qs} &= R_s\,i_{qs} + \omega_r\,L_{ds}\,i_{ds} + \omega_r\,\psi_r \\ &= -6.52 \times 10^{-3} \times 612.24 + 0 + 49.008 \times 8.532 = 414 \text{ V } (peak) \\ V_s &= \sqrt{v_{ds}^2 + v_{qs}^2}/\sqrt{2} = \sqrt{115.5^2 + 414^2}/\sqrt{2} = 304 \text{ V } (rms). \end{aligned}$$

(d) In the generator mode, the stator active power P_s is less than the mechanical input power P_m because of the stator winding copper losses. By using this property, the PMSM stator active is computed, as expressed in the following. The calculation of the apparent power of the generator is also provided.

$$\begin{aligned} P_{cu,s} &= 3\,I_s^2\,R_s = 3 \times 433^2 \times 6.52 \times 10^{-3} = 3.667 \times 10^3 \text{ W} \\ P_s &= P_m + P_{cu,s} = -384 \times 10^3 + 3.667 \times 10^3 = -380.33 \times 10^3 \text{ W} \\ S_s &= 3\,V_s\,I_s = 3 \times 304 \times 433 = 394.9 \times 10^3 \text{ VA}. \end{aligned}$$

(e) The generator PF is

$$\cos\varphi_s = \frac{P_s}{S_s} = \frac{-380.33}{394.9} = -0.9632.$$

The above example summarizes the variable-speed operation of SPMSM. In theory, the wind generators can operate from zero to a rated speed. However, in WECS, the minimum speed corresponds to the cut-in wind speed, which is 0.3 to 0.4 pu of the rated shaft speed. ♦

6.6 SIMULATION OF PERMANENT MAGNET SYNCHRONOUS GENERATOR

The dynamic simulation model of PMSM is shown in Figure 6.10. The model is made up of six subsystems. The input signals to the simulation model are the dq-axis voltages and mechanical torque: v_{ds}, v_{qs}, and T_m. The output signals of the model are i_{ds}, i_{qs}, ψ_{ds}, ψ_{qs}, ω_r, and T_e. The v_{ds}, v_{qs}, and feedback rotor speed ω_r are used in subsystems #1 and #2 to calculate the dq-axis currents. Subsystems #1 and #2 are developed on the basis of (6.14). These subsystems indicate that the i_{ds} and i_{qs} are coupled to each other. Subsystems #3 and #4 use i_{ds} and i_{qs} to calculate ψ_{ds} and ψ_{qs} on the basis of (6.12). Subsystem #5 uses the i_{ds}, i_{qs}, ψ_{ds}, and ψ_{qs} from the previous subsystems to calculate the torque T_e on the basis of (6.22). Subsystem #6 calculates the rotor speed ω_r on the basis of (6.6) and uses T_e, T_m, and feedback rotor speed to estimate the ω_r.

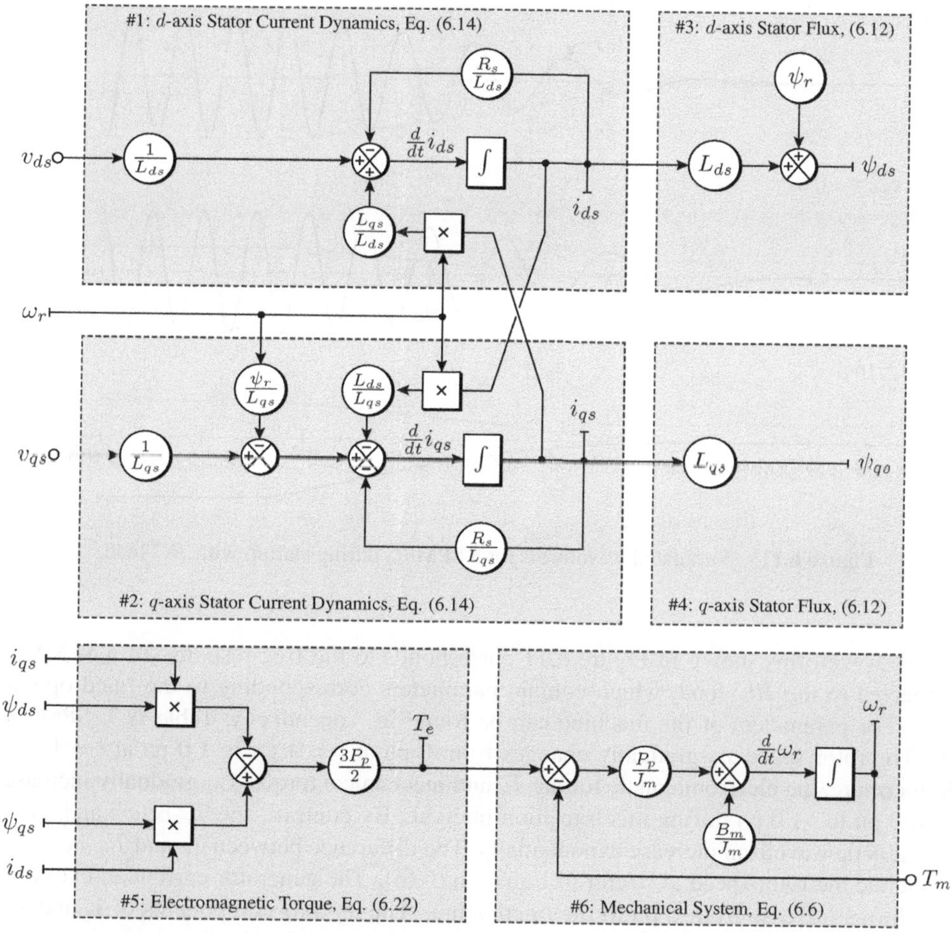

Figure 6.10 Dynamic model of the permanent magnet synchronous machine.

The simulation model shown in Figure 6.10 is different from the model available in the `MATLAB/Simulink` library. However, the performance of the simulation model is similar to that of the `MATLAB/Simulink` model. The performance of a SPMSM during system startup with resistive-capacitive (RC) load is shown in Figure 6.11.

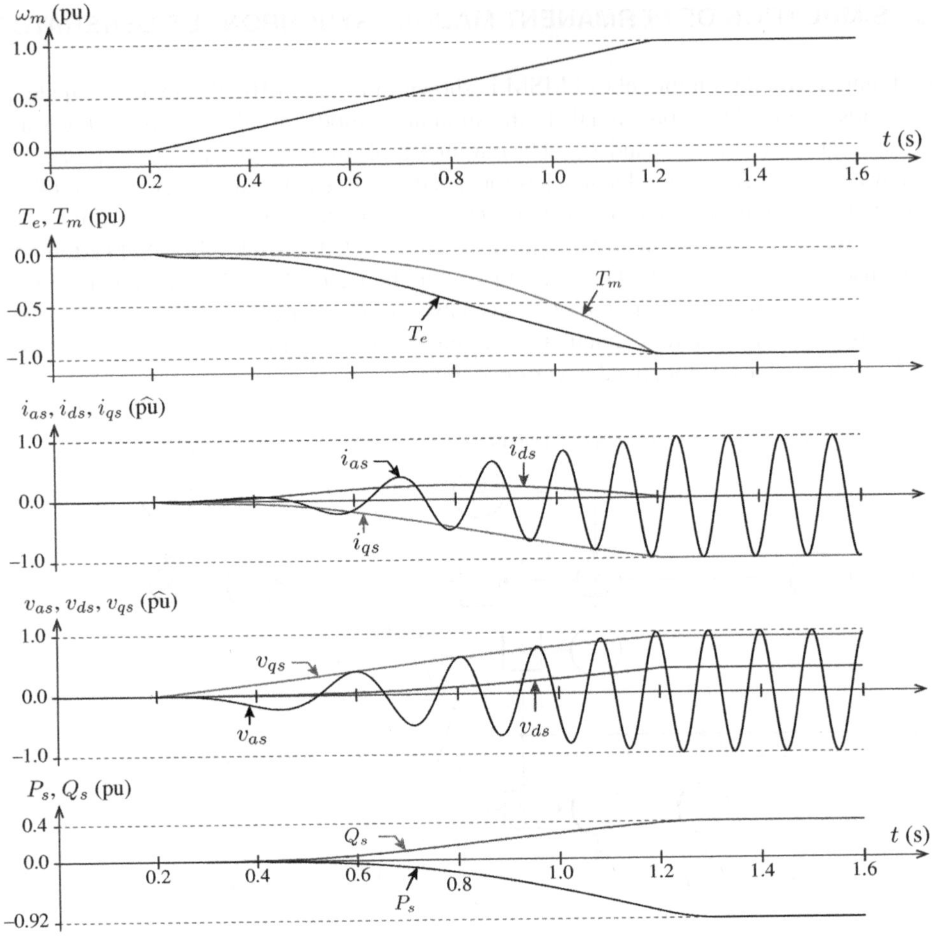

Figure 6.11 Simulated waveforms for a SPMSG during startup with RC load.

The waveforms shown in Figure 6.11 corresponds to the free response of a SPMSM connected to the RC load, which contain parameters corresponding to the rated operation. The parameters of the machine can be found in Appendix A, Table A.3, SPMSG #1. The rotor speed ω_r gradually changes from 0 pu at t = 0.2 s to 1.0 pu at t = 1.2 s. Furthermore, the electromagnetic torque T_e and mechanical torque T_m gradually increase from 0 pu to −1.0 pu during this transition interval. By contrast, the $T_e \propto \omega_r^2$ and electromagnetic waveform increase exponentially. The difference between T_e and T_m helps to accelerate the rotor speed ω_r (refer to Equation (6.6)). The generator current and voltage waveforms increase from 0 pu to the rated value. The dq-axis components of $\mathbf{i_s}$ and $\mathbf{v_s}$ are also shown in Figure 6.11. During the transient interval, the d-axis current i_{ds} is not zero. In the variable-speed WECS, the power converters provide switching actions such that i_{ds} is maintained at a zero value (for SPMSM only) during all operating conditions, including the transient interval. The v_{as} waveform is sinusoidal because of the RC load. However, in variable-speed WECS, the power converters impose pulse-width modulated voltage waveforms across the generator terminals. The active and reactive power reach rated values because of the rated load connected to the generator terminals.

6.7 MODELING OF INDUCTION GENERATOR

In this section, the modeling of an IM is provided in arbitrary, synchronous, and stationary reference frames. The IM models are applicable in both the doubly fed induction machine and squirrel-cage induction machine (SCIM).

6.7.1 Space Vector Model

The induction machine possesses complex dynamic properties. To model and analyze such a machine, space vectors as complex state variables should be used because of their convenience and efficiency. With the help of space phasors, an induction machine can be treated as a third-order dynamic system [19]. To simplify the analysis, the iron losses, slotting effects, and deep bar effects are ignored. Similar to the modeling of PMSM, the phase-a machine winding is aligned with the real axis of the complex plane. The star point of the machine is assumed to be isolated and not connected to the power converter. All the machine physical parameters are normalized. The rotor quantities are referred to the stator with a suitable scaling factor (stator to rotor winding ratio).

A generalized stator space vector is the spatial summation of the three-phase stator voltages, stator currents, or flux-linkages, that is,

$$\mathbf{f_s} = \frac{2}{3}\left(f_{as} + \mathfrak{a}\, f_{bs} + \mathfrak{a}^2\, f_{cs}\right), \quad f \in \{v,\, i,\, \psi\}, \quad \mathbf{f} \in \{\mathbf{v},\, \mathbf{i},\, \boldsymbol{\psi}\} \tag{6.25}$$

where "$\mathfrak{a}$" is a Fortescue operator and its value is given as $e^{j2\pi/3}$. Although f_{as} is shown as a complex variable, it is considered a real value because the phase-a winding is aligned with the real axis of the reference frame. A scaling factor of $2/3$ is used to ensure that variables f_{as}, f_{bs}, and f_{cs} can be constructed from its corresponding space vector $\mathbf{f_s}$.

Similarly, the space vector for the rotor variables is defined as follows:

$$\mathbf{f_r} = \frac{2}{3}\left(f_{ar} + \mathfrak{a}\, f_{br} + \mathfrak{a}^2\, f_{cr}\right), \quad f \in \{v,\, i,\, \psi\}, \quad \mathbf{f} \in \{\mathbf{v},\, \mathbf{i},\, \boldsymbol{\psi}\}. \tag{6.26}$$

To establish the space vector model, the speed of the rotating reference frame is specified as ω_k, which is an arbitrary value. By changing this speed value, the machine models are transformed to different reference frames. The dynamic equivalent circuit of an induction machine in the arbitrary reference frame is shown in Figure 6.12. To simplify the analysis, the equivalent circuit is divided into three parts, as seen in Figure 6.13. The Figures 6.13(a), 6.13(b), and 6.13(c) represent the stator voltage, magnetizing current, and rotor voltage circuits, respectively.

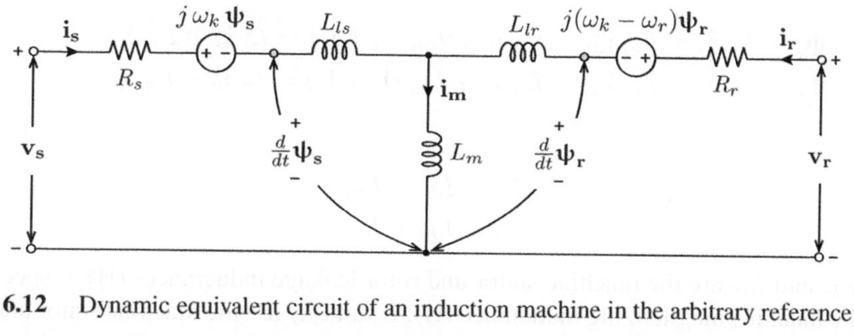

Figure 6.12 Dynamic equivalent circuit of an induction machine in the arbitrary reference frame.

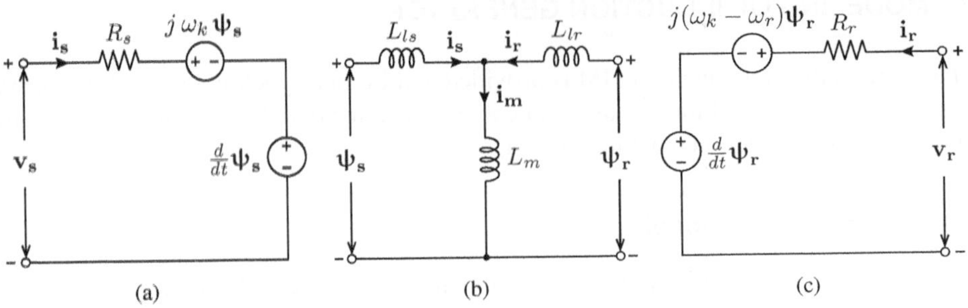

Figure 6.13 Simplified models of an induction machine: (a) stator voltage circuit, (b) magnetizing current circuit, and (c) rotor voltage circuit.

By referring to Figures 6.13(a) and 6.13(c), the stator and rotor voltage space vectors are formulated as follows:

$$\mathbf{v_s} = R_s\,\mathbf{i_s} + \frac{d}{dt}\boldsymbol{\psi}_s + j\,\omega_k\boldsymbol{\psi}_s \tag{6.27}$$

$$\mathbf{v_r} = R_r\,\mathbf{i_r} + \frac{d}{dt}\boldsymbol{\psi}_r + j\,(\omega_k - \omega_r)\,\boldsymbol{\psi}_r \tag{6.28}$$

where

- $\mathbf{v_s}$ and $\mathbf{v_r}$ are the machine stator and rotor voltage vectors (V).
- $\mathbf{i_s}$ and $\mathbf{i_r}$ are the machine stator and rotor current vectors (A).
- $\boldsymbol{\psi}_s$ and $\boldsymbol{\psi}_r$ are the machine stator and rotor flux-linkage vectors (Wb).
- R_s and R_r are the machine stator and rotor winding resistances (Ω).

In (6.27) and (6.28), $R_s\,\mathbf{i_s}$ and $R_r\,\mathbf{i_r}$ represent the stator and rotor voltage drops, respectively. The last terms are introduced by the transformation of the reference frame, which correspond to the induced speed voltages. As mentioned earlier, the induced speed voltages are common to all three-phase systems. For instance, the PMSM model given in (6.11) exhibits a similar phenomenon.

By applying KCL to the Figure 6.13(b), it is observed that the magnetizing current is summation of stator and rotor currents, i.e.,

$$\mathbf{i_m} = \mathbf{i_s} + \mathbf{i_r}. \tag{6.29}$$

Similarly, by applying KVL to the inner loops of Figure 6.13(b), the stator and rotor flux linkages are expressed as follows [20]:

$$\boldsymbol{\psi}_s = L_{ls}\,\mathbf{i_s} + L_m\,\mathbf{i_m} = L_{ls}\,\mathbf{i_s} + L_m\,(\mathbf{i_s}+\mathbf{i_r}) = L_s\,\mathbf{i_s} + L_m\,\mathbf{i_r} \tag{6.30}$$

$$\boldsymbol{\psi}_r = L_{lr}\,\mathbf{i_r} + L_m\,\mathbf{i_m} = L_{lr}\,\mathbf{i_r} + L_m\,(\mathbf{i_s}+\mathbf{i_r}) = L_r\,\mathbf{i_r} + L_m\,\mathbf{i_s} \tag{6.31}$$

with

$$L_s = L_{ls} + L_m \tag{6.32}$$

$$L_r = L_{lr} + L_m \tag{6.33}$$

where L_{ls} and L_{lr} are the machine stator and rotor leakage inductances (H), respectively; L_m is the machine magnetizing inductance (H); L_s and L_r are the machine stator and rotor self-inductances (H), respectively.

6.7.2 Modeling in Arbitrary Reference Frame

To implement various types of predictive control strategies, the dynamics of stator current and stator flux must be obtained. As such, the intended IM model should be in the following state-space format:

$$\begin{aligned}\frac{d}{dt}\left[\mathbf{i_s},\ \boldsymbol{\psi}_\mathbf{r}\right] &= \mathbb{F}\{\ [\mathbf{A}],\ [\mathbf{i_s},\ \boldsymbol{\psi}_\mathbf{r}],\ [\mathbf{B}],\ [\mathbf{v_s},\ \mathbf{v_r}]\} \\ [\mathbf{i_s},\ \boldsymbol{\psi}_\mathbf{s}] &= \mathbb{F}\{\ [\mathbf{C}],\ [\mathbf{i_s},\ \boldsymbol{\psi}_\mathbf{r}]\}.\end{aligned} \tag{6.34}$$

Thus, the objective of this section is to find the parameter matrices **A**, **B**, and **C** for IM modeling. The IM rotor voltages in (6.28), stator flux-linkages in (6.30), and rotor flux-linkages in (6.31) require the rotor current information. However, the measurement of rotor current is not easily obtained, particularly in the case of SCIM. To overcome this, the rotor currents are expressed in terms of stator currents with the help of Equation (6.31):

$$\mathbf{i_r} = \frac{\boldsymbol{\psi}_\mathbf{r} - L_m\,\mathbf{i_s}}{L_r}. \tag{6.35}$$

By substituting (6.35) into (6.30), the stator flux-linkages are expressed as follows:

$$\boldsymbol{\psi}_\mathbf{s} = \sigma\, L_s\, \mathbf{i_s} + k_r\, \boldsymbol{\psi}_\mathbf{r} \tag{6.36}$$

where k_r is the rotor coupling coefficient, $\sigma\, L_s$ is the total leakage inductance, σ is the total leakage coefficient, and k_s is the stator coupling coefficient. These parameters are defined as follows [21]:

$$k_r = \frac{L_m}{L_r},\ k_s = \frac{L_m}{L_s},\ \sigma = 1 - (k_s\, k_r) = 1 - \frac{L_m^2}{L_s\, L_r}. \tag{6.37}$$

Similarly, by substituting (6.35) into (6.28), the rotor flux dynamics are obtained as

$$\frac{d}{dt}\boldsymbol{\psi}_\mathbf{r} = \frac{L_m}{\tau_r}\,\mathbf{i_s} - \left[\frac{1}{\tau_r} + j\,(\omega_k - \omega_r)\right]\boldsymbol{\psi}_\mathbf{r} + \mathbf{v_r} \tag{6.38}$$

where τ_r is the rotor time constant:

$$\tau_r = \frac{L_r}{R_r}. \tag{6.39}$$

The next step is to find the stator current dynamics. By substituting (6.36) into (6.27) and by simplifying the formula, the stator current dynamics are expressed as follows:

$$\frac{d}{dt}\mathbf{i_s} = \frac{1}{\sigma\, L_s}\left[-(R_s + j\,\omega_k\,\sigma\, L_s)\,\mathbf{i_s} - j\,\omega_k\, k_r\,\boldsymbol{\psi}_\mathbf{r} - k_r\,\frac{d}{dt}\boldsymbol{\psi}_\mathbf{r} + \mathbf{v_s}\right]. \tag{6.40}$$

In (6.40), $d\,\boldsymbol{\psi}_\mathbf{r}/dt$ is an unwanted term and must be replaced by its equivalent to achieve the state-space model. By substituting (6.38) into (6.40), we obtain the following [19]:

$$\frac{d}{dt}\mathbf{i_s} = -\left(\frac{1}{\tau_\sigma} + j\,\omega_k\right)\mathbf{i_s} + \frac{k_r}{\sigma\, L_s}\left(\frac{1}{\tau_r} - j\,\omega_r\right)\boldsymbol{\psi}_\mathbf{r} + \frac{1}{\sigma\, L_s}\mathbf{v_s} - \frac{k_r}{\sigma\, L_s}\mathbf{v_r} \tag{6.41}$$

where r_σ is equivalent resistance, and τ_σ is the stator transient time constant; both of these variables are defined as follows:

$$r_\sigma = R_s + k_r^2\, R_r,\quad \tau_\sigma = \frac{\sigma\, L_s}{r_\sigma}. \tag{6.42}$$

By combining (6.36), (6.38), and (6.41), the induction machine CT state-space model in an arbitrary reference frame is obtained by

$$\frac{d}{dt}\begin{bmatrix}\mathbf{i_s}\\ \boldsymbol{\psi}_\mathbf{r}\end{bmatrix} = \mathbf{A}\begin{bmatrix}\mathbf{i_s}\\ \boldsymbol{\psi}_\mathbf{r}\end{bmatrix} + \mathbf{B}\begin{bmatrix}\mathbf{v_s}\\ \mathbf{v_r}\end{bmatrix} \quad \text{and} \quad \boldsymbol{\psi}_\mathbf{s} = \mathbf{C}\begin{bmatrix}\mathbf{i_s}\\ \boldsymbol{\psi}_\mathbf{r}\end{bmatrix} \tag{6.43}$$

and

$$\mathbf{A} = \begin{bmatrix} -\left(\frac{1}{\tau_\sigma} + j\,\omega_k\right) & \frac{k_r}{\sigma L_s}\left(\frac{1}{\tau_r} - j\,\omega_r\right) \\ \frac{L_m}{\tau_r} & -\left(\frac{1}{\tau_r} + j\,(\omega_k - \omega_r)\right) \end{bmatrix}$$
$$\mathbf{B} = \begin{bmatrix} \frac{1}{\sigma L_s} & -\frac{k_r}{\sigma L_s} \\ 0 & 1 \end{bmatrix}, \quad \mathbf{C} = \begin{bmatrix} \sigma L_s & k_r \end{bmatrix}. \tag{6.44}$$

The electromagnetic torque is expressed as follows:

$$T_e = \frac{3\,P_p}{2}\,\Re\mathfrak{e}\,(j\,\boldsymbol{\psi}_\mathbf{s}\,\mathbf{i}_\mathbf{s}^\star) = \frac{3\,P_p}{2}\,\Re\mathfrak{e}\,(j\,\boldsymbol{\psi}_\mathbf{r}\,\mathbf{i}_\mathbf{r}^\star) \tag{6.45}$$

where $\star$ represents the complex conjugate operation.

The stator active and reactive powers are calculated as follows:

$$P_s = \frac{3}{2}\,\Re\mathfrak{e}\,(\mathbf{v_s}\,\mathbf{i}_\mathbf{s}^\star) = \frac{3}{2}\,\Re\mathfrak{e}\,(\mathbf{v_r}\,\mathbf{i}_\mathbf{r}^\star), \quad Q_s = \frac{3}{2}\,\Im\mathfrak{m}\,(\mathbf{v_s}\,\mathbf{i}_\mathbf{s}^\star) = \frac{3}{2}\,\Im\mathfrak{m}\,(\mathbf{v_r}\,\mathbf{i}_\mathbf{r}^\star). \tag{6.46}$$

6.7.3 Modeling in Synchronous Reference Frame

The synchronous (*dq*) reference frame rotates at an angular speed of ω_s. By substituting $\omega_k = \omega_s$, in the arbitrary frame induction machine model, the *dq* model is obtained. The *dq* reference frame equivalent circuit of an induction machine is shown in Figure 6.14. All space vectors are converted to scalar variables via the *d*- and *q*-axis circuits in Figures 6.14(a) and 6.14(b), respectively [22].

The *dq*-frame space phasor is defined as follows:

$$\mathbf{f} = (f_d + j\,f_q), \quad f \in \{v,\, i,\, \psi\}. \tag{6.47}$$

By substituting the above expression into (6.41), we obtain the following:

$$\frac{d}{dt}(i_{ds} + j\,i_{qs}) = -\left(\frac{1}{\tau_\sigma} + j\,\omega_s\right)(i_{ds} + j\,i_{qs}) + \frac{k_r}{\sigma L_s}\left(\frac{1}{\tau_r} - j\,\omega_r\right)$$
$$(\psi_{dr} + j\,\psi_{qr}) + \frac{1}{\sigma L_s}(v_{ds} + j\,v_{qs}) - \frac{k_r}{\sigma L_s}(v_{dr} + j\,v_{qr}). \tag{6.48}$$

By separating real and imaginary components from the above expression, we have [23]:

$$\frac{d\,i_{ds}}{dt} = -\frac{1}{\tau_\sigma}\,i_{ds} + \omega_s\,i_{qs} + \frac{k_r}{\sigma L_s \tau_r}\,\psi_{dr} + \frac{k_r\,\omega_r}{\sigma L_s}\,\psi_{qr} + \frac{1}{\sigma L_s}\,v_{ds} - \frac{k_r}{\sigma L_s}\,v_{dr} \tag{6.49}$$

$$\frac{d\,i_{qs}}{dt} = -\omega_s\,i_{ds} - \frac{1}{\tau_\sigma}\,i_{qs} - \frac{k_r\,\omega_r}{\sigma L_s}\,\psi_{dr} + \frac{k_r}{\sigma L_s \tau_r}\,\psi_{qr} + \frac{1}{\sigma L_s}\,v_{qs} - \frac{k_r}{\sigma L_s}\,v_{qr}. \tag{6.50}$$

Similarly, the *dq*-frame rotor flux dynamics are obtained below by solving (6.38):

$$\frac{d\,\psi_{dr}}{dt} = \frac{L_m}{\tau_r}\,i_{ds} - \frac{1}{\tau_r}\,\psi_{dr} + (\omega_s - \omega_r)\,\psi_{qr} + v_{dr} \tag{6.51}$$

$$\frac{d\,\psi_{qr}}{dt} = \frac{L_m}{\tau_r}\,i_{qs} - (\omega_s - \omega_r)\,\psi_{dr} - \frac{1}{\tau_r}\,\psi_{qr} + v_{qr}. \tag{6.52}$$

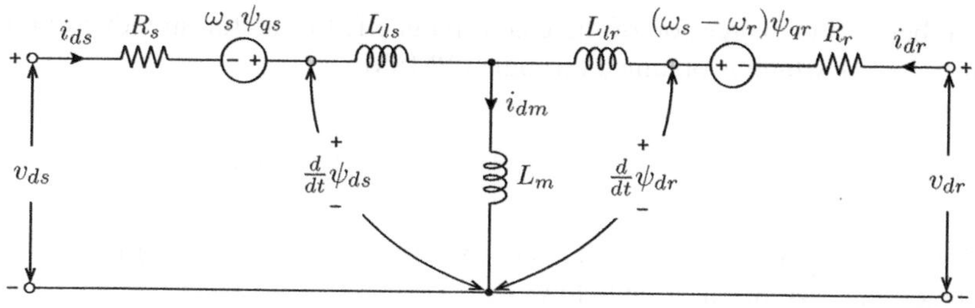

(a) d-axis circuit

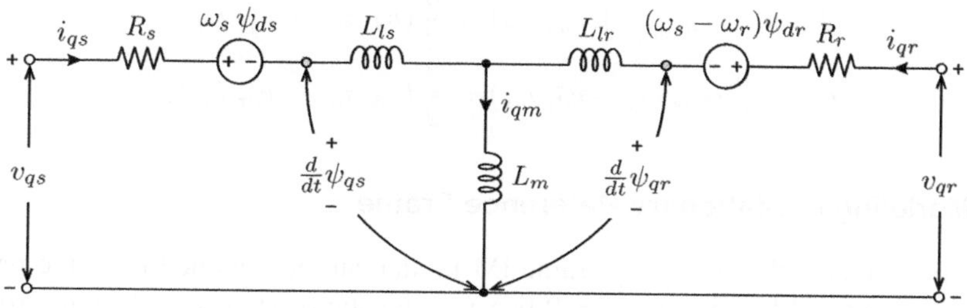

(b) q-axis circuit

Figure 6.14 Equivalent circuit of an induction machine in the dq frame [20].

Finally, dq-frame stator flux-linkages are derived from (6.36):

$$\psi_{ds} = \sigma\, L_s\, i_{ds} + k_r\, \psi_{dr} \tag{6.53}$$

$$\psi_{qs} = \sigma\, L_s\, i_{qs} + k_r\, \psi_{qr}. \tag{6.54}$$

By combining equations (6.49) through (6.54), the CT state-space model in the dq reference frame is expressed as follows [24]:

$$\frac{d}{dt}\begin{bmatrix} i_{ds} \\ i_{qs} \\ \psi_{dr} \\ \psi_{qr} \end{bmatrix} = \mathbf{A}\begin{bmatrix} i_{ds} \\ i_{qs} \\ \psi_{dr} \\ \psi_{qr} \end{bmatrix} + \mathbf{B}\begin{bmatrix} v_{ds} \\ v_{qs} \\ v_{dr} \\ v_{qr} \end{bmatrix} \quad \text{and} \quad \begin{bmatrix} \psi_{ds} \\ \psi_{qs} \end{bmatrix} = \mathbf{C}\begin{bmatrix} i_{ds} \\ i_{qs} \\ \psi_{dr} \\ \psi_{qr} \end{bmatrix} \tag{6.55}$$

and

$$\mathbf{A} = \begin{bmatrix} -\frac{1}{\tau_\sigma} & \omega_s & \frac{k_r}{\sigma\, L_s\, \tau_r} & \frac{k_r\, \omega_r}{\sigma\, L_s} \\ -\omega_s & -\frac{1}{\tau_\sigma} & -\frac{k_r\, \omega_r}{\sigma\, L_s} & \frac{k_r}{\sigma\, L_s\, \tau_r} \\ \frac{L_m}{\tau_r} & 0 & -\frac{1}{\tau_r} & (\omega_s - \omega_r) \\ 0 & \frac{L_m}{\tau_r} & -(\omega_s - \omega_r) & -\frac{1}{\tau_r} \end{bmatrix} \tag{6.56}$$

$$\mathbf{B} = \begin{bmatrix} \frac{1}{\sigma\, L_s} & 0 & -\frac{k_r}{\sigma\, L_s} & 0 \\ 0 & \frac{1}{\sigma\, L_s} & 0 & -\frac{k_r}{\sigma\, L_s} \\ 0 & 0 & 1 & 0 \\ 0 & 0 & 0 & 1 \end{bmatrix} \tag{6.57}$$

$$\mathbf{C} = \begin{bmatrix} \sigma\, L_s & 0 & k_r & 0 \\ 0 & \sigma\, L_s & 0 & k_r \end{bmatrix}. \tag{6.58}$$

As seen below, the expression of the electromagnetic torque for an induction machine is similar to (6.22), which is obtained through a PMSM.

$$T_e = \frac{3\,P_p}{2}\left(\psi_{ds}\, i_{qs} - \psi_{qs}\, i_{ds}\right) = \frac{3\,P_p}{2}\left(\psi_{dr}\, i_{qr} - \psi_{qr}\, i_{dr}\right). \tag{6.59}$$

The stator active and reactive powers are expressed in terms of a dq-frame stator and rotor voltages and currents, as seen below:

$$P_s = \frac{3}{2}\left(v_{ds}\, i_{ds} + v_{qs}\, i_{qs}\right) = \frac{3}{2}\left(v_{dr}\, i_{dr} + v_{qr}\, i_{qr}\right) \tag{6.60}$$

$$Q_s = \frac{3}{2}\left(v_{qs}\, i_{ds} - v_{ds}\, i_{qs}\right) = \frac{3}{2}\left(v_{qr}\, i_{dr} - v_{dr}\, i_{qr}\right). \tag{6.61}$$

6.7.4 Modeling in Stationary Reference Frame

As the name implies, the stationary frame IM model can be obtained from the previous arbitrary frame model by setting $\omega_k = 0$ because the stationary frame does not rotate in space. The equivalent circuit of an induction machine in $\alpha\beta$ reference frame is shown in Figure 6.15. Similar to the dq-frame model shown in Figure 6.14, all space vectors are decomposed into scalar variables through α- and β-axis circuits, as seen in Figures 6.15(a) and 6.15(b), respectively. From comparing Figures 6.14 and 6.15, the following is observed: (1) both circuits are identical, (2) dq variables are replaced by $\alpha\beta$, and (3) the reference frame speed $\omega_s = 0$ in $\alpha\beta$ circuits.

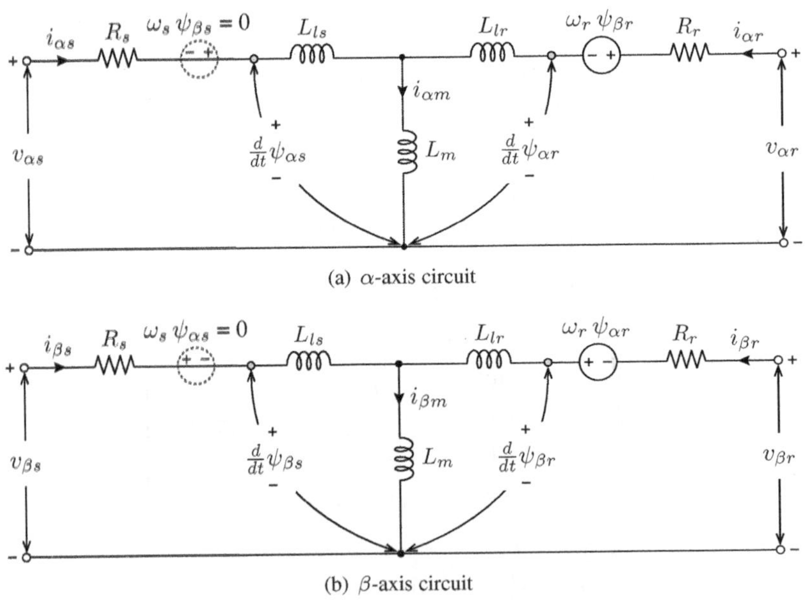

Figure 6.15 The equivalent circuit of an induction machine in the $\alpha\beta$ reference frame.

The stationary frame space phasor is defined as follows:

$$\mathbf{f} = \left(f_\alpha + j\, f_\beta\right), \quad f \in \{v,\, i,\, \psi\}. \tag{6.62}$$

By substituting (6.62) into (6.41) and setting $\omega_k = 0$, obtain the following:

$$\begin{aligned}\frac{d}{dt}(i_{\alpha s} + j\, i_{\beta s}) = -\frac{1}{\tau_\sigma}(i_{\alpha s} + j\, i_{\beta s}) + \frac{k_r}{\sigma L_s}\left(\frac{1}{\tau_r} - j\,\omega_r\right)(\psi_{\alpha r} + j\,\psi_{\beta r}) \\ + \frac{1}{\sigma L_s}(v_{\alpha s} + j\, v_{\beta s}) - \frac{k_r}{\sigma L_s}(v_{\alpha r} + j\, v_{\beta r})\end{aligned} \tag{6.63}$$

from which

$$\frac{d\, i_{\alpha s}}{dt} = -\frac{1}{\tau_\sigma} i_{\alpha s} + \frac{k_r}{\sigma L_s \tau_r}\psi_{\alpha r} + \frac{k_r\,\omega_r}{\sigma L_s}\psi_{\beta r} + \frac{1}{\sigma L_s} v_{\alpha s} - \frac{k_r}{\sigma L_s} v_{\alpha r} \tag{6.64}$$

$$\frac{d\, i_{\beta s}}{dt} = -\frac{1}{\tau_\sigma} i_{\beta s} - \frac{k_r\,\omega_r}{\sigma L_s}\psi_{\alpha r} + \frac{k_r}{\sigma L_s \tau_r}\psi_{\beta r} + \frac{1}{\sigma L_s} v_{\beta s} - \frac{k_r}{\sigma L_s} v_{\beta r}. \tag{6.65}$$

Similarly, the rotor flux dynamics in (6.38) are written in $\alpha\beta$ frame as follows:

$$\frac{d\,\psi_{\alpha r}}{dt} = \frac{L_m}{\tau_r} i_{\alpha s} - \frac{1}{\tau_r}\psi_{\alpha r} - \omega_r\,\psi_{\beta r} + v_{\alpha r} \tag{6.66}$$

$$\frac{d\,\psi_{\beta r}}{dt} = \frac{L_m}{\tau_r} i_{\beta s} + \omega_r\,\psi_{\alpha r} - \frac{1}{\tau_r}\psi_{\beta r} + v_{\beta r}. \tag{6.67}$$

Finally, $\alpha\beta$ frame stator flux-linkages are derived from (6.36) as follows:

$$\psi_{\alpha s} = \sigma L_s\, i_{\alpha s} + k_r\,\psi_{\alpha r} \tag{6.68}$$

$$\psi_{\beta s} = \sigma L_s\, i_{\beta s} + k_r\,\psi_{\beta r}. \tag{6.69}$$

By combining (6.64) through (6.69), the CT state-space model in a stationary reference frame is expressed as follows [25]:

$$\frac{d}{dt}\begin{bmatrix} i_{\alpha s} \\ i_{\beta s} \\ \psi_{\alpha r} \\ \psi_{\beta r}\end{bmatrix} = \mathbf{A}\begin{bmatrix} i_{\alpha s} \\ i_{\beta s} \\ \psi_{\alpha r} \\ \psi_{\beta r}\end{bmatrix} + \mathbf{B}\begin{bmatrix} v_{\alpha s} \\ v_{\beta s} \\ v_{\alpha r} \\ v_{\beta r}\end{bmatrix} \quad \text{and} \quad \begin{bmatrix}\psi_{\alpha s} \\ \psi_{\beta s}\end{bmatrix} = \mathbf{C}\begin{bmatrix} i_{\alpha s} \\ i_{\beta s} \\ \psi_{\alpha r} \\ \psi_{\beta r}\end{bmatrix} \tag{6.70}$$

and

$$\mathbf{A} = \begin{bmatrix} -\frac{1}{\tau_\sigma} & 0 & \frac{k_r}{\sigma L_s \tau_r} & \frac{k_r\,\omega_r}{\sigma L_s} \\ 0 & -\frac{1}{\tau_\sigma} & -\frac{k_r\,\omega_r}{\sigma L_s} & \frac{k_r}{\sigma L_s \tau_r} \\ \frac{L_m}{\tau_r} & 0 & -\frac{1}{\tau_r} & -\omega_r \\ 0 & \frac{L_m}{\tau_r} & \omega_r & -\frac{1}{\tau_r}\end{bmatrix} \tag{6.71}$$

$$\mathbf{B} = \begin{bmatrix} \frac{1}{\sigma L_s} & 0 & -\frac{k_r}{\sigma L_s} & 0 \\ 0 & \frac{1}{\sigma L_s} & 0 & -\frac{k_r}{\sigma L_s} \\ 0 & 0 & 1 & 0 \\ 0 & 0 & 0 & 1\end{bmatrix} \tag{6.72}$$

$$\mathbf{C} = \begin{bmatrix} \sigma L_s & 0 & k_r & 0 \\ 0 & \sigma L_s & 0 & k_r\end{bmatrix}. \tag{6.73}$$

The electromagnetic torque is calculated similar to *dq* frame, as seen below [26]:

$$T_e = \frac{3\,P_p}{2}(\psi_{\alpha s}\, i_{\beta s} - \psi_{\beta s}\, i_{\alpha s}) = \frac{3\,P_p}{2}(\psi_{\alpha r}\, i_{\beta r} - \psi_{\beta r}\, i_{\alpha r}). \tag{6.74}$$

The stator active and reactive powers in a stationary frame are computed as follows:

$$P_s = \frac{3}{2}\left(v_{\alpha s}\, i_{\alpha s} + v_{\beta s}\, i_{\beta s}\right) = \frac{3}{2}\left(v_{\alpha r}\, i_{\alpha r} + v_{\beta r}\, i_{\beta r}\right) \tag{6.75}$$

$$Q_s = \frac{3}{2}\left(v_{\beta s}\, i_{\alpha s} - v_{\alpha s}\, i_{\beta s}\right) = \frac{3}{2}\left(v_{\beta r}\, i_{\alpha r} - v_{\alpha r}\, i_{\beta r}\right). \tag{6.76}$$

The induction machine models in the arbitrary, synchronous, and stationary reference frames are directly applicable for DFIG control. By setting the rotor voltages to zero (i.e., $v_{dr} = v_{qr} = v_{\alpha r} = v_{\beta r} = 0$), these models can be used for SCIG control.

✍ EXAMPLE 6.3 Steady-State Analysis of SCIG

A 3000-kW, 690-V, 60-Hz SCIG has the following rated parameters: $P_p = 2$, $R_s = 0.999$ mΩ, $R_r = 1.132$ mΩ, $L_{ls} = L_{lr} = 0.0431$ mH, $L_m = 1.4372$ mH, and $T_{m,R} = -15.79$ kN.m. The synchronous and rated speeds are $n_s = 1800$ rpm and $n_m = 1814$ rpm. The dq-frame IM model seen in Figure 6.14 is used in this example. During the steady state, the generator stator currents and rotor flux linkages are noted as $i_{ds} = -3496$ A, $i_{qs} = -1798$ A, $\psi_{dr} = 0.31$ (Wb), and $\psi_{qr} = -1.3921$ (Wb). After ignoring the rotational and core losses, the following are calculated: (a) stator flux-linkage values, (b) stator voltages, (c) stator active, reactive and apparent power, (d) electromagnetic torque and mechanical speed, and (e) slip of the generator.

Solution:

(a) The stator flux linkages are

$$\begin{aligned}
L_s &= L_{ls} + L_m = 1.4803 \text{ mH}, \quad L_r = L_{lr} + L_m = 1.4803 \text{ mH} \\
k_s &= L_m / L_s = 1.4372/1.4803 = 0.9709 \\
k_r &= L_m / L_r = 1.4372/1.4803 = 0.9709 \\
\sigma &= 1 - (k_s\, k_r) = 1 - (0.9709 \times 0.9709) = 0.0574 \\
\psi_{ds} &= \sigma\, L_s\, i_{ds} + k_r\, \psi_{dr} = 0.0038 \text{ Wb } (\textit{peak}) \\
\psi_{qs} &= \sigma\, L_s\, i_{qs} + k_r\, \psi_{qr} = -1.5043 \text{ Wb } (\textit{peak}) \\
\Psi_s &= \sqrt{\psi_{ds}^2 + \psi_{qs}^2} / \sqrt{2} = \sqrt{0.0038^2 + (-1.5043)^2} / \sqrt{2} = 1.0632 \text{ Wb } (\textit{rms}).
\end{aligned}$$

(b) The steady-state machine stator voltages are

$$\begin{aligned}
\omega_s &= 2\,\pi\, f_s = 2 \times \pi \times 60 = 377 \text{ rad} \\
v_{ds} &= R_s\, i_{ds} - \omega_s\, \psi_{qs} = -0.999 \times 10^{-3} \times 3496 + 377 \times 1.5043 = 563.6 \text{ V } (\textit{peak}) \\
v_{qs} &= R_s\, i_{qs} + \omega_s\, \psi_{ds} = -0.999 \times 10^{-3} \times 1798 + 377 \times 0.0038 = -0.36 \text{ V } (\textit{peak}) \\
V_s &= \sqrt{v_{ds}^2 + v_{qs}^2} / \sqrt{2} = \sqrt{563.6^2 + (-0.36)^2} / \sqrt{2} = 398.4 \text{ V } (\textit{rms}).
\end{aligned}$$

(c) The stator active, reactive, and apparent powers are

$$\begin{aligned}
P_s &= 1.5\, (v_{ds}\, i_{ds} + v_{qs}\, i_{qs}) \\
&= 1.5 \times (563.6 \times -3496 + 0.36 \times 1798) = -2954 \times 10^3 \text{ W} \\
Q_s &= 1.5\, (v_{qs}\, i_{ds} - v_{ds}\, i_{qs}) \\
&= 1.5 \times (0.36 \times 3496 + 563.6 \times 1798) = 1519 \times 10^3 \text{ VAR} \\
S_s &= \sqrt{P_s^2 + Q_s^2} = \sqrt{(-2954 \times 10^3)^2 + (1519 \times 10^3)^2} = 3321 \times 10^3 \text{ VA}.
\end{aligned}$$

(d) The electromagnetic torque is

$$\begin{aligned} T_e &= \frac{3\,P_p}{2}\,(\psi_{ds}\,i_{qs} - \psi_{qs}\,i_{ds}) \\ &= \frac{3\times 2}{2}\,(0.0038\times -1798 - 1.5043\times 3496) = -15.79\times 10^3\ \text{N.m.} \end{aligned}$$

In steady state, $T_m = T_e = -15.79$ kN.m. Therefore, the mechanical speed is calculated as follows:

$$\begin{aligned} \omega_{m,\text{pu}} &= \sqrt{T_m/T_{m,R}} = \sqrt{-15.79/-15.79} = 1.0 \\ \omega_{m,R} &= 2\,\pi\,n_r/60 = 2\,\pi\,1814/60 = 190\ \text{rad/s} \\ \omega_m &= \omega_{m,R}\times\omega_{m,\text{pu}} = 190\ \text{rad/s.} \end{aligned}$$

(e) The generator slip is

$$\begin{aligned} n_m &= n_{m,R}\times\omega_{m,\text{pu}} = 1814\ \text{rpm} \\ s &= \frac{n_s - n_m}{n_s} = \frac{1800-1814}{1800} = -0.0078. \end{aligned}$$

The negative sign for the slip indicates that the IM is operating in generator mode. All the above values match with the SCIG #2 rated values provided in Appendix A and Table A.7. ♦

6.8 SIMULATION OF INDUCTION GENERATOR

The dynamic simulation model of an induction machine in synchronous reference frame is shown in Figure 6.16. The block diagram seen in the figure can be used to implement the IM model for computer simulations in the `MATLAB/Simulink` environment. The IM model is readily available in the `Simulink` library. The machine model derived in the simulation provides the same transient and steady-state results as those of the standard IM models available in the `Simulink` library.

The simulation model of induction machine consists of eight subsystems. The input signals to the IM simulation model are v_{ds}, v_{qs}, v_{dr}, v_{qr}, ω_s, and T_m. The output signals of the IM model are i_{ds}, i_{qs}, ω_r, and T_e. All input and output signals are similar to the PMSM model shown in Figure 6.10, except that the additional input signals in the IM model are the dq-axis rotor voltages v_{dr} and v_{qr} and synchronous speed ω_s. By setting $\omega_s = 0$, the IM model works in the $\alpha\beta$-frame. Furthermore, by assigning $v_{dr} = v_{qr} = 0$, the IM model works as SCIM. The order of calculations (subsystems) and corresponding dynamic equations are shown in Figure 6.16. Subsystems #1 and #2 calculate the dq-axis stator currents on the basis of v_{ds}, v_{qs}, v_{dr}, v_{qr}, ω_s, and ω_r. Subsystems #3 and #4 calculate the dq-axis rotor flux-linkage values on the basis of i_{ds}, i_{qs}, v_{dr}, v_{qr}, ω_s, and ω_r. Subsystems #5 and #6 use i_{ds}, i_{qs}, ψ_{dr}, ψ_{qr} to compute for stator flux linkages ψ_{ds} and ψ_{qs}. The electromagnetic torque T_e is calculated by subsystem #7 on the basis of the stator currents (i_{ds} and i_{qs}) and stator flux linkages (ψ_{ds} and ψ_{qs}) from the previous subsystems. The rotor speed ω_r is estimated by subsystem #8 on the basis T_e and T_m. Subsystems #7 and #8 of IM and subsystems #5 and #6 of PMSM are essentially the same.

The free-running operation of SCIG with a direct grid connection (Type 1 WT) is shown in Figure 6.17, with the SCIG #2 parameters provided in Appendix A, Table A.7. During the starting procedure, the SCIG speed is raised to a certain value (e.g., 1750 rpm), after which the circuit breaker is closed. Thereafter, the SCIG terminals are directly connected to the three-phase grid (690 V, 60 Hz). The generator output voltages are equal to the grid

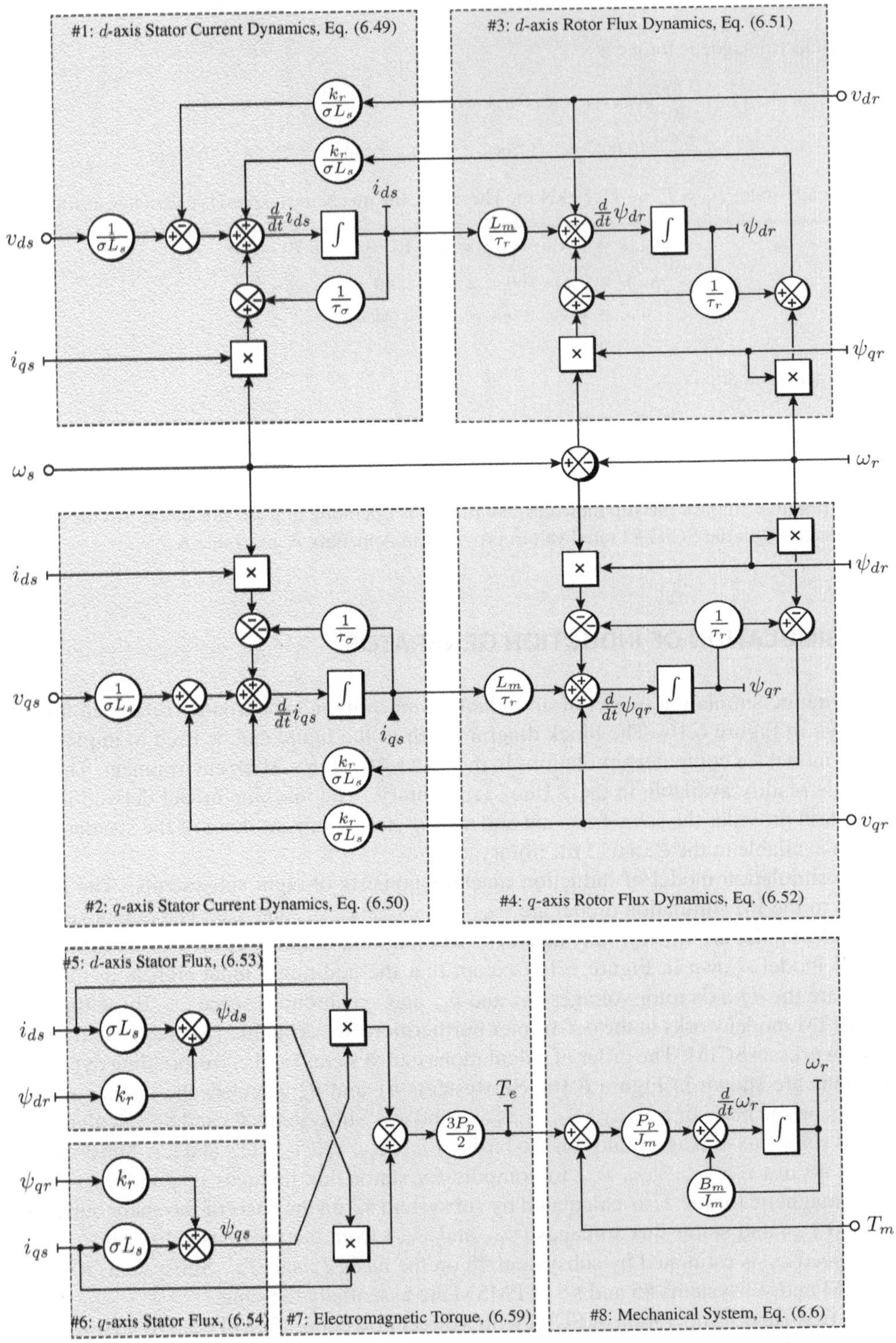

Figure 6.16 Dynamic model of an induction machine in the synchronous reference frame.

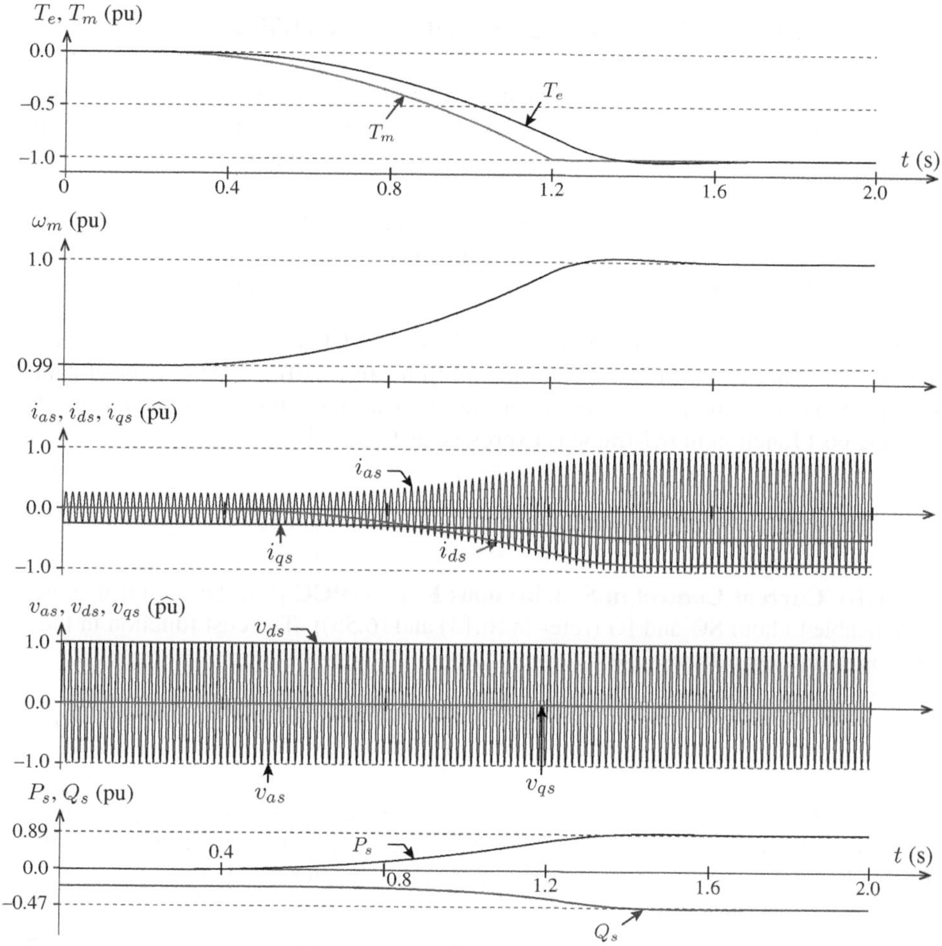

Figure 6.17 Simulated waveforms for a SCIG with direct grid connection.

voltage because the generator is directly connected to the grid. Owing to the startup, a high inrush current flows, thus leading to a large ripple in the electromagnetic torque (not shown in Figure 6.17). To avoid damage to the SCIG, a soft-starter is employed. After the transient period, the SCIG reaches steady-state with the rotor speed settling at synchronous speed (1800 rpm, 0.99 pu) and T_e and T_m approximately equal to zero. The SCIG operates in motoring mode, and the machine produces no active power output ($P_s = 0$). The SCIG draws some reactive power from the grid on the basis of the magnetizing inductance L_m value. The SCIG stator currents correspond to the reactive power consumed by the SCIG.

To produce active power from the SCIG, the mechanical input torque T_m is gradually increased from 0 pu to −1.0 pu from $t = 0.2$ s to 1.2 s. The difference between T_m and T_e helps the SCIG to accelerate from the synchronous speed to rated speed (1814 rpm, 1.0 pu). The absolute value of T_m is greater than the absolute value of T_e because of the generator operation mode ($|T_m| > |T_e|$). The active power output increases from 0 to 2.954 MW (0.89 pu) from $t = 0.2$ s to 1.2 s. In a steady state, the reactive power reaches its rated value of −1.519 MVAR (−0.47 pu). The free-running response obtained is similar to the response produced by the `Simulink` library IM model.

6.9 GENERATOR DYNAMIC MODELS FOR PREDICTIVE CONTROL

The dynamic models developed in this chapter for PMSM and IM can be used to realize various predictive control schemes. The cost functions employed in the different control approaches are given below. In theory, current control can be implemented in the *abc* frame by using the generator models. The predicted currents are compared with the reference currents to identify the power converter switching state that provides the overall minimal error among the three-phase currents. To reduce the number of online calculations in predictive control, the two-axis current control in the $\alpha\beta$ or *dq* frame is recommended.

- **Predictive Current Control in Stationary Frame (PCC-$\alpha\beta$):** This control strategy is applicable only to the $\alpha\beta$-frame IG model (refer to (6.70)). PMSG control cannot be implemented in this frame because of the time-varying nature of inductances. The resulting cost function in $\alpha\beta$-frame is expressed as follows [27]:

$$g(k) = \left[\widehat{i}^*_{\alpha s}(k+1) - i^p_{\alpha s}(k+1)\right]^2 + \left[\widehat{i}^*_{\beta s}(k+1) - i^p_{\beta s}(k+1)\right]^2. \tag{6.77}$$

- **Predictive Current Control in Synchronous Frame (PCC-*dq*):** This control strategy is applicable to both SG and IG (refer to (6.14) and (6.55)). The cost function in the *dq* frame is defined as follows [24]:

$$g(k) = \left[\widehat{i}^*_{ds}(k+1) - i^p_{ds}(k+1)\right]^2 + \left[\widehat{i}^*_{qs}(k+1) - i^p_{qs}(k+1)\right]^2. \tag{6.78}$$

- **Predictive Torque Control:** This strategy controls the generator electromagnetic torque and stator flux through the control of stator voltages and currents [25, 28]. The torque and flux equations for SG and IG are provided in this chapter.

$$g(k) = \left[\widehat{T}^*_e(k+1) - T^p_e(k+1)\right]^2 + \lambda_\psi \left[\widehat{\psi}^*_s(k+1) - \psi^p_s(k+1)\right]^2 \tag{6.79}$$

  where λ_ψ is a weighting factor.

- **Predictive Power Control:** Similar to the PTC, the stator active and reactive power are regulated through the control of stator voltages and currents [29]. The stator active and reactive power models are also provided in this chapter.

$$g(k) = \left[\widehat{P}^*_s(k+1) - P^p_s(k+1)\right]^2 + \left[\widehat{Q}^*_s(k+1) - Q^p_s(k+1)\right]^2. \tag{6.80}$$

- **Predictive Speed Control:** This control strategy regulates the shaft speed through the internal control of *dq*-axis (or $\alpha\beta$-axis) stator currents [30]. The speed dynamics shown in (6.6) is used in conjunction with (6.14) and (6.55) for SG and IG, respectively.

$$\begin{aligned} g(k) = {} & \lambda_\omega \left[\widetilde{\omega}^*_m(k+1) - \omega^p_m(k+1)\right]^2 + \lambda_{id} \left[\widehat{i}^*_{ds}(k+1) - i^p_{ds}(k+1)\right]^2 \\ & + \lambda_{iq} \left[\widehat{i}^*_{qs}(k+1) - i^p_{qs}(k+1)\right]^2 \end{aligned} \tag{6.81}$$

  where λ_ω, λ_{id}, and λ_{iq} are weighting factors.

6.10 CONCLUDING REMARKS

In this chapter, the construction and dynamic modeling of PMSG and IG used in Type 3 and 4 WECS are elucidated. The natural and stationary frame models for PMSG are complex because the generator inductance varies with respect to the rotor position angle. For this reason, the PMSG natural frame model is transformed into the synchronous frame model to simplify the analysis and control. By using the space vector model, the IG model is formulated in the arbitrary reference frame. By setting $\omega_k = \omega_s$ and $\omega_k = 0$, synchronous and stationary frame models are deducted from the arbitrary reference frame model, respectively. The IG model is generic and can be used for both DFIG and SCIG by simply changing the rotor voltage values. Furthermore, the simulation models of SG and IG are presented on the basis of the dynamic generator equations developed in this chapter. Moreover, the steady state and dynamic behavior of SG and IG are studied.

The modeling of various control variables, such as stator currents, stator flux linkages, rotor speed, electromagnetic torque, and stator active and reactive powers, is presented for both SG and IG. In Type 3 and 4 WECS, the power converters are controlled by the MPC scheme, which uses the generator models to precisely regulate the control variables. In this chapter, we learn that the MPC scheme performs “reverse engineering” on the generator models. An accurate generator modeling leads to precise generator control. The predictive controller uses these dynamic models to control the generator; thus, it is more intuitive than linear control.

Link to Next Chapters:

- The SG and IG dynamic models presented in this chapter are transformed to DT for use with MPC in Chapter 7.
- The generator models derived in this chapter are employed in Chapters 9 to 12 to design MPC schemes and to analyze Type 3 and 4 variable-speed WECS performance.

REFERENCES

1. V. Yaramasu, B. Wu, P. C. Sen, S. Kouro, and M. Narimani, “High-power wind energy conversion systems: State-of-the-art and emerging technologies,” *Proceedings of the IEEE*, vol. 103, no. 5, pp. 740–788, May 2015.
2. H. Li and Z. Chen, “Overview of different wind generator systems and their comparisons,” *IET Renewable Power Generation*, vol. 2, no. 2, pp. 123–138, June 2008.
3. H. Polinder, F. van der Pijl, G.-J. de Vilder, and P. Tavner, “Comparison of direct-drive and geared generator concepts for wind turbines,” *IEEE Transactions on Energy Conversion*, vol. 21, no. 3, pp. 725–733, September 2006.
4. I. Boldea, *The Electric Generators Handbook: Variable Speed Generators*. Boca Raton, FL: CRC Press, 2006.
5. X. Yang, D. Patterson, and J. Hudgins, “Permanent magnet generator design and control for large wind turbines,” in *IEEE Symposium on Power Electronics and Machines in Wind Applications (PEMWA)*, Denver, CO , USA, July 2012, pp. 1–5.
6. H. Polinder, J. Ferreira, B. Jensen, A. Abrahamsen, K. Atallah, and R. McMahon, “Trends in wind turbine generator systems,” *IEEE Journal of Emerging and Selected Topics in Power Electronics*, vol. 1, no. 3, pp. 174–185, September 2013.
7. J. Earnest and T. Wizelius, *Wind Power Plants and Project Development*. New Delhi, India: PHI Learning, May 2011.
8. C. Mi, G. Slemon, and R. Bonert, “Modeling of iron losses of permanent-magnet synchronous motors,” *IEEE Transactions on Industry Applications*, vol. 39, no. 3, pp. 734–742, May 2003.
9. S. Kalsi, K. Weeber, H. Takesue, C. Lewis, H.-W. Neumueller, and R. Blaugher, “Development status of rotating machines employing superconducting field windings,” *Proceedings of the IEEE*, vol. 92, no. 10, pp. 1688–1704, October 2004.
10. C. Lewis and J. Muller, “A direct drive wind turbine HTS generator,” in *IEEE Power and Energy Society (PES) General Meeting*, Tampa, FL, USA, June 2007, pp. 1–8.
11. G. Snitchler, B. Gamble, C. King, and P. Winn, “10 MW class superconductor wind turbine generators,” *IEEE Transactions on Applied Superconductivity*, vol. 21, no. 3, pp. 1089–1092, June 2011.

12. S. Bolognani, S. Bolognani, L. Peretti, and M. Zigliotto, "Design and implementation of model predictive control for electrical motor drives," *IEEE Transactions on Industrial Electronics*, vol. 56, no. 6, pp. 1925–1936, June 2009.
13. A. Yazdani and R. Iravani, *Voltage-Sourced Converters in Power Systems*. Hoboken, NJ: Wiley-IEEE Press, January 2010.
14. Z. Chen, M. Tomita, S. Doki, and S. Okuma, "An extended electromotive force model for sensorless control of interior permanent-magnet synchronous motors," *IEEE Transactions on Industrial Electronics*, vol. 50, no. 2, pp. 288–295, April 2003.
15. P. Pillay and R. Krishnan, "Modeling, simulation, and analysis of permanent-magnet motor drives. I. the permanent-magnet synchronous motor drive," *IEEE Transactions on Industry Applications*, vol. 25, no. 2, pp. 265–273, March 1989.
16. L. Springob and J. Holtz, "High-bandwidth current control for torque-ripple compensation in PM synchronous machines," *IEEE Transactions on Industrial Electronics*, vol. 45, no. 5, pp. 713–721, October 1998.
17. T. A. Lipo, *Analysis of Synchronous Machines*, 2nd ed. Boca Raton, FL: CRC Press, 2012.
18. R. Krishnan, *Permanent Magnet Synchronous and Brushless DC Motor Drives*. Boca Raton, FL: CRC Press, 2010.
19. J. Holtz, "The representation of AC machine dynamics by complex signal flow graphs," *IEEE Transactions on Industrial Electronics*, vol. 42, no. 3, pp. 263–271, June 1995.
20. P. Krause and C. Thomas, "Simulation of symmetrical induction machinery," *IEEE Transactions on Power Apparatus and Systems*, vol. 84, no. 11, pp. 1038–1053, November 1965.
21. J. Holtz, "Sensorless control of induction machines: With or without signal injection?" *IEEE Transactions on Industrial Electronics*, vol. 53, no. 1, pp. 7–30, February 2005.
22. B. Bose, *Modern Power Electronics and AC Drives*. Upper Saddle River, NJ: Prentice Hall, 2002.
23. J. Holtz, "Sensorless control of induction motor drives," *Proceedings of the IEEE*, vol. 90, no. 8, pp. 1359–1394, August 2002.
24. C.-S. Lim, E. Levi, M. Jones, N. Rahim, and W.-P. Hew, "A comparative study of synchronous current control schemes based on FCS-MPC and PI-PWM for a two-motor three-phase drive," *IEEE Transactions on Industrial Electronics*, vol. 61, no. 8, pp. 3867–3878, August 2014.
25. H. Miranda, P. Coftes, J. Yuz, and J. Rodríguez, "Predictive torque control of induction machines based on state-space models," *IEEE Transactions on Industrial Electronics*, vol. 56, no. 6, pp. 1916–1924, June 2009.
26. T. Geyer, G. Papafotiou, and M. Morari, "Model predictive direct torque control: Part I: Concept, algorithm, and analysis," *IEEE Transactions on Industrial Electronics*, vol. 56, no. 6, pp. 1894–1905, June 2009.
27. R. Vargas, J. Rodríguez, U. Ammann, and P. Wheeler, "Predictive current control of an induction machine fed by a matrix converter with reactive power control," *IEEE Transactions on Industrial Electronics*, vol. 55, no. 12, pp. 4362–4371, December 2008.
28. M. Preindl and S. Bolognani, "Model predictive direct torque control with finite control set for PMSM drive systems, Part 1: Maximum torque per ampere operation," *IEEE Transactions on Industrial Informatics*, vol. 9, no. 4, pp. 1912–1921, November 2013.
29. J. Hu, J. Zhu, and D. Dorrell, "Predictive direct power control of doubly fed induction generators under unbalanced grid voltage conditions for power quality improvement," *IEEE Transactions on Sustainable Energy*, vol. 6, no. 3, pp. 943–950, July 2015.
30. M. Preindl and S. Bolognani, "Model predictive direct speed control with finite control set of PMSM drive systems," *IEEE Transactions on Power Electronics*, vol. 28, no. 2, pp. 1007–1015, February 2013.

CHAPTER 7

MAPPING OF CONTINUOUS-TIME MODELS TO DISCRETE-TIME MODELS

7.1 INTRODUCTION

From the mathematics and control engineering perspectives, continuous time (CT) and discrete time (DT) are two alternative frameworks that are used in the modeling of variables. Unlike CT variables, which vary continuously over time, DT variables are held constant between any two successive sampling instants [1, 2]. The electrical and mechanical variables in wind energy conversion systems (WECS), such as voltages, currents, wind speeds, and rotor speeds, belong to the CT framework. As mentioned in Chapter 3, digital control has many advantages over analog control: high accuracy, high reliability, user friendliness in changing software and design, and low power consumption, among others [2]. The modern control systems for WECS are realized by digital control platforms, such as microcontrollers, digital signal processors (DSPs), or field-programmable gate arrays (FPGAs). Therefore, the overall WECS is treated as a sampled-data model involving both a CT plant and a DT controller. A simplified block diagram of a WECS sampled-data model is shown in Figure 7.1. The CT plant (WECS) is interfaced with and controlled by the DT controller. The natures of CT and DT variables are also represented graphically. To simplify the block diagram, the disturbance inputs are not shown.

The process of mapping CT functions, models, equations, and systems into discrete counterparts is called "discretization". This process can be thought of transforming continuous differential equations into discrete difference equations. Digital controllers are designed by using direct and indirect approaches [3]. In the direct approach, digital controllers are designed by discretizing the CT plant. In the indirect approach, the controllers

Model Predictive Control of Wind Energy Conversion Systems, First Edition. Venkata Yaramasu and Bin Wu.

Companion Website: www.wiley.com/go/yaramasu_wu_mpc_wecs

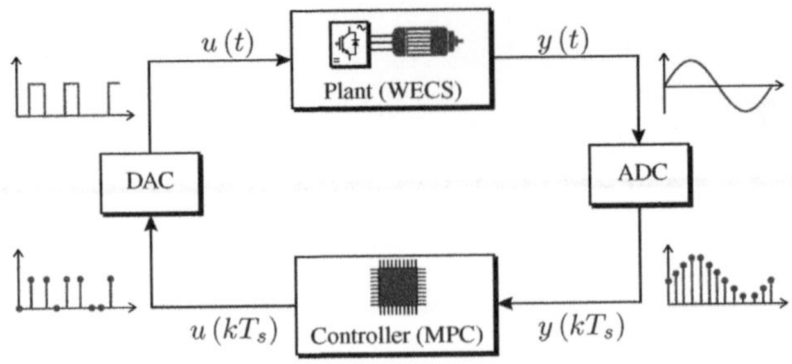

Figure 7.1 Block diagram of a simplified WECS sampled-data model.

designed in CT are discretized for implementation in digital control platforms [4]. For example, in the indirect approach, a proportional-integral (PI) controller designed in CT is converted into DT to be implemented in DSP. Model predictive control (MPC) uses the direct approach, wherein the design and control processes are performed in DT.

To design and implement MPC schemes, CT models should be mapped to DT models. Both frequency and time-domain methods are used for the discretization and design of DT controllers. In this chapter, the state-space CT power converter and wind generator models discussed in Chapters 5 and 6 are mapped to DT state-space models by using several discretization methods. Compared with classical control schemes, MPC involves high computational burden because of the online optimization algorithm [5, 6]. A simple discretization method that does not contribute to additional computational burden is desirable. The MPC performance is greatly influenced by the accuracy of sampled-data models, in addition to other control issues [7]. For a better transient and steady-state performance with the MPC schemes, the CT plant model should be accurately mapped to the DT model but the discretization process should be less complex. Exact, approximate, and quasi-exact discretization methods based on zero-order hold (ZOH), forward Euler, backward Euler, bilinear transformation, matrix factorization, and truncated Taylor series are analyzed in this chapter. Several example problems are designed to support the theory and analysis.

Chapter Overview

- Section 7.2 briefly discusses the properties of the CT plant, DT controller, and interface elements. Section 7.3 provides an overview of the correlation between CT and DT models with respect to the state-space equations and transfer functions.
- Section 7.4 presents an overview of exact, approximate, and quasi-exact discretization methods. The exact discretization of CT state-space models by a step-invariant method (i.e., ZOH) is examined in Section 7.5.
- Section 7.6 analyzes the approximate discretization methods, including forward Euler, backward Euler, and bilinear transformation methods. Section 7.7 discusses the quasi-exact discretization methods based on matrix factorization and truncated Taylor series.
- Section 7.8 presents a comparison between various discretization methods. Section 7.9 shows the offline calculations of DT parameters by MATLAB software with detailed instructions and software code.
- Section 7.10 presents the concluding remarks for this chapter. The main features and challenges of the various discretization methods, as well as their application to the MPC of WECS, are also discussed in this section.

7.2 MODEL PREDICTIVE CONTROL OF WECS

A block diagram of a permanent magnet synchronous generator (PMSG)-based Type 4 WECS is shown in Figure 7.2. All elements are divided into three groups: (1) CT plant, (2) DT controller, and (3) the interface between the CT plant and DT controller.

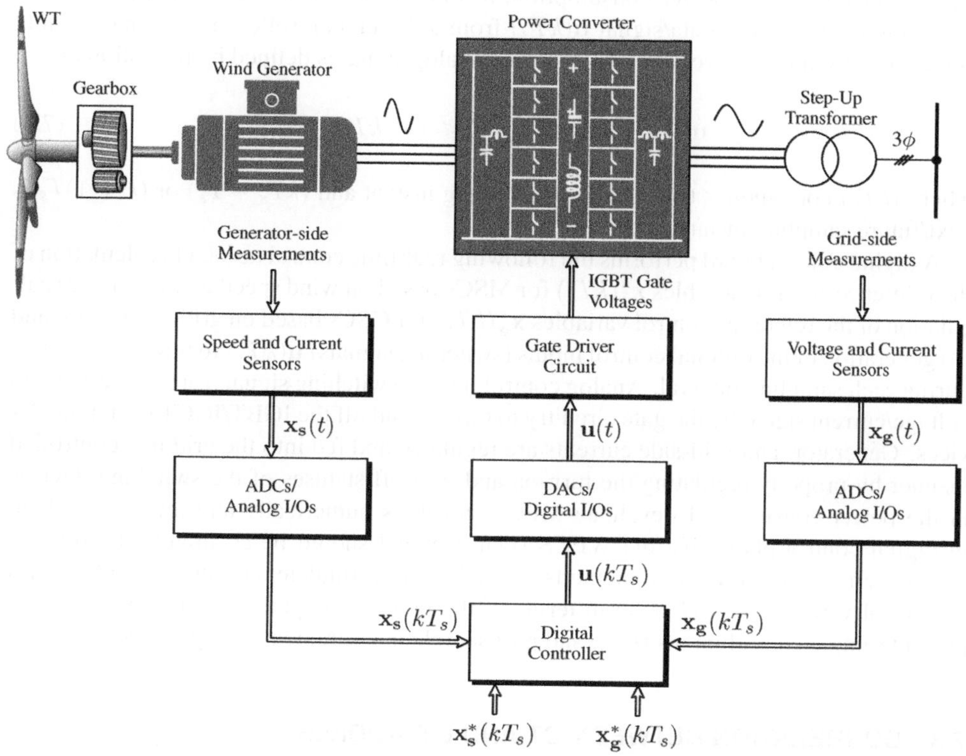

Figure 7.2 Block diagram of digital control system for Type 4 PMSG WECS.

The main elements of a CT plant are a wind turbine, gearbox, wind generator, power converter, and AC and DC filters. Shaft speed and position sensors, generator/grid-side voltage and current sensors, and DC-link voltage sensors are also included in a CT plant. Owing to the decoupling offered by machine-side converters (MSCs) and grid-side converters (GSCs), two digital control platforms are employed in a practical WECS and communicate with each other. DT controller examples include a DSP, an FPGA, and a computer with real-time interface boards. The interface between a CT plant and a DT controller includes analog-to-digital converters (ADCs), digital-to-analog converters (DACs), analog input/output (I/O) channels, and digital I/Os. Commercial DSPs have integrated DACs and ADCs for communication with the CT plant and other DSPs. However, FPGA platforms need an additional interface board containing DAC and ADC channels.

Continuous analog electrical signals are converted into discontinuous digital signals, data, or numbers by an ADC. An ADC (or A/D) uniformly samples a CT signal $\mathbf{x}(t)$ and transforms it into a DT sequence as follows:

$$\mathbf{x}(k) \stackrel{\text{def}}{=} \mathbf{x}(kT_s) \qquad \forall\, k \in 0, \cdots \infty \tag{7.1}$$

where T_s is the sampling period.

The sampling instants k are defined by the synchronization clock of a digital controller. The sampling rate significantly affects the accuracy of the sampled-data models. Sample-and-hold circuits are used in ADCs to sample analog signals and hold them constant for the duration of the sampling time. Bandwidth and signal-to-noise ratio are two primary characteristics that define an ADC. The DAC (or D/A) architectures are selected on the basis of physical size, power consumption, resolution, speed, accuracy, and cost. ZOH is used to convert a DT data/signal $\mathbf{u}(kT_s)$ from a digital controller into an analog input signal $\mathbf{u}(t)$ for a gate driver circuitry [8]. The analog signal is defined by the following:

$$\mathbf{u}(t) = \mathbf{u}(kT_s), \quad kT_s \leqslant t < kT_s + T_s \tag{7.2}$$

where (kT_s) corresponds to the present sampling instant and $(kT_s + T_s)$ or $(k+1)T_s$ is next/future sampling instant.

A digital control board performs the following real-time control tasks: (1) calculation of the reference control variables $\mathbf{x}_\mathbf{s}^*(kT_s)$ for MSCs based on wind speed conditions, (2) calculation of the reference control variables $\mathbf{x}_\mathbf{g}^*(kT_s)$ for GSCs based on grid conditions, and (3) generation of the optimal control inputs (switching signals) $\mathbf{u}(kT_s)$ to power converters during each sampling interval. Analog control inputs (switching signals) are converted into voltage/current signal by the gate circuitry to turn on and off the IGBT/IGCT switching devices. Generator- and grid-side currents are regulated and fed into the grid in a controlled manner by properly regulating the turn-on and turn-off statuses of the switching devices in the power converter. To evaluate MPC algorithms numerically and implement them in digital control platforms, the WECS (plant) model should be accurately discretized. The objective of discretization methods for MPC is to formulate the future sample values of control variables $\mathbf{x}_{\mathbf{s}/\mathbf{g}}(k+1)$ in terms of the present sample measurements $\mathbf{x}_{\mathbf{s}/\mathbf{g}}(k)$, present sample control inputs $\mathbf{u}(k)$, and past sample measurements $\mathbf{x}_{\mathbf{s}/\mathbf{g}}(k-1)$.

7.3 CORRELATION BETWEEN CT AND DT MODELS

To assist in the discussion of various discretization methods, the correlation between CT and DT models is presented. The signals, models, equations, and systems in a CT model have counterparts in a DT model. To simplify the discussion, state-space equations and transfer functions are considered in the analysis.

7.3.1 CT and DT State-Space Equations

In this chapter, the state-space equations are broadly classified into two groups: linear time invariant (LTI) and linear time variant (LTV). In each group, state variables possess different properties and are classified into independent variables, coupled variables, and uncoupled variables. The stationary ($\alpha\beta$) frame state-space model in (7.3) represents a power converter with a resistive-inductive (RL) load. Matrix $\mathbf{A}$ contains time-invariant terms, and the $\alpha\beta$-frame variables are independent (not cross-coupled). Thus, the CT model is called LTI with independent variables.

$$\underbrace{\frac{d}{dt}\begin{bmatrix} i_{\alpha g}(t) \\ i_{\beta g}(t) \end{bmatrix}}_{\dot{\mathbf{x}}(t)} = \underbrace{\begin{bmatrix} -\frac{R_i}{L_i} & 0 \\ 0 & -\frac{R_i}{L_i} \end{bmatrix}}_{\mathbf{A}} \underbrace{\begin{bmatrix} i_{\alpha g}(t) \\ i_{\beta g}(t) \end{bmatrix}}_{\mathbf{x}(t)} + \underbrace{\begin{bmatrix} \frac{1}{L_i} & 0 \\ 0 & \frac{1}{L_i} \end{bmatrix}}_{\mathbf{B}} \underbrace{\begin{bmatrix} v_{\alpha i}(t) \\ v_{\beta i}(t) \end{bmatrix}}_{\mathbf{u}(t)}. \tag{7.3}$$

The state-space model of a GSC in a synchronous (dq) frame is shown in (7.4) [9]. To simplify the analysis, the grid angular frequency, ω_g, is assumed to be constant. The dq-frame currents are coupled to each other because of the ω_g term. From matrix $\mathbf{A}$, the second column in the first row couples the derivative of i_{dg} with the i_{qg}. Similarly, the first column in the second row couples the derivative of i_{qg} with the i_{dg}. Thus, the GSC model in dq-frame is called LTI with cross-coupled state variables.

$$\begin{aligned}\underbrace{\frac{d}{dt}\begin{bmatrix} i_{dg}(t) \\ i_{qg}(t)\end{bmatrix}}_{\dot{\mathbf{x}}(t)} &= \underbrace{\begin{bmatrix} -\frac{R_i}{L_i} & \omega_g \\ -\omega_g & -\frac{R_i}{L_i}\end{bmatrix}}_{\mathbf{A}} \underbrace{\begin{bmatrix} i_{dg}(t) \\ i_{qg}(t)\end{bmatrix}}_{\mathbf{x}(t)} + \underbrace{\begin{bmatrix} \frac{1}{L_i} & 0 \\ 0 & \frac{1}{L_i}\end{bmatrix}}_{\mathbf{B}} \underbrace{\begin{bmatrix} v_{di}(t) \\ v_{qi}(t)\end{bmatrix}}_{\mathbf{u_i}(t)} \\ &+ \underbrace{\begin{bmatrix} -\frac{1}{L_i} & 0 \\ 0 & -\frac{1}{L_i}\end{bmatrix}}_{-\mathbf{B}} \underbrace{\begin{bmatrix} v_{dg}(t) \\ v_{qg}(t)\end{bmatrix}}_{\mathbf{u_g}(t)}.\end{aligned} \tag{7.4}$$

The above two CT LTI state-space models are presented in a generic form as follows:

$$\begin{aligned}\dot{\mathbf{x}}(t) &= \mathbf{A}\,\mathbf{x}(t) + \mathbf{B}\,\mathbf{u}(t) \\ \mathbf{y}(t) &= \mathbf{C}\,\mathbf{x}(t)\end{aligned} \tag{7.5}$$

where $\mathbf{x}(t)$ is the state vector, $\mathbf{u}(t)$ is the input or control or disturbance vector, and $\mathbf{y}(t)$ is the output vector. The matrices are as follows: $\mathbf{A}$ is the state or system matrix, $\mathbf{B}$ is the input or control or disturbance matrix, and $\mathbf{C}$ is the output matrix. The generator and power converter models in this book do not use feedforward matrix $\mathbf{D}$, which connects the input vector $\mathbf{u}(t)$ to the output vector $\mathbf{y}(t)$. To simplify the analysis, matrix $\mathbf{D}$ is set to zero throughout this chapter.

The DT counterpart of the CT LTI state-space model is defined by the following:

$$\begin{aligned}\mathbf{x}(k+1) &= \mathbf{\Phi}\,\mathbf{x}(k) + \mathbf{\Gamma}\,\mathbf{u}(k) \\ \mathbf{y}(k) &= \mathbf{H}\,\mathbf{x}(k)\end{aligned} \tag{7.6}$$

where $\mathbf{\Phi}$, $\mathbf{\Gamma}$, and $\mathbf{H}$ are the DT matrices corresponding to $\mathbf{A}$, $\mathbf{B}$, and $\mathbf{C}$, respectively.

The dq-frame stator current dynamic model of a PMSG from (6.14) in Chapter 6 is a LTV state-space model with cross-coupled terms because $\mathbf{A}(t)$ contains the time-varying term $\omega_r(t)$ and cross-coupled state variables.

$$\begin{aligned}\underbrace{\frac{d}{dt}\begin{bmatrix} i_{ds}(t) \\ i_{qs}(t)\end{bmatrix}}_{\dot{\mathbf{x}}(t)} &= \underbrace{\begin{bmatrix} -\frac{R_s}{L_{ds}} & \frac{\omega_r(t)\,L_{qs}}{L_{ds}} \\ -\frac{\omega_r(t)\,L_{ds}}{L_{qs}} & -\frac{R_s}{L_{qs}}\end{bmatrix}}_{\mathbf{A}(t)} \underbrace{\begin{bmatrix} i_{ds}(t) \\ i_{qs}(t)\end{bmatrix}}_{\mathbf{x}(t)} \\ &+ \underbrace{\begin{bmatrix} \frac{1}{L_{ds}} & 0 \\ 0 & \frac{1}{L_{qs}}\end{bmatrix}}_{\mathbf{B}} \underbrace{\begin{bmatrix} v_{ds}(t) \\ v_{qs}(t)\end{bmatrix}}_{\mathbf{u}(t)} + \underbrace{\begin{bmatrix} 0 \\ -\frac{\omega_r(t)\,\psi_r}{L_{qs}}\end{bmatrix}}_{\mathbf{w}(t)}.\end{aligned} \tag{7.7}$$

The rotor speed dynamic model of PMSG is given in Chapter 6, Equation (6.24). To simplify the analysis, the dynamic model for SPMSG with $L_{ds} = L_{qs}$ is defined below:

$$\frac{d\,\omega_r(t)}{dt} = \frac{3\,P_p^2}{2\,J_m}\,\psi_r\,i_{qs}(t) - \frac{B_m}{J_m}\omega_r(t) - \frac{P_p}{J_m}\,T_m(t). \tag{7.8}$$

When (7.7) and (7.8) are combined, the state-space representation of SPMSG with stator current and rotor speed state variables is obtained as follows:

$$\frac{d}{dt}\begin{bmatrix} i_{ds}(t) \\ i_{qs}(t) \\ \omega_r(t) \end{bmatrix} = \begin{bmatrix} -\frac{R_s}{L_s} & \omega_r(t) & 0 \\ -\omega_r(t) & -\frac{R_s}{L_s} & -\frac{\psi_r}{L_s} \\ 0 & \frac{3}{2}\frac{P_p^2\,\psi_r}{J_m} & -\frac{B_m}{J_m} \end{bmatrix} \begin{bmatrix} i_{ds}(t) \\ i_{qs}(t) \\ \omega_r(t) \end{bmatrix} + \begin{bmatrix} \frac{1}{L_s} & 0 & 0 \\ 0 & \frac{1}{L_s} & 0 \\ 0 & 0 & -\frac{P_p}{J_m} \end{bmatrix} \begin{bmatrix} v_{ds}(t) \\ v_{qs}(t) \\ T_m(t) \end{bmatrix}. \quad (7.9)$$

In (7.9), the state variables i_{ds} and i_{qs} are related to the control input variables v_{ds} and v_{qs}. However, the state variable ω_r is not coupled with the converter/generator voltages v_{ds} and v_{qs}. Therefore, this system is called LTV with uncoupled-state variables. The LTV systems in (7.7) and (7.9) are represented in a generic form as follows:

$$\begin{aligned} \dot{\mathbf{x}}(t) &= \mathbf{A}(t)\,\mathbf{x}(t) + \mathbf{B}(t)\,\mathbf{u}(t) \\ \mathbf{y}(t) &= \mathbf{C}(t)\,\mathbf{x}(t) \end{aligned} \quad (7.10)$$

where matrices $\mathbf{A}(t)$ and/or $\mathbf{B}(t)$ and/or $\mathbf{C}(t)$ contains time-varying coefficients, such as rotor speed and grid angular frequency.

The sampled-data model corresponding to the LTV CT state-space model in (7.10) is

$$\begin{aligned} \mathbf{x}(k+1) &= \mathbf{\Phi}(k)\,\mathbf{x}(k) + \mathbf{\Gamma}(k)\,\mathbf{u}(k) \\ \mathbf{y}(k) &= \mathbf{H}(k)\,\mathbf{x}(k) \end{aligned} \quad (7.11)$$

where $\mathbf{\Phi}(k)$, $\mathbf{\Gamma}(k)$, and $\mathbf{H}(k)$ are time-varying matrices. They are computed and updated online during each sampling interval.

7.3.2 CT and DT Transfer Functions

Transfer function models are not directly useful in MPC design. Instead, a continuous LTI system transfer function in a factored form provides useful information about the stability and natural response of a system. The CT transfer function is defined by the following [10, 11]:

$$\mathbf{G_c}(s) = K\,\frac{(s - z_1)\cdots(s - z_n)}{(s - p_1)(s - p_2)\cdots(s - p_n)} \quad (7.12)$$

where K is the system gain, and z_n and p_n are the nth system zero and pole, respectively. The denominator

$$\lambda(s) = |s\mathbf{I} - \mathbf{A}| \quad (7.13)$$

represents the characteristic equation [12]. The poles of the system (the eigenvalues of matrix $\mathbf{A}$) are the roots of the above polynomial and provide the system stability-related information. The CT eigenvalues $\lambda_1 \cdots \lambda_n$ are mapped to DT as $e^{\lambda_1 T_s} \cdots e^{\lambda_n T_s}$ [1]. Similarly, the CT transfer function is mapped to DT by ZOH method as follows [2]:

$$\mathbf{G_d}(z) = \left(1 - z^{-1}\right) \mathcal{Z}\left\{ \mathcal{L}^{-1}\left[\frac{\mathbf{G_c}}{s}\right]_{t=kT_s} \right\}. \quad (7.14)$$

The correlation between CT and DT models is summarized in Table 7.1 with respect to the state-space representation of LTI and LTV models, time domain and transfer function representations of LTI models, characteristic equations, and eigenvalues [13]. A detailed analysis of various discretization methods to map CT state-space models to DT state-space models are discussed in the succeeding sections.

Table 7.1 Summary of mapping between CT and DT models [13]

	Continuous-Time System	**Discrete-Time System**
LTI State-Space Representation	$\dot{\mathbf{x}}(t) = \mathbf{A}\,\mathbf{x}(t) + \mathbf{B}\,\mathbf{u}(t)$ $\mathbf{y}(t) = \mathbf{C}\,\mathbf{x}(t)$	$\mathbf{x}(k+1) = \boldsymbol{\Phi}\,\mathbf{x}(k) + \boldsymbol{\Gamma}\,\mathbf{u}(k)$ $\mathbf{y}(k) = \mathbf{H}\,\mathbf{x}(k)$
LTV State-Space Representation	$\dot{\mathbf{x}}(t) = \mathbf{A}(t)\,\mathbf{x}(t) + \mathbf{B}(t)\,\mathbf{u}(t)$ $\mathbf{y}(t) = \mathbf{C}\,\mathbf{x}(t)$	$\mathbf{x}(k+1) = \boldsymbol{\Phi}(k)\,\mathbf{x}(k) + \boldsymbol{\Gamma}(k)\,\mathbf{u}(k)$ $\mathbf{y}(k) = \mathbf{H}\,\mathbf{x}(k)$
Time Domain of LTI System	s-Domain Representation: $s\,\mathbf{X}(s) = \mathbf{A}\,\mathbf{X}(s) + \mathbf{B}\,\mathbf{U}(s)$ $\mathbf{Y}(s) = \mathbf{C}\,\mathbf{X}(s)$	z-Domain Representation: $z\,\mathbf{X}(z) = \boldsymbol{\Phi}\,\mathbf{X}(z) + \boldsymbol{\Gamma}\,\mathbf{U}(z)$ $\mathbf{Y}(z) = \mathbf{H}\,\mathbf{X}(z)$
Transfer Function of LTI System	s-Domain Representation: $\frac{\mathbf{Y}(s)}{\mathbf{U}(s)} = \mathbf{C}\,(s\,\mathbf{I} - \mathbf{A})^{-1}\,\mathbf{B}$	z-Domain Representation: $\frac{\mathbf{Y}(z)}{\mathbf{U}(z)} = \mathbf{H}\,(z\,\mathbf{I} - \boldsymbol{\Phi})^{-1}\,\boldsymbol{\Gamma}$
Characteristic Equation	$\lambda(s) = \lvert s\mathbf{I} - \mathbf{A}\rvert$	$\lambda(z) = \lvert z\mathbf{I} - \boldsymbol{\Phi}\rvert$
Eigenvalues	$\lambda_1 \cdots \lambda_n$	$e^{\lambda_1 T_s} \ldots e^{\lambda_n T_s}$

7.4 OVERVIEW OF DISCRETIZATION METHODS

In Chapter 5, power converter terminal voltages are formulated in terms of measured voltages and switching states. Linear models yield the same relationship in CT and DT frameworks. However, DC-link capacitor voltage models contain derivative terms. Similarly, the wind generator models discussed in Chapter 6 employ derivative terms. The mapping of CT differential equations (models) to DT difference equations (models) is performed by various discretization methods, as shown in Figure 7.3. Discretization methods are broadly classified into three groups: (1) exact methods, (2) approximate methods, and (3) quasi-exact methods. In mapping CT models to DT models, the discretization process incurs a discretization error. The objective of discretization methods is to minimize the discretization error to an acceptable minimum value and to involve simple discrete mathematical expressions and calculations.

Exact discretization methods are further classified into pole-zero matching methods and invariant approaches, which include impulse invariant, step invariant (ZOH), and ramp invariant (first-order hold) methods [1, 4, 14]. As the name implies, the poles and zeros in CT and DT models are matched by the pole-zero method. By using the impulse-invariant method, the impulse responses obtained by CT and DT models remain same. Similarly, the step and ramp responses obtained by CT and DT models are the same as those by the step- and ramp-invariant methods, respectively. The DACs used by digital control platforms hold the DT signal/data during each sampling interval and transform it to a CT signal. In other words, the control inputs are piecewise constant over the sample time, T_s, through the ZOH method [7]. First-order hold (FOH) uses piecewise linear approximation to reconstruct the analog signal from the DT signal by a DAC and an integrator. From the MPC perspective, the ZOH method is the most suitable method for exact discretization. For LTI models, DT coefficients can be computed offline by using computer simulation software programs, such as `MATLAB/Simulink`. DT parameters are then included in the digital controller memory for use with MPC schemes.

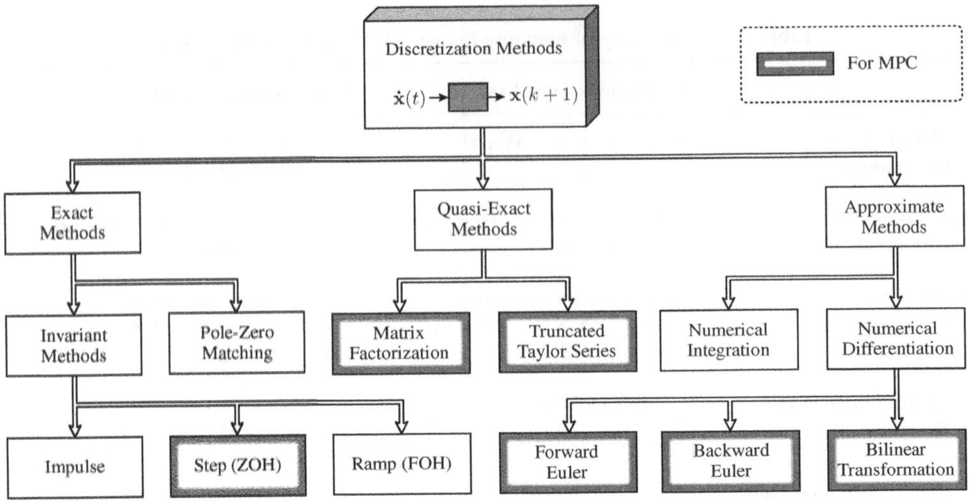

Figure 7.3 Taxonomy of discretization methods for digital control realization.

State matrix **A** in PMSG and induction generator (IG) models is a function of rotor electrical speed $\omega_r(t)$. In such cases, matrices $\mathbf{\Phi}$ and $\mathbf{\Gamma}$ must be computed and updated during each sampling time. For LTV models, obtaining the exact discrete models is often tedious and highly impossible in some cases because of the high-order matrix exponential and integral operations. In such cases, approximate and quasi-exact methods reduce the complexity in discretization for both LTI and LTV systems, assuming that the sampling time T_s is very small. For sampling times greater than the millisecond range, the step response and stability conditions significantly vary across discretization methods. However, in MPC realization, sampling time is in the range of microseconds. Therefore, discretization methods do not exhibit considerable performance variations. Discretization methods are selected according to the specific application and nature of control variables.

Approximate discretization is obtained by numerical integration or numerical differentiation (finite differences). The latter approach is simple and straightforward in mapping CT models to DT models (refer to Chapter 12 in [4]). The models of the various power converters and wind generators discussed in Chapters 5 and 6 demonstrate that the order of derivatives in dynamic models is one. With a given initial value, the *Euler–Maruyama* method, or simply the *Euler* method, uses numerical procedure to solve ordinary differential equations. Forward and backward Euler methods provide derivative approximation by considering different sampling instants. Bilinear transformation combines the operating principles of forward and backward Euler methods to obtain better results [15]. Approximate methods can easily be applied to LTI and LTV models.

For LTV models, exact methods are highly complex and approximate methods are less accurate. Quasi-exact discretization methods provide the best balance between accuracy and complexity needed in real-time implementation. The matrix factorization (or decomposition) method is a modified version of the exact discretization by the ZOH method [16]. The truncated Taylor series considers high-order terms in the matrix exponential, in addition to the first-order terms used by the forward Euler method. Quasi-exact methods contribute to the reduction in computational burden while improving the steady-state and transient performances of MPC schemes.

7.5 EXACT DISCRETIZATION BY ZOH METHOD

An exact discretization by the ZOH method is presented here. The first part of the LTI state-space equation (7.5) corresponding to the state variable $\dot{\mathbf{x}}(t)$ is considered. The output equation $\mathbf{y}(t)$ remains unchanged. Multiplying both sides of (7.5) with $e^{-\mathbf{A}t}$ yields the following:

$$e^{-\mathbf{A}t}\,\dot{\mathbf{x}}(t) = e^{-\mathbf{A}t}\,\mathbf{A}\,\mathbf{x}(t) + e^{-\mathbf{A}t}\,\mathbf{B}\,\mathbf{u}(t). \tag{7.15}$$

From the theory of matrix exponential [17], the following properties are observed:

$$\frac{d}{dt}e^{-\mathbf{A}t} = -\mathbf{A}\,e^{-\mathbf{A}t} = -e^{-\mathbf{A}t}\,\mathbf{A} \tag{7.16}$$

$$\frac{d}{dt}\left[e^{-\mathbf{A}t}\,\mathbf{x}(t)\right] = e^{-\mathbf{A}t}\,\frac{d}{dt}\mathbf{x}(t) - \mathbf{A}\,e^{-\mathbf{A}t}\,\mathbf{x}(t) \tag{7.17}$$

where state matrix $\mathbf{A}$ is nonsingular.

By using the above two properties, the CT model in (7.15) is simplified as follows:

$$\frac{d}{dt}\left[e^{-\mathbf{A}t}\,\mathbf{x}(t)\right] = e^{-\mathbf{A}t}\,\mathbf{B}\,\mathbf{u}(t). \tag{7.18}$$

By integrating both sides of (7.18) and assuming that t_0 and t_1 are two arbitrary sampling points in CT, the following is obtained:

$$e^{-\mathbf{A}t_1}\,\mathbf{x}(t_1) - \mathbf{x}(t_0) = \int_{t_0}^{t_1} e^{-\mathbf{A}\tau}\,\mathbf{B}\,\mathbf{u}(\tau)\,d\tau, \quad t_0 < t_1. \tag{7.19}$$

When both sides of (7.19) are multiplied with $e^{-\mathbf{A}t_1}$,

$$\mathbf{x}(t_1) = e^{\mathbf{A}t_1}\,\mathbf{x}(t_0) = \int_{t_0}^{t_1} e^{\mathbf{A}(t_1-\tau)}\,\mathbf{B}\,\mathbf{u}(\tau)\,d\tau. \tag{7.20}$$

Thus, an analytical solution to the CT state-space model is obtained.

By selecting sampling points corresponding to t_0 and t_1, the CT solution is mapped to the DT solution. When $t_0 = kT_s$ and $t_1 = kT_s + T_s$, and the control input is held constant during each sampling interval, the CT solution given in (7.20) is mapped exactly to DT by the ZOH method as demonstrated in the following [1]:

$$\begin{aligned} \mathbf{x}(kT_s + T_s) &= \mathbf{\Phi}\,\mathbf{x}(kT_s) + \mathbf{\Gamma}\,\mathbf{u}(kT_s) \\ \mathbf{y}(kT_s) &= \mathbf{H}\,\mathbf{x}(kT_s) \end{aligned} \tag{7.21}$$

which is simplified as

$$\begin{aligned} \mathbf{x}(k+1) &= \mathbf{\Phi}\,\mathbf{x}(k) + \mathbf{\Gamma}\,\mathbf{u}(k) \\ \mathbf{y}(k) &= \mathbf{H}\,\mathbf{x}(k). \end{aligned} \tag{7.22}$$

DT matrices $\mathbf{\Phi}$, $\mathbf{\Gamma}$, and $\mathbf{H}$ are defined in terms of CT matrices as [12, 18]:

$$\mathbf{\Phi} = e^{\mathbf{A}T_s}, \quad \mathbf{\Gamma} = \mathbf{A}^{-1}\,(\mathbf{\Phi} - \mathbf{I})\,\mathbf{B}, \quad \mathbf{H} = \mathbf{C} \tag{7.23}$$

where $\mathbf{I}$ is the unity matrix with dimensions equal to the dimension of matrix $\mathbf{A}$.

The DT matrix $\mathbf{\Phi}$ is given by Taylor series expansion method [1, 11]:

$$\mathbf{\Phi} = e^{\mathbf{A}T_s} = \frac{(\mathbf{A}T_s)^0}{0!} + \frac{(\mathbf{A}T_s)^1}{1!} + \frac{(\mathbf{A}T_s)^2}{2!} + \cdots + \frac{(\mathbf{A}T_s)^m}{m!} = \sum_{m=0}^{\infty} \frac{(\mathbf{A}T_s)^m}{m!}$$
$$= \mathbf{I} + \mathbf{A}T_s + \frac{(\mathbf{A}T_s)^2}{2!} + \cdots + \frac{(\mathbf{A}T_s)^m}{m!}. \tag{7.24}$$

The DT matrix $\mathbf{\Gamma}$ is calculated as follows:

$$\mathbf{\Gamma} = \int_0^{T_s} e^{\mathbf{A}\tau}\, d\tau\, \mathbf{B} = \left[\mathbf{I}T_s + \frac{\mathbf{A}T_s^2}{2!} + \frac{\mathbf{A}^2 T_s^3}{3!} + \cdots + \frac{\mathbf{A}^m T_s^{m+1}}{(m+1)!} \right] \mathbf{B}. \tag{7.25}$$

From the comparisons of (7.24) and (7.25), the following is observed:

$$\mathbf{A}\mathbf{\Gamma} = \left(e^{\mathbf{A}T_s} - \mathbf{I}\right)\mathbf{B} \quad \Rightarrow \quad \mathbf{\Gamma} = \mathbf{A}^{-1}\left(e^{\mathbf{A}T_s} - \mathbf{I}\right)\mathbf{B} = \left(e^{\mathbf{A}T_s} - \mathbf{I}\right)\mathbf{A}^{-1}\mathbf{B}. \tag{7.26}$$

The matrix exponential can be computed in 19 different ways as outlined in [19]. The inverse Laplace transformation to calculate matrix exponential is shown below [1, 10]:

$$e^{\mathbf{A}T_s} = \mathcal{L}^{-1}\left\{[s\mathbf{I} - \mathbf{A}]^{-1}\right\}_{t=T_s} \tag{7.27}$$

where s is the Laplace operator.

The sampled-data models obtained by the above procedure are called **explicitly exact models**. In (7.27), the inverse of matrix $[s\mathbf{I} - \mathbf{A}]$ is computed first and the inverse Laplace transform is applied to each element of the resulting Laplace matrix. For second- and third-order state matrices in (7.3) and (7.9), hand calculations by inverse Laplace transformation is a feasible approach. The dimension of state matrix $\mathbf{A}$ for voltage-source converter-fed IG is 4×4, whereas that of state matrix $\mathbf{A}$ for current-source converter-fed IG is 6×6. The Cayley–Hamilton theorem [20] is used to compute $[s\mathbf{I} - \mathbf{A}]^{-1}$ with high-order matrices. As discussed in Section 7.9, computer simulation software programs, such as `MATLAB/Simulink`, can be used as an alternative approach to compute exact DT parameters offline.

7.6 APPROXIMATE DISCRETIZATION METHODS

MPC schemes are realized with sampling times in the range of 10–200 μs. Owing to a small step-size, approximate methods reduce the complexity and computational burden involved in the discretization of LTI and LTV models. In this section, approximate discretization methods based on the forward Euler, backward Euler, and bilinear transformation methods are presented. Approximation methods can be applied to state-space (differential) equations and transfer functions. As a first step in the analysis, the following first-order differential equation is considered:

$$\frac{dx(t)}{dt} = u(t). \tag{7.28}$$

The corresponding transfer function in CT is derived as follows:

$$s\,X(s) = U(s) \quad \Rightarrow \quad \frac{X(s)}{U(s)} = \frac{1}{s}. \tag{7.29}$$

The analytical solution to the differential equation in (7.28) is obtained by numerical integration or numerical differentiation (finite differences). The DT solution with numerical integration is expressed as follows [14]:

$$x(kT_s + T_s) = x(kT_s) + \int_{kT_s}^{kT_s+T_s} u(\tau)\, d\tau. \tag{7.30}$$

The integral term in (7.30) is then approximated to obtain the approximate DT solution. The main difference between various approximation methods, such as the forward Euler, backward Euler, and bilinear transformation methods, lies in the approximation of the integral part. For these methods, the numerical integration approach provides different expressions for both state variable solution $x(k+1)$ and output solution $y(k)$ [14].

The finite difference approach directly considers the derivative term in (7.28) to obtain the approximate sampled-data model. By using this approach, only the state variable solution $x(k+1)$ changes across methods, but the output solution $y(k)$ remains the same. Numerical differentiation is a straightforward approach for obtaining the approximate sampled-data model for MPC design. The forward Euler, backward Euler, and bilinear transformation methods are graphically represented in Figure 7.4. These methods consider different sampling points to approximate derivative and integral terms.

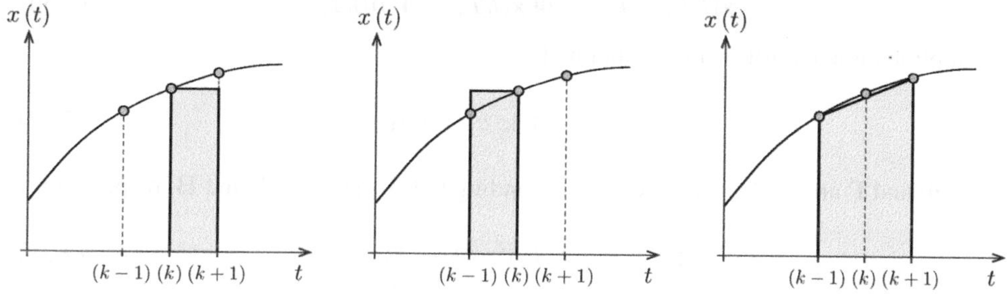

(a) Forward Euler (b) Backward Euler (c) Bilinear Transformation

Figure 7.4 Comparison of approximate discretization methods based on sampling instants [14].

7.6.1 Forward Euler Approximation

The forward Euler method is a simple and straightforward method for obtaining the DT equivalent of CT models. Derivatives are approximated by using the present sampling instant (kT_s) and forward sampling instant $(kT_s + T_s)$; therefore, this Euler method is called forward method or forward difference method. The approximation of derivatives through the forward finite difference method is given by following [21]:

$$\left\{\frac{dx(t)}{dt}\right\}_{t=kT_s} \approx \frac{x(kT_s + T_s) - x(kT_s)}{T_s} \tag{7.31}$$

from which approximate solution is obtained as follows:

$$x(kT_s + T_s) \approx x(kT_s) + T_s \left\{\frac{dx(t)}{dt}\right\}_{t=kT_s}. \tag{7.32}$$

The approximate sampled-data model is obtained by substituting the derivative value in (7.32). For example, combining (7.28) and (7.32) obtains the following solution:

$$x(kT_s + T_s) \approx x(kT_s) + T_s\, u(kT_s). \tag{7.33}$$

From (7.33), the future value of the control variable $x(k+1)$ is defined in terms of the present state value $x(k)$ and present control input $u(k)$. Therefore, the forward Euler method is an explicit method. The following DT transfer function is obtained by applying z-transform to both sides of (7.33):

$$z\,X(z) = X(z) + T_s\, U(z) \quad \Rightarrow \quad \frac{X(z)}{U(z)} = T_s\, \frac{1}{z-1}. \tag{7.34}$$

Application of Forward Euler Approximation to CT State-Space Models:
The substitution of LTI state-space model (7.5) into (7.32) obtains the following approximate sampled-data model:

$$\mathbf{x}(kT_s + T_s) \approx \mathbf{x}(kT_s) + T_s \left\{\mathbf{A}\,\mathbf{x}(kT_s) + \mathbf{B}\,\mathbf{u}(kT_s)\right\}. \tag{7.35}$$

When the state and control variables are rearranged, the sampled-data model is formulated as follows:

$$\mathbf{x}(kT_s + T_s) \approx \mathbf{\Phi}\,\mathbf{x}(kT_s) + \mathbf{\Gamma}\,\mathbf{u}(kT_s) \tag{7.36}$$

by simplification, the following is obtained:

$$\mathbf{x}(k+1) \approx \mathbf{\Phi}\,\mathbf{x}(k) + \mathbf{\Gamma}\,\mathbf{u}(k) \tag{7.37}$$

where $\mathbf{\Phi}$ and $\mathbf{\Gamma}$ are the DT matrices corresponding to CT matrices $\mathbf{A}$ and $\mathbf{B}$, respectively:

$$\mathbf{\Phi} \approx (\mathbf{I} + \mathbf{A}\,T_s), \quad \mathbf{\Gamma} \approx \mathbf{B}\,T_s. \tag{7.38}$$

A comparison of (7.38) and (7.24) shows that the forward Euler method is the same as the Taylor series expansion method if high-order terms are ignored. In other words, the forward Euler method provides the same result as the first-order Taylor series. In case of LTV systems, state matrix $\mathbf{A}$ is replaced by the time-varying matrix $\mathbf{A}(t)$. By using this approach, online matrix exponential and integral operations do not need to be performed. To explain this concept further, the following example is presented.

EXAMPLE 7.1 Approximate PMSG Sampled-Data Model

Consider the PMSG model given in (7.7). The CT model is represented in a simplified form as

$$\frac{d}{dt}\begin{bmatrix} i_{ds}(t) \\ i_{qs}(t) \end{bmatrix} = \mathbf{A}(t)\begin{bmatrix} i_{ds}(t) \\ i_{qs}(t) \end{bmatrix} + \mathbf{B}\begin{bmatrix} v_{ds}(t) \\ v_{qs}(t) \end{bmatrix} + \mathbf{w}(t)$$

and

$$\mathbf{A}(t) = \begin{bmatrix} -\frac{R_s}{L_{ds}} & \frac{\omega_r(t)\,L_{qs}}{L_{ds}} \\ -\frac{\omega_r(t)\,L_{ds}}{L_{qs}} & -\frac{R_s}{L_{qs}} \end{bmatrix}, \quad \mathbf{B} = \begin{bmatrix} \frac{1}{L_{ds}} & 0 \\ 0 & \frac{1}{L_{qs}} \end{bmatrix}, \quad \mathbf{w}(t) = \begin{bmatrix} 0 \\ -\frac{\omega_r(t)\,\psi_r}{L_{qs}} \end{bmatrix}.$$

Calculate the following by using forward Euler approximation method: (a) simplified PMSG sampled-data model, (b) DT matrix $\mathbf{\Phi}(k)$, (c) DT matrix $\mathbf{\Gamma_b}$, and (d) DT matrix $\mathbf{\Gamma_w}(k)$.

Solution:

(a) The sampled-data model for PMSG with forward Euler method is

$$\begin{bmatrix} i_{ds}(k+1) \\ i_{qs}(k+1) \end{bmatrix} \approx \mathbf{\Phi}(k) \begin{bmatrix} i_{ds}(k) \\ i_{qs}(k) \end{bmatrix} + \mathbf{\Gamma_b} \begin{bmatrix} v_{ds}(k) \\ v_{qs}(k) \end{bmatrix} + \mathbf{\Gamma_w}(k).$$

(b) The DT LTV matrix $\mathbf{\Phi}(k)$ corresponding to state matrix $\mathbf{A}(t)$ is

$$\mathbf{\Phi}(k) \approx [\mathbf{I} + \mathbf{A}(k)\, T_s] \approx \begin{bmatrix} 1 - \frac{R_s\, T_s}{L_{ds}} & \frac{\omega_r(k)\, L_{qs}\, T_s}{L_{ds}} \\ -\frac{\omega_r(k)\, L_{ds}\, T_s}{L_{qs}} & 1 - \frac{R_s\, T_s}{L_{qs}} \end{bmatrix}.$$

(c) The DT matrix $\mathbf{\Gamma_b}$ corresponding to $\mathbf{B}$ is

$$\mathbf{\Gamma_b} \approx \mathbf{B}\, T_s \approx \begin{bmatrix} \frac{T_s}{L_{ds}} & 0 \\ 0 & \frac{T_s}{L_{qs}} \end{bmatrix}.$$

(d) The DT LTV matrix $\mathbf{\Gamma_w}(k)$ corresponding to disturbance matrix $\mathbf{w}(t)$ is

$$\mathbf{\Gamma_w}(k) \approx \mathbf{w}(k)\, T_s \approx \begin{bmatrix} 0 \\ -\frac{\omega_r(k)\, \psi_r\, T_s}{L_{qs}} \end{bmatrix}.$$

For the realization of MPC schemes, $\mathbf{\Gamma_b}$ is computed offline. To realize the MPC schemes, two notations are defined for DT LTV matrices: (1)$\widetilde{\mathbf{\Phi}}(k)$ and $\widetilde{\mathbf{\Gamma}}_\mathbf{w}(k)$ are computed offline by excluding $\omega_r(k)$ term; (2) during each sampling interval, the measured rotor speed is $\omega_r(k)$ is used along with offline defined $\widetilde{\mathbf{\Phi}}(k)$ and $\widetilde{\mathbf{\Gamma}}_\mathbf{w}(k)$ coefficients to compute $\mathbf{\Phi}(k)$ and $\mathbf{w}(k)$. The updated DT matrices $\mathbf{\Phi}(k)$ and $\mathbf{w}(k)$ are used to calculate the future values of dq-axes PMSG currents, $i_{ds}(k+1)$ and $i_{qs}(k+1)$. It is the simplest approach to design MPC scheme for both SPMSG with $L_{ds} = L_{qs}$, and IPMSG with $L_{ds} \neq L_{qs}$. ♦

7.6.2 Backward Euler Approximation

As shown below, with the backward Euler method, the derivative is approximated using present (kT_s) and backward $(kT_s - T_s)$ sampling instants [22, 23]. Therefore, this method is also called backward method or backward differences.

$$\left\{ \frac{d\,x(t)}{dt} \right\}_{t=kT_s} \approx \frac{x(kT_s) - x(kT_s - T_s)}{T_s}. \tag{7.39}$$

To perform predictions for control variables by the MPC scheme, the future value of state variable should be modeled in terms of measured state variable. To accomplish this objective, the variables in (7.39) needs to be shifted one sample forward. The resultant approximate solution in DT is expressed as follows:

$$x(kT_s + T_s) \approx x(kT_s) + T_s \left\{ \frac{d\,x}{dt} \right\}_{t=kT_s+T_s}. \tag{7.40}$$

By substituting (7.28) into (7.40), the sampled-data model is obtained as follows:

$$x(kT_s + T_s) \approx x(kT_s) + T_s\, u(kT_s + T_s). \tag{7.41}$$

By using the backward Euler method, the future value $x(k+1)$ is defined in terms of present and future values; hence, this method is an implicit method. When the sampling time T_s is large, the backward Euler provides a stable solution compared with the forward Euler method [24]. The backward Euler is an attractive method in redesigning digital controller from the analog counterpart [15]. A comparison of (7.33) and (7.41) shows that the backward Euler method requires extrapolated input variable $u(k+1)$ if it is not generated from the controller. The extrapolation causes relatively small number of additional calculations compared with the forward Euler method.

With the help of z-transform, the DT transfer function is obtained as follows:

$$z\,X(z) = X(z) + T_s\,z\,U(z) \quad \Rightarrow \quad \frac{X(z)}{U(z)} = T_s\,\frac{z}{z-1}. \tag{7.42}$$

Application of Backward Euler Approximation to CT State-Space Models:
By combining (7.5) and (7.40), the DT state-space model is obtained by the following:

$$\mathbf{x}(kT_s + T_s) \approx \mathbf{x}(kT_s) + T_s\left\{\mathbf{A}\,\mathbf{x}(kT_s + T_s) + \mathbf{B}\,\mathbf{u}(kT_s + T_s)\right\}. \tag{7.43}$$

By simplification, we obtain the following:

$$\mathbf{x}(kT_s + T_s) \approx \mathbf{\Phi}\,\mathbf{x}(kT_s) + \mathbf{\Gamma}\,\mathbf{u}(kT_s + T_s) \tag{7.44}$$

or

$$\mathbf{x}(k+1) \approx \mathbf{\Phi}\,\mathbf{x}(k) + \mathbf{\Gamma}\,\mathbf{u}(k+1) \tag{7.45}$$

where

$$\mathbf{\Phi} \approx (\mathbf{I} - \mathbf{A}\,T_s)^{-1}, \quad \mathbf{\Gamma} \approx (\mathbf{I} - \mathbf{A}\,T_s)^{-1}\,\mathbf{B}\,T_s. \tag{7.46}$$

The backward Euler approximation is applicable to state-space models with only independent state variables. If the state variables are cross-coupled, the above method cannot be used. To explain this concept, an example is presented below.

✍ EXAMPLE 7.2 Comparison of Forward and Backward Euler Methods

Consider the CT $\alpha\beta$ and dq frame dynamic models in (7.3) and (7.4), respectively. These models correspond to LTI models with independent and cross-coupled state variables, respectively.

$$\frac{d}{dt}\begin{bmatrix} i_{\alpha g}(t) \\ i_{\beta g}(t) \end{bmatrix} = \begin{bmatrix} -\frac{R_i}{L_i} & 0 \\ 0 & -\frac{R_i}{L_i} \end{bmatrix}\begin{bmatrix} i_{\alpha g}(t) \\ i_{\beta g}(t) \end{bmatrix} + \underbrace{\begin{bmatrix} \frac{1}{L_i} & 0 \\ 0 & \frac{1}{L_i} \end{bmatrix}}_{\mathbf{B}}\begin{bmatrix} v_{\alpha i}(t) \\ v_{\beta i}(t) \end{bmatrix} \tag{7.47}$$

and

$$\frac{d}{dt}\begin{bmatrix} i_{dg}(t) \\ i_{qg}(t) \end{bmatrix} = \begin{bmatrix} -\frac{R_i}{L_i} & \omega_g \\ -\omega_g & -\frac{R_i}{L_i} \end{bmatrix}\begin{bmatrix} i_{dg}(t) \\ i_{qg}(t) \end{bmatrix} + \mathbf{B}\begin{bmatrix} v_{di}(t) \\ v_{qi}(t) \end{bmatrix} - \mathbf{B}\begin{bmatrix} v_{dg}(t) \\ v_{qg}(t) \end{bmatrix}. \tag{7.48}$$

Calculate the following:
(a) sampled-data model for (7.47) by using the forward Euler method;
(b) sampled-data model for (7.47) by using the backward Euler method;
(c) sampled-data model for (7.48) by using the forward Euler method;
(d) sampled-data model for (7.48) by using the backward Euler method.

Solution:

(a) The sampled-data model corresponding to (7.47) with the forward Euler method is

$$\begin{bmatrix} i_{\alpha g}(k+1) \\ i_{\beta g}(k+1) \end{bmatrix} \approx \begin{bmatrix} 1-\frac{R_i\,T_s}{L_i} & 0 \\ 0 & 1-\frac{R_i\,T_s}{L_i} \end{bmatrix} \begin{bmatrix} i_{\alpha g}(k) \\ i_{\beta g}(k) \end{bmatrix} + \begin{bmatrix} \frac{T_s}{L_i} & 0 \\ 0 & \frac{T_s}{L_i} \end{bmatrix} \begin{bmatrix} v_{\alpha i}(k) \\ v_{\beta i}(k) \end{bmatrix}.$$

(b) To analyze the backward Euler method, the $\alpha\beta$-frame state variables are calculated independently. The DT model corresponding to (7.47) is obtained as follows with the help of (7.40):

$$i_{\alpha g}(k+1) \approx i_{\alpha g}(k) + T_s\left\{-\frac{R_i}{L_i}\,i_{\alpha g}(k+1) + \frac{1}{L_i}\,v_{\alpha i}(k+1)\right\}$$
$$i_{\beta g}(k+1) \approx i_{\beta g}(k) + T_s\left\{-\frac{R_i}{L_i}\,i_{\beta g}(k+1) + \frac{1}{L_i}\,v_{\beta i}(k+1)\right\}.$$

The above two expressions show that α and β variables are independent and uncoupled. By simplification, the above two expressions are given as follows [23]:

$$i_{\alpha g}(k+1) \approx \frac{L_i}{(L_i + R_i\,T_s)}\,i_{\alpha g}(k) + \frac{T_s}{(L_i + R_i\,T_s)}\,v_{\alpha i}(k+1)$$
$$i_{\beta g}(k+1) \approx \frac{L_i}{(L_i + R_i\,T_s)}\,i_{\beta g}(k) + \frac{T_s}{(L_i + R_i\,T_s)}\,v_{\beta i}(k+1).$$

In part (a) and (b) solutions, the $R_i\,T_s$ value is approximately equal to zero. Thus, the DT matrices obtained with the forward and backward Euler methods are almost the same. The $v_{\alpha i}$ and $v_{\beta i}$ are predicted by the digital controller; thus, the (k) and $(k+1)$ state voltages are essentially the same. We observe that, when forward and backward Euler methods are used to control a power converter with an RL load, the outcome is the same; however, the notation of the variables changes. With the forward Euler method, the predicted voltages are represented as $v_{\alpha i}(k)$ and $v_{\beta i}(k)$. The backward method denotes that the predicted converter voltages are $v_{\alpha i}(k+1)$ and $v_{\beta i}(k+1)$.

(c) The sampled-data model for (7.48) with the forward Euler method is expressed as follows:

$$\begin{bmatrix} i_{dg}(k+1) \\ i_{qg}(k+1) \end{bmatrix} \approx \begin{bmatrix} 1-\frac{R_i\,T_s}{L_i} & \omega_g\,T_s \\ -\omega_g\,T_s & 1-\frac{R_i\,T_s}{L_i} \end{bmatrix} \begin{bmatrix} i_{dg}(k) \\ i_{qg}(k) \end{bmatrix} + \begin{bmatrix} \frac{T_s}{L_i} & 0 \\ 0 & \frac{T_s}{L_i} \end{bmatrix} \begin{bmatrix} v_{di}(k) \\ v_{qi}(k) \end{bmatrix}.$$

(d) The analysis with the backward Euler method is given as follows:

$$i_{dg}(k+1) \approx i_{dg}(k) + T_s\left\{-\frac{R_i}{L_i}\,i_{dg}(k+1) + \omega_g\,i_{qg}(k+1)\right\}$$
$$+\,T_s\left\{\frac{1}{L_i}\,v_{di}(k+1) - \frac{1}{L_i}\,v_{dg}(k+1)\right\}$$
$$i_{qg}(k+1) \approx i_{qg}(k) + T_s\left\{-\frac{R_i}{L_i}\,i_{qg}(k+1) - \omega_g\,i_{dg}(k+1)\right\}$$
$$+\left\{\frac{1}{L_i}\,v_{qi}(k+1) - \frac{1}{L_i}\,v_{qg}(k+1)\right\}.$$

In the above representation, $i_{dg}(k+1)$ and $i_{qg}(k+1)$ are cross-coupled. To estimate the future value of i_{dg}, future of i_{qg} is needed and vice versa. Direct estimation of cross-coupled terms is impossible, particularly during small sampling periods. The LTV systems also exhibit the same property for the cross-coupled state variables. Thus, the backward Euler method is not useful for LTI/LTV systems with cross-coupled state variables. ♦

7.6.3 Approximation by Bilinear Transformation

The forward and backward Euler methods provide implicit and explicit solutions, respectively. The combination of these two methods obtains a better solution in DT. The bilinear transformation is also known as the Tustin or trapezoidal method. The derivative is approximated by considering the backward sample $(kT_s - T_s)$ similar to the backward Euler method and forward sample $(kT_s + T_s)$ similar to the forward Euler discretization. The derivative is approximated as follows:

$$\left\{\frac{d\,x(t)}{dt}\right\}_{t=kT_s} \approx \frac{x(kT_s + T_s) - x(kT_s - T_s)}{2\,T_s}. \tag{7.49}$$

The equivalent approximate solution is obtained as follows:

$$x(kT_s + T_s) \approx x(kT_s - T_s) + 2\,T_s \left\{\frac{d\,x(t)}{dt}\right\}_{t=kT_s}. \tag{7.50}$$

The combination of (7.28) and (7.50) becomes the following:

$$x(kT_s + T_s) \approx x(kT_s - T_s) + 2\,T_s\,u(kT_s). \tag{7.51}$$

The DT transfer function is expressed as follows:

$$z\,X(z) = \frac{1}{z}X(z) + 2\,T_s\,U(z) \quad \Rightarrow \quad \frac{X(z)}{U(z)} = \frac{2\,z\,T_s}{z^2 - 1}. \tag{7.52}$$

Application of Bilinear Transformation to CT State-Space Models:
With the help of (7.5) and (7.50), the DT state-space model is established as follows:

$$\mathbf{x}(kT_s + T_s) \approx \mathbf{x}(kT_s - T_s) + 2\,T_s\,\{\mathbf{A}\,\mathbf{x}(kT_s) + \mathbf{B}\,\mathbf{u}(kT_s)\} \tag{7.53}$$

and the simplified representation is

$$\mathbf{x}(kT_s + T_s) \approx \mathbf{x}(kT_s - T_s) + \mathbf{\Phi}\,\mathbf{x}(kT_s) + \mathbf{\Gamma}\,\mathbf{u}(kT_s) \tag{7.54}$$

or

$$\mathbf{x}(k+1) \approx \mathbf{x}(k-1) + \mathbf{\Phi}\,\mathbf{x}(k) + \mathbf{\Gamma}\,\mathbf{u}(k) \tag{7.55}$$

where

$$\mathbf{\Phi} \approx 2\,\mathbf{A}\,T_s, \quad \mathbf{\Gamma} \approx 2\,\mathbf{B}\,T_s. \tag{7.56}$$

Similar to the forward Euler method, the bilinear transformation can be applied to LTI and LTV models with independent and cross-coupled terms.

7.7 QUASI-EXACT DISCRETIZATION METHODS

The ZOH method performs discretization with the highest accuracy; however, this method is applicable for LTI models only. ZOH discretization is highly complex or sometimes impossible for LTV models. The approximation methods are simple, but the discretization error increases with the sampling time T_s. The quasi-exact methods solve this issue by providing moderate error and complexity in the discretization process. In this section, two quasi-exact discretization methods are discussed.

7.7.1 Matrix Factorization

The matrix factorization (or decomposition) method is a modified version of exact discretization method discussed in Section 7.5 [16]. This method is a favorable discretization method for induction and synchronous generator models of the LTV class. The instantaneous values of state matrix $\mathbf{A}(t)$ are assumed to change with respect to time. The matrix $\mathbf{A}(t)$ is then decomposed into two parts, one containing constant parameters and another containing time-varying variables:

$$\mathbf{A}(t) = \mathbf{A_c} + \mathbf{A}_\omega(t). \tag{7.57}$$

By applying the theory of matrix exponential, the model in Equation (7.57) is represented as follows:

$$e^{\mathbf{A}T_s} = e^{(\mathbf{A_c}T_s + \mathbf{A}_\omega T_s)} = e^{\mathbf{A_c}T_s}\, e^{\mathbf{A}_\omega T_s}. \tag{7.58}$$

The above condition is valid if and only if [17]

$$\mathbf{A_c}\,\mathbf{A}_\omega \equiv \mathbf{A}_\omega\,\mathbf{A_c}. \tag{7.59}$$

The first term in (7.58), $e^{\mathbf{A_c}T_s}$ contains time-invariant coefficients and can be computed offline. However, the second term $e^{\mathbf{A}_\omega T_s}$ must be computed online.

By substituting (7.58) into (7.25), the DT matrix $\mathbf{\Gamma}$ is expressed as follows:

$$\mathbf{\Gamma} = \int_0^{T_s} e^{\mathbf{A}\tau}\,\mathbf{B}\,d\tau = \int_0^{T_s} e^{\mathbf{A_c}\tau}\, e^{\mathbf{A}_\omega \tau}\,\mathbf{B}\,d\tau. \tag{7.60}$$

If and only if the generator model and parameters support the following property

$$e^{\mathbf{A}_\omega T_s}\,\mathbf{B} \equiv \mathbf{B} \tag{7.61}$$

then the DT matrix $\mathbf{\Gamma}$ can be computed offline as shown below:

$$\mathbf{\Gamma} = \int_0^{T_s} e^{\mathbf{A_c}\tau}\,\mathbf{B}\,d\tau = \mathbf{A_c}^{-1}\left(e^{\mathbf{A_c}T_s} - \mathbf{I}\right)\mathbf{B}. \tag{7.62}$$

For LTV systems, the DT models obtained by the above procedure are called **explicitly exact models**. If any of the rules in (7.59) and (7.61) are violated, the sampled-data models are called **quasi-exact models** or approximate models with higher accuracy [25].

Application of Matrix Factorization to SPMSG State-Space Models:
For a SPMSG, d-axis inductance is equal to the q-axis inductance. By substituting $L_{ds} = L_{qs} = L_s$ in (7.7), the matrix $\mathbf{A}(t)$ is decomposed as follows:

$$\underbrace{\begin{bmatrix} -\frac{R_s}{L_s} & \omega_r(t) \\ -\omega_r(t) & -\frac{R_s}{L_s} \end{bmatrix}}_{\mathbf{A}(t)} = \underbrace{\begin{bmatrix} -\frac{R_s}{L_s} & 0 \\ 0 & -\frac{R_s}{L_s} \end{bmatrix}}_{\mathbf{A_c}} + \underbrace{\begin{bmatrix} 0 & \omega_r(t) \\ -\omega_r(t) & 0 \end{bmatrix}}_{\mathbf{A}_\omega(t)}. \tag{7.63}$$

For the SPMSG model, it is observed that

$$\mathbf{A_c}\,\mathbf{A}_\omega \equiv \mathbf{A}_\omega\,\mathbf{A_c} \equiv \begin{bmatrix} 0 & -\frac{R_s\,\omega_r(t)}{L_s} \\ \frac{R_s\,\omega_r(t)}{L_s} & 0 \end{bmatrix}. \tag{7.64}$$

Therefore, exact discretization for $\mathbf{\Phi}$ is obtained as follows [26]:

$$\mathbf{\Phi} = e^{\mathbf{A_c} T_s}\, e^{\mathbf{A}_\omega T_s} = \underbrace{\begin{bmatrix} e^{-\frac{R_s\,T_s}{L_s}} & 0 \\ 0 & e^{-\frac{R_s\,T_s}{L_s}} \end{bmatrix}}_{\mathbf{\Phi_c}} \underbrace{\begin{bmatrix} \cos\,(\omega_r\,T_s) & \sin\,(\omega_r\,T_s) \\ -\sin\,(\omega_r\,T_s) & \cos\,(\omega_r\,T_s) \end{bmatrix}}_{\mathbf{\Phi}_\omega} \tag{7.65}$$

For low-speed PMSGs (direct-drive) and high-speed PMSGs (with three-stage gearbox), the product of ω_r and T_s is approximately equal to zero. A low-speed SPMSG is considered in Table A.3 with ω_r = 61.26 rad/s. A high-speed SPMSG is in Table A.5 with ω_r = 377 rad/s. For the sampling time T_s = 20 μs, the DT coefficients are calculated as follows:

$$\begin{aligned} &\text{Low-speed:} \quad \omega_r\,T_s = 0.0012, \quad \cos\,(\omega_r\,T_s) = 0.99999, \quad \sin\,(\omega_r\,T_s) = 0.0012 \\ &\text{High-speed:} \quad \omega_r\,T_s = 0.0075, \quad \cos\,(\omega_r\,T_s) = 0.99997, \quad \sin\,(\omega_r\,T_s) = 0.0075 \end{aligned} \tag{7.66}$$

A similar analysis is given below with T_s = 100 μs:

$$\begin{aligned} &\text{Low-speed:} \quad \omega_r\,T_s = 0.0061, \quad \cos\,(\omega_r\,T_s) = 0.99998, \quad \sin\,(\omega_r\,T_s) = 0.0061 \\ &\text{High-speed:} \quad \omega_r\,T_s = 0.0377, \quad \cos\,(\omega_r\,T_s) = 0.99929, \quad \sin\,(\omega_r\,T_s) = 0.0377 \end{aligned} \tag{7.67}$$

The analysis in (7.66) and (7.67) concludes that, $\mathbf{\Phi}_\omega$ is approximately equal to unity matrix for high-power SPMSG. Therefore, the approximations are expressed as follows:

$$\mathbf{\Phi}_\omega = e^{\mathbf{A}_\omega T_s} \cong \begin{bmatrix} 1 & 0 \\ 0 & 1 \end{bmatrix} \quad \Rightarrow \quad \mathbf{\Phi}_\omega\,\mathbf{B} \cong \mathbf{B}, \quad \mathbf{\Phi}_\omega\,\mathbf{w}(k) \cong \mathbf{w}(k). \tag{7.68}$$

By substituting (7.68) into (7.60), the DT matrices $\mathbf{\Gamma_b}$ and $\mathbf{\Gamma_w}(k)$ are computed as follows:

$$\mathbf{\Gamma_b} \cong \mathbf{A_c^{-1}}\,(e^{\mathbf{A_c} T_s} - \mathbf{I})\,\mathbf{B}, \quad \mathbf{\Gamma_w}(k) \cong \mathbf{A_c^{-1}}\,(e^{\mathbf{A_c} T_s} - \mathbf{I})\,\mathbf{w}(k). \tag{7.69}$$

The matrix $\mathbf{\Gamma_b}$ can be computed offline because $\mathbf{B}$ is constant. However, the matrix $\mathbf{\Gamma_w}(k)$ must be updated online according to the measured rotor speed $\omega_r(k)$. The matrix $\mathbf{\Phi}$ corresponds to exact discretization; however, the calculation of $\mathbf{\Gamma_b}$ and $\mathbf{\Gamma_w}(k)$ involves (acceptable) approximations. Therefore, the overall process is called quasi-exact discretization.

Application of Matrix Factorization to IPMSG State-Space Models:
For an IPMSG, d-axis inductance is unequal to the q-axis inductance ($L_{ds} \neq L_{qs}$). The state matrix $\mathbf{A}(t)$ is factorized into constant and time-varying matrices as follows:

$$\underbrace{\begin{bmatrix} -\frac{R_s}{L_{ds}} & \frac{\omega_r(t)\,L_{qs}}{L_{ds}} \\ -\frac{\omega_r(t)\,L_{ds}}{L_{qs}} & -\frac{R_s}{L_{qs}} \end{bmatrix}}_{\mathbf{A}(t)} = \underbrace{\begin{bmatrix} -\frac{R_s}{L_{ds}} & 0 \\ 0 & -\frac{R_s}{L_{qs}} \end{bmatrix}}_{\mathbf{A_c}} + \underbrace{\begin{bmatrix} 0 & \frac{\omega_r(t)\,L_{qs}}{L_{ds}} \\ -\frac{\omega_r(t)\,L_{ds}}{L_{qs}} & 0 \end{bmatrix}}_{\mathbf{A}_\omega(t)}. \tag{7.70}$$

The factorization above shows that

$$\mathbf{A_c}\,\mathbf{A}_\omega = \begin{bmatrix} 0 & -\frac{R_s\,L_{qs}\,\omega_r(t)}{L_{ds}^2} \\ \frac{R_s\,L_{ds}\,\omega_r(t)}{L_{qs}^2} & 0 \end{bmatrix}, \quad \mathbf{A}_\omega\,\mathbf{A_c} = \begin{bmatrix} 0 & -\frac{R_s\,\omega_r(t)}{L_{ds}} \\ \frac{R_s\,\omega_r(t)}{L_{qs}} & 0 \end{bmatrix}. \tag{7.71}$$

It is observed that $\mathbf{A_c}\,\mathbf{A}_\omega \neq \mathbf{A}_\omega\,\mathbf{A_c}$ in (7.71); hence, the DT matrix $\mathbf{\Phi}$ obtained by the matrix factorization approach corresponds to quasi-exact model:

$$\mathbf{\Phi} \cong e^{\mathbf{A_c} T_s}\, e^{\mathbf{A}_\omega T_s} \cong \underbrace{\begin{bmatrix} e^{-\frac{R_s T_s}{L_{ds}}} & 0 \\ 0 & e^{-\frac{R_s T_s}{L_{qs}}} \end{bmatrix}}_{\mathbf{\Phi_c}} \underbrace{\begin{bmatrix} \cos(\omega_r T_s) & \frac{L_{qs}\sin(\omega_r T_s)}{L_{ds}} \\ -\frac{L_{ds}\sin(\omega_r T_s)}{L_{qs}} & \cos(\omega_r T_s) \end{bmatrix}}_{\mathbf{\Phi}_\omega} \tag{7.72}$$

The coefficients $\mathbf{\Phi}_\omega(1,1)$ and $\mathbf{\Phi}_\omega(2,2)$ are same for SPMSG and IPMSG. However, owing to unequal L_{ds} and L_{qs} values in IPMSG, $\mathbf{\Phi}_\omega(1,2)$ is larger and $\mathbf{\Phi}_\omega(2,1)$ is smaller than the coefficients of SPMSG. The overall $\mathbf{\Phi}_\omega$ is approximately equal to the unity matrix similar to the SPMSG. The approximate DT matrices $\mathbf{\Gamma_b}$ and $\mathbf{\Gamma_w}(k)$ are calculated for IPMSG as follows:

$$\mathbf{\Gamma_b} \cong \mathbf{A_c^{-1}}\left(e^{\mathbf{A_c} T_s} - \mathbf{I}\right)\mathbf{B}, \quad \mathbf{\Gamma_w}(k) \cong \mathbf{A_c^{-1}}\left(e^{\mathbf{A_c} T_s} - \mathbf{I}\right)\mathbf{w}(k). \tag{7.73}$$

The quasi-exact sampled-data models obtained for an IPMSG by the above approach are less accurate than the ZOH method; nevertheless, they produce better steady state and transient performance compared with the forward Euler approximation in (7.31).

Application of Matrix Factorization to IG State-Space Models:
The state matrix $\mathbf{A}(t)$ corresponding to the $\alpha\beta$-frame IG model in (6.71) is separated into two parts as shown below [16]:

$$\underbrace{\begin{bmatrix} -\frac{1}{\tau_\sigma} & 0 & \frac{k_r}{\sigma L_s \tau_r} & \frac{k_r\,\omega_r(t)}{\sigma L_s} \\ 0 & -\frac{1}{\tau_\sigma} & -\frac{k_r\,\omega_r(t)}{\sigma L_s} & \frac{k_r}{\sigma L_s \tau_r} \\ \frac{L_m}{\tau_r} & 0 & -\frac{1}{\tau_r} & -\omega_r(t) \\ 0 & \frac{L_m}{\tau_r} & \omega_r(t) & -\frac{1}{\tau_r} \end{bmatrix}}_{\mathbf{A}(t)} = \underbrace{\begin{bmatrix} -\frac{1}{\tau_\sigma} & 0 & \frac{k_r}{\sigma L_s \tau_r} & 0 \\ 0 & -\frac{1}{\tau_\sigma} & 0 & \frac{k_r}{\sigma L_s \tau_r} \\ \frac{L_m}{\tau_r} & 0 & -\frac{1}{\tau_r} & 0 \\ 0 & \frac{L_m}{\tau_r} & 0 & -\frac{1}{\tau_r} \end{bmatrix}}_{\mathbf{A_c}}$$

$$+ \underbrace{\begin{bmatrix} 0 & 0 & 0 & \frac{k_r\,\omega_r(t)}{\sigma L_s} \\ 0 & 0 & -\frac{k_r\,\omega_r(t)}{\sigma L_s} & 0 \\ 0 & 0 & 0 & -\omega_r(t) \\ 0 & 0 & \omega_r(t) & 0 \end{bmatrix}}_{\mathbf{A}_\omega(t)}. \tag{7.74}$$

The following calculations demonstrate that the IG model exhibits $\mathbf{A_c}\,\mathbf{A}_\omega \neq \mathbf{A}_\omega\,\mathbf{A_c}$ similar to the IPMSG dynamic model [25]:

$$\mathbf{A_c}\,\mathbf{A}_\omega = \begin{bmatrix} 0 & 0 & 0 & -\frac{k_r\,\omega_r}{\sigma L_s \tau_r} - \frac{k_r\,\omega_r}{\sigma L_s \tau_\sigma} \\ 0 & 0 & \frac{k_r\,\omega_r}{\sigma L_s \tau_r} + \frac{k_r\,\omega_r}{\sigma L_s \tau_\sigma} & 0 \\ 0 & 0 & 0 & \frac{\omega_r}{\tau_r} + \frac{L_m\,k_r\,\omega_r}{\sigma L_s \tau_r} \\ 0 & 0 & -\frac{\omega_r}{\tau_r} - \frac{L_m\,k_r\,\omega_r}{\sigma L_s \tau_r} & 0 \end{bmatrix}$$

$$\mathbf{A}_\omega\,\mathbf{A_c} = \begin{bmatrix} 0 & \frac{L_m\,k_r\,\omega_r}{\sigma L_s \tau_r} & 0 & -\frac{k_r\,\omega_r}{\sigma L_s \tau_r} \\ -\frac{L_m\,k_r\,\omega_r}{\sigma L_s \tau_r} & 0 & \frac{k_r\,\omega_r}{\sigma L_s \tau_r} & 0 \\ 0 & -\frac{L_m\,\omega_r}{\tau_r} & 0 & \frac{\omega_r}{\tau_r} \\ \frac{L_m\,\omega_r}{\tau_r} & 0 & -\frac{\omega_r}{\tau_r} & 0 \end{bmatrix} \tag{7.75}$$

The quasi-exact DT matrix $\mathbf{\Phi}$ is obtained by matrix factorization for IG as follows:

$$\mathbf{\Phi} \cong e^{\mathbf{A_c} T_s}\, e^{\mathbf{A}_\omega T_s} \cong \mathbf{\Phi_c} \underbrace{\begin{bmatrix} 1 & 0 & \frac{k_r}{\sigma L_s}\left(1-\cos\left(\omega_r T_s\right)\right) & \frac{k_r}{\sigma L_s}\sin\left(\omega_r T_s\right) \\ 0 & 1 & -\frac{k_r}{\sigma L_s}\sin\left(\omega_r T_s\right) & \frac{k_r}{\sigma L_s}\left(1-\cos\left(\omega_r T_s\right)\right) \\ 0 & 0 & \cos\left(\omega_r T_s\right) & -\sin\left(\omega_r T_s\right) \\ 0 & 0 & \sin\left(\omega_r T_s\right) & \cos\left(\omega_r T_s\right) \end{bmatrix}}_{\mathbf{\Phi}_\omega}. \tag{7.76}$$

The analysis in (7.66) and (7.67) prove that $\cos\left(\omega_r T_s\right) \cong 1$ and $\sin\left(\omega_r T_s\right) \cong 0$. Therefore, $\mathbf{\Phi}_\omega$ in (7.76) is further simplified as

$$\mathbf{\Phi}_\omega \cong \begin{bmatrix} 1 & 0 & \frac{k_r}{\sigma L_s}\left(1-\cos\left(\omega_r T_s\right)\right) & \frac{k_r}{\sigma L_s}\sin\left(\omega_r T_s\right) \\ 0 & 1 & -\frac{k_r}{\sigma L_s}\sin\left(\omega_r T_s\right) & \frac{k_r}{\sigma L_s}\left(1-\cos\left(\omega_r T_s\right)\right) \\ 0 & 0 & \cos\left(\omega_r T_s\right) & -\sin\left(\omega_r T_s\right) \\ 0 & 0 & \sin\left(\omega_r T_s\right) & \cos\left(\omega_r T_s\right) \end{bmatrix} \cong \begin{bmatrix} 1 & 0 & 0 & 0 \\ 0 & 1 & 0 & 0 \\ 0 & 0 & 1 & 0 \\ 0 & 0 & 0 & 1 \end{bmatrix}. \tag{7.77}$$

which is fourth-order unity matrix. Therefore, the assumption $\mathbf{\Phi}_\omega \mathbf{B} \approx \mathbf{B}$ is valid.

The quasi-exact DT matrix $\mathbf{\Gamma_b}$ is defined by

$$\mathbf{\Gamma_b} \cong \mathbf{A_c}^{-1}\left(e^{\mathbf{A_c} T_s} - \mathbf{I}\right)\mathbf{B}. \tag{7.78}$$

The above analysis concludes that both $\mathbf{\Phi}$ and $\mathbf{\Gamma_b}$ are quasi-exact models. The performance obtained by quasi-exact IG models is better than that obtained by the forward Euler approximation (refer to results in [16]). For SPMSG, IPMSG, and IG, the DT matrices $\mathbf{\Phi_c}$ and $\mathbf{\Gamma_b}$ are calculated offline to reduce complexity and computational burden. The offline calculation of $\mathbf{\Gamma_b}$ eliminates integral operations. The high-order matrix exponential calculations in $\mathbf{\Phi}_\omega$ are replaced by simple trigonometrical terms.

7.7.2 Truncated Taylor Series

The main difference between the ZOH and forward Euler method is that the latter ignores high-order terms in the Taylor series expansion. The accuracy of the model can be improved by considering a high number of terms in the Taylor series [27, 28]. As mentioned earlier, in WECS control, the range of sampling times is between 20 μs and 100 μs. For this reason, a second-order Taylor series shown below is sufficient to improve accuracy.

$$x(kT_s + T_s) \cong x(kT_s) + T_s \left\{\frac{d\,x(t)}{dt}\right\}_{t=kT_s} + \frac{T_s^2}{2}\left\{\frac{d^2\,x(t)}{dt^2}\right\}_{t=kT_s}. \tag{7.79}$$

For the particular example considered in (7.28), the truncated Taylor series is expressed as follows:

$$x(kT_s + T_s) \cong x(kT_s) + T_s\left\{u(t)\right\}_{t=kT_s} + \frac{T_s^2}{2}\left\{\frac{d}{dt}u(t)\right\}_{t=kT_s}. \tag{7.80}$$

This solution is the same as the one obtained with the forward Euler because the derivative of $u(t)$ is zero. The corresponding z-transform is also similar to the one shown in Equation (7.34).

Application of Truncated Taylor Series to CT State-Space Models:
By substituting (7.5) into (7.79), the DT state-space model is obtained as follows:

$$\mathbf{x}(kT_s + T_s) \cong \mathbf{x}(kT_s) + T_s \{\mathbf{A}\,\mathbf{x}(kT_s) + \mathbf{B}\,\mathbf{u}(kT_s)\} + \frac{T_s^2}{2}\left\{\frac{d}{dt}[\mathbf{A}\,\mathbf{x}(t) + \mathbf{B}\,\mathbf{u}(t)]\right\}_{t=kT_s}. \tag{7.81}$$

In the above expression, the derivative of $\mathbf{u}(t)$ is zero. The matrix $\mathbf{A}$ is assumed to be constant with time-invariant parameters. By simplification, we obtain the following:

$$\mathbf{x}(kT_s + T_s) \cong (\mathbf{I} + \mathbf{A}\,T_s)\,\mathbf{x}(kT_s) + (\mathbf{B}\,T_s)\,\mathbf{u}(kT_s) + \frac{\mathbf{A}\,T_s^2}{2}\{\mathbf{A}\,\mathbf{x}(kT_s) + \mathbf{B}\,\mathbf{u}(kT_s)\} \tag{7.82}$$

from which the final sampled-data model is deducted as

$$\mathbf{x}(kT_s + T_s) \cong \mathbf{\Phi}\,\mathbf{x}(kT_s) + \mathbf{\Gamma}\,\mathbf{u}(kT_s) \tag{7.83}$$

or

$$\mathbf{x}(k+1) \cong \mathbf{\Phi}\,\mathbf{x}(k) + \mathbf{\Gamma}\,\mathbf{u}(k) \tag{7.84}$$

where

$$\mathbf{\Phi} \cong \mathbf{I} + \mathbf{A}\,T_s + \frac{\mathbf{A}^2\,T_s^2}{2}, \quad \mathbf{\Gamma} \cong \mathbf{B}\,T_s + \frac{\mathbf{A}\,\mathbf{B}\,T_s^2}{2}. \tag{7.85}$$

The above solution is derived on the basis of the assumption that $\mathbf{A}$ is time invariant. In the case of LTV systems, the solution must be obtained according to the CT model. To explain this concept, an example study is presented below.

✍ EXAMPLE 7.3 Comparison of Forward Euler and Truncated Taylor Series

The model in (7.9) is considered in this example with the assumption that $L_{ds} = L_{qs} = L_s$ (SPMSG):

$$\frac{d}{dt}\begin{bmatrix} i_{ds}(t) \\ i_{qs}(t) \\ \omega_r(t) \end{bmatrix} = \mathbf{A}(t)\begin{bmatrix} i_{ds}(t) \\ i_{qs}(t) \\ \omega_r(t) \end{bmatrix} + \mathbf{B}\begin{bmatrix} v_{ds}(t) \\ v_{qs}(t) \\ T_m(t) \end{bmatrix} \tag{7.86}$$

and

$$\mathbf{A}(t) = \begin{bmatrix} -\frac{R_s}{L_s} & \omega_r(t) & 0 \\ -\omega_r(t) & -\frac{R_s}{L_s} & -\frac{\psi_r}{L_s} \\ 0 & \frac{3\,P_p^2\,\psi_r}{2\,J_m} & -\frac{B_m}{J_m} \end{bmatrix}, \quad \mathbf{B} = \begin{bmatrix} \frac{1}{L_s} & 0 & 0 \\ 0 & \frac{1}{L_s} & 0 \\ 0 & 0 & -\frac{P_p}{J_m} \end{bmatrix}. \tag{7.87}$$

The models above show that the rotor electrical speed dynamics is unrelated to the input voltages v_{ds} and v_{qs}; thus, the models are LTV systems with uncoupled-state variables. The following are calculated: (a) sampled-data models for (7.86) by using the forward Euler method, (b) sampled-data model for rotor speed by using the forward Euler method, and (c) sampled-data model for rotor speed by using the truncated Taylor series. Deriving question (b) is unnecessary because it is an integral part of the solution obtained in question (a). Question (b) is introduced to have a smooth discussion.

Solution:

(a) The sampled-data model corresponding to (7.86) with forward Euler method is

$$\begin{bmatrix} i_{ds}(k+1) \\ i_{qs}(k+1) \\ \omega_r(k+1) \end{bmatrix} \approx \begin{bmatrix} 1-\frac{R_s\,T_s}{L_s} & \omega_r(k)\,T_s & 0 \\ -\omega_r(k)\,T_s & 1-\frac{R_s\,T_s}{L_s} & -\frac{\psi_r\,T_s}{L_s} \\ 0 & \frac{3\,P_p^2\,\psi_r\,T_s}{2\,J_m} & 1-\frac{B_m\,T_s}{J_m} \end{bmatrix} \begin{bmatrix} i_{ds}(k) \\ i_{qs}(k) \\ \omega_r(k) \end{bmatrix} + \begin{bmatrix} \frac{T_s}{L_s} & 0 & 0 \\ 0 & \frac{T_s}{L_s} & 0 \\ 0 & 0 & -\frac{P_p\,T_s}{J_m} \end{bmatrix} \begin{bmatrix} v_{ds}(k) \\ v_{qs}(k) \\ T_m(k) \end{bmatrix}. \tag{7.88}$$

(b) To aid the discussion of the rotor speed sampled-data model, the CT dynamic model is represented in a simplified form as follows:

$$\frac{d\,\omega_r(t)}{dt} = K_{\omega r}^{ct}\,\omega_r(t) + K_{iq}^{ct}\,i_{qs}(t) + K_{tm}^{ct}\,T_m(t)$$

with constant CT parameters defined by

$$K_{\omega r}^{ct} = -\frac{B_m}{J_m}, \quad K_{iq}^{ct} = \frac{3\,P_p^2\,\psi_r}{2\,J_m}, \quad K_{tm}^{ct} = -\frac{P_p}{J_m}$$

where superscript ct represents continuous time. The subscripts ωr, iq, and tm are defined in coherence with the variable.

With the help of (7.32), the rotor speed sampled-data model is obtained below:

$$\omega_r(k+1) \approx \omega_r(k) + T_s\left\{K_{iq}^{ct}\,i_{qs}(t) + K_{\omega r}^{ct}\,\omega_r(t) + K_{tm}^{ct}\,T_m(t)\right\}_{t=kT_s}$$

which is simplified as

$$\omega_r(k+1) \approx K_{\omega r}^{dt}\,\omega_r(k) + K_{iq}^{dt}\,i_{qs}(k) + K_{tm}^{dt}\,T_m(k) \tag{7.89}$$

with constant DT parameters defined by

$$K_{\omega r}^{dt} = 1 + K_{\omega r}^{ct}\,T_s, \quad K_{iq}^{dt} = K_{iq}^{ct}\,T_s, \quad K_{tm}^{dt} = K_{tm}^{ct}\,T_s$$

where superscript dt represents discrete time. These values match with the values shown in (7.88). The above analysis shows that the future value of ω_r is unrelated to the predicted voltages v_{ds} and v_{qs} and thus the value remains an uncoupled system in both CT and DT.

(c) The rotor speed sampled-data model with truncated Taylor series method is as follows:

$$\omega_r(k+1) \cong \omega_r(k) + T_s\left\{\frac{d\,\omega_r(t)}{dt}\right\}_{t=kT_s} + \frac{T_s^2}{2}\left\{\frac{d}{dt}\left[\frac{d\,\omega_r(t)}{dt}\right]\right\}_{t=kT_s}.$$

By substituting the derivative of $\omega_r(t)$ from (7.86) and (7.87) into the above expression,

$$\begin{aligned} \omega_r(k+1) \cong\ & \omega_r(k) + T_s\left\{K_{\omega r}^{ct}\,\omega_r(t) + K_{iq}^{ct}\,i_{qs}(t) + K_{tm}^{ct}\,T_m(t)\right\}_{t=kT_s} \\ & + \frac{T_s^2}{2}\left\{\frac{d}{dt}\left[K_{\omega r}^{ct}\,\omega_r(t) + K_{iq}^{ct}\,i_{qs}(t) + K_{tm}^{ct}\,T_m(t)\right]\right\}_{t=kT_s}. \end{aligned}$$

The last term in the above expression indicates that the derivative of $i_{qs}(t)$, $\omega_r(t)$, and $T_m(t)$ are needed to obtain the sampled-data model. The derivative of $T_m(t)$ is zero because it is a constant disturbance input. By substituting the derivative of $i_{qs}(t)$ and $\omega_r(t)$ from (7.86) and (7.87) and by simplification, we obtain the following:

$$\begin{aligned}\omega_r(k+1) \cong\ & K_{\omega r}^{dt}\,\omega_r(k) + K_{id}^{dt}\,\omega_r(k)\,i_{ds}(k) + K_{iq}^{dt}\,i_{qs}(k) \\ & + K_{vq}^{dt}\,v_{qs}(k) + K_{tm}^{dt}\,T_m(k)\end{aligned} \tag{7.90}$$

with the DT constant parameters characterized by

$$\begin{aligned}
K_{\omega r}^{dt} &\cong 1 + K_{\omega r}^{ct}\,T_s + \frac{(K_{\omega r}^{ct}\,T_s)^2}{2} - K_{iq}^{ct}\,\frac{\psi_r\,T_s^2}{2\,L_s} \\
K_{id}^{dt} &\cong K_{iq}^{ct}\,T_s\left(-\frac{T_s}{2}\right) \\
K_{iq}^{dt} &\cong K_{iq}^{ct}\,T_s\left(1 - \frac{R_s\,T_s}{2\,L_s} + K_{\omega r}^{ct}\,\frac{T_s}{2}\right) \\
K_{vq}^{dt} &\cong K_{iq}^{ct}\,T_s\left(-\frac{T_s}{2\,L_s}\right) \\
K_{tm}^{dt} &\cong K_{tm}^{ct}\,T_s\left(1 + K_{\omega r}^{ct}\,\frac{T_s}{2}\right).
\end{aligned}$$

By comparing the sampled-data model in (7.90) with (7.89), it is observed that by truncated Taylor series method, the state variable ω_r becomes a function of v_{qs}. The additional constants introduced by the truncated Taylor series are K_{id}^{dt} and K_{vq}^{dt}, and both of them are a function of previously defined CT constant K_{iq}^{ct}. From this example it is demonstrated that, the uncoupled state variables in CT LTV model becomes coupled state variables in DT by the truncated Taylor series. Similar analysis can be found in [29] for induction machine rotor speed dynamics. ♦

7.8 COMPARISON OF DISCRETIZATION METHODS

In this section, the exact, approximate, and quasi-exact discretization methods are compared. The approximation of derivative with various finite difference methods is summarized in Table 7.2 [4]. Each method approximates the derivative in a different manner. The corresponding z-transformations for derivative are also given.

Table 7.2 Approximation of derivative with different finite difference methods [4]

	Approximate Derivative $d\,\mathbf{x}(t)/dt$	z**-Transform of** s
Forward Euler	$\frac{\mathbf{x}(k+1)-\mathbf{x}(k)}{T_s}$	$\frac{z-1}{T_s}$
Backward Euler	$\frac{\mathbf{x}(k)-\mathbf{x}(k-1)}{T_s}$	$\frac{z-1}{z\,T_s}$
Bilinear	$\frac{\mathbf{x}(k+1)-\mathbf{x}(k-1)}{2\,T_s}$	$\frac{z^2-1}{2\,z\,T_s}$
Truncated Taylor	$\frac{\mathbf{x}(k+1)-\mathbf{x}(k)}{T_s} - \frac{T_s^2}{2}\left\{\frac{d^2\,\mathbf{x}(t)}{dt^2}\right\}_{t=kT_s}$	$\frac{z-1}{T_s}$

The sampled-data models obtained by the exact, approximate, and quasi-exact discretization methods are summarized in Table 7.3. Owing to the finite difference method employed, the output equation is the same for all the methods. The bilinear transformation that is a distinct method among the class of discretization methods uses the present and past state variables, that is, $\mathbf{x}(k-1)$ and $\mathbf{x}(k)$. The backward Euler method is also another distinct method to employ future value of control variable $\mathbf{u}(k+1)$.

The sampled-data models obtained by the ZOH, forward Euler, matrix factorization, and truncated Taylor series are the same. The present-sample state variable $\mathbf{x}(k)$ and control input $\mathbf{u}(k)$ are used to compute the future value of the state variable, $\mathbf{x}(k+1)$. DT matrices $\mathbf{\Phi}$ and $\mathbf{\Gamma}$ are formulated in a different manner with the aforementioned four methods.

Table 7.3 Sampled-data models with different discretization methods

	State Space Equations	$\mathbf{\Phi}$	$\mathbf{\Gamma}$
Exact (ZOH)	$\mathbf{x}(k+1) = \mathbf{\Phi}\,\mathbf{x}(k) + \mathbf{\Gamma}\,\mathbf{u}(k)$	$e^{\mathbf{A}\,T_s}$	$\mathbf{A}^{-1}\,(e^{\mathbf{A}\,T_s} - \mathbf{I})\,\mathbf{B}$
Forward Euler	$\mathbf{x}(k+1) \approx \mathbf{\Phi}\,\mathbf{x}(k) + \mathbf{\Gamma}\,\mathbf{u}(k)$	$\mathbf{I} + \mathbf{A}\,T_s$	$\mathbf{B}\,T_s$
Backward Euler	$\mathbf{x}(k+1) \approx \mathbf{\Phi}\,\mathbf{x}(k) + \mathbf{\Gamma}\,\mathbf{u}(k+1)$	$(\mathbf{I} - \mathbf{A}\,T_s)^{-1}$	$(\mathbf{I} - \mathbf{A}\,T_s)^{-1}\,\mathbf{B}\,T_s$
Bilinear	$\mathbf{x}(k+1) \approx \mathbf{x}(k-1) + \mathbf{\Phi}\,\mathbf{x}(k) + \mathbf{\Gamma}\,\mathbf{u}(k)$	$2\,\mathbf{A}\,T_s$	$2\,\mathbf{B}\,T_s$
Matrix Factorization	$\mathbf{x}(k+1) \cong \mathbf{\Phi}\,\mathbf{x}(k) + \mathbf{\Gamma}\,\mathbf{u}(k)$	$e^{\mathbf{A_c}\,T_s}\,e^{\mathbf{A}_\omega\,T_s}$	$\mathbf{A_c}^{-1}\,(e^{\mathbf{A_c}T_s} - \mathbf{I})\,\mathbf{B}$
Truncated Taylor	$\mathbf{x}(k+1) \cong \mathbf{\Phi}\,\mathbf{x}(k) + \mathbf{\Gamma}\,\mathbf{u}(k)$	$\mathbf{I} + \mathbf{A}\,T_s + \frac{\mathbf{A}^2\,T_s^2}{2}$	$\mathbf{B}\,T_s + \frac{\mathbf{A}\,\mathbf{B}\,T_s^2}{2}$

For all the methods, the output equation is $\mathbf{y}(k) = \mathbf{H}\,\mathbf{x}(k)$, $\mathbf{H} = \mathbf{C}$.

EXAMPLE 7.4 Computation of $\mathbf{\Phi}$ and $\mathbf{\Gamma}$

The following CT LTI parameter matrices from (7.3) are considered as follows:

$$\mathbf{A} = \begin{bmatrix} -\frac{R_i}{L_i} & 0 \\ 0 & -\frac{R_i}{L_i} \end{bmatrix}, \quad \mathbf{B} = \begin{bmatrix} \frac{1}{L_i} & 0 \\ 0 & \frac{1}{L_i} \end{bmatrix}.$$

The parameters are R_i = 10 Ω, L_i = 10 mH, T_s = 100 μs. $\mathbf{\Phi}$ and $\mathbf{\Gamma}$ are calculated with the approximation methods, such as (a) ZOH, (b) forward Euler, (c) backward Euler, and (d) truncated Taylor series.

Solution:

(a) The DT matrices with ZOH method are expressed as follows:

$$\mathbf{\Phi} = e^{\mathbf{A}\,T_s} = \begin{bmatrix} 0.9048 & 0 \\ 0 & 0.9048 \end{bmatrix}, \quad \mathbf{\Gamma} = \mathbf{A}^{-1}(\mathbf{\Phi} - \mathbf{I})\,\mathbf{B} = \begin{bmatrix} 0.0095 & 0 \\ 0 & 0.0095 \end{bmatrix}.$$

(b) The DT matrices by forward Euler approximation are

$$\mathbf{\Phi} \approx \mathbf{I} + \mathbf{A}\,T_s = \begin{bmatrix} 0.9 & 0 \\ 0 & 0.9 \end{bmatrix}, \quad \mathbf{\Gamma} \approx \mathbf{B}\,T_s = \begin{bmatrix} 0.01 & 0 \\ 0 & 0.01 \end{bmatrix}.$$

The values about indicate that the forward Euler method is implicit in nature because of its lower values for $\mathbf{\Phi}$ compared with the ZOH method.

(c) The coefficients in backward Euler method are expressed follows:

$$\mathbf{\Phi} \approx (\mathbf{I} - \mathbf{A}\,T_s)^{-1} = \begin{bmatrix} 0.9091 & 0 \\ 0 & 0.9091 \end{bmatrix}, \quad \mathbf{\Gamma} \approx \mathbf{\Phi}\,\mathbf{B}\,T_s = \begin{bmatrix} 0.0091 & 0 \\ 0 & 0.0091 \end{bmatrix}.$$

The backward Euler approximation method is explicit in nature because the $\mathbf{\Phi}$ value is higher than that in the ZOH method.

(d) The DT coefficients with the truncated Taylor series are

$$\mathbf{\Phi} \cong \mathbf{I} + \mathbf{A}\,T_s + \frac{\mathbf{A}^2\,T_s^2}{2} = \begin{bmatrix} 0.905 & 0 \\ 0 & 0.905 \end{bmatrix}$$

$$\mathbf{\Gamma} \cong \mathbf{B}\,T_s + \frac{\mathbf{A}\,\mathbf{B}T_s^2}{2} = \begin{bmatrix} 0.0095 & 0 \\ 0 & 0.0095 \end{bmatrix}.$$

With T_s = 100 μs, all approximate and quasi-exact DT matrix coefficient values are close to that obtained by the ZOH method. The coefficients with bilinear transformation are excluded because they have different expressions for the sampled-data models. ♦

7.9 OFFLINE CALCULATION OF DT PARAMETERS USING MATLAB

The previous example considers hand calculation of DT parameters for second-order matrices. For higher-order matrices, the calculation of DT matrices $\mathbf{\Phi}$ and $\mathbf{\Gamma}$ is time consuming and cumbersome. In this section, a simple approach to calculate $\mathbf{\Phi}$ and $\mathbf{\Gamma}$ offline by computer simulations softwares such as MATLAB is discussed. The methodologies to calculate DT parameters by exact, approximation, and quasi-exact methods are discussed. For all cases, the CT variables are defined to provide straightforward analysis.

As mentioned in Section 7.5, the Laplace transformations can be used for the manual calculations of $\mathbf{\Phi}$ and $\mathbf{\Gamma}$. The MATLAB code to compute DT parameters by Laplace transformations is shown below by considering the state matrix given in (7.4). The symbol » implies that the code can be executed at MATLAB command window.

```
» syms  s Ri Li wg Ts real             % Define variables
» A     = [-Ri/Li,wg;-wg,-Ri/Li];      % Define matrix A
» IsIA  = inv(s*eye(size(A))-A);       % Compute inverse of [sI − A]
» Phi   = ilaplace(IsIA,Ts)            % Print matrix Φ as L^{-1}[sI − A]^{-1}
Phi =
[  exp(-(Rs*Ts)/Ls)*cosh(Ts*wr*i),        exp(-(Rs*Ts)/Ls)*sin(Ts*wr)]
[     -exp(-(Rs*Ts)/Ls)*sin(Ts*wr),    exp(-(Rs*Ts)/Ls)*cosh(Ts*wr*i)]
```

The above procedure contains complex variables. The `syms` command can be replaced with real values to compute DT parameters numerically. Another approach to compute matrix exponential is to use the in-built math function `expm` of MATLAB which uses a scaling and squaring algorithm with a Pade approximation (methods 2 and 3 among the 19 methods outlined in [19]). The code to obtain $\mathbf{\Phi}$ and $\mathbf{\Gamma}$ is shown below.

```
» syms   Ri Li wg Ts real                   % Define variables
» A      = [-Ri/Li,wg;-wg,-Ri/Li];          % Define matrix A
» B      = [1/Li,0;0,1/Li];                 % Define matrix B
» Phi    = expm(A*Ts)                       % Print matrix Φ which is an exponential of A Ts
Phi =
[   exp(-(Rs*Ts)/Ls)*cos(Ts*wr),   exp(-(Rs*Ts)/Ls)*sin(Ts*wr)]
[  -exp(-(Rs*Ts)/Ls)*sin(Ts*wr),   exp(-(Rs*Ts)/Ls)*cos(Ts*wr)]

» Gamma   = A\(Phi-eye(size(A)))*B   % Print matrix Γ as A^-1 (Φ − I) B
```

$\mathbf{\Gamma}$ goes beyond the display range; thus, it is not shown here. The matrix factorization method discussed in Section 7.7.1 is verified by MATLAB. The stationary frame IG state matrix $\mathbf{A}$ is divided into two parts: $\mathbf{A_c}$ corresponding to time-invariant variables and $\mathbf{A}_\omega$ with rotor speed ω_r.

```
» syms  tau_sigma kr sigma Ls tau_r Lm wr Ts B real
» Ac    = [-1/tau_sigma,0,kr/(sigma*Ls*tau_r),(kr*wr)/(sigma*Ls);
          0,-1/tau_sigma,-(kr*wr)/(sigma*Ls),kr/(sigma*Ls*tau_r);
          Lm/tau_r,0,-1/tau_r,-wr;
          0,Lm/tau_r,wr,-1/tau_r];
» Aw    = [0,0,0,(kr*wr)/(sigma*Ls);0,0,-(kr*wr)/(sigma*Ls),0;
          0,0,0,-wr;0,0,wr,0];
» Phic  = expm(Ac*Ts);
» Phiw  = expm(Aw*Ts)
[  1,  0,  (kr - kr*cos(Ts*wr))/(Ls*sigma),        (kr*sin(Ts*wr))/(Ls*sigma)]
[  0,  1,      -(kr*sin(Ts*wr))/(Ls*sigma),  (kr - kr*cos(Ts*wr))/(Ls*sigma)]
[  0,  0,                      cos(Ts*wr),                      -sin(Ts*wr)]
[  0,  0,                     -sin(Ts*wr),                       cos(Ts*wr)]
```

Matrix $\mathbf{\Phi_c}$ goes beyond the display range; thus, it is not printed here. The MATLAB command `c2d` can also be used to convert CT models to DT models. This command works with numbers only. The second program discussed earlier is considered with `c2d`.

```
» Ri=10;        Li=10e-3;                 % Define parameters
» Ts=100e-6;    wg=2*pi*60;               % Define parameters
» A             = [-Ri/Li,wg;-wg,-Ri/Li]; % Define matrix A
» B             = [1/Li,0;0,1/Li];        % Define matrix B
» C             = eye(size(A));           % Define matrix C
» D             = [0];                    % Define matrix D
» ctss          = ss(A,B,C,D);            % Create continuous-time state space model
» dtss          = c2d(ctss,Ts,'zoh')      % Create sampled-data model with ZOH
```

The above command returns DT parameter matrices. The `c2d` command also supports other methods, such as FOH (`'foh'`), impulse invariant (`'impulse'`), bilinear transformation by numerical integration (`'tustin'`), and pole-zero matching (`'matching'`). Further details are in the MATLAB documentation page in [30].

7.10 CONCLUDING REMARKS

In this chapter, the concept of sampled-data models is introduced. The correlation between the state-space equations and transfer function in CT and DT environment is established. The LTI and LTV models with independent terms, cross-coupled terms, and uncoupled terms are established on the basis of the power converter and wind generator models discussed in Chapters 5 and 6. The mapping of CT models to DT models by various exact, approximate, and quasi-exact methods is analyzed from the MPC perspective. The discretization error produced by various methods is negligible when the sampling time T_s is small (in the order of microseconds).

A comprehensive summary on the usefulness of various discretization methods to energy conversion applications, such as passive loads, motor drives, and WECS is presented in Table 7.4. The discretization methods are compared based on accuracy level (✔✔✔ is highly accurate), application to LTI models (✔ means possible and ✖ means not possible), application to LTV models, application to cross-coupled terms, and application to uncoupled terms. The ZOH method is the highly accurate method among the six listed in Table 7.4. The ZOH method is most preferred for LTI systems because all the DT parameters can be computed offline. Forward Euler method is simplest and the most straightforward method for the discretization of LTI and LTV models. The matrix factorization provides the best compromise between the accuracy and complexity of discretization. By using this approach, the motor drives and wind generators exhibit better control performance than the forward Euler method. The truncated Taylor series is also simple and useful for LTV models with uncoupled terms. We hope that the material presented in this chapter fulfills the gap in MPC literature.

Table 7.4 Summary of the usefulness of discretization methods for energy conversion applications

	Accuracy Level	Linear Time Invariant Systems	Linear Time Variant Systems	Cross-Coupled Variables	Uncoupled Variables
Exact (ZOH)	✔✔✔	✔	✖	✔	✖
Forward Euler	✔	✔	✔	✔	✖
Backward Euler	✔	✔	✔	✖	✖
Bilinear	✔	✔	✔	✔	✖
Matrix Factorization	✔✔	✖	✔	✔	✖
Truncated Taylor	✔✔	✔	✔	✔	✔

Link to Next Chapters:

The discretization methods discussed in this chapter are used in Chapters 8 to 12 to design the MPC schemes for grid-connected converters and variable-speed WECS. To reduce the complexity of analysis and maintain coherence among Chapters 8 to 12, the forward Euler method is employed for discretization. However, adopting a suitable discretization method to improve the performance of overall WECS is advised.

REFERENCES

1. K. J. Åström and B. Wittenmark, *Computer Controlled Systems. Theory And Design*, 3rd ed. Englewood Cliffs, NJ: Prentice Hall, 1997.
2. K. Moudgalya, *Digital Control*. Hoboken, NJ: John Wiley & Sons, 2007.
3. J. Baggio, H. Hey, H. Grundling, H. Pinheiro, and J. Pinheiro, "Discrete control for three-level boost PFC converter," in *IEEE International Telecommunications Energy Conference (INTELEC)*, 2002, pp. 627–633.
4. W. Levine, *The Control Handbook, Second Edition: Control System Fundamentals*, 2nd ed., ser. Electrical Engineering Handbook. Boca Raton, FL: Taylor and Francis, 2010.
5. C. E. García, D. M. Prett, and M. Morari, "Model predictive control: Theory and practice – A survey," *Automatica*, vol. 25, no. 3, pp. 335–348, 1989.
6. M. Morari and J. H. Lee, "Model predictive control: Past, present and future," *International Journal of Computers and Chemical Engineering*, vol. 23, no. 4–5, pp. 667–682, 1999.
7. J. I. Yuz and G. C. Goodwin, *Sampled-Data Models for Linear and Nonlinear Systems*, ser. Communications and Control Engineering. London: Springer Verlag, November 2014.
8. G. Goodwin, J. Agüero, M. Cea Garridos, M. Salgado, and J. Yuz, "Sampling and sampled-data models: The interface between the continuous world and digital algorithms," *IEEE Control Systems Magazine*, vol. 33, no. 5, pp. 34–53, October 2013.
9. V. Yaramasu, B. Wu, and J. Chen, "Model-predictive control of grid-tied four-level diode-clamped inverters for high-power wind energy conversion systems," *IEEE Transactions on Power Electronics*, vol. 29, no. 6, pp. 2861–2873, June 2014.
10. R. C. Dorf and R. H. Bishop, *Modern Control Systems*, 9th ed. Upper Saddle River, NJ: Prentice Hall, 2000.
11. G. Franklin, J. Powell, and A. Emami-Naeini, *Feedback Control of Dynamic Systems*, 7th ed. Pearson Education, Limited, 2011.
12. B. C. Kuo and F. Golnaraghi, *Automatic Control Systems*, 8th ed. Hoboken, NJ: John Wiley & Sons, 2002.
13. B. Kuo, *Discrete-Data Control Systems*, ser. Instrumentation and controls series. Englewood Cliffs, NJ: Prentice Hall, 1970.
14. G. F. Franklin, M. L. Workman, and D. Powell, *Digital Control of Dynamic Systems*, 3rd ed. Boston: Addison-Wesley Longman, 1997.
15. Y. Duan and K. Jin, "Digital controller design for switchmode power converters," in *IEEE Applied Power Electronics Conference and Exposition (APEC)*, vol. 2, March 1999, pp. 967–973.
16. H. Miranda, P. Coftes, J. Yuz, and J. Rodríguez, "Predictive torque control of induction machines based on state-space models," *IEEE Transactions on Industrial Electronics*, vol. 56, no. 6, pp. 1916–1924, June 2009.
17. D. Bernstein, *Matrix Mathematics: Theory, Facts, and Formulas*, 2nd ed., ser. Princeton reference. Princeton, NJ: Princeton University Press, 2009.
18. V. Yaramasu, M. Rivera, M. Narimani, B. Wu, and J. Rodriguez, "Model predictive approach for a simple and effective load voltage control of four-leg inverter with an output LC filter," *IEEE Transactions on Industrial Electronics*, vol. 61, no. 10, pp. 5259–5270, October 2014.
19. C. Moler and C. Van Loan, "Nineteen dubious ways to compute the exponential of a matrix, twenty-five years later," *SIAM review*, vol. 45, no. 1, pp. 3–49, 2003.
20. M. Visser, S. Stramigioli, and C. Heemskerk, "Cayley–Hamilton for roboticists," in *IEEE/RSJ International Conference on Intelligent Robots and Systems*, Beijing, PR China, October 2006, pp. 4187–4192.
21. V. Yaramasu, B. Wu, M. Rivera, M. Narimani, S. Kouro, and J. Rodriguez, "Generalised approach for predictive control with common-mode voltage mitigation in multilevel diode-clamped converters," *IET Power Electronics*, vol. 8, no. 8, p. 1440–1450, August 2015.
22. J. Rodríguez, J. Pontt, C. A. Silva, P. Correa, P. Lezana, P. Coftes, and U. Ammann, "Predictive current control of a voltage source inverter," *IEEE Transactions on Industrial Electronics*, vol. 54, no. 1, pp. 495–503, February 2007.
23. V. Yaramasu, M. Rivera, B. Wu, and J. Rodriguez, "Model predictive current control of two-level four-leg inverters – Part I: Concept, algorithm and simulation analysis," *IEEE Transactions on Power Electronics*, vol. 28, no. 7, pp. 3459–3468, July 2013.
24. J. Jiang, "Preservations of positive realness under discretizations," *Journal of the Franklin Institute*, vol. 330, no. 4, pp. 721–734, 1993.
25. C. Rojas, J. Yuz, C. Silva, and J. Rodríguez, "Comments on predictive torque control of induction machines based on state-space models," *IEEE Transactions on Industrial Electronics*, vol. 61, no. 3, pp. 1635–1638, March 2014.
26. M. Fan, H. Lin, and T. Lan, "Model predictive direct torque control for SPMSM with load angle limitation," *Progress In Electromagnetics Research B*, vol. 58, pp. 245–256, March 2014.
27. C. Silva and J. Yuz, "On sampled-data models for model predictive control," in *IEEE Industrial Electronics Conference (IECON)*, Glendale, AZ, November 2010, pp. 2966–2971.
28. P. Vaclavek and P. Blaha, "PMSM model discretization for model predictive control algorithms," in *IEEE/SICE International Symposium on System Integration (SII)*, Kobe, Japan, December 2013, pp. 13–18.
29. P. Vaclavek and P. Blaha, "Enhanced discrete time model for AC induction machine model predictive control," in *IEEE Industrial Electronics Conference (IECON)*, Montreal, Canada, October 2012, pp. 5043–5048.
30. Mathworks Inc., "R2014a documentation center: c2d," available at: https://www.mathworks.com/help/control/ref/c2d.html, accessed on July 2014.

PART III

CONTROL OF VARIABLE-SPEED WECS

CHAPTER 8

CONTROL OF GRID-SIDE CONVERTERS IN WECS

8.1 INTRODUCTION

As outlined in Chapters 1 and 2, Type 3 and 4 wind turbines (WTs) are connected to the utility grid via power electronic converters. The generic configuration of Type 3 and 4 variable-speed wind energy conversion system (WECS) is shown in Figure 8.1. Type 3 WECS uses a doubly fed induction generator (DFIG) and a partial-scale (30%) power converter in the rotor circuit. Type 4 WECS uses a squirrel-cage induction generator (SCIG), permanent magnet synchronous generator (PMSG) or wound rotor synchronous generator (WRSG), and a full-scale (100%) power converter between the generator and utility grid. The wind generator and power converters can be designed for low voltage (LV) or medium voltage (MV) operations. Among the various possible power converter topologies for variable-speed WECS, the grid-side converter (GSC or DC/AC inverter) is a common and pivotal element not only for delivering the generated power to the utility grid but also for complying with stringent grid codes. Therefore, the digital control of GSCs is important in the successful and efficient operation of Type 3 and 4 WECS [1, 2].

The primary and secondary control objectives for a machine-side converter (MSC or AC/DC rectifier) and a GSC are shown in Figure 8.1. The MSC harvests more energy from wind by performing maximum power point tracking (MPPT) by generator speed, torque, or power control (also called active power control) [3, 4]. The net DC-bus voltage control, grid reactive power control, grid synchronization, and fault-ride through (FRT) operation are performed by the GSC [5]. For high-power WECS, the MSC and GSC should operate with low switching frequency to minimize switching losses. For multilevel power

Model Predictive Control of Wind Energy Conversion Systems, First Edition. Venkata Yaramasu and Bin Wu.

Companion Website: www.wiley.com/go/yaramasu_wu_mpc_wecs

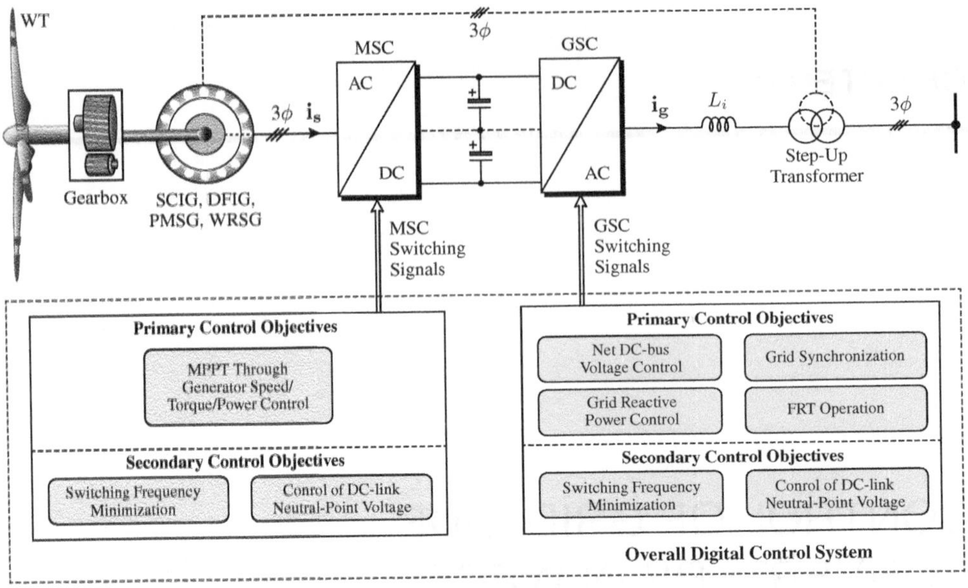

Figure 8.1 Block diagram of Type 3 and 4 WECS with power conversion configuration and primary and secondary control objectives.

converters, the DC-link neutral-point voltage is controlled by both MSC and GSC to ensure low semiconductor device voltage stress. The primary and secondary control objectives are achieved by properly generating switching signals for MSC and GSC during each sampling interval. The control of MSC changes according to the type of wind generator, but the control of GSC is the same for Type 3 and 4 WTs. Therefore, the control of GSC is introduced as a separate chapter in this book to complement the analysis of variable-speed WECS in Chapters 9 to 12. This chapter deals with the model predictive control (MPC) of GSCs in different reference frames. The predictive current control (PCC) and predictive power control (PPC) are discussed for both LV and MV power converters.

Chapter Overview

- The configuration of GSCs in Type 3 and 4 WECS is outlined in Section 8.2. The design and control of GSC is presented in Section 8.3.
- The continuous-time (CT) modeling and discrete-time (DT) modeling of GSCs are presented in Section 8.4 in natural (abc), stationary ($\alpha\beta$), and synchronous (dq) frames. The models of two-level voltage source inverter (2L-VSI) and neutral-point clamped (NPC) inverter output voltages and DC capacitors voltage are also discussed.
- Section 8.5 presents the calculation of reference grid-side variables when MPPT is performed by MSC or GSC. The PCC, which is analogous to the classical voltage-oriented control (VOC), is analyzed in the dq-frame with 2L-VSI in Section 8.6.
- In Section 8.7, the PCC scheme for NPC inverter is developed in $\alpha\beta$-frame. The PPC scheme is discussed in Section 8.8 when MPPT is performed by NPC inverter.
- The real-time implementation issues of MPC schemes for GSCs are investigated in Section 8.9. The concluding remarks of this chapter are given in Section 8.10.

8.2 CONFIGURATION OF GSCS IN TYPE 3 AND 4 WECS

This section deals with various power converter configurations for induction and synchronous generator-based WECS. The 2L back-to-back (BTB) connected voltage source converters (VSCs) are dominantly used in Type 3 WECS for LV operation. For Type 4 WECS, many power converter configurations can be used, including LV/MV matrix converters, LV BTB 2L-VSC, MV BTB NPC converter (3L-VSC), three-phase (3ϕ) diode-bridge rectifier + 2L/3L VSI, and 3ϕ diode-bridge rectifier + 2L/3L boost converter + 2L/3L VSI. These power conversion systems are broadly classified as a single-, two-, and three-stage systems. Given the wide variety of power converter configurations, the corresponding electrical control objectives are also different. The aerodynamic power regulation methods are omitted here to simplify the analysis.

8.2.1 Single-Stage Power Conversion

The generator-side variable voltage and variable frequency is converted to grid compatible fixed voltage and fixed frequency by a single-stage AC/AC power converter, also known as a matrix converter. Compared with BTB 2L-VSCs, matrix converters provide more silicon-based power conversion without any intermediate energy storage components (DC-link capacitors), in addition to low switching losses and low generator-side current distortions [2]. However, given the direct coupling between generator and grid terminals, FRT compliance is complicated. Moreover, matrix converters are not commercialized for high-power WECS yet. The generator- and grid-side currents ($\mathbf{i_s}$ and $\mathbf{i_g}$) are regulated simultaneously to control the MPPT and grid power factor (PF). The control system development is complicated compared with BTB 2L-VSC.

8.2.2 Two-Stage Power Conversion

The two-stage configuration of Type 3 and 4 WECS with BTB 2L-VSC is shown in Figure 8.2. The generator-side 2L voltage source rectifier (2L-VSR) and grid-side 2L-VSI are linked by the DC-link stage (capacitor). The DC-link provides complete decoupling between the generator and grid, thus facilitating decoupled control, particularly during abnormal grid conditions such as grid faults. The grid connection is usually performed by an inductive harmonic filter L_i (or by inductive-capacitive-inductive filter, LCL) to reduce the grid current distortion, and a step-up transformer is used to increase the 2L-VSI output voltage to a wind farm collection-point voltage level.

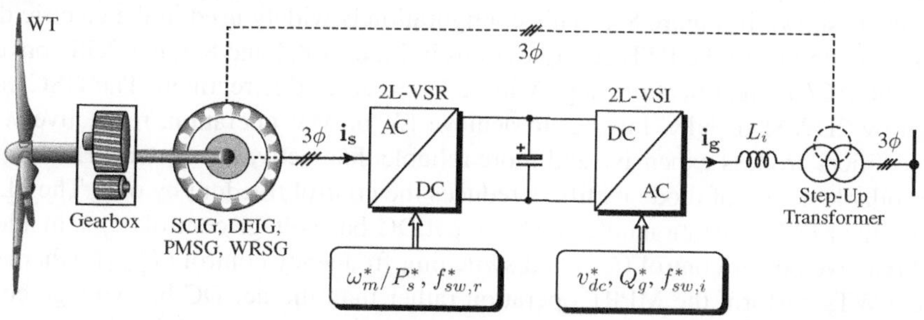

Figure 8.2 Configuration and control objectives of Type 3 and 4 WECS with BTB 2L-VSC.

For Type 4 WECS, BTB 2L-VSC is connected between the generator and grid. For Type 3 WECS, BTB 2L-VSC is connected in the rotor circuit, with the stator terminals of DFIG directly connected to the utility grid via step-up transformer. Given the direct connection of DFIG stator to the utility grid, the FRT operation of Type 3 WECS is complicated compared with Type 4 WECS. The 2L-VSR is used to extract power from wind, whereas the 2L-VSI is employed to make the generated power compatible to the utility grid. On the basis of the measured wind speed, the MPPT algorithm provides reference speed/power (ω_m^* or P_s^*) to the 2L-VSR control system. The net DC-bus voltage reference (v_{dc}^*) and grid reactive power reference (Q_g^*) are provided to the 2L-VSI control system. To accomplish these objectives, the generator and grid currents ($\mathbf{i_s}$ and $\mathbf{i_g}$) are regulated by the 2L-VSR and 2L-VSI digital control systems, respectively. The switching frequencies of VSR and VSI ($f_{sw,r}^*$ and $f_{sw,i}^*$) are minimized to improve the power conversion efficiency.

For high-power Type 4 WECS, the power handling capacity can be increased in two ways: (1) by using multiple BTB 2L-VSCs in parallel and (2) by using MV power converters. The latter approach is proven to be an economical and efficient solution. BTB NPC converters with split DC link are popularly used in Type 4 MV-WECS (Figure 8.3). Compared with 2L-VSCs, NPC converters for high-power WECS offer many benefits such as MV operations with reduced device voltage ratings, increased equivalent switching frequencies, low values for the grid-side filter, grid friendly waveforms, and grid code compliance [6, 7]. Compared with BTB 2L-VSC, the NPC rectifier and inverter perform an additional control function to regulate the DC-link neutral-point voltage ($v_C^* = 0$).

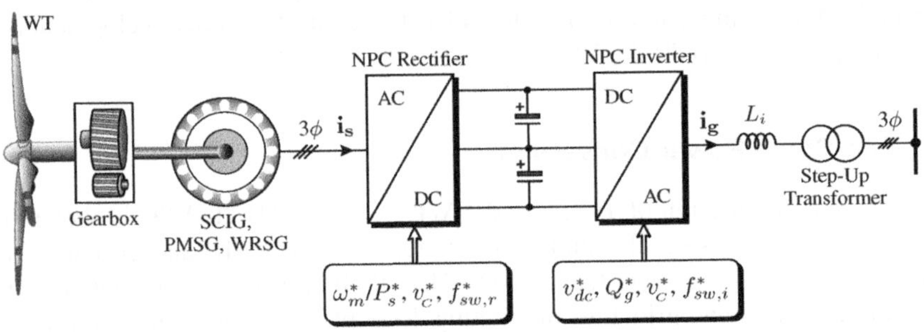

Figure 8.3 Configuration and control objectives of Type 4 WECS with BTB NPC converter.

The two-stage power conversion configuration based on passive generator-side (PGS) converters is shown in Figure 8.4. This configuration is widely used in the electric drives industry. In contrast to the BTB configurations in Figures 8.2 and 8.3, the PGS converters replace the AC/DC rectification stage with a 3ϕ diode-bridge rectifier. The GSC can be realized by 2L-VSI or NPC inverter to achieve LV or MV operation, respectively. The diode rectifiers are less expensive and more reliable than active converters. However, the uncontrollable nature of diode rectifiers reduces the control freedom by one. The 2L-VSI performs the MPPT operation (ω_m^* or P_s^*) or net DC-bus voltage control (v_{dc}^*), in addition to grid reactive power control (Q_g^*) and switching frequency control ($f_{sw,i}^*$), whereas the practical WTs perform the MPPT operation rather than the net DC-bus voltage control. The NPC inverter regulates the DC-link neutral-point voltage, in addition to the control objectives fulfilled by the 2L-VSI.

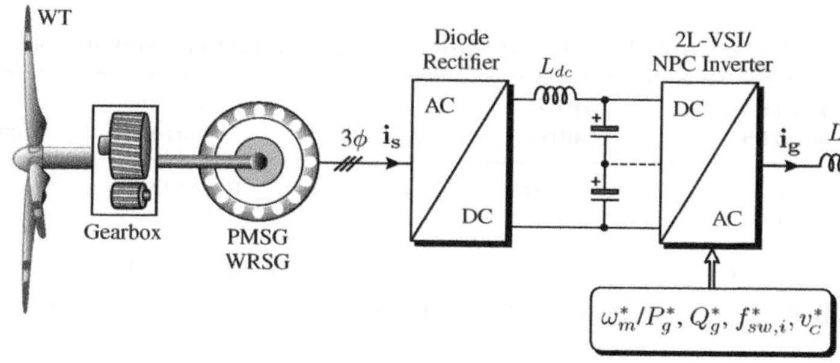

Figure 8.4 Configuration and control objectives of Type 4 WECS with a 3ϕ diode-bridge rectifier and 2L-VSI/ NPC inverter.

8.2.3 Three-Stage Power Conversion

The three-stage power conversion configuration for Type 4 WECS is shown in Figure 8.5. The 2L boost converter is used in conjunction with a 2L-VSI for the LV operation, whereas the 3L boost converter is used with an NPC inverter to enable MV operation. In contrast to the configuration in Figure 8.4, the intermediate 2L boost converter increases the control freedom by one. The 2L boost converter performs MPPT operation (ω_m^* or P_{dc}^*), and GSC is in charge of v_{dc}^* and Q_g^* control. Compared with the 2L boost converter, the 3L boost converter increases the control freedom by one; hence, the 3L boost converter is capable of controlling the DC-link capacitors voltage, in addition to the MPPT. The grid-side NPC inverter does not need to control the DC-link capacitors voltage. Therefore, the grid-side NPC inverter has control flexibility for regulating v_{dc}^* and Q_g^* tightly. This feature leads to better grid power quality than in BTB NPC converters in Figure 8.3.

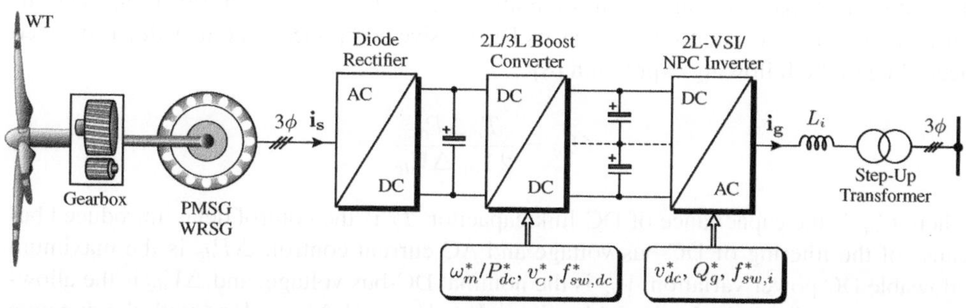

Figure 8.5 Configuration and control objectives of Type 4 WECS with three-phase diode-bridge rectifier, 2L/3L boost converter and 2L-VSI/ NPC inverter.

A detailed summary on the control objectives for various power conversions with respect to the number of power conversion stages, power converter topology, MPPT control, net DC-bus voltage (v_{dc}) control, grid reactive power (Q_g) control, and DC-link neutral-point voltage (v_C) control is given in Table 8.1. The matrix converters and partial-scale BTB 2L-VSCs with DFIG need complicated control structures to implement FRT control. The decoupling between MSC and GSC provides flexibility to Type 4 WECS in performing FRT control. The control objectives can be possibly swapped between MSC and GSC and also between boost converter and GSC. Therefore, GSC can perform MPPT, whereas MSC or boost converter can perform v_{dc} control.

Table 8.1 Summary of control objectives for various power conversion configurations

Power Configuration	Power Converter	MPPT Control	v_{dc} Control	Q_g Control	v_C Control
Single-Stage	Matrix Converters	By AC/AC Stage	Not Needed	By AC/AC Stage	Not Needed
Two-Stage	BTB 2L-VSC	By 2L-VSR	By 2L-VSI	By 2L-VSI	Not Needed
	BTB 3L-VSC	By 3L-VSR	By 3L-VSI	By 3L-VSI	By 3L-VSR and 3L-VSI
	Diode + 2L-VSI	By 2L-VSI	Not Possible	By 2L-VSI	Not Needed
	Diode + 3L-VSI	By 3L-VSI	Not Possible	By 3L-VSI	By 3L-VSI
Three-Stage	Diode + 2L Boost + 2L-VSI	By 2L Boost	By 2L-VSI	By 2L-VSI	Not Needed
	Diode + 3L Boost + 3L-VSI	By 2L Boost	By 3L-VSI	By 3L-VSI	By 3L Boost

8.3 DESIGN AND CONTROL OF GSC

In this section, the design and control of GSC is discussed with respect to the design of passive components (DC-link capacitors and AC-side harmonic filter), design of reference net DC-bus voltage magnitude, definition of grid PF, and grid voltage orientation (grid synchronization).

8.3.1 Design of Passive Components

The GSC needs passive components on both DC and AC sides to perform storage and filtering functions. A capacitor is used as its DC passive component, and its value is decided according to the following expression [8]:

$$C_{dc} = \frac{T_r\,\Delta P_{dc}}{2\,V_{dc}\,\Delta V_{dc}} \tag{8.1}$$

where C_{dc} is the capacitance of DC-link capacitor, T_r is the control delay introduced because of the filtering of DC-bus voltage and AC current control, ΔP_{dc} is the maximum allowable DC power variation, V_{dc} is the nominal DC-bus voltage, and ΔV_{dc} is the allowable DC-bus voltage ripple. Variables T_r, ΔP_{dc}, V_{dc}, and ΔV_{dc}, along with the dynamic response, size, and cost, are used as a trade-off to design the DC-link capacitor value. For high-power applications, the DC-link capacitor value can be simply set to 4.0 pu to obtain a fair trade-off between different design factors.

The AC passive components store only a fraction ($\approx 5\%$) of the total system energy stored. The AC-side harmonic filter can be realized by inductive (L) or inductive-capacitive-inductive (LCL) filter. Compared with the L filter, the LCL filter provides excellent grid current quality; however, this filter involves the critical issue of filter resonance during step change in operating conditions [9]. In this chapter, we mainly deal with the L-filter-based GSC design and control. In commercial WTs based on 2L-VSCs, the harmonic filter L value (combined with the leakage inductance of the step-up transformer, if any) is chosen to be in the range of 0.1–0.25 pu [2]; this value decreases with multilevel converters when the VSI output waveform contains multiple voltage levels.

8.3.2 Design of Reference DC-Bus Voltage

The grid-connected VSI operates with a modulation index m_a between zero and one. To ensure the proper operation of the VSI, the DC-bus voltage v_{dc} should be higher than the *peak* grid line-to-line voltage $V_{g,ll}$. The reference DC-bus voltage v_{dc}^* is selected according to the modulation index and grid *rms* phase voltage V_g, as demonstrated below [10]:

$$v_{dc}^* = \frac{\sqrt{2}\,V_{g,ll}}{m_a} = \frac{\sqrt{6}\,V_g}{m_a}, \quad 0 < m_a \leq 1. \tag{8.2}$$

The *rms* grid phase voltage V_g is equal to the *rms* value of the fundamental frequency component of the VSI phase voltage V_{i1}. The selection of appropriate v_{dc}^* should consider the grid voltage variations and transients during dynamic operating conditions. A modulation index value of 0.8 is reasonable for allowing a 20% margin for adjustment during transients. Therefore, the v_{dc}^* is defined in terms of grid phase voltage as follows:

$$v_{dc}^* = \frac{\sqrt{6}\,V_g}{m_a} = \frac{\sqrt{6}\,V_g}{0.8} = 3.062 \times V_g. \tag{8.3}$$

The control scheme for the VSI should ensure that the v_{dc} is maintained at its reference value during all operating conditions. The net DC-bus voltage control is usually realized by a PI controller in the outer control loop.

8.3.3 Definition of Grid Power Factor

The power flow in GSC is bidirectional: (1) from DC side to AC side, also called an inverter operation; (2) from AC side to DC side, also called a rectifier operation. During normal WECS operation, power flows from the DC-side to AC-side (i.e., inverter operation). However, during system startup, the rectifier operation is used to pre-charge the DC-link capacitors and draw magnetizing current for SCIG and DFIG. The active and reactive powers delivered to (or absorbed from) the grid are defined by the following:

$$\begin{aligned} P_g &= 3\,V_g\,I_g\cos\varphi_g \\ Q_g &= 3\,V_g\,I_g\sin\varphi_g \end{aligned} \tag{8.4}$$

where φ_g is the PF angle, i.e., the angle between the grid voltage and the current vectors.

The four-quadrant operation of GSC is illustrated in Figure 8.6, where quadrants I and IV correspond to inverter operation and quadrants II and III correspond to rectifier operation. The DC link acts as a source during inverter operations or as a load during rectifier operations. The phase-*a* grid voltage v_{ag} and current i_{ag} waveforms are also shown in Figure 8.6. A summary of the PF angle, active power, and reactive power values is given as follows for all four quadrants:

- **Quadrant I:** PF angle is $0° \leq \varphi_g < 90°$, $\cos\varphi_g$ is positive, and $\sin\varphi_g$ is positive. Therefore, P_g is positive and Q_g is positive $\Rightarrow$ "inverter with lagging PF".
- **Quadrant IV:** PF angle is $270° \leq \varphi_g < 360°$, $\cos\varphi_g$ is positive, and $\sin\varphi_g$ is negative. Therefore, P_g is positive and Q_g is negative $\Rightarrow$ "inverter with leading PF".
- **Quadrant II:** PF angle is $90° \leq \varphi_g < 180°$, $\cos\varphi_g$ is negative, and $\sin\varphi_g$ is positive. Therefore, P_g is negative and Q_g is positive $\Rightarrow$ "rectifier with lagging PF".
- **Quadrant III:** PF angle is $180° \leq \varphi_g < 270°$, $\cos\varphi_g$ is negative, and $\sin\varphi_g$ is negative. Therefore, P_g is negative and Q_g is negative $\Rightarrow$ "rectifier with leading PF".

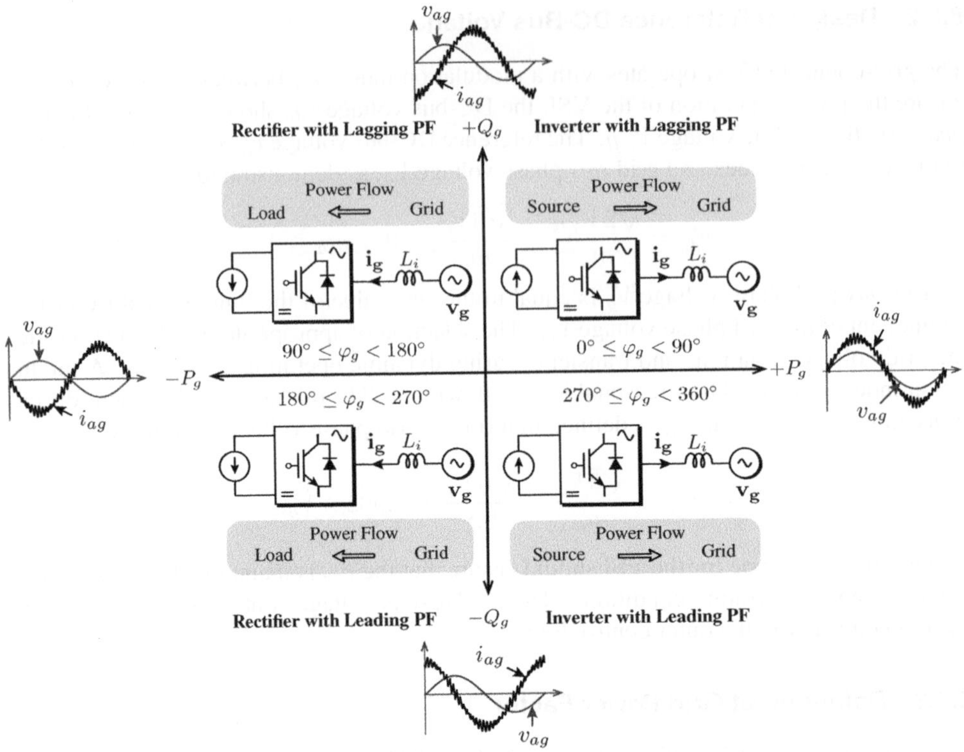

Figure 8.6 Definition of power flow between grid and DC link, and grid PF.

8.3.4 Grid Voltage Orientation

In Chapter 3, various digital control techniques are used to control the load current of three-phase VSI with RL load, including hysteresis control, linear control with PI regulator + pulse width modulation (PWM) or space vector modulation (SVM), sliding mode control, fuzzy logic control, artificial neural networks based control, and predictive control. These digital current control techniques with a slight modification can be applied for GSC control. The grid currents or active and reactive power can be controlled in a decoupled manner by employing grid voltage orientation or grid flux orientation. In summary, four digital control techniques can be utilized to control GSCs [11]:

- VOC, which is similar to the field-oriented control of electric drives, performs decoupled grid current control.
- Voltage-oriented direct power control (V-DPC), which is similar to the direct torque control of electric drives, performs decoupled grid active and reactive power control.
- Virtual flux-oriented control, which is similar to VOC, performs decoupled grid current control.
- Virtual flux-oriented direct power control, which is similar to V-DPC, performs decoupled grid active and reactive power control.

Among the four digital control techniques, the VOC and V-DPC are widely adopted by the wind energy industry, with the former one being very popular. As mentioned in Chapter 3, current control in synchronous (dq) reference frame (SRF) has many advantages including the DC nature of grid currents, accurate reference tracking with less steady-state errors, and easy tuning of PI controller parameters among others. The VOC scheme uses closed-loop current control in the dq-frame. The measured three-phase currents (i_{ag}, i_{bg}, and i_{cg}) and three-phase voltages (v_{ag}, v_{bg}, and v_{cg}) are transformed into SRF currents (i_{dg} and i_{qg}) and voltages (v_{dg} and v_{qg}). Similar to the wind generators modeling (refer to Equations (6.16) and (6.17) in Chapter 6), the grid stator active and reactive powers are computed in SRF as follows:

$$\begin{aligned} P_g &= \mathfrak{Re}\,(S_g) = 1.5\,(v_{dg}\,i_{dg} + v_{qg}\,i_{qg}) \\ Q_g &= \mathfrak{Im}\,(S_g) = 1.5\,(v_{qg}\,i_{dg} - v_{dg}\,i_{qg}) \end{aligned} \tag{8.5}$$

where v_{dg} and v_{qg} are the d- and q-axis grid voltages, respectively; i_{dg} and i_{qg} are the d- and q-axis grid currents, respectively; S_g is the grid apparent power.

From the above expressions, P_g and Q_g are related to both i_{dg} and i_{qg}. The design of control scheme becomes easy if v_{qg} is zero; this situation leads to a linear relationship between P_g and i_{dg} and between Q_g and i_{qg} for a constant v_{dg} value. The process of aligning the d-axis of the SRF with the grid voltage vector $\mathbf{v_g}$ is called voltage orientation. The grid voltage and current vectors ($\mathbf{v_s}$ and $\mathbf{i_s}$) in SRF are shown in Figure 8.7. The grid voltage vector $\mathbf{v_g}$ is aligned with the d-axis of SRF; as a result, v_{qg} becomes zero.

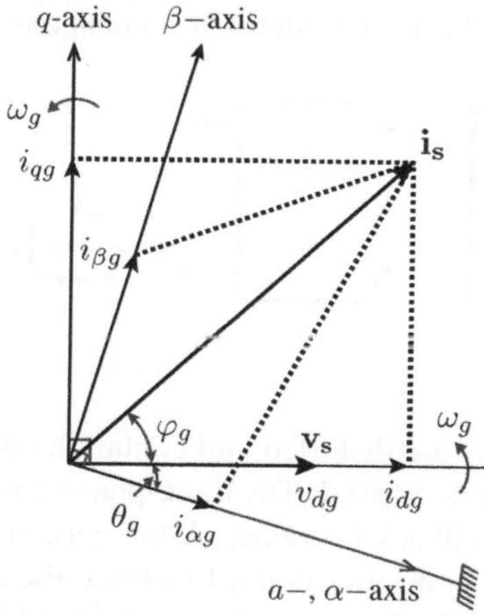

Figure 8.7 Grid voltage and current vectors in dq and $\alpha\beta$ frame.

With $v_{qg} = 0$, the resultant grid active and reactive powers are defined by

$$\left.\begin{aligned} P_g &= +1.5\,v_{dg}\,i_{dg} \\ Q_g &= -1.5\,v_{dg}\,i_{qg} \end{aligned}\right\} \quad \text{for } v_{qg} = 0. \tag{8.6}$$

The grid voltage angle θ_g is used to make v_{dg} equal to $\mathbf{v_s}$ and for grid synchronization, which is one of the most important issues in WECS because it allows the grid and GSC to work in unison. The grid voltage angle θ_g is also important in the transformation of variables between abc and dq frames and in the design of digital control system for GSC. In the literature, various methods such as the zero-crossing method, digital filters, and phase-locked loop (PLL) have been reported for the detection of θ_g [12, 13]. Under ideal grid voltage conditions, θ_g can be detected from $\alpha\beta$-frame voltages as demonstrated below:

$$\theta_g = \tan^{-1}\left(\frac{v_{\beta g}}{v_{\alpha g}}\right) \tag{8.7}$$

where $v_{\alpha g}$ and $v_{\beta g}$ are stationary frame grid voltages, which are estimated from the three-phase grid voltages as follows:

$$\begin{bmatrix} v_{\alpha g} \\ v_{\beta g} \end{bmatrix} = \underbrace{\frac{2}{3}\begin{bmatrix} 1 & -\frac{1}{2} & -\frac{1}{2} \\ 0 & \frac{\sqrt{3}}{2} & -\frac{\sqrt{3}}{2} \end{bmatrix}}_{\mathbf{T}_{abc/\alpha\beta}} \begin{bmatrix} v_{ag} \\ v_{bg} \\ v_{cg} \end{bmatrix} \tag{8.8}$$

where $\mathbf{T}$ denotes a transformation matrix. For a three-phase balanced grid, $v_{ag} + v_{bg} + v_{cg} = 0$; thus, $v_{\alpha g}$ is equal to v_{ag}. The coefficient 2/3 is arbitrarily added to the equation to preserve equal magnitudes for three- and two-phase grid voltages after transformation.

The block diagram for grid voltage angle detection under ideal grid conditions is shown in Figure 8.8. The measured three-phase grid voltages, v_{ag}, v_{bg}, and v_{cg} are transformed to the $\alpha\beta$-frame by using Equation (8.8). The $\alpha\beta$-frame voltages are then transformed from Cartesian coordinates ($x + jy$) to polar coordinates ($r\,e^{j\theta}$). The estimated phase angle θ_g' value changes between $-\pi$ and $+\pi$. To align the peak value of the phase-a grid voltage v_{ag} with the peak phase angle, a modulus function from zero to 2π is used. The resultant grid voltage angle θ_g changes in synchronism with the instantaneous changes in v_{ag} value.

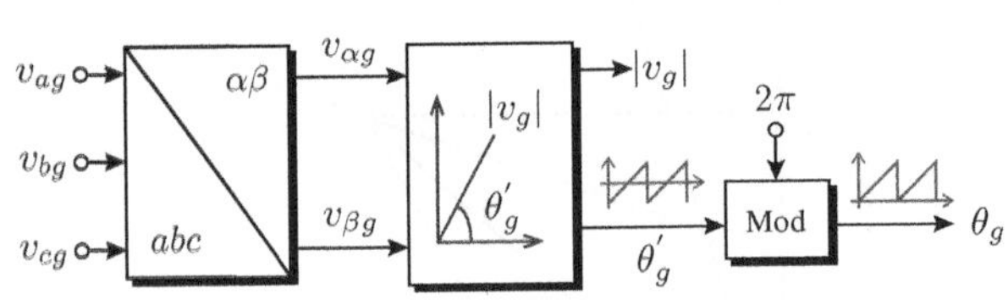

Figure 8.8 Block diagram for grid voltage angle estimation under ideal grid conditions.

In real time, the grid voltages are distorted and contain harmonics. In such cases, the SRF-PLL shown in Figure 8.9 can be used. The three-phase grid voltages are transformed to the dq frame by using the feedback θ_g value. A low-pass filter (LPF) is used to filter the grid voltage error and a PI controller is used to force the q-axis grid voltage v_{qg} to zero value. A feed-forward angular frequency term (ω_o) is added to the output of the PI controller to improve the initial dynamic performance. An integrator is then employed to convert ω_g' to θ_g'. To ensure that θ_g changes between 0 and 2π, a modulus function is used.

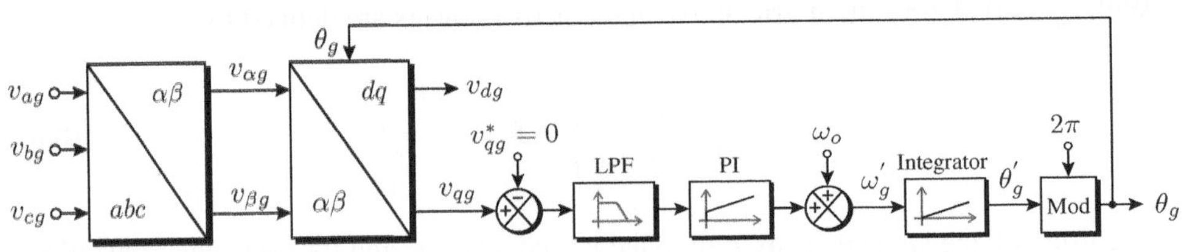

Figure 8.9 Block diagram for grid voltage angle estimation under distorted grid conditions.

The abc, $\alpha\beta$, and dq frame voltages and the grid voltage angle for two fundamental cycles are shown in Figure 8.10. The abc- and $\alpha\beta$-frame voltages are essentially sinusoidal, whereas dq-frame voltages are DC. The instantaneous values of abc- and $\alpha\beta$-frame voltages change as the $\omega_g t$ value changes. The $\theta_g^{'}$ waveform has minimum and maximum values of $-\pi$ and π, respectively. The minimum and maximum values of the θ_g waveform are 0 and 2π, respectively. The peak value of the grid voltage angle θ_g is aligned with the v_{ag} peak value. The dq-axes voltages are estimated on the basis of the value of θ_g. The v_{dg} is equal to 1.0 $\widehat{\text{pu}}$ and v_{qg} is equal to 0.0 $\widehat{\text{pu}}$; hence, the grid voltage orientation is achieved.

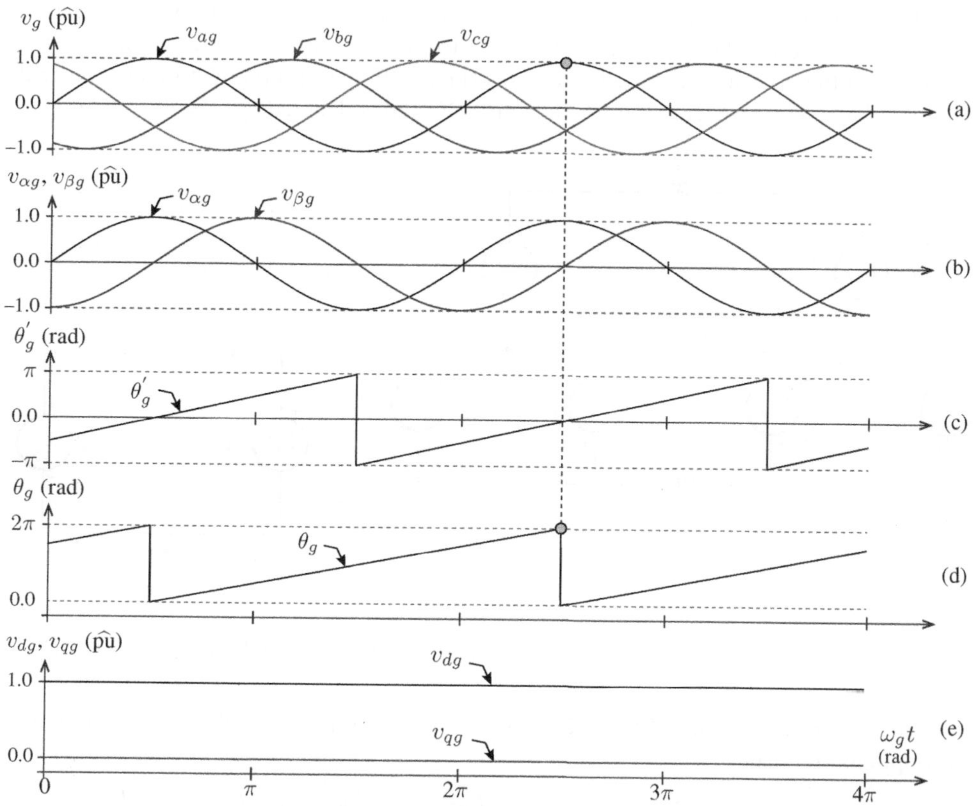

Figure 8.10 Transformation of grid voltages from the abc-frame to $\alpha\beta$- and dq-frames.

8.4 MODELING OF THREE-PHASE GSC

As mentioned in Section 8.2, 2L-VSI and NPC inverter are dominantly used as GSCs in the WECS industry. The primary objective for the GSC includes the regulation of grid active and reactive power directly or through the control of grid currents. In this section, the CT and DT modeling of abc-, $\alpha\beta$-, and dq-frame grid currents and powers is developed. To relate the grid current/power dynamics to power converter switching states, the GSC terminal voltages are formulated in terms of switching states. For an NPC inverter, the secondary control objective is to control the DC-link neutral-point voltage. To accomplish this objective, the DC-link capacitors voltage are defined in terms of NPC inverter switching states. For a classical decoupled VOC, SRF grid currents are controlled for easy PI tuning and low steady-state errors. However, with the MPC scheme, the steady-state current/power error is approximately the same in abc, $\alpha\beta$, and dq reference frames.

8.4.1 Modeling of *abc*-Frame Grid Currents and Powers

A standard and simplified representation of the *abc*-frame GSC is shown in Figure 8.11. To simplify the modeling and analysis, an L-filter is considered. The subscripts a, b, and c represent the three phases. The variables v_{aN}, v_{bN}, and v_{cN} represent the 2L-VSI or NPC inverter terminal voltages with respect to the negative DC-rail (N). v_{ag}, v_{bg}, and v_{cg} are three-phase grid voltages, whereas i_{ag}, i_{bg}, and i_{cg} are three-phase grid currents. L_{ai}, L_{bi}, and L_{ci} correspond to the grid-filter inductance. r_{ai}, r_{bi}, and r_{ci} represent the internal resistance of the grid filter and the leakage resistance of the step-up transformer. The grid neutral voltage with respect to the VSI negative DC-rail is denoted by v_{nN}.

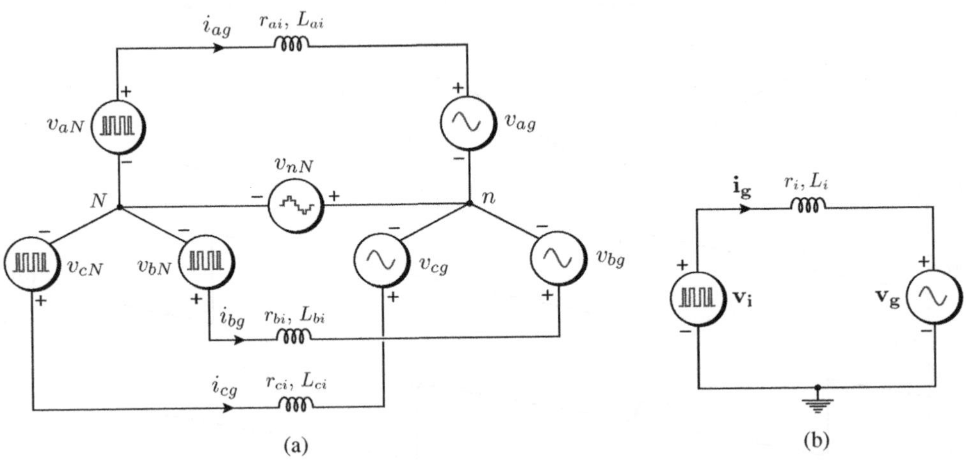

Figure 8.11 Representation of *abc*-frame GSC: (a) standard model and (b) simplified model.

(1) abc-Frame CT Model of Grid Currents: To simplify the analysis, the following assumptions are made:

$$\begin{aligned} r_{ai} &= r_{bi} = r_{ci} = r_i, \quad L_{ai} = L_{bi} = L_{ci} = L_i \\ v_{ag} + v_{bg} + v_{cg} &= 0, \quad i_{ag} + i_{bg} + i_{cg} = 0. \end{aligned} \tag{8.9}$$

By applying Kirchhoff's voltage law to Figure 8.11(a), the VSI AC-side voltages are expressed in terms of grid voltages, grid currents, and filter parameters as follows:

$$\begin{bmatrix} v_{aN} \\ v_{bN} \\ v_{cN} \end{bmatrix} = \begin{bmatrix} r_i & 0 & 0 \\ 0 & r_i & 0 \\ 0 & 0 & r_i \end{bmatrix} \begin{bmatrix} i_{ag} \\ i_{bg} \\ i_{cg} \end{bmatrix} + \begin{bmatrix} L_i & 0 & 0 \\ 0 & L_i & 0 \\ 0 & 0 & L_i \end{bmatrix} \frac{d}{dt} \begin{bmatrix} i_{ag} \\ i_{bg} \\ i_{cg} \end{bmatrix} + \begin{bmatrix} v_{ag} \\ v_{bg} \\ v_{cg} \end{bmatrix} + \begin{bmatrix} v_{nN} \\ v_{nN} \\ v_{nN} \end{bmatrix}. \tag{8.10}$$

For a three-phase balanced grid, the grid neutral voltage v_{nN} is given by the following:

$$v_{nN} = \frac{v_{aN} + v_{bN} + v_{cN}}{3}. \tag{8.11}$$

By substituting (8.11) in (8.10), the VSC output voltages are formulated in matrix notation with respect to the grid neutral n as follows:

$$\begin{bmatrix} v_{an} \\ v_{bn} \\ v_{cn} \end{bmatrix} = \begin{bmatrix} v_{aN} - v_{nN} \\ v_{bN} - v_{nN} \\ v_{cN} - v_{nN} \end{bmatrix} = \begin{bmatrix} r_i & 0 & 0 \\ 0 & r_i & 0 \\ 0 & 0 & r_i \end{bmatrix} \begin{bmatrix} i_{ag} \\ i_{bg} \\ i_{cg} \end{bmatrix} + \begin{bmatrix} L_i & 0 & 0 \\ 0 & L_i & 0 \\ 0 & 0 & L_i \end{bmatrix} \frac{d}{dt} \begin{bmatrix} i_{ag} \\ i_{bg} \\ i_{cg} \end{bmatrix} + \begin{bmatrix} v_{ag} \\ v_{bg} \\ v_{cg} \end{bmatrix} \tag{8.12}$$

which can be simplified as

$$\mathbf{v_i} = r_i\,\mathbf{i_g} + L_i\,\frac{d\,\mathbf{i_g}}{dt} + \mathbf{v_g}. \tag{8.13}$$

The voltage and current vectors in the above expression are defined as follows:

$$\begin{aligned}
\mathbf{v_i} &= \frac{2}{3}\left[v_{an} + \mathfrak{a}\,v_{bn} + \mathfrak{a}^2\,v_{cn}\right] = \left[v_{an}\;\;v_{bn}\;\;v_{cn}\right]^T\\
\mathbf{v_g} &= \frac{2}{3}\left[v_{ag} + \mathfrak{a}\,v_{bg} + \mathfrak{a}^2\,v_{cg}\right] = \left[v_{ag}\;\;v_{bg}\;\;v_{cg}\right]^T\\
\mathbf{i_g} &= \frac{2}{3}\left[i_{ag} + \mathfrak{a}\,i_{bg} + \mathfrak{a}^2\,i_{cg}\right] = \left[i_{ag}\;\;i_{bg}\;\;i_{cg}\right]^T
\end{aligned} \tag{8.14}$$

where $\mathfrak{a} = e^{j\frac{2\pi}{3}}$ is a Fortescue operator and T represents the matrix transpose operator.

From Equation (8.12), the *abc*-frame CT grid current dynamics are expressed as follows:

$$\begin{aligned}
\underbrace{\frac{d}{dt}\begin{bmatrix} i_{ag}(t)\\ i_{bg}(t)\\ i_{cg}(t)\end{bmatrix}}_{\dot{\mathbf{x}}(t)} &= \underbrace{\begin{bmatrix} -\frac{r_i}{L_i} & 0 & 0\\ 0 & -\frac{r_i}{L_i} & 0\\ 0 & 0 & -\frac{r_i}{L_i}\end{bmatrix}}_{\mathbf{A}} \underbrace{\begin{bmatrix} i_{ag}(t)\\ i_{bg}(t)\\ i_{cg}(t)\end{bmatrix}}_{\mathbf{x}(t)} + \underbrace{\begin{bmatrix} \frac{1}{L_i} & 0 & 0\\ 0 & \frac{1}{L_i} & 0\\ 0 & 0 & \frac{1}{L_i}\end{bmatrix}}_{\mathbf{B_i}} \underbrace{\begin{bmatrix} v_{an}(t)\\ v_{bn}(t)\\ v_{cn}(t)\end{bmatrix}}_{\mathbf{v_i}(t)}\\
&+ \underbrace{\begin{bmatrix} -\frac{1}{L_i} & 0 & 0\\ 0 & -\frac{1}{L_i} & 0\\ 0 & 0 & -\frac{1}{L_i}\end{bmatrix}}_{\mathbf{B_g}} \underbrace{\begin{bmatrix} v_{ag}(t)\\ v_{bg}(t)\\ v_{cg}(t)\end{bmatrix}}_{\mathbf{v_g}(t)}
\end{aligned} \tag{8.15}$$

which can be simplified as

$$\frac{d\,\mathbf{i_g}}{dt} = -\frac{r_i}{L_i}\,\mathbf{i_g} + \frac{1}{L_i}\,\mathbf{v_i} - \frac{1}{L_i}\,\mathbf{v_g} = \mathbf{A}\,\mathbf{i_g} + \mathbf{B_i}\,\mathbf{v_i} + \mathbf{B_g}\,\mathbf{v_g}. \tag{8.16}$$

(2) abc-Frame CT Model of Grid Active Reactive Powers: To realize the PPC scheme, the grid active and reactive powers should be related to the three-phase grid voltages and currents. The CT model of grid active and reactive powers is defined by [14]

$$\begin{aligned}
P_g &= v_{ag}\,i_{ag} + v_{bg}\,i_{bg} + v_{cg}\,i_{cg}\\
Q_g &= \frac{1}{\sqrt{3}}\left[(v_{bg} - v_{cg})\,i_{ag} + (v_{cg} - v_{ag})\,i_{bg} + (v_{ag} - v_{bg})\,i_{cg}\right].
\end{aligned} \tag{8.17}$$

(3) abc-Frame DT Model of Grid Currents: The MPC is basically an optimization algorithm. To implement this strategy on digital control platforms, the CT models should be converted to DT equivalents. The CT grid current model in Equation (8.15) is linear time-invariant because of the constant parameter elements in state matrix **A**. For such a model, the exact and approximate DT equivalents can easily be calculated offline. As mentioned in Chapter 7, the discretization error between the exact and forward Euler methods is negligible for small values of r_i and T_s. In this section, we deal with forward Euler approximation only to simplify the analysis. This method considers the future sample $(k+1)$ and present sample (k) as described below:

$$\left\{\frac{d\,x(t)}{dt}\right\}_{t=kT_s} \approx \frac{x(kT_s+T_s)-x(kT_s)}{T_s}, \quad x \in \{i_{ag}, i_{bg}, i_{cg}\} \tag{8.18}$$

which is simplified as

$$x(k+1) \approx x(k) + T_s\left\{\frac{d\,x(t)}{dt}\right\}_{t=k} \tag{8.19}$$

where T_s is the sampling time used in the DT conversion and control.

By substituting (8.19) into (8.15), the DT grid current model is obtained in the *abc*-frame:

$$\begin{bmatrix} i_{ag}(k+1) \\ i_{bg}(k+1) \\ i_{cg}(k+1) \end{bmatrix} = \mathbf{\Phi} \begin{bmatrix} i_{ag}(k) \\ i_{bg}(k) \\ i_{cg}(k) \end{bmatrix} + \mathbf{\Gamma_i} \begin{bmatrix} v_{an}(k) \\ v_{bn}(k) \\ v_{cn}(k) \end{bmatrix} + \mathbf{\Gamma_g} \begin{bmatrix} v_{ag}(k) \\ v_{bg}(k) \\ v_{cg}(k) \end{bmatrix} \tag{8.20}$$

where the DT equivalent matrices are defined by

$$\begin{aligned} \mathbf{\Phi} &\approx [\mathbf{I}+\mathbf{A}\,T_s] \approx \begin{bmatrix} 1-\frac{r_i\,T_s}{L_i} & 0 & 0 \\ 0 & 1-\frac{r_i\,T_s}{L_i} & 0 \\ 0 & 0 & 1-\frac{r_i\,T_s}{L_i} \end{bmatrix} \\ \mathbf{\Gamma_i} &\approx \mathbf{B_i}\,T_s \approx \begin{bmatrix} \frac{T_s}{L_i} & 0 & 0 \\ 0 & \frac{T_s}{L_i} & 0 \\ 0 & 0 & \frac{T_s}{L_i} \end{bmatrix}, \quad \mathbf{\Gamma_g} \approx \mathbf{B_g}\,T_s \approx \begin{bmatrix} -\frac{T_s}{L_i} & 0 & 0 \\ 0 & -\frac{T_s}{L_i} & 0 \\ 0 & 0 & -\frac{T_s}{L_i} \end{bmatrix}. \end{aligned} \tag{8.21}$$

(4) abc-Frame DT Model of Grid Active and Reactive Powers: The calculation of the DT grid active and reactive powers is straightforward because no derivative terms are involved in the CT model. By shifting all the variables to a $(k+1)$ sampling instant, the future behavior of grid active and reactive powers is obtained as described below:

$$\begin{aligned} P_g(k+1) &= v_{ag}(k+1)\,i_{ag}(k+1) + v_{bg}(k+1)\,i_{bg}(k+1) \\ &\quad + v_{cg}(k+1)\,i_{cg}(k+1) \\ Q_g &= \frac{1}{\sqrt{3}}\,[(v_{bg}(k+1)-v_{cg}(k+1))\,i_{ag}(k+1) \\ &\quad + (v_{cg}(k+1)-v_{ag}(k+1))\,i_{bg}(k+1) \\ &\quad + (v_{ag}(k+1)-v_{bg}(k+1))\,i_{cg}(k+1)] \end{aligned} \tag{8.22}$$

where $v_{ag}(k+1)$, $v_{bg}(k+1)$, and $v_{cg}(k+1)$ correspond to the future values of the three-phase grid voltages. These variables can be obtained by the extrapolation methods discussed in Section 4.7 on page 141. For small sampling times, the following is assumed: $v_{ag}(k+1) \simeq v_{ag}(k)$, $v_{bg}(k+1) \simeq v_{bg}(k)$, and $v_{cg}(k+1) \simeq v_{cg}(k)$.

8.4.2 Modeling of $\alpha\beta$-Frame Grid Currents and Powers

A standard and simplified representation of $\alpha\beta$-frame GSC is shown in Figure 8.12. Subscripts α and β correspond to the orthogonal stationary-frame axes. $v_{\alpha i}$ and $v_{\beta i}$ are 2L-VSI or NPC inverter terminal voltages in $\alpha\beta$-frame. $v_{\alpha g}$ and $v_{\beta g}$ are the $\alpha\beta$-frame grid voltages, whereas $i_{\alpha g}$ and $i_{\beta g}$ are $\alpha\beta$-frame grid currents.

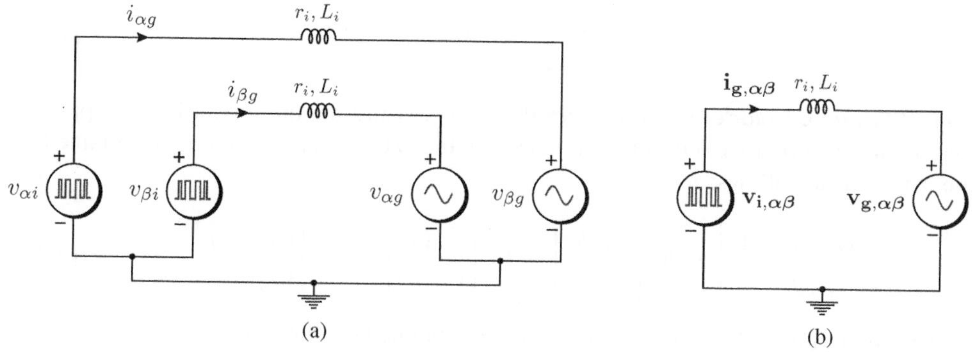

Figure 8.12 Representation of $\alpha\beta$-frame GSC: (a) standard model and (b) simplified model.

(1) $\alpha\beta$-Frame CT Model of Grid Currents: By multiplying both sides of the *abc*-frame VSI voltages in (8.10) with $abc/\alpha\beta$ transformation matrix $\mathbf{T}_{abc/\alpha\beta}$ (refer to Equation (3.6)), the stationary-frame VSI voltages are obtained as follows:

$$\begin{bmatrix} v_{\alpha i} \\ v_{\beta i} \end{bmatrix} = \begin{bmatrix} r_i & 0 \\ 0 & r_i \end{bmatrix} \begin{bmatrix} i_{\alpha g} \\ i_{\beta g} \end{bmatrix} + \begin{bmatrix} L_i & 0 \\ 0 & L_i \end{bmatrix} \frac{d}{dt} \begin{bmatrix} i_{\alpha g} \\ i_{\beta g} \end{bmatrix} + \begin{bmatrix} v_{\alpha g} \\ v_{\beta g} \end{bmatrix}. \tag{8.23}$$

Note that the grid neutral voltage v_{nN} becomes zero during the $abc/\alpha\beta$ transformation. Hence, the above model is simplified as follows:

$$\mathbf{v}_{i,\alpha\beta} = r_i\, \mathbf{i}_{g,\alpha\beta} + L_i \frac{d\, \mathbf{i}_{g,\alpha\beta}}{dt} + \mathbf{v}_{g,\alpha\beta}. \tag{8.24}$$

The stationary-frame voltage and current vectors in (8.24) are defined by the following:

$$\begin{aligned} \mathbf{v}_{i,\alpha\beta} &= [v_{\alpha i} + j\, v_{\beta i}] = [v_{\alpha i} \;\; v_{\beta i}]^T \\ \mathbf{v}_{g,\alpha\beta} &= [v_{\alpha g} + j\, v_{\beta g}] = [v_{\alpha g} \;\; v_{\beta g}]^T \\ \mathbf{i}_{g,\alpha\beta} &= [i_{\alpha g} + j\, i_{\beta g}] = [i_{\alpha g} \;\; i_{\beta g}]^T. \end{aligned} \tag{8.25}$$

The $\alpha\beta$-frame grid current CT model is derived from (8.23) as follows:

$$\begin{aligned} \underbrace{\frac{d}{dt} \begin{bmatrix} i_{\alpha g}(t) \\ i_{\beta g}(t) \end{bmatrix}}_{\dot{\mathbf{x}}(t)} &= \underbrace{\begin{bmatrix} -\frac{r_i}{L_i} & 0 \\ 0 & -\frac{r_i}{L_i} \end{bmatrix}}_{\mathbf{A}} \underbrace{\begin{bmatrix} i_{\alpha g}(t) \\ i_{\beta g}(t) \end{bmatrix}}_{\mathbf{x}(t)} + \underbrace{\begin{bmatrix} \frac{1}{L_i} & 0 \\ 0 & \frac{1}{L_i} \end{bmatrix}}_{\mathbf{B_i}} \underbrace{\begin{bmatrix} v_{\alpha i}(t) \\ v_{\beta i}(t) \end{bmatrix}}_{\mathbf{v_i}(t)} \\ &+ \underbrace{\begin{bmatrix} -\frac{1}{L_i} & 0 \\ 0 & -\frac{1}{L_i} \end{bmatrix}}_{\mathbf{B_g}} \underbrace{\begin{bmatrix} v_{\alpha g}(t) \\ v_{\beta g}(t) \end{bmatrix}}_{\mathbf{v_g}(t)} \end{aligned} \tag{8.26}$$

which can be simplified as

$$\frac{d\,\mathbf{i}_{g,\alpha\beta}}{dt} = -\frac{r_i}{L_i}\,\mathbf{i}_{g,\alpha\beta} + \frac{1}{L_i}\,\mathbf{v}_{i,\alpha\beta} - \frac{1}{L_i}\,\mathbf{v}_{g,\alpha\beta} = \mathbf{A}\,\mathbf{i}_{g,\alpha\beta} + \mathbf{B_i}\,\mathbf{v}_{i,\alpha\beta} + \mathbf{B_g}\,\mathbf{v}_{g,\alpha\beta}. \quad (8.27)$$

(2) $\alpha\beta$-Frame CT Model of Grid Active and Reactive Powers: The CT model of grid active and reactive powers are formulated in terms of $\alpha\beta$-frame grid voltages and currents as follows [15]:

$$\begin{aligned} P_g &= 1.5\,(v_{\alpha g}\,i_{\alpha g} + v_{\beta g}\,i_{\beta g}) \\ Q_g &= 1.5\,(v_{\beta g}\,i_{\alpha g} - v_{\alpha g}\,i_{\beta g})\,. \end{aligned} \quad (8.28)$$

(3) $\alpha\beta$-Frame DT Model of Grid Currents: By substituting the forward Euler approximation in Equation (8.19) into Equation (8.26), the DT grid current model is obtained in the $\alpha\beta$-frame as follows:

$$\begin{bmatrix} i_{\alpha g}(k+1) \\ i_{\beta g}(k+1) \end{bmatrix} = \mathbf{\Phi} \begin{bmatrix} i_{\alpha g}(k) \\ i_{\beta g}(k) \end{bmatrix} + \mathbf{\Gamma_i} \begin{bmatrix} v_{\alpha i}(k) \\ v_{\beta i}(k) \end{bmatrix} + \mathbf{\Gamma_g} \begin{bmatrix} v_{\alpha g}(k) \\ v_{\beta g}(k) \end{bmatrix}. \quad (8.29)$$

The approximate DT equivalent matrices are obtained similar to (8.21) as

$$\begin{aligned} \mathbf{\Phi} &\approx [\mathbf{I} + \mathbf{A}\,T_s] \approx \begin{bmatrix} 1 - \frac{r_i T_s}{L_i} & 0 \\ 0 & 1 - \frac{r_i T_s}{L_i} \end{bmatrix} \\ \mathbf{\Gamma_i} &\approx \mathbf{B_i}\,T_s \approx \begin{bmatrix} \frac{T_s}{L_i} & 0 \\ 0 & \frac{T_s}{L_i} \end{bmatrix}, \; \mathbf{\Gamma_g} \approx \mathbf{B_g}\,T_s \approx \begin{bmatrix} -\frac{T_s}{L_i} & 0 \\ 0 & -\frac{T_s}{L_i} \end{bmatrix}. \end{aligned} \quad (8.30)$$

(4) $\alpha\beta$-Frame DT Model of Grid Active and Reactive Powers: The DT model corresponding to the CT model in Equation (8.28) is obtained as follows:

$$\begin{aligned} P_g(k+1) &= 1.5\,(v_{\alpha g}(k+1)\,i_{\alpha g}(k+1) + v_{\beta g}(k+1)\,i_{\beta g}(k+1)) \\ Q_g(k+1) &= 1.5\,(v_{\beta g}(k+1)\,i_{\alpha g}(k+1) - v_{\alpha g}(k+1)\,i_{\beta g}(k+1)) \end{aligned} \quad (8.31)$$

where $v_{\alpha g}(k+1)$ and $v_{\beta g}(k+1)$ are the future values of the $\alpha\beta$-frame grid voltages.

8.4.3 Modeling of dq-Frame Grid Currents and Powers

A standard and simplified representation of the dq-frame GSC is shown in Figure 8.13. Subscripts d and q correspond to the orthogonal synchronous-frame axes. All the dq-frame variables are DC in nature. v_{di} and v_{qi} are the 2L-VSI or NPC inverter terminal voltages in the dq-frame, respectively. v_{dg} and v_{qg} are the dq-frame grid voltages. i_{dg} and i_{qg} are the dq-frame grid currents. ω_g is the grid angular frequency (rad/s). The terms $-\omega_g L_i i_{qg}$ and $\omega_g L_i i_{dg}$ represent the induced speed voltages caused by the transformation of variables from the *abc*-frame to the dq-frame. The control schemes usually compensate these coupling terms.

(1) dq-Frame CT Model of Grid Currents: Similar to stationary-frame voltages, the dq-frame VSI voltages are obtained by multiplying both sides of Equation (8.10) with the abc/dq transformation matrix given in Equation (3.9), that is,

$$\begin{bmatrix} v_{di} \\ v_{qi} \end{bmatrix} = \begin{bmatrix} r_i & -\omega_g\,L_i \\ \omega_g\,L_i & r_i \end{bmatrix} \begin{bmatrix} i_{dg} \\ i_{qg} \end{bmatrix} + \begin{bmatrix} L_i & 0 \\ 0 & L_i \end{bmatrix} \frac{d}{dt} \begin{bmatrix} i_{dg} \\ i_{qg} \end{bmatrix} + \begin{bmatrix} v_{dg} \\ v_{qg} \end{bmatrix} \quad (8.32)$$

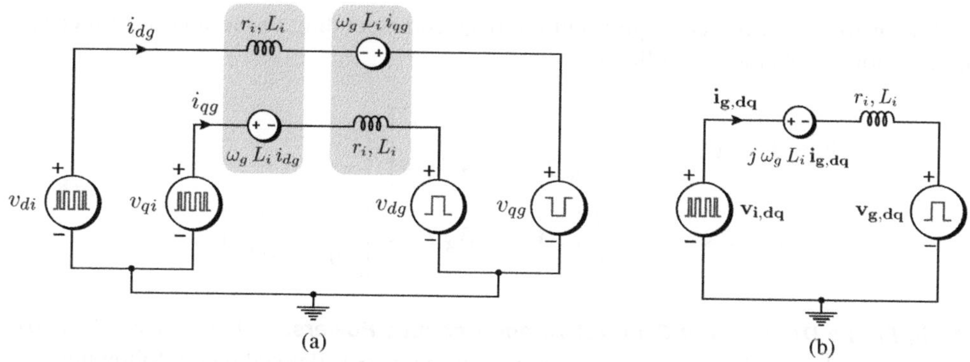

Figure 8.13 Representation of a *dq*-frame GSC: (a) standard model and (b) simplified model.

and can be represented in a simplified form as

$$\mathbf{v_{i,dq}} = (r_i + j\,\omega_g\,L_i)\,\mathbf{i_{g,dq}} + L_i\,\frac{d\,\mathbf{i_{g,dq}}}{dt} + \mathbf{v_{g,dq}}. \tag{8.33}$$

The *dq*-frame voltage and current vectors in (8.33) are defined by the following expressions:

$$\begin{aligned} \mathbf{v_{i,dq}} &= [v_{di} + j\,v_{qi}] = [v_{di}\ \ v_{qi}]^T \\ \mathbf{v_{g,dq}} &= [v_{dg} + j\,v_{qg}] = [v_{dg}\ \ v_{qg}]^T \\ \mathbf{i_{g,dq}} &= [i_{dg} + j\,i_{qg}] = [i_{dg}\ \ i_{qg}]^T. \end{aligned} \tag{8.34}$$

The *dq*-frame grid current dynamic model is obtained from (8.32) as follows:

$$\begin{aligned} \underbrace{\frac{d}{dt}\begin{bmatrix} i_{dg}(t) \\ i_{qg}(t) \end{bmatrix}}_{\dot{\mathbf{x}}(t)} &= \underbrace{\begin{bmatrix} -\frac{r_i}{L_i} & \omega_g \\ -\omega_g & -\frac{r_i}{L_i} \end{bmatrix}}_{\mathbf{A}} \underbrace{\begin{bmatrix} i_{dg}(t) \\ i_{qg}(t) \end{bmatrix}}_{\mathbf{x}(t)} + \underbrace{\begin{bmatrix} \frac{1}{L_i} & 0 \\ 0 & \frac{1}{L_i} \end{bmatrix}}_{\mathbf{B_i}} \underbrace{\begin{bmatrix} v_{di}(t) \\ v_{qi}(t) \end{bmatrix}}_{\mathbf{v_i}(t)} \\ &\quad + \underbrace{\begin{bmatrix} -\frac{1}{L_i} & 0 \\ 0 & -\frac{1}{L_i} \end{bmatrix}}_{\mathbf{B_g}} \underbrace{\begin{bmatrix} v_{dg}(t) \\ v_{qg}(t) \end{bmatrix}}_{\mathbf{v_g}(t)}. \end{aligned} \tag{8.35}$$

(2) dq-Frame CT Model of Grid Active and Reactive Powers: The *dq*-frame CT model of grid active and reactive powers are formulated as follows [16]:

$$\begin{aligned} P_g &= 1.5\,(v_{dg}\,i_{dg} + v_{qg}\,i_{qg}) \\ Q_g &= 1.5\,(v_{qg}\,i_{dg} - v_{dg}\,i_{qg}). \end{aligned} \tag{8.36}$$

(3) dq-Frame DT Model of Grid Currents: By combining (8.19) and (8.35), the *dq*-frame DT grid current model corresponding to the forward Euler approximation is obtained as shown below:

$$\begin{bmatrix} i_{dg}(k+1) \\ i_{qg}(k+1) \end{bmatrix} = \mathbf{\Phi} \begin{bmatrix} i_{dg}(k) \\ i_{qg}(k) \end{bmatrix} + \mathbf{\Gamma_i} \begin{bmatrix} v_{di}(k) \\ v_{qi}(k) \end{bmatrix} + \mathbf{\Gamma_g} \begin{bmatrix} v_{dg}(k) \\ v_{qg}(k) \end{bmatrix}. \tag{8.37}$$

Similar to (8.30), the DT equivalent matrices corresponding to the forward Euler approximation are obtained as follows:

$$\begin{aligned}\mathbf{\Phi} &\approx [\mathbf{I} + \mathbf{A}\,T_s] \approx \begin{bmatrix} 1 - \frac{r_i T_s}{L_i} & \omega_g T_s \\ -\omega_g T_s & 1 - \frac{r_i T_s}{L_i} \end{bmatrix} \\ \mathbf{\Gamma_i} &\approx \mathbf{B_i}\,T_s \approx \begin{bmatrix} \frac{T_s}{L_i} & 0 \\ 0 & \frac{T_s}{L_i} \end{bmatrix}, \; \mathbf{\Gamma_g} \approx \mathbf{B_g}\,T_s \approx \begin{bmatrix} -\frac{T_s}{L_i} & 0 \\ 0 & -\frac{T_s}{L_i} \end{bmatrix}.\end{aligned} \tag{8.38}$$

(4) dq-Frame DT Model of Grid Active and Reactive Powers: On the basis of (8.36), the dq-frame DT model of grid active and reactive powers is defined by the following:

$$\begin{aligned}P_g(k+1) &= 1.5\,(v_{dg}(k+1)\,i_{dg}(k+1) + v_{qg}(k+1)\,i_{qg}(k+1)) \\ Q_g(k+1) &= 1.5\,(v_{qg}(k+1)\,i_{dg}(k+1) - v_{dg}(k+1)\,i_{qg}(k+1))\end{aligned} \tag{8.39}$$

where $v_{dg}(k+1)$ and $v_{qg}(k+1)$ are future value of dq-frame grid voltages. Owing to DC nature of grid voltages in SRF, it can be assumed that $v_{dg}(k+1) \simeq v_{dg}(k)$ and $v_{qg}(k+1) \simeq v_{qg}(k)$.

8.4.4 Modeling of VSI Output Voltages

The DT models of the abc, $\alpha\beta$, and dq frame grid active and reactive powers in Equations (8.22), (8.31), and (8.39) are related to the future values of grid currents $i_{ag}(k+1)$, $i_{bg}(k+1)$, $i_{cg}(k+1)$, $i_{\alpha g}(k+1)$, $i_{\beta g}(k+1)$, $i_{dg}(k+1)$, and $i_{qg}(k+1)$. These future grid current DT models are related to the VSI output voltages $v_{an}(k)$, $v_{bn}(k)$, $v_{cn}(k)$, $v_{\alpha i}(k)$, $v_{\beta i}(k)$, $v_{di}(k)$, and $v_{qi}(k)$, as demonstrated in Equations (8.20), (8.29), and (8.37). To design the PPC and PCC schemes for grid-connected 2L-VSI and NPC inverters, the GSC output voltages should be expressed in abc, $\alpha\beta$, and dq frame switching signals. Thus, the future behavior of the grid currents and powers can be obtained in terms of all the possible switching states of VSI. In this section, the 2L-VSI and NPC inverter output voltages are formulated in terms of measured DC-link capacitors voltage and abc, $\alpha\beta$, and dq frame switching states.

(1) Model of 2L-VSI Output Voltages: The power circuit of grid-connected 2L-VSI with LV insulated gate bipolar transistor (IGBT)-based switching devices is shown in Figure 8.14. Each phase leg is composed of two active switches. The switching signals in each phase operate in a complementary manner, that is, $s_{ai2} = \overline{s}_{ai1}$, $s_{bi2} = \overline{s}_{bi1}$, and $s_{ci2} = \overline{s}_{ci1}$. Grid-side terminals ai, bi, and ci are connected to the positive or negative (P or N) DC-bus on the basis of the switching signal values. The DC-link capacitor voltage is equal to the net DC-bus voltage ($v_{C1} = v_{dc}$). With two switching states '0' and '1' in each phase, the possible number of combinations for a three-phase 2L-VSI is eight (2^3).

The 2L-VSI terminal voltages with respect to the negative DC-bus N are expressed in terms of DC-link capacitor voltage and switching signals as follows [17]:

$$\begin{bmatrix} v_{aN}(k) \\ v_{bN}(k) \\ v_{cN}(k) \end{bmatrix} = v_{C1}(k) \begin{bmatrix} s_{ai1}(k) \\ s_{bi1}(k) \\ s_{ci1}(k) \end{bmatrix}. \tag{8.40}$$

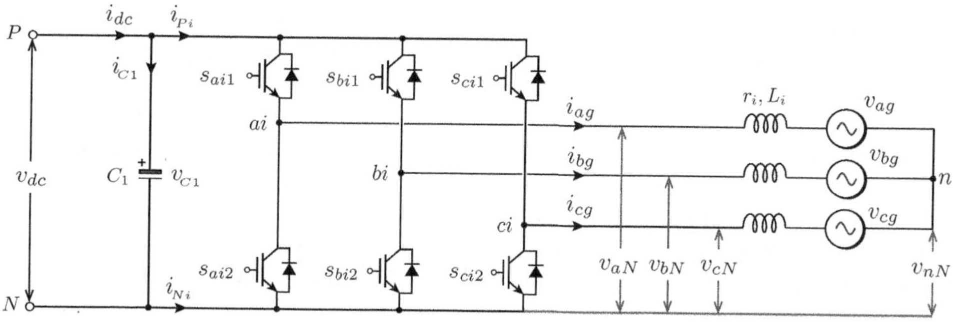

Figure 8.14 Power circuit of three-phase grid-connected 2L-VSI.

To realize the PCC and PPC schemes in the abc-frame, the VSI voltages should be expressed with respect to grid neutral n. To do so, the 2L-VSI terminal voltages in Equation (8.40) are modified as follows:

$$\begin{bmatrix} v_{an}(k) \\ v_{bn}(k) \\ v_{cn}(k) \end{bmatrix} = v_{C1}(k) \underbrace{\begin{bmatrix} \frac{2}{3} & -\frac{1}{3} & -\frac{1}{3} \\ -\frac{1}{3} & \frac{2}{3} & -\frac{1}{3} \\ -\frac{1}{3} & -\frac{1}{3} & \frac{2}{3} \end{bmatrix}}_{\mathbf{T}_{abc/abcn}} \begin{bmatrix} s_{ai1}(k) \\ s_{bi1}(k) \\ s_{ci1}(k) \end{bmatrix} = v_{C1}(k) \begin{bmatrix} s_{ani1}(k) \\ s_{bni1}(k) \\ s_{cni1}(k) \end{bmatrix} \tag{8.41}$$

where s_{ani1}, s_{bni1} and s_{cni1} are the abc-frame switching signals with respect to n. To reduce the computational burden, these calculations are defined offline.

The $\alpha\beta$-frame 2L-VSI terminal voltages are obtained from (8.40) as follows:

$$\begin{bmatrix} v_{\alpha i}(k) \\ v_{\beta i}(k) \end{bmatrix} = v_{C1}(k) \underbrace{\frac{2}{3} \begin{bmatrix} 1 & -\frac{1}{2} & -\frac{1}{2} \\ 0 & \frac{\sqrt{3}}{2} & -\frac{\sqrt{3}}{2} \end{bmatrix}}_{\mathbf{T}_{abc/\alpha\beta}} \begin{bmatrix} s_{ai1}(k) \\ s_{bi1}(k) \\ s_{ci1}(k) \end{bmatrix} = v_{C1}(k) \begin{bmatrix} s_{\alpha i1}(k) \\ s_{\beta i1}(k) \end{bmatrix} \tag{8.42}$$

where $s_{\alpha i1}$ and $s_{\beta i1}$ are the $\alpha\beta$-frame switching signals, which can be defined offline.

Similarly, the dq-frame 2L-VSI terminal voltages are obtained from (8.40) as follows:

$$\begin{aligned} \begin{bmatrix} v_{di}(k) \\ v_{qi}(k) \end{bmatrix} &= v_{C1}(k) \underbrace{\begin{bmatrix} \cos\theta_g(k) & \sin\theta_g(k) \\ -\sin\theta_g(k) & \cos\theta_g(k) \end{bmatrix}}_{\mathbf{T}_{\alpha\beta/dq}} \left[\mathbf{T}_{abc/\alpha\beta}\right] \begin{bmatrix} s_{ai1}(k) \\ s_{bi1}(k) \\ s_{ci1}(k) \end{bmatrix} \\ &= v_{C1}(k) \begin{bmatrix} s_{di1}(k) \\ s_{qi1}(k) \end{bmatrix} \end{aligned} \tag{8.43}$$

where s_{di1} and s_{qi1} are the dq-frame switching signals, which can be computed online by using $s_{\alpha i1}$, $s_{\beta i1}$, and $\theta_g(k)$.

(2) Model of NPC Inverter Output Voltages: The power circuit of grid-connected NPC inverter with MV-IGBT based switching devices is shown in Figure 8.15. Each phase leg is composed of 4 active switches and 2 clamping diodes. In each phase, the top switching signals operate in a complementary manner with the bottom switching signals, that is, $s_{ai3} = \overline{s}_{ai1}$, $s_{ai4} = \overline{s}_{ai2}$, $s_{bi3} = \overline{s}_{bi1}$, $s_{bi4} = \overline{s}_{bi2}$, $s_{ci3} = \overline{s}_{ci1}$, and $s_{ci4} = \overline{s}_{ci2}$. The grid-side terminals ai, bi, and ci are connected to the positive, mid, or negative (P, Z, or N) DC bus on the basis of the switching signal values. The DC-link is composed of 2 identical capacitors, and their mid-point (neutral point) is connected to the mid-point of the clamping diodes in each phase. The net DC-bus voltage is shared equally among the 2 DC-link capacitors (i.e., $v_{C1} = v_{C2} = v_{dc}/2$). With 3 switching states in each phase, the possible number of combinations for a three-phase NPC inverter becomes 27 (3^3).

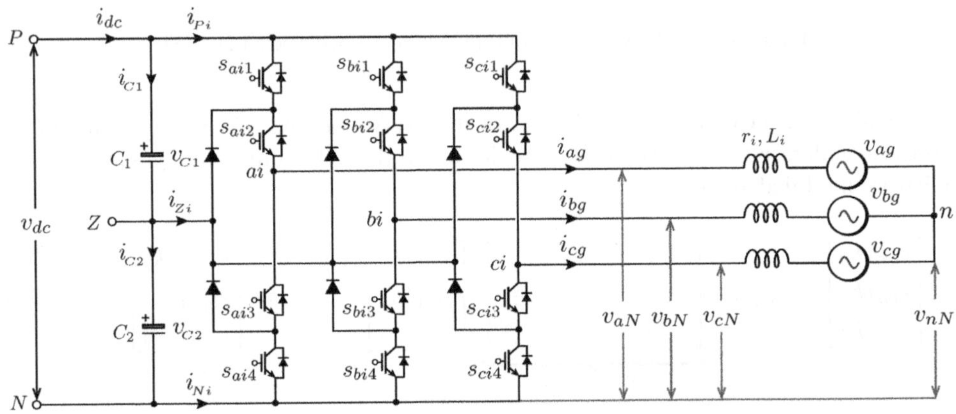

Figure 8.15 Power circuit of three-phase grid-connected NPC inverter.

The NPC inverter terminal voltages with respect to N are expressed in terms of the DC-link capacitors voltage and switching signals as follows [17]:

$$\begin{bmatrix} v_{aN}(k) \\ v_{bN}(k) \\ v_{cN}(k) \end{bmatrix} = v_{C1}(k) \begin{bmatrix} s_{ai1}(k) \\ s_{bi1}(k) \\ s_{ci1}(k) \end{bmatrix} + v_{C2}(k) \begin{bmatrix} s_{ai2}(k) \\ s_{bi2}(k) \\ s_{ci2}(k) \end{bmatrix}. \tag{8.44}$$

The NPC inverter output voltages with respect to the grid neutral n are obtained as follows:

$$\begin{aligned} \begin{bmatrix} v_{an}(k) \\ v_{bn}(k) \\ v_{cn}(k) \end{bmatrix} &= v_{C1}(k) \left[\mathbf{T}_{abc/abcn}\right] \begin{bmatrix} s_{ai1}(k) \\ s_{bi1}(k) \\ s_{ci1}(k) \end{bmatrix} \\ &\quad + v_{C2}(k) \left[\mathbf{T}_{abc/abcn}\right] \begin{bmatrix} s_{ai2}(k) \\ s_{bi2}(k) \\ s_{ci2}(k) \end{bmatrix} \\ &= v_{C1}(k) \begin{bmatrix} s_{ani1}(k) \\ s_{bni1}(k) \\ s_{cni1}(k) \end{bmatrix} + v_{C2}(k) \begin{bmatrix} s_{ani2}(k) \\ s_{bni2}(k) \\ s_{cni2}(k) \end{bmatrix}. \end{aligned} \tag{8.45}$$

The NPC inverter terminal voltages are calculated in the $\alpha\beta$-frame as follows:

$$\begin{aligned}\begin{bmatrix} v_{\alpha i}(k) \\ v_{\beta i}(k) \end{bmatrix} &= v_{C1}(k) \begin{bmatrix} \mathbf{T}_{abc/\alpha\beta} \end{bmatrix} \begin{bmatrix} s_{ai1}(k) \\ s_{bi1}(k) \\ s_{ci1}(k) \end{bmatrix} \\ &+ v_{C2}(k) \begin{bmatrix} \mathbf{T}_{abc/\alpha\beta} \end{bmatrix} \begin{bmatrix} s_{ai2}(k) \\ s_{bi2}(k) \\ s_{ci2}(k) \end{bmatrix} \\ &= v_{C1}(k) \begin{bmatrix} s_{\alpha i1}(k) \\ s_{\beta i1}(k) \end{bmatrix} + v_{C2}(k) \begin{bmatrix} s_{\alpha i2}(k) \\ s_{\beta i2}(k) \end{bmatrix}.\end{aligned} \tag{8.46}$$

Similarly, the NPC inverter terminal voltages are obtained in the dq-frame as follows:

$$\begin{aligned}\begin{bmatrix} v_{di}(k) \\ v_{qi}(k) \end{bmatrix} &= v_{C1}(k) \begin{bmatrix} \mathbf{T}_{\alpha\beta/dq} \end{bmatrix} \begin{bmatrix} \mathbf{T}_{abc/\alpha\beta} \end{bmatrix} \begin{bmatrix} s_{ai1}(k) \\ s_{bi1}(k) \\ s_{ci1}(k) \end{bmatrix} \\ &+ v_{C2}(k) \begin{bmatrix} \mathbf{T}_{\alpha\beta/dq} \end{bmatrix} \begin{bmatrix} \mathbf{T}_{abc/\alpha\beta} \end{bmatrix} \begin{bmatrix} s_{ai2}(k) \\ s_{bi2}(k) \\ s_{ci2}(k) \end{bmatrix} \\ &= v_{C1}(k) \begin{bmatrix} s_{di1}(k) \\ s_{qi1}(k) \end{bmatrix} + v_{C2}(k) \begin{bmatrix} s_{di2}(k) \\ s_{qi2}(k) \end{bmatrix}.\end{aligned} \tag{8.47}$$

8.4.5 Modeling of DC Link Capacitors Voltage in NPC Inverter

As mentioned earlier, an NPC inverter needs to control the DC-link capacitors voltage, in addition to the grid currents/powers. In this section, the DT model of DC-link capacitors voltage is presented in terms of NPC inverter switching states. The modeling given in this section is applicable to the topology given in Figure 8.4 (diode-bridge rectifier + NPC inverter) only. The mid-point Z is considered floating without being connected to the MSC; this situation is also true for the passive front-end + NPC inverter fed electric drive. In the case of BTB NPC converters, the mid-point Z is connected to both the rectifier and inverter mid-points. Therefore, the modeling given here should be modified slightly for BTB NPC converters. This case will be analyzed with PMSG WECS in Chapter 9.

The DC-link capacitors voltage are expressed in terms of the DC-link capacitors current as follows [18]:

$$\begin{aligned}\frac{d}{dt} v_{C1} &= \frac{1}{C_1} i_{C1} \\ \frac{d}{dt} v_{C2} &= \frac{1}{C_2} i_{C2}\end{aligned} \tag{8.48}$$

where i_{C1} and i_{C2} are the currents in DC-link capacitors C_1 and C_2, respectively.

By substituting (8.19) into (8.48), the DT model of the DC-link capacitors voltage is calculated as follows:

$$\begin{aligned} v_{C1}(k+1) &= v_{C1}(k) + \frac{T_s}{C_1} i_{C1}(k) \\ v_{C2}(k+1) &= v_{C2}(k) + \frac{T_s}{C_2} i_{C2}(k). \end{aligned} \tag{8.49}$$

To formulate the DC-link capacitors voltage in terms of NPC inverter switching signals, the correlation between the DC capacitors current and the NPC inverter switching states should be established. By referring to the DC link in Figure 8.15, the DC capacitors current are derived from i_{dc}, i_{Pi}, and i_{Zi} as follows [19]:

$$\begin{aligned} i_{C1} &= i_{dc} - i_{Pi} \\ i_{C2} &= i_{C1} - i_{Zi} = i_{dc} - i_{Pi} - i_{Zi}. \end{aligned} \tag{8.50}$$

The above expressions change in the case of BTB converters because of the neutral-point current flowing from the NPC rectifier to the DC-link mid-point. The objective of the digital control scheme is to maintain equal energy among the DC-link capacitors. With this condition, the net DC current i_{dc} becomes nullified and the expression in (8.50) is modified as follows [20]:

$$\begin{aligned} i_{C1} &= -i_{Pi} \\ i_{C2} &= -i_{Pi} - i_{Zi}. \end{aligned} \tag{8.51}$$

To realize the digital control for NPC inverter, DC branch currents i_{Pi} and i_{Zi} do not need to be measured. As discussed in Chapter 5, the DC branch currents can be expressed in terms of the switching signals and measured three-phase grid currents:

$$\begin{aligned} \begin{bmatrix} i_{Pi} \\ i_{Zi} \end{bmatrix} &= \begin{bmatrix} \mathbf{S_{ai}[2]} & \mathbf{S_{bi}[2]} & \mathbf{S_{ci}[2]} \\ \mathbf{S_{ai}[1]} & \mathbf{S_{bi}[1]} & \mathbf{S_{ci}[1]} \end{bmatrix} \begin{bmatrix} i_{ag} \\ i_{bg} \\ i_{cg} \end{bmatrix} \\ &= \begin{bmatrix} s_{ai1}\, s_{ai2} & s_{bi1}\, s_{bi2} & s_{ci1}\, s_{ci2} \\ s_{ai2}\, s_{ai3} & s_{bi2}\, s_{bi3} & s_{ci2}\, s_{ci3} \end{bmatrix} \begin{bmatrix} i_{ag} \\ i_{bg} \\ i_{cg} \end{bmatrix} \end{aligned} \tag{8.52}$$

where $\mathbf{S_{ai}}$, $\mathbf{S_{bi}}$, and $\mathbf{S_{ci}}$ are switching vectors that represent the group of switching signals in each phase of the NPC inverter. The relationship between the switching vector $\mathbf{S_{xi}}$ and their corresponding switching signals, converter output voltages, and DC branch currents is given in Table 8.2.

By substituting (8.52) in (8.51), the DC-link capacitors current are expressed in terms of the three-phase grid currents and coefficients K_{xi1} and K_{xi2} as follows:

$$\begin{aligned} i_{C1} &= K_{ai1}\, i_{ag} + K_{bi1}\, i_{ag} + K_{ci1}\, i_{cg} \\ i_{C2} &= K_{ai2}\, i_{ag} + K_{bi2}\, i_{ag} + K_{ci2}\, i_{cg}. \end{aligned} \tag{8.53}$$

The coefficients for capacitors current, K_{xi1} and $K_{xi2} \in \{0, -1\}$ can be defined on the basis of NPC inverter switching vectors:

$$\begin{aligned} K_{xi1} &= \text{sgn}\,(0 - \mathbf{S_{xi}}) \\ K_{xi2} &= \text{sgn}\,(1 - \mathbf{S_{xi}}) \cdot \text{sgn}\,(\mathbf{S_{xi}}) \end{aligned} \tag{8.54}$$

Table 8.2 Switching states and coefficients of DC capacitors current $\forall\ x \in \{a, b, c\}$ [1]

Switching Vector	Switching Signals				Output Voltage	Input Currents		DC Current Gains	
$\mathbf{S_{xi}}$	s_{xi1}	s_{xi2}	s_{xi3}	s_{xi4}	v_{xN}	i_{Pi}	i_{Zi}	K_{xi1}	K_{xi2}
[2]	'1'	'1'	'0'	'0'	$v_{C1} + v_{C2}$	i_x	0	-1	-1
[1]	'0'	'1'	'1'	'0'	v_{C2}	0	i_x	-1	0
[0]	'0'	'0'	'1'	'1'	0	0	0	0	0

where sgn(·) is a signum function which is defined as

$$\operatorname{sgn}(x) = \begin{cases} -1 & \text{if } x < 0 \\ 0 & \text{if } x \equiv 0 \\ +1 & \text{if } x > 0. \end{cases} \tag{8.55}$$

The coefficients K_{xi1} and K_{xi2} in Table 8.2 can be defined offline, along with the switching states. Given that coefficients do not need to be calculated online, the modeling given in this section incurs a small number of calculations. Moreover, this modeling can be easily extended to the m-level diode-clamped converter (refer to [21] for details). The dynamics in (8.49), (8.53), and (8.54) suggest that the DC-link capacitors voltage are also a function of switching signals. The proper selection of switching signals lead to an accurate balancing of the capacitors voltage.

8.5 CALCULATION OF REFERENCE GRID-SIDE VARIABLES

As mentioned in Section 8.2, in a complete WECS, the following variables need to be regulated to achieve a high-performance operation for WECS: (1) MPPT control, (2) net DC-bus voltage control, and (3) grid reactive power control. The first two control variables can be exchanged between the MSC and GSC, as shown in Figure 8.16. In the first case (Figure 8.16(a)), the MSC performs MPPT, whereas the GSC performs net DC-bus voltage control and grid reactive power control. The wind generator, along with the AC/DC rectifier, is represented by a variable DC current source, whose magnitude is proportional to the active power extracted from wind. In the second case (Figure 8.16(b)), the GSC performs MPPT and grid reactive power control, whereas the MSC performs net DC-bus voltage control. Thus, the wind generator along with the AC/DC rectifier is represented by a constant DC voltage source.

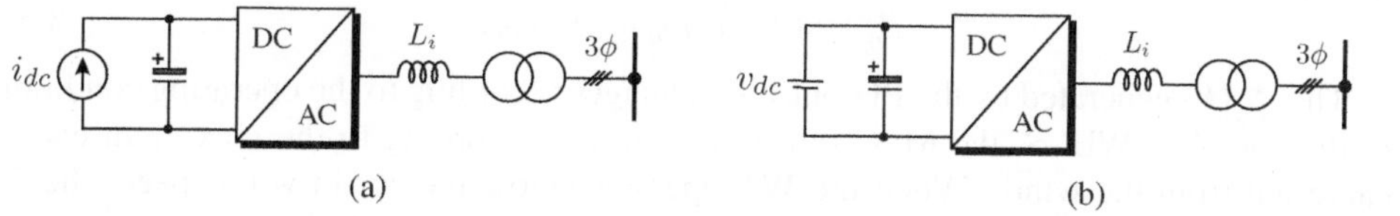

Figure 8.16 Simplified representation of GSC with: (a) generator-side MPPT control and (b) grid-side MPPT control.

To achieve full control over WECS in either control approach, the GSC control objectives are translated into reference currents/powers. The classical VOC uses i^*_{dg} and i^*_{qg}, whereas the classical V-DPC uses P^*_g and Q^*_g as reference control variables for GSC. The abc-frame PCC scheme uses i^*_{ag}, i^*_{bg}, and i^*_{cg}; the $\alpha\beta$-frame PCC scheme uses $i^*_{\alpha g}$ and $i^*_{\beta g}$; the dq-frame PCC scheme uses i^*_{dg} and i^*_{qg}; and the PPC scheme uses P^*_g and Q^*_g as reference control variables for GSC. Therefore, the calculation of reference control variables is common for both classical and predictive control methods. In this section, the calculation of reference currents in the abc, $\alpha\beta$, and dq frames, as well as the reference active and reactive powers, is presented.

8.5.1 Generator-side MPPT

In this case, the GSC regulates net DC-bus voltage v_{dc} at its reference value v^*_{dc} and generates reactive power Q_g according to the grid operator's command Q^*_g. The block diagram for the calculation of the abc-, $\alpha\beta$-, and dq-frame reference grid currents, as well as the grid active and reactive power references considering the generator-side MPPT control, is shown in Figure 8.17.

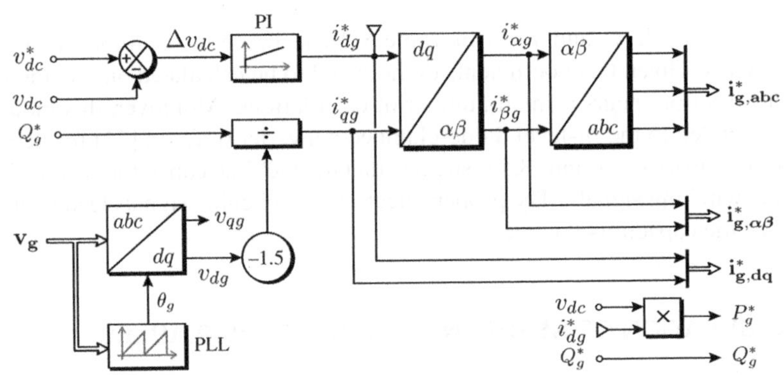

Figure 8.17 Calculation of reference grid-side control variables with generator-side MPPT control.

A PI controller is used to generate the reference d-axis grid current i^*_{dg} such that the measured net DC-bus voltage v_{dc} is maintained at its reference value v^*_{dc}:

$$i^*_{dg} = \left(k_p + \frac{k_i}{s}\right)(v^*_{dc} - v_{dc}) \tag{8.56}$$

where s is the Laplace operator; k_p and k_i are the proportional and integral gains of the PI controller, respectively.

By neglecting losses in GSC, the active power P_g on the AC side will equal the DC power:

$$P_g = 1.5\, v_{dg}\, i_{dg} = v_{dc}\, i_{dc}. \tag{8.57}$$

The $i^*_{dg}(k)$ generated by the PI controller changes according to the operating condition. In the complete WECS, the MSC injects current i_{dc} according to the maximum energy harvested from the wind. When the WT operates below the cut-in wind speed, the i_{dc} supplied by the MSC is zero. Similarly, when the WT operates at the rated wind speed, the i_{dc} is rated. To simulate such a variable-speed WECS, a variable DC current source is considered, as shown in Figure 8.16(a).

The reference reactive power command Q_g^* is provided by the grid operator, which can be set to zero for unity, negative for leading, and positive for the lagging PF. From (8.6), the q-axis reference grid current i_{qg}^* is calculated from Q_g^* as follows:

$$i_{qg}^* = \frac{Q_g^*}{-1.5\, v_{dg}}. \tag{8.58}$$

The dq-frame reference currents are transformed into the $\alpha\beta$ and abc frames with the help of transformation matrices given in (3.12) and (3.7). The reference grid active power P_g^* is obtained by multiplying i_{dg}^* with v_{dc}.

8.5.2 Grid-side MPPT

The block diagram for the calculation of abc-, $\alpha\beta$-, and dq-frame reference grid currents, as well as the grid active and reactive power references considering grid-side MPPT control, is shown in Figure 8.18. The active power reference P_g^* is obtained by the WT power curves based MPPT algorithm (refer to Chapter 1 and Section 1.7.4). This method uses turbine output power versus wind speed v_w curve provided by the WT manufacturer.

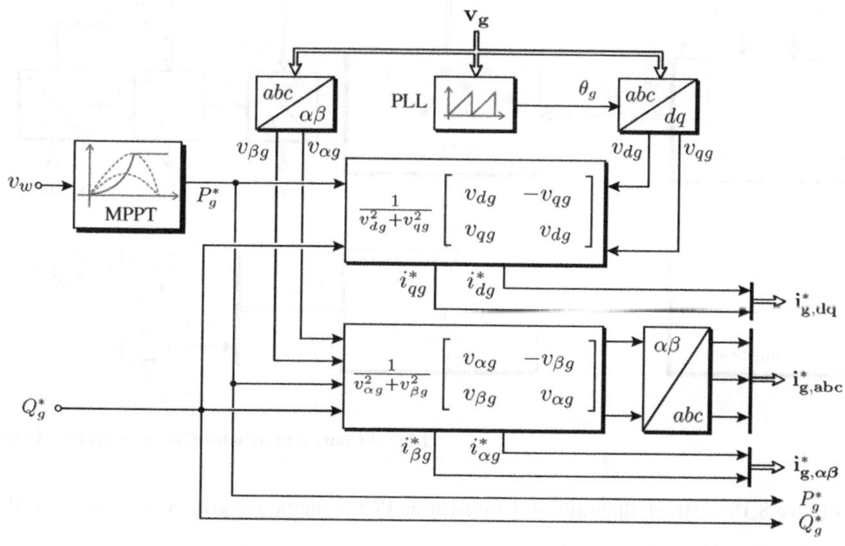

Figure 8.18 Calculation of reference grid-side control variables with grid-side MPPT control.

The measured three-phase grid voltages are transformed into $\alpha\beta$ and dq frames to aid the calculation of reference grid currents. On the basis of the reference active and reactive power commands and the measured $\alpha\beta$-frame grid voltages, the $\alpha\beta$-frame reference grid currents are calculated as follows [2]:

$$\begin{bmatrix} i_{\alpha g}^* \\ i_{\beta g}^* \end{bmatrix} = \frac{1}{v_{\alpha g}^2 + v_{\beta g}^2} \begin{bmatrix} v_{\alpha g} & -v_{\beta g} \\ v_{\beta g} & v_{\alpha g} \end{bmatrix} \begin{bmatrix} P_g^* \\ Q_g^* \end{bmatrix}. \tag{8.59}$$

Similarly, the dq-frame reference grid currents are computed as follows:

$$\begin{bmatrix} i_{dg}^* \\ i_{qg}^* \end{bmatrix} = \frac{1}{v_{dg}^2 + v_{qg}^2} \begin{bmatrix} v_{dg} & -v_{qg} \\ v_{qg} & v_{dg} \end{bmatrix} \begin{bmatrix} P_g^* \\ Q_g^* \end{bmatrix}. \tag{8.60}$$

8.6 PREDICTIVE CURRENT CONTROL OF 2L-VSI IN dq-FRAME

The block diagram of the dq-frame PCC scheme for grid-connected 2L-VSI is shown in Figure 8.19, where the WT, wind generator, and MSC are represented by a variable DC current source [22]. The power flow is considered from the GSC to the grid. Under these conditions, the net DC-bus voltage and reactive power to the grid are regulated by properly generating the gating signals for 2L-VSI.

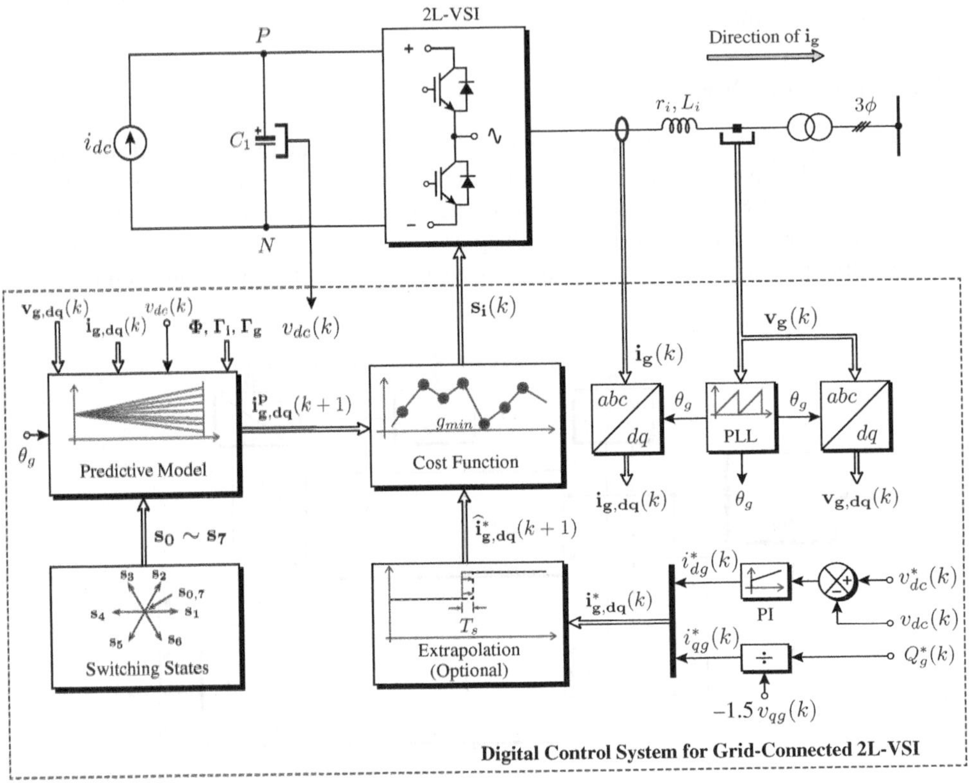

Figure 8.19 Block diagram of the dq-frame PCC scheme for grid-connected 2L-VSI.

8.6.1 Design Procedure

The PCC scheme in Figure 8.19 is very easy and intuitive to understand. It consists of five design steps:

Measurement and Synthesis of Feedback Signals: The three-phase grid voltages $\mathbf{v}_g$, grid currents $\mathbf{i}_g$, and capacitor voltage v_{C1} are measured. For a balanced system, the phase-c grid voltage and current does not need to be measured because $v_{cg} = -v_{ag} - v_{bg}$ and $i_{cg} = -i_{ag} - i_{bg}$. Therefore, to design the PCC scheme, a total of three voltage sensors and two current sensors are required for a balanced system. By using the measured grid voltages, the grid voltage angle θ_g is obtained by SRF-PLL. The abc-frame grid voltages and currents are transformed to $\mathbf{v}_{g,dq}$ and $\mathbf{i}_{g,dq}$ with the help of θ_g. Given the grid voltage orientation, v_{dg} is aligned with the grid voltage vector, and v_{qg} becomes zero.

Calculation of Reference Currents: To realize the PCC scheme with generator-side MPPT, the dq-axes reference currents are obtained from v_{dc}^* and Q_g^* loops. The $i_{dg}^*(k)$ is generated by the DC-link PI controller according to the active power extracted from the wind, thus ensuring that v_{dc} is maintained at its reference value v_{dc}^*. This design step is common for both classical and predictive control schemes, as well as 2L-VSIs and NPC inverters. The outer DC-bus voltage control loop has a slower dynamic response than the inner current control loop. As mentioned in Chapter 4, the PI controller in the outer DC-bus voltage control loop can be avoided by incorporating v_{dc} in the cost function. This control approach is called predictive DC-bus voltage control.

Extrapolation of Reference Currents: Owing to the DC nature of reference dq-axes currents, extrapolation is unnecessary in a steady-state. However, to improve the transient performance, a first-order Lagrange extrapolation can be used with the following generic representation:

$$\widehat{\mathbf{i}}^*_{\mathbf{g,dq}}(k+1) = 2\,\mathbf{i}^*_{\mathbf{g,dq}}(k) - \mathbf{i}^*_{\mathbf{g,dq}}(k-1). \tag{8.61}$$

Prediction of Future Behavior of Grid Currents: The future value of the dq-axes grid currents is assessed by the "predictive model" block, which combines the dq-frame DT model of grid currents in (8.37) with the 2L-VSI model in (8.43). The overall sampled-data model for the prediction of future values of dq-axes grid currents is formulated in terms of 2L-VSI switching signals as follows:

$$\begin{bmatrix} i^p_{dg}(k+1) \\ i^p_{qg}(k+1) \end{bmatrix} = \mathbf{\Phi} \begin{bmatrix} i_{dg}(k) \\ i_{qg}(k) \end{bmatrix} + \mathbf{\Gamma_i} \left\{ v_{C1}(k) \begin{bmatrix} s^p_{di1}(k) \\ s^p_{qi1}(k) \end{bmatrix} \right\} + \mathbf{\Gamma_g} \begin{bmatrix} v_{dg}(k) \\ v_{qg}(k) \end{bmatrix} \tag{8.62}$$

where superscript p denotes the predicted variable. To minimize the number of online calculations, matrices $\mathbf{\Phi}$, $\mathbf{\Gamma_i}$, and $\mathbf{\Gamma_g}$ are defined offline assuming that ω_g is constant.

Cost Function Minimization: The control objectives of the PCC scheme for 2L-VSI are as follows: (1) regulation of active power for which the d-axis current is controlled, (2) regulation of reactive power for which the q-axis current is controlled, and (3) minimization of switching frequency for which the number of switch changes are penalized. All these goals are included into a quadratic cost function for one-sample-ahead prediction as follows:

$$\begin{aligned} g_i(k) = {} & \lambda_{id} \left[\widehat{i}^*_{dg}(k+1) - i^p_{dg}(k+1) \right]^2 \\ & + \lambda_{iq} \left[\widehat{i}^*_{qg}(k+1) - i^p_{qg}(k+1) \right]^2 \\ & + \lambda_{sw,i} \sum_{x=a,\,b,\,c} \left[s^p_{xi1}(k) - s^{\text{op}}_{xi1}(k) \right]^2 \end{aligned} \tag{8.63}$$

where λ_{id}, λ_{iq}, and $\lambda_{sw,i}$ are the weighting factors for i_{dg}, i_{qg}, and switching frequency regulation, respectively. s^{op}_{ai1}, s^{op}_{bi1}, and s^{op}_{ci1} are the optimal switching signals in past sample. By using the per-unit weighting factor method (Section 4.5.2), λ_{id} and λ_{iq} are set to one. The weighting factor $\lambda_{sw,i}$ is selected by using a heuristic approach.

With the above eight combinations for switching signals, eight different predictions are obtained for v_{ds} and v_{qs}, which in turn leads to eight different future values of i_{ds} and i_{qs}. The predicted i_{ds} and i_{qs} are compared with the dq-axes reference currents to select an optimal switching state combination that minimizes the cost function. The selected switching states are applied to the 2L-VSI during $(k+1)$ sampling period.

8.6.2 Control Algorithm

The flowchart of the PCC algorithm for 2L-VSI is shown in Figure 8.20. The DC-link voltage, grid voltages, and grid currents are measured in block ①. The $(k+1)$-state reference dq-axes grid currents are calculated by block ②. In block ③, the PCC algorithm is initialized by setting the switching state combination j to zero, and the optimal cost function (g_{op}) to ∞. In blocks ④ to ⑦, the PCC algorithm enters an iterative loop where the 2L-VSI output voltages and grid currents are predicted and cost function minimization is performed. An optimal vector j_{op} and corresponding switching signals that minimize the cost function are selected and applied directly to the 2L-VSI in blocks ⑧ to ⑩.

The *S-Function Builder* programming in MATLAB environment is shown in Algorithm 8.1. The inputs to the *S-Function* block are $\hat{i}^*_{dg}(k+1)$, $\hat{i}^*_{qg}(k+1)$, $i_{dg}(k)$, $i_{qg}(k)$, $v_{dg}(k)$, $v_{qg}(k)$, $v_{C1}(k)$, cos $\theta_g(k)$, sin $\theta_g(k)$, and $\lambda_{sw,i}$. The output signals from the *S-Function* block are $s^P_{ai1}(k)$, $s^P_{bi1}(k)$, and $s^P_{ci1}(k)$. The external function declarations for the *S-Function* include switching states in $\alpha\beta$ and abc reference frames, control sampling time T_s, and DT parameters $\mathbf{\Phi}$, $\mathbf{\Gamma_i}$, and $\mathbf{\Gamma_g}$. The program lines are explained as follows: Lines `01-02` correspond to PCC initialization; the `for` loop is initialized in lines `03-04`; 2L-VSI voltages are predicted in lines `05-08`; grid currents are predicted in lines `09-10`; the cost function is defined in lines `11-14`; optimal vector `j_op` number is selected in lines `15-19`; `for` loop is terminated in line `20`; optimal three-phase switching states are produced in lines `21-23` on the basis of the `j_op` value; switching states are stored in lines `24-26` for use in the next iteration.

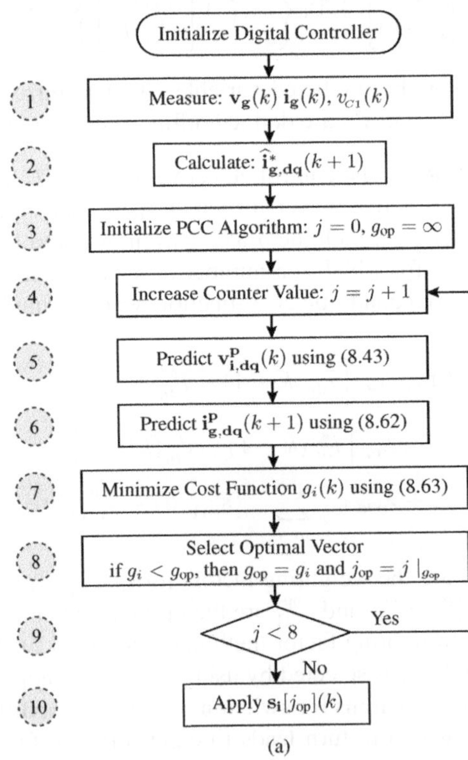

(a)

Figure 8.20 Flowchart of the PCC algorithm for grid-connected 2L-VSI.

ALGORITHM 8.1 ***dq*-Frame PCC Algorithm for Grid-Connected 2L-VSI**

```
\\ P11 = 1-(ri*Ts/Li); P12 = (wg*Ts); P21 = -(wg*Ts); P22 = 1-(ri*Ts/Li);
\\ Gi11 = Gi22 = Ts/Li; Gg11 = Gg22 = -Ts/Li;
j_op = 1000;
g_op = 1000000000;
for(j = 0; j < 8; j++)
{
    vi_ab[0] = vc[0]*vinv[j].s_i_ab[0];
    vi_ab[1] = vc[0]*vinv[j].s_i_ab[1];
    vi_dq[0] =  vi_ab[0]*cstheta_g[0] + vi_ab[1]*cstheta_g[1];
    vi_dq[1] = -vi_ab[0]*cstheta_g[1] + vi_ab[1]*cstheta_g[0];
    ig_dq_k1[0] = P11*ig_dq[0]+P12*ig_dq[1]+Gi11*vi_dq[0]+Gg11*vg_dq[0];
    ig_dq_k1[1] = P21*ig_dq[0]+P22*ig_dq[1]+Gi22*vi_dq[1]+Gg22*vg_dq[1];
    g_id = (ig_ref_k1[0]-ig_dq_k1[0])*(ig_ref_k1[0]-ig_dq_k1[0]);
    g_iq = (ig_ref_k1[1]-ig_dq_k1[1])*(ig_ref_k1[1]-ig_dq_k1[1]);
    g_swi = (vinv[j].s_i_abc[0]-s_i_km1[0])*(vinv[j].s_i_abc[0]-s_i_km1[0])
          + (vinv[j].s_i_abc[1]-s_i_km1[1])*(vinv[j].s_i_abc[1]-s_i_km1[1])
          + (vinv[j].s_i_abc[2]-s_i_km1[2])*(vinv[j].s_i_abc[2]-s_i_km1[2]);
    g_i = g_id + g_iq + alpha[0]*g_swi;
    if(g_i < g_op)
       {
          j_op = j;
          g_op = g_i;
       }
}
s_ai1[0] = vinv[j_op].s_i[0];
s_bi1[0] = vinv[j_op].s_i[1];
s_ci1[0] = vinv[j_op].s_i[2];
s_i_km1[0] = s_ai1[0];
s_i_km1[1] = s_bi1[0];
s_i_km1[2] = s_ci1[0];
```

8.6.3 Comparison of the PCC Design with VOC

In this section, the design procedures for PCC and classical VOC are compared. The block diagram of classical decoupled VOC is shown in Figure 8.21 [10, 23]. With VOC, the measurement of feedback signals $\mathbf{v_g}$, $\mathbf{i_g}$, and v_{C1}, detection of grid voltage angle θ_g, transformation of $\mathbf{v_g}$ and $\mathbf{i_g}$ to $\mathbf{v_{g,dq}}$ and $\mathbf{i_{g,dq}}$, and calculation of i_{dg}^* and i_{qg}^* are similar to the PCC scheme discussed earlier. The internal control loop uses two PI controllers, decoupling terms, $dq/\alpha\beta$ transformation, and space vector modulation. Feedback currents i_{dg} and i_{qg} are compared with reference currents i_{dg}^* and i_{qg}^*, respectively. Current errors Δi_{dg} and Δi_{qg} are processed by two PI controllers. Terms $-\omega_g\, L_i\, i_{qg}$ and $\omega_g\, L_i\, i_{dg}$ are added to the outputs of the PI controllers so as to compensate the coupling between dq-axes currents. The reference inverter voltages v_{di}^* and v_{qi}^* are calculated according to the dynamic model in (8.32). Given that L_i is used in decoupling, the grid impedance variations affect the performance of VOC. The dq-frame reference inverter voltages are transformed into $\alpha\beta$-frame with the help of θ_g. The stationary-frame reference inverter voltages, along with v_{C1}, are used by the modulation stage to generate the gating signals for the 2L-VSI.

A low switching frequency operation is an important requirement at the MW-level. During such a condition, the classical VOC presents several technical and operational related issues such as the following [24]: (1) unsymmetrical performance characteristics because of the predominant nonlinear nature of VSI; (2) significant low-order harmonics, which cause poor power quality and conflict to the grid codes; (3) sluggish transient response caused by the low-bandwidth modulation stage; (4) coupling of control variables such as d- and q-axis grid currents; (5) degraded performance because of the grid voltage harmonics and control delay [25]; (6) complicated approach to include system constraints in the design of controller. The most favorable situation to use the VOC method with PWM/SVM is in fixed switching frequency operations. The MPC is a nonlinear control method and provides an approach that is suitable for controlling power converters while mitigating the aforementioned disadvantages associated with the VOC method. The MPC method also offers a fast dynamic response and the ability to include nonlinearities and constraints in the design of the controller. The classical VOC de facto requirements for internal current control loops and modulation are eliminated by this scheme. The use of MPC entails more computational burden, but modern digital signal processors can perform large amount of calculations with less burden [26].

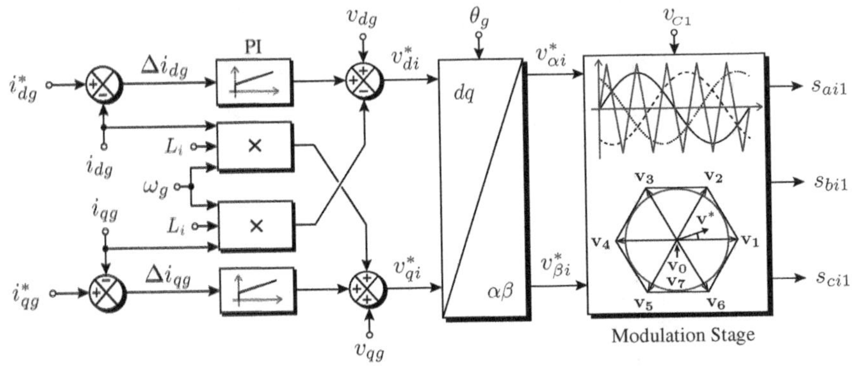

Figure 8.21 Block diagram of decoupled VOC for grid-connected 2L-VSI.

8.6.4 Comparison of GSC Performance with Passive Load Case

The design of PCC scheme for 2L-VSI with passive (RL) load has been discussed in Chapter 4. In this section, the design procedure of PCC and the performance of grid-connected and passive-load cases are compared. By comparing the PCC scheme of the RL load (Figure 4.2) with the PCC scheme of grid-connected 2L-VSI (Figure 8.19), the following observations are made: (1) the grid current DT model replaces the load resistance R with the grid-filter internal resistance r_i; (2) the predictive model in (8.62) uses grid voltages in addition to the DC-link voltage and feedback grid currents. The rest of the control scheme design is the same for grid-connected and RL-load cases.

Figure 8.22 shows the total harmonic distortion (THD) versus the inverter switching frequency ($f_{sw,i}$) performance with grid-connected and passive load cases considering a multilevel converter [22]. The load resistance value is set as the base impedance Z_{B_g} of the system. The THD versus the $f_{sw,i}$ can be attributed to the 2L-VSI with an exception that THD is higher for 2L-VSI compared to multilevel converters. For grid-connected case, the weighting factor $\lambda_{sw,i}$ is changed from 0 to 20,000, the switching frequency reduced from 1338 Hz to 105 Hz, and the THD increased from 2.27% to 14%. For passive load case, the $\lambda_{sw,i}$ is changed from 0 to 10,000, and the $f_{sw,i}$ is reduced from 1333 Hz to 102 Hz with an increase in THD from 2.18% to 14%. The grid voltages have very insignificant impact on the performance of the PCC scheme. The difference between the two cases is also insignificant during the entire switching frequency range. Note that for both cases, a significant reduction in the switching frequency from 1330 Hz to 600 Hz is achieved with a slight increase in the grid current THD from 2.2% to 4.1%.

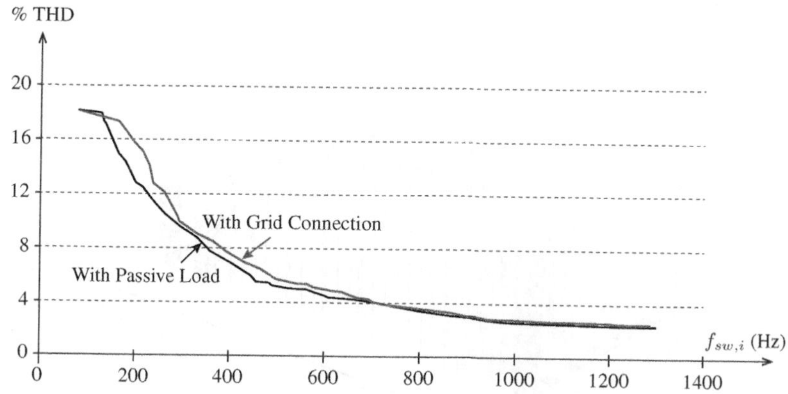

Figure 8.22 Comparison of the % THD versus the $f_{sw,i}$ with grid-connected and resistive load cases.

8.6.5 Switching Frequency Regulation

Owing to the unsynchronized sampling between the measurements and output fundamental reference frequency, the chosen switching states for predicted cost function values within the period of one fundamental cycle vary compared to the selected switching states in the preceding cycle [27]. Thus, the MPC strategy leads to variable switching frequency operation and wide harmonic spectrum. For the grid-connected case, the variable switching frequency operation causes difficulty in the grid-filter design and causes harmonic oscillations in some cases (e.g., with LCL filter) [28]. To solve this issue, the switching frequency can be regulated between two close boundary limits (e.g., 750 and 850 Hz) with the help of a lookup-table-based weighting factor design.

The variable switching frequency nature of MPC is studied for grid-connected multilevel converter with respect to the i^*_{dg} and Q^*_g (Figure 8.23) [22]. The switching frequency increases linearly with the d-axis grid current. The $f_{sw,i}$ increases with negative reactive power compared with positive reactive power. Without the weight factor $\lambda_{sw,i}$, the switching frequency is found to vary in a range of 1800–900Hz, as shown in Figure 8.23(a). To overcome this issue, weighting factors as in Figure 8.23(c) are designed on the basis of the data shown in Figure 8.23(a). By observation, these factors hold a linear relation to the active and reactive powers and are stored in a lookup table for online use. As a result, the switching frequency is maintained within the 750–850 Hz range during all the operating conditions as demonstrated in Figure 8.23(b).

CASE STUDY 8.1 Steady-State Analysis of Grid-Connected 2L-VSI

Case Study Objective: In this case study, the steady-state performance of grid-connected 2L-VSI is analyzed with two operating conditions: (1) pure active power injection (P_g = 1.0 pu, and Q_g = 0.0 pu); (2) pure reactive power injection (P_g = 0.0 pu, and Q_g = 1.0 pu). The first and second cases correspond to the normal operation and FRT operation, respectively. The DC current, grid voltages, currents, and powers are calculated numerically.

Parameters: A 3.0-MVA 2L-VSI is connected to 690 V (line-to-line) three-phase grid. The net DC-bus voltage v_{dc} is designed as 1220 V (3.062 pu).

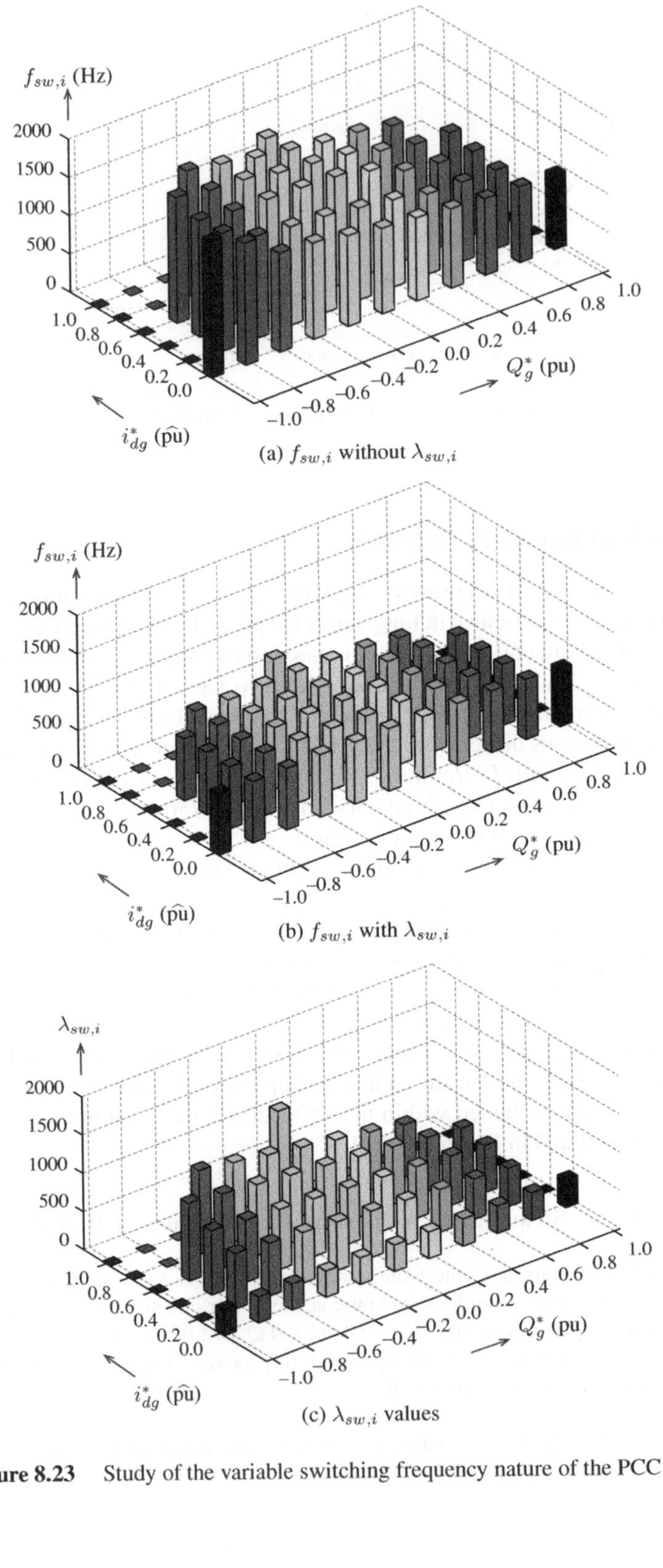

Figure 8.23 Study of the variable switching frequency nature of the PCC scheme.

Analysis:
The steady-state waveforms of grid-connected 2L-VSI are shown in Figure 8.24. The d-axis current, q-axis current, phase-a grid current, 2L-VSI line-to-line (a-b) voltage, and grid active and reactive powers are shown in subplots (a) to (e), respectively. The DC-link and grid-side variables are calculated as follows:

$$v_{dg} = \frac{\sqrt{2}\,V_{g,ll}}{\sqrt{3}} = \frac{\sqrt{2}\times 690}{\sqrt{3}} = 563.4\text{ V }(\textit{peak}) \quad (1.0\ \widehat{\text{pu}})$$

$$i_{dg}^* = \frac{P_g}{1.5\,v_{dg}} = \frac{3.0\times 10^6}{1.5\times 563.4} = 3550\text{ A }(\textit{peak}) \quad (1.0\ \widehat{\text{pu}})$$

$$i_{qg}^* = \frac{Q_g}{-1.5\,v_{dg}} = \frac{0\times 10^6}{-1.5\times 563.4} = 0\text{ A }(\textit{peak}) \quad (0.0\ \widehat{\text{pu}})$$

$$I_g = \sqrt{i_{dg}^2 + i_{qg}^2}/\sqrt{2} = \sqrt{3550^2 + 0^2}/\sqrt{2} = 2510\text{ A }(\textit{rms}) \quad (1.0\text{ pu})$$

$$\varphi_g = \tan^{-1}(i_{qg}/i_{dg}) = \tan^{-1}(0/3550) = 0^\circ$$

$$P_g = P_{dc} = 3.0\text{ MW}, \quad \Rightarrow i_{dc} = P_{dc}/v_{dc} = \frac{3.0\times 10^6}{1220} = 2459\text{ A} \quad (0.98\text{ pu}).$$

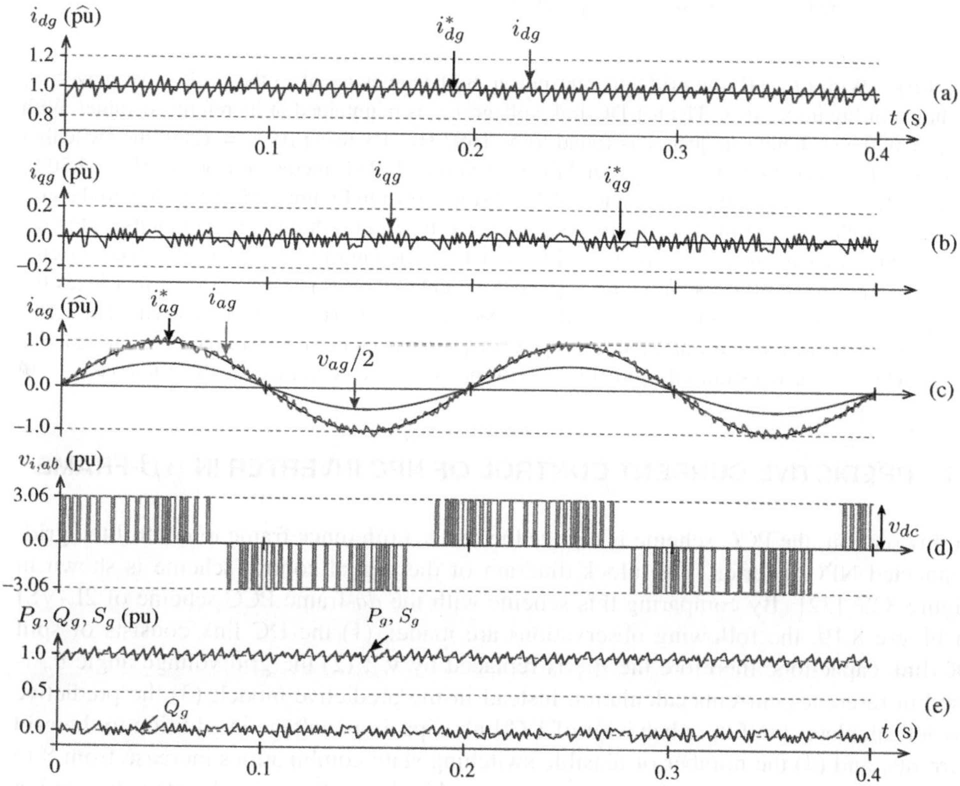

Figure 8.24 Steady-state waveforms for a 2L-VSI with $i_{dg}^* = 1.0\ \widehat{\text{pu}}$, and $i_{qg}^* = 0.0\ \widehat{\text{pu}}$.

This operating condition corresponds to the pure active power injection into the grid, which is the most common scenario in WECS. The active power component of the grid current i_{dg} equals the peak grid current, and the reactive power component of the grid current i_{qg} becomes zero as demonstrated in Figures 8.24(a) and 8.24(b). The grid current is maintained in-phase with the grid voltage as shown in Figure 8.24(c). The 2L-VSI line-to-line voltage contains three-levels: $-v_{dc}$, 0,

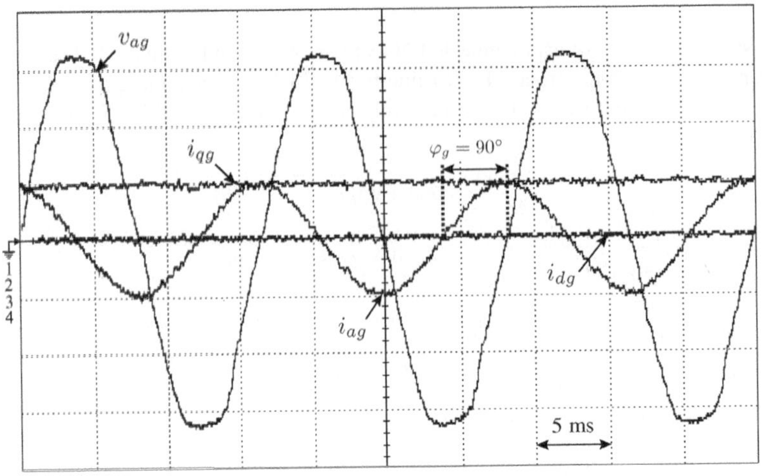

Figure 8.25 Steady-state waveforms for a 2L-VSI with i^*_{dg} = 0.0 $\widehat{\text{pu}}$, and i^*_{qg} = 1.0 $\widehat{\text{pu}}$. v_{ag}: 50 V/div, i_{ag}: 20 A/div, i_{dg}: 20 A/div, and i_{qg}: 20 A/div.

and $+v_{dc}$. With Q_g = 0, the grid apparent power S_g becomes equal to the grid active power P_g as shown in Figure 8.24(e). The net DC-link voltage v_{dc} is maintained at its reference value. With $\lambda_{sw,i}$ = 0, the switching frequency is found to be 1687 Hz. By using $\lambda_{sw,i}$ = 1200, the switching frequency $f_{sw,i}$ is reduced to 1455 Hz, and the grid current THD is increased from 7.05% to 7.30%.

The FRT capability of the grid-connected 2L-VSI is shown in Figure 8.25. According to the grid codes, the high-power WECS should remain connected to the grid during the grid faults and inject pure leading reactive power to help the grid recover from the fault [29–31]. The reference currents corresponding to the FRT operation are i^*_{dg} = 0.0 $\widehat{\text{pu}}$ and i^*_{qg} = 1.0 $\widehat{\text{pu}}$. The grid current leads the voltage by 90°. The i_{dg} becomes zero and the i_{qg} becomes equal to the peak grid current. The active power is maintained at zero, and the grid apparent power becomes equal to the absolute reactive power. This case demonstrates the effectiveness of the PCC scheme in meeting the grid codes. ■

8.7 PREDICTIVE CURRENT CONTROL OF NPC INVERTER IN $\alpha\beta$-FRAME

In this section, the PCC scheme is realized in the $\alpha\beta$ reference frame to control the grid-connected NPC inverter. The block diagram of the digital control scheme is shown in Figure 8.26 [32]. By comparing this scheme with the dq-frame PCC scheme of 2L-VSI in Figure 8.19, the following observations are made: (1) the DC link consists of split DC-link capacitors, therefore the v_{C1} is replaced by $\mathbf{v}_C$; (2) the grid voltage angle θ_g is used in reference currents calculation instead in the predictive model; (3) the predictive model calculates the future behavior of DC-link capacitors voltage in addition to the grid currents; and (4) the number of feasible switching state combinations increase from 8 to 27. For the classical VOC, the grid current control in the dq-frame is advantageous, owing to the DC nature of the dq-axes currents. However, with MPC scheme, the grid current reference tracking error in the dq- and $\alpha\beta$-frames is almost similar. During any sampling interval, the $dq/\alpha\beta$ transformation in the reference currents loop is performed once, but the $\alpha\beta/dq$ transformation in the predictive model is performed 27 times. The corresponding number of online calculations decrease from 27 to 1 because of the elimination of the $\alpha\beta/dq$ transformation in the predictive model, hence, the computational burden on digital signal processor (DSP) becomes lower with the $\alpha\beta$-frame PCC implementation.

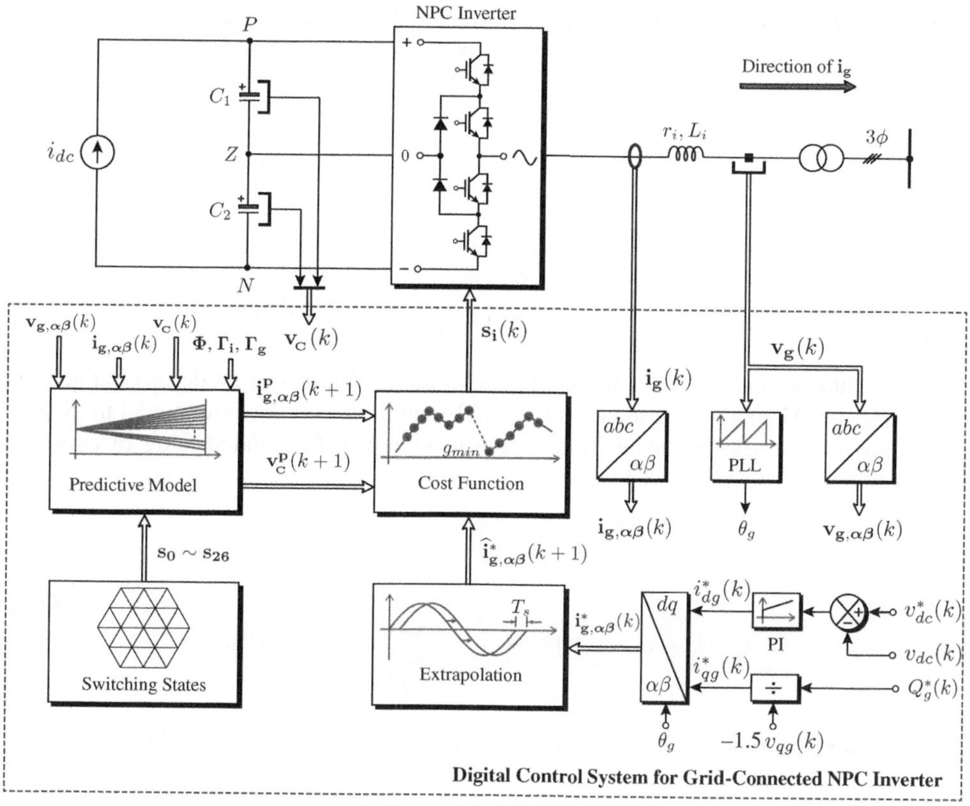

Figure 8.26 Block diagram of the $\alpha\beta$-frame PCC scheme for grid-connected NPC inverter.

8.7.1 Design Procedure

The digital control system implementation for NPC inverter is similar to the 2L-VSI. The control system implementation procedure is given below.

Measurement and Synthesis of Feedback Signals: The grid voltages and currents ($\mathbf{v_g}$ and $\mathbf{i_g}$), and the DC-link capacitors voltage $\mathbf{v_C}$ are measured. The abc-frame grid voltages and currents are transformed to $\mathbf{v}_{g,\alpha\beta}$ and $\mathbf{i}_{g,\alpha\beta}$ with the help of the $abc/\alpha\beta$ transformation in (3.6). The grid voltage angle θ_g is obtained by SRF-PLL for use with the $\alpha\beta$-frame reference currents calculation.

Calculation of Reference Currents: Similar to the control scheme in Figure 8.19, the reference currents i^*_{dg} and i^*_{qg} are obtained from v^*_{dc} and Q^*_g loops, respectively. The SRF reference currents are transformed to stationary frame ($i^*_{\alpha g}$ and $i^*_{\beta g}$) with the help of θ_g.

Extrapolation of Reference Currents: To compute the grid current error in $(k+1)$ sampling instant, the reference currents should be extrapolated to $(k+1)$ state. The vector angle extrapolation is the most suited approach for the stationary-frame variables. The future value of reference grid currents is estimated as shown below [33]:

$$\begin{bmatrix} i^*_{\alpha g}(k+1) \\ i^*_{\beta g}(k+1) \end{bmatrix} = e^{j\omega_g T_s} \begin{bmatrix} i^*_{\alpha g}(k) \\ i^*_{\beta g}(k) \end{bmatrix}. \tag{8.64}$$

Prediction of Future Behavior of Grid Currents and DC Capacitors Voltage: Using the measured quantities and offline computed DT parameters, the future behavior of the $\alpha\beta$-frame grid currents is predicted during each iteration according to following sampled-data model:

$$\begin{aligned}\begin{bmatrix} i^p_{\alpha g}(k+1) \\ i^p_{\beta g}(k+1) \end{bmatrix} &= \mathbf{\Phi} \begin{bmatrix} i_{\alpha g}(k) \\ i_{\beta g}(k) \end{bmatrix} + \mathbf{\Gamma_g} \begin{bmatrix} v_{\alpha g}(k) \\ v_{\beta g}(k) \end{bmatrix} \\ &+ \mathbf{\Gamma_i} \left\{ v_{C1}(k) \begin{bmatrix} s^p_{\alpha i1}(k) \\ s^p_{\beta i1}(k) \end{bmatrix} + v_{C2}(k) \begin{bmatrix} s^p_{\alpha i2}(k) \\ s^p_{\beta i2}(k) \end{bmatrix} \right\}\end{aligned} \tag{8.65}$$

where the superscript p denotes the predicted variable. The above model is obtained by combining the DT model of grid currents in (8.29) with the NPC inverter model in (8.46).

The predictive model of the DC-link capacitors voltage is formulated below by combining the models in (8.49) and (8.53):

$$\begin{aligned} v_{C1}(k+1) &= v_{C1}(k) + \frac{T_s}{C_1} \sum_{x=a,\,b,\,c} K_{xi1}\, i_{xg}(k) \\ v_{C2}(k+1) &= v_{C2}(k) + \frac{T_s}{C_2} \sum_{x=a,\,b,\,c} K_{xi2}\, i_{xg}(k). \end{aligned} \tag{8.66}$$

Cost Function Minimization: The grid-side cost function is defined as follows such that the control goals are met [34]:

$$\begin{aligned} g_i(k) &= \lambda_{i\alpha} \left[\widehat{i}^*_{\alpha g}(k+1) - i^p_{\alpha g}(k+1) \right]^2 \\ &+ \lambda_{i\beta} \left[\widehat{i}^*_{\beta g}(k+1) - i^p_{\beta g}(k+1) \right]^2 \\ &+ \lambda_{dc,i} \left[v^p_{C1}(k+1) - v^p_{C2}(k+1) \right]^2 \\ &+ \lambda_{sw,i} \sum_{x=a,\,b,\,c} \sum_{j=1,\,2} \left[s^p_{xij}(k) - s^{op}_{xij}(k) \right]^2 \end{aligned} \tag{8.67}$$

where $\lambda_{i\alpha}$, $\lambda_{i\beta}$, $\lambda_{dc,i}$, and $\lambda_{sw,i}$ are the weighting factors for $i_{\alpha g}$, $i_{\beta g}$, the DC-link capacitors voltage, and the switching frequency control, respectively. The primary control goal is to force the grid currents to follow their references, and thus $\lambda_{i\alpha} = \lambda_{i\beta} = 1$. The weighting factor $\lambda_{sw,i}$ is selected using a heuristic approach. A weighting factor $\lambda_{dc,i}$ is defined such that the importance for the secondary goal is adjusted according to the desired performance. The value of $\lambda_{dc,i}$ is calculated as follows according to the per-unit weighting factor method in Section 4.5.2:

$$\lambda_{dc,i} = \frac{I_{Bg}}{v^*_{dc}} \tag{8.68}$$

where I_{Bg} is the base *rms* grid current.

When $\lambda_{sw,i} > 0$, the switching frequency reduction is obtained, but at the expense of a higher grid current tracking error and drift in capacitors voltage. During each iteration, the switching state (among 27) which minimizes the cost function is chosen and applied to the NPC inverter gating terminals.

8.7.2 Control Algorithm

Figure 8.27 shows the flowchart of the PCC algorithm for grid-connected NPC inverter, which is identical to the 2L-VSI algorithm in Figure 8.20. The reference grid currents $\widehat{\mathbf{i}}^*_{\mathbf{g},\alpha\beta}(k+1)$ and weighting factor $\lambda_{dc,i}$ are calculated in block ②. The iterative loop is performed between blocks ④ and ⑨. The NPC inverter terminal voltages $\mathbf{v}^{\mathbf{P}}_{\mathbf{i},\alpha\beta}(k)$, grid currents $\mathbf{i}^{\mathbf{P}}_{\mathbf{g},\alpha\beta}(k+1)$, and DC-link capacitors voltage $\mathbf{v}^{\mathbf{P}}_{\mathbf{C}}(k+1)$ are predicted for 27 times in blocks ⑤ and ⑥. The cost function minimization $g_i(k)$ is carried out in block ⑦. The optimal vector j_{op} which provides simultaneous control of grid currents and DC-link capacitors voltage is selected in block ⑧. The optimal switching signals corresponding to j_{op} are applied to the NPC inverter through block ⑩.

The MATLAB *S-Function Builder* programming is shown in Algorithm 8.2. The code for switching frequency minimization is omitted to simplify the analysis. For current control, the external functions are declared similar to Algorithm 8.1. For DC-link capacitors voltage control, the following additional external functions are declared: `C1`, `C2`, `Ki1`, and `Ki2`. The algorithm generates six switching signals (i.e., `s_ai1`, `s_ai2`, `s_bi1`, `s_bi2`, `s_ci1`, and `s_ci2`) for the NPC inverter. The description of program lines is given as follows: Lines `05-06` predict the NPC inverter voltages; lines `07-08` predict the grid currents; lines `09-10` predict the DC-link capacitors current; lines `11-12` predict the DC-link capacitors voltage; lines `13-16` calculate the cost function values; lines `17-21` select the optimal vector `j_op` number; and lines `23-28` select the optimal switching signals based on `j_op` number.

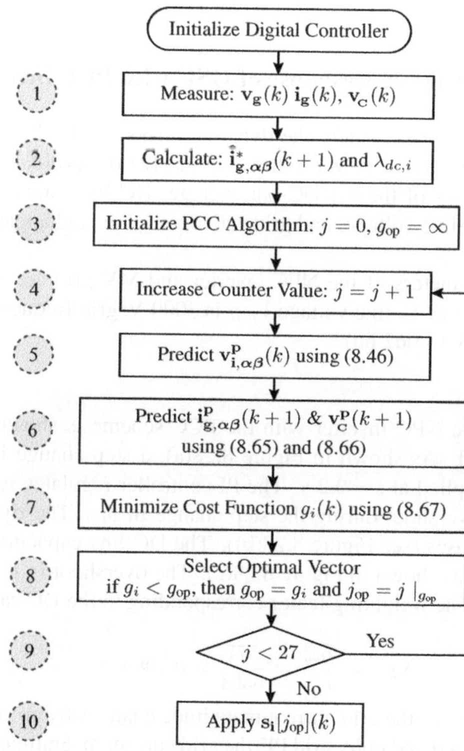

Figure 8.27 Flowchart of the PCC algorithm for grid-connected NPC inverter.

ALGORITHM 8.2 $\alpha\beta$-Frame PCC Algorithm for NPC Inverter

```
\\ P11 = P22 = 1-(ri*Ts/Li); Gi11 = Gi22 = Ts/Li; Gg11 = Gg22 = -Ts/Li;
j_op = 1000;
g_op = 1000000000;
for(j = 0; j < 27; j++)
{
    vi_ab[0] = vc[0]*vinv[j].s_i_ab[0] + vc[1]*vinv[j].s_i_ab[1];
    vi_ab[1] = vc[0]*vinv[j].s_i_ab[2] + vc[1]*vinv[j].s_i_ab[3];
    ig_ab_k1[0] = P11*ig_ab[0]+Gi11*vi_ab[0]+Gg11*vg_ab[0];
    ig_ab_k1[1] = P22*ig_ab[1]+Gi22*vi_ab[1]+Gg22*vg_ab[1];
    ic_k1[0] = vinv[j].K_i1[0]*ig_abc[0] + vinv[j].K_i1[1]*ig_abc[1]
               + vinv[j].K_i1[2]*ig_abc[2];
    ic_k1[1] = vinv[j].K_i2[0]*ig_abc[0] + vinv[j].K_i2[1]*ig_abc[1]
               + vinv[j].K_i2[2]*ig_abc[2];
    vc_k1[0] = vc[0] + (Ts/C1)*ic_k1[0];
    vc_k1[1] = vc[1] + (Ts/C2)*ic_k1[1];
    g_ial = (ig_ref_k1[0]-ig_ab_k1[0])*(ig_ref_k1[0]-ig_ab_k1[0]);
    g_ibe = (ig_ref_k1[1]-ig_ab_k1[1])*(ig_ref_k1[1]-ig_ab_k1[1]);
    g_dc = (vc_k1[0]-vc_k1[1])*(vc_k1[0]-vc_k1[1]);
    g_i = g_ial+ g_ibe + Lambda_dci*g_dc;
    if(g_i < g_op)
       {
          j_op = j;
          g_op = g_i;
       }
}
s_ai1[0] = vinv[j_op].s_i[0];
s_ai2[0] = vinv[j_op].s_i[1];
s_bi1[0] = vinv[j_op].s_i[2];
s_bi2[0] = vinv[j_op].s_i[3];
s_ci1[0] = vinv[j_op].s_i[4];
s_ci2[0] = vinv[j_op].s_i[5];
```

CASE STUDY 8.2 Dynamic Response of GSC with PCC Scheme

Case Study Objective: In this case study, the dynamic response of NPC inverter is studied with respect to: (1) step change in active power extraction and (2) trapezoidal change in grid reactive power reference. The behavior of the net DC-bus voltage, DC-link capacitors voltage, $\alpha\beta$-frame grid currents, inverter line-to-line voltage, and grid active, reactive, and apparent power is studied.

Parameters: The rated parameters of the NPC inverter and MV grid are as follows: the inverter power rating is 3 MVA, grid line-to-line voltage $V_{g,ll}$ is 3000 V, grid frequency f_g is 60 Hz, and net DC-bus voltage v_{dc} is 5304 V (3.062 pu).

Analysis:
The dynamic response of the NPC inverter with the PCC scheme is shown in Figure 8.28 during a step change in wind speed. As shown in Figure 8.28(a), a step change in v_w from 12 m/s (1.0 pu) to 9.6 m/s (0.8 pu) is applied at t = 0.2 s. The PI controller regulates v_{dc} at its reference value of 3.062 pu, with a small overshoot during the step change in v_w. The ripple in v_{dc} decreases as the current magnitude decreases (see Figure 8.28(b)). The DC-link capacitors voltage are regulated equally by the PCC scheme as shown in Figure 8.28(c). The overshoot in v_{dc} waveform is reflected in v_{C1} and v_{C2} waveforms. The weighting factor corresponding to the DC capacitors voltage control is calculated as follows:

$$\lambda_{dc,i} = \frac{I_{Bg}}{v_{dc}^*} = \frac{577}{5304} = 0.1089.$$

During the step change in v_w, the grid current magnitude changes from 1.0 pu to 0.512 pu (Figure 8.28(d)). With rated wind speed and unity grid PF, the grid current magnitude is expressed as follows:

$$I_g = \frac{P_g}{\sqrt{3}\,V_{g,ll}\cos\varphi_g} = \frac{3.0\times10^6}{\sqrt{3}\times3000\times1} = 577\ \text{A}\ (rms)\quad(1.0\ \text{pu}).$$

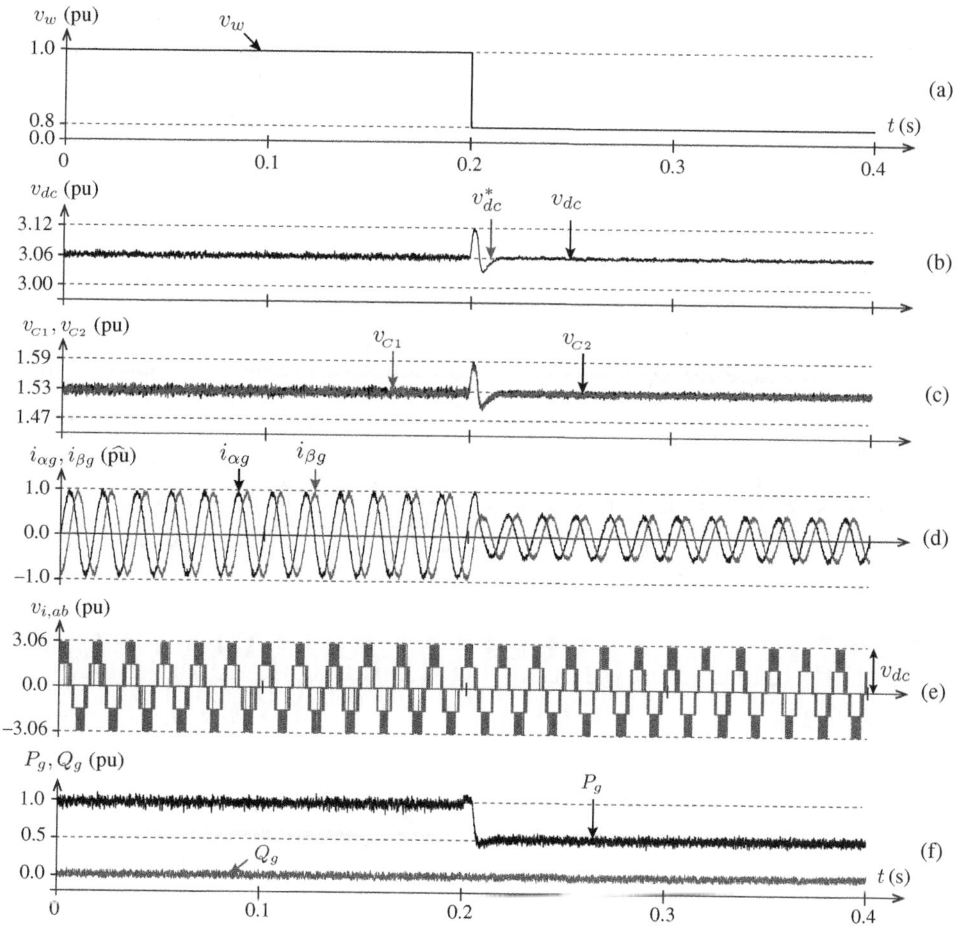

Figure 8.28 Dynamic response of the $\alpha\beta$-frame PCC scheme during a step change in active power extraction (wind speed).

Similarly, the grid current magnitude corresponding to 0.8-pu wind speed is expressed as follows:

$$I_g = \frac{P_g \times v_{w,\,\mathrm{pu}}^3}{\sqrt{3}\, V_{g,ll} \cos \varphi_g} = \frac{3.0 \times 10^6 \times 0.8^3}{\sqrt{3} \times 3000 \times 1} = 295.4\ \mathrm{A}\ (rms) \quad (0.512\ \mathrm{pu}).$$

With grid-tied operation, the NPC inverter contains five discrete voltage levels during all operating conditions (Figure 8.28(e)). During 1.0- and 0.8-pu wind speed conditions, the switching frequency of the NPC inverter $f_{sw,i}$ is noted as 1187 Hz and 1210 Hz, respectively. The grid active power P_g changes from 3.0 MW (1.0 pu) to 1.536 MW (0.512 = 0.8^3 pu) during the step change in v_w. The grid reactive power Q_g is maintained zero by the PCC scheme (Figure 8.28(f)).

The dynamic response of the grid-connected NPC inverter with trapezoidal change in Q_g^* is shown in Figure 8.29. The wind speed v_w is considered to be 0.8 pu and the corresponding grid active power P_g is 1.536 MW (0.512 pu). The Q_g^* follows a trapezoidal pattern with changes from 0 MVAR to 0.858 MVAR to 0 MVAR to –0.858 MVAR to 0 MVAR. The grid reactive power tracks to its reference with fast dynamics and no overshoots (Figure 8.29(a)). A good net DC-bus voltage control and DC-link neutral-point voltage control is achieved during the entire transient interval (Figures 8.29(b) and (c)). The grid current magnitude changes in the order of 1.0, 0.512, 1.0, 0.512, and 1.0 pu. A decoupled control for active and reactive powers is achieved through the control of the $\alpha\beta$-axes grid currents.

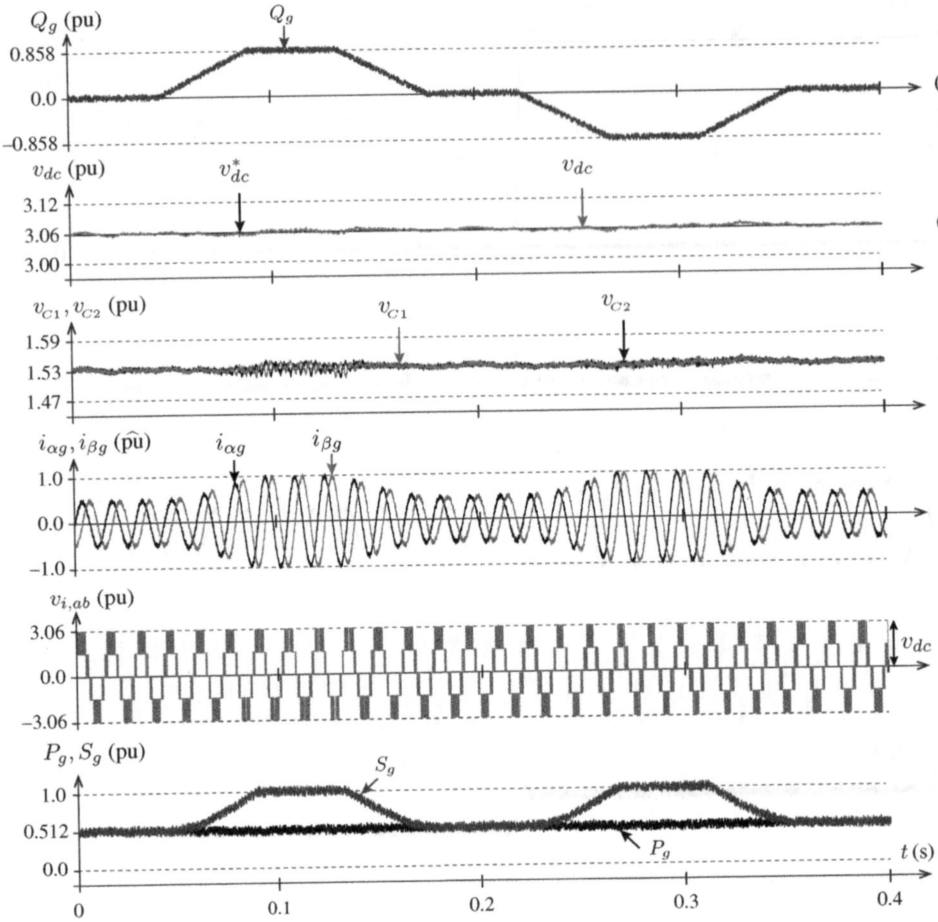

Figure 8.29 Dynamic response of the $\alpha\beta$-frame PCC scheme during trapezoidal changes in reactive power reference.

The grid PF changes in the order of unity, lagging, unity, leading, and then unity. The PFs corresponding to positive and negative Q_g are calculated as follows:

$$\varphi_g = \tan^{-1}(Q_g/P_g) = \tan^{-1}(0.858/0.512) = 59.2^\circ$$
$$\text{PF} = \cos\varphi_g = \cos 59.2^\circ = 0.512 \text{ lagging}$$
$$\varphi_g = \tan^{-1}(Q_g/P_g) = \tan^{-1}(-0.858/0.512) = -59.2^\circ$$
$$\text{PF} = \cos\varphi_g = \cos -59.2^\circ = 0.512 \text{ leading.}$$

The grid apparent power with $Q_g^* = 0$ MVAR and ± 0.858 MVAR is

$$S_g = \sqrt{P_g^2 + Q_g^2} = \sqrt{(1.536 \times 10^6)^2 + (0 \times 10^6)^2} = 1.536 \text{ MVA} \quad (0.512 \text{ pu})$$
$$S_g = \sqrt{P_g^2 + Q_g^2} = \sqrt{(1.536 \times 10^6)^2 + (\pm 0.858 \times 10^6)^2} = 3.0 \text{ MVA} \quad (1.0 \text{ pu}).$$

During all the operating conditions including transient conditions: (1) the net DC-bus v_{dc} is maintained at its reference value, (2) DC-link capacitors voltage v_{C1} and v_{C2} are balanced, (3) $i_{\alpha g}$ and $i_{\beta g}$ tracks their references with fast dynamic response, and (4) reactive power Q_g is supplied according to the grid operator's request. These two operating conditions validate the proper functionality of the NPC inverter and the PCC scheme. ■

8.8 PREDICTIVE POWER CONTROL OF NPC INVERTER WITH GRID-SIDE MPPT

The PPC scheme for grid-tied NPC inverter is shown in Figure 8.30, where the WT, generator, and MSC with net DC-bus voltage control is represented by a constant DC voltage source [16]. The primary objective of the PPC scheme is to achieve the decoupled control of grid active and reactive powers by controlling the dq-axis grid currents. Decoupled control can also be achieved for P_g and Q_g by controlling the $\alpha\beta$-axes grid currents [15].

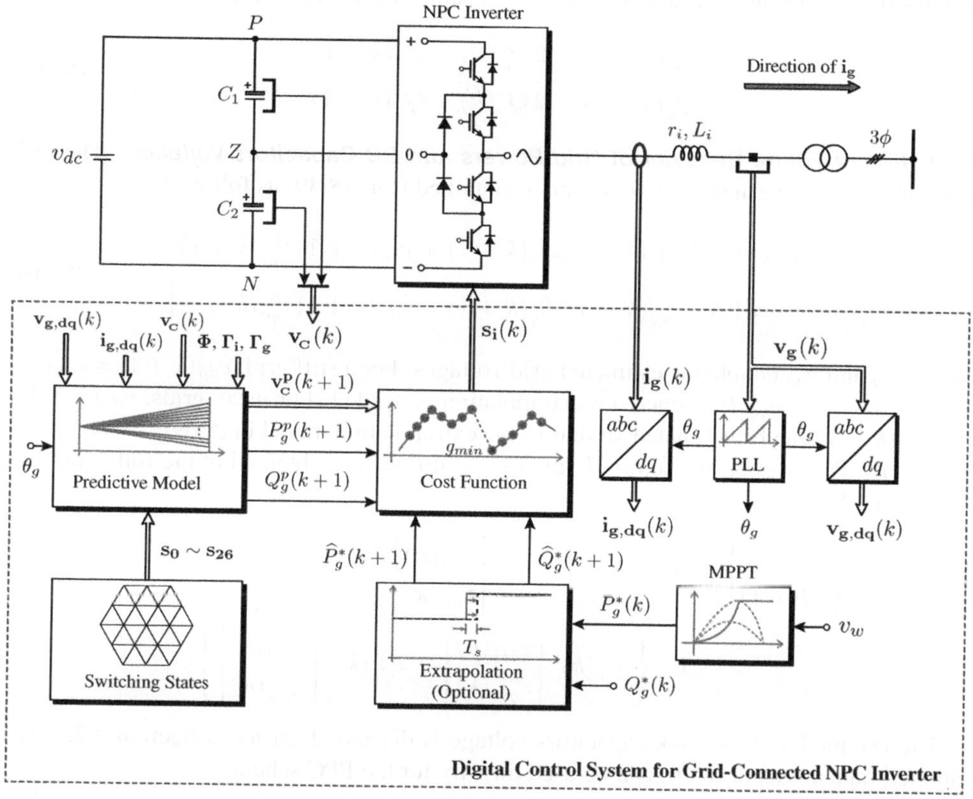

Figure 8.30 Block diagram of the dq-frame PPC scheme for grid-connected NPC inverter.

8.8.1 Design Procedure

The design of the PPC scheme for NPC inverter is similar to the PCC design in Figures 8.19 and 8.26. The major design procedures are explained below.

Measurement and Synthesis of Feedback Signals: The measured grid voltages and currents ($\mathbf{v_g}$ and $\mathbf{i_g}$) are transformed to $\mathbf{v_{g,dq}}$ and $\mathbf{i_{g,dq}}$ with the help of θ_g. The measured DC-link capacitors voltage are used by the predictive model block. The control scheme also requires wind speed v_w information to realize the MPPT algorithm.

Calculation of Reference Powers: The reference reactive power command (Q_g^*) is provided by the grid operator and can be set to zero for unity, negative for leading, and positive for the lagging PF. The active power reference P_g^*, is obtained by the MPPT algorithm. As mentioned earlier, the WT power curves based MPPT algorithm can be employed for GSC. To simplify the analysis, the per-unit P_g^* is assumed to be a cubic of per-unit wind speed [10].

Extrapolation of Reference Powers: Since the reference powers are DC quantities, no extrapolation is needed during the steady-state operation. In practical WECS, the wind speed and thus the active power reference change dynamically, hence, an extrapolation method should be used. Similar to the i_{dg}^* and i_{qg}^* extrapolation in (8.61), the grid active and reactive powers are extrapolated to $(k+1)$ sampling instant as follows:

$$\begin{aligned} \widehat{P}_g^*(k+1) &= 2\,P_g^*(k) - P_g^*(k-1) \\ \widehat{Q}_g^*(k+1) &= 2\,Q_g^*(k) - Q_g^*(k-1). \end{aligned} \tag{8.69}$$

Prediction of Future Behavior of Grid Powers and DC Capacitors Voltage: The DT model for grid active and reactive powers is obtained from (8.39) as follows:

$$\begin{aligned} P_g^p(k+1) &= 1.5\left[\widehat{v}_{dg}(k+1)\,i_{dg}^p(k+1) + \widehat{v}_{qg}(k+1)\,i_{qg}^p(k+1)\right] \\ Q_g^p(k+1) &= 1.5\left[\widehat{v}_{qg}(k+1)\,i_{dg}^p(k+1) - \widehat{v}_{dg}(k+1)\,i_{qg}^p(k+1)\right] \end{aligned} \tag{8.70}$$

where $\widehat{v}_{dg}$ and $\widehat{v}_{qg}$ denote the estimated grid voltages. For a stiff grid, $\widehat{v}_{dg}(k+1) = v_{dg}(k)$ and $\widehat{v}_{qg}(k+1) = v_{qg}(k)$; hence, no extrapolation is needed. For weak grids, $\widehat{v}_{dg}(k+1)$ and $\widehat{v}_{qg}(k+1)$ can be estimated similar to the extrapolation model in (8.69).

For an NPC inverter, the future dq-axes grid currents are obtained in the following by combining (8.37) and (8.47):

$$\begin{aligned} \begin{bmatrix} i_{dg}^p(k+1) \\ i_{qg}^p(k+1) \end{bmatrix} &= \mathbf{\Phi} \begin{bmatrix} i_{dg}(k) \\ i_{qg}(k) \end{bmatrix} + \mathbf{\Gamma_g} \begin{bmatrix} v_{dg}(k) \\ v_{qg}(k) \end{bmatrix} \\ &+ \mathbf{\Gamma_i} \left\{ v_{C1}(k) \begin{bmatrix} s_{di1}^p(k) \\ s_{qi1}^p(k) \end{bmatrix} + v_{C2}(k) \begin{bmatrix} s_{di2}^p(k) \\ s_{qi2}^p(k) \end{bmatrix} \right\}. \end{aligned} \tag{8.71}$$

The DT model of DC-link capacitors voltage is discussed earlier in Section 8.7. The sampled-data model in (8.66) can be used directly for the PPC scheme.

Cost Function Minimization: The control objectives such as regulation of grid active and reactive powers, DC-link capacitors voltage balancing, and switching frequency minimization are included in a cost function as follows:

$$\begin{aligned} g_i(k) &= \lambda_P \left[\widehat{P}_g^*(k+1) - P_g^p(k+1)\right]^2 \\ &+ \lambda_Q \left[\widehat{Q}_g^*(k+1) - Q_g^p(k+1)\right]^2 \\ &+ \lambda_{dc,i} \left[v_{C1}^p(k+1) - v_{C2}^p(k+1)\right]^2 \\ &+ \lambda_{sw,i} \sum_{x=a,\,b,\,c} \sum_{j=1,\,2} \left[s_{xij}^p(k) - s_{xij}^{\text{op}}(k)\right]^2 \end{aligned} \tag{8.72}$$

where λ_P and λ_Q are the weighting factors for active and reactive power control. They are set to 1 as they both represent primary control objective. The weighting factors $\lambda_{dc,i}$ and $\lambda_{sw,i}$ are selected similar to the PCC scheme of NPC inverter in Section 8.7.

8.8.2 Control Algorithm

The flowchart of the PPC algorithm for NPC inverter is shown in Figure 8.31. The DC-link capacitors voltage, along with grid voltages and currents are measured in block ①. The $(k+1)$-state reference active and reactive power, and the weighting factor $\lambda_{dc,i}$ are calculated in block ②. The iterative loop is performed between blocks ④ and ⑨. In blocks ⑤ to ⑦, the NPC inverter output voltages $\mathbf{v}^{\mathrm{P}}_{\mathrm{i,dq}}(k)$, dq-axes grid currents $\mathbf{i}^{\mathrm{P}}_{\mathrm{g,dq}}(k+1)$, active power $P^p_g(k+1)$, reactive power $Q^p_g(k+1)$, and DC-link capacitors voltage $\mathbf{v}^{\mathrm{P}}_{\mathrm{C}}(k+1)$ are predicted. Cost function minimization $g_i(k)$ is performed for all 27 possible switching states. An optimal vector j_{op} and corresponding switching signals are selected and applied directly to the NPC inverter in blocks ⑧ to ⑩. With the PPC algorithm, simultaneous control of grid active and reactive power, and the DC-link capacitors voltage is achieved.

The `MATLAB` *S-Function Builder* programming for the PPC scheme is shown in Algorithm 8.3, which combines Algorithms 8.1 and 8.2 for dq-axes grid currents prediction and uses them for NPC inverter. To simplify the analysis, the code corresponding to the switching frequency minimization is not presented. The program lines are explained as follows: Lines `05-08` predict the dq-axes NPC inverter voltages; lines `09-10` predict the dq-axes grid currents; lines `11-12` predict the grid active and reactive power; lines `13-16` predict the DC-link capacitors current and voltage; lines `17-20` calculate the cost function values; lines `21-25` select the optimal vector `j_op` number; lines `27-32` select the optimal switching signals on the basis of the `j_op` number. The grid is assumed to be stiff, and no extrapolation is used for grid voltages `vg_dq`.

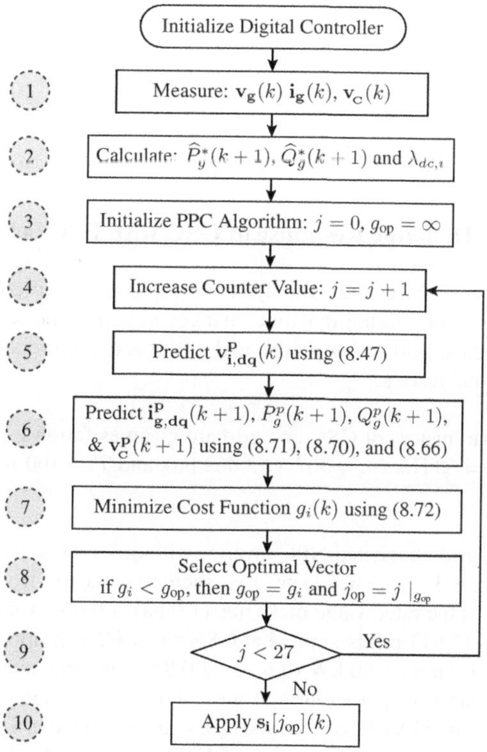

Figure 8.31 Flowchart of the PPC algorithm for grid-connected NPC inverter.

ALGORITHM 8.3 PPC Algorithm for Grid-Connected NPC Inverter

```
\\ P11 = 1-(ri*Ts/Li); P12 = (wg*Ts); P21 = -(wg*Ts); P22 = 1-(ri*Ts/Li);
\\ Gi11 = Gi22 = Ts/Li; Gg11 = Gg22 = -Ts/Li;
j_op = 1000;
g_op = 1000000000;
for(j = 0; j < 27; j++)
{
    vi_ab[0] = vc[0]*vinv[j].s_i_ab[0] + vc[1]*vinv[j].s_i_ab[1];
    vi_ab[1] = vc[0]*vinv[j].s_i_ab[2] + vc[1]*vinv[j].s_i_ab[3];
    vi_dq[0] =  vi_ab[0]*cstheta_g[0] + vi_ab[1]*cstheta_g[1];
    vi_dq[1] = -vi_ab[0]*cstheta_g[1] + vi_ab[1]*cstheta_g[0];
    ig_dq_k1[0] = P11*ig_dq[0]+P12*ig_dq[1]+Gi11*vi_dq[0]+Gg11*vg_dq[0];
    ig_dq_k1[1] = P21*ig_dq[0]+P22*ig_dq[1]+Gi22*vi_dq[1]+Gg22*vg_dq[1];
    pg_k1[0] = 1.5*(vg_dq[0]*ig_dq_k1[0] + vg_dq[1]*ig_dq_k1[1]);
    qg_k1[0] = 1.5*(vg_dq[1]*ig_dq_k1[0] - vg_dq[0]*ig_dq_k1[1]);
    ic_k1[0] = vinv[j].K_i1[0]*ig_abc[0] + vinv[j].K_i1[1]*ig_abc[1]
               + vinv[j].K_i1[2]*ig_abc[2];
    ic_k1[1] = vinv[j].K_i2[0]*ig_abc[0] + vinv[j].K_i2[1]*ig_abc[1]
               + vinv[j].K_i2[2]*ig_abc[2];
    vc_k1[0] = vc[0] + (Ts/C1)*ic_k1[0];
    vc_k1[1] = vc[1] + (Ts/C2)*ic_k1[1];
    g_pg = fabs(pg_ref_k1[0]-pg_k1[0]);
    g_qg = fabs(qg_ref_k1[0]-qg_k1[0]);
    g_dc = (vc_k1[0]-vc_k1[1])*(vc_k1[0]-vc_k1[1]);
    g_i = g_pg + g_qg + Lambda_dci*g_dc;
    if(g_i < g_op)
       {
         j_op = j;
         g_op = g_i;
       }
}
s_ai1[0] = vinv[j_op].s_i[0];
s_ai2[0] = vinv[j_op].s_i[1];
s_bi1[0] = vinv[j_op].s_i[2];
s_bi2[0] = vinv[j_op].s_i[3];
s_ci1[0] = vinv[j_op].s_i[4];
s_ci2[0] = vinv[j_op].s_i[5];
```

CASE STUDY 8.3 Dynamic Response of GSC with PPC Scheme

Case Study Objective: In this case study, the dynamic response of grid-connected NPC inverter is investigated by the PPC scheme during dynamic changes in active and reactive power references. The response of grid currents, grid active power, and grid reactive powers is observed by dSPACE DS1103-based experimental results.

Parameters: The rated parameters of GSC and grid are given as follows: $S_g = 5$ kVA, $V_{g,ll} = 208$ V, $I_g = 13.88$ A (*rms*), $f_g = 60$ Hz, $v_{dc} = 367$ V (3.062 pu), and $T_s = 100$ μs.

Analysis:

The dynamic response of grid-connected NPC inverter with different active and reactive power references is shown in Figure 8.32. The wind speed v_w increases linearly from a cut-in wind speed of 3 m/s (0.25 pu) and reaches the rated value of 12 m/s (1.0 pu) at 0.6 s. A step change in wind speed from 12 m/s to 10.12 m/s (0.843 pu) is applied at 0.8 s. The P_g^* is proportional to the wind speed and changes from 5 kW (1.0 pu) to 3.0 kW (0.6 pu) at 0.8 s. The grid active power P_g tracks to its reference P_g^* with fast dynamic response during transient and steady-state conditions.

A step change in Q_g^* from 0 kVAR to 1.5 kVAR (0.3 pu) is applied at 1.0 s. At $t = 1.5$ s, another step change in Q_g^* is applied from 0 kVAR to –3.5 kVAR (–0.7 pu). During time intervals 1.0 s to 1.3 s and 1.5 s to 1.8 s, the grid apparent power S_g is 3.355 kVA (0.671 pu) and 4.61 kVA (0.922

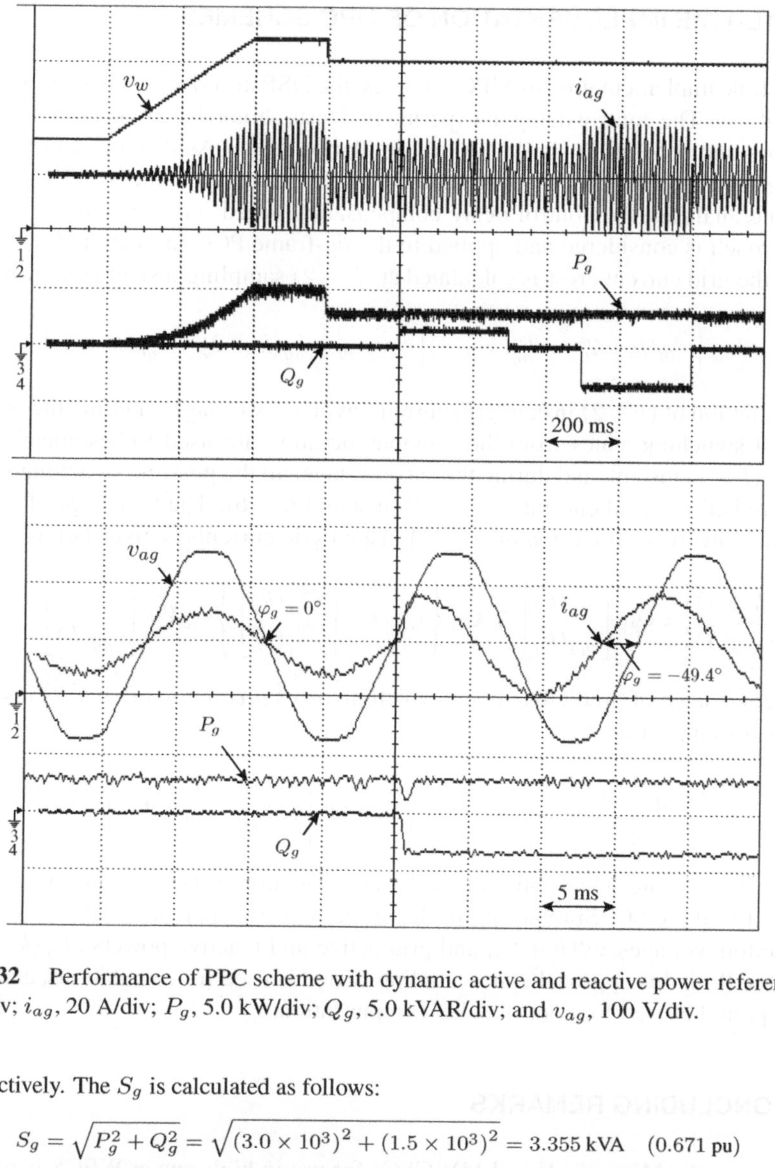

Figure 8.32 Performance of PPC scheme with dynamic active and reactive power references. v_w, 1.0 m/s/div; i_{ag}, 20 A/div; P_g, 5.0 kW/div; Q_g, 5.0 kVAR/div; and v_{ag}, 100 V/div.

pu), respectively. The S_g is calculated as follows:

$$S_g = \sqrt{P_g^2 + Q_g^2} = \sqrt{(3.0 \times 10^3)^2 + (1.5 \times 10^3)^2} = 3.355 \text{ kVA} \quad (0.671 \text{ pu})$$
$$S_g = \sqrt{P_g^2 + Q_g^2} = \sqrt{(3.0 \times 10^3)^2 + (-3.5 \times 10^3)^2} = 4.61 \text{ kVA} \quad (0.922 \text{ pu}).$$

The phase-a grid current i_{ag} magnitude is proportional to the S_g. The DC-link capacitors voltage are very well balanced during the entire operation (not shown in Figure 8.32). The top experimental result in Figure 8.32 is magnified during the step change in Q_g^* from 0 kVAR to –3.5 kVAR (–0.7 pu) at t = 1.5 s. During the step change in Q_g^*, the grid PF changes from unity to 0.65 leading, and the PF angle changes from 0° to –49.4° as demonstrated below:

$$\varphi_g = \tan^{-1}(Q_g/P_g) = \tan^{-1}(-3.5/3.0) = -49.4°$$
$$\text{PF} = \cos\varphi_g = \cos(-49.4°) = 0.65 \text{ leading}.$$

The grid reactive power follows its reference with very fast dynamic response (1.0 ms). These results validate that the grid active and reactive powers can be controlled in a decoupled manner, similar to the classical V-DPC scheme, but with a simpler approach. ■

8.9 REAL-TIME IMPLEMENTATION OF MPC SCHEMES

In the real-time implementation of MPC schemes, the DSP introduces considerable amount of control delay. The control delay in experimental tests should be compensated to achieve high-performance operation similar to the simulation tests. As discussed in Chapter 4, various types of delay compensation methods can be adopted for the MPC schemes. To simplify the analysis, the control delay compensation based on *estimation + prediction* (E+P) approach is considered and applied to the dq-frame PCC in Section 8.6. With this approach, the grid current error is calculated at $(k+2)$ sampling instant as shown below:

$$g_i(k) = \lambda_{id}\left[\hat{i}^*_{dg}(k+2) - i^p_{dg}(k+2)\right]^2 + \lambda_{iq}\left[\hat{i}^*_{qg}(k+2) - i^p_{qg}(k+2)\right]^2. \quad (8.73)$$

The prediction of $(k+2)$ instant grid current involves two stages. During the first stage, the optimal switching states from the previous iteration are used to "estimate" the grid currents at $(k+1)$ instant, and during the second stage, all the possible switching states are used to "predict" the grid currents at $(k+2)$ instant [35, 36]. The first stage of predictive model which involves *estimation* of $(k+1)$ instant grid currents is given below:

$$\begin{bmatrix} \hat{i}_{dg}(k+1) \\ \hat{i}_{qg}(k+1) \end{bmatrix} = \mathbf{\Phi} \begin{bmatrix} i_{dg}(k) \\ i_{qg}(k) \end{bmatrix} + \mathbf{\Gamma_i} \left\{ v_{C1}(k) \begin{bmatrix} s^{\mathrm{op}}_{di1}(k) \\ s^{\mathrm{op}}_{qi1}(k) \end{bmatrix} \right\} + \mathbf{\Gamma_g} \begin{bmatrix} v_{dg}(k) \\ v_{qg}(k) \end{bmatrix}. \quad (8.74)$$

The second stage of predictive model which involves *prediction* of $(k+2)$ instant grid currents is formulated as

$$\begin{bmatrix} i^p_{dg}(k+2) \\ i^p_{qg}(k+2) \end{bmatrix} = \mathbf{\Phi} \begin{bmatrix} \hat{i}_{dg}(k+1) \\ \hat{i}_{qg}(k+1) \end{bmatrix} + \mathbf{\Gamma_i} \left\{ v_{C1}(k) \begin{bmatrix} s^{p}_{di1}(k) \\ s^{p}_{qi1}(k) \end{bmatrix} \right\} + \mathbf{\Gamma_g} \begin{bmatrix} v_{dg}(k) \\ v_{qg}(k) \end{bmatrix}. \quad (8.75)$$

The switching state which minimizes the cost function at $(k+2)$ instant is selected and applied to the GSC. Similar approach can be used to calculate $(k+2)$ instant DC-link capacitors voltages $\mathbf{v}^{\mathrm{p}}_{\mathrm{C}}(k+2)$, and grid active and reactive powers ($P^p_g(k+2)$ and $Q^p_g(k+2)$). With this approach, one sampling period is used for calculations, and another sampling period is used for control delay compensation.

8.10 CONCLUDING REMARKS

In this chapter, the MPC of LV and MV GSCs for use in high-power WECS is presented. The single-stage, two-stage, and three-stage power conversion systems are analyzed along with the control objectives for MSCs and GSCs. The design and control of GSCs are discussed with respect to the design of DC-link capacitor and grid harmonic filter, the definition of reference DC-bus voltage, the notation of grid PF, and the grid voltage orientation. The detection of grid voltage angle during ideal and distorted grid voltage conditions is discussed. CT and DT modeling of three-phase GSC in abc, $\alpha\beta$, and dq frames is presented. The models of grid currents, grid active and reactive power, 2L-VSI and NPC inverter output voltages, and DC-link capacitors voltage are analyzed in detail. The grid-side reference currents and powers are calculated, considering that the MPPT is performed by the MSCs or GSCs. The design of the dq-frame PCC for 2L-VSI, $\alpha\beta$-frame PCC for NPC inverter, and dq-frame PPC for NPC inverter are analyzed in detail with respect to design steps, control algorithm, MATLAB *S-Function Builder* programming, and case studies.

With these digital control schemes, the control objectives such as regulation of net DC-bus voltage, MPPT operation, reactive power generation to meet the grid codes, DC-link capacitors voltage balancing, and switching frequency minimization are included in the cost function. The future behavior of grid currents and/or DC-link voltages are predicted for all the possible switching states using the DT model of the system. The switching state that minimizes the cost function is chosen and applied to the GSC directly. Through a detailed analysis of the simulation and experimental results, the MPC strategy is proved to be a simple and powerful tool in controlling GSCs with excellent performance indices.

Link to Next Chapters:

The material in this chapter will complement the in-depth technical analysis to be conducted in the next four chapters of this book. The PCC and PPC schemes discussed in this chapter are combined with the generator-side digital control schemes in Chapters 9 to 12 such that the overall control of PMSG, SCIG, and DFIG WECS is achieved. To understand the similarities of generator- and grid-side control schemes, the flowcharts in this chapter are reproduced in Chapters 9 to 12 and placed side-by-side with the flowcharts of generator-side control schemes.

REFERENCES

1. V. Yaramasu, "Predictive control of multilevel converters for megawatt wind energy conversion systems," Ph.D. dissertation, Ryerson University, Toronto, ON, Canada, 2014, available at: http://digital.library.ryerson.ca/islandora/object/RULA%3A3459.
2. R. Teodorescu, M. Liserre, and P. Rodriguez, *Grid Converters for Photovoltaic and Wind Power Systems*. Chichester, UK: Wiley-IEEE Press, January 2011.
3. E. Koutroulis and K. Kalaitzakis, "Design of a maximum power tracking system for wind-energy-conversion applications," *IEEE Transactions on Industrial Electronics*, vol. 53, no. 2, pp. 486–494, April 2006.
4. K. Tan and S. Islam, "Optimum control strategies in energy conversion of PMSG wind turbine system without mechanical sensors," *IEEE Transactions on Energy Conversion*, vol. 19, no. 2, pp. 392–399, June 2004.
5. A. Calle-Prado, S. Alepuz, J. Bordonau, J. Nicolas-Apruzzese, P. Cortes, and J. Rodriguez, "Model predictive current control of grid-connected neutral-point-clamped converters to meet low-voltage ride-through requirements," *IEEE Transactions on Industrial Electronics*, vol. 62, no. 3, pp. 1503–1514, March 2015.
6. S. Kouro, M. Malinowski, K. Gopakumar, J. Pou, L. Franquelo, B. Wu, J. Rodríguez, M. Perez, and J. Leon, "Recent advances and industrial applications of multilevel converters," *IEEE Transactions on Industrial Electronics*, vol. 57, no. 8, pp. 2553–2580, August 2010.
7. V. Yaramasu, B. Wu, P. C. Sen, S. Kouro, and M. Narimani, "High-power wind energy conversion systems: State-of-the-art and emerging technologies," *Proceedings of the IEEE*, vol. 103, no. 5, pp. 740–788, May 2015.
8. M. Liserre, F. Blaabjerg, and A. Dell'Aquila, "Step-by-step design procedure for a grid-connected three-phase PWM voltage source converter," *International Journal of Electronics*, vol. 91, no. 8, pp. 445–460, 2004.
9. A. Rockhill, M. Liserre, R. Teodorescu, and P. Rodríguez, "Grid-filter design for a multimegawatt medium-voltage voltage-source inverter," *IEEE Transactions on Industrial Electronics*, vol. 58, no. 4, pp. 1205–1217, April 2011.
10. B. Wu, Y. Lang, N. Zargari, and S. Kouro, *Power Conversion and Control of Wind Energy Systems*, 1st ed., ser. IEEE Press Series on Power Engineering. Hoboken, NJ: Wiley-IEEE Press, July 2011.
11. M. Malinowski, M. Kazmierkowski, and A. Trzynadlowski, "A comparative study of control techniques for PWM rectifiers in AC adjustable speed drives," *IEEE Transactions on Power Electronics*, vol. 18, no. 6, pp. 1390–1396, November 2003.
12. F. Blaabjerg, R. Teodorescu, M. Liserre, and A. Timbus, "Overview of control and grid synchronization for distributed power generation systems," *IEEE Transactions on Industrial Electronics*, vol. 53, no. 5, pp. 1398–1409, October 2006.
13. A. Timbus, M. Liserre, R. Teodorescu, and F. Blaabjerg, "Synchronization methods for three phase distributed power generation systems - an overview and evaluation," in *IEEE Power Electronics Specalists Conference (PESC)*, Recife, Brazil, June 2005, pp. 2474–2481.
14. H. Akagi, E. H. Watanabe, and M. Aredes, *Instantaneous Power Theory and Applications to Power Conditioning*, 1st ed. Hoboken, NJ: Wiley-IEEE Press, March 2007.
15. V. Yaramasu and B. Wu, "Predictive power control of grid-connected four-level inverters in stationary reference frame," in *IEEE International Conference on Circuits, Power and Computing Technologies (ICCPCT)*, Kumaracoil, India, March 2013, pp. 636–641.

16. V. Yaramasu and B. Wu, "Model predictive decoupled active and reactive power control for high-power grid-connected four-level diode-clamped inverters," *IEEE Transactions on Industrial Electronics*, vol. 61, no. 7, pp. 3407–3416, July 2014.

17. J. Rodríguez, J. Pontt, C. A. Silva, P. Correa, P. Lezana, P. Cortés, and U. Ammann, "Predictive current control of a voltage source inverter," *IEEE Transactions on Industrial Electronics*, vol. 54, no. 1, pp. 495–503, February 2007.

18. V. Yaramasu, B. Wu, M. Rivera, and J. Rodriguez, "Predictive current control and DC-link capacitor voltages balancing for four-leg NPC inverters," in *IEEE International Symposium on Industrial Electronics (ISIE)*, Taipei, Taiwan, May 2013, pp. 1–6.

19. M. Saeedifard, R. Iravani, and J. Pou, "A space vector modulation approach for a back-to-back connected four-level converter," in *IEEE Power Electronics Specalists Conference (PESC)*, Orlando, FL, USA, June 2007, pp. 2043–2049.

20. M. Saeedifard, "Space vector modulation of multi-level and multi-module converters for high power applications," Ph.D. dissertation, University of Toronto, Toronto, ON, Canada, 2008.

21. V. Yaramasu, B. Wu, M. Rivera, M. Narimani, S. Kouro, and J. Rodriguez, "Generalised approach for predictive control with common-mode voltage mitigation in multilevel diode-clamped converters," *IET Power Electronics*, vol. 8, no. 8, pp. 1440–1450, August 2015.

22. V. Yaramasu, B. Wu, and J. Chen, "Model-predictive control of grid-tied four-level diode-clamped inverters for high-power wind energy conversion systems," *IEEE Transactions on Power Electronics*, vol. 29, no. 6, pp. 2861–2873, June 2014.

23. V. Yaramasu and B. Wu, "Three-level boost converter based medium voltage megawatt PMSG wind energy conversion systems," in *IEEE Energy Conversion Congress and Exposition (ECCE)*, Phoenix, AZ, USA, September 2011, pp. 561–567.

24. V. Yaramasu, B. Wu, S. Alepuz, and S. Kouro, "Predictive control for low-voltage ride-through enhancement of three-level-boost and NPC-converter-based PMSG wind turbine," *IEEE Transactions on Industrial Electronics*, vol. 61, no. 12, pp. 6832–6843, December 2014.

25. Q. Zeng and L. Chang, "An advanced SVPWM-based predictive current controller for three-phase inverters in distributed generation systems," *IEEE Transactions on Industrial Electronics*, vol. 55, no. 3, pp. 1235–1246, March 2008.

26. S. Kouro, P. Cortés, R. Vargas, U. Ammann, and J. Rodríguez, "Model predictive control-A simple and powerful method to control power converters," *IEEE Transactions on Industrial Electronics*, vol. 56, no. 6, pp. 1826–1838, June 2009.

27. J. Rubinic, B. Wu, V. Yaramasu, and N. Zargari, "Model predictive control of neutral-point clamped inverter with harmonic spectrum shaping and common-mode voltage mitigation," in *IEEE Energy Conversion Congress and Exposition (ECCE)*, Montreal, Canada, September 2015.

28. H. Miranda, R. Teodorescu, P. Rodriguez, and L. Helle, "Model predictive current control for high-power grid-connected converters with output LCL filter," in *IEEE Industrial Electronics Conference (IECON)*, Porto, Portugal, November 2009, pp. 633–638.

29. E.ON Netz Gmbh, "Grid code - high and extra high voltage," April 2006.

30. M. Tsili and S. Papathanassiou, "A review of grid code technical requirements for wind farms," *IET Renewable Power Generation*, vol. 3, no. 3, pp. 308–332, September 2009.

31. M. Kazmierkowski, M. Jasinski, and G. Wrona, "DSP-based control of grid-connected power converters operating under grid distortions," *IEEE Transactions on Industrial Informatics*, vol. 7, no. 2, pp. 204–211, May 2011.

32. V. Yaramasu and B. Wu, "Predictive control of a three-level boost converter and an NPC inverter for high-power PMSG-based medium voltage wind energy conversion systems," *IEEE Transactions on Power Electronics*, vol. 29, no. 10, pp. 5308–5322, October 2014.

33. P. Cortés, F. Quiroz, and J. Rodríguez, "Predictive control of a grid-connected cascaded H-bridge multilevel converter," in *European Conference on Power Electronics and Applications (EPE)*, Birmingham, UK, September 2011, pp. 1–7.

34. R. Vargas, P. Cortés, U. Ammann, J. Rodríguez, and J. Pontt, "Predictive control of a three-phase neutral-point-clamped inverter," *IEEE Transactions on Industrial Electronics*, vol. 54, no. 5, pp. 2697–2705, October 2007.

35. P. Cortés, J. Rodríguez, C. Silva, and A. Flores, "Delay compensation in model predictive current control of a three-phase inverter," *IEEE Transactions on Industrial Electronics*, vol. 59, no. 2, pp. 1323–1325, February 2012.

36. P. Cortés, J. Rodríguez, D. Quevedo, and C. Silva, "Predictive current control strategy with imposed load current spectrum," *IEEE Transactions on Power Electronics*, vol. 23, no. 2, pp. 612–618, March 2008.

CHAPTER 9

CONTROL OF PMSG WECS WITH BACK-TO-BACK CONNECTED CONVERTERS

9.1 INTRODUCTION

Synchronous generators (SGs) and back-to-back (BTB) connected voltage source converters (VSCs) are predominantly used in full-variable speed (Type 4) wind energy conversion systems (WECS). Permanent magnet SG (PMSG), wound rotor SG (WRSG), and high-temperature superconducting SG (HTS-SG) have been employed in Type 4 wind turbines (WTs) [1, 2]. PMSGs are the most popular choice for high-power WT manufacturers. Future WT projects are also expected to use PMSGs and BTB VSCs for power ratings of up to 15 MW. WRSG and HTS-SG WECS are also practical but are not used widely.

With a three-stage (3S) gearbox, PMSGs run with high speed and low torque. The low-torque operation of PMSGs leads to compact design and less weight. However, the gearbox presents several issues such as high audible noise, low reliability, and need for regular maintenance. Direct-drive (DD) technology eliminates the need for the gearbox and its associated disadvantages [3]. By contrast, the low speed and high-torque operations of DD technology significantly increases the stator diameter and weight of PMSG. Single-stage (1S) and two-stage (2S) gearboxes facilitate the medium-speed operation of PMSGs, thus leading to a compromise between 3S-gearbox and DD technology [4, 5]. A few selected SG WTs with BTB VSC are listed in Table 9.1 and distinguished according to gearbox type, generator, power and voltage rating, and power converter configuration. Full-scale two-level VSC (2L-VSC) and three-level VSC (3L-VSC or neutral-point clamped (NPC) converter) respectively enable low-voltage (LV) and medium-voltage (MV) operations for PMSG WECS.

Model Predictive Control of Wind Energy Conversion Systems, First Edition. Venkata Yaramasu and Bin Wu.

Companion Website: www.wiley.com/go/yaramasu_wu_mpc_wecs

Table 9.1 Concise list of commercial SG WECS with BTB VSCs [2, 5, 6]

Manufacturer	Generator	Gen. Rating	Converter	Latest Model #
Areva Wind (Germany)	1S–PMSG	5.0 MW, 3300 V	BTB 3L-VSC	M5000-135
Gamesa (Spain)	2S–PMSG	5.0 MW, 690 V	BTB 2L-VSC	G10X-128
MingYang (PR China)	2S–PMSG	3.0 MW, 850 V	BTB 2L-VSC	SCD3MW
DeWind (USA)	3S–PMSG	3.0 MW, 690 V	BTB 2L-VSC	D9.1
GE Energy (USA)	DD–PMSG	4.1 MW, 690 V	BTB 2L-VSC	GE4.1-113
Darwind (Netherlands)	DD–PMSG	4.5 MW, 3000 V	BTB 3L-VSC	XD115
E.N.O Energy (Germany)	3S–WRSG	3.5 MW, 690 V	BTB 2L-VSC	E.N.O 126
Enercon (Germany)	DD–WRSG	7.5 MW, 690 V	BTB 2L-VSC	E126/7500
Windtec-AMSC (USA)	DD–HTS-SG	10.0 MW, 690 V	BTB 2L-VSC	SeaTitan 10MW

The power conversion unit is composed of a rectifier and an inverter linked by DC-link capacitor(s). This capacitor provides decoupling for machine-side converter (MSC) and grid-side converter (GSC), thus leading to the easy modeling and design of control loops. The control schemes employed for the power converters play an important role in maximizing energy capture and in complying with grid code requirements. Field-oriented control (FOC) and direct torque control (DTC) are two of the most popular control schemes for PMSGs [7]. The model predictive control (MPC) eliminates the need for linear (PI) regulators, lookup tables, and modulation stages, which are commonly employed in classical FOC and DTC schemes [8, 9]. MPC has a simple digital implementation and high dynamic performance. This chapter discusses the predictive control of full-scale LV and MV VSCs employed in PMSG WECS. The MPC scheme uses the discrete-time (DT) model of PMSG WECS and the possible switching states of VSCs to accurately regulate the control variables (generator/grid currents) during the operation.

Chapter Overview

- The PMSG WECS configuration with full-scale LV and MV VSCs is discussed in Section 9.2. The steady-state and dynamic models of PMSGs are presented in Section 9.3.
- The control schemes for PMSGs are outlined in Section 9.4. The overall digital control system development for BTB-converter-based PMSG WECS is given in Section 9.5. The control requirements, maximum power point tracking (MPPT) algorithms, and calculation of reference control variables are also discussed.
- The predictive current control (PCC), which is analogous to the classical FOC, is analyzed with BTB 2L-VSC and BTB NPC converter in Sections 9.6 and 9.7, respectively.
- The predictive torque control (PTC) is developed on the basis of the classical DTC concept and is analyzed in Section 9.8. The design and development of other possible control techniques, such as predictive power control (PPC) and predictive speed control (PSC), are briefly discussed in Section 9.9.
- Practical issues on the real-time implementation of PCC and PTC schemes are investigated, and the solutions are detailed in Section 9.10. Finally, the summary of this chapter and key observations are provided in Section 9.11.

9.2 CONFIGURATION OF PMSG WECS WITH BTB CONVERTERS

As discussed in Chapter 2, BTB configurations such as 2L-VSCs, NPC converters, multilevel active NPC converters, 3L flying capacitor converters, modular multilevel converters, current source converters, and matrix converters can be employed in PMSG WECS. This chapter discusses 2L-VSCs and NPC converters only, but the analysis can be easily extended to other BTB configurations.

9.2.1 PMSG WECS with LV BTB Converters

The block diagram of BTB-connected 2L-VSC for the LV operation of PMSG WECS is shown in Figure 9.1. The MSC and GSC are labeled as voltage source rectifier (VSR) and voltage source inverter (VSI), respectively. The 2L-VSR and 2L-VSI are linked by DC-link capacitor C_1. The wind speed is not constant; thus, the generator output voltage and frequency changes with respect to wind speed. The variable voltage and frequency output of PMSG is first rectified to DC by using the generator-side 2L-VSR. DC is then inverted to a fixed voltage and frequency by the grid-side 2L-VSI.

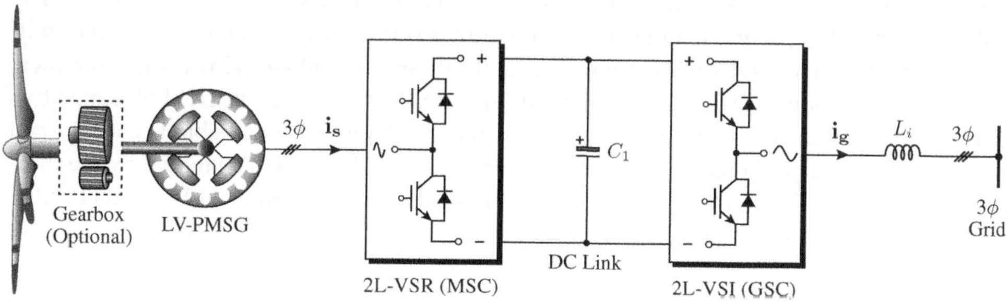

Figure 9.1 Block diagram of PMSG WECS with BTB 2L-VSC.

In terms of hardware, the construction and assembly of VSR and VSI are the same. As the name implies, the output phase voltage waveform of VSR and VSI contains two discrete voltage levels. Two-level VSC is composed of six LV-insulated gate bipolar transistors (IGBTs) with two IGBTs per phase. For high-power applications, VSI operates at a few kHz to make a compromise between the switching losses and grid current total harmonic distortion (THD). BTB 2L-VSCs are widely used for power ratings of up to 750 kW. To increase power ratings, BTB converters are connected in parallel [10]. The most common voltages for LV WTs are 575 and 690 V.

9.2.2 PMSG WECS with MV BTB Converters

The configuration of BTB connected NPC converter for the MV operation of PMSG WECS is shown in Figure 9.2 [11–13]. NPC converters are also known in literature as 3L diode-clamped converters. The NPC leg is composed of four switching devices and two clamping diodes. The DC link is composed of two split DC-link capacitors with the mid-point (neutral point) connected to the neutral point of the clamping diodes [14]. Under ideal operating conditions, the split DC-link capacitors equally share the net DC-bus voltage. The NPC inverter is connected to the MV grid through an L or LCL filter.

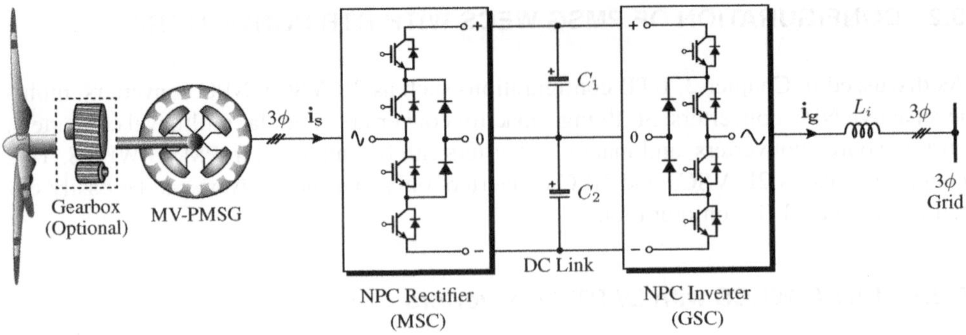

Figure 9.2 Block diagram of PMSG WECS with BTB NPC converter.

The NPC converters' output phase voltage waveform contains three levels that lead to small voltage steps, low *dv/dt*, low electromagnetic interference, low current ripple, small grid-side filter, and good power quality [15]. Switching devices are realized by the MV-IGBT or integrated gate-commutated thyristor (IGCT), and clamping devices comprise fast-recovery diodes. MV-IGBTs are more popularly used in WTs than the IGCTs. The voltage rating of semiconductor switches and diodes is merely half of the net DC-bus voltage, assuming that the DC-link capacitors equally share this voltage. However, in practice, the switching actions of both the rectifier and inverter lead to a drift in DC-link capacitors voltage. An imbalance between DC-link capacitors voltage is undesirable because it can cause high stress and damage to semiconductor devices. This issue is handled by the digital control schemes employed for NPC rectifier and inverter. In practical WTs, BTB NPC converters has a 6.0-MW power rating and 3-4-kV operating voltage [2, 5, 16].

9.2.3 Power Flow in PMSG WECS

The Sankey diagram in Figure 9.3 illustrates the power flow in BTB-converter-based WECS with detailed power conversion stages and losses between the stages. The power flows from the PMSG to the three-phase grid for the WECS application and in the reverse direction for the electric motor drive application. Among all these losses, the gearbox and generator losses are the most significant. Power converters, along with harmonic filters, operate with high efficiency (98% to 99%) for high-power applications.

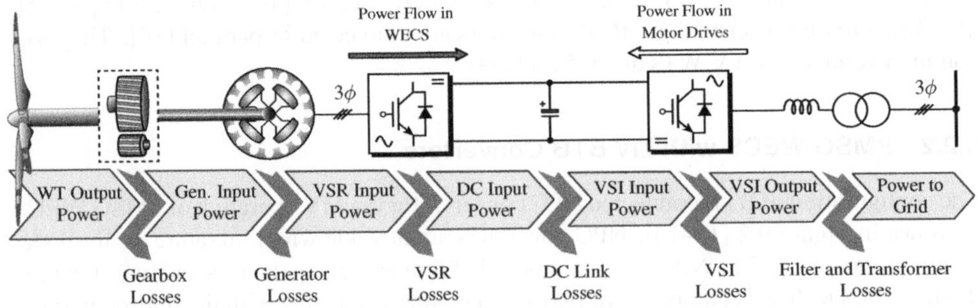

Figure 9.3 Sankey diagram for power flow in BTB-converter-based PMSG WECS.

9.3 MODELING OF PERMANENT MAGNET SYNCHRONOUS GENERATOR

The PMSG model discussed previously in Chapters 6 and 7 is briefly reviewed here. PMSG models are usually used for motor operations but can also be used for WECS operations with a negative sign for the mechanical input torque T_m. Steady-state and dynamic models are discussed to design and analyze the control techniques for PMSG.

9.3.1 Steady-State Models of PMSG

In the synchronous (dq) reference frame, the dq-axis inductances and rotor flux linkage exhibit time invariance, thus leading to simplified PMS machine (PMSM) modeling. The steady-state equivalent dq-axis circuits and the space vector diagram of PMSM, along with an MSC, are shown in Figure 9.4. Steady-state machine models ignore the derivative terms for stator currents and flux linkages. The rotor flux linkage is aligned with the d-axis. The dq-axes and space vectors corresponding to stator currents, voltages, and stator flux linkages rotate in space at synchronous speed ω_r (electrical speed of PMSG).

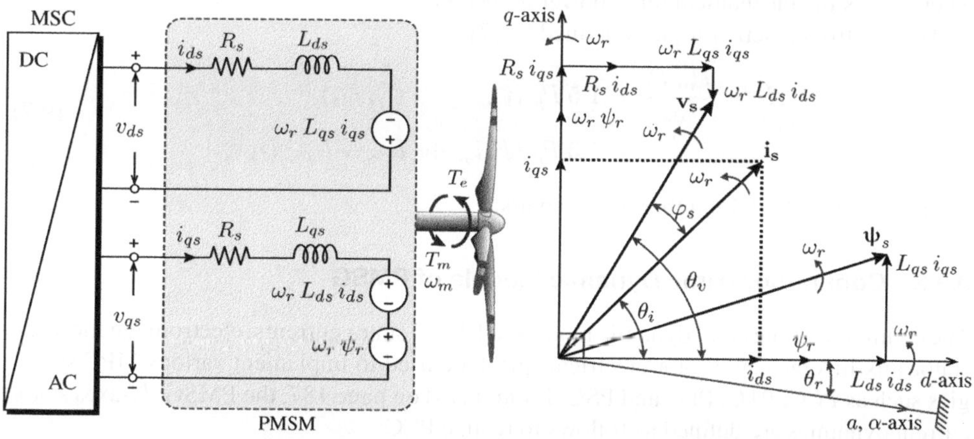

Figure 9.4 Equivalent steady-state circuits and space vector diagram of PMSM.

The steady-state dq-axis stator flux linkages are obtained from (6.12) on page 187 as

$$\psi_{ds} = L_{ds}\, i_{ds} + \psi_r, \qquad \psi_{qs} = L_{qs}\, i_{qs} \tag{9.1}$$

where i_{ds} and i_{qs} are the stator currents in the dq frame, L_{ds} and L_{qs} are the dq-axis inductances, and ψ_r is the peak value of the flux linkages produced by the PMs.

Similarly, the dq-axis stator voltages are reproduced from (6.11) and (6.13):

$$\begin{aligned} v_{ds} &= R_s\, i_{ds} - \omega_r\, \psi_{qs} = R_s\, i_{ds} - \omega_r\, L_{qs}\, i_{qs} \\ v_{qs} &= R_s\, i_{qs} + \omega_r\, \psi_{ds} = R_s\, i_{qs} + \omega_r\, L_{ds}\, i_{ds} + \omega_r\, \psi_r \end{aligned} \tag{9.2}$$

where R_s is the stator winding resistance and ω_r is the rotor electrical speed.

The peak and root-mean-square (*rms*) stator flux linkages, voltages, and currents are expressed as follows:

$$x_s = \sqrt{x_{ds}^2 + x_{qs}^2}, \quad X_s = \sqrt{x_{ds}^2 + x_{qs}^2}/\sqrt{2}, \quad x \in \{v,\, i,\, \psi\},\ X \in \{V,\, I,\, \Psi\}. \tag{9.3}$$

The stator output active, reactive, and apparent power are expressed as follows:

$$\begin{aligned} P_s &= 3\,V_s\,I_s\cos\varphi_s = 1.5\,(v_{ds}\,i_{ds} + v_{qs}\,i_{qs}) \\ Q_s &= 3\,V_s\,I_s\sin\varphi_s = 1.5\,(v_{qs}\,i_{ds} - v_{ds}\,i_{qs}) \\ S_s &= 3\,V_s\,I_s = \sqrt{P_s^2 + Q_s^2} \end{aligned} \tag{9.4}$$

where φ_s is the power factor (PF) angle.

The angle between $\mathbf{v_s}$ and $\mathbf{i_s}$ is the stator PF angle and is defined as follows:

$$\varphi_s = \theta_v - \theta_i, \quad \theta_v = \tan^{-1}\frac{v_{qs}}{v_{ds}}, \quad \theta_i = \tan^{-1}\frac{i_{qs}}{i_{ds}} \tag{9.5}$$

where θ_v and θ_i are the angles of the stator voltage and current vectors with respect to the d-axis.

The stator winding copper losses and stator PF are expressed as follows:

$$P_{cu,s} = P_m - P_s = 3\,I_s^2\,R_s, \quad \cos\varphi_s = \cos(\theta_v - \theta_i) = \frac{P_s}{S_s} \tag{9.6}$$

where P_m is the mechanical input power to PMSG.

The electromagnetic torque is defined by [7]:

$$\begin{aligned} T_e = \frac{P_m\,P_p}{\omega_r} &= 1.5\,P_p\,(\psi_{ds}\,i_{qs} - \psi_{qs}\,i_{ds}) \\ &= 1.5\,P_p\,[\psi_r\,i_{qs} + (L_{ds} - L_{qs})\,i_{ds}\,i_{qs}] \end{aligned} \tag{9.7}$$

where P_p is the PMSG number of pole pairs.

9.3.2 Continuous-Time Dynamic Models of PMSG

The continuous-time (CT) dynamic models of PMSG stator currents, electromagnetic torque, stator flux linkages, and rotor electrical speed are used to implement various MPC strategies such as PCC, PTC, PPC and PSC. From (6.14) on page 187, the PMSG dq-axes stator current dynamics are defined as follows to realize PCC:

$$\frac{d}{dt}\begin{bmatrix} i_{ds}(t) \\ i_{qs}(t) \end{bmatrix} = \mathbf{A}(t)\begin{bmatrix} i_{ds}(t) \\ i_{qs}(t) \end{bmatrix} + \mathbf{B}\begin{bmatrix} v_{ds}(t) \\ v_{qs}(t) \end{bmatrix} + \mathbf{w}(t) \tag{9.8}$$

where the CT matrices $\mathbf{A}(t)$, $\mathbf{B}$, and $\mathbf{w}(t)$ are defined in terms of PMSG parameters and rotor electrical speed [17]:

$$\mathbf{A}(t) = \begin{bmatrix} -\frac{R_s}{L_{ds}} & \frac{\omega_r(t)\,L_{qs}}{L_{ds}} \\ -\frac{\omega_r(t)\,L_{ds}}{L_{qs}} & -\frac{R_s}{L_{qs}} \end{bmatrix}, \quad \mathbf{B} = \begin{bmatrix} \frac{1}{L_{ds}} & 0 \\ 0 & \frac{1}{L_{qs}} \end{bmatrix}, \quad \mathbf{w}(t) = \begin{bmatrix} 0 \\ -\frac{\omega_r(t)\,\psi_r}{L_{qs}} \end{bmatrix}. \tag{9.9}$$

From (6.11), the stator flux dynamics for the PTC scheme are defined as follows:

$$\frac{d}{dt}\begin{bmatrix} \psi_{ds}(t) \\ \psi_{qs}(t) \end{bmatrix} = \mathbf{A_s}(t)\begin{bmatrix} \psi_{ds}(t) \\ \psi_{qs}(t) \end{bmatrix} + \mathbf{B_v}\begin{bmatrix} v_{ds}(t) \\ v_{qs}(t) \end{bmatrix} + \mathbf{B_i}\begin{bmatrix} i_{ds}(t) \\ i_{qs}(t) \end{bmatrix} \tag{9.10}$$

where

$$\mathbf{A_s}(t) = \begin{bmatrix} 0 & \omega_r(t) \\ -\omega_r(t) & 0 \end{bmatrix}, \quad \mathbf{B_v} = \mathbf{I} = \begin{bmatrix} 1 & 0 \\ 0 & 1 \end{bmatrix}, \quad \mathbf{B_i} = \begin{bmatrix} -R_s & 0 \\ 0 & -R_s \end{bmatrix}. \tag{9.11}$$

9.3.3 Discrete-Time Dynamic Models of PMSG

The CT models expressed as (9.8) through (9.11) are linear time variant (LTV) because of the $\omega_r(t)$ element in the state matrices $\mathbf{A}(t)$ and $\mathbf{A_s}(t)$. The exact discretization of such LTV models is complicated and cumbersome. To obtain the approximate DT equivalent of LTV CT models, the forward Euler method is implemented. The analysis of the complete WECS by the forward Euler approximation method is simple and straightforward. However, the discretization accuracy becomes slightly lower than that achieved by exact and quasi-exact discretization approaches. For high-power PMSGs, the product of R_s and T_s is approximately zero and the discretization error introduced by the approximation is negligible. Other discretization methods such as bilinear transformation, matrix factorization, and truncated Taylor series, which are discussed earlier in Chapter 7, can also be used to improve the accuracy of DT models.

The approximation of the first-order derivative by the forward Euler method is

$$\left\{\frac{dx(t)}{dt}\right\}_{t=kT_s} \approx \frac{x(kT_s+T_s)-x(kT_s)}{T_s}, \qquad x \in \{i_{ds},\, i_{qs},\, \psi_{ds},\, \psi_{qs}\} \tag{9.12}$$

which is simplified as

$$x(k+1) \approx x(k) + T_s \left\{\frac{dx(t)}{dt}\right\}_{t=k}. \tag{9.13}$$

By substituting (9.13) into (9.8), the stator current DT dynamic model is derived in the following. For a detailed analysis, refer to Example 7.1 in Chapter 7.

$$\begin{bmatrix} i_{ds}(k+1) \\ i_{qs}(k+1) \end{bmatrix} = \mathbf{\Phi}(k) \begin{bmatrix} i_{ds}(k) \\ i_{qs}(k) \end{bmatrix} + \mathbf{\Gamma_b} \begin{bmatrix} v_{ds}(k) \\ v_{qs}(k) \end{bmatrix} + \mathbf{\Gamma_w}(k) \tag{9.14}$$

where the DT equivalent matrices are obtained as,

$$\begin{aligned} \mathbf{\Phi}(k) &\approx [\mathbf{I} + \mathbf{A}(k)\,T_s] \approx \begin{bmatrix} 1-\frac{R_s\,T_s}{L_{ds}} & \frac{\omega_r(k)\,L_{qs}\,T_s}{L_{ds}} \\ -\frac{\omega_r(k)\,L_{ds}\,T_s}{L_{qs}} & 1-\frac{R_s\,T_s}{L_{qs}} \end{bmatrix} \\ \mathbf{\Gamma_b} &\approx \mathbf{B}\,T_s \approx \begin{bmatrix} \frac{T_s}{L_{ds}} & 0 \\ 0 & \frac{T_s}{L_{qs}} \end{bmatrix}, \quad \mathbf{\Gamma_w}(k) \approx \mathbf{w}(k)\,T_s \approx \begin{bmatrix} 0 \\ -\frac{\omega_r(k)\,\psi_r\,T_s}{L_{qs}} \end{bmatrix}. \end{aligned} \tag{9.15}$$

The DT dynamic model corresponding to the stator flux is obtained by substituting Equation (9.13) into Equation (9.10):

$$\begin{bmatrix} \psi_{ds}(k+1) \\ \psi_{qs}(k+1) \end{bmatrix} = \mathbf{\Phi_s}(k) \begin{bmatrix} \psi_{ds}(k) \\ \psi_{qs}(k) \end{bmatrix} + \mathbf{\Gamma_v} \begin{bmatrix} v_{ds}(k) \\ v_{qs}(k) \end{bmatrix} + \mathbf{\Gamma_i} \begin{bmatrix} i_{ds}(k) \\ i_{qs}(k) \end{bmatrix} \tag{9.16}$$

and

$$\begin{aligned} \mathbf{\Phi_s}(k) &\approx [\mathbf{I} + \mathbf{A_s}(k)\,T_s] \approx \begin{bmatrix} 1 & \omega_r(k)\,T_s \\ -\omega_r(k)\,T_s & 1 \end{bmatrix} \\ \mathbf{\Gamma_v} &\approx \mathbf{B_v}\,T_s \approx \begin{bmatrix} T_s & 0 \\ 0 & T_s \end{bmatrix}, \quad \mathbf{\Gamma_i} \approx \mathbf{B_i}\,T_s \approx \begin{bmatrix} -R_s\,T_s & 0 \\ 0 & -R_s\,T_s \end{bmatrix}. \end{aligned} \tag{9.17}$$

The above stator current and stator flux discrete-models are combined with the power converter models to realize the MPC schemes for MSCs in PMSG WECS. This topic is discussed in detail in the succeeding sections.

9.4 CONTROL OF PERMANENT MAGNET SYNCHRONOUS GENERATOR

To obtain high dynamic performance and decoupled torque and flux control similar to the separately excited DC motor, vector control strategies are adopted for PMSGs. By reference frame transformations, the magnitude, frequency, and instantaneous phase angle of the space vector are controlled by vector control. As discussed in Chapter 1, the most popular control strategies for wind generators are FOC and DTC [18]. FOC is a linear control scheme, whereas DTC is a nonlinear hysteresis control scheme. Other possible strategies include direct power control, feedback linearization, and passivity-based control [19]. The PMSG in Type 4 variable-speed WECS is controlled by vector control strategies to obtain high performance during steady state and transient conditions. The degree of freedom for PMSG allows the optimization of different performance indices such as electromagnetic torque, PF, power losses, and efficiency. The main vector control strategies for PMSG are listed as follows [20–23]:

- Zero d-axis current (ZDC) control
- Maximum torque per ampere (MTPA) control
- Unity power factor (UPF) control
- Maximum efficiency control
- Constant stator flux control

The above control strategies have their own merits and demerits with respect to performance indices such as torque versus current, torque versus speed, and PF [21, 24]. For a surface mount PMSG (SPMSG), the d-axis inductance is equal to the q-axis inductance, that is, $L_{ds} = L_{qs}$. For interior PMSG (IPMSG), the d-axis inductance is lower than the q-axis inductance, that is, $L_{ds} < L_{qs}$. The ZDC method produces a linear relationship between the stator currents and electromagnetic torque. Optimal control performance is also attained for SPMSG in the entire operational boundary. For IPMSG, MTPA control produces maximum electromagnetic torque with a minimum value of stator current, thereby reducing winding copper losses. The UPF control maintains a constant phase angle φ_s between the stator voltage and current vectors in all operating conditions such that PMSG operates with a unity PF. However, with this method, both SPMSG and IPMSG will not be able to attain their rated output power. The other control strategies listed above are also examined to improve PMSG performance. ZDC and MTPA control strategies are widely used in the industry for SPMSG and IPMSG, respectively. The ZDC and MTPA control strategies are discussed as follows.

9.4.1 Zero d-axis Current Control

Given $L_{ds} = L_{qs}$, the electromagnetic torque expression in (9.7) is simplified as follows:

$$T_e = 1.5\, P_p\, \psi_r\, i_{qs}. \tag{9.18}$$

In PMSG, the rotor flux linkage value ψ_r is constant. To achieve a linear relationship between the torque and stator current, the d-axis current is set to zero, that is,

$$i_{ds} = 0, \qquad i_s = \sqrt{i_{ds}^2 + i_{qs}^2} = i_{qs}. \tag{9.19}$$

To realize the ZDC strategy, the three-phase stator currents are transformed to the dq-frame, and a controller is used to maintain the d-axis current at zero [25]. The stator current vector $\mathbf{i_s}$ becomes perpendicular to the rotor flux linkage ψ_r (Figure 9.4). Thus, the stator current angle θ_i becomes the following:

$$\theta_i = \tan^{-1}\frac{i_{qs}}{i_{ds}} = -\frac{\pi}{2}\ \text{rad}\ (-90°). \tag{9.20}$$

In the motor drive application, both T_e and i_{qs} are positive; therefore, θ_i is +90°. In the case of the generator, θ_i is –90° and T_e and i_{qs} are negative values. The PF angle φ_s yields $0° \leq \varphi_s < 90°$ for the motor drive and $90° \leq \varphi_s < 180°$ for the generator. The stator PF is positive for the motor and negative for the generator.

The torque angle θ_T is the angle between the electromagnetic torque and rotor flux linkage ψ_r. Given that T_e is aligned with i_{qs}, and ψ_r is aligned with the d-axis current, θ_T becomes equal to θ_i. The above notation implies that despite the changes in i_{qs}, the θ_i and θ_T are constant at 90°. Therefore, ZDC control is also called constant torque angle control. With ZDC, the steady-state stator flux linkages and voltages become the following:

$$\begin{aligned} \psi_{ds} &= \psi_r, & \psi_{qs} &= L_{qs}\, i_{qs} \\ v_{ds} &= -\omega_r\, L_{qs}\, i_{qs}, & v_{qs} &= R_s\, i_{qs} + \omega_r\, \psi_r. \end{aligned} \tag{9.21}$$

9.4.2 Maximum Torque per Ampere Control

By referring to (9.7), the electromagnetic torque in IMPSG can be understood as the torque produced by both d- and q-axis currents as $L_{ds} \neq L_{qs}$. The MPTA strategy produces the maximum torque using the minimum values of i_{ds} and i_{qs} by adjusting the ratio between them [17, 26]. To simplify the analysis, the d-axis current is represented in terms of i_s and i_{qs} by using the notation given in (9.19). The resulting electromagnetic torque expression is as follows [23]:

$$T_e = 1.5\, P_p \left[\psi_r\, i_{qs} + (L_{ds} - L_{qs})\left(\sqrt{i_s^2 - i_{qs}^2}\right) i_{qs}\right]. \tag{9.22}$$

To obtain the value of i_{ds} corresponding to the MTPA operation, the above expression is differentiated with respect to i_{qs} and set equal to zero:

$$\frac{d\,T_e}{d\,i_{qs}} = 1.5\, P_p \left[\psi_r + (L_{ds} - L_{qs})\, i_{ds} + (L_{ds} - L_{qs})\, i_{qs}^2 \frac{1}{\sqrt{i_s^2 - i_{qs}^2}}\right] = 0 \tag{9.23}$$

which is deducted to a quadratic expression in the form of $a\,x^2 + b\,x + c = 0$:

$$(L_{ds} - L_{qs})\, i_{ds}^2 + \psi_r\, i_{ds} + (L_{ds} - L_{qs})\, i_{qs}^2 = 0 \tag{9.24}$$

and its solution is obtained as follows [27]:

$$i_{ds} = -\frac{\psi_r}{2\,(L_{ds} - L_{qs})} \pm \sqrt{\frac{\psi_r^2}{4\,(L_{ds} - L_{qs})^2} + i_{qs}^2}, \quad L_{ds} \neq L_{qs} \tag{9.25}$$

To obtain the minimum value for i_{ds}, the negative term is used in the right side of the expression. For the generator, both i_{ds} and i_{qs} are negative with the MTPA control. The steady-state values corresponding to the MTPA control are calculated by using Equations (9.1) to (9.7).

9.5 DIGITAL CONTROL OF BTB CONVERTER-BASED PMSG WECS

Digital control systems play an important role in the power conversion of WECS. They enable the safe, reliable, and efficient operation of power converters to ensure maximum energy capture and compatibility to the grid integration requirements. The information presented in this section can be applied in a broad context and can be used for LV and MV power converters.

9.5.1 Block Diagram of the Digital Control System

The block diagram of the digital control system for BTB-VSC-based PMSG WECS is shown in Figure 9.5. As discussed in Chapter 1, the overall WECS control system includes Level I to Level VI control loops (Figure 1.22). To simplify the analysis, only Level 1 to Level III control loops are considered, as shown in the block diagram in Figure 9.5. The decoupling between the MSC and GSC enables the use of two independent digital control systems. The generator side, DC link, and grid-side feedback signals are used by these digital control systems to generate optimal gating signals for the MSC and GSC. The generator currents (flowing from PMSG to MSC) are multiplied by -1 after considering the motor equations in the previously performed modeling. This process is the only difference between the MPC of the PMS motor and that of the generator.

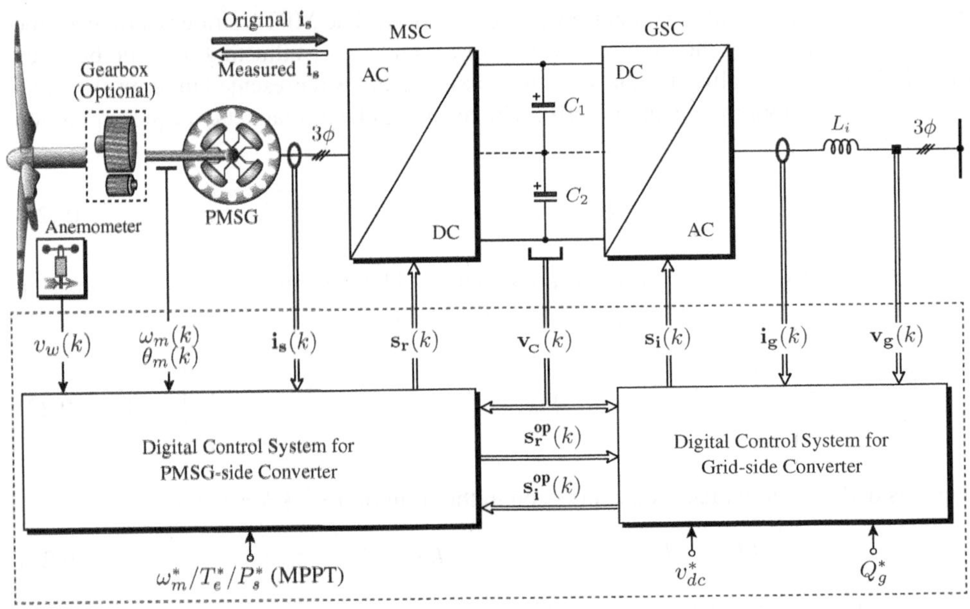

Figure 9.5 Block diagram of digital control scheme for BTB-converter-based PMSG WECS.

9.5.2 Control Requirements

The technical, operational, and legal requirements for the MW-WECS operation are previously discussed in Chapters 1 and 2. The primary control objectives for high-power WECS are listed as follows [5, 16]:

- Improvement of wind energy conversion efficiency in all wind speed conditions by employing MPPT algorithm
- Net DC-bus voltage control to ensure proper operation for the GSC
- Reactive power generation to meet the grid code requirements
- Synchronization of GSC to the three-phase grid to enable grid-tied operation of WTs

The first objective is achieved by the MSC, while the last three objectives are achieved by the GSC. For the power converters, the following additional variables should be controlled by both the MSC and GSC:

- Balancing of the DC-link capacitors voltage in NPC converters to reduce the device voltage stress and improve power quality
- Lower switching frequency operation to reduce converter switching losses
- Mitigation of common-mode voltage to safeguard generator insulation
- Limitation of control variables within the set boundaries (e.g., the generator current should not exceed the prescribed limit during the transient operation)

9.5.3 Notation of Variables

The notation of variables employed in the digital control system is defined as follows:

- Three-phase grid voltages and currents, $\mathbf{v_g} = [v_{ag}\ \ v_{bg}\ \ v_{cg}]^T$, $\mathbf{i_g} = [i_{ag}\ \ i_{bg}\ \ i_{cg}]^T$
- dq-axis PMSG stator currents, $\mathbf{i_{s,dq}} = [i_{ds}\ \ i_{qs}]^T$, $\mathbf{i^*_{s,dq}} = [i^*_{ds}\ \ i^*_{qs}]^T$
- 2L and 3L DC-link capacitor(s) voltage, $\mathbf{v_C} = [v_{C1}]^T$, $\mathbf{v_C} = [v_{C1}\ \ v_{C2}]^T$
- 2L VSR and VSI switching signals, $\mathbf{s_r} = [s_{ar1}\ \ s_{br1}\ \ s_{cr1}]^T$, $\mathbf{s_i} = [s_{ai1}\ \ s_{bi1}\ \ s_{ci1}]^T$
- NPC rectifier switching signals, $\mathbf{s_r} = [s_{ar1}\ \ s_{ar2}\ \ s_{br1}\ \ s_{br2}\ \ s_{cr1}\ \ s_{cr2}]^T$
- NPC inverter switching signals, $\mathbf{s_i} = [s_{ai1}\ \ s_{ai2}\ \ s_{bi1}\ \ s_{bi2}\ \ s_{ci1}\ \ s_{ci2}]^T$

9.5.4 Calculation of Reference Control Variables

On the basis of the generator construction and operating condition to be achieved, the reference control variables, such as rotor speed ω^*_m, dq-axis currents i^*_{ds} and i^*_{qs}, electromagnetic torque T^*_e, stator flux linkage ψ^*_s, and stator active and reactive power (P^*_s and Q^*_s), are calculated in conjunction with the MPPT algorithms. The PCC scheme uses i^*_{ds} and i^*_{qs}, PTC uses T^*_e and ψ^*_s, PPC scheme uses P^*_s and Q^*_s, and PSC scheme uses ω^*_m.

The optimal tip-speed ratio (OTSR) and optimal torque (OT) MPPT algorithms are simple and most commonly used in WECS [28]. The block diagram for the calculation of reference control variables along with the MPPT algorithms is shown in Figure 9.6. The other MPPT algorithms discussed in Section 1.7.4 can also be used to calculate the reference speed ω^*_m or electromagnetic torque reference T^*_e. By using the reference control variables, as shown in Figure 9.6, the generator speed and torque are regulated, achieving MPPT in varying wind speed conditions.

The OTSR MPPT algorithm uses wind speed v_w to calculate the reference mechanical rotor speed [29]. The coefficient K_1^{op} is calculated such that the WT operates at an optimal tip-speed ratio λ_T^{op} at varying wind speed conditions, that is,

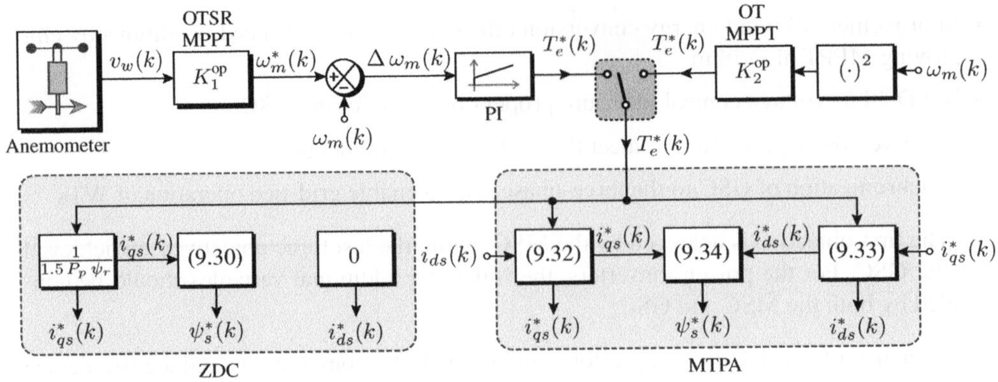

Figure 9.6 Calculation of reference control variables for ZDC and MTPA control.

$$\omega_m^* = K_1^{\mathrm{op}}\, v_w, \qquad K_1^{\mathrm{op}} = \frac{\lambda_T^{\mathrm{op}}\, r_{gb}}{r_T} \tag{9.26}$$

where r_{gb} is the gear ratio and r_T is the blade radius (m). For a particular WT design, K_1^{op} is constant; therefore, ω_m^* holds a linear relationship with the measured v_w.

A PI controller acts on the rotor speed error $\Delta\,\omega_m(k)$ to calculate the reference torque T_e^*, as demonstrated below:

$$T_e^*(k) = \left(k_p + \frac{k_i}{s}\right)\left[\omega_m^*(k) - \omega_m(k)\right] \tag{9.27}$$

where s is the Laplace operator; k_p and k_i are the proportional and integral gains.

The OT MPPT algorithm eliminates the need for the wind speed sensor [30]. This algorithm uses rotor speed feedback $\omega_m(k)$ and rated WT parameters to calculate T_e^*:

$$T_e^* = K_2^{\mathrm{op}}\, \omega_m^2, \qquad K_2^{\mathrm{op}} = \frac{1}{2}\,\rho\,\pi\, r_T^5\, \frac{C_P^{\mathrm{op}}}{\lambda_T^{\mathrm{op}\,3}\, r_{gb}^3} \tag{9.28}$$

where ρ is the air density (kg/m^3) and C_P^{op} is the optimal power coefficient of the WT blades. The coefficient K_2^{op} is also calculated on the basis of the rated WT parameters. Thus, the coefficient is constant for all wind speed values. The wind speed variations are not instantaneously reflected by the OT MPPT; therefore, the efficiency of the algorithm is slightly lower than that of the OTSR algorithm [28]. By changing the toggle switch position in Figure 9.6, the WECS can be operated by using the OTSR or OT MPPT algorithm.

The ZDC and MTPA schemes use the reference torque T_e^* to calculate the reference currents and stator flux linkage. On the basis of the steady-state model in (9.18), the *dq*-axis reference currents are calculated as follows to achieve the ZDC operation:

$$i_{ds}^*(k) = 0, \qquad i_{qs}^*(k) = \frac{T_e^*(k)}{1.5\,P_p\,\psi_r}. \tag{9.29}$$

On the basis of the models in Equations (9.1) and (9.3), the stator flux reference is calculated as follows [25]:

$$\psi_s^*(k) = \sqrt{\psi_r^2 + \left[L_{qs}\, i_{qs}^*(k)\right]^2}. \tag{9.30}$$

The stator flux reference ψ_s^* comprises ψ_r and i_{qs}^*, but the former is the dominant part. For the PMSG, the rotor flux linkage produced by the permanent magnets is constant. The variations in wind speed are reflected by i_{qs}^*. Therefore, ψ_s^* changes with respect to the operating condition (wind speed). The stator flux reference is calculated online to improve the efficiency of PMSG [22, 31, 32]. By substituting (9.29) into (9.30), the stator flux reference is expressed in terms of reference torque T_e^* as follows [33]:

$$\psi_s^*(k) = \sqrt{\psi_r^2 + \left[\frac{L_{qs}\, T_e^*(k)}{1.5\, P_p\, \psi_r}\right]^2}. \tag{9.31}$$

By using the models expressed in (9.7) and (9.25) and the measured $i_{ds}(k)$, the reference dq-axes currents to achieve the MTPA operation are defined by the following:

$$i_{qs}^*(k) = \frac{T_e^*(k)}{1.5\, P_p\, [\psi_r + (L_{ds} - L_{qs})\, i_{ds}(k)]} \tag{9.32}$$

$$i_{ds}^*(k) = -\frac{\psi_r}{2\,(L_{ds} - L_{qs})} - \sqrt{\frac{\psi_r^2}{4\,(L_{ds} - L_{qs})^2} + [i_{qs}^*(k)]^2}. \tag{9.33}$$

On the basis of (9.1) and (9.3), the reference stator flux corresponding to MTPA operation is defined in terms of i_{ds}^* and i_{qs}^*:

$$\psi_s^*(k) = \sqrt{[\psi_r + L_{ds}\, i_{ds}^*(k)]^2 + \left[L_{qs}\, i_{qs}^*(k)\right]^2}. \tag{9.34}$$

With respect to the variations in i_{ds}^* and i_{qs}^*, the variation in ψ_s^* is very small. The analysis for the calculation of ω_m^*, T_e^*, i_{ds}^*, i_{qs}^*, and ψ_s^* can be applied for both classical and MPC schemes. Chapter 8 presents more details about the calculation of the grid-side reference variables, $v_{dc}^*(k)$ and $Q_g^*(k)$.

CASE STUDY 9.1 Calculation of Reference Variables for SPMSG and IPMSG

Case Study Objective: This case study deals with the calculation of the reference control variables for SPMSG and IPMSG by using the OTSR MPPT algorithm. The mathematical expressions in Section 9.5.4 are verified by the calculations and MATLAB simulation results.

Parameters: The 750-kW, 690-V, 9.75-Hz DD SPMSG #1 and IPMSG #1 parameters are presented in Table A.3. The 750-kW WT parameters in Table A.1 are considered. During the rated operating condition, the d-axis current of IPMSG #1 is measured as –213 A (*peak*).

Analysis:
Figure 9.7 shows the reference control variables during the WECS startup operation. The wind speed, v_w, changes from 0 m/s at t = 0.2 s to 12 m/s at t = 1.2 s (Figure 9.7(a)). The mechanical shaft speed reference ω_m^* increases linearly with respect to the wind speed profile (Figure 9.7(b)). The electromagnetic torque reference T_e^* exhibits a quadratic increase with respect to v_w (Figure 9.7(c)). For SPMSG, the d-axis current reference i_{ds}^* is zero for the entire operating range, whereas the q-axis current reference i_{qs}^* follows the T_e^* waveform (Figure 9.7(d)). The stator flux reference ψ_s^* increases from 8.53 Wb to 9.29 Wb (*peak*) (Figure 9.7(e)). The minimum and maximum values of ψ_s^* correspond to ψ_r and the rated stator flux, respectively. For IPMSG, the dq-axes stator current references (i_{ds}^* and i_{qs}^*) increase from zero to the respective rated values (Figure 9.7(f)). The ψ_s^* increases from the ψ_r value to the rated stator flux value (Figure 9.7(g)).

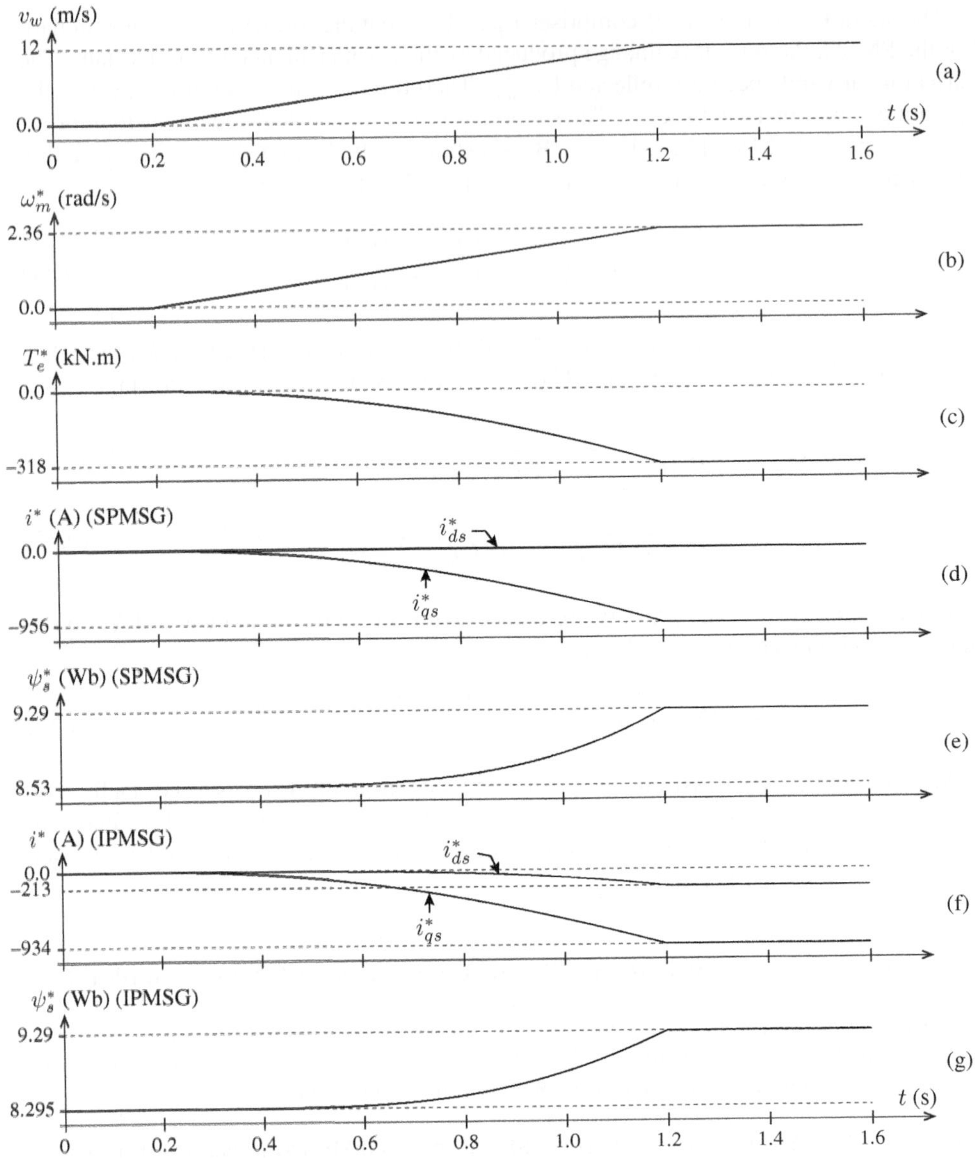

Figure 9.7 Reference control variables for SPMSG and IPMSG during startup.

Under the rated wind speed (12 m/s), the reference control variables for SPMSG are

$$\omega_m^* = \frac{\lambda_T^{\text{op}}\, r_{gb}}{r_T} v_w = \frac{4.4971 \times 1}{22.9} \times 12 = 2.36 \text{ rad/s} \quad (1.0 \text{ pu})$$

$$T_e^* = T_{e,R} \times \left(\frac{v_w}{v_{w,R}}\right)^2 = -318 \times 10^3 \times \left(\frac{12}{12}\right)^2 = -318 \text{ kN.m} \quad (-1.0 \text{ pu})$$

$$i_{ds}^* = 0 \text{ A} \quad (0 \,\widehat{\text{pu}}) \quad (\text{ZDC scheme})$$

$$i_{qs}^* = \frac{T_e^*}{1.5\, P_p\, \psi_r} = \frac{-318 \times 10^3}{1.5 \times 26 \times 8.53} = -956 \text{ A} \quad (-1.0 \,\widehat{\text{pu}})$$

$$\psi_s^* = \sqrt{\psi_r^2 + (L_{qs}\, i_{qs}^*)^2} = \sqrt{8.53^2 + (3.85 \times 10^{-3} \times -956)^2} = 9.29 \text{ Wb} \quad (1.01 \,\widehat{\text{pu}}).$$

Similarly, the reference control variables for IPMSG under the rated wind speed are

$$
\begin{aligned}
i_{qs}^* &= \frac{T_e^*}{1.5\, P_p\, [\psi_r + (L_{ds} - L_{qs})\, i_{ds}]} \\
&= \frac{-318 \times 10^3}{1.5 \times 26 \times [8.53 + (3.74 - 5.88) \times 10^{-3} \times -213]} = -934 \text{ A} \quad (-0.98 \,\widehat{\text{pu}}) \\
i_{ds}^* &= -\frac{\psi_r}{2\,(L_{ds} - L_{qs})} - \sqrt{\frac{\psi_r^2}{4\,(L_{ds} - L_{qs})^2} + \left(i_{qs}^*\right)^2} \\
&= -\frac{8.53}{2 \times (3.74 - 5.88) \times 10^{-3}} - \sqrt{\frac{8.53^2}{4 \times ((3.74 - 5.88) \times 10^{-3})^2} + (-934)^2} \\
&= -213 \text{ A} \quad (-0.22 \,\widehat{\text{pu}}) \\
\psi_s^* &= \sqrt{\left(\psi_r + L_{ds}\, i_{ds}^*\right)^2 + \left(L_{qs}\, i_{qs}^*\right)^2} \\
&= \sqrt{(8.53 + 3.74 \times 10^{-3} \times -213)^2 + (5.88 \times 10^{-3} \times -934)^2} = 9.29 \text{ Wb} \quad (1.01 \,\widehat{\text{pu}}).
\end{aligned}
$$

Given the steady-state operating condition, the calculated i_{ds}^* is equal to the measured i_{ds}. However, during variable-speed operation, i_{ds}^* becomes different from i_{ds}. Following the reference control variables in Figure 9.7, the PCC, PTC, and PSC schemes, along with the MPPT operation, can be realized for SPMSG and IPMSG WECS. ■

9.6 PREDICTIVE CURRENT CONTROL OF BTB 2L-VSC-BASED PMSG WECS

The overall digital control of PMSG WECS with BTB 2L-VSC is presented in this section. Along with 2L-VSR, the PMSG is controlled by the PCC scheme, whereas the grid-side 2L-VSI is assumed to be controlled by the PCC or PPC scheme. The generator-side PCC scheme controls the 2L-VSR such that the stator currents are tracked to their reference currents. The control scheme for 2L-VSI is not discussed in detail in this section and can be implemented as a separate subsystem by using the guidelines in Chapter 8.

9.6.1 Generator-side Control Scheme

Figure 9.8 shows the block diagram of the PCC scheme for an SPMSG with 2L-VSR. By conducting slight modifications on the block diagram, particularly the reference currents calculation, the PCC scheme can be adopted for IPMSG WECS. The generator-side control scheme aims to perform MPPT, wherein the generator currents are controlled in the *dq*-frame. The control scheme mainly comprises five design steps as explained below.

Measurement and Synthesis of Feedback Signals: The feedback signals, DC-link capacitor voltage $v_{C1}(k)$, three-phase generator/converter currents $\mathbf{i_s}(k)$, and generator rotor mechanical speed $\omega_m(k)$ and position $\theta_m(k)$ are used to design PCC. The generator currents are measured in the reverse direction (VSR to PMSG) to adopt the motor model for the PCC scheme. The wind speed $v_w(k)$ information is also needed to calculate the reference currents. By multiplying $\omega_m(k)$ and $\theta_m(k)$ with the generator pole pairs, the equivalent electrical speed $\omega_r(k)$ and position $\theta_r(k)$ are obtained. The electrical rotor position $\omega_r(k)$ is then used to convert the three-phase currents to the *dq*-frame.

Calculation of Reference Currents: The reference currents for the considered control scheme are calculated by combining the OTSR MPPT, speed control loop, and ZDC scheme (Figure 9.6). By changing the reference currents, the PCC scheme can also be applied to 2L-VSR-based IPMSG.

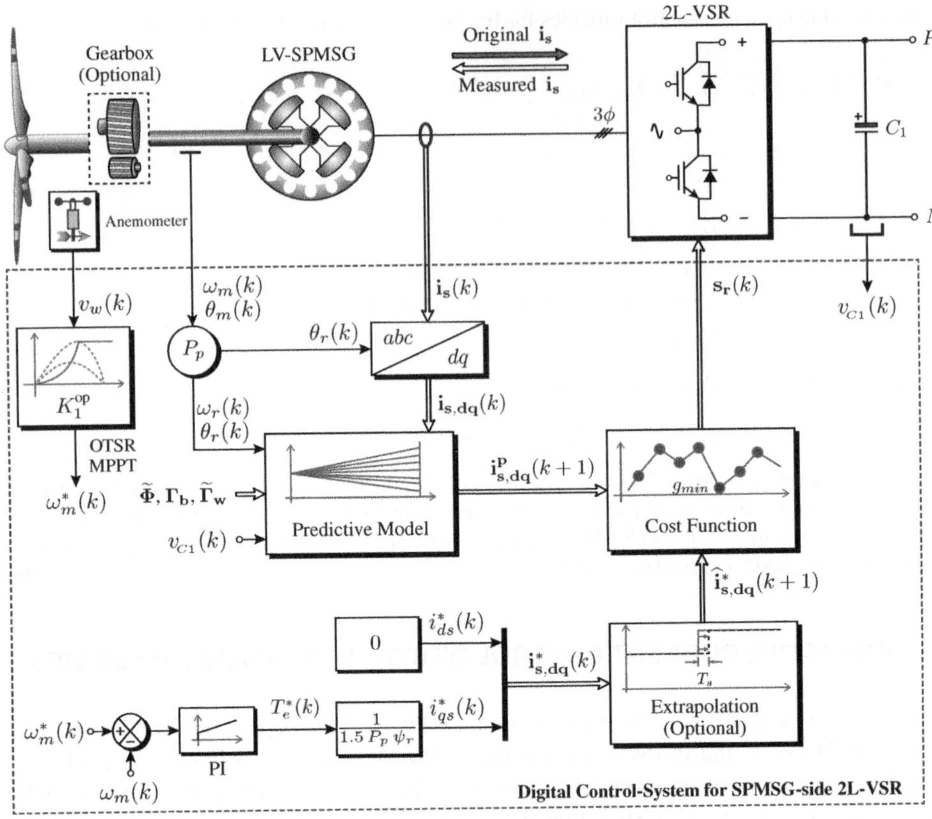

Figure 9.8 Block diagram of the PCC scheme for ZDC control of the SPMSG with 2L-VSR.

Extrapolation of Reference Currents: To compute the PMSG current error in $(k+1)$ sampling instant, the reference currents are extrapolated to the $(k+1)$ state. Given that i^*_{ds} is always zero in ZDC control, no extrapolation is needed. In the steady state, no extrapolation is needed for i^*_{qs}. To improve the transient performance, a first-order Lagrange extrapolation is performed by using the following generic representation:

$$\mathbf{i}^*_{\mathbf{s,dq}}(k+1) = 2\,\mathbf{i}^*_{\mathbf{s,dq}}(k) - \mathbf{i}^*_{\mathbf{s,dq}}(k-1). \tag{9.35}$$

Prediction of Future Behavior of PMSG Currents: The "predictive model" block calculates the future value of the dq-axis currents by using the DT model of PMSG and 2L-VSR. The dq-axes currents are predicted as follows:

- The 2L-VSR terminal voltages are equal to generator voltages v_{ds} and v_{qs} (Figure 9.8). By substituting the 2L-VSR dq-axes voltages (Equation (5.12)) into the PMSG sampled-data model (Equation (9.14)), the DT model is defined as follows:

$$\begin{bmatrix} i^p_{ds}(k+1) \\ i^p_{qs}(k+1) \end{bmatrix} = \mathbf{\Phi}(k) \begin{bmatrix} i_{ds}(k) \\ i_{qs}(k) \end{bmatrix} + \mathbf{\Gamma_b} \underbrace{\begin{bmatrix} v^p_{ds}(k) \\ v^p_{qs}(k) \end{bmatrix}}_{\text{2L-VSR Model}} + \mathbf{\Gamma_w}(k) \tag{9.36}$$

where superscript p denotes the predicted variable.

As mentioned in Chapter 5, eight switching states (predictions) are available for a 2L-VSR. To minimize the number of online calculations, matrices $\widetilde{\mathbf{\Phi}}$ and $\widetilde{\mathbf{\Gamma}}_{\mathbf{w}}$ (in addition to $\mathbf{\Gamma_b}$) are defined offline by excluding the $\omega_r(k)$ term.

- The predictions for v_{ds} and v_{qs} are obtained in terms of the dq-axes switching signals and measured DC-link capacitor voltage. The model is defined as follows:

$$\begin{bmatrix} v_{ds}^p(k) \\ v_{qs}^p(k) \end{bmatrix} = v_{C1}(k) \begin{bmatrix} s_{dr1}^p(k) \\ s_{qr1}^p(k) \end{bmatrix}. \tag{9.37}$$

The dq-axis switching signals are computed from the natural switching signals as

$$\begin{bmatrix} s_{dr1}^p(k) \\ s_{qr1}^p(k) \end{bmatrix} = \underbrace{\begin{bmatrix} \cos\theta_r(k) & \sin\theta_r(k) \\ -\sin\theta_r(k) & \cos\theta_r(k) \end{bmatrix}}_{\mathbf{T}_{\alpha\beta/dq}} \underbrace{\frac{2}{3}\begin{bmatrix} 1 & -\frac{1}{2} & -\frac{1}{2} \\ 0 & \frac{\sqrt{3}}{2} & -\frac{\sqrt{3}}{2} \end{bmatrix}}_{\mathbf{T}_{abc/\alpha\beta}} \begin{bmatrix} s_{ar1}^p(k) \\ s_{br1}^p(k) \\ s_{cr1}^p(k) \end{bmatrix}. \tag{9.38}$$

In the above expression, the natural-frame switching signals are initially converted to the $\alpha\beta$-frame and then to the dq-frame by using $\theta_r(k)$. As mentioned in Chapter 5, the $\alpha\beta$-frame switching signals can be defined offline to reduce the number of online calculations.

- Lastly, by combining (9.36) and (9.37), the overall sampled-data model for predicting the future dq-axis current values is formulated as follows in terms of the 2L-VSR switching signals:

$$\begin{bmatrix} i_{ds}^p(k+1) \\ i_{qs}^p(k+1) \end{bmatrix} = \mathbf{\Phi}(k) \begin{bmatrix} i_{ds}(k) \\ i_{qs}(k) \end{bmatrix} + \mathbf{\Gamma_b} \left\{ v_{C1}(k) \begin{bmatrix} s_{dr1}^p(k) \\ s_{qr1}^p(k) \end{bmatrix} \right\} + \mathbf{\Gamma_w}(k). \tag{9.39}$$

The switching signals can have eight different combinations as shown below:

$$\begin{aligned} \mathbf{s_r^p}(k) \in [s_{ar1}^p(k), s_{br1}^p(k), s_{cr1}^p(k)] \in & [\text{'0', '0', '0'}], [\text{'1', '0', '0'}], [\text{'1', '1', '0'}], \\ & [\text{'0', '1', '0'}], [\text{'0', '1', '1'}], [\text{'0', '0', '1'}], [\text{'1', '0', '1'}], [\text{'1', '1', '1'}]. \end{aligned} \tag{9.40}$$

Cost Function Minimization: The PCC scheme primarily aims to regulate the dq-axes PMSG currents, which are defined in the form of two sub-cost functions as follows:

$$\begin{aligned} g_{id}(k) &= \left[\widehat{i}_{ds}^*(k+1) - i_{ds}^p(k+1) \right]^2 \\ g_{iq}(k) &= \left[\widehat{i}_{qs}^*(k+1) - i_{qs}^p(k+1) \right]^2. \end{aligned} \tag{9.41}$$

The objective of the switching frequency minimization for 2L-VSR is defined by

$$g_{sw,r}(k) = \left[s_{ar1}^p(k) - s_{ar1}^{\text{op}}(k)\right]^2 + \left[s_{br1}^p(k) - s_{br1}^{\text{op}}(k)\right]^2 + \left[s_{cr1}^p(k) - s_{cr1}^{\text{op}}(k)\right]^2 \tag{9.42}$$

where s_{ar1}^{op}, s_{br1}^{op}, and s_{cr1}^{op} are the optimal switching signals in past sampling instant.

As demonstrated below, the global generator-side cost function combines the above sub-cost functions through the following weighting factors:

$$g_r(k) = \lambda_{id}\, g_{id}(k) + \lambda_{iq}\, g_{iq}(k) + \lambda_{sw,r}\, g_{sw,r}(k) \tag{9.43}$$

where λ_{id}, λ_{iq}, and $\lambda_{sw,r}$ are the weighting factors. By using the per-unit weighting factor selection method (Section 4.5.2), λ_{id} and λ_{iq} are calculated as follows:

$$\lambda_{id} = \lambda_{iq} = 1. \tag{9.44}$$

The weighting factor $\lambda_{sw,r}$ is selected by heuristic approach. By using the above eight combinations for switching signals, eight different predictions are obtained for v_{ds} and v_{qs}, thus leading to eight different future values of i_{ds} and i_{qs}. The predicted i_{ds} and i_{qs} are compared to the reference dq-axes currents to select an optimal switching state combination that minimizes the cost function. The selected switching states are applied to the 2L-VSR during the $(k+1)$ sampling period.

9.6.2 Grid-side Control Scheme

For the grid-side 2L-VSI, the $\alpha\beta$-frame reference currents are obtained by regulating the net DC-bus voltage v_{dc} and the grid-side reactive power Q_g at their reference values. The predictive model calculates the future values of the $\alpha\beta$-frame grid currents. The grid-side cost function evaluates the error between the reference and predicted currents as follows:

$$g_i(k) = \lambda_{i\alpha}\left[\widehat{i}^{*}_{\alpha g}(k+1) - i^{p}_{\alpha g}(k+1)\right]^2 + \lambda_{i\beta}\left[\widehat{i}^{*}_{\beta g}(k+1) - i^{p}_{\beta g}(k+1)\right]^2 \tag{9.45}$$
$$+ \lambda_{sw,i}\left(\left[s^{p}_{ai1}(k) - s^{\text{op}}_{ai1}(k)\right]^2 + \left[s^{p}_{bi1}(k) - s^{\text{op}}_{bi1}(k)\right]^2 + \left[s^{p}_{ci1}(k) - s^{\text{op}}_{ci1}(k)\right]^2\right)$$

Similar to the generator-side cost function, the switching states that minimize the cost function g_i are selected and directly applied to the 2L-VSI. The $\alpha\beta$-frame PCC for GSC is explained further in Chapter 8.

9.6.3 Control Algorithm

The flowcharts of the PCC algorithm for 2L-VSR and 2L-VSI are shown in Figures 9.9(a) and 9.9(b), respectively. Both the VSR and VSI algorithms are identical and have the same digital implementation. Block ⑥ uses a different predictive model from that of the PCC algorithm, as illustrated in Figure 4.3. After initializing the digital controller and calculating the reference generator and grid currents, the PCC algorithm enters an iterative loop where the generator and grid currents are predicted and cost function minimization is performed. An optimal vector j_{op} and the corresponding switching signals that minimize the cost function are selected and directly applied to the VSR and VSI.

Algorithm 9.1 shows the *S-Function Builder* programming in `MATLAB` environment that is used to realize the PCC scheme for 2L-VSR. The inputs to the *S-Function* block include $\widehat{i}^{*}_{ds}(k+1)$, $\widehat{i}^{*}_{qs}(k+1)$, $i_{ds}(k)$, $i_{qs}(k)$, $v_{C1}(k)$, $\omega_r(k)$, cos $\theta_r(k)$, sin $\theta_r(k)$, and $\lambda_{sw,r}$, whereas the output signals from the *S-Function* block include $s_{ar1}(k)$, $s_{br1}(k)$, and $s_{cr1}(k)$. The external function declarations for the *S-Function* includes switching states in the $\alpha\beta$ and abc reference frames and DT parameters $\widetilde{\mathbf{\Phi}}$, $\mathbf{\Gamma}_\mathbf{b}$, and $\widetilde{\mathbf{\Gamma}}_\mathbf{w}$. The *S-Function* is executed at the control sampling time T_s irrespective of the simulation fundamental sample time (fixed step-size). The program lines are explained as follows: Lines `01` and `02` correspond to PCC initialization, the `for` loop is initialized in lines `03` and `04`, the VSR voltages are predicted in lines `05` to `08`, the PMSG currents are predicted in lines `09` and `10`, the cost function is defined in lines `11` to `14`, the optimal vector `j_op` number is selected in lines `15` to `19`, the `for` loop is terminated in line `20`, the optimal three-phase switching states are produced in lines `21` to `23` according to the `j_op` value, and the switching states are stored in lines `24` to `26` for the next iteration.

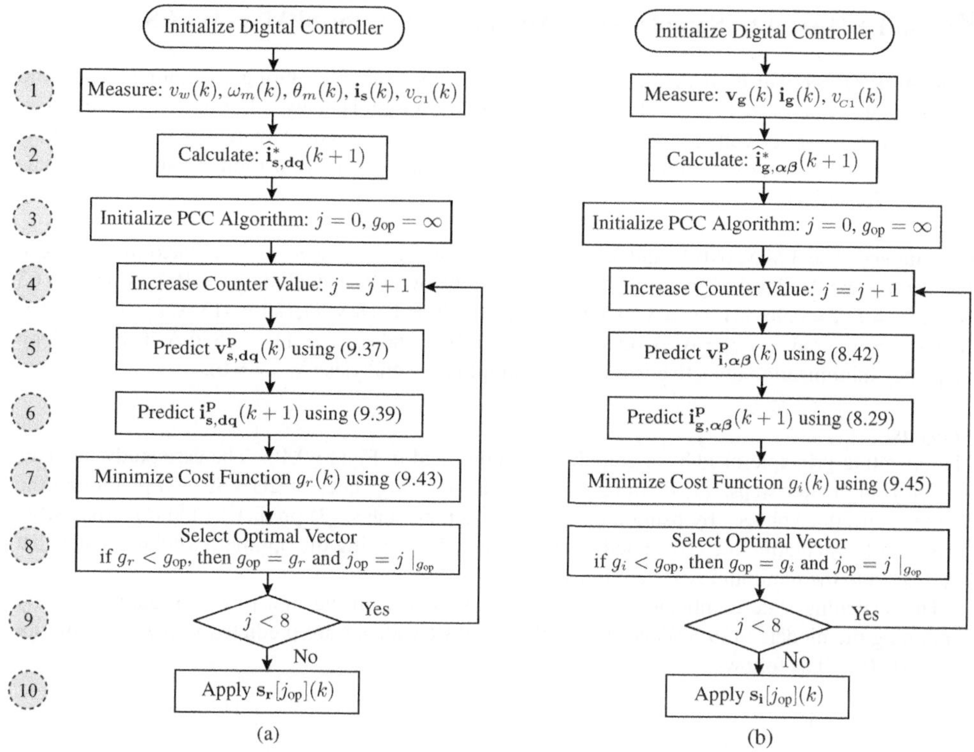

Figure 9.9 Flowchart of the PCC algorithm for: (a) PMSG-side 2L-VSR and (b) grid-side 2L-VSI.

ALGORITHM 9.1 PCC Algorithm for the 2L-VSR-Based PMSG WECS

```
\\ P11 = 1-(Rs*Ts/Lds); P12 = (Lqs*Ts)/Lds; P21 = -(Lds*Ts)/Lqs;
\\ P22 = 1-(Rs*Ts/Lqs); G11 = Ts/Lds; G22 = Ts/Lqs; W21 = -(Psir*Ts)/Lqs;
j_op = 1000;
g_op = 1000000000;
for(j = 0; j < 8; j++)
{
    vr_ab[0] = vc[0]*vrec[j].s_r_ab[0];
    vr_ab[1] = vc[0]*vrec[j].s_r_ab[1];
    vr_dq[0] =  vr_ab[0]*cstheta_r[0] + vr_ab[1]*cstheta_r[1];
    vr_dq[1] = -vr_ab[0]*cstheta_r[1] + vr_ab[1]*cstheta_r[0];
    is_dq_k1[0] = P11*is_dq[0]+P12*wr[0]*is_dq[1]+G11*vr_dq[0];
    is_dq_k1[1] = P21*wr[0]*is_dq[0]+P22*is_dq[1]+G22*vr_dq[1]+W21*wr[0];
    g_id = (is_ref_k1[0]-is_dq_k1[0])*(is_ref_k1[0]-is_dq_k1[0]);
    g_iq = (is_ref_k1[1]-is_dq_k1[1])*(is_ref_k1[1]-is_dq_k1[1]);
    g_swr = (vrec[j].s_r_abc[0]-s_r_km1[0])*(vrec[j].s_r_abc[0]-s_r_km1[0])
          + (vrec[j].s_r_abc[1]-s_r_km1[1])*(vrec[j].s_r_abc[1]-s_r_km1[1])
          + (vrec[j].s_r_abc[2]-s_r_km1[2])*(vrec[j].s_r_abc[2]-s_r_km1[2]);
    g_r = g_id + g_iq + alpha[0]*g_swr;
    if(g_r < g_op)
       {
          j_op = j;
          g_op = g_r;
       }
}
s_ar1[0] = vrec[j_op].s_r[0];
s_br1[0] = vrec[j_op].s_r[1];
s_cr1[0] = vrec[j_op].s_r[2];
s_r_km1[0] = s_ar1[0];
s_r_km1[1] = s_br1[0];
s_r_km1[2] = s_cr1[0];
```

CASE STUDY 9.2 Sample-Based Analysis of PCC for SPMSG WECS

Case Study Objective: By using the feedback and reference variables during a sampling instant, the operating principle of PCC for a 2L-VSR-based PMSG WECS is analyzed in this case study. The prediction of converter voltages and generator currents with all possible switching states as well as the selection of optimal switching signals through cost function minimization are analyzed by numerical calculations.

Parameters: The 3 MW, 690 V, and 9.75 Hz DD SPMSG #2 parameters are presented in Table A.3. The parameters, feedback signals, and reference currents with ZDC include $T_s = 100\ \mu$s, $R_s = 1.63$ mΩ, $L_{ds} = L_{qs} = 0.96$ mH, $\psi_r = 8.53$ Wb (*peak*), $v_{C1}(k) = 1218$ V, $i_{ds}(k) = -13$ A, $i_{qs}(k) = -3856$ A, $\theta_r(k) = 3.13$ rad, $\omega_r(k) = 61.3$ rad/s, $\widehat{i}^*_{ds}(k+1) = 0$ A, and $\widehat{i}^*_{qs}(k+1) = -3826$ A. The switching frequency minimization objective is ignored to simplify the analysis ($\lambda_{sw,r} = 0$).

Analysis:

The operating principle of PCC is graphically represented in Figure 9.10. The case study involves the following major steps: (1) defining all possible switching states of 2L-VSR, (2) predicting the *dq*-axis converter voltages by using all possible switching states, (3) predicting the *dq*-axis PMSG currents, (4) calculating the cost function values, and (5) selecting the optimal switching states that correspond to the minimum cost function value.

The switching state combinations ($\mathbf{s_0} \sim \mathbf{s_7}$) are shown in the first row of Figure 9.10. By combining the models in (9.37) and (9.38), the 2L-VSR voltages are predicted in the *dq*-frame for $\mathbf{s_5} \in$ ['0','0','1'] as follows:

$$\begin{bmatrix} v^p_{ds}(k) \\ v^p_{qs}(k) \end{bmatrix} = v_{C1}(k) \begin{bmatrix} \cos\theta_r(k) & \sin\theta_r(k) \\ -\sin\theta_r(k) & \cos\theta_r(k) \end{bmatrix} \frac{2}{3} \begin{bmatrix} 1 & -\frac{1}{2} & -\frac{1}{2} \\ 0 & \frac{\sqrt{3}}{2} & -\frac{\sqrt{3}}{2} \end{bmatrix} \begin{bmatrix} s^p_{ar1}(k) \\ s^p_{br1}(k) \\ s^p_{cr1}(k) \end{bmatrix}$$

$$= 1218 \begin{bmatrix} \cos(3.13) & \sin(3.13) \\ -\sin(3.13) & \cos(3.13) \end{bmatrix} \frac{2}{3} \begin{bmatrix} 1 & -\frac{1}{2} & -\frac{1}{2} \\ 0 & \frac{\sqrt{3}}{2} & -\frac{\sqrt{3}}{2} \end{bmatrix} \begin{bmatrix} 0 \\ 0 \\ 1 \end{bmatrix} = \begin{bmatrix} 397.8 \\ 707.9 \end{bmatrix}.$$

By performing the same analysis with other switching state combinations, the converter voltages are predicted and represented in the second row of Figure 9.10. The DT matrices $\mathbf{\Phi}(k)$, $\mathbf{\Gamma_b}$, and $\mathbf{\Gamma_w}$ are needed to calculate the future values of the PMSG currents. By using the PMSG parameters and rotor electrical speed, the DT matrices for $T_s = 100\ \mu$s are calculated as follows:

$$\mathbf{\Phi}(k) \approx \begin{bmatrix} 1 - \frac{R_s T_s}{L_{ds}} & \frac{\omega_r(k) L_{qs} T_s}{L_{ds}} \\ -\frac{\omega_r(k) L_{ds} T_s}{L_{qs}} & 1 - \frac{R_s T_s}{L_{qs}} \end{bmatrix} \approx \begin{bmatrix} 0.99983 & 0.00613 \\ -0.00613 & 0.99983 \end{bmatrix}$$

$$\mathbf{\Gamma_b} \approx \begin{bmatrix} \frac{T_s}{L_{ds}} & 0 \\ 0 & \frac{T_s}{L_{qs}} \end{bmatrix} \approx \begin{bmatrix} 0.10417 & 0 \\ 0 & 0.10417 \end{bmatrix}$$

$$\mathbf{\Gamma_w}(k) \approx \begin{bmatrix} 0 \\ -\frac{\omega_r(k)\psi_r T_s}{L_{qs}} \end{bmatrix} \approx \begin{bmatrix} 0 \\ -54.468 \end{bmatrix}.$$

By using the DT parameters and VSR predicted voltages, the PMSG *dq*-axes future currents for $\mathbf{s_5} \in$ ['0','0','1'] are calculated as follows:

$$\begin{bmatrix} i^p_{ds}(k+1) \\ i^p_{qs}(k+1) \end{bmatrix} = \mathbf{\Phi}(k) \begin{bmatrix} i_{ds}(k) \\ i_{qs}(k) \end{bmatrix} + \mathbf{\Gamma_b} \begin{bmatrix} v^p_{ds}(k) \\ v^p_{qs}(k) \end{bmatrix} + \mathbf{\Gamma_w}(k)$$

$$= \begin{bmatrix} 0.99983 & 0.00613 \\ -0.00613 & 0.99983 \end{bmatrix} \begin{bmatrix} -13 \\ -3856 \end{bmatrix} + \begin{bmatrix} 0.10417 & 0 \\ 0 & 0.10417 \end{bmatrix} \begin{bmatrix} 397.8 \\ 707.9 \end{bmatrix}$$

$$+ \begin{bmatrix} 0 \\ -54.468 \end{bmatrix} = \begin{bmatrix} 4.8 \\ -3836 \end{bmatrix}.$$

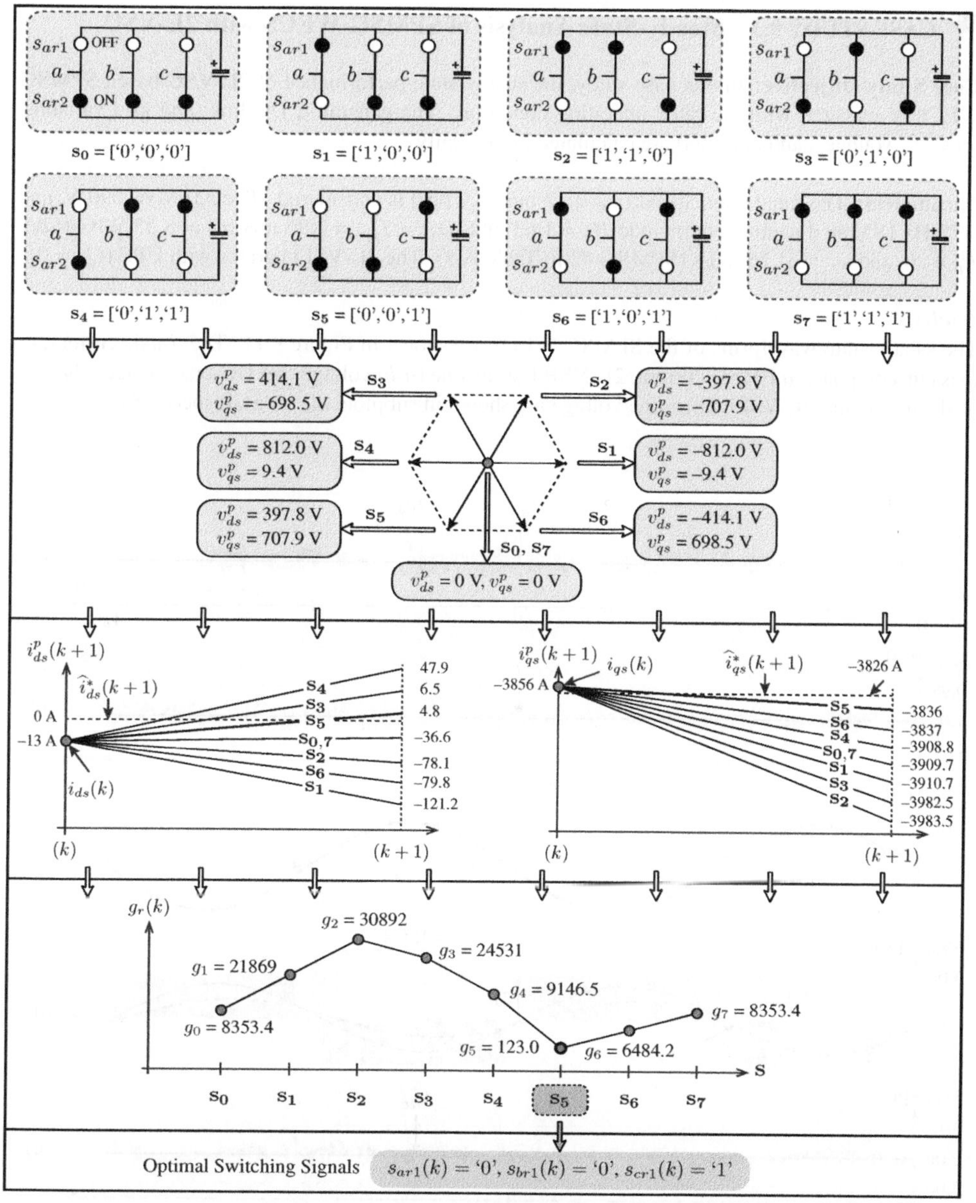

Figure 9.10 Graphical representation of the sample-based analysis of 2L-VSR with SPMSG.

The third row in Figure 9.10 represents the predicted dq-axes currents for all possible switching states. By using the predicted currents, the cost function is calculated as follows:

$$g_r(k) = \left[\widehat{i}^*_{ds}(k+1) - i^p_{ds}(k+1)\right]^2 + \left[\widehat{i}^*_{qs}(k+1) - i^p_{qs}(k+1)\right]^2$$
$$= [0 - 4.8]^2 + [-3826 + 3836]^2 = 123.$$

As shown in the fourth row of Figure 9.10, the switching state combination $\mathbf{s}_5$ produces a minimum cost function value. The corresponding optimal switching signals are $s_{ar1}(k) = $ '0', $s_{br1}(k) = $ '0', and $s_{cr1}(k) = $ '1'. The above procedure is repeated during each sampling interval such that the PMSG dq-axes currents follow the reference currents. ■

CASE STUDY 9.3 Steady-State Analysis of SPMSG WECS with 2L-VSC

Case Study Objective: In this case study, the steady-state performance of 2L-VSC-based SPMSG WECS is analyzed with the rated operating condition. The generator, DC link, and grid variables (powers, voltages, and currents) are calculated numerically.

Parameters: The rated wind speed (v_w = 12 m/s (1.0 pu)) is considered. The 3.0-MW, 690-V, and 9.75-Hz DD rated parameters include R_s = 1.63 mΩ, $L_{ds} = L_{qs}$ = 0.96 mH, ψ_r = 8.53 Wb (*peak*), P_p = 26, and ω_r = 61.26 rad/s (SPMSG #2 in Table A.3). The 2L-VSI operates with UPF ($Q_g = 0$).

Analysis:
The steady-state waveforms of the SPMSG WECS are shown in Figure 9.11. The d-axis current, q-axis current, phase-a PMSG current, 2L-VSR line-to-line (a-b) voltage, net DC-bus voltage, phase-a grid current, and 2L-VSI line-to-line voltage are shown in subplots (a) to (g), respectively.

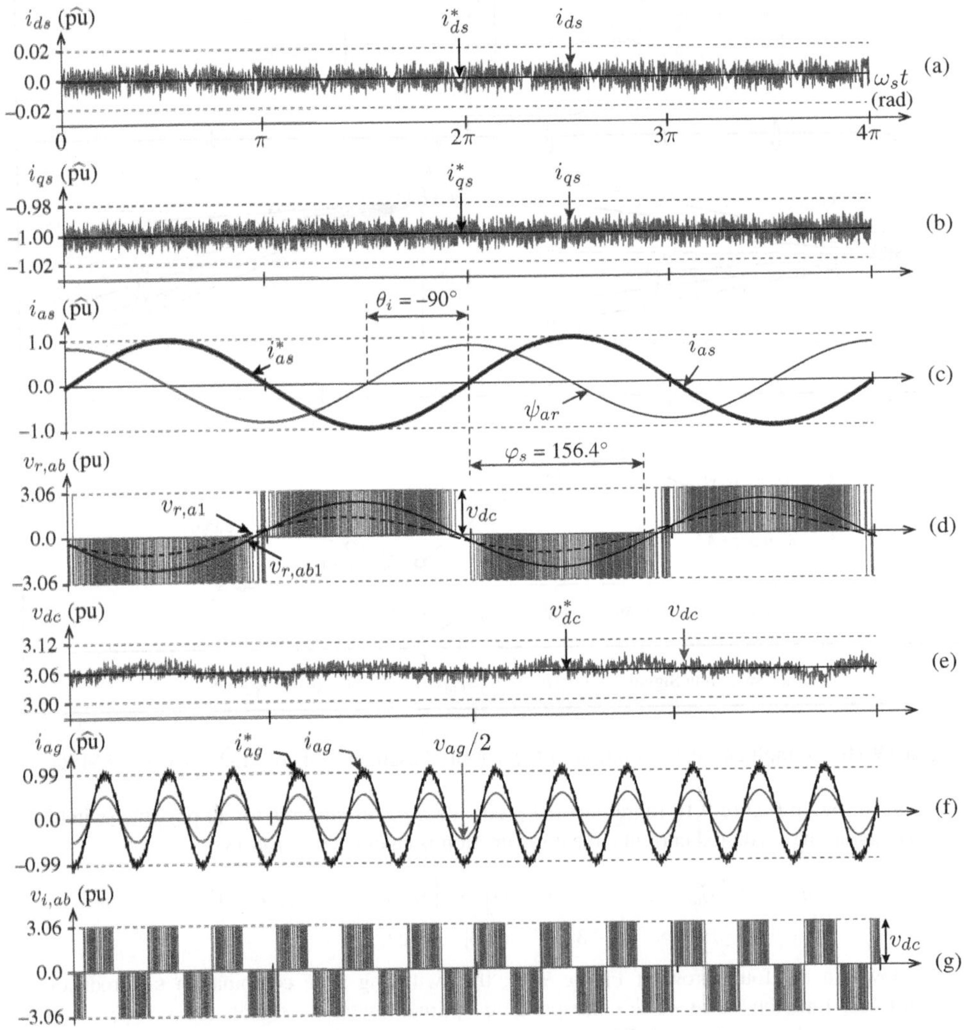

Figure 9.11 Steady-state waveforms with the PCC scheme for 2L-VSC-based SPMSG WECS.

The rotor electrical speed, mechanical input power, and electromagnetic torque are calculated under the rated operating condition:

$$\omega_r = \omega_{r,R} \times v_{w,\,\text{pu}} = 61.26 \times 1.0 = 61.26 \text{ rad/s} \quad (1.0 \text{ pu})$$

$$T_e = \frac{P_m\, P_p}{\omega_r} = \frac{-3000 \times 10^3 \times 26}{61.26} = -318 \times 10^3 \text{ N.m} \quad (-1.0 \text{ pu}).$$

Along with the speed control loop and ZDC scheme, the MPPT operation produces the reference dq-axes currents. The q-axis current is calculated as follows:

$$i_{qs} = \frac{T_e}{1.5\, P_p\, \psi_r} = \frac{-318 \times 10^3}{1.5 \times 26 \times 8.53} = -3826.4 \text{ A } (peak) \quad (-1.0 \text{ pu})$$

$$I_s = \sqrt{i_{ds}^2 + i_{qs}^2}/\sqrt{2} = \sqrt{0^2 + (-3826.4)^2}/\sqrt{2} = 2705.7 \text{ A } (rms) \quad (1.0 \text{ pu}).$$

As shown in Figures 9.11(a) and 9.11(b), the PCC scheme selects the optimal switching states during each sampling interval such that the PMSG dq-axes currents follow i_{ds}^* and i_{qs}^* with $\pm 1\%$ peak current ripple. The THD of the phase-a PMSG current is negligibly low (0.67%) because of the filtering effect of the PMSG dq-axes inductances (Figure 9.11(c)). The fundamental frequency of the PMSG currents (9.75 Hz) is approximately six times lower than the grid frequency (60 Hz). The steady-state SPMSG dq-axes voltages are obtained as follows:

$$v_{ds} = R_s\, i_{ds} - \omega_r\, L_{qs}\, i_{qs} = 225.5 \text{ V } (peak) \quad (0.4 \text{ pu})$$

$$v_{qs} = R_s\, i_{qs} + \omega_r\, L_{ds}\, i_{ds} + \omega_r\, \psi_r = 516.4 \text{ V } (peak) \quad (0.92 \text{ pu})$$

$$V_s = \sqrt{v_{ds}^2 + v_{qs}^2}/\sqrt{2} = \sqrt{225.5^2 + 516.4^2}/\sqrt{2} = 398.4 \text{ V } (rms) \quad (1.0 \text{ pu}).$$

The switching frequency ($f_{sw,r}$) of 2L-VSR is noted as 1800 Hz. The 2L-VSR line-to-line voltage and its fundamental $v_{r,ab1}$, as well as the fundamental component of the phase-a VSR voltage $v_{r,a1}$ ($= v_{s,a1}$), are shown in Figure 9.11(d). The stator PF is calculated as follows:

$$\theta_i = \tan^{-1}\frac{i_{qs}}{i_{ds}} = \tan^{-1}\frac{-3826.4}{0} = -1.5708 \text{ rad} \quad (-90^\circ)$$

$$\theta_v = \tan^{-1}\frac{v_{qs}}{v_{ds}} = \tan^{-1}\frac{516.4}{225.5} = 1.1591 \text{ rad} \quad (66.4^\circ)$$

$$\varphi_s = \theta_v - \theta_i = 1.1591 + 1.5708 = 2.7299 \quad (156.4^\circ)$$

$$\cos\varphi_s = \cos(2.7299) = -0.92.$$

As indicated in Figures 9.11(c) and 9.11(d), the PMSG phase-a current is lagging the phase-a rotor flux ψ_{ar} by 90° and leading the $v_{r,a1}$ waveform by 156.4°. The PMSG stator output power P_s is calculated using the following approaches:

$$P_s = 3\, V_s\, I_s \cos\varphi_s = 3 \times 398.4 \times 2705.7 \times -0.92 = -2.964 \text{ MW} \quad (-0.92 \text{ pu})$$

$$P_s = 1.5\,(v_{ds}\, i_{ds} + v_{qs}\, i_{qs}) = 1.5\,(0 + 516.4 \times -3826.4) = -2.964 \text{ MW} \quad (-0.92 \text{ pu}).$$

The PI controller that is employed in the outer DC control loop of the GSC maintains the net DC-bus voltage at its reference value (Figure 9.11(e)). By neglecting the losses in the power converters, the PMSG output power is equal to the grid-side power ($|\, P_s \,| = P_g$). The phase-a grid current and its reference current are shown in Figure 9.11(f). The grid currents track their references well even during the peaks and valleys of the waveform. The phase angle between the grid voltage and current is equal to zero because of the UPF operation. The grid current magnitude is calculated as follows:

$$i_g = \frac{P_g}{1.5\, v_g\, \cos\varphi_g} = \frac{2.964 \times 10^6}{1.5 \times 563.4 \times 1} = 3507.4 \text{ A } (peak) \quad (0.99 \text{ pu}).$$

The 2L-VSI line-to-line waveform is shown in Figure 9.11(g). The 2L-VSI operates with an average switching frequency $f_{sw,i}$ of 1700 Hz. The grid current THD is 7.13%. The generator and grid currents follow their respective references with a minimum reference tracking error to satisfy the overall control objectives of the PMSG WECS.

■

9.7 PREDICTIVE CURRENT CONTROL OF BTB-NPC-CONVERTER-BASED PMSG WECS

The concept of PCC is extended to BTB-NPC-converter-based IPMSG WECS in this section. In addition to the MPPT operation (through the regulation of stator currents), the DC-link capacitors voltage are balanced at half the net DC-bus voltage by applying the PCC scheme to the NPC rectifier and inverter [34].

9.7.1 Generator-side Control Scheme

The block diagram of the PCC scheme for the NPC rectifier is shown in Figure 9.12. This block diagram is similar to that discussed earlier for 2L-VSC (Figure 9.8), except that an additional "estimator" subsystem is used to meet the control objective of the DC-link neutral-point voltage control. A step-by-step procedure for the PMSG-side control scheme is presented below.

Measurement and Synthesis of Feedback Signals: The DC-link measurement includes two capacitors voltage. The grid-side currents and NPC inverter optimal switching states are used to estimate the DC-branch currents. All other measurements are similar to the control system employed for BTB 2L-VSC.

Calculation and Extrapolation of Reference Currents: The reference dq-axes currents $\mathbf{i}^*_{\mathbf{s,dq}}(k)$ are calculated by combining OTSR MPPT, speed control loop, and MTPA operation (Section 9.5.4). The reference currents are extrapolated to the $(k+1)$ sampling instant using the Lagrange extrapolation that is presented in (9.35).

Prediction of Future Behavior of PMSG Currents: The future behavior of PMSG currents must be predicted to satisfy the basic requirement of MPPT operation. The predictions for i_{ds} and i_{qs} with an NPC rectifier are obtained following the procedures presented below. Chapter 5 presents additional details about the NPC converter modeling.

- The NPC rectifier terminal voltages are predicted in terms of measured DC-link capacitors voltage and dq-frame switching signals:

$$\begin{bmatrix} v^p_{ds}(k) \\ v^p_{qs}(k) \end{bmatrix} = v_{C1}(k) \begin{bmatrix} s^p_{dr1}(k) \\ s^p_{qr1}(k) \end{bmatrix} + v_{C2}(k) \begin{bmatrix} s^p_{dr2}(k) \\ s^p_{qr2}(k) \end{bmatrix} \tag{9.46}$$

  where the dq-frame switching signals are related to the natural frame signals:

$$\begin{bmatrix} s^p_{drj}(k) \\ s^p_{qrj}(k) \end{bmatrix} = \begin{bmatrix} \mathbf{T}_{\alpha\beta/dq} \end{bmatrix} \begin{bmatrix} \mathbf{T}_{abc/\alpha\beta} \end{bmatrix} \begin{bmatrix} s^p_{arj}(k) \\ s^p_{brj}(k) \\ s^p_{crj}(k) \end{bmatrix}, \quad j \in \{1,2\}. \tag{9.47}$$

- By replacing the 2L-VSR model with the NPC rectifier model in (9.36), the sampled-data model for the dq-axes currents is computed as follows:

$$\begin{aligned} \begin{bmatrix} i^p_{ds}(k+1) \\ i^p_{qs}(k+1) \end{bmatrix} &= \mathbf{\Phi}(k) \begin{bmatrix} i_{ds}(k) \\ i_{qs}(k) \end{bmatrix} + \mathbf{\Gamma_w}(k) \\ &+ \mathbf{\Gamma_b} \left\{ v_{C1}(k) \begin{bmatrix} s^p_{dr1}(k) \\ s^p_{qr1}(k) \end{bmatrix} + v_{C2}(k) \begin{bmatrix} s^p_{dr2}(k) \\ s^p_{qr2}(k) \end{bmatrix} \right\}. \end{aligned} \tag{9.48}$$

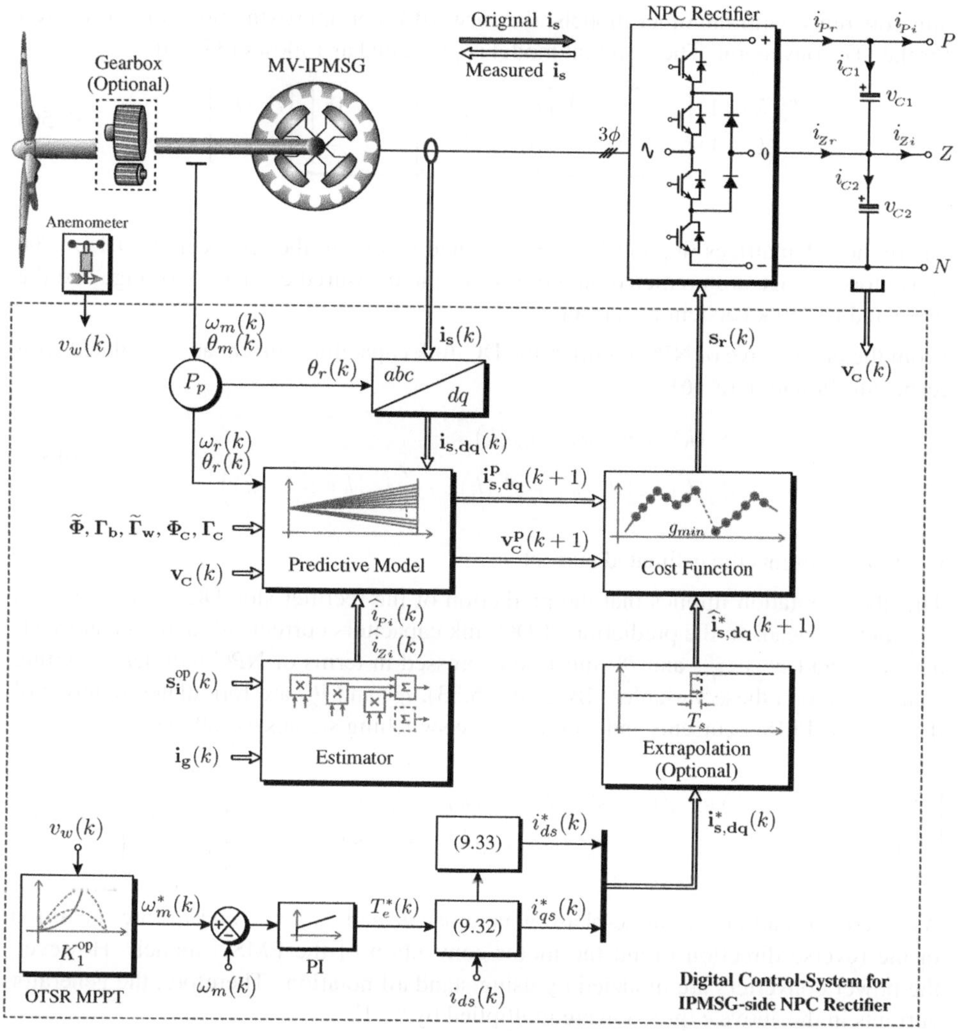

Figure 9.12 PCC scheme for the MTPA control of IPMSG with NPC rectifier.

Prediction of Future Behavior of Capacitors Voltage: The NPC rectifier and inverter employ cooperative switching actions to fulfill the control objective of the regulating DC-link capacitors voltage. The regulation of the DC-link capacitors voltage (v_{C1} and v_{C2}) from the perspective of NPC rectifier is presented as follows:

- From (5.35), the CT model of DC-link capacitors voltage is expressed in state-space form as follows:

$$\frac{d}{dt}\begin{bmatrix} v_{C1}(t) \\ v_{C2}(t) \end{bmatrix} = \underbrace{\begin{bmatrix} 0 & 0 \\ 0 & 0 \end{bmatrix}}_{\mathbf{A_C}} \begin{bmatrix} v_{C1}(t) \\ v_{C2}(t) \end{bmatrix} + \underbrace{\begin{bmatrix} \frac{1}{C_1} & 0 \\ 0 & \frac{1}{C_2} \end{bmatrix}}_{\mathbf{B_C}} \begin{bmatrix} i_{C1}(t) \\ i_{C2}(t) \end{bmatrix}. \quad (9.49)$$

- The above model implies that the CT model is linear time-invariant. Therefore, the exact discretization method can be employed to map the CT model to the DT model. To maintain coherence with the other models, the forward Euler approximation is reconsidered for the DT conversion. The resultant model is presented as follows [35, 36]:

$$\begin{bmatrix} v_{C1}^p(k+1) \\ v_{C2}^p(k+1) \end{bmatrix} = \underbrace{\begin{bmatrix} 1 & 0 \\ 0 & 1 \end{bmatrix}}_{\mathbf{\Phi_C}} \begin{bmatrix} v_{C1}(k) \\ v_{C2}(k) \end{bmatrix} + \underbrace{\begin{bmatrix} \frac{T_s}{C_1} & 0 \\ 0 & \frac{T_s}{C_2} \end{bmatrix}}_{\mathbf{\Gamma_C}} \begin{bmatrix} i_{C1}^p(k) \\ i_{C2}^p(k) \end{bmatrix} \tag{9.50}$$

where the DT matrices $\mathbf{\Phi_C}$ and $\mathbf{\Gamma_C}$ are computed similar to the approach in (9.15). The capacitors voltage are predicted on the basis of the measured capacitors voltage and the predicted DC-link capacitors current.

- From the perspective of NPC rectifier, the DC-link capacitors current are calculated with respect to Equation (5.36):

$$\begin{aligned} i_{C1}^p(k) &= i_{Pr}^p(k) - \widehat{i}_{Pi}(k), \\ i_{C2}^p(k) &= \left(i_{Pr}^p(k) + i_{Zr}^p(k)\right) - \left(\widehat{i}_{Pi}(k) + \widehat{i}_{Zi}(k)\right) \end{aligned} \tag{9.51}$$

where $\widehat{i}$ represents the estimated current.

- The above notation implies that the prediction of the rectifier-side DC-branch currents (i_{Pr}^p and i_{Zr}^p) leads to the prediction of DC-link capacitors current (i_{C1}^p and i_{C2}^p) and voltages (v_{C1}^p and v_{C2}^p). i_{Pr}^p and i_{Zr}^p must be expressed in terms of NPC rectifier switching states to predict these variables. By using (5.33), i_{Pr}^p and i_{Zr}^p are formulated in terms of three-phase PMSG currents and natural-frame switching signals as follows:

$$\begin{bmatrix} i_{Pr}^p(k) \\ i_{Zr}^p(k) \end{bmatrix} = \begin{bmatrix} s_{ar1}^p(k)\, s_{ar2}^p(k) & s_{br1}^p(k)\, s_{br2}^p(k) & s_{cr1}^p(k)\, s_{cr2}^p(k) \\ s_{ar2}^p(k)\, s_{ar3}^p(k) & s_{br2}^p(k)\, s_{br3}^p(k) & s_{cr2}^p(k)\, s_{cr3}^p(k) \end{bmatrix} \begin{bmatrix} -i_{as}(k) \\ -i_{bs}(k) \\ -i_{cs}(k) \end{bmatrix}. \tag{9.52}$$

As discussed earlier, the three-phase generator currents i_{as}, i_{bs}, and i_{cs} are measured in the reverse direction to aid the motor convention of the PMSG model. However, the power converters are modeled by using standard notation. Therefore, the generator currents in the above expression are multiplied by "−1".

- The DC-branch currents $\widehat{i}_{Pi}$ and $\widehat{i}_{Zi}$ can be measured to aid the model in (9.51). This procedure increases sensor costs and system complexity. Currents $\widehat{i}_{Pi}$ and $\widehat{i}_{Zi}$ can be estimated by using the grid-side currents and the NPC inverter optimal switching states (Equation (5.34)). The "estimator" subsystem in Figure 9.12 uses the following model:

$$\begin{bmatrix} \widehat{i}_{Pi}(k) \\ \widehat{i}_{Zi}(k) \end{bmatrix} = \begin{bmatrix} s_{ai1}^{\text{op}}(k)\, s_{ai2}^{\text{op}}(k) & s_{bi1}^{\text{op}}(k)\, s_{bi2}^{\text{op}}(k) & s_{ci1}^{\text{op}}(k)\, s_{ci2}^{\text{op}}(k) \\ s_{ai2}^{\text{op}}(k)\, s_{ai3}^{\text{op}}(k) & s_{bi2}^{\text{op}}(k)\, s_{bi3}^{\text{op}}(k) & s_{ci2}^{\text{op}}(k)\, s_{ci3}^{\text{op}}(k) \end{bmatrix} \begin{bmatrix} i_{ag}(k) \\ i_{bg}(k) \\ i_{cg}(k) \end{bmatrix}. \tag{9.53}$$

- As discussed in Chapter 5, 27 (3^3) switching combinations are available for an NPC converter. These 27 switching states lead to 27 predictions for the dq-axes converter output voltages (Equation (9.46)) and dq-axes stator currents (Equation (9.48)). The 27 switching states also produce 27 different values for the DC-branch currents i_{Pr}^p and i_{Zr}^p (Equation (9.52)). These currents, along with the estimated inverter branch currents, lead to 27 different possible values for i_{C1}^p and i_{C2}^p (Equation (9.51)). Consequently, the DC-link capacitors current lead to 27 predictions for v_{C1}^p and v_{C2}^p (Equation (9.50)).

Cost Function Minimization: The sub-cost functions that deal with the regulation of the dq-axis currents are similar to the one discussed earlier with 2L-VSC. The sub-cost function for balancing the DC-link capacitors voltage is expressed as follows:

$$g_{dc,r}(k) = \left[v_{C1}^{p}(k+1) - v_{C2}^{p}(k+1)\right]^2. \tag{9.54}$$

For the NPC rectifier, the sub-cost function for switching frequency minimization is defined in terms of upper-leg switching signals:

$$\begin{aligned} g_{sw,r}(k) = {} & \left[s_{ar1}^{p}(k) - s_{ar1}^{\mathrm{op}}(k)\right]^2 + \left[s_{ar2}^{p}(k) - s_{ar2}^{\mathrm{op}}(k)\right]^2 + \left[s_{br1}^{p}(k) - s_{br1}^{\mathrm{op}}(k)\right]^2 \\ & + \left[s_{br2}^{p}(k) - s_{br2}^{\mathrm{op}}(k)\right]^2 + \left[s_{cr1}^{p}(k) - s_{cr1}^{\mathrm{op}}(k)\right]^2 + \left[s_{cr2}^{p}(k) - s_{cr2}^{\mathrm{op}}(k)\right]^2. \end{aligned} \tag{9.55}$$

The global generator-side cost function combines the sub-cost functions through the weighting factors [34]:

$$g_r(k) = \lambda_{id}\, g_{id}(k) + \lambda_{iq}\, g_{iq}(k) + \lambda_{dc,r}\, g_{dc,r}(k) + \lambda_{sw,r}\, g_{sw,r}(k) \tag{9.56}$$

where λ_{id}, λ_{iq}, $\lambda_{dc,r}$, and $\lambda_{sw,r}$ are the weighting factors.

The weighting factor values specify the importance of DC-link capacitors voltage and switching frequency minimization over current control. The weighting factor $\lambda_{sw,r}$ is selected by using the heuristic approach. The weighting factors λ_{id}, λ_{iq}, and $\lambda_{dc,r}$ are calculated by the per-unit method. By considering the generator-rated *rms* current as the base value (I_{Bs}), the *approximate weighting factor values* are calculated as follows:

$$\lambda_{id} = \lambda_{iq} = \frac{I_{Bs}}{I_{Bs}} = 1, \qquad \lambda_{dc,r} = \frac{I_{Bs}}{v_{dc}^{*}}. \tag{9.57}$$

9.7.2 Grid-side Control Scheme

Figure 9.13 shows the block diagram of the grid-side PCC scheme for simultaneous grid current control and DC-link capacitors voltage control. The calculation and extrapolation of reference grid currents, as well as the prediction of the future behavior of grid currents, have been performed by using the guidelines in Chapter 8. The predictive model and cost function corresponding to the DC-link capacitors voltage are discussed as follows.

Prediction of Future Behavior of Capacitors Voltage: The grid-side NPC inverter collaborates with the NPC rectifier to balance the DC-link capacitors voltage, in addition to the basic requirement of grid currents regulation [36]. The prediction of DC-link capacitors voltage v_{C1} and v_{C2} from the perspective of NPC inverter is discussed in this section.

- Using the notation in (5.36), the DC-link capacitors current are calculated as follows:

$$\begin{aligned} i_{C1}^{p}(k) &= \widehat{i}_{Pr}(k) - i_{Pi}^{p}(k) \\ i_{C2}^{p}(k) &= \left(\widehat{i}_{Pr}(k) + \widehat{i}_{Zr}(k)\right) - \left(i_{Pi}^{p}(k) + i_{Zi}^{p}(k)\right). \end{aligned} \tag{9.58}$$

- The inverter-side DC branch currents i_{Pi}^{p} and i_{Zi}^{p} are formulated in terms of the three-phase grid currents and the NPC inverter switching signals as follows:

$$\begin{bmatrix} i_{Pi}^{p}(k) \\ i_{Zi}^{p}(k) \end{bmatrix} = \begin{bmatrix} s_{ai1}^{p}(k)\, s_{ai2}^{p}(k) & s_{bi1}^{p}(k)\, s_{bi2}^{p}(k) & s_{ci1}^{p}(k)\, s_{ci2}^{p}(k) \\ s_{ai2}^{p}(k)\, s_{ai3}^{p}(k) & s_{bi2}^{p}(k)\, s_{bi3}^{p}(k) & s_{ci2}^{p}(k)\, s_{ci3}^{p}(k) \end{bmatrix} \begin{bmatrix} i_{ag}(k) \\ i_{bg}(k) \\ i_{cg}(k) \end{bmatrix}. \tag{9.59}$$

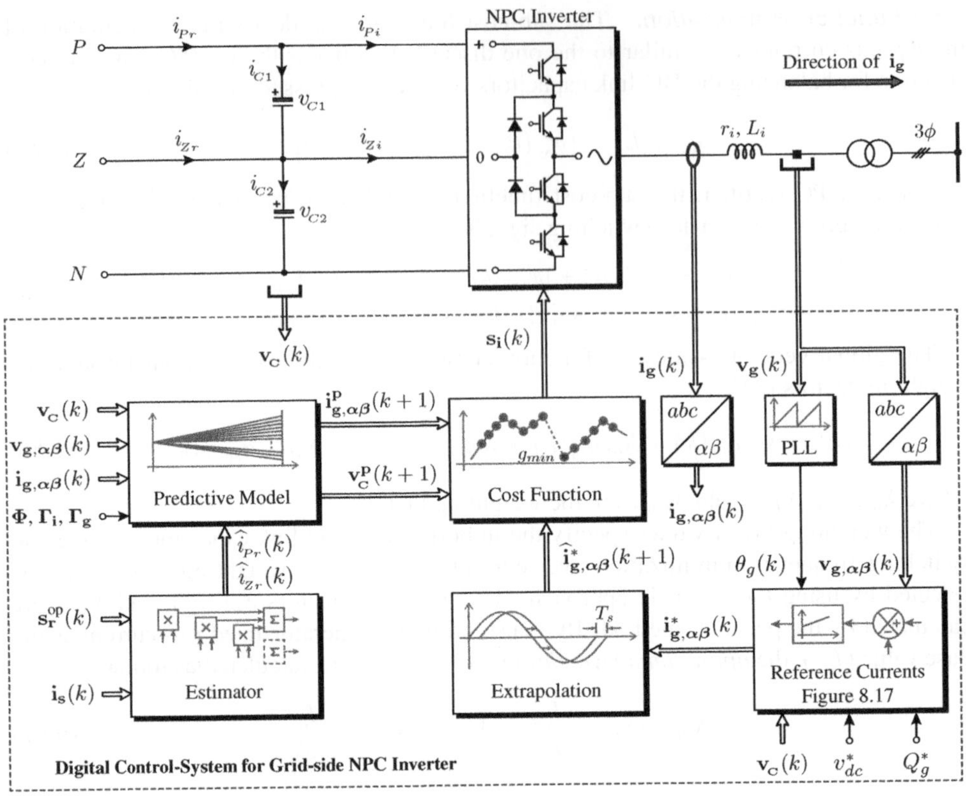

Figure 9.13 Block diagram of the PCC scheme for the grid-connected NPC inverter.

- The "estimator" subsystem in Figure 9.13 is similar to the one discussed with generator-side control system. This subsystem uses the following model to calculate $\widehat{i}_{Pr}$ and $\widehat{i}_{Zr}$:

$$\begin{bmatrix} \widehat{i}_{Pr}(k) \\ \widehat{i}_{Zr}(k) \end{bmatrix} = \begin{bmatrix} s^{op}_{ar1}(k)\, s^{op}_{ar2}(k) & s^{op}_{br1}(k)\, s^{op}_{br2}(k) & s^{op}_{cr1}(k)\, s^{op}_{cr2}(k) \\ s^{op}_{ar2}(k)\, s^{op}_{ar3}(k) & s^{op}_{br2}(k)\, s^{op}_{br3}(k) & s^{op}_{cr2}(k)\, s^{op}_{cr3}(k) \end{bmatrix} \begin{bmatrix} -i_{as}(k) \\ -i_{bs}(k) \\ -i_{cs}(k) \end{bmatrix} \tag{9.60}$$

- The 27 switching states in NPC inverter produces 27 different possible values for: (a) i^p_{Pi} and i^p_{Zi}, (b) i^p_{C1} and i^p_{C2}, (c) v^p_{C1} and v^p_{C2}, and (d) $i^p_{\alpha g}$ and $i^p_{\beta g}$.

Cost Function Minimization: The global grid-side cost function with current control, capacitors voltage control, and switching frequency minimization is given as follows:

$$g_i(k) = \lambda_{i\alpha}\, g_{i\alpha}(k) + \lambda_{i\beta}\, g_{i\beta}(k) + \lambda_{dc,i}\, g_{dc,i}(k) + \lambda_{sw,i}\, g_{sw,i}(k) \tag{9.61}$$

where $\lambda_{i\alpha}$, $\lambda_{i\beta}$, $\lambda_{dc,i}$, and $\lambda_{sw,i}$ are the weighting factors. The cost functions $g_{i\alpha}$ and $g_{i\beta}$ calculate the current error values in the $\alpha\beta$-frame. The cost functions $g_{dc,i}$ and $g_{sw,i}$ are defined similar to (9.54) and (9.55), respectively.

The grid-side weighting factors are calculated as follows:

$$\lambda_{i\alpha} = \lambda_{i\beta} = \frac{I_{Bg}}{I_{Bg}} = 1, \qquad \lambda_{dc,i} = \frac{I_{Bg}}{v^*_{dc}}. \tag{9.62}$$

9.7.3 Control Algorithm

Figures 9.14(a) and 9.14(b) show the generator- and grid-side PCC algorithms, respectively. These algorithms are identical to the 2L-VSC PCC algorithms in Figure 9.9. In block ②, the DC-branch currents and weighting factors are calculated, in addition to the reference currents $\widehat{\mathbf{i}}^*_{\mathbf{s,dq}}(k+1)$ and $\widehat{\mathbf{i}}^*_{\mathbf{g},\alpha\beta}(k+1)$. In block ⑤, the DC-link capacitors current ($\mathbf{i}^{\mathrm{P}}_{\mathrm{C}}$) are predicted, in addition to the NPC converter output voltages ($\mathbf{v}^{\mathrm{P}}_{\mathbf{s,dq}}$ and $\mathbf{v}^{\mathrm{P}}_{\mathbf{i},\alpha\beta}$). The generator/grid currents and DC-link capacitors voltage are predicted in block ⑥. The cost function values are calculated in block ⑦. The iteration is performed 27 times during each sampling interval. The optimal vector that provides the optimal control of generator/grid currents and DC-link capacitors voltage is selected; the gating signals corresponding to this vector are directly applied to the NPC rectifier and inverter.

Algorithm 9.2 for the NPC rectifier PCC scheme shows the *S-Function Builder* programming for implementation by using `MATLAB` software. The code for switching frequency minimization is omitted to simplify the analysis. The external function declarations for *S-Function* are similar to Algorithm 9.1 The program line description is given as follows: Lines `05` to `08`, prediction of the NPC rectifier voltages; lines `09` to `10`, prediction of the PMSG currents; lines `11` to `12`, prediction of the DC-branch currents; lines `13` to `14`, prediction of the DC-link capacitors current; lines `15` to `16`, prediction of the DC-link capacitors voltage; lines `17` to `20`, calculation of cost function values; lines `21` to `25`, selection of the optimal vector `j_op` number; and lines `27` to `32`, selection of the optimal switching signals based on the `j_op` number.

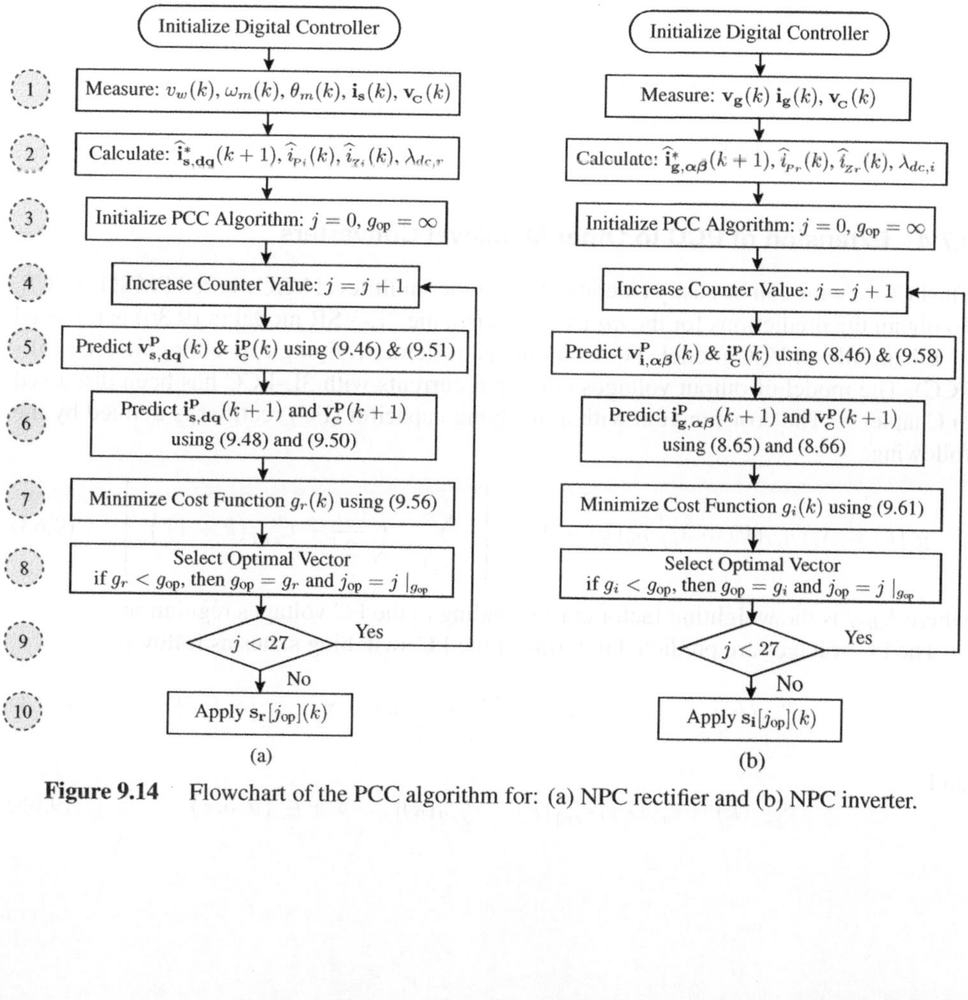

Figure 9.14 Flowchart of the PCC algorithm for: (a) NPC rectifier and (b) NPC inverter.

ALGORITHM 9.2 PCC Algorithm for NPC Rectifier-Based PMSG WECS

```
j_op = 1000;
g_op = 1000000000;
for(j = 0; j < 27; j++)
{
    vr_ab[0] = vc[0]*vrec[j].s_r_ab[0] + vc[1]*vrec[j].s_r_ab[1];
    vr_ab[1] = vc[0]*vrec[j].s_r_ab[2] + vc[1]*vrec[j].s_r_ab[3];
    vr_dq[0] =  vr_ab[0]*cstheta_r[0] + vr_ab[1]*cstheta_r[1];
    vr_dq[1] = -vr_ab[0]*cstheta_r[1] + vr_ab[1]*cstheta_r[0];
    is_dq_k1[0] = P11*is_dq[0] + P12*wr[0]*is_dq[1] + G11*vr_dq[0];
    is_dq_k1[1] = P21*wr[0]*is_dq[0] + P22*is_dq[1]
                  + G22*vr_dq[1] + W21*wr[0];
    ipzr_k1[0] = - vrec[j].s_r[0]*vrec[j].s_r[1]*is[0]
                   - vrec[j].s_r[2]*vrec[j].s_r[3]*is[1]
                   - vrec[j].s_r[4]*vrec[j].s_r[5]*is[2];
    ipzr_k1[1] = - !vrec[j].s_r[0]*vrec[j].s_r[1]*is[0]
                  - !vrec[j].s_r[2]*vrec[j].s_r[3]*is[1]
                  - !vrec[j].s_r[4]*vrec[j].s_r[5]*is[2];
    ic_k1[0] = ipzr_k1[0] - ipzi[0];
    ic_k1[1] = ipzr_k1[0] + ipzr_k1[1] - ipzi[0] - ipzi[1];
    vc_k1[0] = vc[0] + (Ts/C1)*ic_k1[0];
    vc_k1[1] = vc[1] + (Ts/C2)*ic_k1[1];
    g_id = (is_ref_k1[0]-is_dq_k1[0])*(is_ref_k1[0]-is_dq_k1[0]);
    g_id = (is_ref_k1[1]-is_dq_k1[1])*(is_ref_k1[1]-is_dq_k1[1]);
    g_dc = (vc_k1[0]-vc_k1[1])*(vc_k1[0]-vc_k1[1]);
    g_r = g_id + g_iq + Lambda_dcr*g_dc;
    if(g_r < g_op)
        {
          j_op = j;
          g_op = g_r;
        }
}
s_ar1[0] = vrec[j_op].s_r[0];
s_ar2[0] = vrec[j_op].s_r[1];
s_br1[0] = vrec[j_op].s_r[2];
s_br2[0] = vrec[j_op].s_r[3];
s_cr1[0] = vrec[j_op].s_r[4];
s_cr2[0] = vrec[j_op].s_r[5];
```

9.7.4 Extension of PCC to Other Multilevel Converters

The PCC scheme can be easily extended to the other multilevel VSCs-based PMSG WECS. To obtain the predictions for the dq-axis currents, the 2L-VSR model in (9.36) is replaced by the multilevel VSC model. For example, consider a 3L flying capacitor converter (3L-FCC). The model of output voltages and input currents with 3L-FCC has been discussed in Chapter 5. The cost function with a 3L flying capacitor (FC) rectifier is defined by the following:

$$g_r(k) = \lambda_{id}\, g_{id}(k) + \lambda_{iq}\, g_{iq}(k) + \lambda_{dc,r} \left[\sum_{x=a,\, b,\, c} \left(\frac{v_{dc}^*}{2} - v_{xrf}^p(k+1) \right)^2 \right] \tag{9.63}$$

where $\lambda_{dc,r}$ is the weighting factor corresponding to the FC voltages regulation.

The FC voltages are predicted in terms of the FC switching states as follows:

$$v_{xrf}^p(k+1) = v_{xrf}(k) + \frac{T_s}{C_{xrf}}\, i_{xrf}^p(k), \quad \forall\ x \in \{a, b, c\} \tag{9.64}$$

and

$$i_{xrf}^p(k) = i_{xs}(k)\left[s_{xr1}^p(k) - s_{xr2}^p(k)\right], \quad \forall\ x \in \{a, b, c\}. \tag{9.65}$$

CASE STUDY 9.4 Startup Transient Analysis of IPMSG WECS

Case Study Objective: The startup transient analysis of IPMSG WECS with the BTB NPC converter and PCC scheme is studied in this case study. The performance of the PMSG-side, DC-link, and grid-side control variables during a ramp wind speed profile are likewise analyzed.

Parameters: The rated parameters of a 3.0-MW, 3000-V, 9.75-Hz DD generator are as follows: R_s = 30.52 mΩ, L_{ds} = 17.68 mH, L_{qs} = 27.79 mH, ψ_r = 36.06 Wb (*peak*), P_p = 26, and ω_r = 61.26 rad/s (IPMSG #3 in Table A.3). The NPC inverter operates with UPF (cos φ_g = 1).

Analysis:

Figure 9.15 shows the startup transient waveforms of IPMSG WECS with the BTB NPC converter. The reference generator-side control variables are calculated on the basis of Case Study 9.1. During the first phase of the startup procedure (t = 0 to 0.2 s), the inverter is synchronized to the grid by regulating the net DC-bus voltage at its reference value (Figure 9.15(f)) and the grid reactive power (Q_g) at zero (Figure 9.15(j)). The dq-axes grid currents corresponding to v_{dc} and Q_g are approximately zero (Figure 9.15(h)). Once the inverter is tied to the three-phase grid, WT drives PMSG to produce electric power. The NPC rectifier converts variable voltage and frequency to the DC voltage, and the NPC inverter feeds the power to the three-phase grid.

In practice, the yaw position is adjusted such that the WT blades are subjected to a maximum-possible wind speed. The mechanical brakes are released such that the WT speed increases gradually to the rated speed. To simplify the analysis and to avoid the WT dynamics, the wind speed profile is defined such that it closely represents the WT startup. From t = 0.2 to 1.2 s, the wind speed v_w increases from 3 m/s (0.25 pu) to 12 m/s (1.0 pu). OTSR MPPT produces the reference speed ω_m^* on the basis of the measured wind speed and coefficient K_1^{op}. Reference speed ω_m^* increases linearly with respect to v_w and reaches its rated value at t = 1.2 s. The PI controller employed in the speed control loop regulates the PMSG speed at its reference value (Figure 9.15(a)). The PI controller also produces reference torque T_e^*, and the power converters adjust the switching actions such that the PMSG torque T_e follows its reference T_e^* and then ramps up to the rated value (Figure 9.15(b)). The difference between mechanical input torque T_m and T_e aids the PMSG to accelerate the rotor speed ω_m. In steady-state, $\omega_m^* = \omega_m$; thus, T_e becomes equal to T_m.

The magnitude and frequency of the PMSG stator currents increase with respect to the rotor speed (Figure 9.15(c)). The THD of the PMSG stator current is noted as 0.48%. The d-axis current is measured as –195.7 A (*peak*, –0.22 $\widehat{\text{pu}}$). The q-axis current i_{qs} is proportional to the PMSG torque T_e. The magnitude of the q-axis current is calculated by the following:

$$i_{qs} = \frac{T_e}{1.5\,P_p\,[\psi_r + (L_{ds} - L_{qs})\,i_{ds}]}$$

$$= \frac{-1273 \times 10^3}{1.5 \times 26 \times [36.06 + (17.68 - 27.79) \times 10^{-3} \times -195.7]} = -858.3\text{ A} \quad (-0.98\ \widehat{\text{pu}})$$

$$i_s = \sqrt{i_{ds}^2 + i_{qs}^2} = \sqrt{(-195.7)^2 + (-858.3)^2} = 880\text{ A}\,(\textit{peak}) \quad (1.0\ \widehat{\text{pu}}).$$

The dq-axes stator voltages are noted as follows: v_{ds} = 980 V (*peak*, 0.4 $\widehat{\text{pu}}$) and v_{qs} = 2254 V (*peak*, 0.92 $\widehat{\text{pu}}$). The rectifier line-to-line voltage $v_{r,ab}$ waveform shows three and five distinct voltage levels during low and high wind speeds, respectively. At low wind speeds, the NPC rectifier operates as a 2L converter. At high wind speeds, the NPC rectifier operates as a 3L converter. The switching frequency of the NPC rectifier $f_{sw,r}$ is noted as 1260 Hz. The PMSG active, reactive, and apparent output powers at the rated wind speed are expressed as follows:

$$P_s = 1.5\,(v_{ds}\,i_{ds} + v_{qs}\,i_{qs}) = -2.964\text{ MW} \quad (-0.92\text{ pu})$$

$$Q_s = 1.5\,(v_{qs}\,i_{ds} - v_{ds}\,i_{qs}) = 1.294\text{ MVAR} \quad (0.4\text{ pu})$$

$$S_s = \sqrt{P_s^2 + Q_s^2} = 3.234\text{ MVA} \quad (1.0\text{ pu}).$$

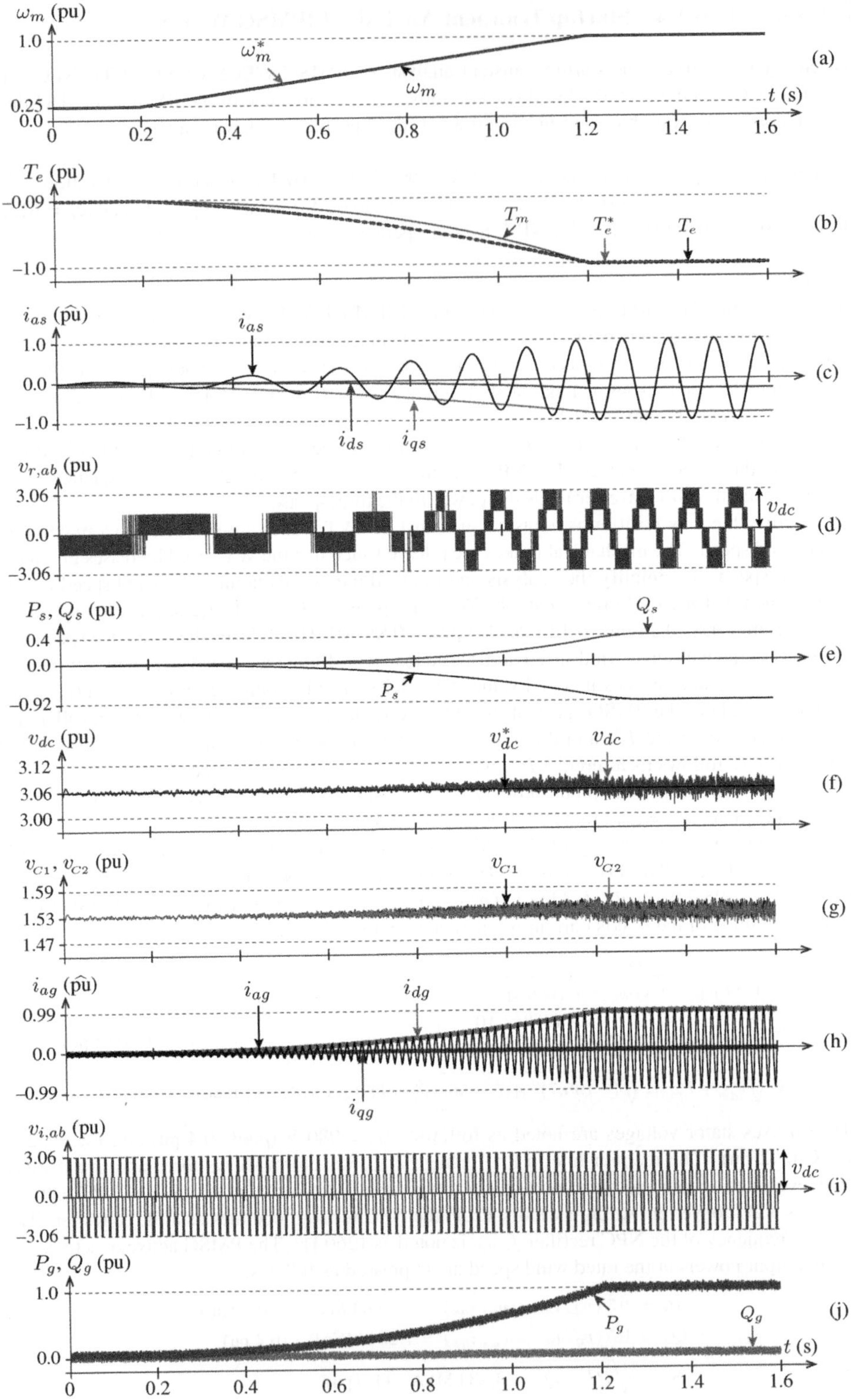

Figure 9.15 Simulated waveforms with the PCC scheme for IPMSG WECS during startup.

The rated PMSG PF is –0.92 (P_s/S_s). The net DC-bus voltage v_{dc} is maintained constant at its reference value of 3.062 pu by a PI controller employed in the DC control loop (Figure 9.15(f)). The ripple in v_{dc} increases with respect to the generator and grid current magnitude. The DC-link capacitors voltage are well-regulated by the PCC schemes employed for the NPC rectifier and inverter (Figure 9.15(g)). The weighting factors corresponding to the DC capacitors voltage control are calculated for the generator- and grid-side PCC schemes as follows:

$$\lambda_{dc,r} = \frac{I_{Bs}}{v_{dc}^*} = \frac{622}{5304} = 0.1174$$
$$\lambda_{dc,i} = \frac{I_{Bg}}{v_{dc}^*} = \frac{577}{5304} = 0.1089.$$

The grid current magnitude increases proportionally to the PMSG active power output. The d-axis grid current follows the grid current waveform, whereas the q-axis current is maintained zero because of the grid-side UPF operation [13]. The grid current THD is noted as 4.5%. With the grid-tied operation, the NPC inverter contains five discrete voltage levels during all operating conditions. During the rated condition, the switching frequency of the NPC inverter $f_{sw,i}$ is noted as 1200 Hz. The grid active power P_g follows the P_s spectrum, and the grid reactive power Q_g is maintained zero by the grid-side PCC scheme (Figure 9.15(j)). As discussed in Chapter 8, GSC can operate with unity, leading or lagging PF by changing Q_g^*.

The weighting factors designed earlier ensure the balancing of the DC-link capacitors voltage during all operating conditions. Figure 9.16 shows the performance of IPMSG WECS with the BTB NPC converter during a step change in wind speed. The wind speed is considered to change from 12 m/s (1.0 pu) to 9 m/s (0.75 pu) at $t = 0.1$ s and to 6 m/s (0.5 pu) at $t = 0.2$ s. The PI controller in the GSC DC control loop exhibits transient oscillations for the net DC-bus voltage v_{dc} because of the step change in wind speed. However, the DC-link capacitors voltage are regulated at equal voltages, including transient interval (Figure 9.16(b)). The ripple in the DC-link capacitors voltage decreases with respect to the generator and grid current magnitude. The PMSG and grid currents track to their references during both steady state and transient intervals.

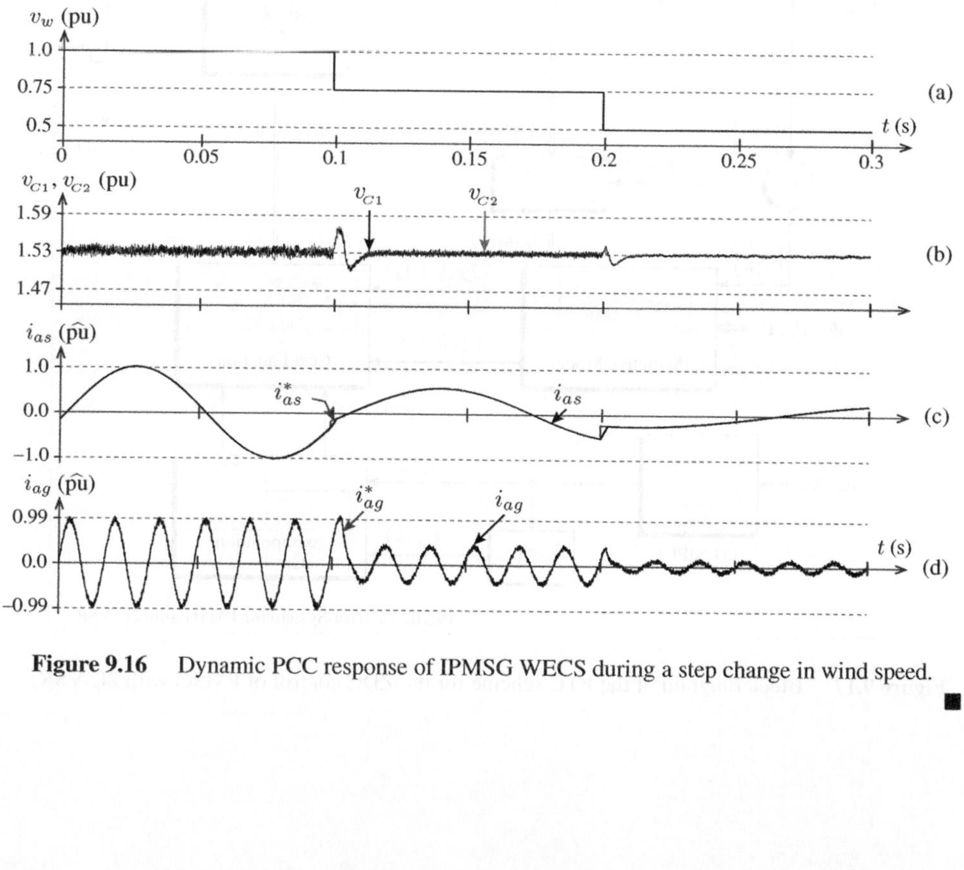

Figure 9.16 Dynamic PCC response of IPMSG WECS during a step change in wind speed.

■

9.8 PREDICTIVE TORQUE CONTROL OF BTB 2L-VSC-BASED PMSG WECS

DTC provides good dynamic response, robustness against generator parameter variations, and low complexity in real-time implementation compared with FOC. The PTC scheme is analogous to the classical DTC to achieve variable-speed operations for PMSG WECS. The PTC scheme eliminates the hysteresis regulators and lookup table used in the classical DTC [31, 33]. In the PTC structure, the PMSG electromagnetic torque and stator flux are controlled directly with rapid dynamic response, thus leading to the indirect control of generator speed and stator currents [7, 37].

9.8.1 Generator-side Control Scheme

Figure 9.17 shows the block diagram for the PTC scheme for the 2L-VSR-based SPMSG WECS. The block diagram configuration and the measurement and synthesis of feedback signals is similar to the PCC of the 2L-VSR-based PMSG WECS given in Figure 9.8. The reference torque, T_e^* and stator flux ψ_s^* are calculated based on OT MPPT and ZDC control (Section 9.5.4). The torque and flux are extrapolated to the $(k+1)$ sampling instant similar to the dq-axes reference currents in the PCC scheme. The Lagrange extrapolation model in (9.35) is used to calculate $\widehat{T}_e^*(k+1)$ and $\widehat{\psi}_s^*(k+1)$. The predictive model and cost function design with the PTC scheme are discussed as follows.

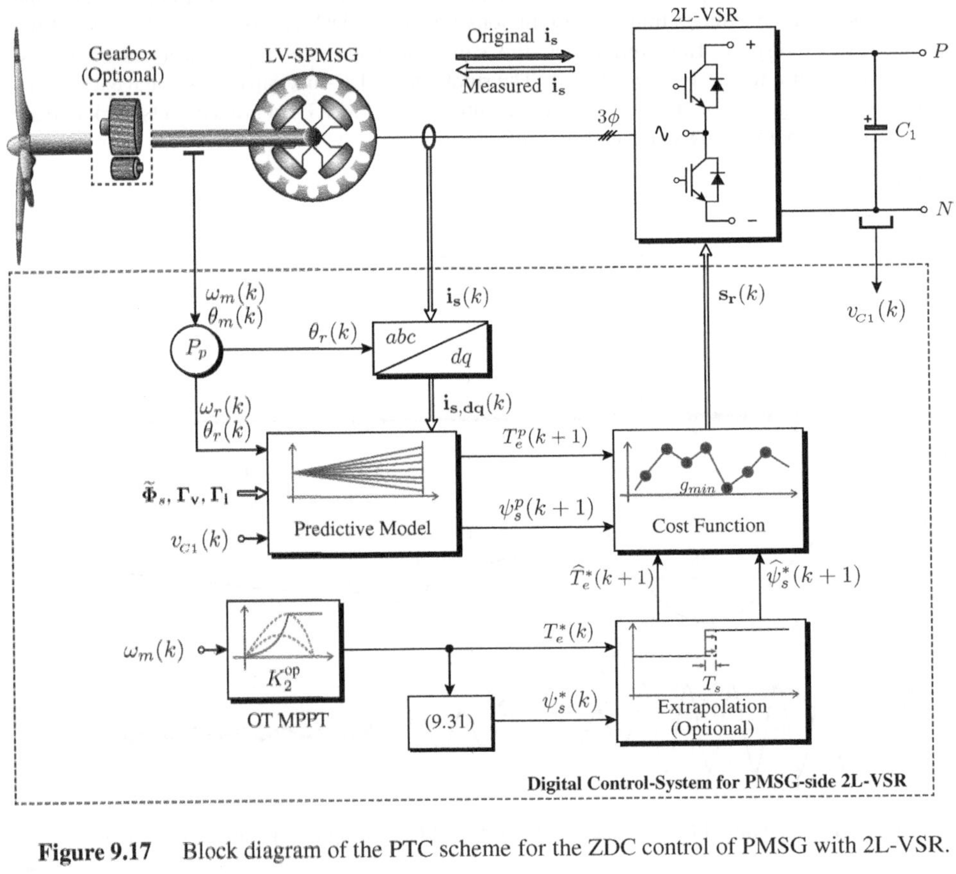

Figure 9.17 Block diagram of the PTC scheme for the ZDC control of PMSG with 2L-VSR.

Prediction of Future Behavior of PMSG Torque and Flux: The step-by-step procedure to predict the future values of stator flux and electromagnetic torque is provided as follows:

- The DT model in (9.16) is used to predict the future values of the dq-axes stator flux linkages, that is,

$$\begin{bmatrix} \psi^p_{ds}(k+1) \\ \psi^p_{qs}(k+1) \end{bmatrix} = \mathbf{\Phi_s}(k) \begin{bmatrix} \psi_{ds}(k) \\ \psi_{qs}(k) \end{bmatrix} + \mathbf{\Gamma_v} \underbrace{\begin{bmatrix} v^p_{ds}(k) \\ v^p_{qs}(k) \end{bmatrix}}_{\text{2L-VSR Model}} + \mathbf{\Gamma_i} \begin{bmatrix} i_{ds}(k) \\ i_{qs}(k) \end{bmatrix}. \tag{9.66}$$

  The dq-axes stator flux linkages $\psi_{ds}(k)$ and $\psi_{qs}(k)$ are calculated on the basis of dq-axes stator currents $i_{ds}(k)$ and $i_{qs}(k)$ and the PMSG steady-state model in (9.1). The DT matrices $\mathbf{\Phi_s}(k)$, $\mathbf{\Gamma_v}$, and $\mathbf{\Gamma_i}$ are defined in (9.17) with respect to the PMSG parameters. The 2L-VSR voltages are predicted by using the DT model in (9.37).

- The model in (9.66) is used to calculate the peak value of the stator flux linkage:

$$\psi^p_s(k+1) = \sqrt{[\psi^p_{ds}(k+1)]^2 + [\psi^p_{qs}(k+1)]^2}. \tag{9.67}$$

- The steady-state model indicates the relationship of the electromagnetic torque to the dq-axis stator flux linkages and stator currents (refer to (9.7)). By moving the variables one-sample forward, the predictive model of electromagnetic torque is obtained as follows:

$$T^p_e(k+1) = 1.5\, P_p \left[\psi^p_{ds}(k+1)\, i^p_{qs}(k+1) - \psi^p_{qs}(k+1)\, i^p_{ds}(k+1)\right]. \tag{9.68}$$

  From the preceding model, $T^p_e(k+1)$ is assessed on the basis of the future values of the stator flux linkages and stator currents. The stator flux model in (9.66) and stator currents model in (9.39) are used to obtain predictions for T_e.

Cost Function Minimization: The simultaneous control of the PMSG electromagnetic torque and stator flux is defined by a cost function:

$$g_r(k) = \lambda_T \left[\widehat{T}^*_e(k+1) - T^p_e(k+1)\right]^2 + \lambda_\psi \left[\widehat{\psi}^*_s(k+1) - \psi^p_s(k+1)\right]^2 \tag{9.69}$$

where λ_T and λ_ψ are the weighting factors for torque and flux control, respectively. Similar to the NPC rectifier cost function in (9.56), which involves the regulation of currents and capacitors voltage, the PTC cost function comprises two terms. The weighting factors are designed as follows by using the per-unit method discussed in Chapter 4.

$$\lambda_T = \frac{T_{e,R}}{T_{e,R}} = 1, \qquad \lambda_\psi = \frac{T_{e,R}}{\psi_{s,R}} \tag{9.70}$$

where $T_{e,R}$ and $\psi_{s,R}$ are the rated values of torque and stator flux, respectively. The multiobjective ranking algorithm can also be employed to eliminate the weighting factors employed in the PTC scheme. In contrast to the classical DTC method, which involves hysteresis controllers and lookup tables, the PTC method discussed here is intuitive and uses only generator and converter model to simultaneously control T_e and ψ_s.

The eight possible switching state combinations are used to predict the following: (1) 2L-VSR voltages $v^p_{ds}(k+1)$ and $v^p_{qs}(k+1)$, (2) dq-axes stator currents $i^p_{ds}(k+1)$ and $i^p_{qs}(k+1)$, (3) stator flux linkage $\psi^p_s(k+1)$, and (4) electromagnetic torque $T^p_e(k+1)$. The selection of optimal switching signals leads to the decoupled control of both the electromagnetic torque and stator flux.

9.8.2 Control Algorithm

Figure 9.18 shows the flowchart for the realization of the PTC algorithm for the 2L-VSR-based SPMSG WECS. The implementation procedure of PTC is similar to the PCC algorithm in Figure 9.9(a), except that blocks②, ⑥, and ⑦ are different. In block ②, the reference torque and flux ($\widehat{T}_e^*(k+1)$ and $\widehat{\psi}_s^*(k+1)$) are calculated replacing the reference currents ($\widehat{i}_{ds}^*$ and $\widehat{i}_{qs}^*$) in the PCC algorithm. In block ⑤, the 2L-VSR voltages are predicted similar to the PCC algorithm. Both torque and flux are predicted in block ⑥; this prediction replaces i_{ds}^p and i_{qs}^p prediction in the PCC flowchart. The cost function in block ⑦ calculates the error in the torque and flux, whereas the PCC algorithm calculates the error in the dq-axis currents. The optimal switching signals for 2L-VSR are obtained in block ⑩ on the basis of the j_{op} value that shows a minimum cost function value.

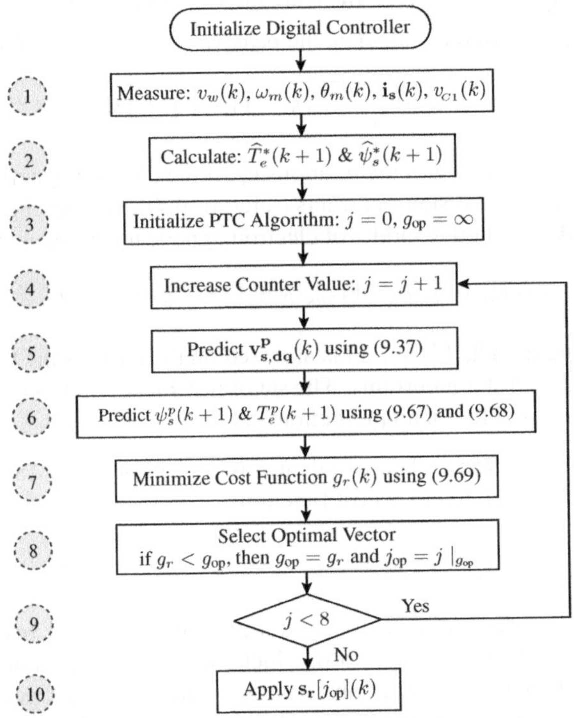

Figure 9.18 Flowchart of the PTC algorithm for the 2L-VSR-based PMSG WECS.

9.8.3 Extension of PTC to BTB NPC Converter

The PTC scheme can be easily extended to IPMSG WECS by just changing the calculation of the reference stator flux in Figure 9.17. The PTC scheme can also be applied to the NPC-converter-based PMSG WECS by simply changing the VSR model in (9.36) and (9.66). The error between the predicted DC-link capacitors voltage is also included in the cost function. The model of the NPC rectifier results in new prediction values for $i_{ds}^p(k+1)$, $i_{qs}^p(k+1)$, $\psi_{ds}^p(k+1)$, and $\psi_{qs}^p(k+1)$. Consequently, these values produce a high number of predictions for $\psi_s^p(k+1)$ and $T_e^p(k+1)$. Developing complicated lookup tables and space vector analysis is not required to realize PTC for the NPC converters.

CASE STUDY 9.5 Dynamic Response of SPMSG WECS with the PTC Scheme

Case Study Objective: The objective of this case study is to investigate the dynamic response of PMSG WECS with the PTC scheme during a step change in wind speed. The behavior of the PMSG torque and flux is studied, in addition to mechanical speed, stator currents, stator output active power, and grid currents. BTB 2L-VSC is considered to realize the power conversion for PMSG WECS.

Parameters: The rated parameters of a medium-speed (400 rpm), 750-kW, 690-V, 40-Hz generator are as follows: R_s = 6.52 mΩ, $L_{ds} = L_{qs}$ = 0.938 mH, ψ_r = 2.08 Wb (*peak*), P_p = 6, ω_m = 41.89 rad/s, and ω_r = 251.3 rad/s (SPMSG #4 in Table A.4). The original moment of inertia J_m is reduced to 14.73 kg.m^2 to reduce the simulation runtime.

Analysis:

Figure 9.19 shows the dynamic response of SPMSG WECS with the PTC scheme and 2L-VSC. A step change in wind speed from 8.4 m/s (0.7 pu) to 12 m/s (1.0 pu) is applied at t = 0.2 s. The PMSG rotor speed, electromagnetic torque, stator currents, and stator output power during a 0.7-pu wind speed are calculated as follows:

$$T_e = T_{e,R} \times v_{w,\,\text{pu}}^2 = -17.91 \times 10^3 \times 0.7^2 = -8.78 \text{ kN.m} \quad (-0.49 \text{ pu})$$

$$\psi_s = \sqrt{\psi_r^2 + \left[\frac{L_{qs}\,T_e}{1.5\,P_p\,\psi_r}\right]^2} = \sqrt{2.08^2 + \left[\frac{0.938 \times 10^{-3} \times 8780}{1.5 \times 6 \times 2.08}\right]^2} = 2.126 \text{ Wb} \quad (0.95\ \widehat{\text{pu}})$$

$$i_{qs} = \frac{T_e}{1.5\,P_p\,\psi_r} = \frac{-8.78 \times 10^3}{1.5 \times 6 \times 2.08} = -468.7 \text{ A} \quad (-0.49\ \widehat{\text{pu}})$$

$$I_s = \sqrt{i_{ds}^2 + i_{qs}^2}/\sqrt{2} = \sqrt{0^2 + (-468.7)^2}/\sqrt{2} = 331.5 \text{ A } (rms) \quad (0.49 \text{ pu})$$

$$P_m = P_{m,R} \times v_{w,\,\text{pu}}^3 = -750 \times 10^3 \times 0.7^3 = -257.3 \text{ kW} \quad (-0.34 \text{ pu})$$

$$P_{cu,s} = 3\,I_s^2\,R_s = 3 \times 331.5^2 \times 6.52 \times 10^{-3} = 2.15 \text{ kW}$$

$$P_s = P_m + P_{cu,s} = -257.3 + 2.15 = -255.1 \text{ kW} \quad (-0.32 \text{ pu}).$$

With step change in wind speed, the corresponding ω_m^* changes from 0.7 pu to 1.0 pu. High wind speed causes the ω_m to increase to the rated value (Figure 9.19(a)). In high-power WECS, the dynamic response of ω_m is considerably slow because of the large moment of inertia. With the step change in wind speed from 0.7 pu to 1.0 pu, the T_m increases immediately from –0.49 (–0.7^2) pu to –1.0 pu. The difference between T_m and T_e causes the rotor sped ω_m to accelerate to the new operating condition. As ω_m increases, OT MPPT produces new T_e^* reference for the PTC scheme. The optimal switching signals produced by the PTC scheme ensures that T_e follows T_e^* closely with less reference tracking error (Figure 9.19(b)). As the electromagnetic torque increases, the stator flux magnitude ψ_s increases from 0.95 $\widehat{\text{pu}}$ to 1.01 $\widehat{\text{pu}}$ (Figure 9.19(c)). The PMSG *dq*-axes currents are internally controlled by the PTC scheme. The *d*-axis current is maintained at zero, and the *q*-axis current follows the T_e waveform (Figure 9.19(d)). The stator active power output P_s increases from –0.32 pu to –0.92 pu (Figure 9.19(e)). The grid current magnitude increases from 0.485 pu to 0.99 pu (Figure 9.19(f)).

Figure 9.20 shows the dynamic response of the PTC scheme with step change in T_e^*. A step change in T_e^* is possible for high-speed or low-power PMSG with low value of moment of inertia J_m. Wind speed changes from 0.7 pu to 1.0 pu at t = 0.1 s and from 1.0 pu to 0.5 pu at t = 0.2 s. The corresponding T_e^* changes from –0.49 pu to –1.0 pu at t = 0.1 s, and from –1.0 pu to –0.25 pu at t = 0.2 s. The stator flux ψ_s corresponds to –0.49 pu, –1.0 pu, and –0.25 pu torque are 2.126 (0.95 $\widehat{\text{pu}}$), 2.265 (1.01 $\widehat{\text{pu}}$) and 2.092 (0.93 $\widehat{\text{pu}}$), respectively. The generator current magnitude changes in proportion to the torque. The waveforms for the PMSG electromagnetic torque, stator flux, and stator current validate the rapid dynamic response and decoupled control of the PTC scheme. The results presented through this case study demonstrate that both the PCC and PTC schemes achieve similar control performance for PMSG WECS.

■

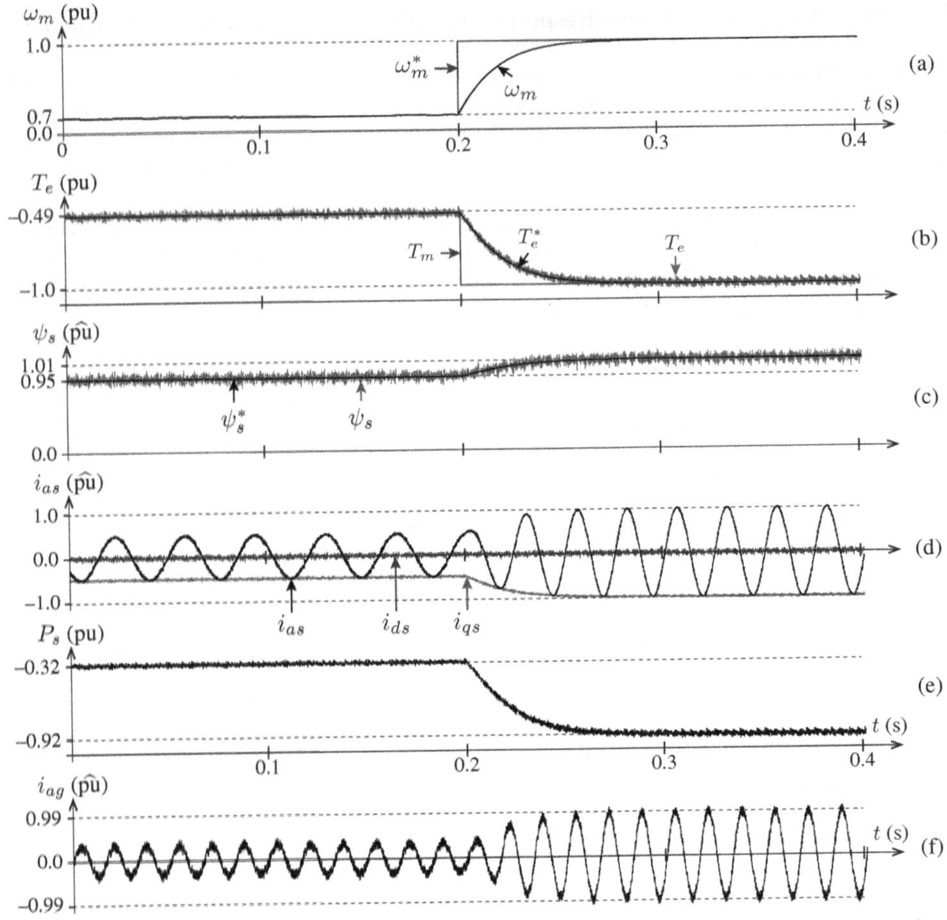

Figure 9.19 Dynamic PTC response of SPMSG WECS during the step change in wind speed.

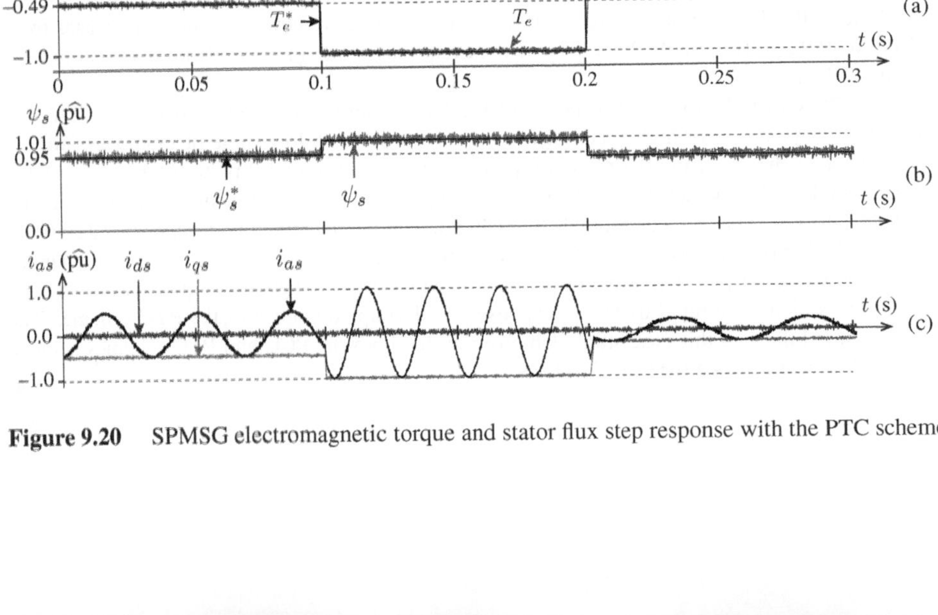

Figure 9.20 SPMSG electromagnetic torque and stator flux step response with the PTC scheme.

9.9 OTHER MPC SCHEMES FOR PMSG WECS

The control objectives of PMSG WECS can be achieved by regulating the stator currents, torque and flux, active and reactive powers, and rotor speed. The modeling and design of the PCC and PTC algorithms can be easily extended to other MPC designs as outlined below.

9.9.1 Predictive Power Control

PPC was previously analyzed in Chapter 8 for the decoupled active and reactive power of GSCs. This strategy can be extended to PMSG WECS, where the generator active and reactive powers are controlled directly instead of the dq-axis currents or torque/flux [38]. Similar to the control of the dq-axes currents, the importance of active and reactive control is the same. The cost function for the PPC scheme is defined by the following:

$$g_r(k) = \left[\widehat{P}_s^*(k+1) - P_s^p(k+1) \right]^2 + \left[\widehat{Q}_s^*(k+1) - Q_s^p(k+1) \right]^2 \tag{9.71}$$

where $\widehat{P}_s^*(k+1)$ and $\widehat{Q}_s^*(k+1)$ are extrapolated reference active and reactive powers, respectively. The references for the PPC scheme are calculated as follows:

$$\begin{aligned} P_s^*(k) &= 1.5 \left[v_{ds}(k)\, i_{ds}^*(k) + v_{qs}(k)\, i_{qs}^*(k) \right] \\ Q_s^*(k) &= 1.5 \left[v_{qs}(k)\, i_{ds}^*(k) - v_{ds}(k)\, i_{qs}^*(k) \right]. \end{aligned} \tag{9.72}$$

The reference currents are calculated on the basis of the ZDC/MTPA control as outlined in Section 9.5.4. The active and reactive powers are predicted as follows:

$$\begin{aligned} P_s^p(k+1) &= 1.5 \left[v_{ds}^p(k)\, i_{ds}^p(k+1) + v_{qs}^p(k)\, i_{qs}^p(k+1) \right] \\ Q_s^p(k+1) &= 1.5 \left[v_{qs}^p(k)\, i_{ds}^p(k+1) - v_{ds}^p(k)\, i_{qs}^p(k+1) \right] \end{aligned} \tag{9.73}$$

where $v_{ds}^p(k)$ and $v_{qs}^p(k)$ are predicted converter voltages; $i_{ds}^p(k+1)$ and $i_{qs}^p(k+1)$ are predicted PMSG currents similar to the PPC scheme.

9.9.2 Predictive Speed Control

To realize the PCC and PTC schemes, a PI controller is employed in the outer speed control loop. The speed PI controller can be eliminated, thereby leading to a cascade-free control of the complete PMSG WECS. PSC directly controls the rotor electrical speed by internally generating the dq-axis currents [39, 40]. The PSC cost function is defined by the following:

$$\begin{aligned} g_r(k) &= \left[\widetilde{\omega}_r^*(k+1) - \omega_r^p(k+1) \right]^2 \\ &+ \left[\widehat{i}_{ds}^*(k+1) - i_{ds}^p(k+1) \right]^2 + \left[\widehat{i}_{qs}^*(k+1) - i_{qs}^p(k+1) \right]^2 \end{aligned} \tag{9.74}$$

where $\widehat{i}_{ds}^*$ and $\widehat{i}_{qs}^*$ are the reference currents produced internally by the PSC algorithm.

With the cost function in (9.74), the rotor speed is regulated through the internal control of the dq-axis currents; thus, this method can also be called the PSC-PCC method. The aforementioned cost function can be modified to enable the decoupled control of torque/flux or active/reactive power. For example, the PSC strategy can be implemented through PTC (leading to a new PSC-PTC method) by replacing the i_{ds} and i_{qs} terms by T_e and ψ_s, respectively.

9.10 REAL-TIME IMPLEMENTATION OF MPC SCHEMES

As discussed in Chapter 4, control delay is an inherent property of the digital signal processor-based real-time implementation of the MPC schemes. The control delay results in the deterioration of the MPC schemes performance. To obtain a control performance similar to the simulation results, the control delay can be compensated by: (1) changing the DT models with delay incorporated or (2) employing delay compensation methods on top of the standard DT models [41]. In this section, the latter approach is investigated for PMSG WECS through the *prediction + prediction* approach.

Figure 9.21 shows the block diagram of two-samples-ahead (2H) PCC for the 2L-VSR-based PMSG WECS. The reference *dq*-axis PMSG currents are calculated on the basis of OTSR MPPT, speed control loop, and ZDC/MTPA operation (Figure 9.6). Compared with the one-sample-ahead (1H) PCC in Figure 9.8, 2H-PCC uses two-samples-ahead prediction for PMSG currents, and cost function minimization at $(k + 2)$ sampling instant. Therefore, one sampling period is used for predictions and another sampling period is used for control delay compensation.

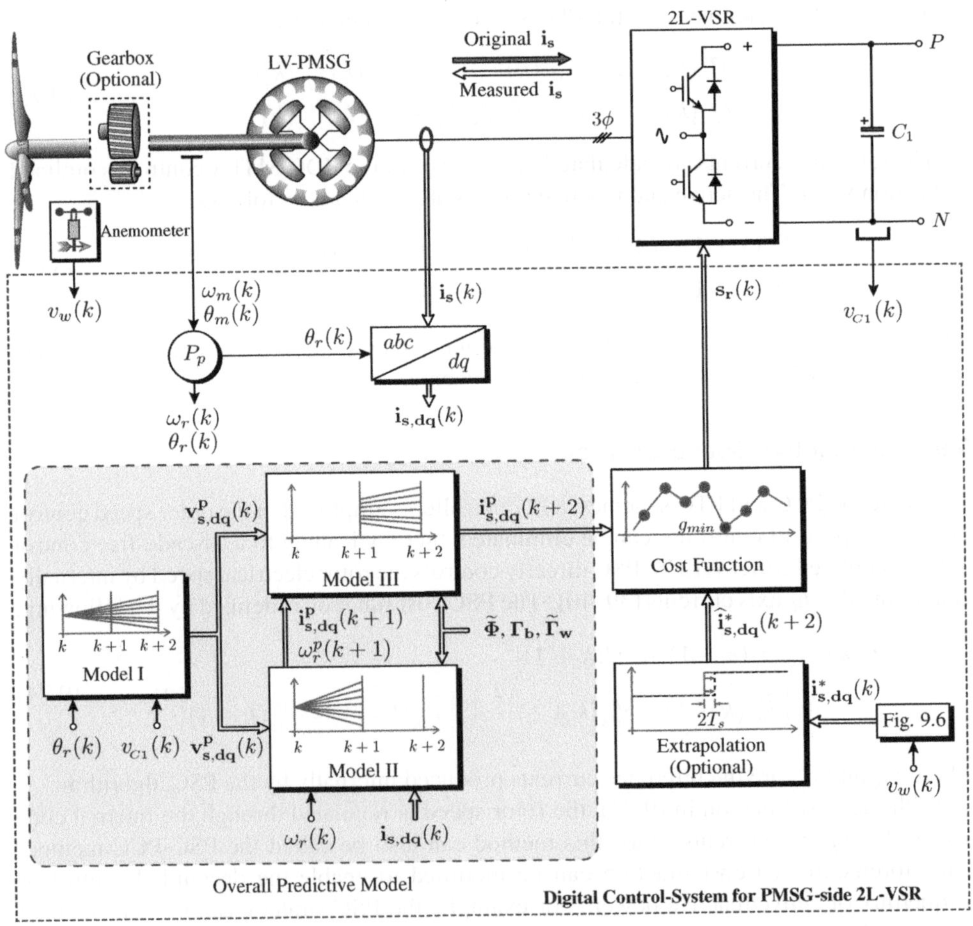

Figure 9.21 Block diagram of the 2H-PCC for ZDC/MTPA control of PMSG with 2L-VSR.

The cost function with 2H-PCC is defined by

$$g_r(k) = \left[\widehat{i}^*_{ds}(k+2) - i^p_{ds}(k+2)\right]^2 + \left[\widehat{i}^*_{qs}(k+2) - i^p_{qs}(k+2)\right]^2 \tag{9.75}$$

where $\widehat{i}^*_{ds}(k+2)$ and $\widehat{i}^*_{qs}(k+2)$ are the extrapolated dq-axes reference currents, respectively. They are calculated according to the first-order Lagrange extrapolation:

$$\mathbf{i}^*_{\mathbf{s,dq}}(k+2) = 3\,\mathbf{i}^*_{\mathbf{s,dq}}(k) - 2\,\mathbf{i}^*_{\mathbf{s,dq}}(k-1). \tag{9.76}$$

The dq-axes PMSG currents are predicted up to $(k+2)$ sampling interval using the following DT model:

$$\begin{aligned}\begin{bmatrix} i^p_{ds}(k+2) \\ i^p_{qs}(k+2) \end{bmatrix} &= \begin{bmatrix} 1 - \frac{R_s\,T_s}{L_{ds}} & \frac{\omega^p_r(k+1)\,L_{qs}\,T_s}{L_{ds}} \\ -\frac{\omega^p_r(k+1)\,L_{ds}\,T_s}{L_{qs}} & 1 - \frac{R_s\,T_s}{L_{qs}} \end{bmatrix} \begin{bmatrix} i^p_{ds}(k+1) \\ i^p_{qs}(k+1) \end{bmatrix} \\ &\quad + \boldsymbol{\Gamma}_{\mathbf{b}} \begin{bmatrix} v^p_{ds}(k) \\ v^p_{qs}(k) \end{bmatrix} + \begin{bmatrix} 0 \\ -\frac{\omega^p_r(k+1)\,\psi_r\,T_s}{L_{qs}} \end{bmatrix}\end{aligned} \tag{9.77}$$

where $\omega^p_r(k+1)$ is the predicted rotor speed.

The DT matrices $\widetilde{\boldsymbol{\Phi}}$, $\boldsymbol{\Gamma}_{\mathbf{b}}$, and $\widetilde{\boldsymbol{\Gamma}}_{\mathbf{w}}$ are commonly used by models II and III. With *prediction + prediction* delay compensation approach, $i^p_{ds}(k+2)$ and $i^p_{qs}(k+2)$ are obtained by the following design procedure:

- **Model I:** Similar to standard 1H-PCC, the 2L-VSR output voltages $v^p_{ds}(k)$ and $v^p_{qs}(k)$ are predicted using the model in (9.37). These voltages are used in both models II and III to predict future behavior of PMSG currents, therefore the computation burden caused by the delay compensation method becomes lower [42].
- **Model II:** Using $v^p_{ds}(k)$ and $v^p_{qs}(k)$, the PMSG currents and rotor speed ($i^p_{ds}(k+1)$, $i^p_{qs}(k+1)$, and $\omega^p_r(k+1)$) are predicted. The PMSG DT model which combines the rotor speed dynamics with the dq-axes currents can be found in example 7.3 in Chapter 7. As discussed in this case study, the truncated Taylor series discretization approach provides better performance compared to the forward Euler method.
- **Model III:** The two-samples-ahead PMSG currents $i^p_{ds}(k+2)$ and $i^p_{qs}(k+2)$ are predicted by using $v^p_{ds}(k)$, $v^p_{qs}(k)$, $i^p_{ds}(k+1)$, $i^p_{qs}(k+1)$, and $\omega^p_r(k+1)$ from the above two design steps. The rotor speed prediction $\omega^p_r(k+2)$ is not performed by model III because it is no longer necessary for cost function calculation.

Control delay can be similarly compensated by using the BTB NPC converters-based PMSG WECS. With 2H-PCC, the DC-link capacitors voltage $v^p_{C1}(k+2)$ and $v^p_{C2}(k+2)$ are predicted, in addition to $i^p_{ds}(k+2)$ and $i^p_{qs}(k+2)$ [34]. In two-samples-ahead PTC scheme, $i^p_{ds}(k+2)$ and $i^p_{qs}(k+2)$ are replaced by $T^p_e(k+2)$ and $\psi^p_s(k+2)$.

Figure 9.22 shows the flowchart for the implementation of 2H-PCC for the 2L-VSR-based PMSG WECS. This process is similar to the one-sample-ahead PCC algorithm in Figure 9.9 except that blocks ②, ⑥, and ⑦ are different. The $(k+2)$ sample reference currents $\widehat{\mathbf{i}}^*_{\mathbf{s,dq}}(k+2)$ are calculated in block ②. Block ⑥ is split into two components: the first component corresponds to the $(k+1)$ prediction of the dq-axes currents and rotor speed, whereas the second component predicts the $(k+2)$-sample dq-axes currents. Cost function is calculated in block ⑦ with $(k+2)$ state reference and predicted currents. In block ⑩, the switching states that minimizes the cost function at the $(k+2)$ instant are selected and applied to 2L-VSR directly.

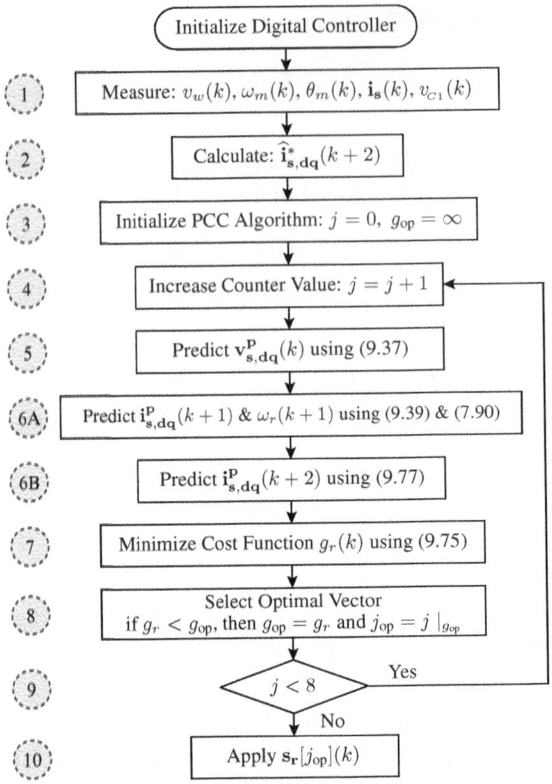

Figure 9.22 Flowchart of the 2H-PCC algorithm for the 2L-VSR-based PMSG WECS.

9.11 CONCLUDING REMARKS

This chapter discusses the MPC of SPMSG and IPMSG WECS comprising BTB VSCs. The DT model of the complete WECS is developed by combining the PMSG and VSC models. LV and MV PMSG WECS are analyzed through the predictive control of BTB 2L-VSC and BTB NPC converter. The steady state, CT and DT models of PMSG are developed. The calculation of reference control variables based on OTSR and OT MPPT algorithms is also analyzed. The PCC and PTC schemes are analyzed in detail with the aid of control block diagrams, flowcharts, and case studies. The other control schemes, such as the PPC and PSC schemes, are also discussed briefly. With the four types of MPC schemes, the dq-axis currents, electromagnetic torque and stator flux, stator active and reactive powers, and rotor speed follow their respective references with less tracking error. The per-unit method to calculate the weighting factors for the DC-link capacitors voltage balancing, as well as the decoupled torque and flux control, is discussed. The delay compensation based on two-samples-ahead PCC is discussed in detail. The current PI controllers, hysteresis controllers, lookup tables, and modulation stage in the internal control loop are eliminated by using the MPC scheme. With a detailed analysis of various power converters and MPC schemes, MPC has been demonstrated to be simple and intuitive. This realization of MPC schemes is similar for both generator- and grid-side power conversion units.

Link to Next Chapters:

The digital control of BTB converters is common for both the SG and IG WECS. The LV and MV converters, block diagrams of the overall digital control system, notation of control variables, calculation of reference control variables, the PCC and PTC schemes, weighting factors calculation, and control-delay compensation method discussed in this chapter are recalled in Chapters 11 and 12 to develop the MPC schemes for SCIG and DFIG WECS.

REFERENCES

1. H. Polinder, J. Ferreira, B. Jensen, A. Abrahamsen, K. Atallah, and R. McMahon, "Trends in wind turbine generator systems," *IEEE Journal of Emerging and Selected Topics in Power Electronics*, vol. 1, no. 3, pp. 174–185, September 2013.
2. V. Yaramasu, B. Wu, P. C. Sen, S. Kouro, and M. Narimani, "High-power wind energy conversion systems: State-of-the-art and emerging technologies," *Proceedings of the IEEE*, vol. 103, no. 5, pp. 740–788, May 2015.
3. H. Polinder, F. van der Pijl, G.-J. de Vilder, and P. Tavner, "Comparison of direct-drive and geared generator concepts for wind turbines," *IEEE Transactions on Energy Conversion*, vol. 21, no. 3, pp. 725–733, September 2006.
4. H. Li, Z. Chen, and H. Polinder, "Optimization of multibrid permanent-magnet wind generator systems," *IEEE Transactions on Energy Conversion*, vol. 24, no. 1, pp. 82–92, March 2009.
5. M. Liserre, R. Cardenas, M. Molinas, and J. Rodríguez, "Overview of multi-MW wind turbines and wind parks," *IEEE Transactions on Industrial Electronics*, vol. 58, no. 4, pp. 1081–1095, April 2011.
6. X. Yang, D. Patterson, and J. Hudgins, "Permanent magnet generator design and control for large wind turbines," in *IEEE Symposium on Power Electronics and Machines in Wind Applications (PEMWA)*, Denver, CO , USA, July 2012, pp. 1–5.
7. L. Zhong, M. Rahman, W. Y. Hu, and K. W. Lim, "Analysis of direct torque control in permanent magnet synchronous motor drives," *IEEE Transactions on Power Electronics*, vol. 12, no. 3, pp. 528–536, May 1997.
8. T. Geyer, "A comparison of control and modulation schemes for medium-voltage drives: Emerging predictive control concepts versus PWM-based schemes," *IEEE Transactions on Industry Applications*, vol. 47, no. 3, pp. 1380–1389, May/June 2011.
9. J. Rodríguez, R. Kennel, J. Espinoza, M. Trincado, C. Silva, and C. Rojas, "High-performance control strategies for electrical drives: An experimental assessment," *IEEE Transactions on Industrial Electronics*, vol. 59, no. 2, pp. 812–820, February 2012.
10. J. Chivite-Zabalza, C. Girones, A. Carcar, I. Larrazabal, E. Olea, and M. Zabaleta, "Comparison of power conversion topologies for a multi-megawatt off-shore wind turbine, based on commercial power electronic building blocks," in *IEEE Industrial Electronics Conference (IECON)*, Vienna, Austria, November 2013, pp. 5242–5247.
11. A. Yazdani and R. Iravani, "A neutral-point clamped converter system for direct-drive variable-speed wind power unit," *IEEE Transactions on Energy Conversion*, vol. 21, no. 2, pp. 596–607, June 2006.
12. A. Faulstich, J. Stinke, and F. Wittwer, "Medium voltage converter for permanent magnet wind power generators up to 5 MW," in *European Conference on Power Electronics and Applications (EPE)*, Dresden, Germany, 2005, pp. 9–P.9.
13. S. Alepuz, A. Calle, S. Busquets-Monge, S. Kouro, and B. Wu, "Use of stored energy in PMSG rotor inertia for low-voltage ride-through in back-to-back NPC converter-based wind power systems," *IEEE Transactions on Industrial Electronics*, vol. 60, no. 5, pp. 1787–1796, May 2013.
14. J. Rodríguez, S. Bernet, P. Steimer, and I. Lizama, "A survey on neutral-point-clamped inverters," *IEEE Transactions on Industrial Electronics*, vol. 57, no. 7, pp. 2219–2230, July 2010.
15. S. Kouro, M. Malinowski, K. Gopakumar, J. Pou, L. Franquelo, B. Wu, J. Rodríguez, M. Perez, and J. Leon, "Recent advances and industrial applications of multilevel converters," *IEEE Transactions on Industrial Electronics*, vol. 57, no. 8, pp. 2553–2580, August 2010.
16. F. Blaabjerg and K. Ma, "Future on power electronics for wind turbine systems," *IEEE Journal of Emerging and Selected Topics in Power Electronics*, vol. 1, no. 3, pp. 139–152, September 2013.
17. M. Preindl and S. Bolognani, "Model predictive direct torque control with finite control set for PMSM drive systems, Part 1: Maximum torque per ampere operation," *IEEE Transactions on Industrial Informatics*, vol. 9, no. 4, pp. 1912–1921, November 2013.
18. D. Casadei, F. Profumo, G. Serra, and A. Tani, "FOC and DTC: two viable schemes for induction motors torque control," *IEEE Transactions on Power Electronics*, vol. 17, no. 5, pp. 779–787, September 2002.
19. G. Buja and M. Kazmierkowski, "Direct torque control of PWM inverter-fed AC motors - A survey," *IEEE Transactions on Industrial Electronics*, vol. 51, no. 4, pp. 744–757, August 2004.
20. R. Krishnan, *Permanent Magnet Synchronous and Brushless DC Motor Drives*. Boca Raton, FL: CRC Press, 2010.
21. M. Kazmierkowski, R. Krishnan, and F. Blaabjerg, *Control in Power Electronics: Selected Problems*. San Diego, CA: Academic Press, 2002.

22. B. Bose, *Power Electronics and Motor Drives: Recent Advances and Trends*. Orlando, FL: Academic Press, 2006.
23. B. Wu, Y. Lang, N. Zargari, and S. Kouro, *Power Conversion and Control of Wind Energy Systems*, 1st ed., ser. IEEE Press Series on Power Engineering. Hoboken, NJ: Wiley-IEEE Press, July 2011.
24. S. Morimoto, Y. Takeda, and T. Hirasa, "Current phase control methods for permanent magnet synchronous motors," *IEEE Transactions on Power Electronics*, vol. 5, no. 2, pp. 133–139, April 1990.
25. M. Chinchilla, S. Arnaltes, and J. Burgos, "Control of permanent-magnet generators applied to variable-speed wind-energy systems connected to the grid," *IEEE Transactions on Energy Conversion*, vol. 21, no. 1, pp. 130–135, March 2006.
26. S. Morimoto, K. Hatanaka, Y. Tong, Y. Takeda, and T. Hirasa, "Servo drive system and control characteristics of salient pole permanent magnet synchronous motor," *IEEE Transactions on Industry Applications*, vol. 29, no. 2, pp. 338–343, March 1993.
27. Y.-R. Mohamed and T. Lee, "Adaptive self-tuning MTPA vector controller for IPMSM drive system," *IEEE Transactions on Energy Conversion*, vol. 21, no. 3, pp. 636–644, Sept 2006.
28. M. Abdullah, A. Yatim, C. Tan, and R. Saidur, "A review of maximum power point tracking algorithms for wind energy systems," *Renewable and Sustainable Energy Reviews*, vol. 16, no. 5, pp. 3220–3227, 2012.
29. E. Koutroulis and K. Kalaitzakis, "Design of a maximum power tracking system for wind-energy-conversion applications," *IEEE Transactions on Industrial Electronics*, vol. 53, no. 2, pp. 486–494, April 2006.
30. S. Morimoto, H. Nakayama, M. Sanada, and Y. Takeda, "Sensorless output maximization control for variable-speed wind generation system using IPMSG," *IEEE Transactions on Industry Applications*, vol. 41, no. 1, pp. 60–67, January 2005.
31. F. Niu, B. Wang, A. Babel, K. Li, and E. Strangas, "Comparative evaluation of direct torque control strategies for permanent magnet synchronous machines," *IEEE Transactions on Power Electronics*, vol. 31, no. 2, pp. 1408–1424, February 2016.
32. Y. Cho, K.-B. Lee, J.-H. Song, and Y. Lee, "Torque-ripple minimization and fast dynamic scheme for torque predictive control of permanent-magnet synchronous motors," *IEEE Transactions on Power Electronics*, vol. 30, no. 4, pp. 2182–2190, April 2015.
33. M. Pacas and J. Weber, "Predictive direct torque control for the PM synchronous machine," *IEEE Transactions on Industrial Electronics*, vol. 52, no. 5, pp. 1350–1356, October 2005.
34. A. Calle, S. Alepuz, J. Bordonau, P. Cortes, and J. Rodriguez, "Predictive control of a back-to-back NPC converter-based wind power system," *IEEE Transactions on Industrial Electronics*, early access article, 2016, doi: 10.1109/TIE.2016.2529564.
35. R. Vargas, P. Cortes, U. Ammann, J. Rodríguez, and J. Pontt, "Predictive control of a three-phase neutral-point-clamped inverter," *IEEE Transactions on Industrial Electronics*, vol. 54, no. 5, pp. 2697–2705, October 2007.
36. A. Calle-Prado, S. Alepuz, J. Bordonau, J. Nicolas-Apruzzese, P. Cortes, and J. Rodriguez, "Model predictive current control of grid-connected neutral-point-clamped converters to meet low-voltage ride-through requirements," *IEEE Transactions on Industrial Electronics*, vol. 62, no. 3, pp. 1503–1514, March 2015.
37. J. Scoltock, T. Geyer, and U. Madawala, "A comparison of model predictive control schemes for MV induction motor drives," *IEEE Transactions on Industrial Informatics*, vol. 9, no. 2, pp. 909–919, May 2013.
38. L. Xiao, S. Huang, K. Huang, Z. Chen, S. Xiong, and J. Tan, "Space vector modulation based on constant switching frequency direct power control for direct-drive permanent magnet synchronous generator," in *IEEE International Conference on Electrical Machines and Systems (ICEMS)*, October 2010, pp. 574–577.
39. E. Fuentes, J. Rodríguez, C. Silva, S. Diaz, and D. Quevedo, "Speed control of a permanent magnet synchronous motor using predictive current control," in *IEEE International Conference on Power Electronics and Motion Control (IPEMC)*, Wuhan, China, May 2009, pp. 390–395.
40. M. Preindl and S. Bolognani, "Model predictive direct speed control with finite control set of PMSM drive systems," *IEEE Transactions on Power Electronics*, vol. 28, no. 2, pp. 1007–1015, February 2013.
41. P. Cortes, J. Rodríguez, C. Silva, and A. Flores, "Delay compensation in model predictive current control of a three-phase inverter," *IEEE Transactions on Industrial Electronics*, vol. 59, no. 2, pp. 1323–1325, February 2012.
42. V. Yaramasu, M. Rivera, M. Narimani, B. Wu, and J. Rodriguez, "High performance operation for a four-leg NPC inverter with two-sample-ahead predictive control strategy," *International Journal of Electric Power Systems Research*, vol. 123, pp. 31–39, June 2015.

CHAPTER 10

CONTROL OF PMSG WECS WITH PASSIVE GENERATOR-SIDE CONVERTERS

10.1 INTRODUCTION

The back-to-back (BTB) connected voltage source converters (VSCs) discussed in Chapter 9 allow bidirectional power flow for permanent magnet synchronous generator (PMSG) based Type 4 wind turbines (WTs). However, in wind energy conversion systems (WECS), power flow is unidirectional, that is, the power flows from the wind generator to the utility grid. Thus, passive (diode bridge) converters can be employed on the generator side instead of pulse width modulated (PWM) active converters. Diode-bridge rectifiers are less expensive and inherently more reliable than PWM converters. In the PMSG and wound rotor synchronous generator (WRSG), the rotor flux is generated by permanent magnets and rotor field excitation, respectively. Passive converters can realize generator-side power conversion because no power is required from the utility grid to excite the PMSG and WRSG [1]. During system startup, squirrel cage and doubly fed induction generators require magnetizing current (reactive power) from the utility grid. Therefore, bidirectional power flow BTB converters are mandatory for the realization of power conversion systems with induction generator-based WECS.

The output voltage of the diode rectifier varies according to the generator speed. During low wind speeds, the generator output voltage and the diode rectifier output voltage significantly decreases. To transfer the generated power to the grid, the DC-link voltage must be higher than the *peak* value of grid line-to-line voltage. To increase the output voltage of the diode rectifier, a boost converter is employed as an intermediate stage. The configuration based on a diode rectifier, boost converter, and voltage source inverter (VSI) is referred to

Model Predictive Control of Wind Energy Conversion Systems, First Edition. Venkata Yaramasu and Bin Wu.

Companion Website: www.wiley.com/go/yaramasu_wu_mpc_wecs

as a three-stage power conversion system. The boost converter also enables variable-speed operation for the SG WECS by performing maximum power point tracking (MPPT). The wind energy conversion efficiency improves with this scenario, particularly during low wind speeds. Passive generator-side (PGS) converters have been used in practical WECS, ranging from a few kilowatts (kWs) to megawatts (MWs). A few examples of MW WTs with PGS converters are given in Table 10.1 with respect to gearbox type, generator, power and voltage rating, and power converter configuration. The low-voltage (LV) three-stage power converter configuration is commonly employed by present WT manufacturers. To increase power handling capacity, a number of two-level (2L) boost converters and 2L-VSI modules are connected in parallel [2]. Medium-voltage (MV) PGS converters are promising candidates for future development.

Table 10.1 List of commercial SG WECS with PGS converters

Manufacturer	Generator	Gen. Rating	Converter	Model #
Clipper (USA)	2S–4x PMSG	2.5 MW, 690 V	Diode + 2L-VSI	Liberty C99
Enercon (Germany)	DD–WRSG	2.0 MW, 450 V	Diode + 2L-Boost + 2L-VSI	E82/2000
Goldwind (PR China)	DD–6ϕ PMSG	1.5 MW, 690 V	Diode + 2L-Boost + 2L-VSI	GW70/1500
Vensys (Germany)	DD–6ϕ PMSG	1.5 MW, 690 V	Diode + 2L-Boost + 2L-VSI	V70/1500

4x = quantum drivetrain, 2S = two-stage gearbox, DD = direct-drive (no gearbox), 2L = two-level

In addition to the power converter configuration, digital control schemes play an important role in achieving many technical and operational requirements. The power converters in WECS are controlled by digital control systems for efficient and reliable power conversion. The digital control schemes are referred to as both classical and model predictive control (MPC) techniques. This chapter deals with the LV and MV PGS-converters-based Type 4 PMSG WECS and the digital control using MPC. The information given in this chapter is also applicable for WRSG WECS and the classical control of SG WECS.

Chapter Overview

- The configuration of PMSG WECS with LV and MV PGS converters is given in Section 10.2. BTB and PGS converters are also compared and tabulated.
- The continuous-time (CT) and discrete-time (DT) modeling of 2L and three-level (3L) boost converters is presented in Sections 10.3 and 10.4, respectively.
- The digital control system development for the PGS-converter-based PMSG WECS is given in Section 10.5. The MPPT algorithms and calculations of reference control variables are also discussed to realize various MPC schemes.
- The predictive current control (PCC) schemes for PMSG WECS with 2L and 3L PGS converters are analyzed in Sections 10.6 and 10.7, respectively. The performance of PMSG WECS with 3L-PGS converters is analyzed through case studies in Section 10.8.
- Other predictive control schemes such as predictive power control (PPC) and predictive speed control (PSC) are briefly discussed in Section 10.9. Section 10.10 addresses the real-time implementation issues such as computational burden, and digital control delay compensation by two-samples-ahead predictive control. The concluding remarks of this chapter are given in Section 10.11.

10.2 CONFIGURATION OF PMSG WECS WITH PGS CONVERTERS

In this section, two basic configurations of PMSG WECS with LV and MV PGS power converters are presented. The complete list of power converter configurations, including parallel converters and current source converters, can be found in Chapter 2, Section 2.5. The gearbox is optional for these configurations because PMSG can operate at low-shaft speeds (usually 8–30 rpm) by employing a large number of poles.

10.2.1 PMSG WECS with LV PGS Converters

The configuration of PMSG WECS with 2L PGS converters is shown in Figure 10.1 [3, 4]. The generator and primary side of the step-up transformer (not shown in Figure 10.1) are rated for LV operation (usually 690 or 575 V). The AC/DC, DC/DC, and DC/AC power conversion stages are implemented by a three-phase (3ϕ) diode-bridge rectifier, 2L boost converter, and 2L-VSI, respectively. The output of the diode rectifier remains unregulated but is limited by the rated speed of the PMSG. The voltage ripple in the diode rectifier output has a minimal effect on the power conversion and control scheme; thus, the capacitor used in the first DC link can have a smaller value than the capacitor used in the second DC link or can be completely eliminated [5]. The boost converter active switch is realized by using an LV-insulated gate bipolar transistor (IGBT) and its voltage rating is decided according to the net DC-bus voltage. The voltage in the second DC link is always higher than the first DC link because of the boost nature of the DC/DC converter. The 2L-VSI is connected to the three-phase collection point through a harmonic filter (L or LCL) and a step-up transformer. The 2L converters shown in Figure 10.1 can operate up to 0.75 MW without connecting the switching devices in series or parallel.

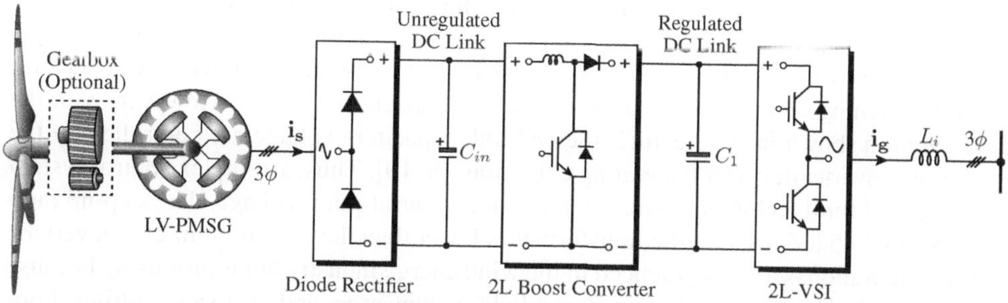

Figure 10.1 Block diagram of PMSG WECS with diode rectifier, 2L boost converter, and 2L-VSI.

10.2.2 PMSG WECS with MV PGS Converters

The PMSG WECS with 3L converters is shown in Figure 10.2, where the PMSG and primary side of the transformer are rated for MV (normally 3-4-kV) operations. The 3ϕ diode rectifier features series-connected diodes owing to the MV generator. The active switching devices in this configuration are realized by using an MV-IGBT or integrated gate-commutated thyristor (IGCT) and are operated at a few hundred Hz to decrease switching losses [6]. The second DC link consists of two capacitors that share equal voltages. The output of the 3L boost converter directly fits the two DC-link capacitors of the grid-side neutral-point clamped (NPC) inverter.

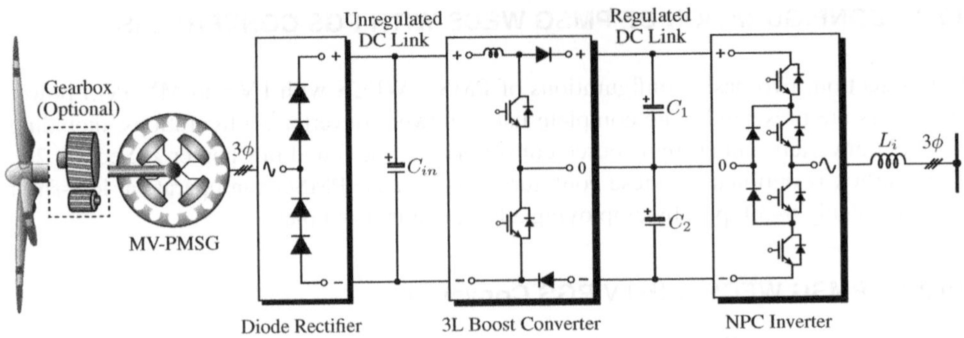

Figure 10.2 PMSG WECS with diode rectifier, 3L boost converter, and NPC inverter.

The 3L-boost converter enables the MV operation for the DC/DC stage. This converter is developed by using two MV-IGBTs/IGCTs, two fast recovery diodes, and a DC inductor. This converter offers many benefits, including reduced switching losses and reduced reverse recovery losses, compared with the 2L-boost converter shown in Figure 10.1 [7, 8]. The rating for active switches and diodes is half the DC-bus voltage. The 2L-boost converter uses one active switch and operates in two modes only. With the two active switches, the 3L-boost converter can operate in four modes. The increase in the number of modes provides additional freedom to charge and discharge the DC-link capacitors independently, in addition to the basic feature of MPPT operation. The step-up voltage ratio offered by the 3L-boost converter is the same as the 2L-boost converter.

Compared with the BTB-NPC converter-based configuration discussed in the previous chapter, the number of active switches in the machine-side converter (MSC) decreases from 12 to 2. Thus, the overall cost and complexity of the power converter is reduced. In the BTB-NPC converter configuration, both the MSC and grid-side converter (GSC) perform precise switching actions, thus balancing the DC capacitors voltage. However, with the configuration shown in Figure 10.2, the DC-link capacitors voltage are controlled by the 3L-boost converter during all operating conditions [9, 10]. Thus, a complex control system is not needed for the GSC to regulate the DC-link neutral-point voltage. The step-up ratio (3/4 kV: 33/34.5 kV) used in the transformer is lower than the ratio used in 2L converters. This configuration is not yet practiced in the wind energy industry but is promising because it combines the advantages of generator-side PGS converters and grid-side multilevel operation.

10.2.3 Comparison Between BTB and PGS Converters

The LV PGS converters shown in Figure 10.1 and the BTB 2L-VSC shown in Chapter 9 and Figure 9.1 are considered here for comparison. The grid-side 2L-VSI is a common element in both configurations. The BTB converters constitute a two-stage (AC/DC + DC/AC) power conversion, whereas the PGS converters constitute a three-stage (AC/DC + DC/DC + DC/AC) conversion. In addition to low cost and reliable operation, PGS converters simplify the complexity of the control system for both classical and predictive control schemes. The control system uses only feedback signals and does not require the model and parameters of PMSG. Given the reduced control complexity, a low-cost control platform can be used with PGS converters.

The use of PGS converters is associated with a few disadvantages. PMSM currents, voltages, and electromagnetic torque, as well their harmonic analyses, are given in Figure 10.3 under the rated operating condition [11]. The stator currents and voltages are plotted according to the machine convention introduced in Chapter 6.

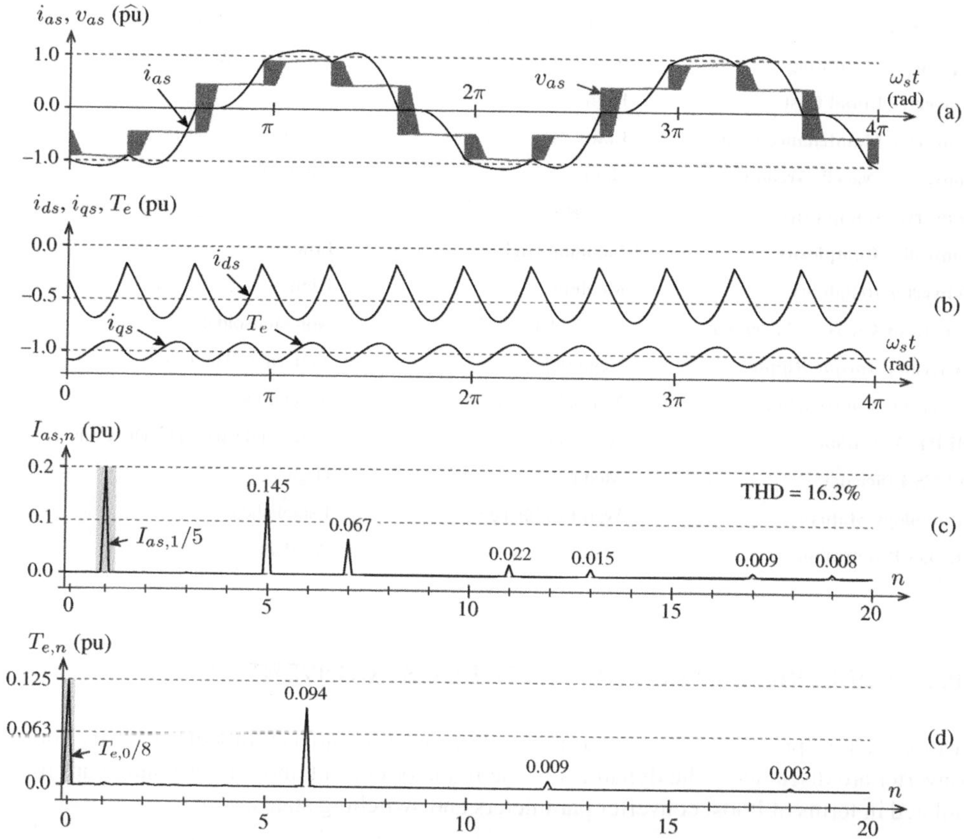

Figure 10.3 Three-phase PMSG phase voltage, stator current and electromagnetic torque waveforms, and harmonic spectrum under rated operating condition with PGS converters.

As shown in the harmonic spectrum plots in Figure 10.3, the generator stator currents contain significant 5th (14.5%) and 7th (6.7%) harmonics, which leads to 6th harmonic distortion (9.4%) in the electromagnetic torque. However, because of the decoupling offered by the second DC link, these torque ripples and generator current distortion do not cause conflicts to the grid codes. To solve the problem of high torque ripple, a 6-phase PMSG with a 30° phase shift and a 12-pulse diode-bridge rectifier can be used. The phase displacement cancels the 5th and 7th harmonics, thus decreasing the torque ripple. Commercial WTs, `Vensys` V70/77, and `Goldwind` GW70/77 feature six-phase PMSG configurations. The PMSG operates close to the unity power factor (UPF) because of the diode-bridge rectifier. As discussed in the previous chapter, the UPF operation of PMSG leads to stator currents higher than their rated values. To solve this issue, the PMSG must be designed and overrated (approximately 5% higher) compared with the one employed with the BTB converters. The summary of comparisons between BTB and PGS power converter configurations is given in Table 10.2.

Table 10.2 Comparison of BTB and PGS converters for LV PMSG-WECS [1]

	BTB Power Converters	**PGS Power Converters**
Typical Power Converter	Figure 9.1	Figure 10.1
Power Stages	AC/DC + DC/AC	AC/DC + DC/DC + DC/AC
Typical Power Rating	0.75–6.0 MW	0.75–3.0 MW
Converter Initial Cost	High	Low
Converter Maintenance Cost	Low	Very Low
Converter/ Nacelle Weight	High	Medium
Converter Complexity	Very High	Low
Controller Complexity	Medium–High	Low
Converter Reliability	Medium	High
Generator Current Waveform	Sinusoidal	Non-sinusoidal
Generator Torque Ripple	Negligible	Noticeably High
Generator Construction	Normal Design	Special Design
MPPT Operation	Achievable	Achievable with DC/DC Converter
WECS Efficiency	Medium	High
Technology Status	Well Established	Established
Market Penetration	Mature	Available

10.3 MODELING OF THE TWO-LEVEL BOOST CONVERTER

In this section, power circuit, operating modes, and CT and DT models of a 2L boost converter are discussed. The dynamics of the inductor current and output voltage are formulated in terms of boost converter parameters and switching states.

10.3.1 Power Circuit

The PMSG WECS configuration is presented in Figure 10.4(a) to facilitate the modeling of a 2L boost converter. The corresponding simplified configuration using generic switch and diode models is shown in Figure 10.4(b). The PMSG, along with the 3ϕ-diode bridge rectifier and capacitive filter C_{in}, is represented as a variable DC voltage source v_{in}. The grid-side 2L-VSI is represented by virtual resistive load R_1. The virtual resistive loads are introduced to facilitate the discussion about the modeling of the boost converter, assuming that the grid-side 2L-VSI operates with a high-power factor. For the given DC-link voltage, the current (power) delivered to the grid can be changed by varying the R_1 value.

The 2L boost converter is realized by using an active switch T_{dc1}, fast recovery diode D_1, DC-inductor L_{dc}, and DC capacitive filter C_1. The internal resistance of DC inductor L_{dc} is represented as r_{dc}. The equivalent series resistance of C_1 is ignored to simplify the analysis. The gating (switching) signal applied to active switch T_{dc1} is denoted as s_{dc1}. The current through the DC inductor and virtual load are represented as i_{dc} and i_{R1}, respectively. i_{Pi} and i_{Ni} corresponds to the grid-side 2L-VSI positive and negative DC branch currents, respectively.

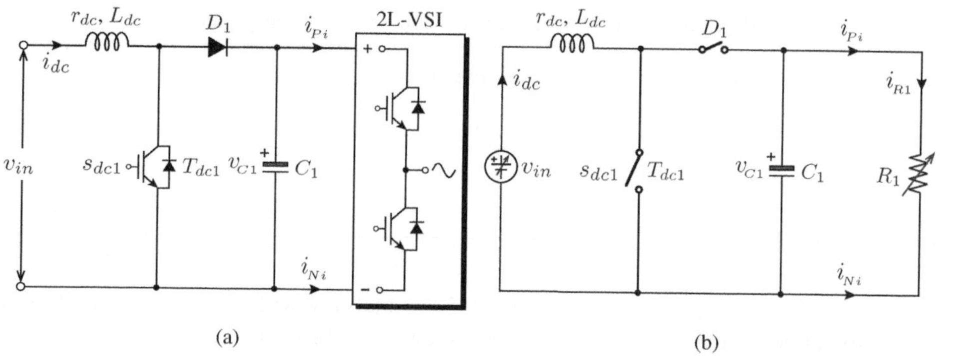

Figure 10.4 Representation of a 2L boost converter power circuit: (a) standard configuration and (b) simplified configuration.

10.3.2 Operating Modes

The one active switch T_{dc1} in a 2L-boost converter leads to two operating modes, as shown in Figure 10.5 [12]. Table 10.3 summarizes the turn *on/off* status of the active switch and diode, as well as the charging/discharging status of the capacitor. The analysis during these two operating modes is given below:

- **Mode I Operation:** As demonstrated in Figure 10.5(a), during this operating mode, the semiconductor switch T_{dc1} is closed and fast recovery diode D_1 is disabled by applying s_{dc1} = '1'. The DC current i_{dc} flows through L_{dc} and T_{dc1}. Thus, energy is stored in the DC inductor. The energy that is already stored in the capacitors C_1 is discharged to the virtual load R_1.
- **Mode II Operation:** The switching signal is disabled, that is, s_{dc1} = '0' during this mode of operation. The semiconductor switch T_{dc1} becomes open and the diode D_1 becomes closed as shown in Figure 10.5(b). The DC-current flows through the L_{dc}, D_1, C_1, and R_1. This condition leads to the charging of capacitor C_1 and current flow through the virtual load R_1.

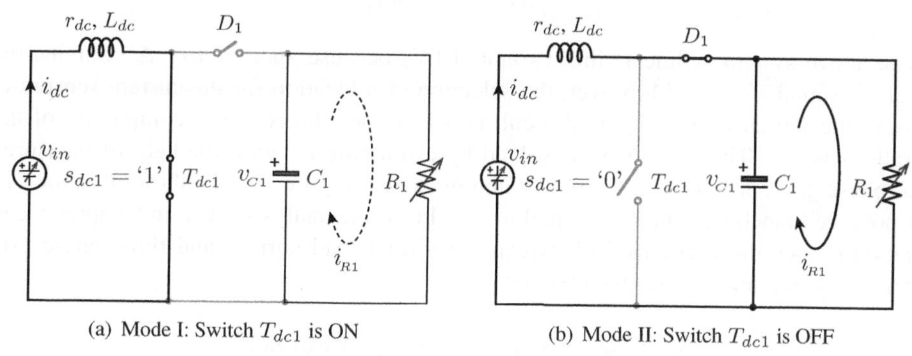

Figure 10.5 Representation of operating modes for a 2L boost converter.

Table 10.3 Summary of 2L-boost converter switching states

Mode	Switching States	Switching Device	Diode	Capacitor
	s_{dc1}	T_{dc1}	D_1	C_1
I	'1'	ON	OFF	Discharge
II	'0'	OFF	ON	Charge

The operating modes presented in Figure 10.5 are analyzed with the help of Kirchhoff's law. The dynamics of the inductor current are given below [13, 14]:

$$\frac{d\,i_{dc}}{dt} = \begin{cases} \frac{v_{in} - r_{dc}\,i_{dc}}{L_{dc}} & \text{if } T_{dc1} \text{ is on} \quad (s_{dc1} = \text{'1'}) \\ \frac{v_{in} - r_{dc}\,i_{dc} - v_{C1}}{L_{dc}} & \text{if } T_{dc1} \text{ is off} \quad (s_{dc1} = \text{'0'}). \end{cases} \tag{10.1}$$

The capacitor voltage dynamics are obtained as follows:

$$\frac{d\,v_{C1}}{dt} = \begin{cases} -\frac{i_{R1}}{C_1} & \text{if } T_{dc1} \text{ is on} \quad (s_{dc1} = \text{'1'}) \\ \frac{i_{dc} - i_{R1}}{C_1} & \text{if } T_{dc1} \text{ is off} \quad (s_{dc1} = \text{'0'}). \end{cases} \tag{10.2}$$

10.3.3 Continuous-Time Model

By combining the dynamic equations in (10.1) and (10.2), the CT model describing the inductor current and DC-link capacitor voltage is formulated as follows:

$$\underbrace{\frac{d}{dt}\begin{bmatrix} i_{dc}(t) \\ v_{C1}(t) \end{bmatrix}}_{\dot{\mathbf{x}}(t)} = \underbrace{\begin{bmatrix} -\frac{r_{dc}}{L_{dc}} & -\frac{(1 - s_{dc1}(t))}{L_{dc}} \\ \frac{(1 - s_{dc1}(t))}{C_1} & 0 \end{bmatrix}}_{\mathbf{A}(t)} \underbrace{\begin{bmatrix} i_{dc}(t) \\ v_{C1}(t) \end{bmatrix}}_{\mathbf{x}(t)} + \underbrace{\begin{bmatrix} \frac{1}{L_{dc}} & 0 \\ 0 & -\frac{1}{C_1} \end{bmatrix}}_{\mathbf{B}} \underbrace{\begin{bmatrix} v_{in}(t) \\ i_{R1}(t) \end{bmatrix}}_{\mathbf{u}(t)}. \tag{10.3}$$

The above system is linear time variant (LTV) because state matrix **A** contains the switching signal $s_{dc1}(t)$. Moreover, the selection of a location for the current sensors to measure the virtual current i_{R1} is difficult and increases the cost and complexity of the overall converter. This issue can be solved by estimating i_{R1} with the help of measured three-phase grid currents and 2L-VSI switching states. Figure 10.4(b) shows that the 2L-VSI positive branch current i_{Pi} is equal to i_{R1}. From the analysis given in Chapter 5 and Equation (5.14), the relationship between the virtual load current and three-phase grid currents (i_{ag}, i_{bg}, and i_{cg}) is given below:

$$i_{R1} = i_{Pi} = s_{ai1}\,i_{ag} + s_{bi1}\,i_{bg} + s_{ci1}\,i_{cg} \tag{10.4}$$

where s_{ai1}, s_{bi1}, and s_{ci1} are the switching signals applied to the top switching devices of the grid-side 2L-VSI.

10.3.4 Discrete-Time Model

The calculation of the exact DT model corresponding to the CT model in (10.3) is not straightforward and contains complex mathematical expressions because of the time-varying nature of state matrix $\mathbf{A}$. The forward Euler method discussed in Chapter 7 is used to obtain an approximate DT model as shown below:

$$\underbrace{\begin{bmatrix} i_{dc}(k+1) \\ v_{C1}(k+1) \end{bmatrix}}_{\mathbf{x}(k+1)} = \underbrace{\begin{bmatrix} 1 - \frac{r_{dc}\,T_s}{L_{dc}} & -\frac{T_s\,(1-s_{dc1}(k))}{L_{dc}} \\ \frac{T_s\,(1-s_{dc1}(k))}{C_1} & 1 \end{bmatrix}}_{\mathbf{\Phi}(k)} \underbrace{\begin{bmatrix} i_{dc}(k) \\ v_{C1}(k) \end{bmatrix}}_{\mathbf{x}(k)} + \underbrace{\begin{bmatrix} \frac{T_s}{L_{dc}} & 0 \\ 0 & -\frac{T_s}{C_1} \end{bmatrix}}_{\mathbf{\Gamma}} \underbrace{\begin{bmatrix} v_{in}(k) \\ i_{R1}(k) \end{bmatrix}}_{\mathbf{u}(k)} \tag{10.5}$$

where T_s is the controller sampling time.

The calculation of DT matrices $\mathbf{\Phi}$ and $\mathbf{\Gamma}$ from the corresponding CT matrices $\mathbf{A}$ and $\mathbf{B}$ is shown below:

$$\mathbf{\Phi}(k) \approx (\mathbf{I} + \mathbf{A}(k)\,T_s), \quad \mathbf{\Gamma} \approx \mathbf{B}\,T_s \tag{10.6}$$

where $\mathbf{I}$ is second-order identity matrix.

The above notation shows that the future behavior of the inductor current and DC-link capacitor voltage can be obtained with the help of measurements and switching signals of the 2L boost converter. However, the 2L boost converter is capable of controlling either the inductor current or DC-link capacitor only. In verifying the control algorithm through simulation studies, the calculation of $i_{dc}(k+1)$ is sufficient. However, in the experimental implementation, both $i_{dc}(k+1)$ and $v_{C1}(k+1)$ should be calculated to compensate for the control delay caused by the digital signal processor (DSP). This calculation is explained in more detail in Section 10.10.

In the classical control scheme, the closed-loop PI controller generates the duty cycle D according to the operating condition [15]. The D value is compared with the carrier signal to generate the gating signal s_{dc1}. The switching frequency of the converter is equal to the carrier frequency. The modeling presented here uses a different approach to generate the gating signal s_{dc1} and does not involve the concept of duty cycle. By using the classical and MPC schemes, the voltage step-up ratio remains unchanged. In other words, the predictive controller cannot step up the voltage beyond the maximum value. Although this process is unnecessary with the predictive controller, one can estimate the duty cycle of the boost converter during any sampling instant as follows:

$$D(k) = 1 - \frac{v_{in}(k)}{v_{dc}(k)} = 1 - \frac{v_{in}(k)}{v_{C1}(k)}. \tag{10.7}$$

The theoretical minimum and maximum values for the duty cycle D are zero and one, respectively. The above calculation is also applicable to multilevel boost converters, which is discussed in the next section.

10.4 MODELING OF THE THREE-LEVEL BOOST CONVERTER

The 3L boost converter is used together with the NPC inverter to enable the MV operation of PMSG WECS. As discussed earlier, the 3L boost converter can regulate the DC-link capacitors voltage, in addition to the regulation of the inductor current (MPPT operation). In this section, the model of inductor current and DC-link capacitors voltage is presented in terms of boost converter parameters and switching states.

10.4.1 Power Circuit

The standard and simplified configurations of PMSG WECS with a 3L boost converter are shown in Figures 10.6(a) and 10.6(b), respectively. Two active switches (T_{dc1}, T_{dc2}), two diodes (D_1, D_2), two capacitors (C_1, C_2), and a DC-inductor (L_{dc}) are used to realize the 3L boost converter [7, 16]. The gating signals applied to active switches T_{dc1} and T_{dc2} are denoted as s_{dc1} and s_{dc2}, respectively. The current through virtual loads R_1 and R_2 are represented as i_{R1} and i_{R2}, respectively. The grid-side NPC inverter branch currents are as follows: i_{Pi}, i_{Zi} and i_{Ni}.

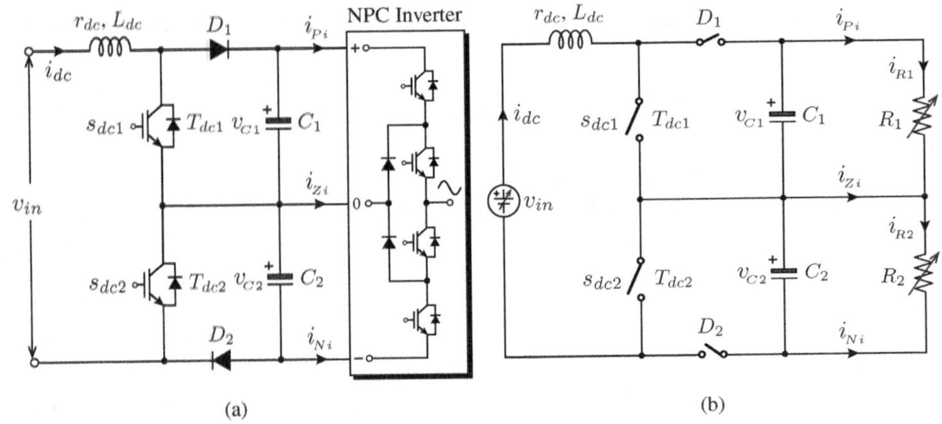

Figure 10.6 Representation of a 3L boost converter power circuit: (a) standard configuration and (b) simplified configuration.

10.4.2 Operating Modes

Active switches T_{dc1} and T_{dc2} produce four operating modes, as shown in Figure 10.7 [17, 18]. The increase in switching states leads to additional control freedom to charge and discharge the DC-link capacitors independently. The turn *on* and *off* status of active switches, diodes, and charging/discharging condition of the capacitors are summarized in Table 10.4. The 3L boost converter operating modes are analyzed as follows:

- **Mode I Operation:** As shown in Figure 10.7(a), semiconductor switches T_{dc1} and T_{dc2} are closed, whereas diodes D_1 and D_2 are open. The DC current i_{dc} flows through L_{dc}, T_{dc1}, and T_{dc2}. Thus, energy is stored in the inductor only and the DC-link capacitors are not charged. The energy stored in capacitors C_1 and C_2 is discharged to virtual loads R_1 and R_2, respectively.

- **Mode II Operation:** Semiconductor devices T_{dc1} and D_2 are ON, and T_{dc2} and D_1 are OFF during this operating mode, as depicted in Figure 10.7(b). The DC-current flows through L_{dc}, T_{dc1}, C_2, R_2, and D_2. The stored inductor energy is transferred to C_2 and R_2. The energy in capacitor C_1 is discharged to virtual load R_1.
- **Mode III Operation:** In the third operating mode, T_{dc2} and D_1 are ON, and T_{dc1} and D_2 are OFF (refer to Figure 10.7(c)). The DC-current flows through L_{dc}, D_1, C_1, R_1, and T_{dc2}. Thus, stored inductor energy is transferred to C_1 and R_1. The energy in capacitor C_2 is discharged to R_2.
- **Mode IV Operation:** Diodes D_1 and D_2 are closed, whereas switches T_{dc1} and T_{dc2} are open during the last operating mode, as shown in Figure 10.7(d). The DC current flows through L_{dc}, D_1, C_1, C_2, R_1, R_2, and D_2. The inductor energy is transferred to capacitors C_1 and C_2, as well as virtual loads R_1 and R_2.

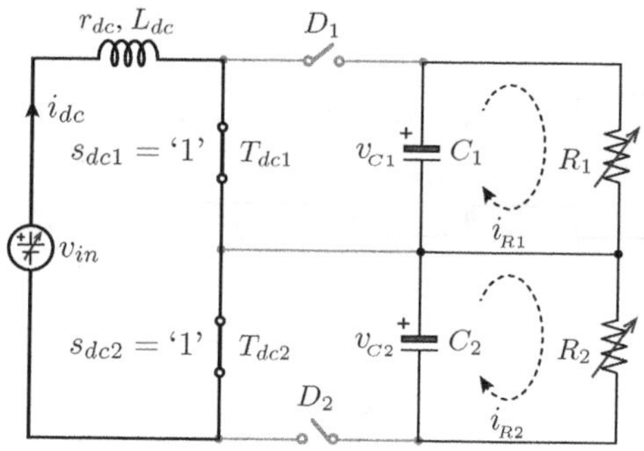

(a) Mode I: Switches T_{dc1} and T_{dc2} are ON

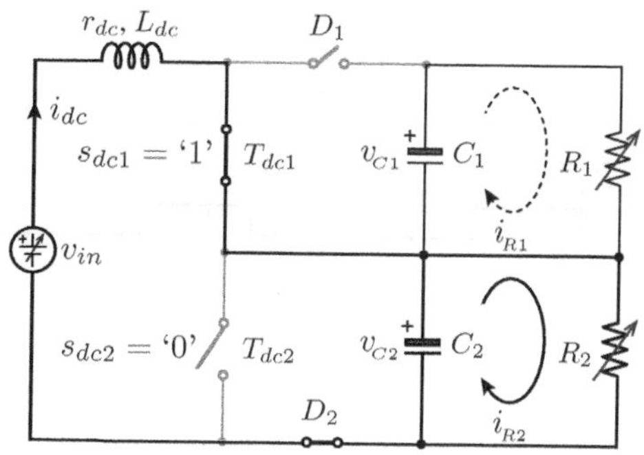

(b) Mode II: Switch T_{dc1} is ON and T_{dc2} is OFF

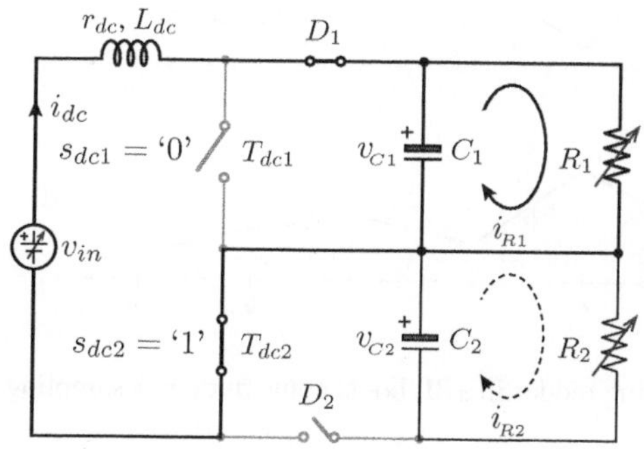

(c) Mode III: Switch T_{dc1} is OFF and T_{dc2} is ON

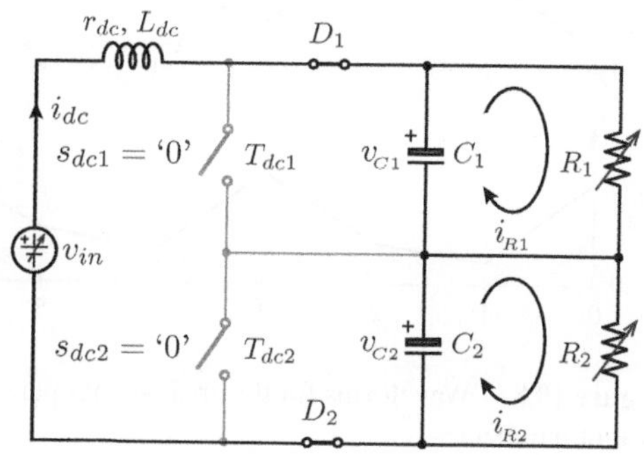

(d) Mode IV: Switches T_{dc1} and T_{dc2} are OFF

Figure 10.7 Representation of operating modes for a 3L boost converter [17, 18].

The graphical representation of the above analysis is given in Figure 10.8. The behavior of the inductor current and DC-link capacitors voltage during these four operating modes are highlighted. In Mode I, i_{dc} increase, whereas v_{C1} and v_{C2} decrease. In Mode II, i_{dc} and v_{C2} increase, and v_{C1} decrease. In Mode III, i_{dc} and v_{C1} increase, whereas v_{C2} decrease. Finally in Mode IV, v_{C1} and v_{C2} increase, and i_{dc} decrease.

Table 10.4 Summary of 3L-boost converter switching states

Mode	Switching States		Switching Devices		Diodes		Capacitors	
	s_{dc1}	s_{dc2}	T_{dc1}	T_{dc2}	D_1	D_2	C_1	C_2
I	'1'	'1'	ON	ON	OFF	OFF	Discharge	Discharge
II	'1'	'0'	ON	OFF	OFF	ON	Discharge	Charge
III	'0'	'1'	OFF	ON	ON	OFF	Charge	Discharge
IV	'0'	'0'	OFF	OFF	ON	ON	Charge	Charge

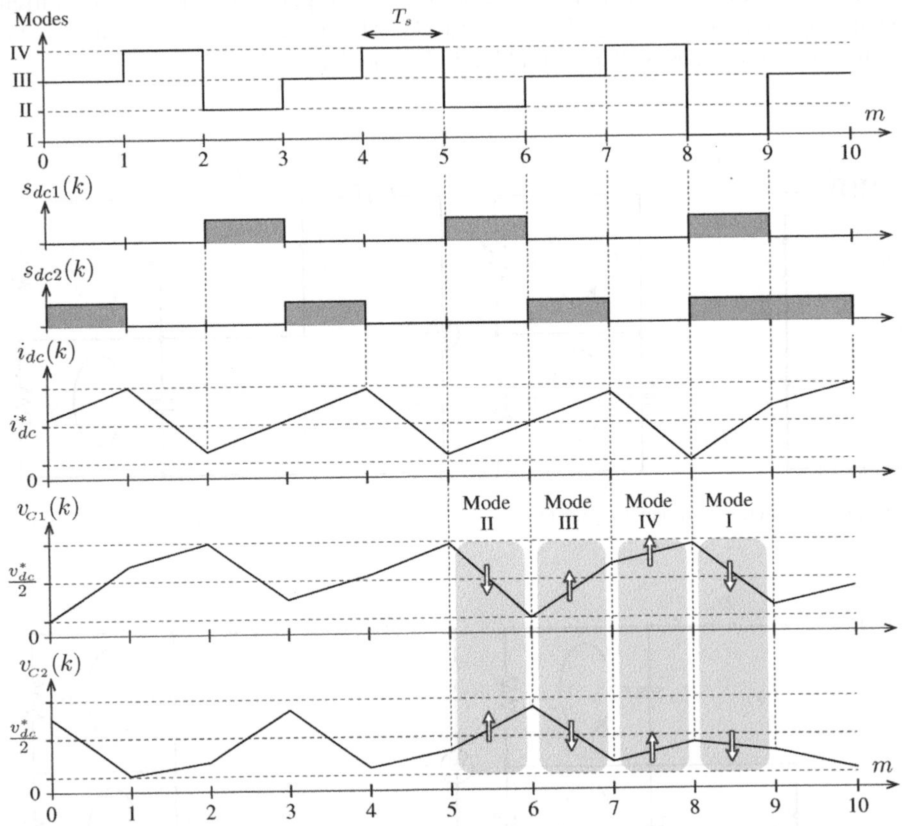

Figure 10.8 Waveforms for the analysis of operating modes in a 3L boost converter (m is sampling instant number).

With the help of Figure 10.7, the inductor current dynamics during the four operating modes are summarized as follows [9]:

$$\frac{di_{dc}}{dt} = \begin{cases} \frac{v_{in} - r_{dc}\, i_{dc}}{L_{dc}} & \text{if } T_{dc1} \text{ is on and } T_{dc2} \text{ is on} \quad (s_{dc1} = \text{`1'},\ s_{dc2} = \text{`1'}) \\ \frac{v_{in} - r_{dc}\, i_{dc} - v_{C2}}{L_{dc}} & \text{if } T_{dc1} \text{ is on and } T_{dc2} \text{ is off} \quad (s_{dc1} = \text{`1'},\ s_{dc2} = \text{`0'}) \\ \frac{v_{in} - r_{dc}\, i_{dc} - v_{C1}}{L_{dc}} & \text{if } T_{dc1} \text{ is off and } T_{dc2} \text{ is on} \quad (s_{dc1} = \text{`0'},\ s_{dc2} = \text{`1'}) \\ \frac{v_{in} - r_{dc}\, i_{dc} - v_{dc}}{L_{dc}} & \text{if } T_{dc1} \text{ is off and } T_{dc2} \text{ is off} \quad (s_{dc1} = \text{`0'},\ s_{dc2} = \text{`0'}) \end{cases} \tag{10.8}$$

The model for the first DC-link capacitor voltage is formulated as follows:

$$\frac{d\,v_{C1}}{dt} = \begin{cases} -\frac{i_{R1}}{C_1} & \text{if } T_{dc1} \text{ is on and } T_{dc2} \text{ is on} \quad (s_{dc1} = \text{`1'},\ s_{dc2} = \text{`1'}) \\ -\frac{i_{R1}}{C_1} & \text{if } T_{dc1} \text{ is on and } T_{dc2} \text{ is off} \quad (s_{dc1} = \text{`1'},\ s_{dc2} = \text{`0'}) \\ \frac{i_{dc}-i_{R1}}{C_1} & \text{if } T_{dc1} \text{ is off and } T_{dc2} \text{ is on} \quad (s_{dc1} = \text{`0'},\ s_{dc2} = \text{`1'}) \\ \frac{i_{dc}-i_{R1}}{C_1} & \text{if } T_{dc1} \text{ is off and } T_{dc2} \text{ is off} \quad (s_{dc1} = \text{`0'},\ s_{dc2} = \text{`0'}) \end{cases} \tag{10.9}$$

Similarly, the model of the second DC-link capacitor voltage is expressed as follows:

$$\frac{d\,v_{C2}}{dt} = \begin{cases} -\frac{i_{R2}}{C_2} & \text{if } T_{dc1} \text{ is on and } T_{dc2} \text{ is on} \quad (s_{dc1} = \text{`1'},\ s_{dc2} = \text{`1'}) \\ \frac{i_{dc}-i_{R2}}{C_2} & \text{if } T_{dc1} \text{ is on and } T_{dc2} \text{ is off} \quad (s_{dc1} = \text{`1'},\ s_{dc2} = \text{`0'}) \\ -\frac{i_{R2}}{C_2} & \text{if } T_{dc1} \text{ is off and } T_{dc2} \text{ is on} \quad (s_{dc1} = \text{`0'},\ s_{dc2} = \text{`1'}) \\ \frac{i_{dc}-i_{R2}}{C_2} & \text{if } T_{dc1} \text{ is off and } T_{dc2} \text{ is off} \quad (s_{dc1} = \text{`0'},\ s_{dc2} = \text{`0'}) \end{cases} \tag{10.10}$$

10.4.3 Continuous-Time Model

The state-space CT model for the 3L boost converter is obtained as follows by combining equations (10.8) through (10.10) [10]:

$$\underbrace{\frac{d}{dt}\begin{bmatrix} i_{dc}(t) \\ v_{C1}(t) \\ v_{C2}(t) \end{bmatrix}}_{\dot{\mathbf{x}}(t)} = \underbrace{\begin{bmatrix} -\frac{r_{dc}}{L_{dc}} & -\frac{(1-s_{dc1}(t))}{L_{dc}} & -\frac{(1-s_{dc2}(t))}{L_{dc}} \\ \frac{(1-s_{dc1}(t))}{C_1} & 0 & 0 \\ \frac{(1-s_{dc2}(t))}{C_2} & 0 & 0 \end{bmatrix}}_{\mathbf{A}(t)} \underbrace{\begin{bmatrix} i_{dc}(t) \\ v_{C1}(t) \\ v_{C2}(t) \end{bmatrix}}_{\mathbf{x}(t)}$$

$$+ \underbrace{\begin{bmatrix} \frac{1}{L_{dc}} & 0 & 0 \\ 0 & -\frac{1}{C_1} & 0 \\ 0 & 0 & -\frac{1}{C_2} \end{bmatrix}}_{\mathbf{B}} \underbrace{\begin{bmatrix} v_{in}(t) \\ i_{R1}(t) \\ i_{R2}(t) \end{bmatrix}}_{\mathbf{u}(t)}. \tag{10.11}$$

Virtual load currents i_{R1} and i_{R2} and DC branch currents i_{Pi}, i_{Zi}, and i_{Ni} are shown in Figure 10.6(b). The mathematical relationship between these currents is formulated as follows:

$$i_{R1} = i_{Pi}, \quad i_{R2} = i_{R1} + i_{Zi}. \tag{10.12}$$

As discussed in Chapter 5, the NPC inverter branch currents i_{Pi} and i_{Zi} can be estimated by using the measured grid currents and switching signals:

$$\begin{aligned} i_{Pi} &= s_{ai1}\, s_{ai2}\, i_{ag} + s_{bi1}\, s_{bi2}\, i_{bg} + s_{ci1}\, s_{ci2}\, i_{cg} \\ i_{Zi} &= s_{ai2}\, s_{ai3}\, i_{ag} + s_{bi2}\, s_{bi3}\, i_{bg} + s_{ci2}\, s_{ci3}\, i_{cg} \end{aligned} \tag{10.13}$$

where s_{xi1}, s_{xi2}, and s_{xi3} for all $x \in \{a, b, c\}$ are NPC inverter switching signals (refer to Figure 5.7 on page 162 for details). By substituting Equation (10.13) into (10.12), the required virtual load currents can be estimated by using the grid currents and switching signals information.

10.4.4 Discrete-Time Model

With the help of the forward Euler method, the approximate DT model for the 3L boost converter is derived from Equation (10.11) as follows:

$$\underbrace{\begin{bmatrix} i_{dc}(k+1) \\ v_{C1}(k+1) \\ v_{C2}(k+1) \end{bmatrix}}_{\mathbf{x}(k+1)} = \underbrace{\begin{bmatrix} 1-\frac{r_{dc}T_s}{L_{dc}} & -\frac{T_s(1-s_{dc1}(k))}{L_{dc}} & -\frac{T_s(1-s_{dc2}(k))}{L_{dc}} \\ \frac{T_s(1-s_{dc1}(k))}{C_1} & 1 & 0 \\ \frac{T_s(1-s_{dc2}(k))}{C_2} & 0 & 1 \end{bmatrix}}_{\mathbf{\Phi}(k)} \underbrace{\begin{bmatrix} i_{dc}(k) \\ v_{C1}(k) \\ v_{C2}(k) \end{bmatrix}}_{\mathbf{x}(k)} + \underbrace{\begin{bmatrix} \frac{T_s}{L_{dc}} & 0 & 0 \\ 0 & -\frac{T_s}{C_1} & 0 \\ 0 & 0 & -\frac{T_s}{C_2} \end{bmatrix}}_{\mathbf{\Gamma}} \underbrace{\begin{bmatrix} v_{in}(k) \\ i_{R1}(k) \\ i_{R2}(k) \end{bmatrix}}_{\mathbf{u}(k)}. \tag{10.14}$$

The above model correlates the inductor current and DC-link capacitors voltage in terms of boost converter switching states. This model is similar to the 2L boost converter model, except that the dimensions of $\mathbf{x}(k)$, $\mathbf{u}(k)$, $\mathbf{\Phi}(k)$, and $\mathbf{\Gamma}$ are increased by one to accommodate the additional (second) DC-link capacitor voltage.

10.4.5 Extension of Modeling to Multilevel Boost Converters

The multilevel boost converters are used in conjunction with multilevel diode-clamped inverters to increase WECS operating voltages. In the literature, boost converters are reported until four levels [19]. As shown below, the modeling presented earlier can be easily extended to four-level (4L) boost converter without involving operating modes and conducting detailed analyses [20]:

$$\underbrace{\begin{bmatrix} i_{dc}(k+1) \\ v_{C1}(k+1) \\ v_{C2}(k+1) \\ v_{C3}(k+1) \end{bmatrix}}_{\mathbf{x}(k+1)} = \underbrace{\begin{bmatrix} 1-\frac{r_{dc}T_s}{L_{dc}} & -\frac{T_s(1-s_{dc1}(k))}{L_{dc}} & -\frac{T_s(1-s_{dc2}(k))}{L_{dc}} & -\frac{T_s(1-s_{dc3}(k))}{L_{dc}} \\ \frac{T_s(1-s_{dc1}(k))}{C_1} & 1 & 0 & 0 \\ \frac{T_s(1-s_{dc2}(k))}{C_2} & 0 & 1 & 0 \\ \frac{T_s(1-s_{dc3}(k))}{C_3} & 0 & 0 & 1 \end{bmatrix}}_{\mathbf{\Phi}(k)} \underbrace{\begin{bmatrix} i_{dc}(k) \\ v_{C1}(k) \\ v_{C2}(k) \\ v_{C3}(k) \end{bmatrix}}_{\mathbf{x}(k)} + \underbrace{\begin{bmatrix} \frac{T_s}{L_{dc}} & 0 & 0 & 0 \\ 0 & -\frac{T_s}{C_1} & 0 & 0 \\ 0 & 0 & -\frac{T_s}{C_2} & 0 \\ 0 & 0 & 0 & -\frac{T_s}{C_3} \end{bmatrix}}_{\mathbf{\Gamma}} \underbrace{\begin{bmatrix} v_{in}(k) \\ i_{R1}(k) \\ i_{R2}(k) \\ i_{R3}(k) \end{bmatrix}}_{\mathbf{u}(k)}. \tag{10.15}$$

The above model shows that the fourth row and fourth column are new entities compared with the 3L boost converter model presented in (10.14). A 4L-boost converter is composed of three active switches, four diodes, and an inductor. The voltage rating of semiconductor devices is one-third of those used in a 2L-boost converter. For this converter, five operating modes are available. The DC-link capacitors can be charged or discharged independently through these operating modes, thus leading to effective control over the DC-link capacitors voltage. As demonstrated in Sections 10.3 and 10.4, PGS converters do not require the dynamic model and parameters of the PMSG. Moreover, the PMSG rotor position is not required for modeling, thus greatly simplifying the complexity of the control system compared with active generator-side converters.

10.5 DIGITAL CONTROL OF PGS CONVERTER-BASED PMSG WECS

The digital control system for PMSG WECS with PGS converters is similar to the system discussed in the previous chapter with BTB converters. The control of GSC is the same for BTB and PGS converters. The information given is here is common for both LV and MV converters; the dashed line in second DC-link corresponds to 3L converters.

10.5.1 Block Diagram of Digital Control System

The block diagram of a digital control system for the PMSG WECS with PGS converters is shown in Figure 10.9. The feedback signals (measured variables) from the generator side, DC link, and grid side are provided to the intermediate DC/DC converter and GSC control systems. Two individual control loops can be developed because of the decoupled nature of the MSC and GSC.

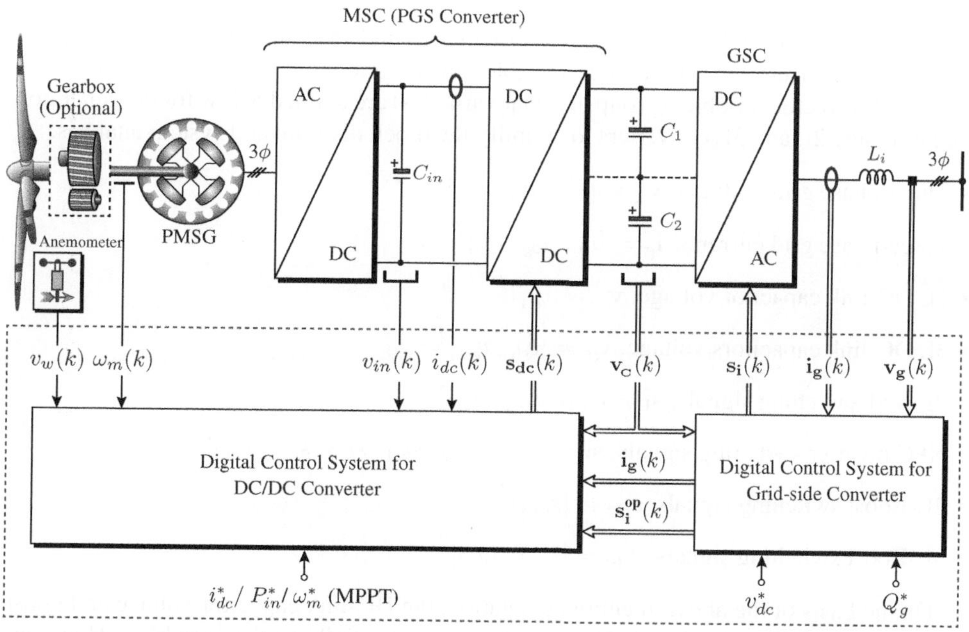

Figure 10.9 Block diagram of digital control scheme for PGS-converter-based PMSG WECS.

10.5.2 Control Requirements

The control requirements for the PMSG WECS with BTB converters are discussed in Chapter 9. Although the power converter architecture is changed in this chapter to PGS converters, the control objectives remain the same. The Level I to III control objectives are briefly reviewed here again:

- Maximum energy capture through MPPT operation.
- Regulation of net DC-bus voltage at its reference value.
- Generation of grid reactive power whenever required.
- Grid synchronization.

The MSC accomplishes the first variable, and the GSC handles the last three variables. In BTB power converters, MPPT operation is achieved through the control of generator currents. By contrast, in PGS converters, the inductor current is controlled to achieve MPPT operation. The control objectives can be interchanged between the MSC and GSC. The DC/DC boost converter can handle the regulation of the net DC-bus voltage, whereas the GSC can perform MPPT operation. In addition to the above basic requirements, some specific converter-related requirements are as follows: balancing of split DC-link capacitors voltage in multilevel converters, low switching frequency operation, common-mode voltage mitigation, and minimization of generator torque ripple. With PGS converters, the balancing of DC-link capacitors voltage can be solely handled by the 3L boost converter, which can provide high control flexibility for the grid-side NPC inverter to manage the last three basic objectives.

10.5.3 Notation of Variables

The scalar variables in the digital control system are wind speed v_w, input DC voltage v_{in}, and inductor current i_{dc}. These three variables are common to both 2L and 3L boost converters. The vector variables (group of scalar variables) are defined below for the generator- and grid-side 2L and 3L converters to simplify the block diagram and system analysis:

- Three-phase grid voltages, $\mathbf{v_g} = [v_{ag} \;\; v_{bg} \;\; v_{cg}]^T$
- Three-phase grid currents, $\mathbf{i_g} = [i_{ag} \;\; i_{bg} \;\; i_{cg}]^T$
- 2L DC-link capacitor voltage, $\mathbf{v_C} = [v_{C1}]^T$
- 3L DC-link capacitors voltage, $\mathbf{v_C} = [v_{C1} \;\; v_{C2}]^T$
- 2L-VSI switching signals, $\mathbf{s_i} = [s_{ai1} \;\; s_{bi1} \;\; s_{ci1}]^T$
- NPC inverter switching signals, $\mathbf{s_i} = [s_{ai1} \;\; s_{ai2} \;\; s_{bi1} \;\; s_{bi2} \;\; s_{ci1} \;\; s_{ci2}]^T$
- 2L-Boost switching signals, $\mathbf{s_{dc}} = [s_{dc1}]^T$
- 3L-Boost switching signals, $\mathbf{s_{dc}} = [s_{dc1} \;\; s_{dc2}]^T$

On the basis of the above-mentioned notation, the DC-link capacitor voltage and boost converter switching signal with 2L converters are essentially scalar variables. However, to generalize the control block diagram for both 2L and 3L power converters, they are represented as vector variables.

10.5.4 Calculation of Reference Control Variables

By using PGS converters, variable-speed operation for the PMSG WECS can be obtained by PCC, PPC, or PSC scheme. The reference control variables that correspond to the PCC, PPC, and PSC are reference inductor current i_{dc}^*, reference input power P_{in}^* and reference generator speed ω_m^*, respectively. With the notation in Section 1.7.4, the MPPT algorithms based on optimal tip-speed ratio (OTSR) and power signal feedback (PSF) are presented in Figures 10.10(a) and 10.10(b), respectively.

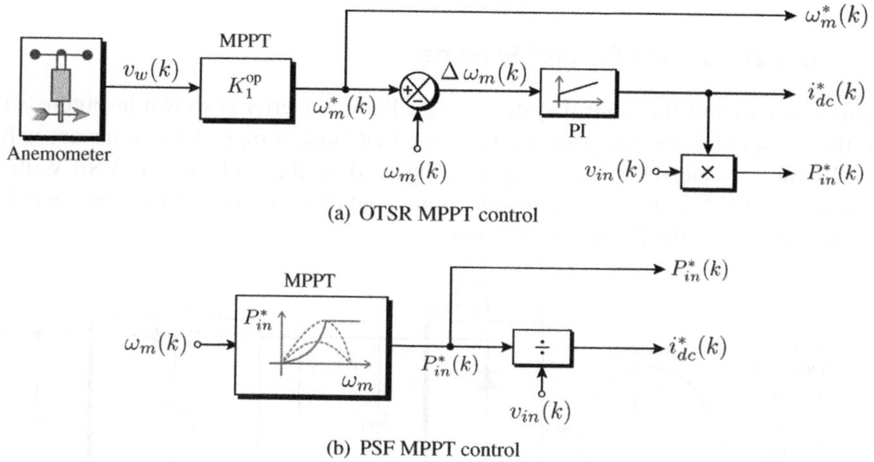

Figure 10.10 Calculation of reference control variables for PCC, PPC and PSC schemes.

The OTSR MPPT algorithm (Figure 10.10(a)) provides reference generator speed ω_m^* on the basis of the measured wind speed v_w and other WT parameters [15, 21]. The coefficient K_1^{op} is calculated as follows:

$$\omega_m^* = \frac{\lambda_T^{op}\, v_w\, r_{gb}}{r_T} = K_1^{op}\, v_w \tag{10.16}$$

where λ_T^{op} is optimal tip-speed ratio, r_{gb} is the gear ratio, r_T is the blade radius (m), and v_w is the wind speed (m/s).

A PI controller is used to generate the reference inductor current $i_{dc}^*(k)$ such that the generator speed ω_m is maintained at its reference value ω_m^*:

$$i_{dc}^*(k) = \left(k_p + \frac{k_i}{s}\right)(\omega_m^*(k) - \omega_m(k)) \tag{10.17}$$

where s is the Laplace operator. k_p and k_i are the proportional and integral gains of the PI controller, respectively. With the varying wind-speed conditions, this reference inductor current magnitude changes and the MPPT operation can be achieved by tracking this quantity. The reference power P_{in}^* is calculated by multiplying $i_{dc}^*(k)$ with $v_{in}(k)$.

The PSF MPPT algorithm (Figure 10.10(b)) replaces the wind-speed sensor with a generator speed sensor [22]. With the initial tests, P_{in}^* versus ω_m curves are computed and stored in memory. The PSF control calculates reference power P_{in}^* on the basis of the measured generator speed ω_m. The reference inductor current $i_{dc}^*(k)$ is calculated by dividing $P_{in}^*(k)$ with the measured input voltage $v_{in}(k)$.

10.6 PREDICTIVE CURRENT CONTROL OF 2L-PGS-CONVERTER-BASED PMSG WECS

As mentioned in the previous section, the control algorithms for the MSC and GSC are independent of each other because of the decoupling offered by the second DC link. A detailed analysis on the control of MSC comprised of 3ϕ diode bridge rectifier and 2L boost converter is presented in this section. The control system for the grid-side 2L-VSI can be found in Chapter 8.

10.6.1 Generator-side Control Scheme

The block diagram of the control system for 2L PGS converter is shown in Figure 10.11. From the boost converter perspective, the second DC link is treated as a constant voltage source because the net DC-bus voltage is controlled by the grid-side 2L-VSI. With this condition, the PMSG stator current and electromagnetic torque are regulated by controlling the gating signals for the 2L boost converter.

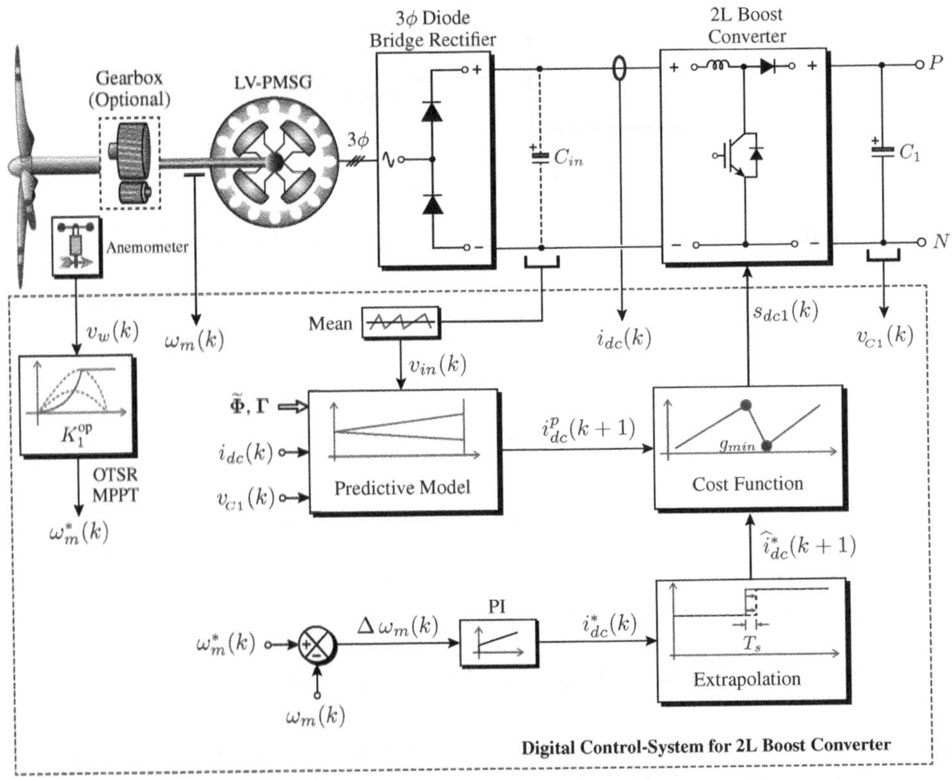

Figure 10.11 Block diagram of the PCC scheme for a 2L boost converter-based PMSG WECS.

To generate gating signals through the predictive control, five major tasks are outlined: (1) measurement of variables, (2) calculation of reference inductor current, (3) extrapolation of reference inductor current, (4) prediction of future behavior of inductor current, and (5) generation of gating signals through the cost function minimization. These major tasks are analyzed below.

Measurement and Synthesis of Feedback Signals: The first step in the implementation is to measure the required feedback signals: shaft speed $\omega_m(k)$, filtered input DC voltage $v_{in}(k)$, DC current $i_{dc}(k)$, and capacitor voltage $v_{C1}(k)$. As shown by dashed lines, C_{in} is an optional component. The $v_{in}(k)$ contains significant ripple (due to small capacitance value), and its mean value must be extracted by using the proper filtering approach.

Calculation of Reference Inductor Current: The main objective for the 2L boost converter is to enable the variable-speed operation by performing MPPT. To achieve this operation, the OTSR MPPT algorithm produces generator reference speed $\omega_m^*(k)$ in proportion to $v_w(k)$. The reference inductor current $i_{dc}^*(k)$ is calculated by the closed-loop PI controller such that the generator speed $\omega_m(k)$ is maintained at its rated value $\omega_m^*(k)$.

Extrapolation of Reference Inductor Current: The PCC anticipates that errors will occur in the future sample. This algorithm then generates the switching signal such that largest error terms are eliminated. Given that the error values must be determined in the future sample, the reference current is extrapolated to the $(k+1)$ sample. The estimation of $\widehat{i_{dc}^*}(k+1)$ by using the first-order Lagrange extrapolation is shown below:

$$\widehat{i_{dc}^*}(k+1) = 2\, i_{dc}^*(k) - i_{dc}^*(k-1). \tag{10.18}$$

Prediction of Future Behavior of Inductor Current: The next step is to calculate the possible future values of the inductor current based on converter parameters and measured variables. To reduce the number of online calculations, the DT matrices $\widetilde{\mathbf{\Phi}}$ (by excluding the "$(1 - s_{dc1}(k))$" term) and $\mathbf{\Gamma}$ are defined offline. By using the DT model of 2L-boost converter in Equation (10.5), two possible future values of the inductor current are calculated with $s_{dc1}^p(k)$ = '1' and $s_{dc1}^p(k)$ = '0'. To verify the control algorithm through simulations, estimating the virtual load current $i_{R1}(k)$ is unnecessary because it is not connected to the inductor current dynamics.

Cost Function Minimization: The cost function deals with the control objectives of the system. The predicted variables are used by the cost function, and the switching signals that minimize the cost function are selected and applied to the boost converter. The primary objective of inductor current regulation is incorporated in the sub-cost function:

$$g_{idc}(k) = \left[\widehat{i_{dc}^*}(k+1) - i_{dc}^p(k+1)\right]^2. \tag{10.19}$$

Another requirement for the high-power boost converter is to operate at lower switching frequency. This requirement is embedded into another sub-cost function as follows:

$$g_{sw,dc}(k) = \left[s_{dc1}^p(k) - s_{dc1}^{\text{op}}(k)\right]^2 \tag{10.20}$$

where $s_{dc1}^{\text{op}}(k)$ is the optimal switching signal in the previous sample.

The final cost function combines the above two objectives, that is,

$$g_{dc}(k) = \lambda_{idc}\, g_{idc}(k) + \lambda_{sw,dc}\, g_{sw,dc}(k). \tag{10.21}$$

The weighting factor λ_{idc} associated with the primary objective is set to one. With $\lambda_{sw,dc} > 0$, the switching frequency reduction is obtained but at the expense of high inductor current tracking error. Given that the cost function includes more than one control goal, the quadratic cost function performs better than the absolute cost function, as discussed in Chapter 4. During each sampling instant, the switching signal $s_{dc1}(k)$, which minimizes the cost function (10.21), is chosen and applied to the 2L-boost converter directly. Employing internal PI controllers and modulator is not required.

10.6.2 Control Algorithm

The flowchart for the digital implementation of the PCC algorithm for a 2L boost converter is shown in Figure 10.12. This algorithm is identical to the 2L-VSI PCC algorithm illustrated in Figure 4.3. The PCC algorithm mainly consists of nine blocks/steps. By using the feedback measurements in ①, the extrapolated reference inductor current is calculated in ②. The blocks in ③ and ④ correspond to the PCC initialization and counter operation. The future value of inductor current is predicted in ⑤, and the current error is computed in ⑥ through a cost function. The switching state number j, which produces minimum g_{dc} value, is stored as $j_{\rm op}$ in ⑦. The optimal $j_{\rm op}$ number is produced by block ⑧ after the control loop runs twice ($0 \sim 1$). On the basis of the $j_{\rm op}$ value, the block ⑨ produces the optimal switching state for the 2L boost converter.

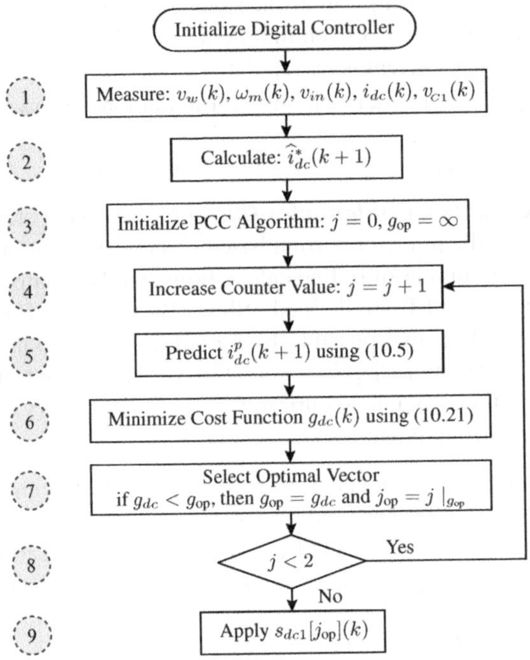

Figure 10.12 Flowchart of the PCC algorithm for 2L boost converter-based PMSG WECS.

Algorithm 10.1 shows the *S-Function Builder* code to realize PCC in a `MATLAB` environment. The input ports of the *S-Function Builder* are extrapolated reference current `idc_ref_k1`, weighting factor `alpha`, filtered input voltage `vin`, inductor current `idc`, and DC-link capacitor voltage `vc`. The output port is switching signal `s_dc` whose value corresponds to '`1`' or '`0`'. The DT parameters `P11`, `P12`, and `G11`, are computed offline and included in the *library* of *S-Function Builder*.

The PCC algorithm initialization and counter operation is performed in the first three lines. The PCC algorithm `for` loop is depicted in lines `04-14`. The inductor current prediction is conducted in line `05`. The cost function calculation is performed in lines `06-08`. The selection of optimal switching state number and generation of optimal switching signals for the boost converter is performed in lines `09-15`. The optimal switching state value is stored in line `16`. It is used in the next iteration to calculate `g_swdc` in line `07`. The quadratic operation in line `07` eliminates the negative cost function values.

ALGORITHM 10.1 PCC Algorithm for 2L Boost Converter

```
\\ P11 = 1-(rdc*Ts/Ldc); P12 = -Ts/Ldc; G11 = Ts/Ldc;
j_op = 1000;
g_op = 1000000000;
for(j = 0; j < 2; j++)
{
    idc_k1[0] = P11*idc[0]+P12*(1-vboost[j].s_dc[0])*vc[0]+G11*vin[0];
    g_idc = (idc_ref_k1[0]-idc_k1[0])*(idc_ref_k1[0]-idc_k1[0]);
    g_swdc = (vboost[j].s_dc[0]-s_dc1_km1[0])
            *(vboost[j].s_dc[0]-s_dc1_km1[0]);
    g_dc = g_idc + alpha[0]*g_swdc;
    if(g_dc < g_op)
        {
            j_op = j;
            g_op = g_dc;
        }
}
s_dc1[0] = vboost[j_op].s_dc[0];
s_dc1_km1[0] = s_dc1[0];
```

10.7 PREDICTIVE CURRENT CONTROL OF 3L-PGS-CONVERTER-BASED PMSG WECS

The PCC of the 3L boost converter is presented in this section as an extension to the PCC of the 2L boost converter discussed in the previous section. The 3L boost converter controls the DC-link neutral-point voltage, in addition to the inductor current. The grid-side NPC inverter controls the reactive power and net DC-bus voltage through the regulation of decomposed grid currents. The NPC inverter need not to include the task of DC-link capacitors voltage balancing because it is solely performed by the 3L boost converter.

10.7.1 Generator-side Control Scheme

The PCC scheme for the 3L boost converter-based PMSG WECS is shown in Figure 10.13. The additional feedback measurements required in comparison with the PCC of the 2L boost converter are as follows: (1) $\mathbf{i}_\mathrm{g}(k)$ and NPC inverter optimal switching signals $\mathrm{s}_\mathrm{i}^\mathrm{op}(k)$ to estimate virtual load currents, $\widehat{\mathbf{i}}_\mathrm{R}(k)$, and (2) two DC-link capacitors voltage, $\mathbf{v}_\mathrm{C}(k)$. The OTSR in Figure 10.11 is replaced by PSF to show the feasibility of applying different MPPT algorithms to the PGS converter-based PMSG WECS. With the help of $P_{in}^*(k)$ and $v_{in}(k)$, the reference inductor current, $i_{dc}^*(k)$, is computed. The predictive model uses the DT model of the 3L boost converter provided in (10.14) to estimate the future behavior of both inductor current and DC-link capacitors voltage.

The final cost function for a 3L boost converter is formulated as follows:

$$\begin{aligned} g_{dc}(k) = {} & \lambda_{idc}\left[\widehat{i}_{dc}^*(k+1) - i_{dc}^p(k+1)\right]^2 \\ & + \lambda_{dc,dc}\left[v_{C1}^p(k+1) - v_{C2}^p(k+1)\right]^2 \\ & + \lambda_{sw,dc}\left(\left[s_{dc1}^p(k) - s_{dc1}^{\mathrm{op}}(k)\right]^2 + \left[s_{dc2}^p(k) - s_{dc2}^{\mathrm{op}}(k)\right]^2\right) \end{aligned} \tag{10.22}$$

where λ_{idc}, $\lambda_{dc,dc}$, and $\lambda_{sw,dc}$ are weighting factors for the inductor current control, balancing of DC-link capacitors voltage, and switching frequency minimization, respectively.

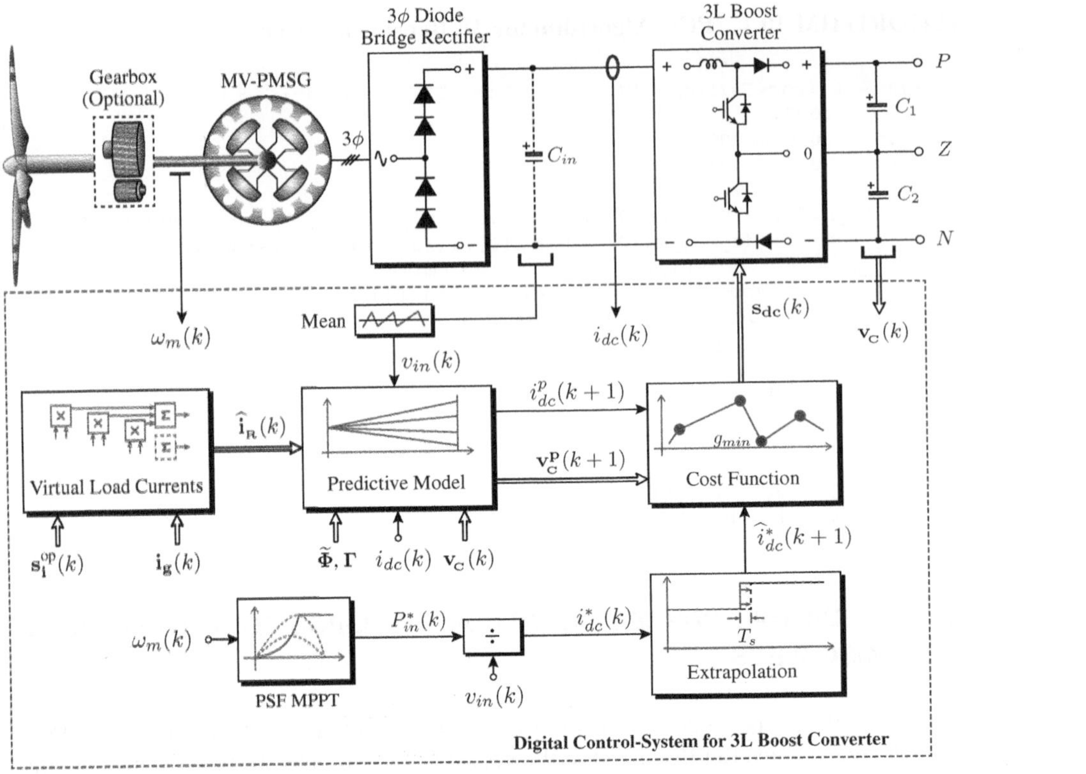

Figure 10.13 Block diagram of the PCC scheme for a 3L boost converter-based PMSG WECS.

By using the per-unit weighting factor method introduced in Chapter 4, the weighting factors λ_{idc} and $\lambda_{dc,dc}$ are calculated as follows:

$$\lambda_{idc} = \frac{i_{dc,R}}{i_{dc,R}} = 1, \quad \lambda_{dc,dc} = \frac{i_{dc,R}}{v_{\mathrm{dc}}^*}. \tag{10.23}$$

The rated inductor current $i_{dc,R}$ and rated DC-bus voltage v_{dc}^* for a 3.0-MW, 3000-V PMSG WECS are 850 A and 5304 V, respectively. With these values, $\lambda_{idc} = 1$ and $\lambda_{dc,dc} = 850/5304 = 0.16$. These weighting factor values ensure excellent reference tracking for inductor current and perfect balancing of DC-link capacitors voltage during steady state and transient operating conditions. In 3L converters, $\lambda_{sw,dc}$ affects both inductor current tracking and capacitors voltage balancing. Switching signals $s_{dc1}(k)$ and $s_{dc2}(k)$, which produce the lowest cost function value, are applied to the 3L boost converter directly.

10.7.2 Control Algorithm

The flowchart for the implementation of the PCC algorithm for a 3L boost converter-based PMSG is shown in Figure 10.14. The block in ① uses additional measurements to accommodate the calculation of $\hat{\mathbf{i}}_{\mathrm{R}}(k)$ in block ②. The predictive model in ⑤ calculates the future values of inductor current and DC-link capacitors voltage. Block ⑨ derives the optimal switching signals for the 3L boost converter-based on the j_{op} number. The rest of the flowchart and its realization are similar to those of the 2L boost converter in Figure 10.12. The corresponding *S-Function Builder* code is presented in Algorithm 10.2.

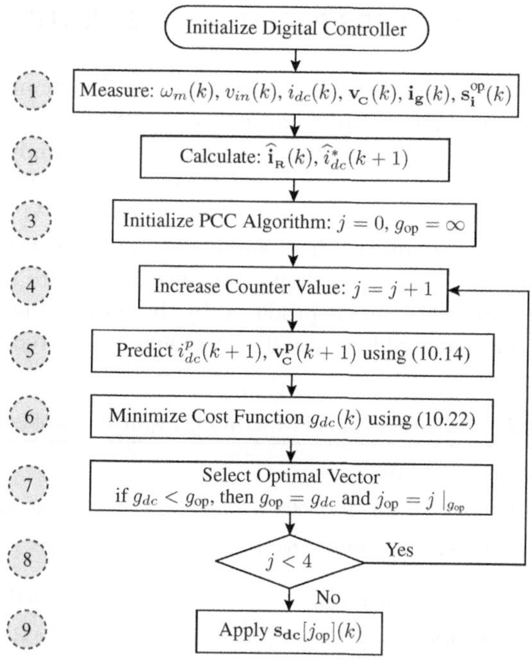

Figure 10.14 Flowchart of the PCC algorithm for a 3L boost converter-based PMSG WECS.

ALGORITHM 10.2 PCC Algorithm for 3L Boost Converter

```
\\ P11 = 1-(rdc*Ts/Ldc); P12 = P13 = -Ts/Ldc; P21 = Ts/C1;
\\ P31 = Ts/C2; G11 = Ts/Ldc; G22 = -Ts/C1; G33 = -Ts/C2;
j_op = 1000;
g_op = 1000000000;
for(j = 0; j < 4; j++)
{
    idc_k1[0] = P11*idc[0]+P12*(1-vboost[j].s_dc[0])*vc[0]
                  +P13*(1-vboost[j].s_dc[1])*vc[1]+G11*vin[0];
    vc_k1[0] = P21*(1-vboost[j].s_dc[0])*idc[0]+vc[0]+G22*ir[0];
    vc_k1[1] = P31*(1-vboost[j].s_dc[1])*idc[0]+vc[1]+G33*ir[1];
    g_idc = (idc_ref_k1[0]-idc_k1[0])*(idc_ref_k1[0]-idc_k1[0]);
    g_dcdc = (vc_k1[0]-vc_k1[1])*(vc_k1[0]-vc_k1[1]);
    g_swdc = ((vboost[j].s_dc[0]-s_dc_km1[0])
             *(vboost[j].s_dc[0]-s_dc_km1[0]))
             + ((vboost[j].s_dc[1]-s_dc_km1[1])
             *(vboost[j].s_dc[1]-s_dc_km1[1]));
    g_dc = g_idc + alpha[1]*g_dcdc + alpha[0]*g_swdc ;
    if(g_dc < g_op)
        {
          j_op = j;
          g_op = g_dc;
        }
}
s_dc1[0] = vboost[j_op].s_dc[0];
s_dc2[0] = vboost[j_op].s_dc[1];
s_dc_km1[0] = s_dc1[0];
s_dc_km1[1] = s_dc2[0];
```

10.8 ANALYSIS OF PMSG WECS PERFORMANCE WITH PGS CONVERTERS

The PMSG WECS performance with PGS converters and MPC scheme is analyzed through `MATLAB/Simulink` simulation results and `dSPACE` DS1103-based experimental results. The analysis with 2L boost converters is omitted in this study because its performance is similar to that of the 3L converters except that controlling the DC-link capacitors voltage is unnecessary. The parameters of the 3L converter-based PMSG WECS are shown in Table 10.5. The PMSG is assumed to have surface mount magnets by which $L_{ds} = L_{qs}$. The generator-side, DC link, and grid-side performance indices are analyzed through six case studies. The switching harmonics produced by the power converters, losses in the converters, and core and rotational losses of the generator are ignored in the analysis.

Table 10.5 Parameters for a PMSG WECS with 3L boost converter

Wind turbine power rating: 3.0 MW	Table A.1 (Appendix A)
Wind generator: DD-PMSG, 3ϕ, 3.0 MW, 3000 V, 9.75 Hz	Table A.6 (Appendix A)
Grid and base values: 3ϕ, 3.0 MVA, 3000 V, 60 Hz	Table A.2 (Appendix A)
Boost converter DC choke inductance, L_{dc}	2.5 mH (0.31 pu)
Internal resistance of DC choke, r_{dc}	0.03 Ω (0.01 pu)
Boost converter input capacitance, C_{in}	3537 μF (4.0 pu)
DC-link capacitor values, C_1 and C_2	3537 μF (4.0 pu)
Grid-side filter inductance, L_i	1.59 mH (0.2 pu)
Internal resistance of grid-side filter, r_i	0.03 Ω (0.01 pu)
Reference net DC-bus voltage, v_{dc}^*	5304 V (3.062 pu)
Predictive controller sampling time, T_s	100 μs
Weighting factor, $\lambda_{dc,dc}$	0.16

CASE STUDY 10.1 Selection of Optimal Gating Signals for a 3L Boost Converter

Case Study Objective: This case study provides in-depth analysis related to the DT model, prediction of future behavior of inductor current and DC-link capacitors voltage, cost function minimization, and selection of weighting factors. The mathematical expressions of a 3L boost converter are translated into operating principles through this case study.

Parameters: The parameters for the 3L boost converter are shown in Table 10.5. The measured and estimated values are as follows: $v_{C1}(k) = 2640.2$ V, $v_{C2}(k) = 2660.6$ V, $\widehat{i}_{R1}(k) = 811.64$ A, $\widehat{i}_{R2}(k) = 433.91$ A, $v_{in}(k) = 3472.8$ V, and $i_{dc}(k) = 842.24$ A. The value of the extrapolated reference current $\widehat{i}^*_{dc}(k+1)$ is 852.92 A. Three different weighting factor values, $\lambda_{dc,dc} = 0, 0.16$, and 1, are considered to analyze the balancing issue of DC-link capacitors voltage. The weighting factor associated with the switching frequency reduction, $\lambda_{sw,dc}$, is set to zero to simplify the analysis.

Analysis:
The graphical representation of this case study is shown in Figure 10.15. It is composed of four main parts: (1) identification of switching combinations during the four possible operating modes, (2) prediction of inductor current and capacitors voltage, (3) minimization of cost function considering different weighting factor values, and (4) selection of optimal switching signals to apply to the 3L boost converter during the next sampling interval.

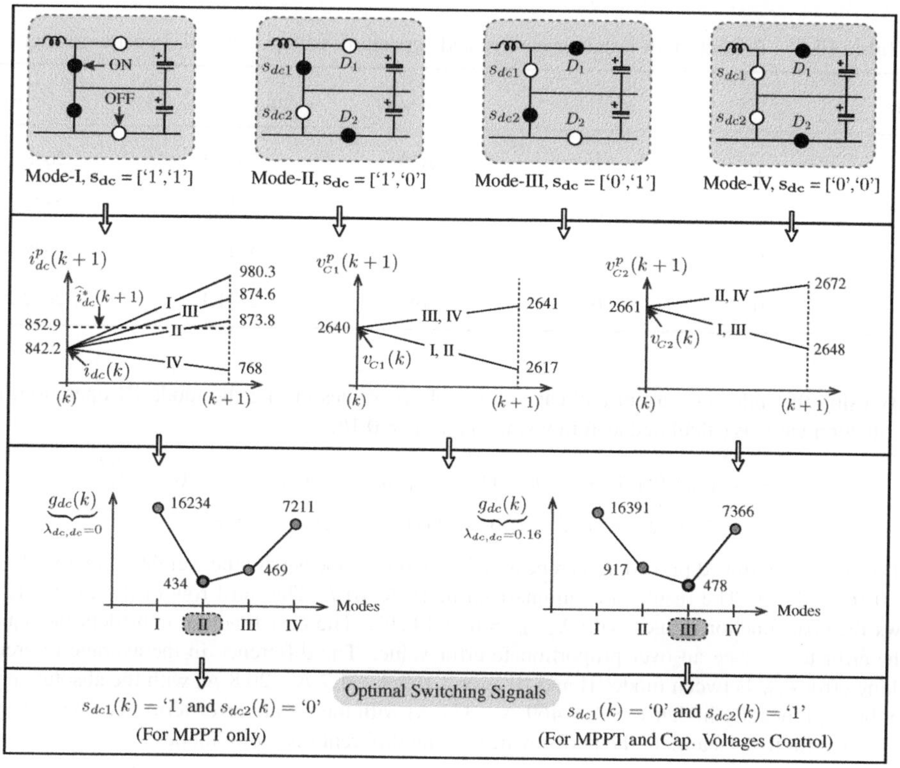

Figure 10.15 Graphical representation of sample-based analysis for a 3L boost converter.

As a first step in the analysis, the switching combinations during the operating modes I to IV are identified as $\mathbf{s_{dc}} = [s_{dc1}, s_{dc2}]$ = ['1', '1'], ['1', '0'], ['0', '1'], ['0', '0'], respectively. These switching combinations are shown in the first row of Figure 10.15. To perform predictions for the variables to be controlled, the DT state matrix $\mathbf{\Phi}$ and input matrix $\mathbf{\Gamma}$ are estimated as follows:

$$\mathbf{\Phi}(k) = \begin{bmatrix} 0.9988 & -0.04\,(1 - s_{dc1}(k)) & -0.04\,(1 - s_{dc2}(k)) \\ 0.028\,(1 - s_{dc1}(k)) & 1 & 0 \\ 0.028\,(1 - s_{dc2}(k)) & 0 & 1 \end{bmatrix}$$

$$\mathbf{\Gamma} = \begin{bmatrix} \frac{T_s}{L_{dc}} & 0 & 0 \\ 0 & -\frac{T_s}{C_1} & 0 \\ 0 & 0 & -\frac{T_s}{C_2} \end{bmatrix} = \begin{bmatrix} 0.04 & 0 & 0 \\ 0 & -0.028 & 0 \\ 0 & 0 & -0.028 \end{bmatrix}.$$

By using the model in (10.14), the required variables are predicted as follows:

$$\begin{bmatrix} i_{dc}(k+1) \\ v_{C1}(k+1) \\ v_{C2}(k+1) \end{bmatrix} = \mathbf{\Phi}(k) \begin{bmatrix} i_{dc}(k) \\ v_{C1}(k) \\ v_{C2}(k) \end{bmatrix} + \mathbf{\Gamma} \begin{bmatrix} v_{in}(k) \\ i_{R1}(k) \\ i_{R2}(k) \end{bmatrix}.$$

Four different $\mathbf{\Phi}(k)$ values are obtained by substituting four switching state combinations. Again four predictions are obtained for the inductor current and capacitors voltage by substituting the $\mathbf{\Phi}(k)$ values in DT model. The summary of these values is shown in Table 10.6 and in the second row of Figure 10.15. These values show that $i_{dc}(k+1)$ has four different values, whereas $v_{C1}(k+1)$ and $v_{C2}(k+1)$ have two different values. During modes I and II, the $s_{dc1}(k) = $ '1', and thus $v_{C1}(k+1)$ value remains unchanged. Similarly, the $v_{C2}(k+1)$ value is the same during modes I and III because $s_{dc2}(k) = $ '1' in both modes.

Table 10.6 Prediction of inductor current and capacitors voltage for a 3L boost converter

Mode	$s_{dc1}(k)$	$s_{dc2}(k)$	$i_{dc}(k+1)$	$v_{C1}(k+1)$	$v_{C2}(k+1)$
I	'1'	'1'	980.3	2617.3	2648.3
II	'1'	'0'	873.8	2617.3	2672.2
III	'0'	'1'	874.6	2641.1	2648.3
IV	'0'	'0'	768	2641.1	2672.2

By using the inductor current and capacitors voltage values during the mode III operation, the cost function value is calculated as follows, with $\lambda_{dc,dc} = 0.16$:

$$g_{dc}(k) = \left[\hat{i}^{*}_{dc}(k+1) - i^{p}_{dc}(k+1)\right]^2 + \lambda_{dc,dc}\left[v^{p}_{C1}(k+1) - v^{p}_{C2}(k+1)\right]^2$$
$$= [852.92 - 874.6]^2 + 0.16 \times [2641.1 - 2648.3]^2 = 478.$$

The above-mentioned procedure is repeated during other modes with the weighting factors $\lambda_{dc,dc}$ = 0, 0.16, and 1.0. The results are summarized in Table 10.7. The third row of the Figure 10.15 shows the cost function values with $\lambda_{dc,dc} = 0$ and 0.16. The cost function considers the square of the error to produce an over proportionate error value. The difference in the average reference tracking error e_{idc} between modes II and III is only 0.9 A (21.7 A – 20.8 A) with the absolute error term, but significantly higher (35 A = 469 A – 434 A) with the square error term. The higher error value is helpful to substantiate the performance during different operating modes.

Table 10.7 Cost function values for a 3L boost converter with different weighting factors

Mode	$s_{dc1}(k)$	$s_{dc2}(k)$	e_{idc}	e_{vc}	$\underbrace{g_{dc}(k)}_{\lambda_{dc,dc}=0}$	$\underbrace{g_{dc}(k)}_{\lambda_{dc,dc}=0.16}$	$\underbrace{g_{dc}(k)}_{\lambda_{dc,dc}=1}$
I	'1'	'1'	127.4	31.1	16234	16391	17203
II	'1'	'0'	**20.8**	54.9	**434**	917	3448
III	'0'	'1'	21.7	**7.3**	469	**478**	**522**
IV	'0'	'0'	84.9	31.1	7211	7366	8177

The results in Table 10.7 indicate that during mode II, a minimum value of the tracking error (20.8 A) can be obtained if the capacitors voltage are not regulated. The secondary objective with the 3L boost converter is to regulate the DC-link capacitors voltage. When $\lambda_{dc,dc} = 0.16$ is applied to the above-mentioned cost function, mode III provides a minimum value for the cost function ($g_{dc}(k) = 478$). This condition means that during mode III, the error in the capacitors voltage, e_{vc}, becomes low (7.3 V). However, the error e_{idc} increases by 0.9 A. The switching states in mode III, $s_{dc1}(k) =$ '0' and $s_{dc2}(k) =$ '1', are selected by the controller and applied to the boost converter. The third and fourth rows in Figure 10.15 support the above-mentioned explanation.

The weighting factor $\lambda_{dc,dc} = 1$ also produces the same switching signals for the particular sampling instant considered in this case study. The use of a high value for $\lambda_{dc,dc}$ is not recommended because it will produce higher ripple (error) in the inductor current. The weighting factor $\lambda_{dc,dc}$, which provides an error of 2% in capacitors voltage, should be sufficient to ensure safe and efficient operation for the power converters. ■

CASE STUDY 10.2 Balancing of Capacitors Voltage with Predictive Control

Case Study Objective: The performance of PCC scheme with the unbalanced DC-link capacitors voltage is investigated in this case study. The unbalanced conditions that arise from perturbation in the DC-link model, failure of components in the DC link, loss of controller for a few sampling cycles, or transient operating conditions cause the diversion of the capacitors voltage from their nominal values. The voltage waveforms of the experimental DC-link capacitors are shown in Figure 10.16 with the absence of controller for a few sampling instants. The objective of this case study is to analyze the effectiveness of MPC in converting unbalanced condition to balanced condition.

Parameters: Case study 10.1 is considered again in this study to simplify the analysis. The v_{C1} value is increased by 1000 V, and the v_{C2} value is decreased by 1000 V to assume an unbalanced condition. The resulting new capacitors voltage are as follows: v_{C1} = 3640.2 V and v_{C2} = 1660.6 V. The difference in the capacitors voltage, e_{vc}, is 1979.6 V.

Analysis:
To balance the capacitors voltage, v_{C1} must be discharged to the virtual load R_1 rapidly, while charging capacitor C_2. Mode II in Figure 10.7 is the most suitable mode to discharge and charge C_1 and C_2, respectively. The new predicted inductor current and capacitors voltage, error magnitudes, and cost function values are shown in Table 10.8.

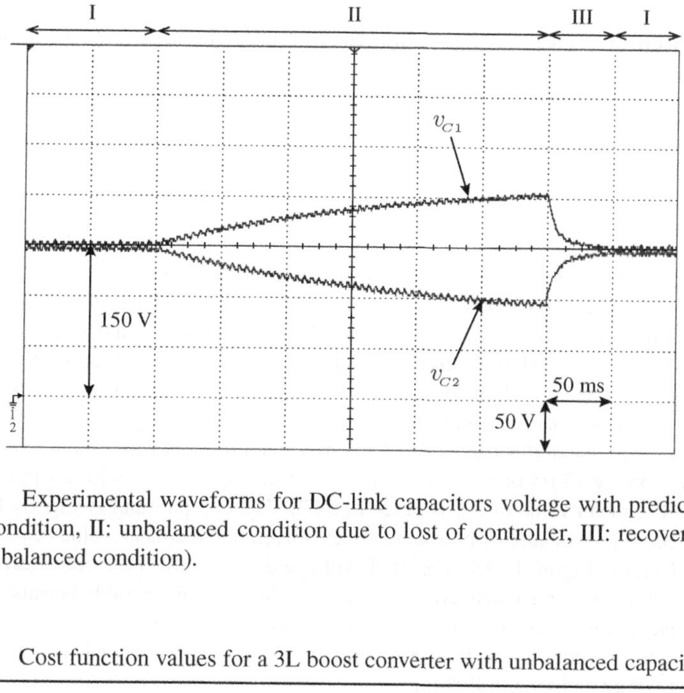

Figure 10.16 Experimental waveforms for DC-link capacitors voltage with predictive controller (I: balanced condition, II: unbalanced condition due to lost of controller, III: recovery period from unbalanced to balanced condition).

Table 10.8 Cost function values for a 3L boost converter with unbalanced capacitors voltage

Mode	$s_{dc1}(k)$	$s_{dc2}(k)$	$i_{dc}(k+1)$	$v_{C1}(k+1)$	$v_{C2}(k+1)$	e_{idc}	e_{vc}	$\underbrace{g_{dc}(k)}_{\lambda_{dc,dc}=0.16}$
I	'1'	'1'	980.3	3617.3	1648.3	127.4	1968.9	636499
II	'1'	'0'	913.8	**3617.3** ⇓	**1672.1** ⇑	60.9	**1945.1**	**609059**
III	'0'	'1'	834.5	3641.1	1648.3	**18.4**	1992.7	635697
IV	'0'	'0'	768	3641.1	1672.1	84.9	1968.9	627475

The predictions show that only mode II operation produces less e_{vc} value than the measured error value of 1979.6 V. Although the e_{idc} is lower in mode III operation, the cost function that deals with inductor current and capacitors voltage provides the minimum value for mode II operation. This operating mode decreases the v_{C1} by 22.9 V and increases the v_{C2} by 11.5 V, whereas the net error e_{vc} decreases by 34.5 V (1979.6 V – 1945.1 V). Although the capacitors voltage do not become balanced in one sampling instant, this case study demonstrates that the capacitors voltage follow the convergence direction. Within a few milliseconds, the v_{C1} and v_{C2} would reach their nominal values.

The experimental waveforms that correspond to this case study are presented in Figure 10.16. These test results are obtained with low-power IGBT-based boost converter. The net DC-bus voltage is 300 V and is equally shared among the DC-link capacitors. At t = 0.1 s, the weighting factor associated with the capacitors voltage balancing ($\lambda_{dc,dc}$) is set to zero to imitate the loss of controller and create an unbalanced condition in the capacitors voltage. The voltage v_{C1} across the top capacitor starts to increase, whereas the bottom capacitor voltage v_{C2} gradually decreases. At t = 0.4 s, the weighting factor is activated again ($\lambda_{dc,dc}$ = 0.16), and the DC-link capacitors voltage become balanced within 50 ms. For the particular low-power test, the predictive controller operating with a sampling time T_s of 100 μs took approximately 500 sampling cycles (50 ms/100 μs) to bring the capacitors voltage back to their rated value. The high-power converter also exhibits similar performance with a recovery period of 50 ms. ■

CASE STUDY 10.3 Steady-State Analysis of PMSG with 3L PGS Converter

Case Study Objective: The steady-state performance of the 3L PGS converter-based PMSG WECS is analyzed in this case study under rated operating conditions.

Parameters: The wind speed is considered to be 12 m/s (1.0 pu). The grid-side NPC inverter operates with UPF.

Analysis:

The steady-state waveforms (in per-unit values) of a 3 MW, 3000 V PMSG WECS operating under rated conditions are shown in Figure 10.17. The measured generator stator current and line-to-line voltages are 650 A and 2700 V, respectively. The generator output voltage is lower than the rated value because of the voltage drop across stator winding resistance R_s and synchronous reactance X_s. The generator current is slightly higher than the rated value to produce a rated output power. Therefore, the generator used with the diode rectifiers should have extra insulation levels.

Figure 10.17(a) shows that the input DC voltage v_{in} contains a large ripple because of the low value of the input filter capacitor C_{in} = 3537 μF. The maximum and minimum values for the v_{in} are measured as 3551 V (2.05 pu) and 3378 V (1.95 pu), respectively, with a ripple frequency of 58.5 Hz (6 pu). The mean value of v_{in} is noted as 3472 V (2.005 pu), and is slightly lower than the ideal value of $1.35 \times V_{s,ll}$ because of harmonic losses. Given that the mean value of v_{in} is used in the control scheme (refer to Figure 10.13), the ripple voltage does not affect the operation and control of the boost converter. However, a considerably large ripple in v_{in} is undesirable because it will impose low-frequency oscillations in the DC-link capacitors voltage v_{C1} and v_{C2}.

At the rated wind speed, the mechanical input power P_m, stator copper losses $P_{cu,s}$, and stator output power P_s are given by the following:

$$P_{cu,s} = 3\, I_s^2\, R_s = 3 \times 650^2 \times 37.521 \times 10^{-3} = 47.6\,\text{kW} \quad (0.016\,\text{pu})$$

$$P_s = P_m + P_{cu,s} = -3.0 \times 10^6 + 47.6 \times 10^3 = -2.95\,\text{MW} \quad (0.98\,\text{pu}).$$

The generator output power becomes equal to the DC-link power and grid-side power ($|\, P_s \,| = P_{in} = P_g$) by neglecting the losses in the power converters. By using this assumption, the inductor current is calculated as follows:

$$i_{dc}^* = i_{dc} = \frac{P_{in}}{v_{in}} = \frac{2.95 \times 10^6}{3472} = 850\ \text{A} \quad (1.47\ \text{pu}).$$

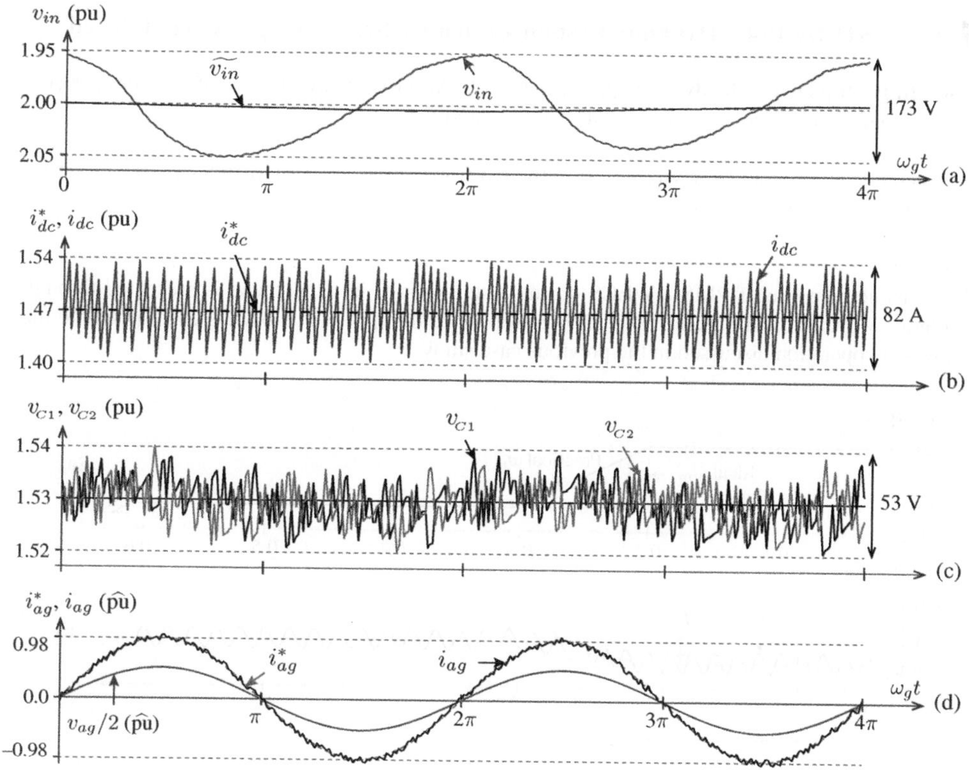

Figure 10.17 Steady-state waveforms for a 3L boost converter-based PMSG WECS.

The DC-current waveform in Figure 10.17(b) conforms to the above-mentioned pu value and analysis. The ripple in the inductor current is noted as 82 A (0.14 pu). The average switching frequency ($f_{sw,dc}$) of the boost converter is measured as 2934 Hz with $\lambda_{sw,dc} = 0$. By using the measured input and output voltages of the boost converter, the ripple in the inductor current is estimated below [23]:

$$D = 1 - \frac{v_{in}}{v_{dc}} = 1 - \frac{3472}{5304} = 0.345$$

$$\Delta i_{dc} = D\,(1 - D)\,\frac{v_{dc}}{2\,f_{sw,dc}\,L_{dc}} = 82\text{ A} \quad (0.14\text{ pu}).$$

The switching frequency of the boost converter can be decreased by simply increasing the weighting factor $\lambda_{sw,dc}$. With $\lambda_{sw,dc} = 100$, the $f_{sw,dc}$ is decreased from 2934 Hz to 1652 Hz with a slight increase in e_{idc} and e_{vc}. The net DC-bus voltage v_{dc} is maintained at its reference value by the grid-side NPC inverter, and the ripple in v_{dc} is found to be ±26.5 V. As demonstrated in Figure 10.17(c), v_{C1} and v_{C2} are accurately balanced by the 3L boost converter with a maximum difference value of 53 V (0.02 pu or 2% of base voltage). The grid current magnitude is calculated as follows:

$$I_g = \frac{P_g}{\sqrt{3}\,V_{g,ll}\,\cos\varphi_g} = \frac{2.95 \times 10^6}{\sqrt{3} \times 3000 \times 1} = 568\text{ A }(rms) \quad (0.98\text{ pu}).$$

The phase-a grid current and its reference current are shown in Figure 10.17(d). The grid currents track their references well even during the peaks and valleys of the waveform. The phase angle between the grid voltage and current is zero because of the UPF operation. The grid-side inverter operates with an average switching frequency ($f_{sw,i}$) of 1190 Hz. The grid current total harmonic distortion (THD) is noted as 4.15%. ■

CASE STUDY 10.4 Dynamic Response during a Step-Change in Wind Speed

Case Study Objective: The dynamic performance of PMSG WECS with 3L PGS converter during a step change in wind speed is analyzed in this case study.

Parameters: A step change in wind speed is applied from 12 m/s (1.0 pu) to 10.8 m/s (0.9 pu) at $t =$ 0.2 s. The grid-side power factor is assumed to be unity.

Analysis:
The dynamic response of PMSG WECS during a step change in wind speed is presented through simulation waveforms in Figure 10.18. The performance during $t = 0$ s to 0.2 s corresponds to the steady-state operation discussed in the previous case study.

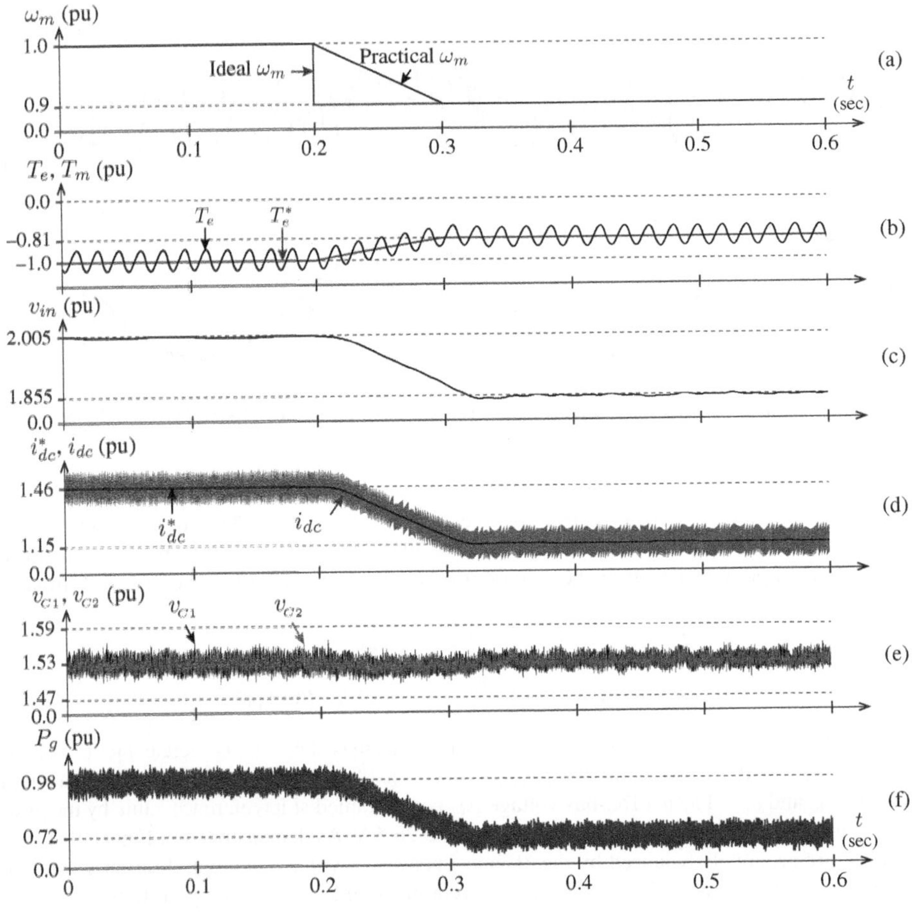

Figure 10.18 Simulated waveforms for a 3L boost converter-based PMSG WECS during step change in wind speed.

The generator-side mechanical system response in MW WECS is considerably slower than the grid-side electrical response because of large moment of inertia in the turbine and generator shafts [10, 24]. Given this condition, the generator speed cannot be instantaneously changed even though the wind speed suddenly changes. The generator mechanical speed is considered to change gradually from 1.0 pu to 0.9 pu, as shown in Figure 10.18. The generator current decreases from 657 A to 506.3

A with decreasing generator speed. The generator output voltage also decreases; thus, v_{in} decreases from 3472 V to 3220 V. The electromagnetic torque T_e also decreases in proportion to the generator speed ω_m. With the changes in the generator speed, the MPPT algorithm generates a new i^*_{dc} value and the generator-side predictive controller forces the DC inductor current to track to its reference. Thus, MPPT is obtained with varying wind speed conditions. The active power injected into the grid changes on the basis of the generator output power. The DC-link capacitors voltage are well balanced by the boost converter even during transient interval (t = 0.2 s to 0.3 s).

The generator speed, electromagnetic torque, mechanical input power, generator output power, and DC reference current after the transient interval are obtained as follows:

$$\omega_m = \omega_{m,R} \times \omega_{m,\,\mathrm{pu}} = 22.5 \times \frac{2\,\pi}{60} \times 0.9 = 2.12 \text{ rad/s} \quad (0.9 \text{ pu})$$

$$T_e = T_{m,R} \times \omega^2_{m,\,\mathrm{pu}} = 1.273 \times 10^6 \times 0.9^2 = -1.031 \text{ MN.m} \quad (-0.81 \text{ pu})$$

$$P_m = P_{m,R} \times \omega^3_{m,\,\mathrm{pu}} = -3.0 \times 10^6 \times 0.9^3 = -2.187 \text{ MW} \quad (0.729 \text{ pu})$$

$$P_{cu,s} = 3 \times I_s^2 \times R_s = 3 \times 506.3^2 \times 37.521 \times 10^{-3} = 28.85 \text{ kW} \quad (0.01 \text{ pu})$$

$$P_s = P_m + P_{cu,s} = -2.187 \times 10^6 + 28.85 \times 10^3 = -2.158 \text{ MW} \quad (0.72 \text{ pu})$$

$$i^*_{dc} = i_{dc} = \frac{|\,P_s\,|}{v_{in}} = \frac{P_{in}}{v_{in}} = \frac{2.158 \times 10^6}{3220} = 670 \text{ A} \quad (1.15 \text{ pu}).$$

The active power injected into the grid and the corresponding amplitude of grid currents is calculated as follows:

$$P_g = P_{in} = 2.158 \text{ MW} \quad (0.72 \text{ pu})$$

$$I_g = \frac{i_{dg}}{\sqrt{2}} = \frac{P_g}{\sqrt{2} \times 1.5 \times v_{dg}} = \frac{2.158 \times 10^6}{\sqrt{2} \times 1.5 \times 2450} = 415 \text{ A } (rms) \quad (0.719 \text{ pu}).$$

During t = 0.4 s to 0.6 s, the switching frequencies of the 3L boost converter and the NPC inverter are found to be 3046 and 1216 Hz, respectively (with $\lambda_{sw,dc} = \lambda_{sw,i} = 0$). These values are slightly higher than the rated operating condition. However, with $\lambda_{sw,dc} = \lambda_{sw,i} = 100$, the switching frequencies become 1022 and 890 Hz, respectively for 3L boost converter and the NPC inverter. The grid current THD is noted as 5.31% according to FFT analysis. ■

CASE STUDY 10.5 DC Link Capacitors Voltage with External Resistor R_x

Case Study Objective: The model of the DC link can be changed by connecting a resistor across any DC-link capacitor. The objective of this case study is to analyze the robustness of the predictive controller in balancing the DC-link capacitors voltage and grid-side currents during the perturbation in the DC-link model. This test resembles one of the worst-case scenarios in the operation of power converters in PMSG WECS.

Parameters: The analysis has been conducted by using the rated wind speed condition at the generator side and the UPF operation at the grid side. A low-power (3.6 kW), 208 V, 20 Hz DD-PMSG is used in the experimental tests. The parameters of the experimental test are given as follows: DC power P_{dc} = 3242 W, grid-phase voltage V_g = 120 V (*rms*), grid frequency f_g = 60 Hz, and disturbance resistance R_x = 100 Ω.

Analysis:
The block diagram of the experimental setup with external (disturbance) resistor R_x is shown in Figure 10.19(a). The perturbation in the DC-link model is created by connecting the resistor R_x across the top DC-link capacitor C_1 through manual switch S_{Rx}. The corresponding experimental results are shown in Figure 10.19(b). The switch S_{Rx} is suddenly closed at 42.5 ms. Even with the step-connection of R_x, the net DC-bus voltage v_{dc} is maintained at its reference value v^*_{dc}. This voltage is equally shared among the DC-link capacitors C_1 and C_2. The DC-link capacitors voltage v_{C1} and v_{C2} are tightly regulated, and the difference between v_{C1} and v_{C2} is negligible.

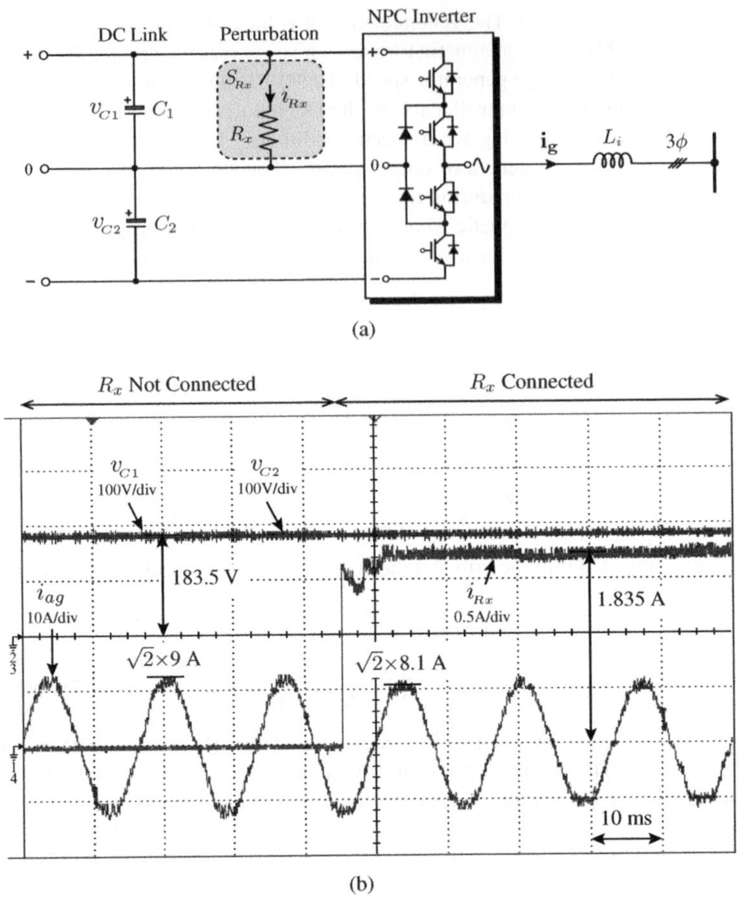

Figure 10.19 DC-link perturbation with R_x across C_1: (a) block diagram of test setup and (b) experimental waveforms.

The following analysis is conducted to estimate the current i_{Rx} flowing through the resistor R_x and the corresponding power loss (dissipation):

$$v_{dc} = v_{dc}^{*} = 3.062\, V_g = 3.062 \times 120 = 367 \text{ V}$$

$$v_{C1} = v_{C2} = \frac{v_{dc}}{2} = 183.5 \text{ V}$$

$$i_{Rx} = \frac{v_{C1}}{R_x} = \frac{183.5}{100} = 1.835 \text{ A}$$

$$P_{Rx} = i_{Rx}^2\, R_x = 1.835^2 \times 100 = 336.7 \text{ W}.$$

The assumption that $P_{dc} = P_g$ is made by neglecting the losses in power converters. The reference grid currents before the DC-link perturbation are calculated below:

$$i_{dg}^{*} = \frac{P_{dc}}{1.5\, v_{dg}} = \frac{3242}{1.5 \times \sqrt{2} \times 120} = 12.73 \text{ A}, \quad I_g^{*} = \frac{i_{dg}^{*}}{\sqrt{2}} = 9 \text{ A } (rms).$$

Before connecting the R_x across C_1, the current delivered to the grid closely follows its reference of 9 A. With this condition, the active power injected to the grid is as follows [3]:

$$P_g = 3\, V_g\, I_g \cos \varphi_g = 3 \times 120 \times 9 \times 1 = 3242 \text{ W}.$$

This value matches the statement provided earlier.

The reference grid currents after the DC-link perturbation are as follows:

$$i_{dg}^* = \frac{P_{dc} - P_{Rx}}{1.5\, v_{dg}} = \frac{3242 - 336.7}{1.5 \times \sqrt{2} \times 120} = 11.41 \text{ A}, \quad I_g^* = \frac{i_{dg}^*}{\sqrt{2}} = 8.065 \text{ A } (rms)$$

$$P_g = 3\, V_g\, I_g \cos \varphi_g = 3 \times 120 \times 8.065 \times 1 = 2906 \text{ W}.$$

The above-mentioned analysis indicates that the power converters and control scheme assume that the power dissipated in R_x as a low wind speed (low active power extraction) operation. Figure 10.19(b) shows that the phase-a grid current tracks to its reference before and after the DC-link perturbation, and the capacitors voltage are balanced with the help of a predictive controller. During the step connection of R_x, the transient duration is noted to be 5 ms. This case study verifies that the 3L boost converter and predictive control scheme perform well in controlling the neutral-point voltage even with the changes in the DC-link model. ■

CASE STUDY 10.6 Switching Frequency Minimization of NPC Inverter

Case Study Objective: The objective of this case study is to analyze the grid-side NPC inverter performance during switching frequency minimization. Two cases are considered to study the performance: (1) the NPC inverter employed in the BTB converters (refer to Figure 9.2) and (2) the NPC inverter connected to the 3L boost converter (refer to Figure 10.2).

Parameters: The predictive control scheme is applied for the above-mentioned topologies under rated wind speed and zero grid-side reactive power.

Analysis:
The simulation results with the BTB-NPC topology are presented in Figure 10.20. As discussed in Chapter 9, the NPC inverter in the BTB-NPC topology performs three tasks: (1) controlling the net DC-bus voltage, (2) balancing neutral-point voltage, and (3) regulating the reactive power.

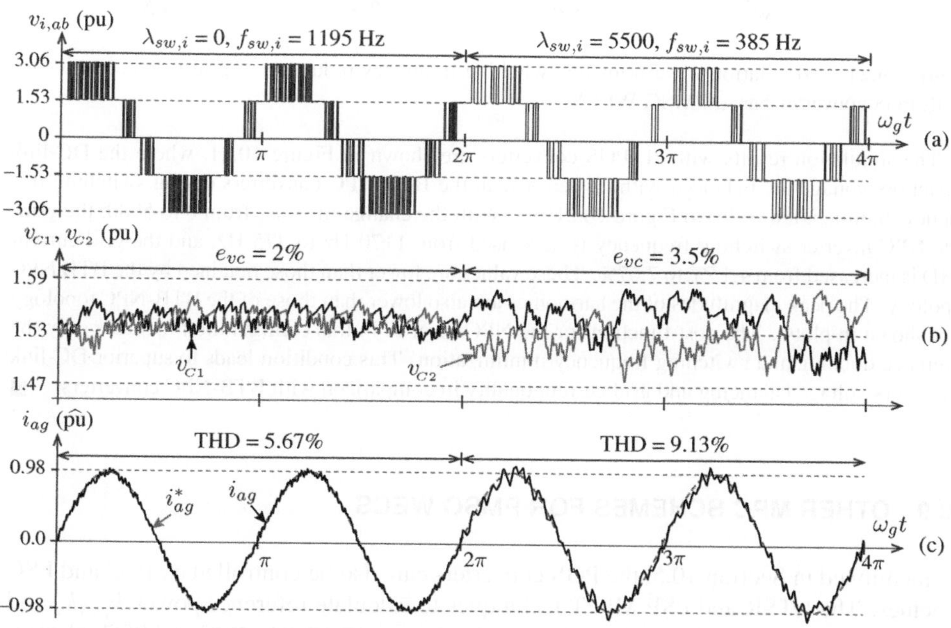

Figure 10.20 Simulation waveforms for switching frequency reduction of grid-side NPC inverter in BTB NPC converter-based PMSG WECS.

For the waveforms with BTB-NPC converters shown in Figure 10.20, at time $\omega_g t = 2\pi$, the weighting factor for the switching frequency minimization ($\lambda_{sw,i}$) is changed from 0 to 5500. With this condition, the $f_{sw,i}$ is changed from 1195 Hz to 385 Hz. This condition can be observed by examining the number of switch changes in the inverter line-to-line voltage waveform presented in Figure 10.20(a). With the introduction of the switching frequency minimization term to the cost function, two other variables (neutral-point voltage and grid current reference tracking) are affected, as shown in Figures 10.20(b) and 10.20(c). The grid current THD with $\lambda_{sw,i} = 0$ and $\lambda_{sw,i} = 5500$ are 5.67% and 9.13%, respectively.

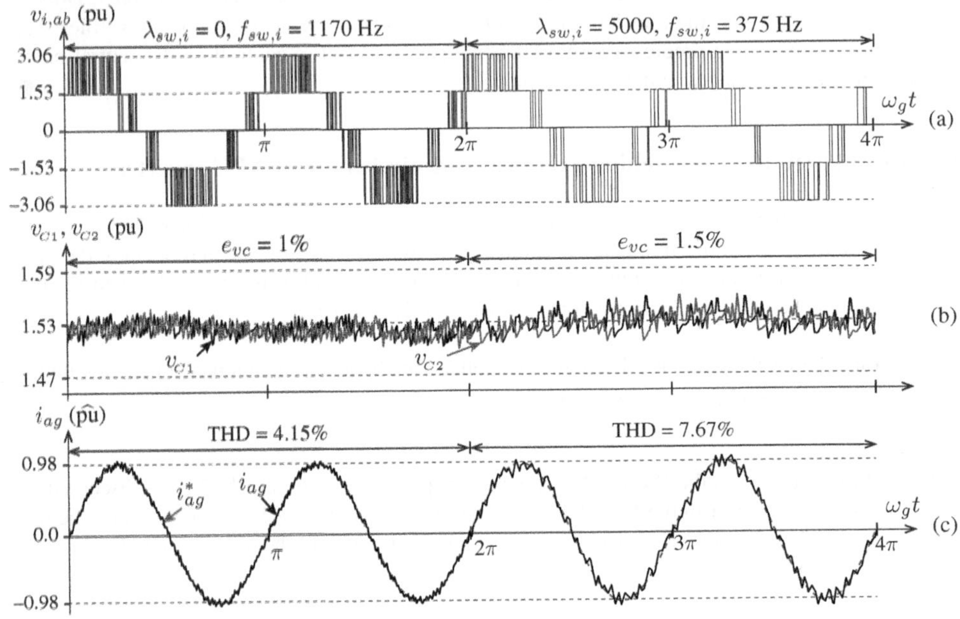

Figure 10.21 Simulation waveforms for switching frequency reduction of grid-side NPC inverter in 3L PGS converter-based PMSG WECS.

The simulation results with 3L PGS converters are shown in Figure 10.21, where the DC-link capacitors voltage are balanced with less error than the BTB-NPC converters during switching frequency minimization (refer to Figure 10.21(b)). With the change in $\lambda_{sw,i}$ from 0 to 5000, the grid-side NPC inverter switching frequency is decreased from 1170 Hz to 375 Hz, and the grid current THD is increased from 4.15% to 7.67%. These values are lower than those obtained by the BTB NPC topology. The peak magnitudes of the harmonics are also lower than those of the BTB-NPC topology (not shown in plots). The cost function for the NPC inverter in this configuration includes only the reference tracking and switching frequency minimization. This condition leads to superior DC-link capacitors voltage balancing and grid current quality in comparison with BTB-NPC converters. ■

10.9 OTHER MPC SCHEMES FOR PMSG WECS

As mentioned in Section 10.5, the PGS converters can also be controlled by PPC and PSC schemes. The OTSR and PSF MPPT techniques to calculate reference power $P_{in}^*(k)$ and reference generator speed $\omega_m^*(k)$ can be found in Figure 10.10. The PPC and PSC schemes allow the integration of predictive control with different MPPT techniques. The PPC and PSC schemes are briefly analyzed in this section.

10.9.1 Predictive Power Control

With the PPC scheme, the predictive controller forces the boost converter to track to the extrapolated reference input power, $\widehat{P}^*_{in}(k+1)$. The generator speed, current, and electromagnetic torque are regulated by tracking to the reference input power. The cost function definition for the PPC scheme is defined as follows:

$$g_{dc}(k) = \left[\widehat{P}^*_{in}(k+1) - P^p_{in}(k+1) \right]^2 . \tag{10.24}$$

The DT model for the prediction of input power is depicted as follows:

$$P^p_{in}(k+1) = \widehat{v}_{in}(k+1)\, i^p_{dc}(k+1) \tag{10.25}$$

where $\widehat{v}_{in}(k+1)$ is the extrapolated input voltage, and $i^p_{dc}(k+1)$ is the predicted DC-link current. The DT models in (10.5) and (10.14) are used to calculate $i^p_{dc}(k+1)$ for 2L and 3L boost converters, respectively.

10.9.2 Predictive Speed Control

Figure 10.10 shows that a PI controller is used to calculate the reference inductor current i^*_{dc}. The PSC scheme eliminates the requirement for a PI controller in the outer speed control loop. This approach leads to the cascade structure-free operation of PMSG WECS. The cost function definition for the PSC scheme is defined as follows:

$$g_{dc}(k) = \left[\widetilde{\omega}^*_m(k+1) - \omega^p_m(k+1) \right]^2 + \left[\widehat{i}^*_{dc}(k+1) - i^p_{dc}(k+1) \right]^2 \tag{10.26}$$

where $\widetilde{\omega}^*_m(k+1)$ is the filtered reference speed. The PSC scheme produces reference inductor current $\widehat{i}^*_{dc}$ internally. The dynamics of generator speed should be formulated in terms of boost converter switching states to realize the PSC scheme.

10.10 REAL-TIME IMPLEMENTATION OF MPC SCHEMES

The number of feasible switching states for 2L and 3L boost converters are two and four, respectively. Given the few numbers of switching states, the computational burden on DSP is lower with PGS converters than that with BTB converters. In other words, PGS converters require much lower sampling time (T_s) to execute the PCC algorithm than BTB power converters. The MPC schemes discussed earlier considers cost function minimization at $(k+1)$ sampling instant, which is suitable for computer simulations. However, in the real-time implementation of MPC schemes, the computational delay caused by the DSP deteriorates the performance of the predictive controller. To overcome this issue, the cost function minimization can be performed at the $(k+2)$ sampling instant. This approach allows one sampling period for calculations and another sampling period for control delay compensation. Given the few numbers of switching states, *prediction + prediction* (P+P) or *prediction + double prediction* (P+P^2) delay compensation methods can be employed without involving much computational burden [20]. The PCC algorithm with P+P delay compensation is shown in Figure 10.22 for the 2L boost converter-based PMSG WECS. It is identical to the PCC algorithm in Figure 10.11 with additional measurements to calculate virtual load current $\widehat{i}_{R1}(k)$. The predicted and extrapolated variables correspond to the $(k+2)$ sampling instant.

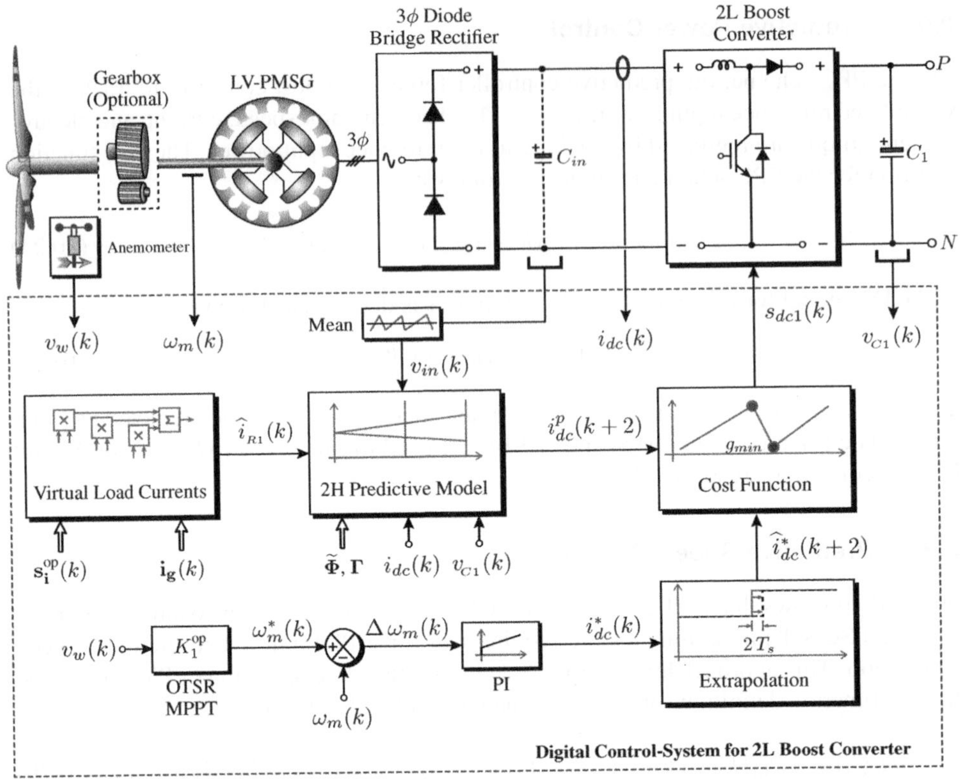

Figure 10.22 Block diagram of the PCC scheme with P+P delay compensation for 2L boost converter-based PMSG WECS.

The inductor current error is computed at the $(k+2)$ sampling instant by defining a cost function as follows:

$$g_{idc}(k) = \left[\widehat{i^*_{dc}}(k+2) - i^p_{dc}(k+2)\right]^2 \tag{10.27}$$

where $\widehat{i^*_{dc}}(k+2)$ is the extrapolated reference inductor current. A first-order Lagrange extrapolation for the two-samples-ahead extrapolation is as follows:

$$\widehat{i^*_{dc}}(k+2) = 3\,i^*_{dc}(k) - 2\,i^*_{dc}(k-1). \tag{10.28}$$

The $(k+2)$ sample inductor current prediction involves two steps. First, $i^p_{dc}(k+1)$ and $v^p_{C1}(k+1)$ are calculated by using the DT model in (10.5) and all possible switching states. Finally, $i^p_{dc}(k+2)$ is calculated by using the following DT model:

$$\begin{bmatrix} i^p_{dc}(k+2) \\ v^p_{C1}(k+2) \end{bmatrix} = \begin{bmatrix} 1 - \frac{r_{dc}\,T_s}{L_{dc}} & -\frac{T_s\,(1-s_{dc1}(k))}{L_{dc}} \\ \frac{T_s\,(1-s_{dc1}(k))}{C_1} & 1 \end{bmatrix} \begin{bmatrix} i^p_{dc}(k+1) \\ v^p_{C1}(k+1) \end{bmatrix} + \Gamma \begin{bmatrix} v_{in}(k) \\ i_{R1}(k) \end{bmatrix}. \tag{10.29}$$

As mentioned in Chapter 4, with the P + P delay compensation method, the same switching state combinations are used in $(k+1)$ and $(k+2)$ predictions. The switching state that minimizes the cost function at the $(k+2)$ sampling instant is selected and applied at sampling instant (k). The flowchart for the PCC algorithm with P+P delay compensation is shown in Figure 10.23. Blocks ⑤ and ⑥ indicate that the `for` loop uses the same switching state combination number `j` to calculate $i^p_{dc}(k+1)$ and $i^p_{dc}(k+2)$.

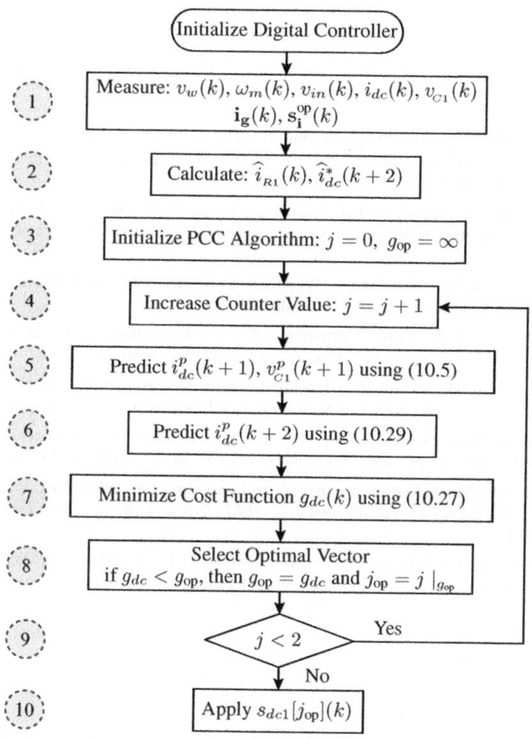

Figure 10.23 Flowchart for the PCC algorithm with P+P delay compensation for the 2L boost converter-based PMSG WECS.

10.11 CONCLUDING REMARKS

This chapter provides a comprehensive analysis on PGS converters for LV and MV PMSG-based variable-speed WECS. The CT and DT models of 2L and 3L boost converters are developed. The generator-side control variables, such as inductor current and DC-link capacitors voltage, are formulated in terms of boost converter switching states. The models provided for 2L and 3L converters are intuitive, and they can be easily extended to multilevel boost converters. The FCS-MPC digital control system for 2L and 3L PGS converter-based PMSG WECS is presented. The design steps indicate that the MPC scheme can be developed in a generalized way for the 2L and 3L converters with a few changes to control objectives and cost function definition. The WT control requirements, such as MPPT, balancing of DC-link capacitors voltage (in multilevel converters), regulation of the net DC-bus voltage, and grid reactive power control, are analyzed with PGS converters and PCC schemes. The other predictive control schemes and real-time implementation issues are also discussed. The operating principles of predictive control, performance of PMSG WECS, and important concepts are presented through case studies.

REFERENCES

1. V. Yaramasu, B. Wu, P. C. Sen, S. Kouro, and M. Narimani, "High-power wind energy conversion systems: State-of-the-art and emerging technologies," *Proceedings of the IEEE*, vol. 103, no. 5, pp. 740–788, May 2015.
2. M. Aguirre, S. Kouro, J. Rodriguez, and H. Abu-Rub, "Model predictive control of interleaved boost converters for synchronous generator wind energy conversion systems," in *IEEE International Conference on Industrial Technology (ICIT)*, Seville, Spain, March 2015, pp. 2295–2301.
3. K. Amei, Y. Takayasu, T. Ohji, and M. Sakui, "A maximum power control of wind generator system using a permanent magnet synchronous generator and a boost chopper circuit," in *Power Conversion Conference (PCC)*, Osaka, Japan, vol. 3, 2002, pp. 1447–1452.
4. S.-H. Song, S. il Kang, and N. kun Hahm, "Implementation and control of grid connected AC-DC-AC power converter for variable speed wind energy conversion system," in *IEEE Applied Power Electronics Conference and Exposition (APEC)*, Miami Beach, FL, USA, vol. 1, February 2003, pp. 154–158.
5. Y. Xia, J. Fletcher, S. Finney, K. Ahmed, and B. Williams, "Torque ripple analysis and reduction for wind energy conversion systems using uncontrolled rectifier and boost converter," *IET Renewable Power Generation*, vol. 5, no. 5, pp. 377–386, September 2011.
6. J. Rodríguez, S. Bernet, P. Steimer, and I. Lizama, "A survey on neutral-point-clamped inverters," *IEEE Transactions on Industrial Electronics*, vol. 57, no. 7, pp. 2219–2230, July 2010.
7. J.-M. Kwon, B.-H. Kwon, and K.-H. Nam, "Three-phase photovoltaic system with three-level boosting MPPT control," *IEEE Transactions on Power Electronics*, vol. 23, no. 5, pp. 2319–2327, September 2008.
8. H.-C. Chen and W.-J. Lin, "MPPT and voltage balancing control with sensing only inductor current for photovoltaic-fed, three-level, boost-type converters," *IEEE Transactions on Power Electronics*, vol. 29, no. 1, pp. 29–35, January 2014.
9. C. Xia, X. Gu, T. Shi, and Y. Yan, "Neutral-point potential balancing of three-level inverters in direct-driven wind energy conversion system," *IEEE Transactions on Energy Conversion*, vol. 26, no. 1, pp. 18–29, March 2011.
10. V. Yaramasu and B. Wu, "Predictive control of a three-level boost converter and an NPC inverter for high power PMSG-based medium voltage wind energy conversion systems," *IEEE Transactions on Power Electronics*, vol. 29, no. 10, pp. 5308–5322, October 2014.
11. A. Di Gerlando, G. Foglia, M. Iacchetti, and R. Perini, "Analysis and test of diode rectifier solutions in grid-connected wind energy conversion systems employing modular permanent-magnet synchronous generators," *IEEE Transactions on Industrial Electronics*, vol. 59, no. 5, pp. 2135–2146, May 2012.
12. W. Zhang, G. Feng, Y.-F. Liu, and B. Wu, "A digital power factor correction (PFC) control strategy optimized for DSP," *IEEE Transactions on Power Electronics*, vol. 19, no. 6, pp. 1474–1485, November 2004.
13. J. Chen, A. Prodic, R. Erickson, and D. Maksimovic, "Predictive digital current programmed control," *IEEE Transactions on Power Electronics*, vol. 18, no. 1, pp. 411–419, January 2003.
14. M. Perez, J. Rodríguez, and A. Coccia, "Predictive current control in a single phase PFC boost rectifier," in *IEEE International Conference on Industrial Technology (ICIT)*, Gippsland, Australia, February 2009, pp. 1–6.
15. M. Haque, M. Negnevitsky, and K. Muttaqi, "A novel control strategy for a variable-speed wind turbine with a permanent-magnet synchronous generator," *IEEE Transactions on Industry Applications*, vol. 46, no. 1, pp. 331–339, January/February 2010.
16. X. Ruan, B. Li, Q. Chen, S.-C. Tan, and C. Tse, "Fundamental considerations of three-level DC-DC converters: Topologies, analyses, and control," *IEEE Transactions on Circuits and Systems—Part I: Fundamental Theory and Applications*, vol. 55, no. 11, pp. 3733–3743, December 2008.
17. J. Baggio, H. Hey, H. Grundling, H. Pinheiro, and J. Pinheiro, "Discrete control for three-level boost PFC converter," in *IEEE International Telecommunications Energy Conference (INTELEC)*, Montreal, QC, Canada, 2002, pp. 627–633.
18. V. Yaramasu and B. Wu, "Three-level boost converter based medium voltage megawatt PMSG wind energy conversion systems," in *IEEE Energy Conversion Congress and Exposition (ECCE)*, Phoenix, AZ, USA, September 2011, pp. 561–567.
19. K. Corzine and S. Majeethia, "Analysis of a novel four-level dc/dc boost converter," *IEEE Transactions on Industry Applications*, vol. 36, no. 5, pp. 1342–1350, September/October 2000.
20. V. Yaramasu, B. Wu, M. Rivera, and J. Rodriguez, "A new power conversion system for megawatt PMSG wind turbines using four-level converters and a simple control scheme based on two-step model predictive strategy – Part I: Modeling and theoretical analysis," *IEEE Journal of Emerging and Selected Topics in Power Electronics*, vol. 2, no. 1, pp. 3–13, March 2014.
21. E. Koutroulis and K. Kalaitzakis, "Design of a maximum power tracking system for wind-energy-conversion applications," *IEEE Transactions on Industrial Electronics*, vol. 53, no. 2, pp. 486–494, April 2006.
22. Z. Chen, J. Guerrero, and F. Blaabjerg, "A review of the state of the art of power electronics for wind turbines," *IEEE Transactions on Power Electronics*, vol. 24, no. 8, pp. 1859–1875, August 2009.
23. B. Wu, Y. Lang, N. Zargari, and S. Kouro, *Power Conversion and Control of Wind Energy Systems*, 1st ed., ser. IEEE Press Series on Power Engineering. Hoboken, NJ: Wiley-IEEE Press, July 2011.
24. S. Alepuz, A. Calle, S. Busquets-Monge, S. Kouro, and B. Wu, "Use of stored energy in PMSG rotor inertia for low-voltage ride-through in back-to-back NPC converter-based wind power systems," *IEEE Transactions on Industrial Electronics*, vol. 60, no. 5, pp. 1787–1796, May 2013.

CHAPTER 11

CONTROL OF SCIG WECS WITH VOLTAGE SOURCE CONVERTERS

11.1 INTRODUCTION

During the 1980s, Type 1 fixed-speed wind energy conversion systems (WECS) were developed by using a multi-stage gearbox and a squirrel-cage induction generator (SCIG). The SCIG was a popular choice because of its off-the-shelf availability for high-power applications, significant cost-effectiveness, simple and rugged construction, and maintenance-free operation [1, 2]. The SCIG is directly connected to a three-phase grid through a soft starter and a step-up transformer. The wind energy conversion efficiency is low with this type of configuration because SCIG operates in a narrow range around synchronous speed. Type 1 wind turbines (WTs) have been commercialized by `NEG Micon` (currently `Vestas`), `Bonus` (currently `Siemens`), `Made` and `Nordex` [3]. During the 1990s, Type 4 variable-speed WECS were introduced with synchronous generator (SG) and full-scale power converter based on back-to-back (BTB)-connected two-level (2L) voltage source converter (VSC). Type 4 WECS have high wind energy conversion efficiency, require no soft starter, operate without external reactive power compensation, have less mechanical stress on mechanical drivetrain, and are robust against power system grid faults [4, 5]. To combine the advantages of SCIG and Type 4 WECS, a few manufacturers developed variable-speed WTs with SCIG and BTB VSC.

Table 11.1 highlights the list of megawatt (MW)-level Type-4 SCIG WTs with respect to power and voltage rating, gearbox steps and gear ratio, and power converter configuration [1]. Low-voltage (LV) operations with BTB 2L-VSC are popularly used with SCIG WECS. Given the increasing power levels of SCIGs, medium-voltage (MV) op-

Model Predictive Control of Wind Energy Conversion Systems, First Edition. Venkata Yaramasu and Bin Wu.

Companion Website: www.wiley.com/go/yaramasu_wu_mpc_wecs

Table 11.1 List of commercial SCIG WECS with BTB VSCs [1]

Manufacturer	Generator	Power Rating	Gear Steps & Ratio	Converter	Latest Model #
RRB Energy (India)	LV–SCIG	1.8 MW, 690 V	3, 1:61	BTB 2L-VSC	PS-1800
Siemens (Germany)	LV–SCIG	3.6 MW, 690 V	3, 1:119	BTB 2L-VSC	SWT 3.6-120
Windtec-AMSC (USA)	LV–SCIG	5.5 MW, 690 V	3, –	BTB 2L-VSC	WT5500
Condor Ltd.	MV–SCIG	6.6 MW, 3300 V	2.5, 1:34.81	BTB 3L-VSC	Condor 6

eration with BTB three-level (3L) VSC is gaining popularity in recent years. During the system startup, SCIG draws magnetizing current from the utility grid; therefore, power converters must have bidirectional power flow capability. SCIGs are equipped with four or six poles and employ a three-stage (3S) gearbox to enable high-speed operation. A recent project by `Condor` uses a 2.5-stage gearbox along with a multi-pole SCIG to reduce the weight of the gearbox and the rotational speed of SCIG. The major drawbacks for Type 4 SCIG WECS are high initial costs owing to the full-scale converter and mandatory gearbox requirement. Variable-speed SCIGs will have high application potential owing to the continuous innovations in cost-effective power electronics, and gear-step reduction.

Similar to SG, field-oriented control (FOC) and direct torque control (DTC) are the two most popular control schemes for SCIG [6, 7]. Compared with SG, the calculation of reference control variables and speed control is relatively complicated with SCIGs. Given the increasing computational power of microprocessors, model predictive control (MPC) is becoming an attractive alternative to classical FOC and DTC [8–10]. The speed, torque, and flux of SCIG are controlled accurately by the MPC scheme by generating optimal switching signals for the machine-side converter (MSC) via the use of discrete-time (DT) model of a complete power conversion system. Moreover, the MPC scheme eliminates the need for proportional-integral (PI) controllers and modulation stage needed by FOC, as well as hysteresis regulators and lookup table needed by DTC. In addition to simple digital implementation, MPC schemes for SCIG offer high dynamic performance throughout the operational speed range [11, 12]. This chapter deals with the predictive control of full-scale LV and MV VSCs employed in SCIG WECS.

Chapter Overview

- The SCIG WECS configuration with BTB LV and MV VSCs is discussed in Section 11.2. The equivalent circuit and steady-state model of SCIG is presented in Section 11.3. The FOC and DTC schemes for SCIG are outlined in Section 11.4.
- The overall digital control system development for BTB converter-based SCIG WECS, including maximum power point tracking (MPPT) and calculation of reference control variables for MPC design, is given in Section 11.5.
- Predictive current control (PCC), which is analogous to classical FOC, is analyzed with BTB 2L-VSC in Section 11.6.
- Predictive torque control (PTC) is developed on the basis of the classical DTC concept and is analyzed with BTB neutral-point clamped (NPC) converter in Section 11.7.
- The delay compensation for the real-time implementation of MPC schemes is addressed in Section 11.8. Finally, the concluding remarks of this chapter are given in Section 11.9.

11.2 CONFIGURATION OF SCIG WECS WITH BTB CONVERTERS

The configuration of SCIG WECS is similar to the permanent magnet synchronous generator (PMSG)-based WECS discussed in Chapter 9, except that the multi-stage gearbox is mandatory because of the high-speed operation. The analysis given in this chapter with 2L-VSCs and NPC converters can be attributed to other promising power converter topologies such as active NPC converters, flying capacitor converters, modular multilevel converters, current source converters, and matrix converters discussed in Chapter 2 [1, 13, 14]. Passive generator-side converters cannot be used with SCIGs because the former provides only unidirectional power flow (from wind generator to the utility grid).

11.2.1 SCIG WECS with LV BTB Converters

The configuration of grid-connected Type 4 WECS with three-stage (3S) gearbox-based SCIG and BTB 2L-VSC is shown in Figure 11.1 [15, 16]. The generated power is instantaneously injected to a utility grid via the full-scale converter connected between the generator stator and grid terminals. The generator-side 2L voltage source rectifier (2L-VSR) enables full variable-speed operation (low-speed during weak wind speed conditions to high-speed during rated wind speed) by controlling the output voltage amplitude and frequency. The variable-speed operation leads to high efficiency and energy yield, in addition to low stress on mechanical components. The grid-side converter (GSC) based on 2L voltage source inverter (2L-VSI) performs smooth grid connection, reactive power generation, and fault ride-through (FRT) operation without affecting the normal SCIG operation. The DC link provides complete decoupling between the MSC and GSC. Thus, a 50-Hz SCIG can be employed for 60-Hz grid voltage operation or vice versa.

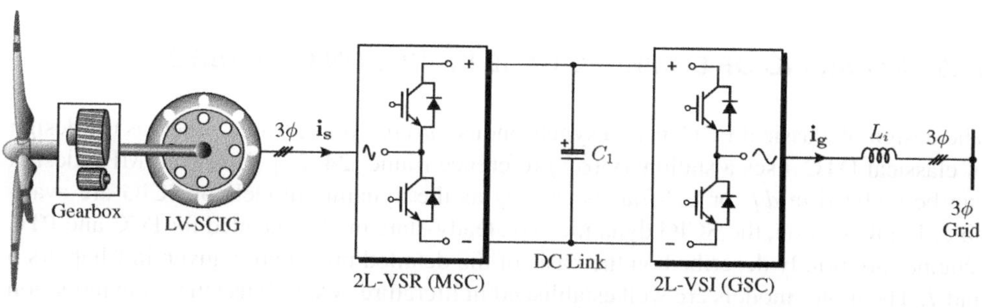

Figure 11.1 Block diagram of SCIG WECS with BTB 2L-VSC.

The configuration in Figure 11.1 is used for power ratings of up to 0.75 MW and 575 or 690 V. For higher power ratings, more converters need to be connected in parallel to increase the current handling capacity of SCIG WECS (similar to SG WECS) [17]. System redundancy increases because of the large number of parallel converters, but at the expense of increased converter cost, weight, and losses.

11.2.2 SCIG WECS with MV BTB Converters

Parallel converters with LV operations are inefficient and cost-ineffective for WT power ratings above 3 MW. The MV operation of WTs is an attractive alternative to increase the power density of Type 4 WTs [18, 19]. Figure 11.2 shows the configuration of Type 4 MV-WECS; in this configuration, the BTB 2L-VSC in the previous configuration is replaced

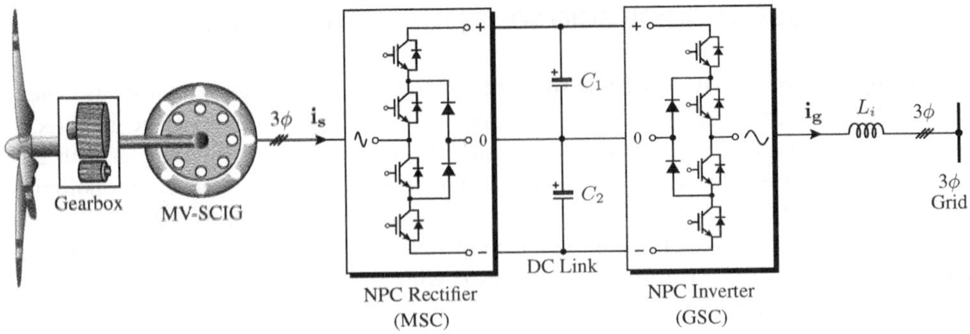

Figure 11.2 Block diagram of SCIG WECS with BTB NPC converter.

by a BTB NPC converter [20]. The performance of MV induction motor drives are proven in the electric drive industry [21]. The wind energy industry has started adopting MV technology for SCIG-based Type 4 WECS. All of the advantages discussed earlier with BTB 2L-VSC are preserved by a BTB NPC converter. Furthermore, the MV operation on the generator side decreases the generator losses and current total harmonic distortion (THD). On the grid-side, the cable size and losses are reduced by the MV operation of NPC inverter, in addition to excellent grid power quality and low harmonic filter requirement. The BTB NPC converter in Figure 11.2 can be used for power and voltage ratings up to 6.0 MW and 3-4 kV without connecting switching devices in series or parallel; these converters operate at about 98% efficiency [22]. The balancing of DC capacitors voltage and equal distribution of switching losses are two well-known challenges for NPC converters [23].

11.3 MODELING OF SQUIRREL-CAGE INDUCTION GENERATOR

The design of classical FOC uses a synchronous (dq) reference frame, whereas the design of classical DTC uses a stationary ($\alpha\beta$) reference frame [24, 25]. However, MPC design can be realized in dq- or $\alpha\beta$-frames as long as the dynamic models of SCIG are available. In this section, the SCIG dynamic and steady-state models needed for PCC and PTC schemes are briefly described on the basis of the detailed information given in Chapters 6 and 7. The motor models are well established in literature. We will use the same induction motor models by only changing the sign of mechanical input torque T_m to negative.

11.3.1 Equivalent Circuit of SCIG

The overall SCIG model includes both electrical and mechanical variables. The electrical model includes stator voltages, currents and flux linkages, and rotor flux linkages, whereas the mechanical model includes the dynamics of rotor speed, moment of inertia, and electromagnetic torque. In Chapters 6 and 9, dynamic models and MPC schemes are considered for PMSG in the dq-frame because inductances exhibit a time-invariant nature. For an SCIG, no issues arise with the inductance variations in dq- and $\alpha\beta$-frames. Therefore, PCC and PTC schemes can be realized in any reference frame. Similar to the control of a GSC in Chapter 8, the $\alpha\beta$-frame control decreases design complexity, number of online calculations, and computational burden. In this chapter, design and control aspects are performed in the $\alpha\beta$-frame for both PCC and PTC schemes.

The dynamic equivalent circuits in the $\alpha\beta$-axes of SCIG are shown in Figure 11.3, along with an MSC, gearbox, and WT. The $\alpha\beta$-frame circuit is deducted from the arbitrary frame IG model by setting the arbitrary frame speed ω_k to zero; thus, the stationary frame does not rotate in space [24]. The squirrel-cage rotor consists of a steel cylinder and aluminum or copper conductors embedded in its surface. The rotor conductors are permanently shorted by end rings; hence, rotor-induced voltages $v_{\alpha r}$ and $v_{\beta r}$ become zero.

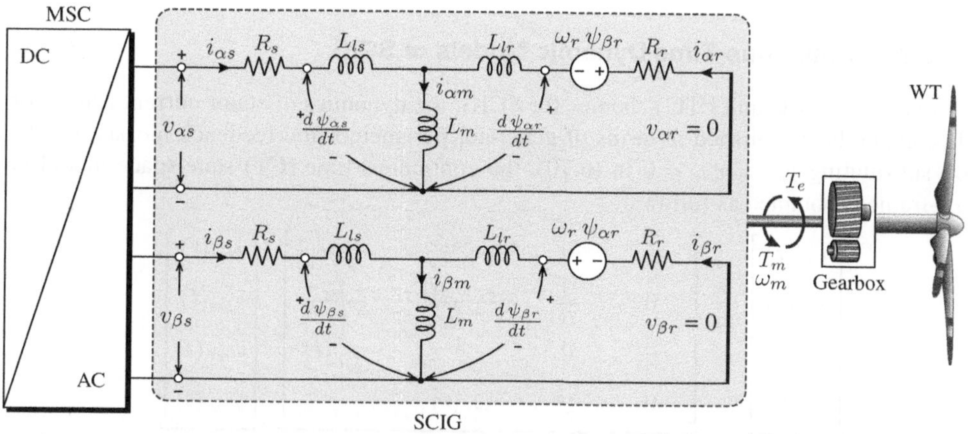

Figure 11.3 Equivalent circuit of an induction machine with 2L/3L MSC and WT.

In the $\alpha\beta$-frame, the stator and rotor voltage vectors are defined by the following:

$$\mathbf{v}_{\mathbf{s},\alpha\beta} = R_s\,\mathbf{i}_{\mathbf{s},\alpha\beta} + \frac{d}{dt}\boldsymbol{\psi}_{\mathbf{s},\alpha\beta} \tag{11.1}$$

$$\mathbf{v}_{\mathbf{r},\alpha\beta} = R_r\,\mathbf{i}_{\mathbf{r},\alpha\beta} + \frac{d}{dt}\boldsymbol{\psi}_{\mathbf{r},\alpha\beta} - j\,\omega_r\,\boldsymbol{\psi}_{\mathbf{r},\alpha\beta} = 0 \tag{11.2}$$

where

- $\mathbf{v}_{\mathbf{s},\alpha\beta} = [v_{\alpha s}\ v_{\beta s}]^T$ and $\mathbf{v}_{\mathbf{r},\alpha\beta} = [v_{\alpha r}\ v_{\beta r}]^T$ are the stator and rotor voltage vectors (V), respectively.
- $\mathbf{i}_{\mathbf{s},\alpha\beta} = [i_{\alpha s}\ i_{\beta s}]^T$ and $\mathbf{i}_{\mathbf{r},\alpha\beta} = [i_{\alpha r}\ i_{\beta r}]^T$ are the stator and rotor current vectors (A), respectively.
- $\boldsymbol{\psi}_{\mathbf{s},\alpha\beta} = [\psi_{\alpha s}\ \psi_{\beta s}]^T$ and $\boldsymbol{\psi}_{\mathbf{r},\alpha\beta} = [\psi_{\alpha r}\ \psi_{\beta r}]^T$ are the stator and rotor flux vectors (Wb), respectively.
- R_s and R_r are the stator and rotor winding resistances (Ω).
- ω_r is the electrical angular speed of SCIG (rad/s).

The vectors of stator and rotor flux linkages are obtained as follows [26]:

$$\boldsymbol{\psi}_{\mathbf{s},\alpha\beta} = L_s\,\mathbf{i}_{\mathbf{s},\alpha\beta} + L_m\,\mathbf{i}_{\mathbf{r},\alpha\beta}, \qquad \boldsymbol{\psi}_{\mathbf{r},\alpha\beta} = L_r\,\mathbf{i}_{\mathbf{r},\alpha\beta} + L_m\,\mathbf{i}_{\mathbf{s},\alpha\beta} \tag{11.3}$$

with

$$L_s = L_{ls} + L_m, \qquad L_r = L_{lr} + L_m \tag{11.4}$$

where L_{ls} and L_{lr} are the stator and rotor leakage inductances (H). L_m is the magnetizing inductance (H). L_s and L_r are the stator and rotor self-inductances (H).

The electromagnetic torque, and stator active and reactive powers are defined by [27]:

$$T_e = 1.5\, P_p\, \Re\mathfrak{e}\left(j\, \boldsymbol{\psi}_{s,\alpha\beta}\, \mathbf{i}^{\star}_{s,\alpha\beta}\right) = 1.5\, P_p\, (\psi_{\alpha s}\, i_{\beta s} - \psi_{\beta s}\, i_{\alpha s}) \tag{11.5}$$

$$P_s = 1.5\, \Re\mathfrak{e}\left(\mathbf{v}_{s,\alpha\beta}\, \mathbf{i}^{\star}_{s,\alpha\beta}\right) = 1.5\, (v_{\alpha s}\, i_{\alpha s} + v_{\beta s}\, i_{\beta s}) \tag{11.6}$$

$$Q_s = 1.5\, \Im\mathfrak{m}\left(\mathbf{v}_{s,\alpha\beta}\, \mathbf{i}^{\star}_{s,\alpha\beta}\right) = 1.5\, (v_{\beta s}\, i_{\alpha s} - v_{\alpha s}\, i_{\beta s}) \tag{11.7}$$

where $\star$ represents the complex conjugate operation; P_p denotes the SCIG pole pairs.

11.3.2 Continuous-Time Dynamic Models of SCIG

To realize the PCC and PTC schemes for SCIG, the dynamics of stator current and stator flux should be established in terms of generator parameters and feedback measurements. By substituting $v_{\alpha r} = v_{\beta r} = 0$ in (6.70), the continuous-time (CT) state-space model in $\alpha\beta$-frame is obtained as follows:

$$\underbrace{\frac{d}{dt}\begin{bmatrix} i_{\alpha s}(t) \\ i_{\beta s}(t) \\ \psi_{\alpha r}(t) \\ \psi_{\beta r}(t) \end{bmatrix}}_{\dot{\mathbf{x}}(t)} = \underbrace{\begin{bmatrix} -\frac{1}{\tau_\sigma} & 0 & \frac{k_r}{\sigma\, L_s\, \tau_r} & \frac{k_r\, \omega_r(t)}{\sigma\, L_s} \\ 0 & -\frac{1}{\tau_\sigma} & -\frac{k_r\, \omega_r(t)}{\sigma\, L_s} & \frac{k_r}{\sigma\, L_s\, \tau_r} \\ \frac{L_m}{\tau_r} & 0 & -\frac{1}{\tau_r} & -\omega_r(t) \\ 0 & \frac{L_m}{\tau_r} & \omega_r(t) & -\frac{1}{\tau_r} \end{bmatrix}}_{\mathbf{A}(t)} \underbrace{\begin{bmatrix} i_{\alpha s}(t) \\ i_{\beta s}(t) \\ \psi_{\alpha r}(t) \\ \psi_{\beta r}(t) \end{bmatrix}}_{\mathbf{x}(t)} + \underbrace{\begin{bmatrix} \frac{1}{\sigma\, L_s} & 0 \\ 0 & \frac{1}{\sigma\, L_s} \\ 0 & 0 \\ 0 & 0 \end{bmatrix}}_{\mathbf{B}} \underbrace{\begin{bmatrix} v_{\alpha s}(t) \\ v_{\beta s}(t) \end{bmatrix}}_{\mathbf{u}(t)} \tag{11.8}$$

$$\underbrace{\begin{bmatrix} \psi_{\alpha s}(t) \\ \psi_{\beta s}(t) \end{bmatrix}}_{\mathbf{y}(t)} = \underbrace{\begin{bmatrix} \sigma\, L_s & 0 & k_r & 0 \\ 0 & \sigma\, L_s & 0 & k_r \end{bmatrix}}_{\mathbf{C}} \underbrace{\begin{bmatrix} i_{\alpha s}(t) \\ i_{\beta s}(t) \\ \psi_{\alpha r}(t) \\ \psi_{\beta r}(t) \end{bmatrix}}_{\mathbf{x}(t)} \tag{11.9}$$

where $\mathbf{A}(t)$ is the state matrix, $\mathbf{B}$ is the input matrix, and $\mathbf{C}$ is the output matrix. The SCIG variables are defined as follows [28, 29]:

- Stator coupling coefficient, $k_s = \frac{L_m}{L_s}$.
- Rotor coupling coefficient, $k_r = \frac{L_m}{L_r}$.
- Total leakage coefficient, $\sigma = 1 - k_s\, k_r = 1 - \frac{L_m^2}{L_s\, L_r}$.
- Equivalent resistance, $r_\sigma = R_s + k_r^2\, R_r$ (Ω).
- Stator transient time constant, $\tau_\sigma = \frac{\sigma\, L_s}{r_\sigma}$ (s).
- Stator time constant, $\tau_s = \frac{L_s}{R_s}$ (s).
- Rotor time constant, $\tau_r = \frac{L_r}{R_r}$ (s).
- Total leakage inductance, $\sigma\, L_s = \frac{L_s\, L_r - L_m^2}{L_r}$.

11.3.3 Discrete-Time Dynamic Models of SCIG

The MPC performance is significantly influenced by the accuracy of sampled-data models, in addition to other control issues [30, 31]. As mentioned in Chapter 7, the DT matrices can be calculated by exact, quasi-exact, and approximate discretization methods. The state matrix $\mathbf{A}(t)$ in Equation (11.8) contains time-varying rotor speed term $\omega_r(t)$. Therefore, this model is a linear time variant (LTV) model. Given the LTV and complex nature of the SCIG state-space model, exact discretion is impossible because it involves high-order matrix exponential and integral operations in the online computation of $\mathbf{\Phi}(k)$ and $\mathbf{\Gamma}$. The approximate discretization by the forward Euler method and the quasi-exact discretization by matrix factorization are recommended for SCIG state-space models. The latter approach provides an excellent compromise between discretization error and computational burden [32]. To simplify the analysis and maintain coherence with other chapters, forward Euler discretization is employed in the present chapter. To obtain better transient and steady-state performance with the PCC and PTC schemes, the $\mathbf{\Phi}(k)$ and $\mathbf{\Gamma}$ corresponding to the matrix factorization method can be calculated by using the guidelines given in Section 7.7.1.

The SCIG sampled-data model with forward Euler discretization is given by

$$\underbrace{\begin{bmatrix} i_{\alpha s}(k+1) \\ i_{\beta s}(k+1) \\ \psi_{\alpha r}(k+1) \\ \psi_{\beta r}(k+1) \end{bmatrix}}_{\mathbf{x}(k+1)} = \underbrace{\begin{bmatrix} 1-\frac{T_s}{\tau_\sigma} & 0 & \frac{k_r\,T_s}{\sigma\,L_s\,\tau_r} & \frac{k_r\,\omega_r(k)\,T_s}{\sigma\,L_s} \\ 0 & 1-\frac{T_s}{\tau_\sigma} & -\frac{k_r\,\omega_r(k)\,T_s}{\sigma\,L_s} & \frac{k_r\,T_s}{\sigma\,L_s\,\tau_r} \\ \frac{L_m\,T_s}{\tau_r} & 0 & 1-\frac{T_s}{\tau_r} & -\omega_r(k)\,T_s \\ 0 & \frac{L_m\,T_s}{\tau_r} & \omega_r(k)\,T_s & 1-\frac{T_s}{\tau_r} \end{bmatrix}}_{\mathbf{\Phi}(k)} \underbrace{\begin{bmatrix} i_{\alpha s}(k) \\ i_{\beta s}(k) \\ \psi_{\alpha r}(k) \\ \psi_{\beta r}(k) \end{bmatrix}}_{\mathbf{x}(k)} + \underbrace{\begin{bmatrix} \frac{T_s}{\sigma\,L_s} & 0 \\ 0 & \frac{T_s}{\sigma\,L_s} \\ 0 & 0 \\ 0 & 0 \end{bmatrix}}_{\mathbf{\Gamma_b}} \underbrace{\begin{bmatrix} v_{\alpha s}(k) \\ v_{\beta s}(k) \end{bmatrix}}_{\mathbf{u}(k)} \qquad (11.10)$$

with

$$\underbrace{\begin{bmatrix} \psi_{\alpha s}(k) \\ \psi_{\beta s}(k) \end{bmatrix}}_{\mathbf{y}(k)} = \underbrace{\begin{bmatrix} \sigma\,L_s & 0 & k_r & 0 \\ 0 & \sigma\,L_s & 0 & k_r \end{bmatrix}}_{\mathbf{H}} \underbrace{\begin{bmatrix} i_{\alpha s}(k) \\ i_{\beta s}(k) \\ \psi_{\alpha r}(k) \\ \psi_{\beta r}(k) \end{bmatrix}}_{\mathbf{x}(k)} \qquad (11.11)$$

where T_s is the sampling time used in discretization. $\mathbf{\Phi}(k)$, $\mathbf{\Gamma_b}$, and $\mathbf{H}$ are DT matrices corresponding to $\mathbf{A}(t)$, $\mathbf{B}$, and $\mathbf{C}$, respectively. They are defined as follows:

$$\mathbf{\Phi}(k) \approx (\mathbf{I} + \mathbf{A}(t)\,T_s), \quad \mathbf{\Gamma_b} \approx \mathbf{B}\,T_s, \quad \mathbf{H} = \mathbf{C} \qquad (11.12)$$

where $\mathbf{I}$ is a fourth-order unity matrix.

11.4 CONTROL OF SQUIRREL-CAGE INDUCTION GENERATOR

Vector control techniques such as FOC and DTC were developed during the 1970s and 1980s to enable adjustable-speed electric drives to obtain high dynamic performance, including fast acceleration and deceleration over a full speed range [7, 24, 25, 33]. The PCC and PTC schemes are analogous to the classical FOC and DTC, respectively [34]. Therefore, understanding the operating principle of the classical FOC and DTC schemes is important.

11.4.1 Field-Oriented Control

FOC is an important invention in the field of electric drives. This scheme enables AC motors to achieve decoupled torque and flux control similar to separately excited DC motors. FOC has been used in many other applications with different names. For example, this scheme is used as voltage-oriented control (VOC) for GSC (Chapter 8) and as zero d-axis current (ZDC) control for PMSG (Chapter 9).

For GSCs, the grid active and reactive powers are formulated in terms of dq-axes grid voltages and currents as shown below (obtained from (8.5) in page 245):

$$\begin{aligned} P_g &= 1.5\,(v_{dg}\,i_{dg} + v_{qg}\,i_{qg}) \\ Q_g &= 1.5\,(v_{qg}\,i_{dg} - v_{dg}\,i_{qg}) \end{aligned} \tag{11.13}$$

where P_g and Q_g are the respective functions of i_{dg} and i_{qg} and causes coupling issues in control. To achieve decoupled control, the VOC scheme aligns grid voltage vector $\mathbf{v_g}$ along the d-axis; thus, v_{qg} is forced to become zero. As a result, a linear relationship between P_g and i_{dg} and between Q_g and i_{qg} is obtained for all operating conditions.

For PMSGs, the electromagnetic torque is a function of both i_{ds} and i_{qs}, as shown in the following equation (obtained from Equation (9.7) in page 290):

$$T_e = 1.5\,P_p\,(\psi_{ds}\,i_{qs} - \psi_{qs}\,i_{ds}), \quad \psi_{ds} = L_{ds}\,i_{ds} + \psi_r, \quad \psi_{qs} = L_{qs}\,i_{qs}. \tag{11.14}$$

The ZDC scheme forces the d-axis current i_{ds} to become zero; hence, ψ_{ds} is equal to the flux linkage produced by the permanent magnets (ψ_r). Consequently, T_e becomes proportional to i_{qs}. Similarly, for an SCIG, the electromagnetic torque expression is related to both i_{ds} and i_{qs}, as demonstrated in the following equation [13]:

$$T_e = \frac{3\,P_p\,L_m}{2\,L_r}\,(\psi_{dr}\,i_{qs} - \psi_{qr}\,i_{ds}). \tag{11.15}$$

The FOC scheme aligns the rotor flux vector $\boldsymbol{\psi}_\mathbf{r}$ along the d-axis of the synchronous reference frame (see Figure 11.4). The speed of the synchronous frame ω_s changes with respect to the stator frequency (f_s) of generator. As a result, ψ_{dr} becomes equal to ψ_r and ψ_{qr} becomes zero. Therefore, the T_e expression in (11.15) is simplified as follows:

$$T_e = K_T\,\psi_r\,i_{qs} \quad (\text{for } \psi_{dr} = \psi_r \text{ and } \psi_{qr} = 0), \quad K_T = \frac{3\,P_p\,L_m}{2\,L_r}. \tag{11.16}$$

The flux linkage component of current i_{ds} is aligned with the d-axis, and the torque component of current i_{qs} is aligned along the q-axis. By controlling the currents of dq-axes by MSC, the generator rotor flux and electromagnetic torque are controlled in a decoupled manner. In addition to the rotor flux orientation discussed above, FOC can be realized by using stator flux and air-gap flux orientations. However, the rotor flux orientation is most commonly used in electric drives and wind generators because of its simplicity and flexibility in real-time implementations.

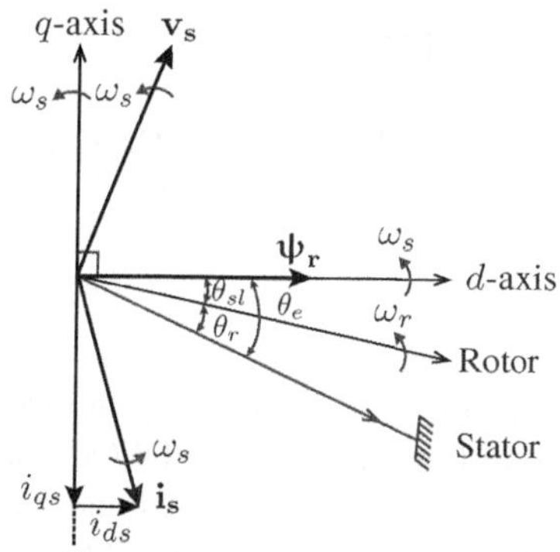

Figure 11.4 Space vector diagram of SCIG for rotor flux-oriented control.

(1) Block Diagram of FOC: The block diagram of the FOC scheme for an SCIG is shown in Figure 11.5. The FOC uses a cascaded control structure: (1) the outer speed control loop defines rotor flux and torque references; (2) the rotor flux and torque references are transformed to orthogonal stator current references (i^*_{ds} and i^*_{qs}); (3) the inner control loop employs PI controllers to force currents i_{ds} and i_{qs} to be close to their corresponding references and to generate suitable rectifier voltages references (v^*_{ds} and v^*_{qs}); (4) the dq-frame voltages are transformed to the $\alpha\beta$- or abc-frame with the help of rotor flux angle θ_e; (5) finally a modulation stage is used to transform $v^*_{\alpha s}$ and $v^*_{\beta s}$ to corresponding three-phase switching signals s_{ar1}, s_{br1}, and s_{cr1}. The sinusoidal pulse-width modulation and space vector modulation are commonly used in FOC. The rotor flux magnitude and angle (ψ_r and θ_e) are estimated by an additional subsystem.

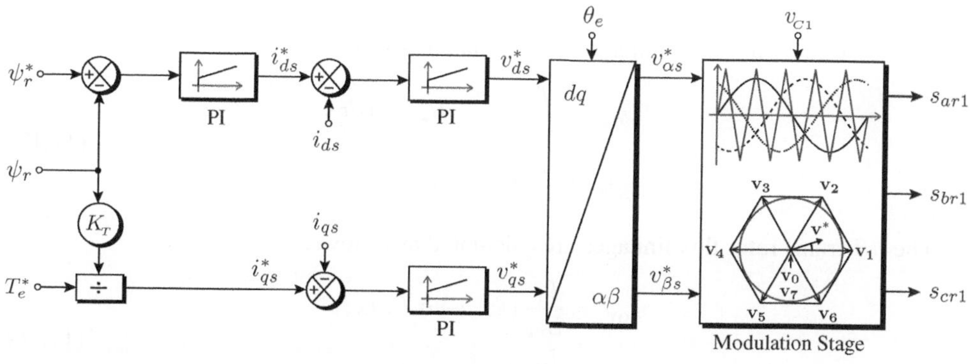

Figure 11.5 Block diagram of the FOC scheme for 2L-VSR-connected SCIG.

The FOC can be realized by two design approaches: (1) feedback or direct FOC (DFOC) developed by *F. Blaschke* and (2) feedforward or indirect FOC (IFOC) developed by *K. Hasse*. The FOC block diagram shown in Figure 11.5 is applicable for both DFOC and IFOC. However, they use different methods to estimate ψ_r and θ_e (Figure 11.6). The DFOC uses three-phase stator voltages and currents ($\mathbf{v_s}$ and $\mathbf{i_s}$) to directly estimate ψ_r and θ_e. The IFOC estimates ψ_r and θ_e indirectly with the help of three-phase stator currents $\mathbf{i_s}$ and rotor electrical speed ω_r. IFOC is more popular than DFOC because of its easier implementation and better performance during the entire speed range.

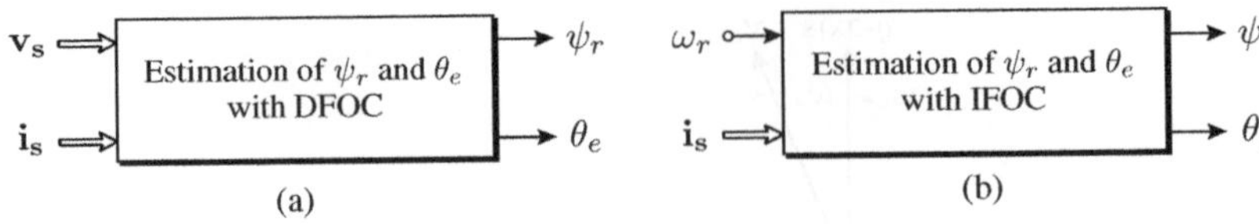

Figure 11.6 Input requirements for ψ_r and θ_e estimation with (a) DFOC and (b) IFOC.

((2) Direct Field-Oriented Control: The block diagram for the estimation of ψ_r and θ_e with DFOC scheme is shown in Figure 11.7. The DFOC uses the $\alpha\beta$-frame IG model to estimate ψ_r and θ_e. The three-phase stator voltages and currents are transformed to the $\alpha\beta$-frame with the help of $abc/\alpha\beta$ transformation in (3.6).

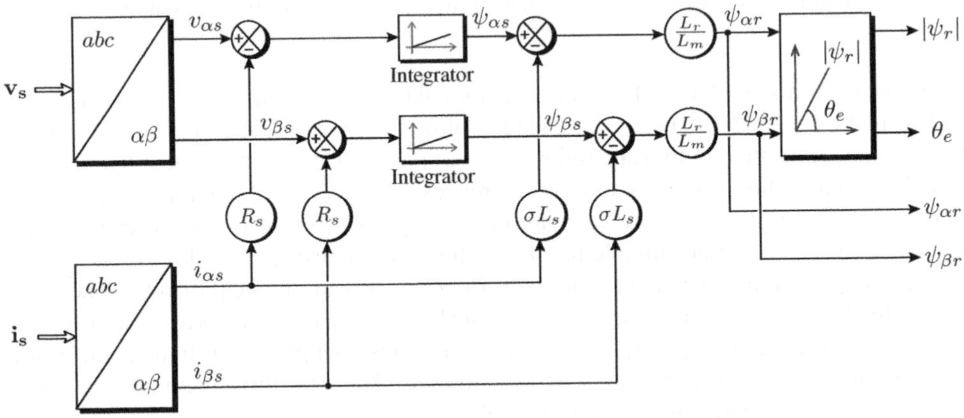

Figure 11.7 Estimation of ψ_r and θ_e with DFOC scheme.

The $\alpha\beta$-frame stator flux linkages are estimated as follows:

$$\begin{aligned} \psi_{\alpha s} &= \int (v_{\alpha s} - R_s\, i_{\alpha s})\, dt \\ \psi_{\beta s} &= \int (v_{\beta s} - R_s\, i_{\beta s})\, dt. \end{aligned} \tag{11.17}$$

The $\alpha\beta$-frame rotor flux linkages are calculated as follows:

$$\begin{aligned} \psi_{\alpha r} &= \frac{L_r}{L_m}\,(\psi_{\alpha s} - \sigma\, L_s\, i_{\alpha s}) \\ \psi_{\beta r} &= \frac{L_r}{L_m}\,(\psi_{\beta s} - \sigma\, L_s\, i_{\beta s}). \end{aligned} \tag{11.18}$$

Finally, the magnitude and angle of rotor flux vector are obtained by the following:

$$\begin{aligned} \psi_r &= \sqrt{\psi_{\alpha r}^2 + \psi_{\beta r}^2} \\ \theta_e &= \tan^{-1}\left(\frac{\psi_{\beta r}}{\psi_{\alpha r}}\right). \end{aligned} \tag{11.19}$$

The sinusoidal nature of the $\alpha\beta$-frame variables leads to problems in the integrator. This issue is usually solved by using low-pass filters in place of integrators [13]. The estimated ψ_r and θ_e are used in the FOC block for the independent control of torque and flux.

(3) Indirect Field-Oriented Control: Figure 11.8 shows the IFOC scheme in the dq-frame for the estimation of ψ_r and θ_e via the measured $\mathbf{i_s}$ and ω_r. The three-phase currents are transformed to the dq-frame with the help of the feedback rotor flux angle θ_e. The stator angular frequency ω_s is obtained by adding the rotor speed ω_r with the estimated slip frequency ω_{sl}. The rotor flux angle θ_e is then obtained by integrating ω_s:

$$\theta_e = \int \omega_s \, dt = \int (\omega_r + \omega_{sl}) \, dt = \theta_r + \theta_{sl}. \tag{11.20}$$

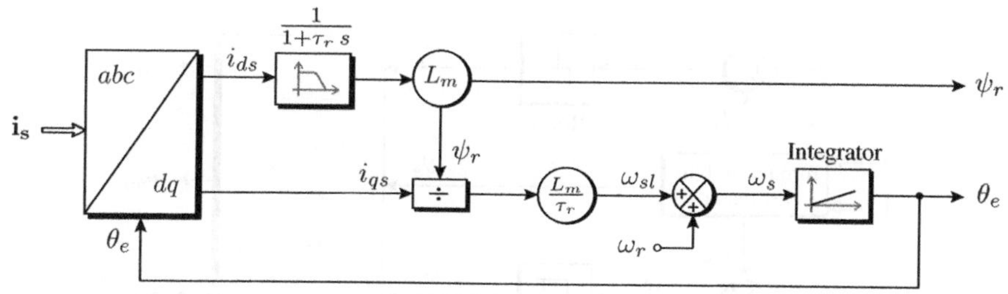

Figure 11.8 Estimation of ψ_r and θ_e with IFOC scheme.

By using the detailed IG model given in Chapter 6, the slip angular frequency ω_{sl} is expressed in terms of generator parameters and q-axis current as follows [13]:

$$\omega_{sl} = \left(\frac{L_m}{\tau_r \, \psi_r} \right) i_{qs}. \tag{11.21}$$

The rotor flux magnitude ψ_r needed in the above expression is estimated from d-axis current, as shown in the following expression:

$$\psi_r = \frac{L_m}{(1 + \tau_r \, s)} i_{ds} \tag{11.22}$$

where s is the Laplace operator.

The rotor flux magnitude ψ_r and angle θ_e estimated by the above feedforward approach are used in Figure 11.5 to realize the FOC scheme.

11.4.2 Direct Torque Control

Manfred Depenbrock, *Isao Takahashi* and *Toshihiko Noguchi* are credited for the innovation of DTC in the 1980s [35, 36]. In DTC scheme, the stator flux and electromagnetic torque are controlled directly without involving internal current control loop. The electromagnetic torque of SCIG is expressed as follows [7]:

$$T_e = \frac{1.5 \, P_p \, L_m}{\sigma \, L_s \, L_r} \, \psi_s \, \psi_r \sin \theta_T \tag{11.23}$$

where θ_T is called the torque angle.

During different operating conditions, the rotor flux magnitude ψ_r changes slightly. However, the variation in ψ_r is significantly small; thus, ψ_r is assumed constant. The objective of DTC is to maintain a linear relationship between T_e and θ_T by maintaining stator flux magnitude ψ_s constant [37]. Figure 11.9 shows the block diagram of the DTC scheme for a 2L-VSR-connected SCIG. The FOC scheme uses rotor field coordinates, whereas the DTC scheme uses stator field coordinates. Similar to FOC, the DTC scheme also uses a cascaded structure: (1) the outer speed control loop defines stator flux and torque references (ψ_s^* and T_e^*); (2) stator flux linkage ψ_s and electromagnetic torque T_e are estimated with the help of an additional subsystem; (3) estimated ψ_s and T_e are compared with the corresponding references and error terms ΔT_e and $\Delta \psi_s$ are passed to 3L and 2L hysteresis comparators (relays); and (4) finally, a lookup table is used to define the switching signals for 2L-VSR on the basis of the output logic of relays (x_T and x_ψ) and stator flux angle θ_s such that the torque and flux errors are bounded to the hysteresis bandwidths.

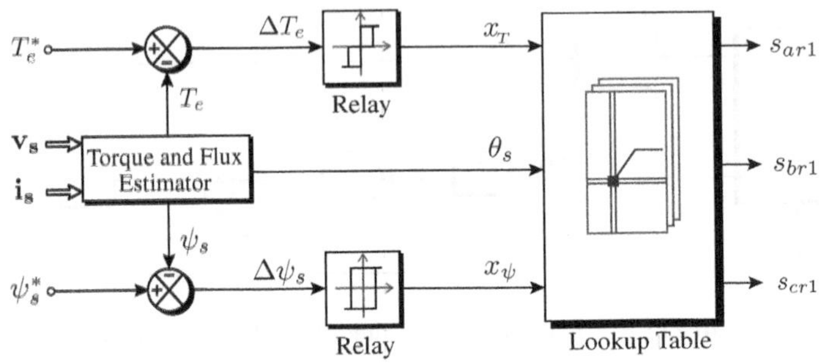

Figure 11.9 Block diagram of the DTC scheme for a 2L-VSR-connected SCIG.

Similar to the DFOC in Figure 11.7, the stator flux linkage is estimated with the help of $\mathbf{v_s}$ and $\mathbf{i_s}$. The electromagnetic torque is computed by the cross product of estimated stator flux vector and measured stator current vector. The perturbations in stator voltage measurement and stator winding resistance lead to inaccurate flux estimation, especially at low-speed operation. The switching frequency varies with the DTC scheme because no modulation stage is employed. The DTC uses no PI controllers and modulation stage; thus, its design is simple compared with that of the FOC. The DTC scheme produces a fast dynamic response for torque and flux compared with the FOC method.

11.5 DIGITAL CONTROL OF BTB CONVERTER-BASED SCIG WECS

The digital control system development for SCIG WECS is similar to that for PMSG WECS (Chapter 9). In this section, the digital control system development for SCIG WECS is briefly discussed in terms of the block diagram of the control scheme and the calculation of reference control variables.

11.5.1 Block Diagram of Digital Control System

The block diagram of the digital control system for BTB VSC-based SCIG WECS is shown in Figure 11.10. The MSC and GSC are realized by 2L or 3L VSC. The DC-link consists of one capacitor with 2L-VSC and two capacitors with 3L-VSC. The Level I to III control loops are used for generator torque and flux control, net DC-bus voltage control, and grid reactive power control. Compared with the block diagram of PMSG WECS, the SCIG WECS additionally uses generator voltages to allow the DFOC method to estimate the rotor flux magnitude and angle. To utilize the motor models for generator operation, the stator currents are measured in the opposite direction, that is, from MSC to SCIG.

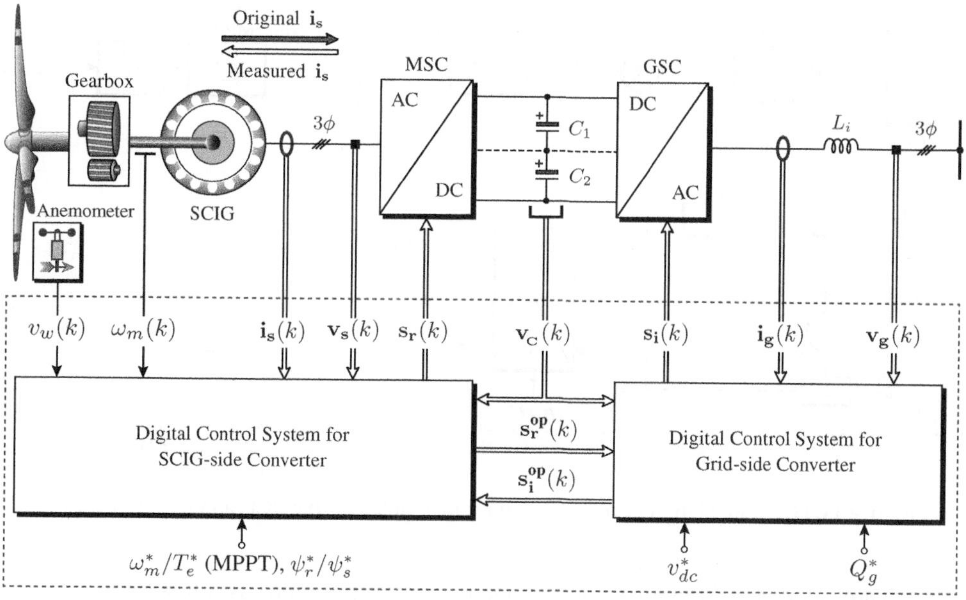

Figure 11.10 Block diagram of digital control scheme for BTB VSC-based SCIG WECS.

The technical, operational, and legal requirements for the operation of MW-WECS are fulfilled by the digital control systems of MSC and GSC. The MPPT operation of SCIG WECS is achieved by controlling the generator torque and flux. The torque and flux commands are transformed to reference generator currents. The MSC control system regulates feedback currents at their reference values to achieve variable-speed operation. On the grid-side, the net DC-bus voltage and grid reactive power control requirements are transformed to reference grid currents. The GSC control system generates optimal switching signals such that grid currents follow the respective reference currents. To reduce the switching losses at the MW level, the MSC and GSC switching frequencies are also minimized by the digital control systems. For BTB NPC converters, the DC-link capacitors voltage are maintained at half the net DC-bus voltage to reduce stress on semiconductor devices. The nation of variables is nearly the same as in PMSG WECS. Refer to Section 9.5.3 for further details.

11.5.2 Calculation of Reference Control Variables

Figure 11.11 shows the block diagram for the calculation of reference control variables. The references corresponding to PCC and PTC are shown, along with the optimal tip-speed ratio (OTSR) and optimal torque (OT) MPPT algorithms. The electromagnetic torque reference T_e^* generated by the OTSR and OT algorithms are used by both the PCC and PTC methods. The PCC method uses $i_{\alpha s}^*$ and $i_{\beta s}^*$ and the PTC method uses T_e^* and ψ_s^*.

The OTSR MPPT algorithm uses wind speed v_w and coefficient K_1^{op} and calculates reference speed ω_m^*. The OT algorithm uses rotor speed feedback $\omega_m(k)$ and rated WT parameters and calculates reference torque T_e^*. The rotor and stator flux references (ψ_r^* and ψ_s^*) are defined according to the rated parameters of SCIG. The reference currents of

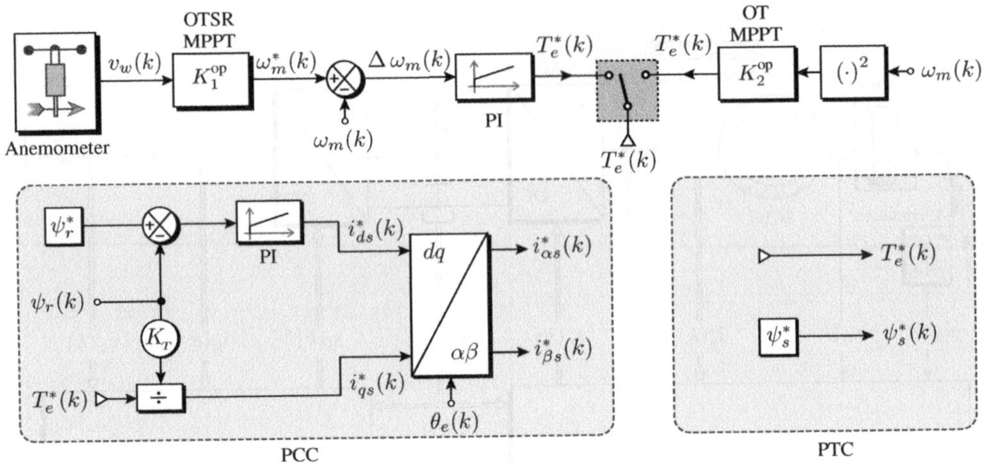

Figure 11.11 Calculation of reference control variables for the PCC and PTC schemes.

dq-axes are calculated from the rotor flux and torque control loops as follows:

$$i_{ds}^*(k) = \left(k_p + \frac{k_i}{s}\right)[\psi_r^*(k) - \psi_r(k)] \tag{11.24}$$

$$i_{qs}^*(k) = \frac{T_e^*(k)}{K_T\,\psi_r(k)}. \tag{11.25}$$

The reference currents of dq-axes are transformed to the $\alpha\beta$-frame with the help of rotor flux angle θ_e. The rotor flux magnitude $\psi_r(k)$ needed in the above two calculations and θ_e are obtained by DFOC or IFOC methods discussed earlier (Figures 11.7 and 11.8).

CASE STUDY 11.1 Calculation of Reference Variables for SCIG

Case Study Objective: This case study extends the analysis of reference control variables for SCIG WECS. The reference control variables based on the OTSR MPPT algorithm are analyzed on the basis of the MATLAB simulation results. This case study also provides a comparison between reference control variables for SCIG WECS and PMSG WECS (refer to case study 9.1).

Parameters: The parameters for a 750-kW, 690-V, 60-Hz SCIG #1, and a 750-kW WT are given in Tables A.7 and A.1, respectively. The rated wind speed of 12 m/s is considered.

Analysis:
Figure 11.12 shows the reference control variables with respect to wind speed during the startup operation of SCIG WECS. The wind speed v_w is programmed to change from 0 m/s at t = 1.0 s to 12 m/s at t = 4.0 s (Figure 11.12(a)). The OTSR MPPT formulates the mechanical shaft speed reference as a linear function of wind speed. Therefore, ω_m^* increases in proportion to v_w (Figure 11.12(b)). The electromagnetic torque reference T_e^* increases two-fold with respect to v_w (Figure 11.12(c)). The ω_m^* and T_e^* waveforms are similar to those in case study 9.1. The d-axis reference current i_{ds}^* is constant because ψ_r^* is maintained constant during the entire wind speed range. The transients in i_{ds}^* waveform are not shown. During the system startup, the estimated ψ_r value is brought to ψ_r^* by a PI controller (Figure 11.11). During this transient interval, the PI controller saturates and generates 1.0 $\widehat{\text{pu}}$ i_{ds}^* to enable fast system startup. Considering that the rotor flux magnitude is

maintained constant, the q-axis reference current i_{qs}^* shows a linear relationship with T_e^* (refer to Equation (11.25)). Figure 11.12(d) shows the i_{ds}^* and i_{qs}^* waveforms with respect to v_w. The rotor flux reference ψ_r^* for PCC scheme and stator flux reference ψ_s^* for PTC scheme are maintained constant according to the rated parameters of SCIG (Figure 11.12(e)).

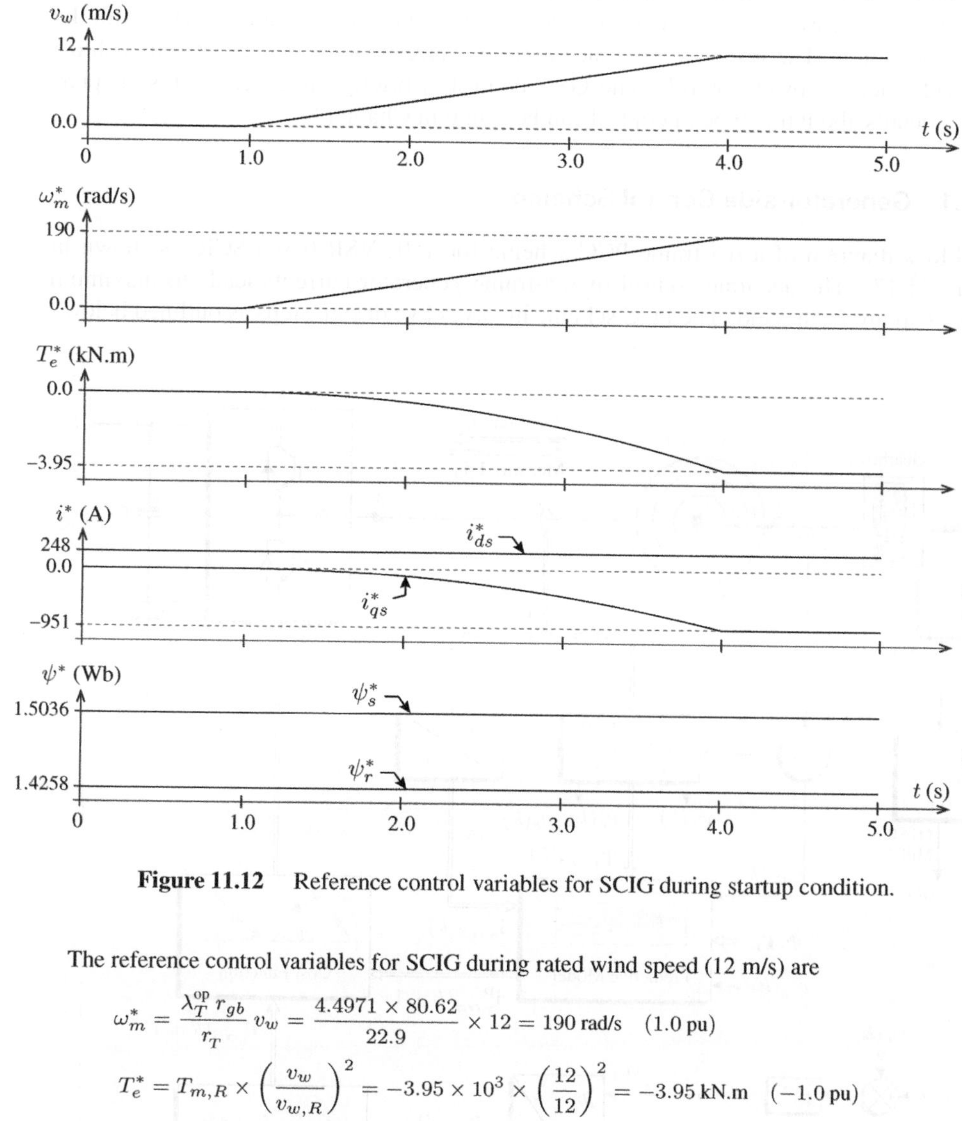

Figure 11.12 Reference control variables for SCIG during startup condition.

The reference control variables for SCIG during rated wind speed (12 m/s) are

$$\omega_m^* = \frac{\lambda_T^{\rm op}\, r_{gb}}{r_T}\, v_w = \frac{4.4971 \times 80.62}{22.9} \times 12 = 190 \text{ rad/s} \quad (1.0 \text{ pu})$$

$$T_e^* = T_{m,R} \times \left(\frac{v_w}{v_{w,R}}\right)^2 = -3.95 \times 10^3 \times \left(\frac{12}{12}\right)^2 = -3.95 \text{ kN.m} \quad (-1.0 \text{ pu})$$

$$i_{ds}^* = \frac{\psi_r^*}{L_m} = \frac{1.4258}{5.75 \times 10^{-3}} = 248 \text{ A} \quad (0.25\ \widehat{\rm pu})$$

$$i_{qs}^* = \frac{T_e^*}{K_T\, \psi_r} = \frac{T_e^*\, 2\, L_r}{3\, P_p\, L_m\, \psi_r} = \frac{-3.95 \times 10^3 \times 2 \times 0.00592}{3 \times 2 \times 0.00575 \times 1.4258} = -951 \text{ A} \quad (-0.97\ \widehat{\rm pu}).$$

The variable-speed operation of SCIG WECS is achieved by following the above reference control variables. The PCC scheme forces the measured currents to track i_{ds}^* and i_{qs}^*, whereas the PTC scheme considers T_e^* and ψ_s^* to accomplish the MPPT operation.

■

11.6 PREDICTIVE CURRENT CONTROL OF BTB 2L-VSC-BASED SCIG WECS

The overall digital control of BTB 2L-VSC-based SCIG WECS is analyzed in this section. The SCIG-side 2L-VSR and grid-side 2L-VSI are controlled by $\alpha\beta$-frame PCC schemes. The main control objective for VSR and VSI control schemes is tracking the respective $\alpha\beta$-frame reference currents. The generator-side reference currents are calculated to fulfill the MPPT operation. The grid-side reference currents correspond to the net DC-bus voltage and grid reactive power control. The GSC control is briefly discussed in this chapter; further details about this type of control can be found in Chapter 8.

11.6.1 Generator-side Control Scheme

The block diagram of a $\alpha\beta$-frame PCC scheme for a 2L-VSR-based SCIG is shown in Figure 11.13. The accurate control of $\alpha\beta$-frame generator currents leads to maximum energy capture from wind. The PCC scheme involves five design steps as outlined below.

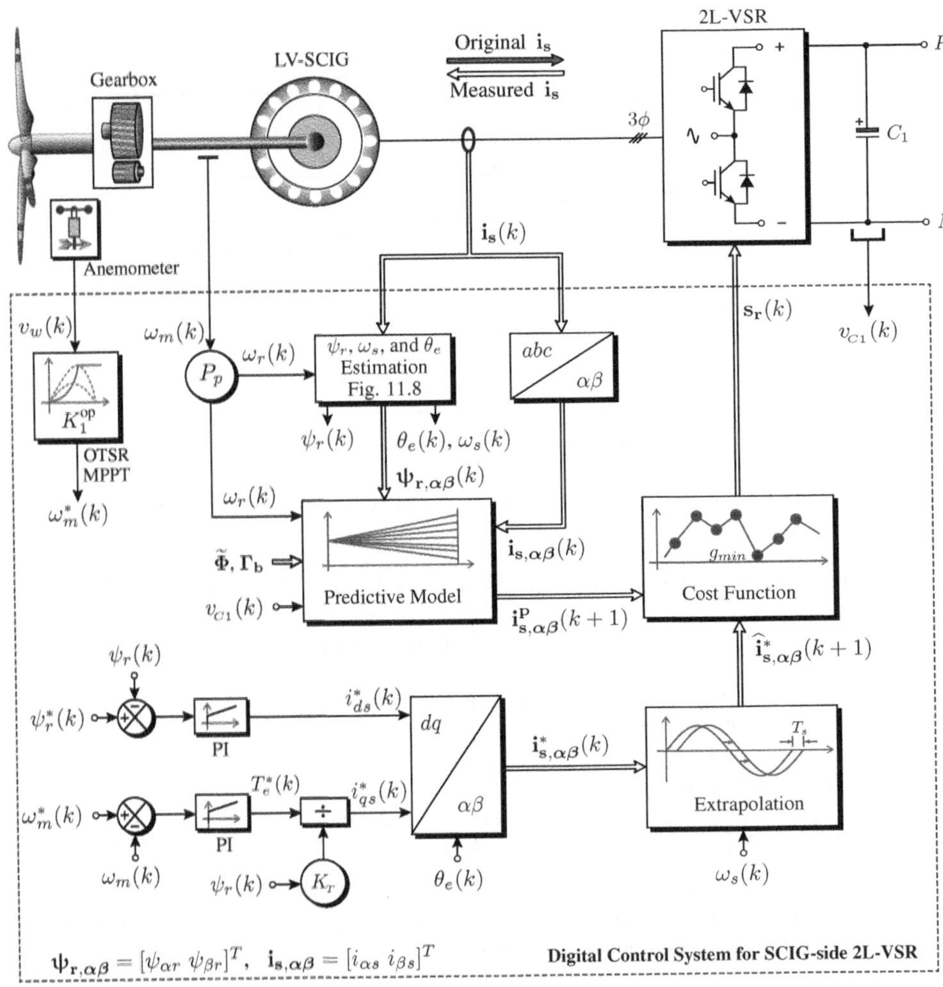

Figure 11.13 Block diagram of IFOC-based PCC scheme for SCIG with 2L-VSR.

Measurement and Synthesis of Feedback Signals: The wind speed v_w, DC- link capacitor voltage $v_{C1}(k)$, three-phase generator/converter currents $\mathbf{i_s}(k)$, and generator rotor mechanical speed $\omega_m(k)$ are measured as a first step in the design of PCC scheme. To adopt the motor model for generator control, the three-phase stator currents $\mathbf{i_s}$ are measured in opposite direction to the original flow. The rotor electrical speed $\omega_r(k)$ is obtained by multiplying $\omega_m(k)$ with generator pole pairs, P_p. The rotor flux magnitude ψ_r and angle θ_e, as well as the speed of synchronous frame ω_s are calculated by the IFOC scheme by utilizing the dq-frame model of SCIG (Figure 11.8). The $\alpha\beta$-frame rotor flux linkages $\psi_{\alpha r}$ and $\psi_{\beta r}$ are also computed for use in the "predictive model" subsystem.

Calculation of Reference Currents: The calculation of reference currents involves both the dq and $\alpha\beta$ reference frames. $i^*_{ds}(k)$ and $i^*_{qs}(k)$ are calculated by combining the OTSR MPPT algorithm and IFOC method in Figure 11.11. $i^*_{ds}(k)$ is calculated from the rotor flux control loop, and $i^*_{qs}(k)$ is obtained from the torque control loop. The dq-frame reference currents are transformed to the $\alpha\beta$-frame with the help of the estimated $\theta_e(k)$.

Extrapolation of Reference Currents: The cost function subsystem evaluates the error between predicted and reference currents during the $(k+1)$ sampling instant. To accurately estimate the future values of the generator current error, the reference currents are extrapolated to the $(k+1)$ sampling instant with the help of vector angle extrapolation:

$$\begin{bmatrix} \widehat{i}^*_{\alpha s}(k+1) \\ \widehat{i}^*_{\beta s}(k+1) \end{bmatrix} = e^{j\omega_s(k)\,T_s} \begin{bmatrix} i^*_{\alpha s}(k) \\ i^*_{\beta s}(k) \end{bmatrix}. \tag{11.26}$$

The speed of synchronous frame $\omega_s(k)$ is updated by the ψ_r and θ_e estimation subsystem during each sampling interval.

Prediction of Future Behavior of SCIG Currents: The PCC scheme can be implemented in the dq-frame, where $i^*_{ds}(k)$ and $i^*_{qs}(k)$ are directly used in the cost function. However, the prediction of dq-frame currents involves $\alpha\beta/dq$ transformation in iterative loop leading to more number of calculations during each sampling interval. The $\alpha\beta$-frame PCC uses $dq/\alpha\beta$ transformation only once during each sampling interval to transform $i^*_{ds}(k)$ and $i^*_{qs}(k)$ to $i^*_{\alpha s}(k)$ and $i^*_{\beta s}(k)$. The "predictive model" subsystem uses the following DT model of SCIG and 2L-VSR to predict the future behavior of $\alpha\beta$-frame currents.

- For one-sample-ahead PCC, only the prediction of stator currents is needed and rotor flux linkages need not be predicted. Therefore, the DT model in Equation (11.10) is modified as follows to suit one-sample-ahead PCC applications:

$$\begin{bmatrix} i^p_{\alpha s}(k+1) \\ i^p_{\beta s}(k+1) \end{bmatrix} = \underbrace{\begin{bmatrix} 1-\frac{T_s}{\tau_\sigma} & 0 & \frac{k_r\,T_s}{\sigma\,L_s\,\tau_r} & \frac{k_r\,\omega_r(k)\,T_s}{\sigma\,L_s} \\ 0 & 1-\frac{T_s}{\tau_\sigma} & -\frac{k_r\,\omega_r(k)\,T_s}{\sigma\,L_s} & \frac{k_r\,T_s}{\sigma\,L_s\,\tau_r} \end{bmatrix}}_{\mathbf{\Phi}(k)} \begin{bmatrix} i_{\alpha s}(k) \\ i_{\beta s}(k) \\ \psi_{\alpha r}(k) \\ \psi_{\beta r}(k) \end{bmatrix} + \underbrace{\begin{bmatrix} \frac{T_s}{\sigma\,L_s} & 0 \\ 0 & \frac{T_s}{\sigma\,L_s} \end{bmatrix}}_{\mathbf{\Gamma_b}} \underbrace{\begin{bmatrix} v^p_{\alpha s}(k) \\ v^p_{\beta s}(k) \end{bmatrix}}_{\text{2L-VSR Model}} \tag{11.27}$$

where superscript p denotes the predicted variable.

- The next step is to define the 2L-VSR output voltages in terms of $\alpha\beta$-frame switching signals and measured DC-link capacitor voltage. Without any harmonic filter between SCIG and 2L-VSR, the rectifier output voltages become equal to the generator voltages (Figure 11.13). As shown below, the model from (5.11) is adopted for the rectifier case.

$$\begin{bmatrix} v^p_{\alpha r}(k) \\ v^p_{\beta r}(k) \end{bmatrix} = v_{C1}(k)\left[\mathbf{T}_{abc/\alpha\beta}\right]\begin{bmatrix} s_{ar1}(k) \\ s_{br1}(k) \\ s_{cr1}(k) \end{bmatrix} = v_{C1}(k)\begin{bmatrix} s_{\alpha r1}(k) \\ s_{\beta r1}(k) \end{bmatrix} \tag{11.28}$$

where $s_{\alpha r1}$ and $s_{\beta r1}$ are $\alpha\beta$-frame switching signals that can be calculated offline.

- Finally, by combining Equations (11.27) and (11.28), the complete sampled-data model for the prediction of future behavior of $\alpha\beta$-frame generator currents is obtained in terms of 2L-VSR switching signals:

$$\begin{bmatrix} i^p_{\alpha s}(k+1) \\ i^p_{\beta s}(k+1) \end{bmatrix} = \mathbf{\Phi}(k)\begin{bmatrix} i_{\alpha s}(k) \\ i_{\beta s}(k) \\ \psi_{\alpha r}(k) \\ \psi_{\beta r}(k) \end{bmatrix} + \mathbf{\Gamma_b}\left\{ v_{C1}(k)\begin{bmatrix} s^p_{\alpha r1}(k) \\ s^p_{\beta r1}(k) \end{bmatrix}\right\}. \tag{11.29}$$

For a 2L-VSR, eight switching state combinations (predictions) are available, thus leading to eight different predictions are obtained for $v^p_{\alpha r}$ and $v^p_{\beta r}$, and $i^p_{\alpha s}$ and $i^p_{\beta s}$. The DT matrix $\widetilde{\mathbf{\Phi}}$ is computed offline (by excluding $\omega_r(k)$ term), in addition to $\mathbf{\Gamma_b}$. This approach leads to a minimal number of online calculations.

Cost Function Minimization: The control objectives for the generator-side 2L-VSR are the regulation of $\alpha\beta$-frame generator currents and the minimization of switching frequency. As demonstrated below, these control objectives are defined by a cost function with three sub-control terms:

$$\begin{aligned} g_r(k) = {} & \lambda_{i\alpha}\left[\widehat{i}^*_{\alpha s}(k+1) - i^p_{\alpha s}(k+1)\right]^2 \\ & + \lambda_{i\beta}\left[\widehat{i}^*_{\beta s}(k+1) - i^p_{\beta s}(k+1)\right]^2 \\ & + \lambda_{sw,r}\sum_{x=a,\,b,\,c}\left[s^p_{xr1}(k) - s^{\text{op}}_{xr1}(k)\right]^2 \end{aligned} \tag{11.30}$$

where $\lambda_{i\alpha}$, $\lambda_{i\beta}$, and $\lambda_{sw,r}$ are weighting factors. s^{op}_{ar1}, s^{op}_{br1}, and s^{op}_{cr1} are the previous sample optimal switching signals of 2L-VSR. Weighting factors $\lambda_{i\alpha}$ and $\lambda_{i\beta}$ correspond to the regulation of the $\alpha\beta$-frame generator currents. Considering that the generator current control is the primary control objective, $\lambda_{i\alpha}$ and $\lambda_{i\beta}$ are set to one. The heuristic approach is commonly used to select the $\lambda_{sw,r}$ value to minimize the 2L-VSR switching frequency to desired value.

Among the eight combinations, the switching signal combination that produces the minimum value of $g_r(k)$ is selected from the pool of eight and is applied directly to the 2L-VSR. As demonstrated through the step-by-step design, with the PCC scheme, no PI controllers are needed for the regulation of generator currents, and no modulation stage is needed for the generation of switching signals. Therefore, PCC implementation is simple compared with classical FOC.

11.6.2 Grid-side Control Scheme

On the basis of the v_{dc} and Q_g control loops, the $\alpha\beta$-frame reference grid currents are obtained. Similar to the generator-side control scheme, the grid-side control scheme regulates the $\alpha\beta$-frame grid currents and switching frequency of 2L-VSI. The grid-side cost function defined below is similar to Equation (11.30), except that subscripts s and r are replaced by g and i, respectively.

$$\begin{aligned} g_i(k) &= \lambda_{i\alpha}\left[\widehat{i}^*_{\alpha g}(k+1) - i^p_{\alpha g}(k+1)\right]^2 \\ &+ \lambda_{i\beta}\left[\widehat{i}^*_{\beta g}(k+1) - i^p_{\beta g}(k+1)\right]^2 \\ &+ \lambda_{sw,i}\sum_{x=a,\,b,\,c}\left[s^p_{xi1}(k) - s^{\text{op}}_{xi1}(k)\right]^2. \end{aligned} \tag{11.31}$$

The analysis on the selection of weighting factors for generator-side cost function can also be attributed to the grid-side cost function. A detailed discussion on GSC control has already been presented in Chapter 8.

11.6.3 Control Algorithm

To demonstrate the easy digital implementation of the PCC scheme, the flowchart and MATLAB *S-Function Builder* programming are discussed in this section. The flowchart of GSC control is also given to exhibit the identical nature of PCC algorithms for 2L-VSR and 2L-VSI. Figures 11.14(a) and 11.14(b) show the flowcharts of the PCC algorithm for 2L-VSR and 2L-VSI, respectively. The feedback measurements needed are obtained in block ①. The $\alpha\beta$-frame extrapolated reference currents are calculated in block ②. The PCC algorithm is initiated in block ③, which means that the previous two blocks are outside the iterative loop. The PCC algorithm enters iterative loop in block ④ and ends in block ⑨. The $\alpha\beta$-frame 2L-VSR and 2L-VSI output voltages are predicted in block ⑤. The prediction of generator and grid currents is conducted by block ⑥. Block ⑦ evaluates the generator-side and grid-side cost functions. An optimal vector j_{op} corresponding to the minimum (optimal) cost function value is stored in block ⑧. After the completion of an iterative loop for eight counter values, the switching signals corresponding to j_{op} are produced by block ⑩ and are applied directly to 2L-VSR and 2L-VSI.

The MATLAB *S-Function Builder* programming for 2L-VSR PCC scheme is given in Algorithm 11.1. The inputs to the *S-Function* block are $\widehat{i}^*_{\alpha s}(k+1)$, $\widehat{i}^*_{\beta s}(k+1)$, $i_{\alpha s}(k)$, $i_{\beta s}(k)$, $\psi_{\alpha r}(k)$, $\psi_{\beta r}(k)$, $v_{C1}(k)$, and $\lambda_{sw,r}$. The *S-Function* block output signals are $s^p_{ar1}(k)$, $s^p_{br1}(k)$, and $s^p_{cr1}(k)$. The switching states in the $\alpha\beta$ and abc reference frames, as well as DT parameters $\mathbf{\Phi}$ and $\mathbf{\Gamma}_\text{b}$, are defined as the external functions for the *S-Function*. The *S-Function* is executed at the control sampling time T_s; thus, `RateTransition` blocks should be connected to the input and output ports of *S-Function Builder*. The port sampling time for input and output `RateTransition` blocks are T_s and fundamental simulation sampling time T_{sim}, respectively. The program lines are explained as follows: PCC initialization is shown in lines `01-02`; the `for` loop is initialized in lines `03-04`; The VSR voltages are predicted in lines `05-06`; SCIG currents are predicted in lines `07-08`; the cost function is calculated in lines `09-12`; the optimal vector `j_op` number is selected in lines `13-17`; the `for` loop is terminated in line `18`; the optimal three-phase switching states are produced in lines `19-21` on the basis of the `j_op` value; the switching states are stored in lines `22-24` for use in the next iteration.

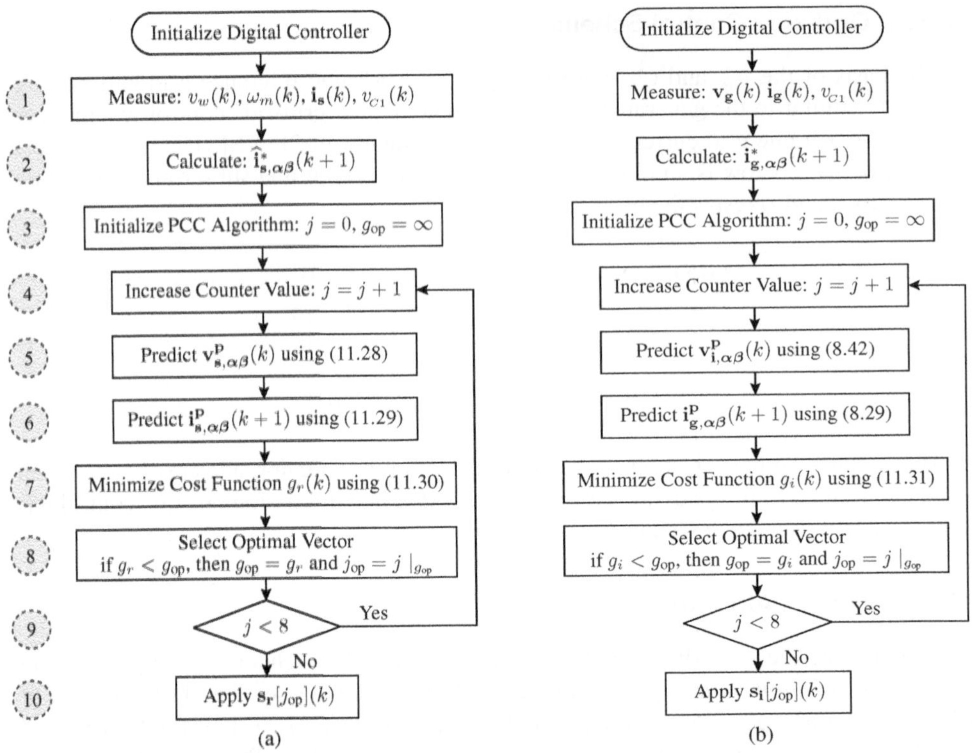

Figure 11.14 Flowchart of the PCC algorithm for the (a) 2L-VSR and (b) 2L-VSI.

ALGORITHM 11.1 PCC Algorithm for 2L-VSR-Based SCIG WECS

```
\\ P11 = P22 = 1-(Ts/tau_sigma); P13 = P24 = (kr*Ts)/(sigma*Ls*tau_r);
\\ P14 = -P23 = (kr*Ts)/(sigma*Ls); G11 = G22 = Ts/(sigma*Ls);
j_op = 1000;
g_op = 1000000000;
for(j = 0; j < 8; j++)
{
    vr_ab[0] = vc[0]*vrec[j].s_r_ab[0];
    vr_ab[1] = vc[0]*vrec[j].s_r_ab[1];
    is_ab_k1[0] = P11*is_ab[0] + P13*psir_ab[0] + P14*wr[0]*psir_ab[1]
                  + G11*vr_ab[0];
    is_ab_k1[1] = P22*is_ab[1] + P23*wr[0]*psir_ab[0] + P24*psir_ab[1]
                  + G22*vr_ab[1];
    g_ial = (is_ref_k1[0]-is_ab_k1[0])*(is_ref_k1[0]-is_ab_k1[0]);
    g_ibe = (is_ref_k1[1]-is_ab_k1[1])*(is_ref_k1[1]-is_ab_k1[1]);
    g_swr = (vrec[j].s_r_abc[0]-s_r_km1[0])*(vrec[j].s_r_abc[0]-s_r_km1[0])
          + (vrec[j].s_r_abc[1]-s_r_km1[1])*(vrec[j].s_r_abc[1]-s_r_km1[1])
          + (vrec[j].s_r_abc[2]-s_r_km1[2])*(vrec[j].s_r_abc[2]-s_r_km1[2]);
    g_r = g_ial + g_ibe + alpha[0]*g_swr;
    if(g_r < g_op)
        {
          j_op = j;
          g_op = g_r;
        }
}
s_ar1[0] = vrec[j_op].s_r[0];
s_br1[0] = vrec[j_op].s_r[1];
s_cr1[0] = vrec[j_op].s_r[2];
s_r_km1[0] = s_ar1[0];
s_r_km1[1] = s_br1[0];
s_r_km1[2] = s_cr1[0];
```

CASE STUDY 11.2 Steady-State Analysis of 2L-VSC-Based SCIG WECS

Case Study Objective: This case study analyzes the steady-state performance of the 2L-VSC-based SCIG WECS during the rated operating condition. The generator- and grid-side variables are calculated numerically to assist the theoretical analysis given earlier.

Parameters: The parameters of a 750-kW, 690-V, 60-Hz SCIG are $R_s = 3.996$ mΩ, $R_r = 4.529$ mΩ, $L_{ls} = L_{lr} = 0.173$ mH, $L_m = 5.745$ mH, $\psi_r = 1.4258$ Wb (*peak*), $P_p = 2$, and $n_m = 1814$ rpm (SCIG #1 in Table A.7). The parameters of GSC are $r_i = 6.3$ mΩ (0.01 pu), $L_i = 0.337$ mH (0.2 pu), and $v_{dc}^* = 1220$ V (3.062 pu). The reference reactive power of GSC is zero ($Q_g^* = 0$). The wind speed is assumed to be rated with $v_w = 12$ m/s (1.0 pu). The PCC scheme is implemented with $T_s = 100$ μs.

Analysis:
The steady-state waveforms of SCIG WECS during the rated wind speed are shown in Figure 11.15. The generator currents of $\alpha\beta$-axes, 2L-VSR phase and line-to-line voltages, rotor flux linkages of $\alpha\beta$-axes, rotor flux angle, and grid currents of $\alpha\beta$-axes are shown in subplots (a) to (e), respectively.

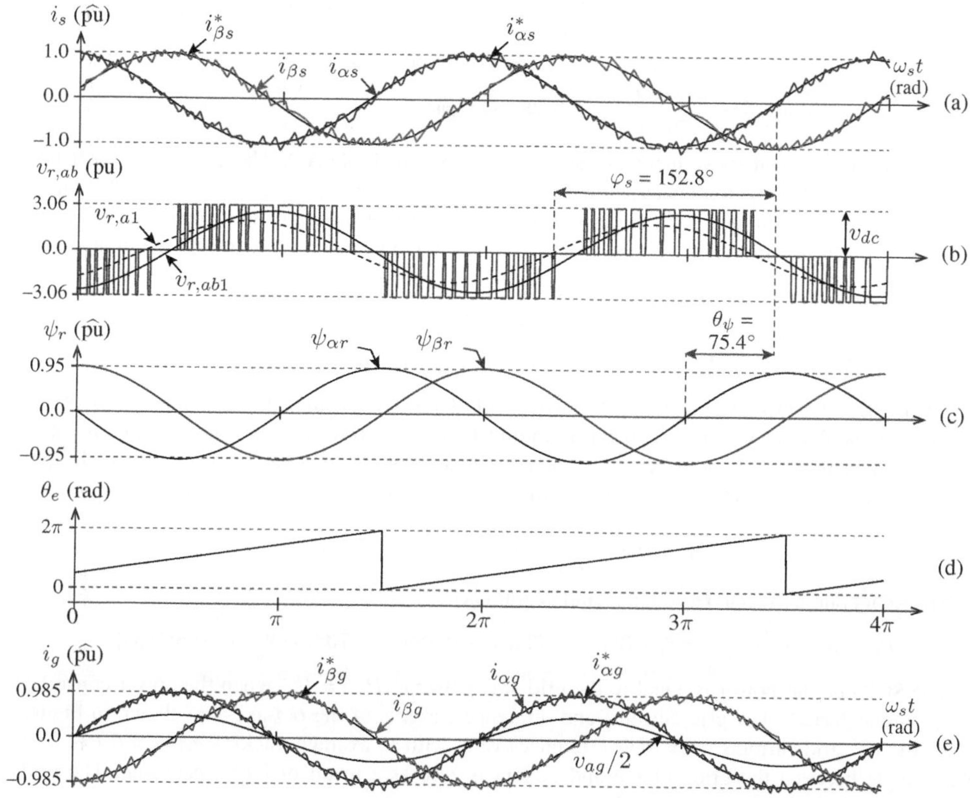

Figure 11.15 Steady-state waveforms with PCC scheme for 2L-VSC-based SCIG WECS.

The rotor electrical speed and electromagnetic torque with $v_w = 1.0$ pu are expressed as follows:

$$\omega_r = \omega_{m,R} \times P_p \times v_{w,\text{pu}} = 1814 \times 2 \times (2 \times \pi/60) \times 1.0 = 379.9 \text{ rad/s} \quad (1.0 \text{ pu})$$

$$T_e = \frac{P_m P_p}{\omega_r} = \frac{-750 \times 10^3 \times 2}{379.9} = -3.948 \times 10^3 \text{ N.m} \quad (-1.0 \text{ pu}).$$

The rotor flux ψ_r and electromagnetic torque are maintained at their reference values. The corresponding dq-axes currents during the steady-state are expressed as follows:

$$i_{ds} = \frac{\psi_r}{L_m} = \frac{1.4258}{5.75 \times 10^{-3}} = 248 \text{ A } (peak) \quad (0.25 \widehat{\text{pu}})$$

$$i_{qs} = \frac{T_e^* \, 2 \, L_r}{3 \, P_p \, L_m \, \psi_r} = \frac{-3.95 \times 10^3 \times 2 \times 0.00592}{3 \times 2 \times 0.00575 \times 1.4258} = -951 \text{ A } (peak) \quad (-0.97 \widehat{\text{pu}})$$

$$i_s = \sqrt{i_{ds}^2 + i_{qs}^2} = \sqrt{248^2 + (-951)^2} = 983 \text{ A } (peak) \quad (1.0 \widehat{\text{pu}})$$

$$\overline{I}_s = i_s/\sqrt{2} = 695\angle 0^\circ \text{ A } (rms) \quad \text{(reference phasor)}.$$

Figure 11.15(a) shows that the $\alpha\beta$-frame currents corresponding to the dq-frame currents are well-regulated at their reference values by the PCC scheme with a 4.7% average tracking error. The THD of the phase-a SCIG current is considerably high (9.09%) compared with that of the PMSG (0.67%, as shown in Figure 9.11(c)). The switching frequency ($f_{sw,r}$) of 2L-VSR is noted as 1360 Hz with a weighting factor of $\lambda_{sw,r} = 0$. The 2L-VSR line-to-line voltage and its fundamental $v_{r,ab1}$, as well as the fundamental component of phase-a VSR voltage $v_{r,a1}$ ($= v_{s,a1}$), are shown in Figure 11.15(b). The SCIG slip is calculated with the help of slip and stator frequencies:

$$\omega_{sl} = \left(\frac{L_m}{\tau_r \, \psi_r}\right) i_{qs} = \frac{5.75 \times 10^{-3} \times -951}{1.307 \times 1.4258} = -2.932 \text{ rad/s} \quad (1.0 \text{ pu})$$

$$\omega_s = \omega_r + \omega_{sl} = 379.9 - 2.932 = 376.99 \text{ rad/s} \quad (1.0 \text{ pu})$$

$$s = \frac{\omega_{sl}}{\omega_s} = \frac{-2.932}{376.99} = -0.0078 \quad (1.0 \text{ pu}).$$

The slip value shown above matches the rated slip given in Table A.7. The generator impedance corresponding to the slip frequency shown above is calculated first; the power factor (PF) angle, stator PF, and stator voltage vector are calculated on the basis of this generator impedance [13]. To assist the analysis, the dq-frame equivalent circuit of the IG in Figure 6.14 is considered in this part.

$$\overline{Z}_s = |Z_s| \angle\varphi_s = (R_s + j\,X_{ls} + j\,X_m) \parallel (R_r/s + j\,X_{lr}) = 0.57\angle 152.8^\circ \ \Omega$$

$$\text{PF} = \cos\varphi_s = \cos(152.8^\circ) = -0.889$$

$$\overline{V}_s = \overline{I}_s \, \overline{Z}_s = |V_s| \angle\varphi_s = 398.4\angle 152.8^\circ \text{ V } (rms) \quad (1.0 \text{ pu}).$$

The PF derived above also matches the rated PF of SCIG. The PF angle (φ_s) is measured between the voltage and current vectors, as shown in Figure 11.15. The rotor flux is maintained constant at its reference value by a PI controller; thus, the flux linkages of $\alpha\beta$-axes have a magnitude of 1.0 $\widehat{\text{pu}}$. The angle of rotor flux vector with respect to the stator currents is calculated as follows:

$$\theta_\psi = \cos^{-1}\frac{i_{qs}}{i_{ds}} = \cos^{-1}\frac{-951}{248} = 75.4^\circ.$$

The SCIG output power P_s is calculated as follows:

$$P_s = 3\,V_s\,I_s\cos\varphi_s = 3 \times 398.4 \times 695 \times -0.889 = -738.5 \text{ kW} \quad (-0.889 \text{ pu}).$$

The SCIG output power is equal to the grid-side power ($|\ P_s\ | = P_g$) when the power converter losses are neglected. The grid currents and reference currents of the $\alpha\beta$-axes are shown in Figure 11.15(e). The grid currents track their references well, with an average tracking error of 6.1%. The phase angle between the grid voltage and current is zero as a result of UPF operation. The grid current magnitude is calculated as follows:

$$i_g = \frac{P_g}{1.5\,v_g\cos\varphi_g} = \frac{738.5 \times 10^6}{1.5 \times 563.4 \times 1} = 873.9 \text{ A } (peak) \quad (0.985 \widehat{\text{pu}}).$$

The 2L-VSI operates with an average switching frequency $f_{sw,i}$ of 1580 Hz. The grid current THD is noted as 8.21%. The generator and grid currents follow their respective references to satisfy the overall control objectives of SCIG WECS. ■

CASE STUDY 11.3 Startup Transient Analysis of SCIG WECS

Case Study Objective: In this case study, the transient response of SCIG WECS is investigated during a ramp change in wind speed. The BTB 2L-VSC is controlled by PCC scheme. This case study closely resembles the startup procedure for SCIG WECS. The waveforms on SCIG side, DC link, and grid side are investigated with respect to the varying wind speed condition. The PCC scheme discussed in Section 11.6 is verified by this case study.

Parameters: The rated parameters of 750-kW, 690-V, 60-Hz SCIG given in case study 11.2 are considered again in this case study with cos $\varphi_g = 1$. To reduce the simulation runtime, the original moment of inertia J_m is changed to 10 kg.m^2.

Analysis:
The transient waveforms corresponding to SCIG WECS startup are shown in Figure 11.16. The reference control variables for the generator-side PCC are calculated according to the block diagram given in Figure 11.13. During $t = 0$ s to 0.5 s, the GSC is synchronized to the grid and the PI controller in the DC-bus voltage control loop maintains measured v_{dc} at its reference value v_{dc}^* (Figure 11.16(g)). The output of DC voltage PI controller is i_{dg}^*, and it changes dynamically according to the energy captured from wind (Figure 11.16(h)). Once the GSC is tied to the grid, rotor flux reference ψ_r^* is applied to the generator-side control scheme. The rotor flux ψ_r starts to build up and reaches the rated value after a short interval. To simplify the plots, simulations are changed such that rotor flux transients occur before $t = 0$ s. The PI controller in rotor flux control loop ensures that ψ_r is maintained constant all the time.

To bring the speed of WT and SCIG from zero to the rated value, a ramp profile for wind speed is considered. The v_w gradually increases from cut-in wind speed, 3 m/s (0.25 pu) to rated speed 12 m/s (1.0 pu) during time interval from $t = 0.5$ s to 3.5 s. This wind speed profile resembles the release of mechanical brakes in practical WTs. Using the measured wind speed v_w and offline computed coefficient $K_1^{\rm op}$, the OTSR MPPT produces the reference speed ω_m^* for the generator-side control system. Given that the coefficient $K_1^{\rm op}$ is constant for the particular WT design, the ω_m^* exhibits a linear relationship with respect to v_w. The speed loop PI controller regulates ω_m at its reference value (Figure 11.16(a)). The output of the PI controller is the reference electromagnetic torque T_e^*. The reference torque, along with ψ_r^*, are transformed to the generator reference currents. The output of the PI controller in ψ_r loop is i_{ds}^*, and is maintained constant during the entire speed range (Figure 11.16(b)). The q-axis reference current is proportional to T_e^*. The PCC scheme maintains the measured generator currents at their reference values, by which the generator speed and torque are controlled (Figures 11.16(c) and 11.16(d)). The difference between the mechanical input torque T_m and T_e helps the SCIG to accelerate the rotor speed ω_m.

The magnitude and frequency of SCIG stator currents increase with respect to the rotor speed (Figure 11.16(d)). The currents of dq-axes and *rms* stator current during rated wind speed are already calculated in the previous case study. The 2L-VSR line-to-line voltage $v_{r,ab}$ waveform shows three distinct voltage levels during the entire wind speed range (Figure 11.16(e)). The stator active and reactive powers (P_s and Q_s) during rated wind speed are –738 kW (–0.889 pu) and –423.5 kVAR (–0.51 pu) (Figure 11.16(f)). The negative signs for P_s and Q_s represent active power production and reactive power consumption, respectively. These values match the rated apparent power of SCIG, that is, $S_s = 830.4$ kVA (1.0 pu). The SCIG PF is measured as –0.889 (P_s/S_s) with $v_w = 1.0$ pu. The net DC-bus voltage v_{dc} is maintained constant at its reference value of 3.06 pu by a PI controller employed in the DC control loop (Figure 11.16(g)). The ripple in v_{dc} increases with respect to the generator and grid current magnitude.

The grid current magnitude increases with increasing SCIG active power output. The d-axis grid current follows the grid current waveform, whereas the q-axis current is maintained zero because of grid-side UPF operation (Figure 11.16(h)). With grid-tied operation, the 2L-VSI contains three discrete voltage levels during all operating conditions (Figure 11.16(i)). The GSC can be operated with unity, leading or lagging PF by changing Q_g^*. ■

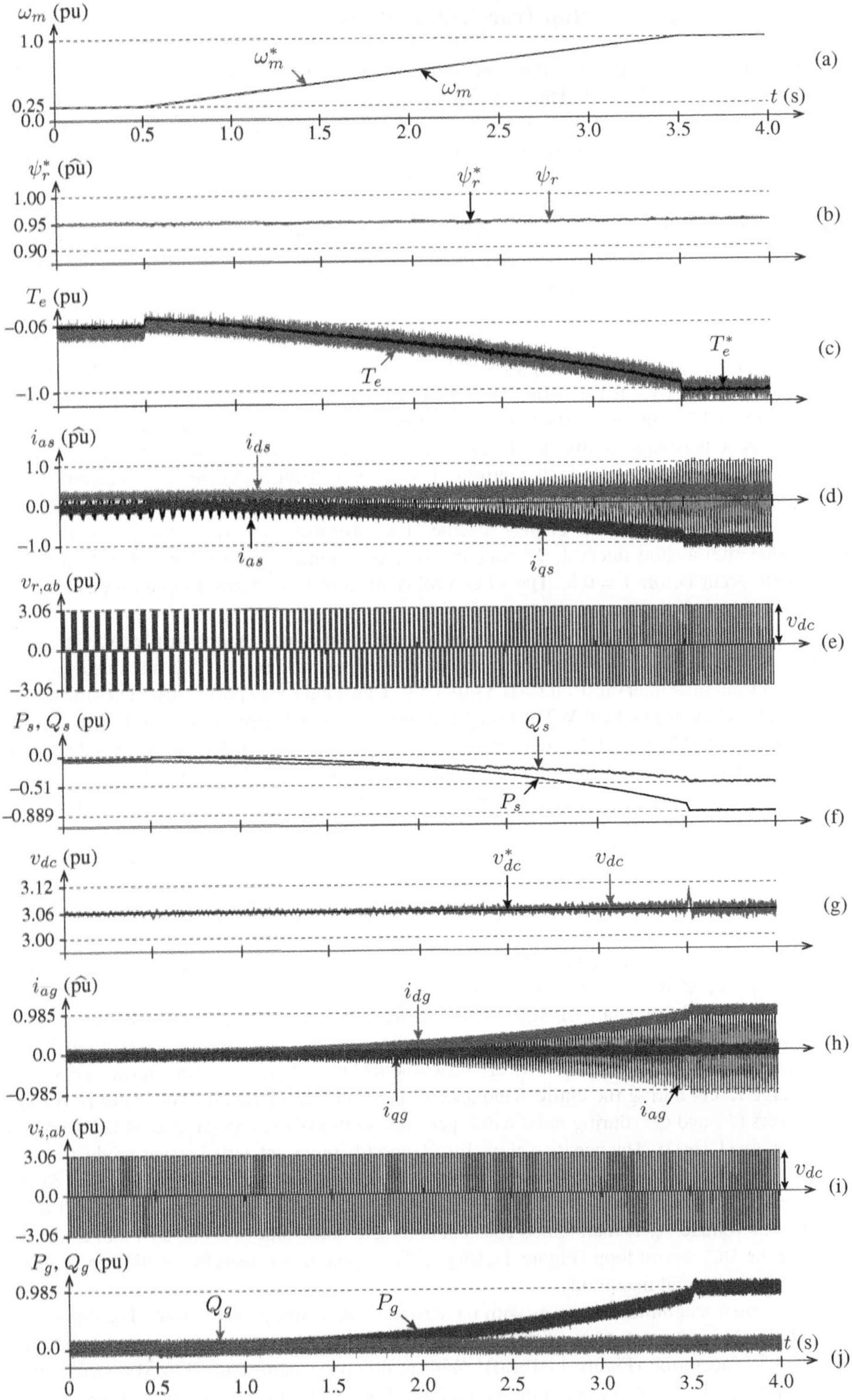

Figure 11.16 Simulated waveforms with PCC scheme for SCIG WECS during startup.

11.7 PREDICTIVE TORQUE CONTROL OF BTB NPC CONVERTER-BASED SCIG WECS

In this section, the PTC scheme for BTB NPC converter-based SCIG WECS is discussed by combining: (1) PCC of 2L-VSC-based SCIG WECS (Section 11.6) and (2) PCC of BTB NPC converter-based PMSG WECS (Section 9.7). The electromagnetic torque and stator flux of SCIG are controlled directly by the indirect control of rotor speed and stator currents [38]. The control systems for the NPC rectifier and inverter balance the DC-link capacitors voltage and switching frequency minimization, in addition to common-mode voltage elimination.

11.7.1 Generator-side Control Scheme

Figure 11.17 shows the block diagram of PTC for NPC rectifier-based SCIG WECS. This control system is similar to the PCC discussed earlier in Section 11.6, except (1) the 2L-VSR is replaced by NPC rectifier, (2) the DC-link consists of two DC-link capacitors, (3) the wind speed sensor-based OTSR MPPT is replaced by rotor speed sensor-based OT MPPT, (4) the synchronous frame speed $\omega_s(k)$ is not needed by extrapolation block, and (5) the "estimator" subsystem calculates the grid-side DC branch currents ($\widehat{i}_{Pi}$ and $\widehat{i}_{Zi}$). The reference torque $T_e^*(k)$ is calculated according to OT MPPT control, and stator flux reference $\psi_s^*(k)$ is defined constant (Figure 11.11). For the reference stator flux extrapolation $\widehat{\psi}_s^*(k+1)$ becomes equal to $\psi_s^*(k)$. The $T_e^*(k)$ is DC in a steady-state; thus, no extrapolation is needed. However, to improve transient response, the following first-order Lagrange extrapolation can be used to extrapolate the $T_e^*(k)$ to $(k+1)$ sampling instant:

$$\widehat{T}_e^*(k+1) = 2T_e^*(k) - T_e^*(k-1). \tag{11.32}$$

The other significant design steps in the PTC scheme are predictions of DC-link capacitors voltage, electromagnetic torque, and stator flux, as well as the definition of cost function [39]. These design steps are elaborated below.

Prediction of Future Behavior of SCIG Torque and Flux and DC Capacitors Voltage: The future values of stator flux, electromagnetic torque, and DC-link capacitors voltage are calculated by the following design steps:

- The DT model in Equation (11.10) is used to predict the future values of the $\alpha\beta$-frame stator currents and rotor flux linkages. In this model, $v_{\alpha s}(k)$ and $v_{\beta s}(k)$ are the SCIG (or NPC rectifier) output voltages. They are formulated in terms of DC-link capacitors voltage and $\alpha\beta$-frame switching signals as follows:

$$\begin{bmatrix} v_{\alpha s}^p(k) \\ v_{\beta s}^p(k) \end{bmatrix} = v_{C1}(k) \begin{bmatrix} s_{\alpha r1}^p(k) \\ s_{\beta r1}^p(k) \end{bmatrix} + v_{C2}(k) \begin{bmatrix} s_{\alpha r2}^p(k) \\ s_{\beta r2}^p(k) \end{bmatrix}. \tag{11.33}$$

- By substituting (11.33) into (11.10), the overall sampled-data model for the stator currents and rotor flux linkages is obtained as follows:

$$\begin{bmatrix} i_{\alpha s}^p(k+1) \\ i_{\beta s}^p(k+1) \\ \psi_{\alpha r}^p(k+1) \\ \psi_{\beta r}^p(k+1) \end{bmatrix} = \mathbf{\Phi}(k) \begin{bmatrix} i_{\alpha s}(k) \\ i_{\beta s}(k) \\ \psi_{\alpha r}(k) \\ \psi_{\beta r}(k) \end{bmatrix} + \mathbf{\Gamma}_{\rm b} \left\{ v_{C1}(k) \begin{bmatrix} s_{\alpha r1}^p(k) \\ s_{\beta r1}^p(k) \end{bmatrix} + v_{C2}(k) \begin{bmatrix} s_{\alpha r2}^p(k) \\ s_{\beta r2}^p(k) \end{bmatrix} \right\} \tag{11.34}$$

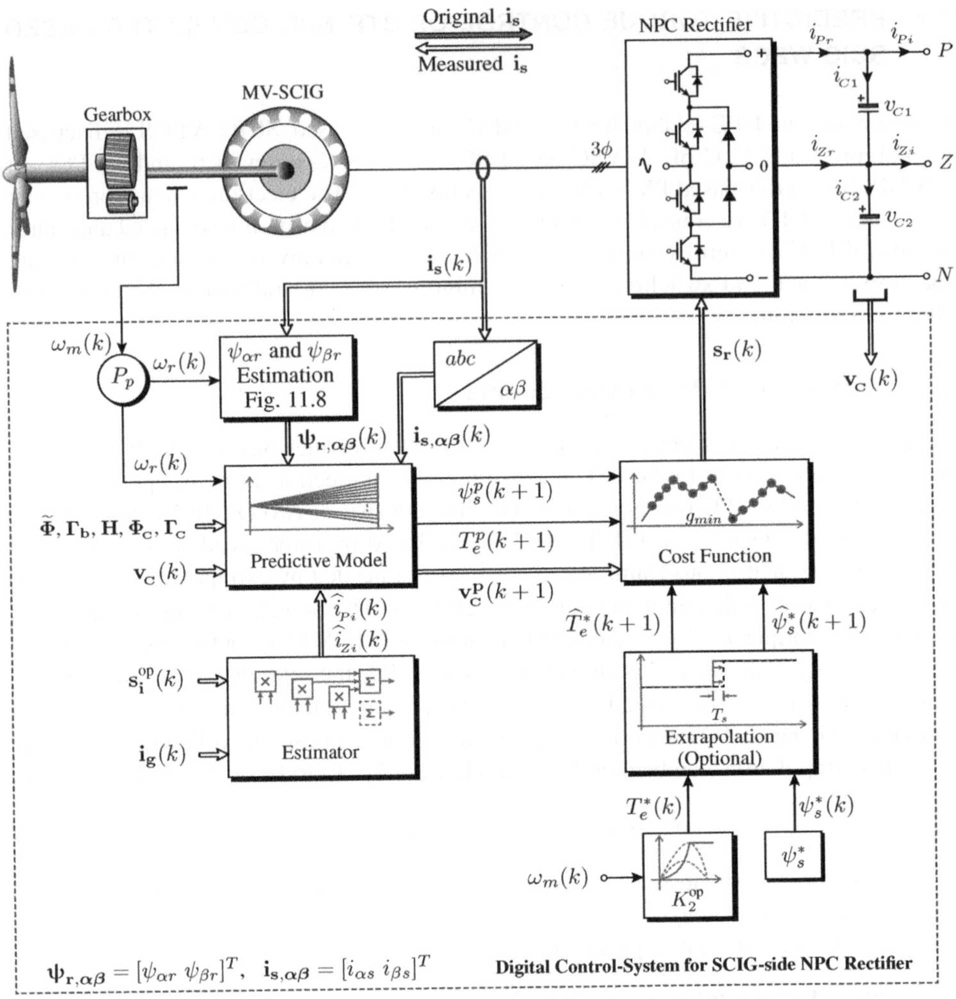

Figure 11.17 Block diagram of the PTC scheme for NPC rectifier-based SCIG.

- The DT model in Equation (11.11) defines the $\alpha\beta$-frame stator flux linkages in terms of stator currents and rotor flux linkages. By moving the variables in Equation (11.11) one-sample forward, the future value of stator flux linkages is obtained in terms of the predicted stator currents and rotor flux linkages in (11.34), that is,

$$\begin{bmatrix} \psi_{\alpha s}^p(k+1) \\ \psi_{\beta s}^p(k+1) \end{bmatrix} = \underbrace{\begin{bmatrix} \sigma L_s & 0 & k_r & 0 \\ 0 & \sigma L_s & 0 & k_r \end{bmatrix}}_{\mathbf{H}} \begin{bmatrix} i_{\alpha s}^p(k+1) \\ i_{\beta s}^p(k+1) \\ \psi_{\alpha r}^p(k+1) \\ \psi_{\beta r}^p(k+1) \end{bmatrix}. \tag{11.35}$$

- The peak value of stator flux linkage is deduced from the above model as

$$\psi_s^p(k+1) = \sqrt{[\psi_{\alpha s}^p(k+1)]^2 + [\psi_{\beta s}^p(k+1)]^2}. \tag{11.36}$$

- The future value of electromagnetic torque is obtained by the cross product of predicted $\alpha\beta$-frame stator currents and stator flux linkages in (11.34) and (11.35), respectively. The model of $T_e^p(k+1)$ is defined by the following:

$$T_e^p(k+1) = 1.5\, P_p \left[\psi_{\alpha s}^p(k+1)\, i_{\beta s}^p(k+1) - \psi_{\beta s}^p(k+1)\, i_{\alpha s}^p(k+1)\right]. \tag{11.37}$$

- The DT model for DC-link capacitors voltage is reconsidered from (9.50) [40, 41]:

$$\begin{bmatrix} v_{C1}^p(k+1) \\ v_{C2}^p(k+1) \end{bmatrix} = \underbrace{\begin{bmatrix} 1 & 0 \\ 0 & 1 \end{bmatrix}}_{\mathbf{\Phi_C}} \begin{bmatrix} v_{C1}(k) \\ v_{C2}(k) \end{bmatrix} + \underbrace{\begin{bmatrix} \frac{T_s}{C_1} & 0 \\ 0 & \frac{T_s}{C_2} \end{bmatrix}}_{\mathbf{\Gamma_C}} \begin{bmatrix} i_{C1}^p(k) \\ i_{C2}^p(k) \end{bmatrix} \tag{11.38}$$

where $\mathbf{\Phi_C}$ and $\mathbf{\Gamma_C}$ are DT matrices and can be defined offline to reduce the computational burden; $i_{C1}^p(k)$ and $i_{C2}^p(k)$ are the predicted DC-link capacitors current and are calculated on the basis of the NPC rectifier switching states and estimated $\widehat{i}_{P_i}$ and $\widehat{i}_{Z_i}$. Further details about the balancing of DC-link capacitors voltage can be found in Section 9.7.

According to the comparison of the DT models in Equations (11.33) to (11.38), the predictive models of the stator flux, electromagnetic torque, and DC-link capacitors voltage are related to the NPC rectifier switching signals.

Cost Function Minimization: The stator flux and electromagnetic torque of SCIG, DC link capacitors voltage, and switching frequency are controlled simultaneously by defining a cost function as follows [42, 43]:

$$\begin{aligned} g_r(k) &= \lambda_T \left[\widehat{T}_e^*(k+1) - T_e^p(k+1)\right]^2 \\ &+ \lambda_\psi \left[\widehat{\psi}_s^*(k+1) - \psi_s^p(k+1)\right]^2 \\ &+ \lambda_{dc,r} \left[v_{C1}^p(k+1) - v_{C2}^p(k+1)\right]^2 \\ &+ \lambda_{sw,r} \sum_{x=a,b,c} \sum_{j=1,2} \left[s_{xrj}^p(k) - s_{xrj}^{\text{op}}(k)\right]^2 \end{aligned} \tag{11.39}$$

where λ_T, λ_ψ, $\lambda_{dc,r}$, and $\lambda_{sw,r}$ are the weighting factors for torque control, flux control, balancing of DC-link capacitors voltage, and switching frequency minimization, respectively. By using the per-unit method in Chapter 4, the weighting factors are defined by the following [32]:

$$\lambda_T = \frac{T_{e,R}}{T_{e,R}} = 1, \qquad \lambda_\psi = \frac{T_{e,R}}{\psi_{s,R}}, \qquad \lambda_{dc,r} = \frac{I_{Bs}}{v_{dc}^*} \tag{11.40}$$

where $T_{e,R}$ and $\psi_{s,R}$ are rated values of torque and stator flux, respectively. I_{Bs} is the base *rms* generator current. The weighting factor requirement can be eliminated by employing multiobjective ranking algorithm in the PTC cost function [44–46].

The PTC scheme is analogous to the classical DTC scheme. This scheme preserves the fast dynamic response of torque and flux but with less complexity in digital implementation. The classical DTC de facto requirements for hysteresis controllers and lookup tables are eliminated in the PTC scheme. The generator and converter models along with 27 possible switching state combinations are used to predict (1) NPC rectifier voltages

$v^p_{\alpha s}(k+1)$ and $v^p_{\beta s}(k+1)$, (2) stator currents $i^p_{\alpha s}(k+1)$ and $i^p_{\beta s}(k+1)$, (3) rotor flux linkages $\psi^p_{\alpha r}(k+1)$ and $\psi^p_{\beta r}(k+1)$, (4) $\alpha\beta$-axes' stator flux linkages $\psi^p_{\alpha s}(k+1)$ and $\psi^p_{\beta s}(k+1)$, (5) peak value of stator flux linkage $\psi^p_s(k+1)$, (6) electromagnetic torque $T^p_e(k+1)$, and (7) DC-link capacitors voltage $v^p_{C1}(k+1)$ and $v^p_{C2}(k+1)$. The optimal switching signals for NPC rectifier are selected such that the stator flux, electromagnetic torque, and DC-link capacitors voltage are controlled simultaneously.

11.7.2 Grid-side Control Scheme

The grid-side cost function shown below considers simultaneous control of the $\alpha\beta$-frame grid currents, DC-link capacitors voltage, and switching frequency minimization.

$$
\begin{aligned}
g_i(k) &= \lambda_{i\alpha}\left[\widehat{i}^*_{\alpha g}(k+1) - i^p_{\alpha g}(k+1)\right]^2 \\
&+ \lambda_{i\beta}\left[\widehat{i}^*_{\beta g}(k+1) - i^p_{\beta g}(k+1)\right]^2 \\
&+ \lambda_{dc,i}\left[v^p_{C1}(k+1) - v^p_{C2}(k+1)\right]^2 \\
&+ \lambda_{sw,i}\sum_{x=a,b,c}\sum_{j=1,2}\left[s^p_{xij}(k) - s^{\text{op}}_{xij}(k)\right]^2
\end{aligned}
\tag{11.41}
$$

where $\lambda_{i\alpha}$, $\lambda_{i\beta}$, $\lambda_{dc,i}$, and $\lambda_{sw,i}$ are weighting factors. More details about NPC inverter PCC scheme can be found in Section 9.7.

11.7.3 Control Algorithm

The flowchart and `MATLAB` *S-Function Builder* programming for PTC scheme are discussed in this section. Figures 11.18(a) and 11.18(b) show the flowcharts of PTC for an NPC rectifier and PCC for an NPC inverter, respectively. The feedback measurements needed are obtained in block ①, and reference control variables are calculated in block ②. The PTC and PCC algorithms enter an iterative loop in block ④ and end in block ⑨. The NPC rectifier and inverter output voltages and capacitors current are predicted in block ⑤. Block ⑥ calculates the future values of the torque and flux for the PTC scheme, the $\alpha\beta$-frame grid currents for the PCC scheme, and DC-link capacitors voltage of both schemes. Block ⑦ evaluates the generator-side and grid-side cost functions. An optimal vector j_{op} corresponding to the minimum (optimal) cost function value is stored in block ⑧. The optimal switching signals for NPC rectifier and inverter are obtained in block ⑩ on the basis of j_{op}, which produces the minimum cost function value.

Algorithm 11.2 shows the *S-Function* programming for the PTC of the NPC rectifier-based SCIG. The external function declarations for the *S-Function* can be found on the left-hand side of code (before "=" notation). The description of program lines is given as follows: Lines `05-06` involve the prediction of NPC rectifier voltages; lines `07-08` involve the prediction of SCIG stator currents; lines `09-10` involve the prediction of SCIG rotor flux linkages; lines `11-12` involve the prediction of SCIG stator flux linkages; line `13` is the prediction of peak stator flux linkage; line `14` is the prediction of electromagnetic torque; lines `15-16` involve the prediction of DC branch currents; lines `17-18` involve the prediction of DC-link capacitors current; lines `19-20` involve the prediction of DC-link capacitors voltage; lines `21-24` involve the calculation of cost function values; lines `25-29` involve the selection of optimal vector `j_op` number; lines `31-36` involve the selection of optimal switching signals on the basis of the `j_op` number.

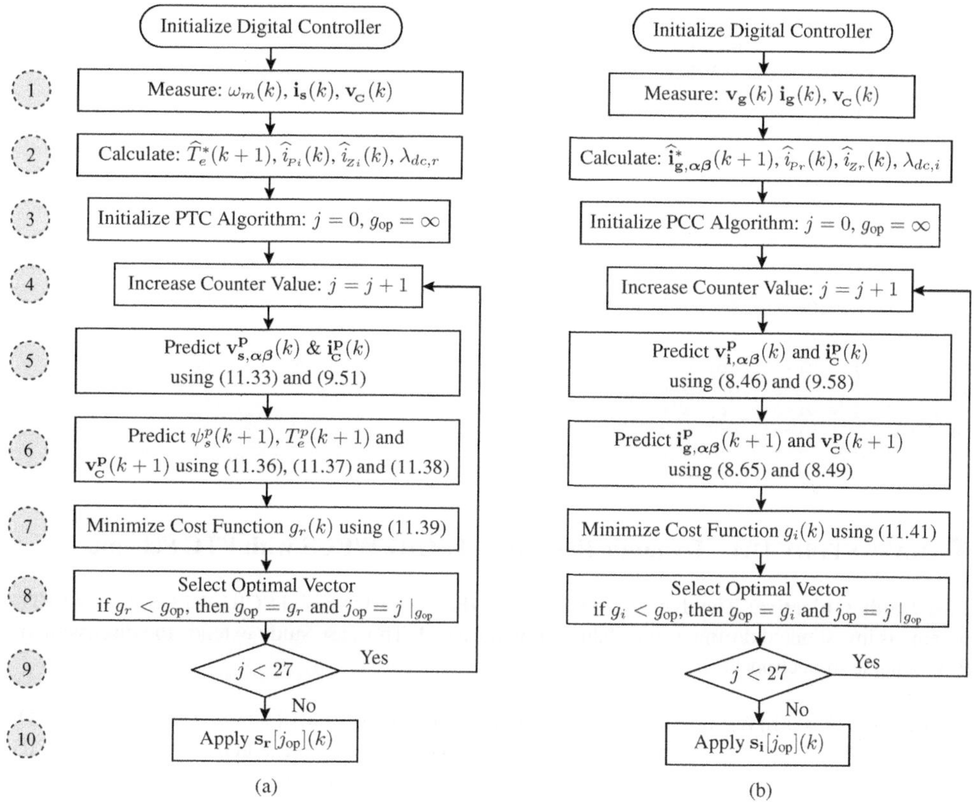

Figure 11.18 Flowchart of (a) PTC algorithm NPC rectifier and (b) PCC algorithm NPC inverter.

ALGORITHM 11.2 PTC Algorithm for NPC Rectifier-Based SCIG WECS

```
\\ P11 = P22 = P33 = P44 = 1-(Ts/tau_sigma); P31 = P42 = Lm*Ts/tau_r;
\\ P13 = P24 = (kr*Ts)/(sigma*Ls*tau_r); P43 = -P34 = Ts;
\\ P14 = -P23 = (kr*Ts)/(sigma*Ls); G11 = G22 = Ts/(sigma*Ls);
\\ H11 = H22 = sigma*Ls; H13 = H24 = kr; kTe = 1.5*PP;
j_op = 1000;
g_op = 1000000000;
for(j = 0; j < 27; j++)
{
    vr_ab[0] = vc[0]*vrec[j].s_r_ab[0] + vc[1]*vrec[j].s_r_ab[1];
    vr_ab[1] = vc[0]*vrec[j].s_r_ab[2] + vc[1]*vrec[j].s_r_ab[3];
    is_ab_k1[0] = P11*is_ab[0] + P13*psir_ab[0] + P14*wr[0]*psir_ab[1]
                + G11*vr_ab[0];
    is_ab_k1[1] = P22*is_ab[1] + P23*wr[0]*psir_ab[0] + P24*psir_ab[1]
                + G22*vr_ab[1];
    psir_ab_k1[0] = P31*is_ab[0] + P33*psir_ab[0] + P34*wr[0]*psir_ab[1];
    psir_ab_k1[1] = P42*is_ab[1] + P43*wr[0]*psir_ab[0] + P44*psir_ab[1];
    psis_ab_k1[0] = H11*is_ab_k1[0] + H13*psir_ab_k1[0];
    psis_ab_k1[1] = H22*is_ab_k1[1] + H24*psir_ab_k1[1];
    psis_k1[0] = sqrt((psis_ab_k1[0]*psis_ab_k1[0])+
                    (psis_ab_k1[1]*psis_ab_k1[1]));
    te_k1[0] = kTe*((psis_ab_k1[0]*is_ab_k1[1])-
                    (psis_ab_k1[1]*is_ab_k1[0]));
    ipzr_k1[0] = - vrec[j].s_r[0]*vrec[j].s_r[1]*is[0]
                 - vrec[j].s_r[2]*vrec[j].s_r[3]*is[1]
                 - vrec[j].s_r[4]*vrec[j].s_r[5]*is[2];
    ipzr_k1[1] = - !vrec[j].s_r[0]*vrec[j].s_r[1]*is[0]
```

```
                         - !vrec[j].s_r[2]*vrec[j].s_r[3]*is[1]
                         - !vrec[j].s_r[4]*vrec[j].s_r[5]*is[2];
    ic_k1[0] = ipzr_k1[0] - ipzi[0];
    ic_k1[1] = ipzr_k1[0] + ipzr_k1[1] - ipzi[0] - ipzi[1];
    vc_k1[0] = vc[0] + (Ts/C1)*ic_k1[0];
    vc_k1[1] = vc[1] + (Ts/C2)*ic_k1[1];
    g_te = fabs(te_ref_k1[0]-te_k1[0]);
    g_psis = fabs(psis_ref_k1[0]-psis_k1[0]);
    g_dc = (vc_k1[0]-vc_k1[1])*(vc_k1[0]-vc_k1[1]);
    g_r = g_te + Lambda_psis*g_psis + Lambda_dcr*g_dc;
    if(g_r < g_op)
        {
            j_op = j;
            g_op = g_r;
        }
}
s_ar1[0] = vrec[j_op].s_r[0];
s_ar2[0] = vrec[j_op].s_r[1];
s_br1[0] = vrec[j_op].s_r[2];
s_br2[0] = vrec[j_op].s_r[3];
s_cr1[0] = vrec[j_op].s_r[4];
s_cr2[0] = vrec[j_op].s_r[5];
```

CASE STUDY 11.4 Dynamic Response of SCIG WECS with PTC Scheme

Case Study Objective: The dynamic response of NPC converter-based SCIG WECS with the PTC scheme is investigated during a step change in wind speed. This case study extends the discussion of PTC scheme in Section 11.7.

Parameters: A 3.0-MW, 3000-V, and 60-Hz SCIG WECS with R_s = 18.885 mΩ, R_r = 21.404 mΩ, $L_{ls} = L_{lr}$ = 0.815 mH, L_m = 27.168 mH, ψ_r = 6.198 Wb (*peak*), P_p = 2, and n_m = 1814 rpm (SCIG #3 in Table A.7) is considered. To avoid a long simulation runtime, the original moment of inertia J_m is reduced to 10 kg.m^2. The GSC and control parameters are r_i = 30 mΩ (0.01 pu), L_i = 1.6 mH (0.2 pu), v_{dc}^* = 5304 V (3.062 pu), Q_g^* = 0 MVAR, and T_s = 100 μs.

Analysis:
Figure 11.19 shows the dynamic response of the SCIG WECS with PTC scheme during a step change in wind speed from 12 m/s (1.0 pu) to 7.2 m/s (0.6 pu). The behavior of SCIG torque and flux is studied, in addition to the mechanical speed, stator currents, stator output active power, net DC-bus voltage, DC-link capacitors voltage, rectifier and inverter line-to-line voltages, and grid currents. The reference generator speed ω_m^* changes from 1.0 pu to 0.6 pu, which reflects the step change in v_w. A lower wind speed causes the generator speed ω_m to decelerate to lower value (Figure 11.19(a)). Given the large moment of inertia in MW-WECS, the outer speed control loop possesses a low dynamic response compared with internal torque and flux control loop. The speed PI controller changes T_e^* from –1.0 pu to –0.36 (–0.6^2) pu. The optimal switching signals produced by the PTC scheme ensures that T_e follows T_e^* closely with less reference tracking error (Figure 11.19(b)). The stator flux is maintained constant during the entire speed range by the PTC scheme (Figure 11.19(c)). The d-axis current is maintained nearly constant, and the q-axis current follows the T_e waveform (Figure 11.19(d)). The THD of the SCIG stator current is 6.81%. The NPC rectifier line-to-line voltage $v_{r,ab}$ waveform shows five and three distinct voltage levels during high and low wind speeds, respectively (Figure 11.19(e)). At high and low wind speeds, the NPC rectifier operates as 3L and 2L VSC. During rated operating conditions, the switching frequency of the NPC rectifier $f_{sw,r}$ is 1160 Hz. The stator active power P_s decreases from –0.889 pu to –0.22 pu, and the stator reactive power Q_s decreases from –0.512 pu to –0.22 pu (Figure 11.19(f)).

Considering the step change in wind speed, the PI controller in the GSC DC control loop exhibits transient oscillations for net DC-bus voltage v_{dc} (Figure 11.19(g)). With the help of the weighting factors in Equation (11.40), the DC-link capacitors voltage are regulated at equal voltages including transient interval (Figure 11.19(h)). The ripple in DC-link capacitors voltage decrease with respect

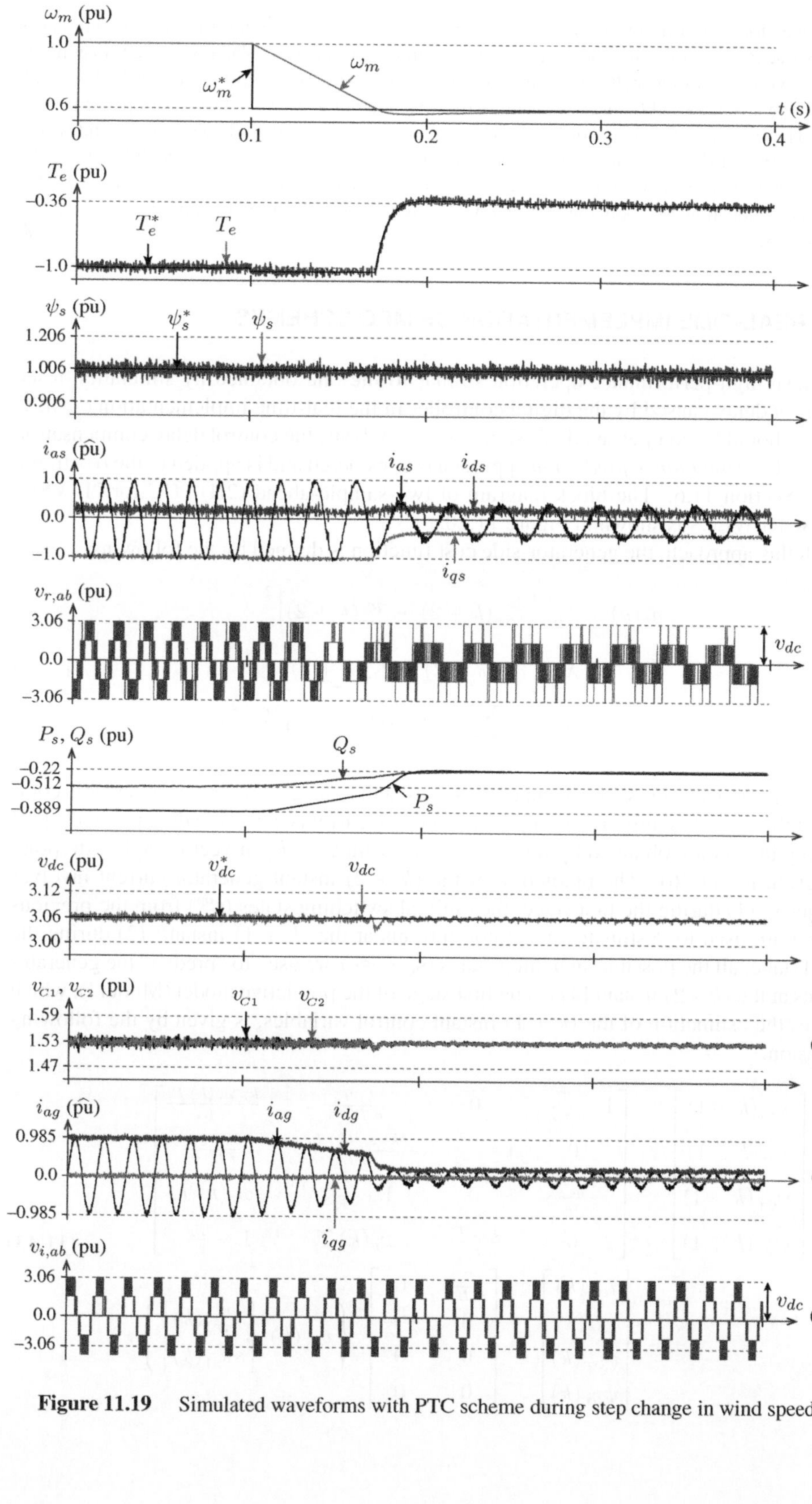

Figure 11.19 Simulated waveforms with PTC scheme during step change in wind speed.

to the generator and grid current magnitude. The SCIG and grid currents follow their references with less tracking error during both steady-state and transient intervals. The grid current magnitude decreases with decreasing SCIG active power output. The THD of the grid current is 4.74%. With grid-tied operation, the NPC inverter contains five discrete voltage levels during all operating conditions. The switching frequency of NPC inverter $f_{sw,i}$ is noted as 1060 Hz with rated wind speed. The waveforms of the electromagnetic torque, stator flux, and stator currents of SCIG validate the fast dynamic response and decoupled control of the PTC scheme. The results presented through this case study demonstrate that both the PCC and PTC schemes achieve similar control performance for SCIG WECS. ■

11.8 REAL-TIME IMPLEMENTATION OF MPC SCHEMES

To achieve high-performance operation similar to the one obtained by simulation tests, the control delay caused by the digital controller in the real-time implementation of MPC schemes should be compensated. To simplify the analysis, the control delay compensation based on the *estimation + prediction* approach is considered and is applied to the $\alpha\beta$-frame PCC in Section 11.6. The block diagram of two-sample-ahead (2H) PCC for 2L-VSR-based SCIG WECS is shown in Figure 11.20.

With this approach, the generator-side cost function is defined by the following:

$$\begin{aligned} g_r(k) = {} & \lambda_{i\alpha}\left[\widehat{i}^{*}_{\alpha s}(k+2) - i^{p}_{\alpha s}(k+2)\right]^2 \\ & + \lambda_{i\beta}\left[\widehat{i}^{*}_{\beta s}(k+2) - i^{p}_{\beta s}(k+2)\right]^2 \\ & + \lambda_{sw,r} \sum_{x=a,\,b,\,c} \left[s^{p}_{xr1}(k) - s^{\text{op}}_{xr1}(k)\right]^2. \end{aligned} \tag{11.42}$$

From the above expression, it can be understood that the generator current error is computed at the $(k+2)$ sampling instant. The reference currents corresponding to the $(k+2)$ sampling instant are obtained by replacing $\omega_s(k)$ with $2\,\omega_s(k)$ in vector angle extrapolation given in (11.26). The prediction of the $(k+2)$ instant generator current involves two stages: (1) during the first stage, the optimal switching states ($\mathbf{s}^{\text{op}}$) from the previous iteration are used to "estimate" generator currents at the $(k+1)$ instant; (2) during the second stage, all the possible switching states ($\mathbf{s}_0 \sim \mathbf{s}_7$) are used to "predict" the generator currents at the $(k+2)$ instant [47]. The first stage of the predictive model (Model I), which involves the estimation of the $(k+1)$ instant control variables, is given by the following expression:

$$\begin{aligned} \begin{bmatrix} \widehat{i}_{\alpha s}(k+1) \\ \widehat{i}_{\beta s}(k+1) \\ \widehat{\psi}_{\alpha r}(k+1) \\ \widehat{\psi}_{\beta r}(k+1) \end{bmatrix} = {} & \begin{bmatrix} 1-\frac{T_s}{\tau_\sigma} & 0 & \frac{k_r\,T_s}{\sigma\,L_s\,\tau_r} & \frac{k_r\,\omega_r(k)\,T_s}{\sigma\,L_s} \\ 0 & 1-\frac{T_s}{\tau_\sigma} & -\frac{k_r\,\omega_r(k)\,T_s}{\sigma\,L_s} & \frac{k_r\,T_s}{\sigma\,L_s\,\tau_r} \\ \frac{L_m\,T_s}{\tau_r} & 0 & 1-\frac{T_s}{\tau_r} & -\omega_r(k)\,T_s \\ 0 & \frac{L_m\,T_s}{\tau_r} & \omega_r(k)\,T_s & 1-\frac{T_s}{\tau_r} \end{bmatrix} \\ & \begin{bmatrix} i_{\alpha s}(k) \\ i_{\beta s}(k) \\ \psi_{\alpha r}(k) \\ \psi_{\beta r}(k) \end{bmatrix} + \begin{bmatrix} \frac{T_s}{\sigma\,L_s} & 0 \\ 0 & \frac{T_s}{\sigma\,L_s} \\ 0 & 0 \\ 0 & 0 \end{bmatrix} \left\{ v_{C1}(k) \begin{bmatrix} s^{\text{op}}_{\alpha r1}(k) \\ s^{\text{op}}_{\beta r1}(k) \end{bmatrix} \right\}. \end{aligned} \tag{11.43}$$

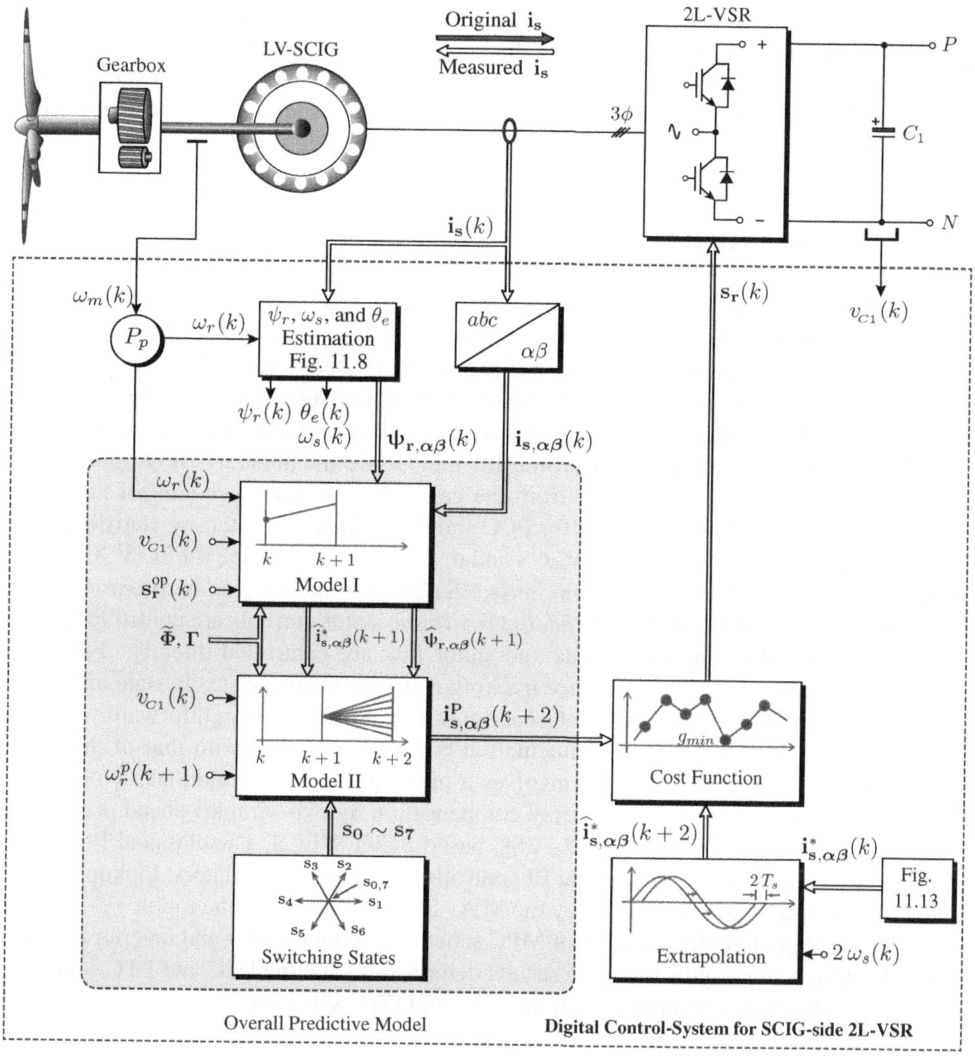

Figure 11.20 Block diagram of the 2H PCC scheme for SCIG with 2L-VSR.

Compared with the one-sample-ahead PCC sampled-data model in Equation (11.29), the above model estimates the future behavior of both stator currents and rotor fluxes. The second stage of predictive model (Model II) involves the prediction of the $(k+2)$ instant generator currents only as shown in the following:

$$\begin{bmatrix} i^p_{\alpha s}(k+2) \\ i^p_{\beta s}(k+2) \end{bmatrix} = \begin{bmatrix} 1-\frac{T_s}{\tau_\sigma} & 0 & \frac{k_r\,T_s}{\sigma\,L_s\,\tau_r} & \frac{k_r\,\omega_r^p(k+1)\,T_s}{\sigma\,L_s} \\ 0 & 1-\frac{T_s}{\tau_\sigma} & -\frac{k_r\,\omega_r^p(k+1)\,T_s}{\sigma\,L_s} & \frac{k_r\,T_s}{\sigma\,L_s\,\tau_r} \end{bmatrix} \begin{bmatrix} \widehat{i}_{\alpha s}(k+1) \\ \widehat{i}_{\beta s}(k+1) \\ \widehat{\psi}_{\alpha r}(k+1) \\ \widehat{\psi}_{\beta r}(k+1) \end{bmatrix} + \begin{bmatrix} \frac{T_s}{\sigma\,L_s} & 0 \\ 0 & \frac{T_s}{\sigma\,L_s} \end{bmatrix} \left\{ v_{C1}(k) \begin{bmatrix} s^p_{\alpha r1}(k) \\ s^p_{\beta r1}(k) \end{bmatrix} \right\} \tag{11.44}$$

where $\omega_r^p(k+1)$ is the predicted rotor speed and can be obtained according to the rotor speed dynamics. Refer to example 7.3 in Chapter 7 for further details.

The switching state, which minimizes the cost function at the $(k+2)$ instant, is selected and applied to the 2L-VSR. With this approach, one sampling period is used for calculations and another sampling period is used for control delay compensation.

11.9 CONCLUDING REMARKS

A detailed analysis on the modeling and control of BTB VSC-based SCIG WECS is presented in this chapter. The CT modeling of the stator currents, rotor flux linkages, and stator flux linkages of SCIG is developed. The DT model of a complete WECS is developed by combining the SCIG models with 2L-VSC and NPC converter models. The control of SCIG with DFOC, IFOC, and DTC schemes is analyzed on the basis of the developed PCC and PTC schemes. The calculation of rotor flux magnitude and angle by DFOC and IFOC methods is discussed in detail, apart from the calculation of reference control variables with OTSR and OT MPPT algorithms for PCC and PTC schemes. A step-by-step design of PCC scheme for the 2L-VSC-based WECS and that of the PTC scheme for 3L-VSC-based WECS are presented in addition to flowcharts, MATLAB *S-Function Builder* programming, and case studies. With the PCC scheme, the $\alpha\beta$-frame stator currents are controlled; with the PTC scheme, the generator torque and stator flux are controlled directly. For both the control schemes, excellent reference tracking is observed during steady state and transient conditions. The calculation of reference control variables is straightforward with the PTC scheme, and its real-time implementation is easier compared with that of the PCC scheme. However, the PTC scheme involves a high number of calculations, which lead to high computational burden. The delay compensation by two-samples-ahead predictive control is discussed by considering 2L-VSC-based SCIG WECS. The classical FOC and DTC de facto requirements for current PI controllers, hysteresis regulators, lookup tables, and modulation stage are eliminated by the MPC scheme. Similar to the topology of MSC and GSC, the digital implementation of MPC schemes for the rectifier and inverter is identical. The design steps and simulation results demonstrate that the PCC and PTC schemes are simple and intuitive compared with the FOC and DTC schemes.

Link to Next Chapters:

The modeling of SCIG, block diagram of overall digital control system, and calculation of reference control variables are further discussed in the next chapter (Chapter 12) to develop MPC schemes for DFIG WECS.

REFERENCES

1. V. Yaramasu, B. Wu, P. C. Sen, S. Kouro, and M. Narimani, "High-power wind energy conversion systems: State-of-the-art and emerging technologies," *Proceedings of the IEEE*, vol. 103, no. 5, pp. 740–788, May 2015.

2. L. H. Hansen, L. Helle, F. Blaabjerg, E. Ritchie, S. Munk-Nielsen, H. Bindner, P. Sorensen, and B. Bak-Jensen, "Conceptual survey of generators and power electronics for wind turbines," Riso National Laboratory, Roskilde, Denmark, Riso-R-1205(EN), December 2001, available at: http://orbit.dtu.dk.

3. H. Li and Z. Chen, "Overview of different wind generator systems and their comparisons," *IET Renewable Power Generation*, vol. 2, no. 2, pp. 123–138, June 2008.

4. Z. Chen, J. Guerrero, and F. Blaabjerg, "A review of the state of the art of power electronics for wind turbines," *IEEE Transactions on Power Electronics*, vol. 24, no. 8, pp. 1859–1875, August 2009.

5. J. Conroy and R. Watson, "Frequency response capability of full converter wind turbine generators in comparison to conventional generation," *IEEE Transactions on Power Systems*, vol. 23, no. 2, pp. 649–656, May 2008.

6. D. Casadei, F. Profumo, G. Serra, and A. Tani, "FOC and DTC: two viable schemes for induction motors torque control," *IEEE Transactions on Power Electronics*, vol. 17, no. 5, pp. 779–787, September 2002.

7. G. Buja and M. Kazmierkowski, "Direct torque control of PWM inverter-fed AC motors - A survey," *IEEE Transactions on Industrial Electronics*, vol. 51, no. 4, pp. 744–757, August 2004.

8. F. Wang, S. Li, X. Mei, W. Xie, J. Rodriguez, and R. Kennel, "Model-based predictive direct control strategies for electrical drives: An experimental evaluation of PTC and PCC methods," *IEEE Transactions on Industrial Informatics*, vol. 11, no. 3, pp. 671–681, June 2015.

9. C. Bordons and C. Montero, "Basic principles of MPC for power converters: Bridging the gap between theory and practice," *IEEE Industrial Electronics Magazine*, vol. 9, no. 3, pp. 31–43, September 2015.

10. S. Vazquez, J. Leon, L. Franquelo, J. Rodriguez, H. Young, A. Marquez, and P. Zanchetta, "Model predictive control: A review of its applications in power electronics," *IEEE Industrial Electronics Magazine*, vol. 8, no. 1, pp. 16–31, March 2014.

11. T. Geyer, "A comparison of control and modulation schemes for medium-voltage drives: Emerging predictive control concepts versus PWM-based schemes," *IEEE Transactions on Industry Applications*, vol. 47, no. 3, pp. 1380–1389, May/June 2011.

12. J. Rodríguez, R. Kennel, J. Espinoza, M. Trincado, C. Silva, and C. Rojas, "High-performance control strategies for electrical drives: An experimental assessment," *IEEE Transactions on Industrial Electronics*, vol. 59, no. 2, pp. 812–820, February 2012.

13. B. Wu, Y. Lang, N. Zargari, and S. Kouro, *Power Conversion and Control of Wind Energy Systems*, 1st ed., ser. IEEE Press Series on Power Engineering. Hoboken, NJ: Wiley-IEEE Press, July 2011.

14. R. Cardenas, R. Pena, P. Wheeler, J. Clare, and G. Asher, "Control of the reactive power supplied by a WECS based on an induction generator fed by a matrix converter," *IEEE Transactions on Industrial Electronics*, vol. 56, no. 2, pp. 429–438, February 2009.

15. R. Pena, R. Cardenas, R. Blasco, G. Asher, and J. Clare, "A cage induction generator using back to back PWM converters for variable speed grid connected wind energy system," in *IEEE Industrial Electronics Conference (IECON)*, Denver, CO, USA, vol. 2, 2001, pp. 1376–1381.

16. M. Simoes, B. Bose, and R. J. Spiegel, "Fuzzy logic based intelligent control of a variable speed cage machine wind generation system," *IEEE Transactions on Power Electronics*, vol. 12, no. 1, pp. 87–95, January 1997.

17. B. Andresen and J. Birk, "A high power density converter system for the Gamesa G10x 4,5 MW wind turbine," in *European Conference on Power Electronics and Applications (EPE)*, Aalborg, Denmark, September 2007, pp. 1–8.

18. V. Yaramasu and B. Wu, "Predictive control of a three-level boost converter and an NPC inverter for high-power PMSG-based medium voltage wind energy conversion systems," *IEEE Transactions on Power Electronics*, vol. 29, no. 10, pp. 5308–5322, October 2014.

19. M. Liserre, R. Cardenas, M. Molinas, and J. Rodríguez, "Overview of multi-MW wind turbines and wind parks," *IEEE Transactions on Industrial Electronics*, vol. 58, no. 4, pp. 1081–1095, April 2011.

20. E. Bueno, S. Cobreces, F. Rodriguez, A. Hernandez, and F. Espinosa, "Design of a back-to-back NPC converter interface for wind turbines with squirrel-cage induction generator," *IEEE Transactions on Energy Conversion*, vol. 23, no. 3, pp. 932–945, September 2008.

21. S. Kouro, M. Malinowski, K. Gopakumar, J. Pou, L. Franquelo, B. Wu, J. Rodríguez, M. Perez, and J. Leon, "Recent advances and industrial applications of multilevel converters," *IEEE Transactions on Industrial Electronics*, vol. 57, no. 8, pp. 2553–2580, August 2010.

22. P. Maibach, A. Faulstich, M. Eichler, and S. Dewar, "Full-scale medium-voltage converters for wind power generators up to 7 MVA," 2010, available at: http://www.abb.com.

23. Z. Zhu and J. Hu, "Electrical machines and power-electronic systems for high-power wind energy generation applications: Part II – power electronics and control systems," *COMPEL: The International Journal for Computation and Mathematics in Electrical and Electronic Engineering*, vol. 32, no. 1, pp. 34–71, 2013.

24. B. Bose, *Modern Power Electronics and AC drives*. Upper Saddle River, NJ: Prentice Hall, 2002.

25. B. Wu, *High-Power Converters and AC Drives*, 1st ed. Hoboken, NJ: Wiley-IEEE Press, March 2006.

26. P. Krause and C. Thomas, "Simulation of symmetrical induction machinery," *IEEE Transactions on Power Apparatus and Systems*, vol. 84, no. 11, pp. 1038–1053, November 1965.

27. T. Geyer, G. Papafotiou, and M. Morari, "Model predictive direct torque control: Part I: Concept, algorithm, and analysis," *IEEE Transactions on Industrial Electronics*, vol. 56, no. 6, pp. 1894–1905, June 2009.

28. J. Holtz, "The representation of AC machine dynamics by complex signal flow graphs," *IEEE Transactions on Industrial Electronics*, vol. 42, no. 3, pp. 263–271, June 1995.

29. J. Holtz, "Sensorless control of induction motor drives," *Proceedings of the IEEE*, vol. 90, no. 8, pp. 1359–1394, August 2002.

30. J. I. Yuz and G. C. Goodwin, *Sampled-Data Models for Linear and Nonlinear Systems*, ser. Communications and Control Engineering. London: Springer Verlag, November 2014.

31. P. Vaclavek and P. Blaha, "Enhanced discrete time model for AC induction machine model predictive control," in *IEEE Industrial Electronics Conference (IECON)*, Montreal, QC, Canada, October 2012, pp. 5043–5048.

32. H. Miranda, P. Coftes, J. Yuz, and J. Rodríguez, "Predictive torque control of induction machines based on state-space models," *IEEE Transactions on Industrial Electronics*, vol. 56, no. 6, pp. 1916–1924, June 2009.

33. B. Bose, "The past, present, and future of power electronics," *IEEE Industrial Electronics Magazine*, vol. 3, no. 2, pp. 7–11, 14, June 2009.

34. J. Guzinski and H. Abu-Rub, "Speed sensorless induction motor drive with predictive current controller," *IEEE Transactions on Industrial Electronics*, vol. 60, no. 2, pp. 699–709, February 2013.

35. M. Depenbrock, "Direct self-control of the flux and rotary moment of a rotary-field machine," July 1987, United States Patent, US 4,678,248.

36. T. Noguchi, and I. Takahashi, "Quick torque response control of an induction motor based on a new concept," in *IEEJ Technical Meeting on Rotating Machines*, vol. RM84-76, pp.61–70, September 1984.

37. S. Kouro, R. Bernal, H. Miranda, C. Silva, and J. Rodríguez, "High-performance torque and flux control for multilevel inverter fed induction motors," *IEEE Transactions on Power Electronics*, vol. 22, no. 6, pp. 2116–2123, November 2007.

38. J. Scoltock, T. Geyer, and U. Madawala, "A comparison of model predictive control schemes for MV induction motor drives," *IEEE Transactions on Industrial Informatics*, vol. 9, no. 2, pp. 909–919, May 2013.

39. P. Correa, M. Pacas, and J. Rodríguez, "Predictive torque control for inverter-fed induction machines," *IEEE Transactions on Industrial Electronics*, vol. 54, no. 2, pp. 1073–1079, April 2007.

40. R. Vargas, P. Coftes, U. Ammann, J. Rodríguez, and J. Pontt, "Predictive control of a three-phase neutral-point-clamped inverter," *IEEE Transactions on Industrial Electronics*, vol. 54, no. 5, pp. 2697–2705, October 2007.

41. V. Yaramasu, M. Rivera, M. Narimani, B. Wu, and J. Rodriguez, "High performance operation for a four-leg NPC inverter with two-sample-ahead predictive control strategy," *International Journal of Electric Power Systems Research*, vol. 123, pp. 31–39, June 2015.

42. P. Urrejola, M. Perez, J. Rodríguez, and M. Trincado, "Direct torque control of an 3L-NPC inverter-fed induction machine: A model predictive approach," in *IEEE Industrial Electronics Conference (IECON)*, November 2010, pp. 2947–2952, Glendale, AZ, USA.

43. R. Vargas, U. Ammann, B. Hudoffsky, J. Rodríguez, and P. Wheeler, "Predictive torque control of an induction machine fed by a matrix converter with reactive input power control," *IEEE Transactions on Power Electronics*, vol. 25, no. 6, pp. 1426–1438, June 2010.

44. C. Rojas, J. Rodríguez, F. Villarroel, J. Espinoza, C. Silva, and M. Trincado, "Predictive torque and flux control without weighting factors," *IEEE Transactions on Industrial Electronics*, vol. 60, no. 2, pp. 681–690, February 2013.

45. S. Davari, D. Khaburi, and R. Kennel, "Using a weighting factor table for FCS-MPC of induction motors with extended prediction horizon," in *IEEE Industrial Electronics Conference (IECON)*, Montreal, QC, Canada, October 2012, pp. 2086–2091.

46. S. Davari, D. Khaburi, and R. Kennel, "An improved FCS-MPC algorithm for an induction motor with an imposed optimized weighting factor," *IEEE Transactions on Power Electronics*, vol. 27, no. 3, pp. 1540–1551, March 2012.

47. P. Coftes, J. Rodríguez, C. Silva, and A. Flores, "Delay compensation in model predictive current control of a three-phase inverter," *IEEE Transactions on Industrial Electronics*, vol. 59, no. 2, pp. 1323–1325, February 2012.

CHAPTER 12

CONTROL OF DFIG WECS WITH VOLTAGE SOURCE CONVERTERS

12.1 INTRODUCTION

Since the beginning of grid-connected operations in the 1980s, various combinations of electric generators and power electronic converters have been developed in commercial wind turbines (WTs). In first-generation WTs based on squirrel-cage induction generator (SCIG), the power converter was used in the form of a soft starter for a smooth connection of the generator to the utility grid. The major challenges imposed by this configuration include mandatory soft-starter requirements, reactive power compensation, and limited speed range (only around 1% of the synchronous speed). The limited speed range has many drawbacks, such as low energy yield and high stress on the turbine drivetrain during varying wind speed conditions. To solve these issues, second-generation WTs based on wound rotor induction generators (WRIGs) were developed. A small power converter is used in the rotor circuit to change the rotor winding resistance and increase the speed range to approximately 10% around the synchronous speed. The power converter does not process the generator output power; thus, a soft starter and reactive power compensation are required. In a continuous effort to solve the aforementioned issues, third-generation WTs were developed where a power converter is used between the WRIG rotor winding and utility grid. The power converter in a rotor circuit processes the slip power such that: (1) the variable-speed operation is extended to 30% above or below the synchronous speed, (2) a generator is smoothly connected to a utility grid without employing a soft starter, and (3) the power factor (PF) at the wind farm collection point (WFCP, also called as point of common coupling) is adjusted without any external reactive power compensation [1–3].

Model Predictive Control of Wind Energy Conversion Systems, First Edition. Venkata Yaramasu and Bin Wu.

Companion Website: www.wiley.com/go/yaramasu_wu_mpc_wecs

Third-generation WTs are innovative in processing generator output power and in solving the issues of former WT types. The power is fed to a utility grid through both stator and rotor windings, thus leading to a convenient nomenclature of a doubly fed induction generator (DFIG) even though it has the same construction as a WRIG. DFIG WTs are dominantly used by the wind energy industry (market share of more than 50%) [4, 5]. Table 12.1 highlights the list of megawatt (MW)-level DFIG WTs with respect to power and voltage rating, speed range, and gearbox gear ratio [1].

Table 12.1 Concise list of commercial DFIG WECS (Type 3 WECS) [1]

Manufacturer	Power, Stator /Rotor Voltages	Speed Range (rpm)	Gear Ratio	Latest Model #
Acciona Wind (Spain)	3.0 MW, 12000/690 V	770 – 1320	1:77	AW125/3000
Alstom (France)	3.0 MW, 1000/760 V	1025 – 1800	1:131.64	ECO110
BARD GmbH (Germany)	5.0 MW, 690/– V	679 – 1358	1:96.97	Bard 5.0
CCWE (PR China)	3.0 MW, 10000/690 V	775 – 1381	1:84.21	3000-100DF
Gamesa (Spain)	2.0 MW, 690/– V	905 – 1910	1:100.5	G90-2.0
GE Energy (USA)	3.0 MW, 690/– V	1003 – 1805	1:118	GE3.6S
MingYang (PR China)	1.5 MW, 690/– V	975 – 1960	1:100.5	MY1500-77
Nordex (Germany)	2.5 MW, 660/– V	890 – 1560	1:93.2	N100/2500
Senvion (Germany)	6.15 MW, 6600/660 V	747 – 1174	1:97	6.2M126
Suzlon (India)	2.1 MW, 690/– V	1440 – 1830	1:99	S97-2100
Vestas (Denmark)	2.0 MW, 690/– V	1083 – 1916	1:100.33	V80

Given the success of DFIG wind energy conversion systems (WECS), digital control techniques have been highly researched in the industry and academia [5–7]. Similar to the SCIGs in Chapter 11, the DFIG can be controlled by field-oriented control (FOC), direct torque control (DTC), or direct power control with the reference frame fixed to a stator flux or voltage [8]. To improve control performance, a model predictive control (MPC) can be used in place of classical control schemes. This chapter deals with the MPC of power converters employed in DFIG WECS.

Chapter Overview

- The DFIG WECS power conversion configuration and power flow during different wind speeds are discussed in Section 12.2.
- The control of a DFIG is outlined in Section 12.3. The steady state modeling and dynamic modeling of DFIG are presented in Section 12.4.
- The overall digital control system development for DFIG WECS is given in Section 12.5, including the maximum power point tracking (MPPT) and calculation of reference rotor currents for MPC design.
- The predictive current control (PCC), which is analogous to the classical FOC, is analyzed in a synchronous reference frame in Sections 12.6 and 12.7 with the help of indirect and direct rotor current dynamic models, respectively.
- Finally, the concluding remarks of this chapter are given in Section 12.8.

12.2 CONFIGURATION OF DFIG WECS AND POWER FLOW

12.2.1 Power Conversion Configuration

The block diagram of DFIG WECS with back-to-back (BTB) connected two-level voltage source converter (2L-VSC) is shown in Figure 12.1 [9, 10]. This configuration is also referred to as the Scherbius topology in the electric drives industry. A three-winding step-up transformer is used to connect the DFIG stator and rotor terminals to the WFCP. The stator terminals are connected to the WFCP through a step-up transformer. By using a medium voltage stator design (Table 12.1), the stator terminals are connected directly to the WFCP. The rotor-side converter (RSC) is connected to rotor windings via slip rings and brushes. The generator-side converter (GSC) is connected to the WFCP through a harmonic filter and a step-up transformer. The power converter in a rotor circuit processes the slip power, which is approximately 30% of the rated generator power. Therefore, the power rating of BTB VSC is only 30% of the generator rating. For example, a 2.5 MW DFIG only requires a 0.75 MW power converter. BTB 2L-VSCs are commonly used for the commercial DFIG WTs given in Table 12.1.

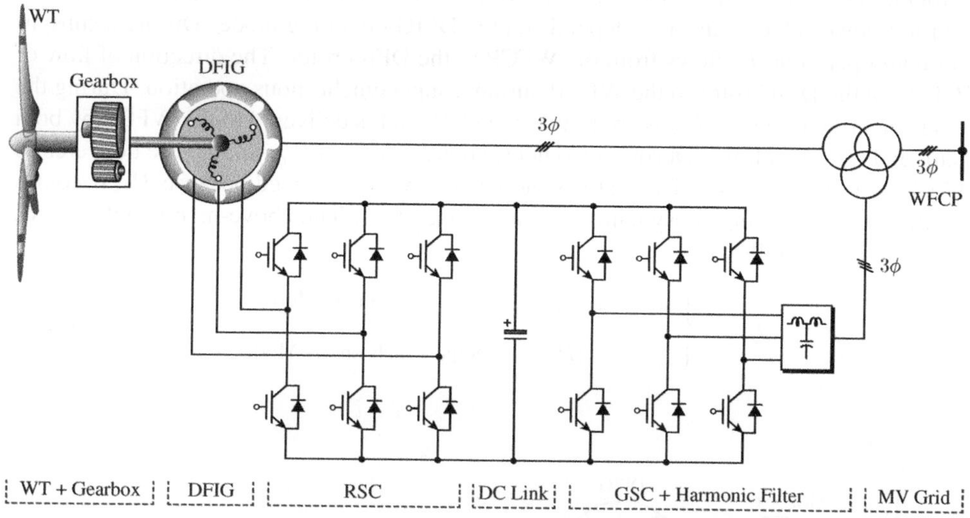

Figure 12.1 Block diagram of DFIG WECS with BTB 2L-VSC.

The RSC and GSC are linked by a DC-link capacitor. The BTB VSC enables bidirectional power flow in a rotor circuit (power can flow from rotor to grid or vice versa), thus leading to a high speed range (30% above or below the synchronous speed), low stress on the drivetrain during wind speed gusts, and robustness against power system disturbances compared with SCIG and WRIG WECS [11]. By controlling RSC, the generator stator active and reactive powers are controlled in a decoupled manner such that the MPPT operation can be obtained while maintaining unity, lagging, or leading PF at the WFCP. The DFIG is popularly used in modern WTs because of its low converter cost, reduced power losses, and full controllability over active and reactive power [3].

DFIGs are usually equipped with two or three pole pairs; therefore, a three-stage gearbox is most commonly used (Table 12.1). A low-speed operation is not possible as the construction of a multi-pole DFIG is not technically feasible. A major drawback is the mandatory requirement of a gearbox, but the other advantages possessed by the DFIG WECS make it a dominant technology in today's wind energy industry.

The direct and indirect matrix converters have been proposed in literature as an alternative to BTB VSCs [12, 13]. Given the partial power converter capacity needed, the matrix converters are promising for Type 3 WECS. The matrix converters provide more silicon-based AC to AC power conversion with sinusoidal input and output current waveforms. However, their performance during grid faults is inferior compared with BTB VSCs because no energy storage elements are used.

12.2.2 Power Flow in DFIG WECS

The power flow in BTB VSC-based DFIG WECS is shown by a Sankey diagram in Figure 12.2. The gearbox, generator and power converter losses, and block diagram of a harmonic filter are ignored to simplify the analysis. With this assumption, the WT output power becomes equal to the generator mechanical input power P_m. According to wind speed, the DFIG operates below the synchronous speed ω_s (subsynchronous mode) or above the synchronous speed ω_s (supersynchronous mode). During an entire wind speed range, the stator output power P_s flows from the DFIG stator to the WFCP. However, the direction of the rotor power P_r (or currents) depends on the DFIG operating mode. During a subsynchronous operation, P_r flows from the WFCP to the DFIG rotor. The direction of flow of P_r is from the DFIG rotor to the WFCP during a supersynchronous operation. During the supersynchronous mode, P_m is split into P_s and P_r and is delivered to the WFCP by both stator and rotor windings. During subsynchronous mode, both P_m and P_r are delivered to WFCP by stator winding. The stator windings are not overloaded because P_m becomes significantly low during subsynchronous mode operation. The above-mentioned analysis is summarized as follows:

$$|P_s| = \begin{cases} |P_m| + |P_r| & \text{(Subsynchronous Mode)} \\ |P_m| - |P_r| & \text{(Supersynchronous Mode).} \end{cases} \tag{12.1}$$

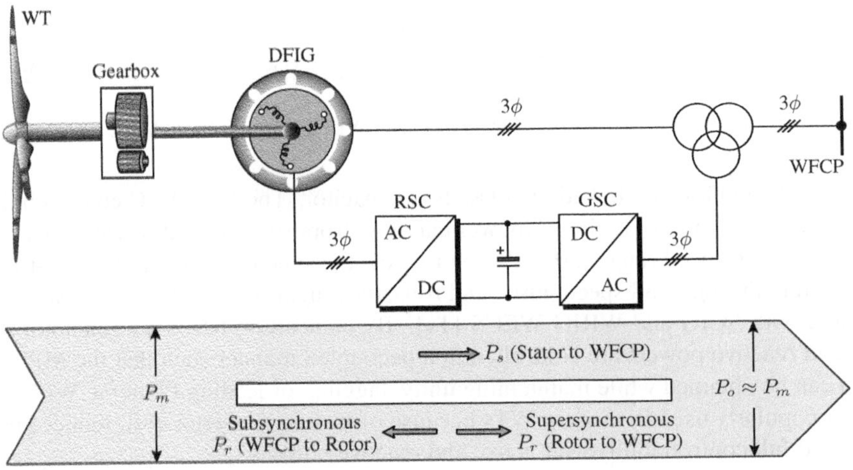

Figure 12.2 Sankey diagram for power flow in BTB VSC-based DFIG WECS.

12.3 CONTROL OF DOUBLY FED INDUCTION GENERATOR

In the previous chapter, the FOC scheme for SCIG has been discussed by using the synchronous (dq) reference frame (SRF) theory (Section 11.4.1). The three-phase stator currents are decomposed into dq-axes with the help of the rotor flux angle θ_e. The rotor flux orientation is used to align the flux linkage component of stator current i_{ds} with d-axis and the torque component of stator current i_{qs} with q-axis. The FOC then uses a cascaded control structure with an outer speed control loop, an inner current control loop, and a modulation stage to produce switching signals for a generator-side rectifier (Figure 11.5). The dq-axes stator current control through the rectifier leads to a decoupled control of rotor flux and electromagnetic torque similar to a separately excited DC motor. The rotor flux magnitude and angle (ψ_r and θ_e) are estimated according to direct FOC (DFOC) or indirect FOC (IFOC). The DFOC uses three-phase stator voltages and currents, whereas the IFOC employs three-phase stator currents and a rotor electrical speed ω_r for the estimation of ψ_r and θ_e (refer to Figure 11.6).

The FOC scheme can be used in DFIG to achieve high dynamic performance during variable-speed operation [2]. Three reference frames are commonly used to design and analyze FOC schemes for DFIG: (1) a stationary frame fixed to the stator ($\alpha\beta$), (2) a stationary frame fixed to the rotor (xy), and (3) a synchronous frame (dq) fixed to either the stator flux or stator voltage [14]. The SRF is commonly employed to decouple the rotor currents into active power (or torque) and reactive power (or flux) producing components [14]. The dq-axes rotor currents are then controlled independently by using the FOC block diagram given in Figure 11.5. The stator angle θ_s and slip angle θ_{sl} are needed to realize FOC for DFIG. Similar to SCIG, two methods are commonly employed for the estimation of θ_s and θ_{sl}: (1) stator flux-oriented control (SFOC); (2) stator voltage-oriented control (SVOC).

12.3.1 Stator Flux-Oriented Control

Figure 12.3 shows the block diagram of SFOC for the estimation of θ_s and θ_{sl}. The SFOC has been employed in the first commercial developments of DFIG WTs. The SFOC scheme is a simplified version of the DFOC scheme discussed in Chapter 11 and is analogous to the virtual flux-oriented control of GSCs discussed in Chapter 8 and the DTC torque and stator flux estimator discussed in Chapter 11. The three-phase stator voltages and currents ($\mathbf{v_s}$ and $\mathbf{i_s}$), measured rotor electrical position angle θ_r, and $\alpha\beta$-frame IG model are used by the SFOC to estimate θ_s and θ_{sl}. The $\mathbf{v_s}$ and $\mathbf{i_s}$ are transformed to the $\alpha\beta$-frame with the help of the $abc/\alpha\beta$ transformation in Equation (3.6).

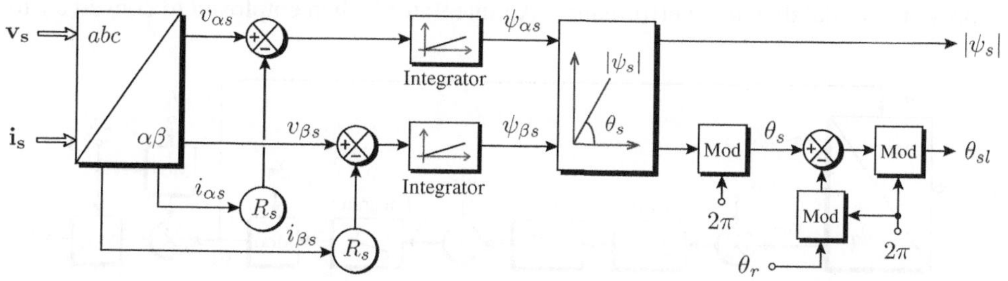

Figure 12.3 Estimation of stator flux angle and slip angle with the SFOC scheme.

With the help of estimated $\alpha\beta$-axes stator voltages and currents and stator winding resistance R_s, the $\alpha\beta$-frame stator flux linkages are obtained by the following [8, 9]:

$$\begin{aligned} \psi_{\alpha s} &= \int (v_{\alpha s} - R_s\, i_{\alpha s})\, dt \\ \psi_{\beta s} &= \int (v_{\beta s} - R_s\, i_{\beta s})\, dt. \end{aligned} \tag{12.2}$$

The angle of stator flux vector is then calculated as follows:

$$\theta_s = \tan^{-1}\left(\frac{\psi_{\beta s}}{\psi_{\alpha s}}\right). \tag{12.3}$$

The slip angle is then obtained by subtracting θ_r from θ_s, that is,

$$\theta_{sl} = \theta_s - \theta_r. \tag{12.4}$$

The stator flux angle θ_s, rotor electrical position angle θ_r, and slip angle θ_{sl} are transformed to "0 to 2π" range with the help of a modulus function. The modulus function makes the stator flux angle θ_s change from 0 to 2π when the stator voltage vector $\mathbf{v_s}$ makes one revolution in space.

12.3.2 Stator Voltage-Oriented Control

As mentioned earlier, the stator terminals of DFIG are connected to WFCP directly or through a step-up transformer. Under ideal operating conditions, the stator voltage angle θ_s is obtained from $\alpha\beta$-axis grid voltages as follows:

$$\theta_s = \tan^{-1}\left(\frac{v_{\beta s}}{v_{\alpha s}}\right). \tag{12.5}$$

However, to compensate the stator voltage harmonics and distortions in practical operating conditions, a SRF phase-locked loop (SRF-PLL) (discussed in Chapter 8) can be employed to estimate θ_s [15]. The SVOC scheme uses three-phase stator voltages $\mathbf{v_s}$ and θ_r to estimate θ_s and the slip angle θ_{sl} (Figure 12.4). The block diagram of SVOC is similar to the SRF-PLL in Figure 8.9, except that the subscript g (grid) is now replaced by s (stator). The three-phase stator voltages $\mathbf{v_s}$ are transformed to the dq-frame by using the feedback θ_s value. A low-pass filter (LPF) is used to filter the stator voltage error. A proportional-integral (PI) controller is used to force the q-axis stator voltage v_{qs} to zero. A feed-forward angular frequency term ω_o is added to the output of the PI controller to improve the initial dynamic performance. An integrator is then employed to convert ω_s to

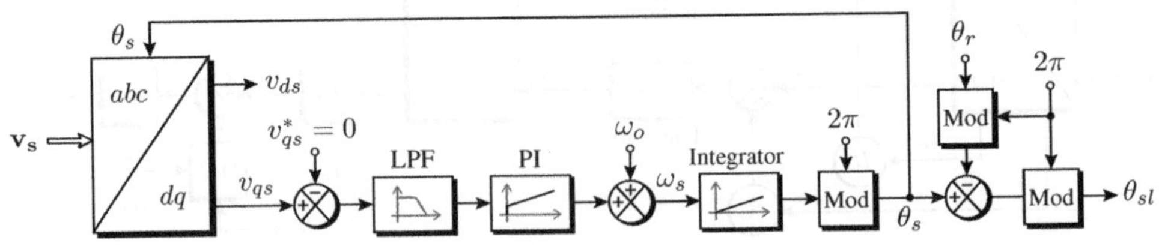

Figure 12.4 Estimation of stator voltage angle and slip angle with the SVOC scheme.

θ_s. To ensure that the θ_s changes between 0 and 2π, a modulus function is used. The slip angle θ_{sl} is then obtained by subtracting θ_r from θ_s (similar to Equation (12.4)).

A space vector diagram for the SVOC scheme is shown in Figure 12.5 when the DFIG is operating with a unity power factor (UPF) in supersynchronous mode [16]. The process of aligning the d-axis of the SRF with the stator voltage vector $\mathbf{v_s}$ is called stator voltage orientation. With SVOC, the stator voltage vector $\mathbf{v_s}$ is aligned with the d-axis of SRF. As a result, the q-axis grid voltage becomes zero, that is,

$$v_{qs} = 0, \qquad v_{ds} = \sqrt{v_s^2 - v_{qs}^2} = v_s \tag{12.6}$$

where v_s represents the length of the stator voltage vector or the peak stator voltage.

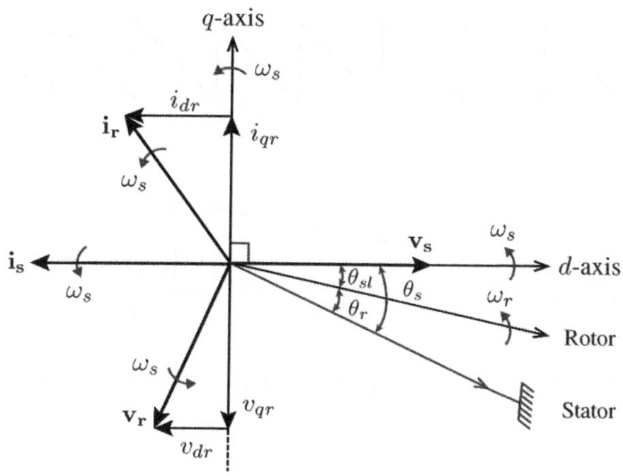

Figure 12.5 Space vector diagram of DFIG with SVOC in supersynchronous mode [16].

The rotating speed of the SRF is given by the following:

$$\omega_s = 2\,\pi\,f_s \tag{12.7}$$

where f_s is the stator or WFCP frequency.

All the stator-related space phasors in Figure 12.5 rotate at the ω_s value. The rotor moves in space at the ω_r value. Both the stator voltage angle θ_s and rotor position angle θ_r are referenced to the stator. The stator current vector $\mathbf{i_s}$ is aligned with stator voltage vector $\mathbf{v_s}$ but is 180° out of phase because of the UPF and generator mode operation. The rotor voltage and current vectors $\mathbf{v_r}$ and $\mathbf{i_r}$ are also decomposed into dq-axes with the help of the slip angle θ_{sl}. The digital control system for the RSC controls the dq-axis rotor voltages and currents such that the control objectives for the DFIG, such as the MPPT operation and WFCP PF control, are met.

The correlation between stator voltages, θ_s, θ_r, and θ_{sl} is shown in Figures 12.6 and 12.7 during subsynchronous and supersynchronous modes of operation for the two fundamental cycles of slip angular frequency ω_{sl}, respectively. In both cases, the peak value of θ_s is aligned with the peak value of the phase-a stator voltage v_{as}. By subtracting θ_r from θ_s, the slip angle θ_{sl} is obtained. The slip angle θ_{sl} monotonically increases from 0 to 2π in subsynchronous mode and monotonically decreases from 2π to 0 in supersynchronous mode. All three angle types vary from 0 to 2π because of the modulus function employed for them. To maintain coherence with Chapter 8, the SVOC scheme is used in the further analysis of DFIG WECS.

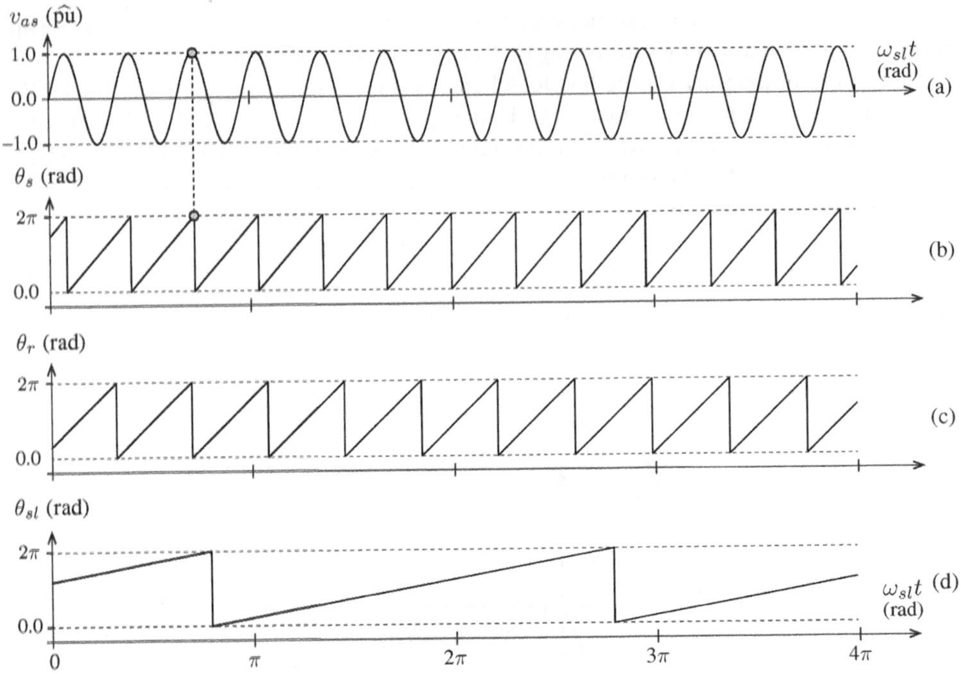

Figure 12.6 Stator voltage angle, rotor position angle and slip angle in subsynchronous mode.

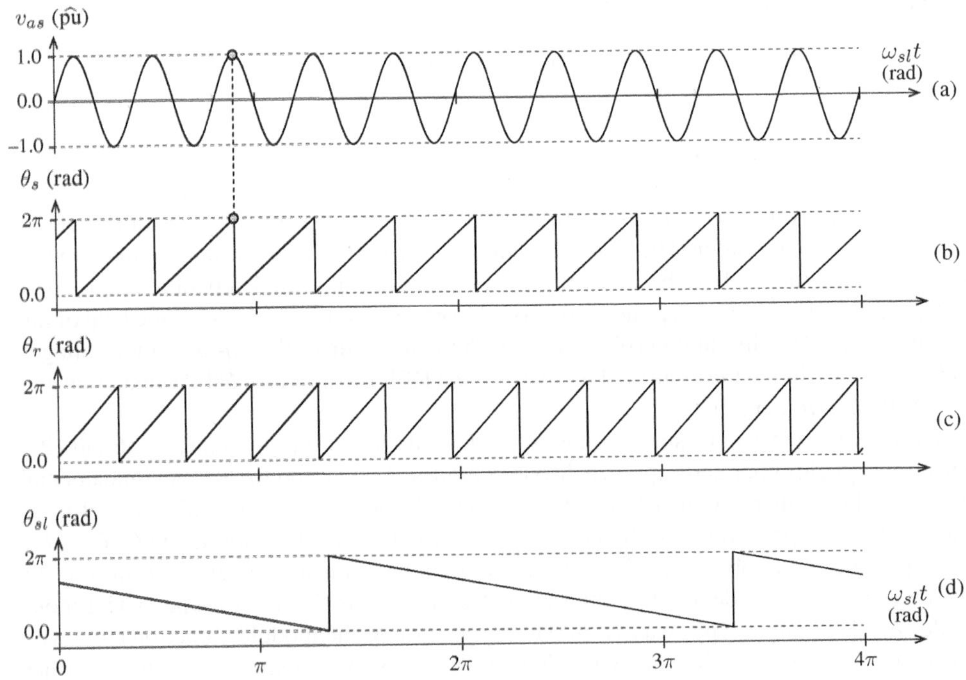

Figure 12.7 Stator voltage angle, rotor position angle and slip angle in supersynchronous mode.

12.4 MODELING OF DOUBLY FED INDUCTION GENERATOR

As mentioned in the previous section, the DFIG can be analyzed in the $\alpha\beta$, xy, or dq frames. In SCIG and permanent magnet synchronous generator (PMSG), all stator variables vary at one angular frequency. In DFIG, the stator and rotor variables have two different angular frequencies (i.e., ω_s and ω_{sl}). The PCC design has a flexibility to be realized in the $\alpha\beta$, xy, or dq frames as long as the dynamic models of DFIG are available. However, the $\alpha\beta$ and xy frames lead to complicated analyses and coupling issues in digital control. By using dq-frame control, all stator and rotor variables become DC, thus leading to a simplified analysis and decoupled control. In this section, the modeling of DFIG is analyzed in SRF and is fixed to the stator voltage vector (SVOC scheme). The steady state and dynamic models of the DFIG are presented for PCC design. The correlation between rotor currents and stator active and reactive powers is also established. Further details about IG modeling can be found in Chapters 6 and 7.

12.4.1 Equivalent Circuit of DFIG

The dynamic dq-axes equivalent circuits of DFIG are shown in Figure 12.8, along with an RSC and WFCP. The dq-frame circuit is deducted from the arbitrary frame IG model by setting the arbitrary frame speed ω_k to ω_s, thus indicating that SRF rotates at a speed of ω_s. The equivalent circuit refers to motor models and can be employed for the generator mode of the operation by changing the mechanical input torque T_m sign to negative.

By referring to Figure 12.8, the stator and rotor voltage vectors are defined by [9]

$$\mathbf{v_{s,dq}} = R_s\,\mathbf{i_{s,dq}} + \frac{d}{dt}\boldsymbol{\psi}_{\mathbf{s,dq}} + j\,\omega_s \boldsymbol{\psi}_{\mathbf{s,dq}} \tag{12.8}$$

$$\mathbf{v_{r,dq}} = R_r\,\mathbf{i_{r,dq}} + \frac{d}{dt}\boldsymbol{\psi}_{\mathbf{r,dq}} + j\,(\omega_s - \omega_r)\,\boldsymbol{\psi}_{\mathbf{r,dq}} \tag{12.9}$$

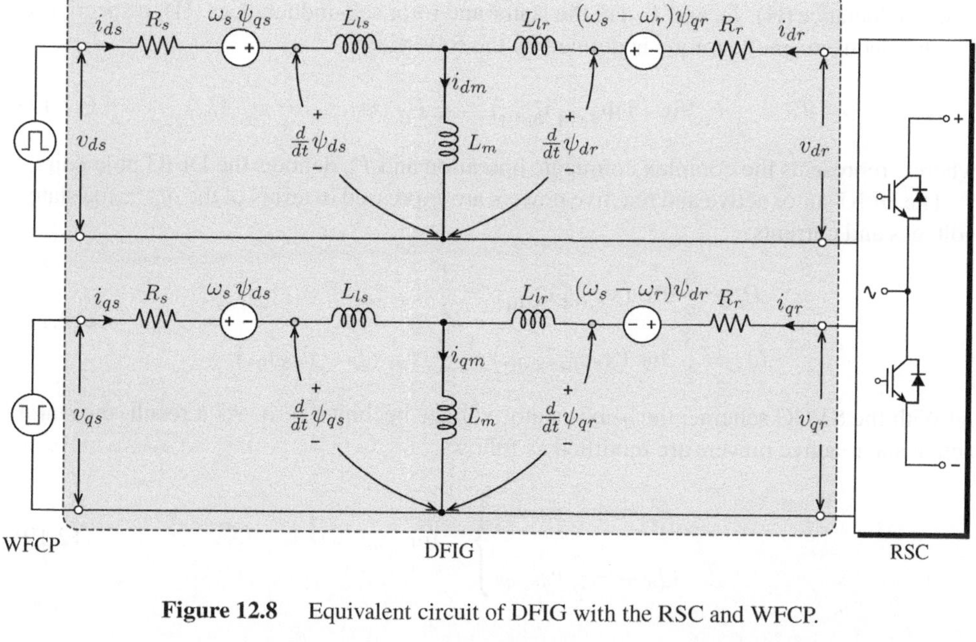

Figure 12.8 Equivalent circuit of DFIG with the RSC and WFCP.

where

- $\mathbf{v_{s,dq}} = [v_{ds}\ v_{qs}]^T$ and $\mathbf{v_{r,dq}} = [v_{dr}\ v_{qr}]^T$ are the stator and rotor voltage vectors (V).
- $\mathbf{i_{s,dq}} = [i_{ds}\ i_{qs}]^T$ and $\mathbf{i_{r,dq}} = [i_{dr}\ i_{qr}]^T$ are the stator and rotor current vectors (A).
- $\boldsymbol{\psi}_{\mathbf{s,dq}} = [\psi_{ds}\ \psi_{qs}]^T$ and $\boldsymbol{\psi}_{\mathbf{r,dq}} = [\psi_{dr}\ \psi_{qr}]^T$ are the stator and rotor flux vectors (Wb).
- R_s and R_r are the stator and rotor winding resistances (Ω).
- $\omega_{sl} = (\omega_s - \omega_r)$ is the slip angular speed (rad/s).

During the steady state operation, the derivative terms in Equations (12.8) and (12.9) become zero. The stator flux vector is obtained from Equation (12.8) as follows:

$$\boldsymbol{\psi}_{\mathbf{s,dq}} = \frac{\mathbf{v_{s,dq}} - R_s\,\mathbf{i_{s,dq}}}{j\,\omega_s} \tag{12.10}$$

which can be decomposed into dq-axes as

$$\psi_{ds} = \frac{v_{qs} - R_s\,i_{qs}}{\omega_s}, \qquad \psi_{qs} = -\frac{v_{ds} - R_s\,i_{ds}}{\omega_s}. \tag{12.11}$$

Alternatively, the stator and rotor flux linkage vectors are defined by the following:

$$\boldsymbol{\psi}_{\mathbf{s,dq}} = L_s\,\mathbf{i_{s,dq}} + L_m\,\mathbf{i_{r,dq}} \tag{12.12}$$

$$\boldsymbol{\psi}_{\mathbf{r,dq}} = L_r\,\mathbf{i_{r,dq}} + L_m\,\mathbf{i_{s,dq}} \tag{12.13}$$

which are decomposed into dq-axes as follows:

$$\psi_{ds} = L_s\,i_{ds} + L_m\,i_{dr}, \qquad \psi_{qs} = L_s\,i_{qs} + L_m\,i_{qr} \tag{12.14}$$

$$\psi_{dr} = L_r\,i_{dr} + L_m\,i_{ds}, \qquad \psi_{qr} = L_r\,i_{qr} + L_m\,i_{qs} \tag{12.15}$$

with

$$L_s = L_{ls} + L_m, \qquad L_r = L_{lr} + L_m \tag{12.16}$$

where L_{ls} and L_{lr} are the stator and rotor leakage inductances (H) and L_m is the magnetizing inductance (H). L_s and L_r are the stator and rotor self-inductances (H), respectively.

The electromagnetic torque is defined by the following:

$$T_e = \frac{3}{2}\,P_p\,\mathfrak{Re}\left(j\,\boldsymbol{\psi}_{\mathbf{s,dq}}\,\mathbf{i}^\star_{\mathbf{s,dq}}\right) = \frac{3}{2}\,P_p\,(\psi_{ds}\,i_{qs} - \psi_{qs}\,i_{ds}) \tag{12.17}$$

where $\star$ represents the complex conjugate operation and P_p denotes the DFIG pole pairs.

The DFIG stator active and reactive powers are expressed in terms of the dq-frame stator voltages and currents:

$$\begin{aligned} P_s &= \frac{3}{2}\,\mathfrak{Re}\left(\mathbf{v_{s,dq}}\,\mathbf{i}^\star_{\mathbf{s,dq}}\right) = \frac{3}{2}\,(v_{ds}\,i_{ds} + v_{qs}\,i_{qs}) \\ Q_s &= \frac{3}{2}\,\mathfrak{Im}\left(\mathbf{v_{s,dq}}\,\mathbf{i}^\star_{\mathbf{s,dq}}\right) = \frac{3}{2}\,(v_{qs}\,i_{ds} - v_{ds}\,i_{qs})\,. \end{aligned} \tag{12.18}$$

With the SVOC scheme, the q-axis stator voltage becomes zero. As a result, the stator active and reactive powers are modified as follows:

$$\left.\begin{aligned} P_s &= +\frac{3}{2}\,v_{ds}\,i_{ds} \\ Q_s &= -\frac{3}{2}\,v_{ds}\,i_{qs} \end{aligned}\right\} \quad \text{for } v_{qs} = 0. \tag{12.19}$$

12.4.2 Correlation Between Rotor Currents and Control Requirements

In DFIG WECS, the MPPT operation is achieved by controlling P_s (electromagnetic torque control loop) and the WFCP PF is adjusted by controlling Q_s. The control objectives are fulfilled by regulating the rotor currents through the digital control of RSC. Therefore, the correlation between rotor currents and control variables T_e, P_s, and Q_s should be understood.

By solving Equations (12.11), (12.14), and (12.17), the electromagnetic torque is expressed in terms of the dq-axes rotor currents:

$$T_e = \frac{3\,P_p\,L_m}{2\,\omega_s\,L_s}\,(-i_{qr}\,v_{qs} + R_s\,i_{qs}\,i_{qr} + R_s\,i_{ds}\,i_{dr} - i_{dr}\,v_{ds})\,. \tag{12.20}$$

By using the SVOC scheme, the q-axis stator voltage is maintained at zero. For a high-power DFIG, the stator winding resistance R_s is approximately equal to zero. Therefore, the electromagnetic torque expression in Equation (12.20) is simplified as follows:

$$T_e = -\frac{3\,P_p\,L_m}{2\,\omega_s\,L_s}\,i_{dr}\,v_{ds}. \tag{12.21}$$

By solving Equations (12.11), (12.14), and (12.19), the rotor currents are expressed in terms of stator active and reactive powers [17, 18]:

$$\begin{aligned} i_{dr} &= -\left(\frac{2\,L_s}{3\,v_{ds}\,L_m}\right)P_s - \left(\frac{R_s}{\omega_s\,L_m}\right)i_{qs} \\ i_{qr} &= +\left(\frac{2\,L_s}{3\,v_{ds}\,L_m}\right)Q_s + \left(\frac{R_s}{\omega_s\,L_m}\right)i_{ds} - \left(\frac{1}{\omega_s\,L_m}\right)v_{ds}. \end{aligned} \tag{12.22}$$

By neglecting stator winding resistance, the aforementioned expressions are simplified to the following:

$$\begin{aligned} i_{dr} &= -\left(\frac{2\,L_s}{3\,v_{ds}\,L_m}\right)P_s \\ i_{qr} &= \left(\frac{2\,L_s}{3\,v_{ds}\,L_m}\right)Q_s - \left(\frac{1}{\omega_s\,L_m}\right)v_{ds}. \end{aligned} \tag{12.23}$$

As shown in Equations (12.21) and (12.23), T_e is directly proportional to i_{dr}, P_s is directly proportional to i_{dr}, and Q_s is directly proportional to i_{qr}. By controlling the dq-axis rotor currents, the electromagnetic torque, stator active power, and stator reactive power of DFIG are controlled in a decoupled manner.

12.4.3 Continuous-Time Dynamic Models of DFIG

The PCC scheme uses the mathematical model of DFIG and RSC to predict the behavior of rotor currents at the $(k+1)$ sampling instant. To design the PCC scheme, the dynamic model of rotor currents is needed. The continuous-time (CT) dq-frame dynamic model of IG given in Equation (6.55) is reproduced below to analyze the rotor current dynamic model:

$$\frac{d}{dt}\begin{bmatrix} i_{ds} \\ i_{qs} \\ \psi_{dr} \\ \psi_{qr} \end{bmatrix} = \underbrace{\begin{bmatrix} -\frac{1}{\tau_\sigma} & \omega_s & \frac{k_r}{\sigma L_s \tau_r} & \frac{k_r \omega_r}{\sigma L_s} \\ -\omega_s & -\frac{1}{\tau_\sigma} & -\frac{k_r \omega_r}{\sigma L_s} & \frac{k_r}{\sigma L_s \tau_r} \\ \frac{L_m}{\tau_r} & 0 & -\frac{1}{\tau_r} & (\omega_s - \omega_r) \\ 0 & \frac{L_m}{\tau_r} & -(\omega_s - \omega_r) & -\frac{1}{\tau_r} \end{bmatrix}}_{\mathbf{A}(t)} \begin{bmatrix} i_{ds} \\ i_{qs} \\ \psi_{dr} \\ \psi_{qr} \end{bmatrix}$$
$$+ \underbrace{\begin{bmatrix} \frac{1}{\sigma L_s} & 0 & -\frac{k_r}{\sigma L_s} & 0 \\ 0 & \frac{1}{\sigma L_s} & 0 & -\frac{k_r}{\sigma L_s} \\ 0 & 0 & 1 & 0 \\ 0 & 0 & 0 & 1 \end{bmatrix}}_{\mathbf{B}} \begin{bmatrix} v_{ds} \\ v_{qs} \\ v_{dr} \\ v_{qr} \end{bmatrix} \tag{12.24}$$

where $\mathbf{A}(t)$ and $\mathbf{B}$ are the state and input matrices in CT, respectively. The DFIG variables are defined by the following [19, 20]:

- Stator and rotor coupling coefficients: $k_s = \frac{L_m}{L_s}$ and $k_r = \frac{L_m}{L_r}$.
- Total leakage coefficient, $\sigma = 1 - k_s\, k_r = 1 - \frac{L_m^2}{L_s\, L_r}$.
- Equivalent resistance, $r_\sigma = R_s + k_r^2\, R_r$ (Ω).
- Stator transient time constant, $\tau_\sigma = \frac{\sigma L_s}{r_\sigma}$ (s).
- Stator and rotor time constants: $\tau_s = \frac{L_s}{R_s}$ (s) and $\tau_r = \frac{L_r}{R_r}$ (s).
- Total leakage inductance, $\sigma L_s = \frac{L_s\, L_r - L_m^2}{L_r}$.

From the aforementioned model, the dq-axes stator currents and rotor flux linkages are state variables. Therefore, an output CT model should be established to relate the dq-axis rotor currents to the state variables in Equation (12.24). The expression in (6.35) is reproduced below to support the analysis:

$$\mathbf{i_r} = \frac{\boldsymbol{\psi}_\mathbf{r} - L_m\, \mathbf{i_s}}{L_r} \tag{12.25}$$

which can be expressed in the form of a state-space equation as

$$\begin{bmatrix} i_{dr} \\ i_{qr} \end{bmatrix} = \underbrace{\begin{bmatrix} -\frac{L_m}{L_r} & 0 & \frac{1}{L_r} & 0 \\ 0 & -\frac{L_m}{L_r} & 0 & \frac{1}{L_r} \end{bmatrix}}_{\mathbf{C}} \begin{bmatrix} i_{ds} \\ i_{qs} \\ \psi_{dr} \\ \psi_{qr} \end{bmatrix} \tag{12.26}$$

where $\mathbf{C}$ is the output matrix in CT.

The dq-axes rotor voltages v_{dr} and v_{qr} in (12.24) are controlled by RSC, which in turn leads to the direct control of stator currents and rotor flux linkages and the indirect control of rotor currents. Therefore, the state-space models in Equations (12.24) and (12.26) are referred to as the indirect dynamic models of rotor currents.

The direct dynamic model of rotor currents can be obtained, as shown below (Equation (4.45) in [21]):

$$\frac{d}{dt}\begin{bmatrix} i_{ds} \\ i_{qs} \\ i_{dr} \\ i_{qr} \end{bmatrix} = k_\sigma \underbrace{\begin{bmatrix} -R_s\,L_r & \omega_r\,L_m^2 + k_\omega & R_r\,L_m & \omega_r\,L_m\,L_r \\ -\omega_r\,L_m^2 - k_\omega & -R_s\,L_r & -\omega_r\,L_m\,L_r & R_r\,L_m \\ R_s\,L_m & -\omega_r\,L_s\,L_m & -R_r\,L_s & -\omega_r\,L_r\,L_s + k_\omega \\ \omega_r\,L_s\,L_m & R_s\,L_m & \omega_r\,L_r\,L_s - k_\omega & -R_r\,L_s \end{bmatrix}}_{\mathbf{A}(t)} \begin{bmatrix} i_{ds} \\ i_{qs} \\ i_{dr} \\ i_{qr} \end{bmatrix}$$

$$+\,k_\sigma \underbrace{\begin{bmatrix} L_r & 0 & -L_m & 0 \\ 0 & L_r & 0 & -L_m \\ -L_m & 0 & L_s & 0 \\ 0 & -L_m & 0 & L_s \end{bmatrix}}_{\mathbf{B}} \begin{bmatrix} v_{ds} \\ v_{qs} \\ v_{dr} \\ v_{qr} \end{bmatrix} \tag{12.27}$$

with

$$k_\sigma = \left(\frac{1}{\sigma\,L_s\,L_r}\right), \quad k_\omega = \omega_s\,\sigma\,L_s\,L_r. \tag{12.28}$$

12.4.4 Discrete-Time Dynamic Models of DFIG

The indirect and direct dynamic models of rotor currents are linear time varying because of $\omega_r(t)$ in state matrix $\mathbf{A}(t)$. To reduce the complexity of discretization, sampled data models are obtained by the forward Euler approximation. The approximation of the first-order derivative with the forward Euler method is defined by the following:

$$x(k+1) \approx x(k) + T_s \left\{\frac{d\,x(t)}{dt}\right\}_{t=k}, \quad x \in \{i_{ds},\, i_{qs},\, \psi_{dr},\, \psi_{qr},\, i_{dr},\, i_{qr}\}. \tag{12.29}$$

By combining Equations (12.24), (12.26), and (12.29), the discrete-time (DT) indirect rotor current dynamic model is obtained as follows:

$$\begin{bmatrix} i_{ds}(k+1) \\ i_{qs}(k+1) \\ \psi_{dr}(k+1) \\ \psi_{qr}(k+1) \end{bmatrix} \approx \underbrace{\begin{bmatrix} 1 - \frac{T_s}{\tau_\sigma} & \omega_s(k)\,T_s & \frac{k_r\,T_s}{\sigma\,L_s\,\tau_r} & \frac{k_r\,T_s\,\omega_r(k)}{\sigma\,L_s} \\ -\omega_s\,T_s & 1 - \frac{T_s}{\tau_\sigma} & -\frac{k_r\,T_s\,\omega_r(k)}{\sigma\,L_s} & \frac{k_r\,T_s}{\sigma\,L_s\,\tau_r} \\ \frac{L_m\,T_s}{\tau_r} & 0 & 1 - \frac{T_s}{\tau_\sigma} & \omega_{sl}(k)\,T_s \\ 0 & \frac{L_m\,T_s}{\tau_r} & -\omega_{sl}(k)\,T_s & 1 - \frac{T_s}{\tau_\sigma} \end{bmatrix}}_{\mathbf{\Phi}(k)} \begin{bmatrix} i_{ds}(k) \\ i_{qs}(k) \\ \psi_{dr}(k) \\ \psi_{qr}(k) \end{bmatrix}$$

$$+ \underbrace{\begin{bmatrix} \frac{T_s}{\sigma\,L_s} & 0 & -\frac{k_r\,T_s}{\sigma\,L_s} & 0 \\ 0 & \frac{T_s}{\sigma\,L_s} & 0 & -\frac{k_r\,T_s}{\sigma\,L_s} \\ 0 & 0 & T_s & 0 \\ 0 & 0 & 0 & T_s \end{bmatrix}}_{\mathbf{\Gamma_b}} \begin{bmatrix} v_{ds}(k) \\ v_{qs}(k) \\ v_{dr}(k) \\ v_{qr}(k) \end{bmatrix} \tag{12.30}$$

with

$$\begin{bmatrix} i_{dr}(k) \\ i_{qr}(k) \end{bmatrix} = \underbrace{\begin{bmatrix} -\frac{L_m}{L_r} & 0 & \frac{1}{L_r} & 0 \\ 0 & -\frac{L_m}{L_r} & 0 & \frac{1}{L_r} \end{bmatrix}}_{\mathbf{H}} \begin{bmatrix} i_{ds}(k) \\ i_{qs}(k) \\ \psi_{dr}(k) \\ \psi_{qr}(k) \end{bmatrix} \tag{12.31}$$

where T_s is the control sampling time. The correlation between the CT and DT matrices is defined by the following:

$$\mathbf{\Phi}(k) \approx (\mathbf{I} + \mathbf{A}(t)\, T_s), \quad \mathbf{\Gamma_b} \approx \mathbf{B}\, T_s, \quad \mathbf{H} = \mathbf{C} \tag{12.32}$$

where $\mathbf{I}$ is the unity matrix.

The rotor current direct DT dynamic model is obtained below by substituting (12.29) into (12.27):

$$\begin{bmatrix} i_{ds}(k+1) \\ i_{qs}(k+1) \\ i_{dr}(k+1) \\ i_{qr}(k+1) \end{bmatrix} = \mathbf{\Phi}(k) \begin{bmatrix} i_{ds}(k) \\ i_{qs}(k) \\ i_{dr}(k) \\ i_{qr}(k) \end{bmatrix} + \mathbf{\Gamma_b} \begin{bmatrix} v_{ds}(k) \\ v_{qs}(k) \\ v_{dr}(k) \\ v_{qr}(k) \end{bmatrix} \tag{12.33}$$

where

$$\mathbf{\Phi}(k) \approx k_\sigma \begin{bmatrix} 1 - R_s\, L_r\, T_s & \omega_r(k)\, L_m^2\, T_s + k_\phi & R_r\, L_m\, T_s & \omega_r(k)\, L_m\, L_r\, T_s \\ -\omega_r(k)\, L_m^2\, T_s - k_\phi & 1 - R_s\, L_r\, T_s & -\omega_r(k)\, L_m\, L_r & R_r\, L_m\, T_s \\ R_s\, L_m\, T_s & -\omega_r(k)\, L_s\, L_m\, T_s & 1 - R_r\, L_s\, T_s & -\omega_r(k)\, L_r\, L_s\, T_s + k_\phi \\ \omega_r(k)\, L_s\, L_m\, T_s & R_s\, L_m\, T_s & \omega_r(k)\, L_r\, L_s\, T_s - k_\phi & 1 - R_r\, L_s\, T_s \end{bmatrix}$$

$$\mathbf{\Gamma_b} \approx k_\sigma \begin{bmatrix} L_r\, T_s & 0 & -L_m\, T_s & 0 \\ 0 & L_r\, T_s & 0 & -L_m\, T_s \\ -L_m\, T_s & 0 & L_s\, T_s & 0 \\ 0 & -L_m\, T_s & 0 & L_s\, T_s \end{bmatrix} \tag{12.34}$$

with

$$k_\phi \approx k_\omega\, T_s \approx \omega_s\, \sigma\, L_s\, L_r\, T_s. \tag{12.35}$$

The aforementioned DT matrices $\mathbf{\Phi}(k)$ and $\mathbf{\Gamma_b}$ employ the model in (12.32). The DT indirect and direct rotor current models discussed in the preceding paragraphs involve low complexity in the discretization and real-time implementation. To improve the control performance with a better compromise between discretization error and computational burden, other discretization methods such as bilinear transformation, matrix factorization, and truncated Taylor series (discussed earlier in Chapter 7) can also be used.

12.5 DIGITAL CONTROL OF BTB CONVERTER-BASED DFIG WECS

Digital control system development for the classical or predictive control of DFIG WECS requires a brief understanding of the block diagram of control scheme, control requirements, and calculation of reference control variables. In this section, the digital control system development for DFIG WECS is discussed with respect to the digital control systems analyzed for PMSG and SCIG WECS in Chapters 9 to 11. The analysis has been conducted by targeting BTB VSCs and PCC schemes but the information presented in this section can be used for classical SVOC, other power converters (matrix converters), and other MPC schemes (predictive torque control and predictive power control) [22, 23].

12.5.1 Block Diagram of Digital Control System

Figure 12.9 shows the block diagram of a digital control system with Level I to III control loops for BTB VSC-based DFIG WECS. To simplify the analysis, the DFIG stator terminals are assumed to be directly connected to the WFCP and the rotor terminals are connected to the WFCP by BTB VSC, a harmonic filter, and a step-up transformer with a step-up ratio of 1:V_s/V_r. V_s and V_r are the rated *rms* stator and rotor voltages, respectively. For the configuration shown in Figure 12.9, the DFIG stator voltage $\mathbf{v_s}$ is equal to the WFCP (or WT output) voltage $\mathbf{v_o}$. The WFCP (or WT output) current $\mathbf{i_o}$ is equal to the summation of the DFIG stator current $\mathbf{i_s}$ and grid (GSC) current $\mathbf{i_g}$. The currents in the rotor circuit can have a bidirectional flow, depending upon the rotor speed of DFIG. The stator currents are unidirectional, that is, they always flow from the stator to the WFCP.

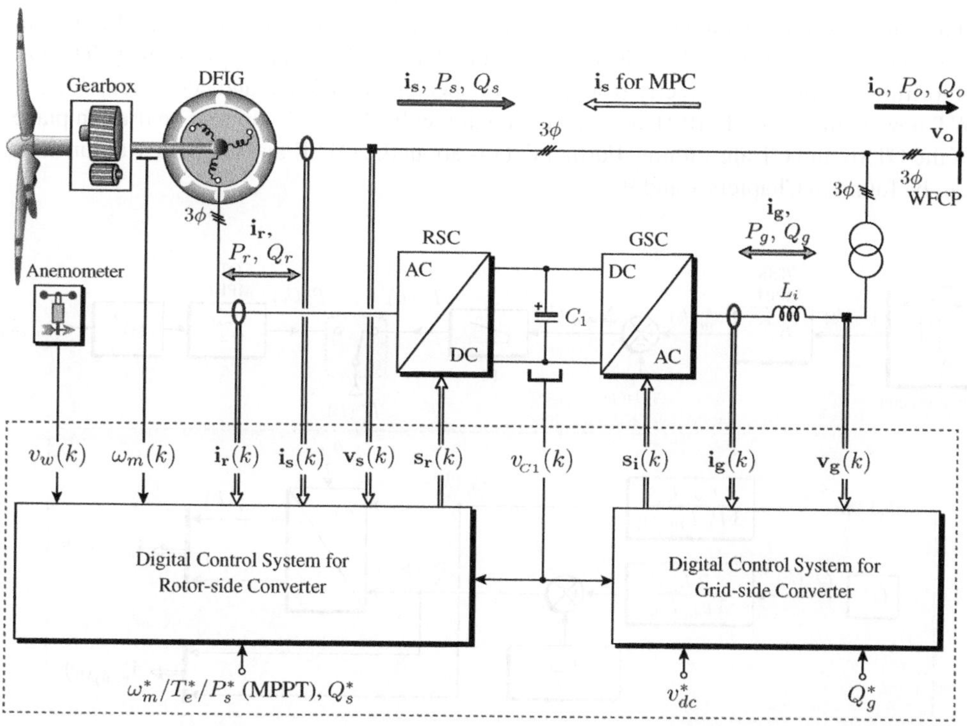

Figure 12.9 Block diagram of the digital control scheme for BTB VSC-based DFIG WECS.

The decoupling between the RSC and GSC enables the use of two independent digital control systems. The stator-, rotor-, and grid-side feedback signals, as well as the DC-link, are used by the RSC and GSC digital control systems. The stator currents are measured in the opposite direction to enable the use of motor models for generator operation. The GSC reactive power reference Q_g^* is set to zero, thus indicating that GSC is not responsible for the overall WECS PF control. The stator reactive power reference Q_s^* is defined such that the RSC controls the overall WECS reactive power Q_o. Owing to generator operation, the notation of reactive power is opposite to the notation discussed with Figure 8.6 in Chapter 8. The reference stator reactive power command (Q_s^*) can be set to zero for unity, positive for leading, and negative for the lagging PF of the WFCP.

The technical and operational requirements for DFIG WECS are fulfilled through the digital control systems for RSC and GSC. The MPPT operation and WFCP PF are controlled through the regulation of rotor currents. The net DC-bus voltage is maintained at its reference through the control of GSC-side currents. To reduce the switching losses at MW level, the RSC and GSC switching frequencies are minimized to a few kHz.

12.5.2 Calculation of Reference Control Variables

The block diagram in Figure 12.10 shows the calculation of the reference rotor currents in the $\alpha\beta$ and dq frames. In the next section, the dq-frame reference rotor currents are only conducted for PCC scheme realization. The electromagnetic torque reference T_e^* is calculated by the optimal tip-speed ratio (OTSR) or optimal torque (OT) MPPT algorithms [24–26]. With the help of the measured wind speed v_w and offline computed coefficient K_1^{op}, the OTSR MPPT algorithm calculates the reference speed ω_m^*. The measured mechanical speed is compared with the reference speed. A PI controller processes the error in rotor mechanical speed and produces T_e^* such that ω_m follows the ω_m^* trajectory. The OT algorithm computes T_e^* directly from the rotor speed feedback ω_m and coefficient K_2^{op}. A WT power curve-based MPPT or a power signal feedback MPPT can also be used in place of the OTSR and OT algorithms. Further details about the OTSR and OT MPPT algorithms can be found in Chapters 1 and 9.

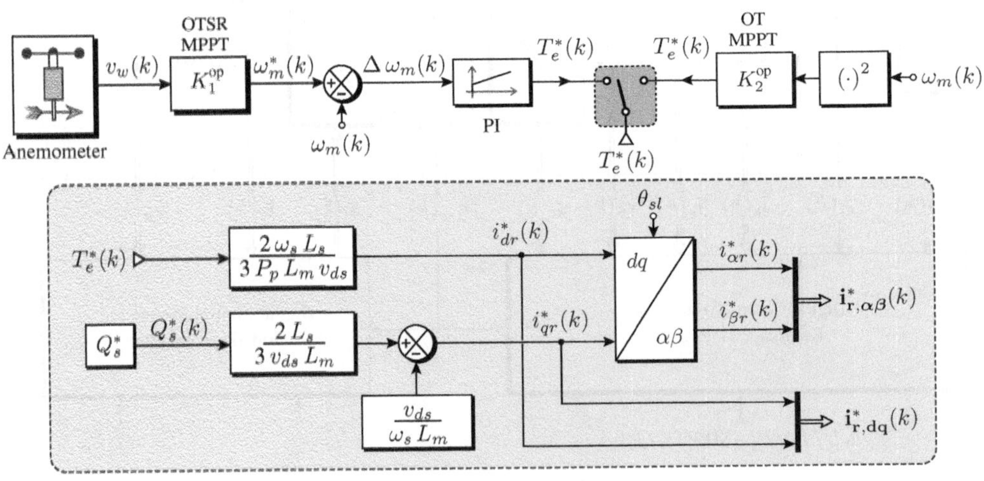

Figure 12.10 Calculation of reference control variables for PCC of DFIG WECS.

The dq-axes reference rotor currents are calculated dynamically from the electromagnetic torque reference T_e^* (or stator active reference) and stator reactive power reference Q_s^* [16]:

$$i_{dr}^*(k) = T_e^*(k) \left(\frac{2\,\omega_s\,L_s}{3\,P_p\,L_m\,v_{ds}} \right) \tag{12.36}$$

$$i_{qr}^*(k) = Q_s^*(k) \left(\frac{2\,L_s}{3\,v_{ds}\,L_m} \right) - \left(\frac{v_{ds}}{\omega_s\,L_m} \right). \tag{12.37}$$

To obtain the UPF operation at WFCP, the Q_s^* is set to zero. The corresponding q-axis reference rotor current becomes the following:

$$i_{qr}^*(k) = -\left(\frac{v_{ds}}{\omega_s\,L_m} \right) \qquad \text{for UPF operation.} \tag{12.38}$$

The dq-axes reference rotor currents are transformed to the $\alpha\beta$-frame with the help of the slip angle θ_{sl} estimated by the SVOC scheme [22]. The voltage in the rotor circuit is only a fraction of the WFCP (or DFIG stator) voltage. Therefore, the GSC in DFIG WECS operates with considerably lower grid voltages than the Type 4 WECS. The net DC-bus voltage reference for GSC in DFIG WECS is calculated as follows:

$$v_{dc}^*(k) = 3.062 \times V_g \times \frac{V_r}{V_s} = 3.062 \times V_r. \tag{12.39}$$

12.6 INDIRECT PREDICTIVE CURRENT CONTROL OF DFIG WECS

In this section, the digital control of RSC and GSC is analyzed such that a complete control of DFIG WECS is achieved. The main control objective of the RSC and GSC control schemes is to track the respective dq-frame reference currents during all operating conditions and to minimize the switch frequency.

12.6.1 Generator-side Control Scheme

The block diagram of the dq-frame PCC scheme for a DFIG with RSC is shown in Figure 12.11. With optimal gating signals for RSC, the DFIG stator active and reactive powers are controlled independently. The MPPT operation is limited to a $\pm 30\%$ speed range. The dq-axes rotor currents are utilized indirectly by the control of stator currents and rotor flux linkages. The indirect PCC scheme eliminates the need for a rotor current sensor, thus leading to low costs and low complexity for hardware configuration. This approach is similar to the PTC scheme of PMSG and SCIG WECS, where the electromagnetic torque and stator flux are regulated by the indirect control of stator currents and rotor flux linkages. The PCC scheme involves five intuitive design steps, which are outlined below.

Measurement and Synthesis of Feedback Signals: The wind speed v_w, DC-link capacitor voltage $v_{C1}(k)$, stator currents $\mathbf{i_s}(k)$, stator voltages $\mathbf{v_s}(k)$, and generator rotor mechanical speed $\omega_m(k)$ are measured to design the PCC scheme. The three-phase stator currents $\mathbf{i_s}(k)$ are measured in the opposite direction of the original flow, because the "predictive model" subsystem uses motor corresponding dynamic equations. The mechanical speed $\omega_m(k)$ is multiplied with DFIG pole pairs P_p to obtain the equivalent rotor electrical speed $\omega_r(k)$. The stator voltage angle θ_s, slip angle θ_{sl}, and stator angular frequency

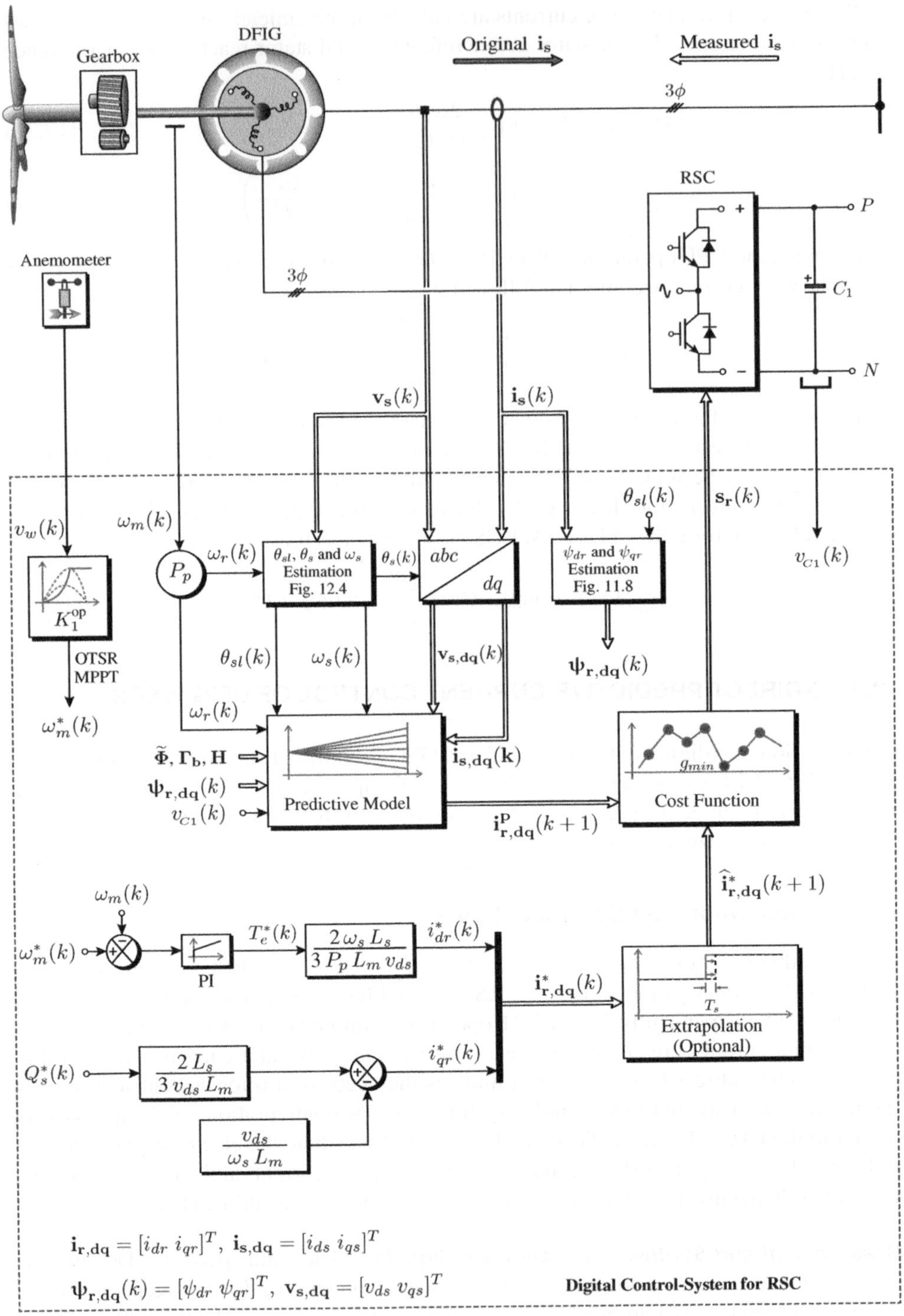

Figure 12.11 Block diagram of the indirect PCC scheme for DFIG with RSC.

ω_s are estimated by the SVOC scheme discussed in Section 12.3.2. The dq-axes rotor flux linkages are estimated, which is similar to the IFOC scheme of SCIG (Figure 11.8). The stator voltage angle θ_s is used to transform three-phase stator voltages and currents to equivalent dq-frame voltages and currents ($\mathbf{v_{s,dq}}$ and $\mathbf{i_{s,dq}}$). Owing to the stator voltage orientation, $v_{ds} = v_s$, and $v_{qs} = 0$. The slip angle θ_{sl} is used in the estimation of dq-axes rotor flux linkages ψ_{dr} and ψ_{qr}.

Calculation of Reference Currents: The dq-axes reference rotor currents are calculated by combining the OTSR MPPT, speed control loop, and SVOC scheme (Figure 12.10). The $i^*_{dr}(k)$ is calculated from the speed control loop and $i^*_{qr}(k)$ is obtained from the stator reactive power control loop.

Extrapolation of Reference Currents: To compute for the future value of rotor current error, the reference rotor currents are extrapolated from the $(k+1)$ sampling instant with the help of the first-order Lagrange extrapolation:

$$\widehat{\mathbf{i}}^*_{\mathbf{r,dq}}(k+1) = 2\,\mathbf{i}^*_{\mathbf{r,dq}}(k) - \mathbf{i}^*_{\mathbf{r,dq}}(k-1). \tag{12.40}$$

Prediction of Future Behavior of DFIG Rotor Currents: The future values of the dq-axes rotor currents are calculated by the "predictive model" block with the help of: (1) a DT model of DFIG and RSC, (2) dq-axes-estimated rotor flux linkages $\boldsymbol{\psi}_{\mathbf{r,dq}}(k)$, (3) dq-axes stator voltages and currents, $\mathbf{v_{s,dq}}(k)$ and $\mathbf{i_{s,dq}}(k)$, (4) rotor angular speed $\omega_r(k)$ and stator angular frequency $\omega_s(k)$, (5) slip angle $\theta_{sl}(k)$, (6) measured DC-link voltage $v_{C1}(k)$, and (7) all possible switching state combinations of RSC. A step-by-step procedure for the prediction of dq-axes rotor currents is given below:

- The RSC output voltages are equal to the DFIG rotor voltages v_{dr} and v_{qr} (Figure 12.11). By predicting the RSC voltages in a dq-frame, future values of stator currents and rotor flux linkages can be obtained. To illustrate this concept, the DT model in (12.30) is simplified and represented as follows:

$$\begin{bmatrix} i^p_{ds}(k+1) \\ i^p_{qs}(k+1) \\ \psi^p_{dr}(k+1) \\ \psi^p_{qr}(k+1) \end{bmatrix} = \widetilde{\mathbf{\Phi}}(k) \begin{bmatrix} i_{ds}(k) \\ i_{qs}(k) \\ \psi_{dr}(k) \\ \psi_{qr}(k) \end{bmatrix} + \mathbf{\Gamma_b} \begin{bmatrix} v_{ds}(k) \\ v_{qs}(k) \\ v^p_{dr}(k) \\ v^p_{qr}(k) \end{bmatrix} \tag{12.41}$$

  where superscript p denotes the predicted variable. To minimize the number of online calculations, $\widetilde{\mathbf{\Phi}}$ and $\mathbf{\Gamma_b}$ are defined offline by excluding the $\omega_r(k)$ and $\omega_{sl}(k)$ terms.

- The rotor voltages in the aforementioned expression are predicted with the help of dq-axes RSC switching signals and measured DC-link capacitor voltage. The model is defined by the following:

$$\begin{bmatrix} v^p_{dr}(k) \\ v^p_{qr}(k) \end{bmatrix} = v_{C1}(k) \begin{bmatrix} s^p_{dr1}(k) \\ s^p_{qr1}(k) \end{bmatrix} \tag{12.42}$$

  where the dq-axes switching signals are computed from the natural frame switching signals and slip angle $\theta_{sl}(k)$, which are shown below:

$$\begin{bmatrix} s^p_{dr1}(k) \\ s^p_{qr1}(k) \end{bmatrix} = \underbrace{\begin{bmatrix} \cos\theta_{sl}(k) & \sin\theta_{sl}(k) \\ -\sin\theta_{sl}(k) & \cos\theta_{sl}(k) \end{bmatrix}}_{\mathbf{T}_{\alpha\beta/dq}} \underbrace{\frac{2}{3}\begin{bmatrix} 1 & -\frac{1}{2} & -\frac{1}{2} \\ 0 & \frac{\sqrt{3}}{2} & -\frac{\sqrt{3}}{2} \end{bmatrix}}_{\mathbf{T}_{abc/\alpha\beta}} \begin{bmatrix} s^p_{ar1}(k) \\ s^p_{br1}(k) \\ s^p_{cr1}(k) \end{bmatrix}. \tag{12.43}$$

The natural frame switching signals in the aforementioned expression are first converted to the $\alpha\beta$-frame and then to the dq-frame by using $\theta_{sl}(k)$. This approach leads to a low number of online calculations.

- By combining Equations (12.41) and (12.42), the overall sampled data model for the prediction of future values of dq-axes stator currents and rotor flux linkages is formulated in terms of RSC switching signals as follows:

$$\begin{bmatrix} i_{ds}^{p}(k+1) \\ i_{qs}^{p}(k+1) \\ \psi_{dr}^{p}(k+1) \\ \psi_{qr}^{p}(k+1) \end{bmatrix} = \mathbf{\Phi}(k) \begin{bmatrix} i_{ds}(k) \\ i_{qs}(k) \\ \psi_{dr}(k) \\ \psi_{qr}(k) \end{bmatrix} + \mathbf{\Gamma_b} \begin{bmatrix} v_{ds}(k) \\ v_{qs}(k) \\ v_{C1}(k)\, s_{dr1}^{p}(k) \\ v_{C1}(k)\, s_{qr1}^{p}(k) \end{bmatrix}. \tag{12.44}$$

- By moving the variables in Equation (12.31) one-sample forward, the future value of rotor currents is predicted in terms of stator currents and rotor flux linkages in Equation (12.44), that is,

$$\begin{bmatrix} i_{dr}^{p}(k+1) \\ i_{qr}^{p}(k+1) \end{bmatrix} = \mathbf{H} \begin{bmatrix} i_{ds}^{p}(k+1) \\ i_{qs}^{p}(k+1) \\ \psi_{dr}^{p}(k+1) \\ \psi_{qr}^{p}(k+1) \end{bmatrix}. \tag{12.45}$$

As seen from the aforementioned prediction procedure, the rotor currents does not need to be measured.

Cost Function Minimization: The control objectives for RSC include the regulation of dq-frame rotor currents and the minimization of switching frequency. These control objectives are defined by a cost function with three sub-control terms:

$$\begin{aligned} g_r(k) = {} & \lambda_{id} \left[\widehat{i_{dr}^{*}}(k+1) - i_{dr}^{p}(k+1) \right]^2 \\ & + \lambda_{iq} \left[\widehat{i_{qr}^{*}}(k+1) - i_{qr}^{p}(k+1) \right]^2 \\ & + \lambda_{sw,r} \sum_{x=a,\,b,\,c} \left[s_{xr1}^{p}(k) - s_{xr1}^{\text{op}}(k) \right]^2 \end{aligned} \tag{12.46}$$

where λ_{id}, λ_{iq}, and $\lambda_{sw,r}$ are the weighting factors. s_{ar1}^{op}, s_{br1}^{op}, and s_{cr1}^{op} are the optimal switching signals of RSC in past sampling instant. Weighting factors λ_{id} and λ_{iq} correspond to the primary objective of the regulation of dq-frame rotor currents. Therefore, λ_{id} and λ_{iq} are set to one. The $\lambda_{sw,r}$ value is heuristically selected according to the desired switching frequency minimization.

By using the eight aforementioned combinations for switching signals, eight different predictions are obtained for i_{ds}, i_{qs}, ψ_{dr}, and ψ_{qr}, which leads to eight different future values of i_{dr} and i_{qr}. During each sampling interval, the best suited switching states for RSC are selected on the basis of the minimum cost function value, which ensures the accurate tracking of rotor currents to their respective reference currents. The selected switching states are applied to the RSC during the $(k+1)$ sampling period.

12.6.2 Grid-side Control Scheme

The cost function for GSC is analogous to the RSC cost function in Equation (12.46). The GSC cost function includes the dq-axes grid currents regulation and switching frequency minimization, which is shown below:

$$\begin{aligned} g_i(k) = {} & \lambda_{id}\left[\hat{i}^*_{dg}(k+1) - i^p_{dg}(k+1)\right]^2 + \lambda_{iq}\left[\hat{i}^*_{qg}(k+1) - i^p_{qg}(k+1)\right]^2 \\ & + \lambda_{sw,i}\sum_{x=a,\,b,\,c}\left[s^p_{xi1}(k) - s^{\mathrm{op}}_{xi1}(k)\right]^2. \end{aligned} \tag{12.47}$$

A detailed discussion on GSC control can be found in Chapter 8.

12.6.3 Control Algorithm

The flowcharts of the dq-frame PCC algorithms for RSC and GSC are shown in Figures 12.12(a) and 12.12(b), respectively. The design and digital implementation of both algorithms are identical. Both PCC algorithms consist of 10 major blocks/steps. The feedback measurements are obtained in block ①. The dq-frame extrapolated reference currents are calculated in block ②, in addition to the rotor flux linkages, stator voltage angle, slip angle, and stator angular frequency needed for RSC PCC algorithm. The PCC algorithm is initiated in block ③, and enters an iterative loop in block ④, and ends in block ⑨. The dq-frame

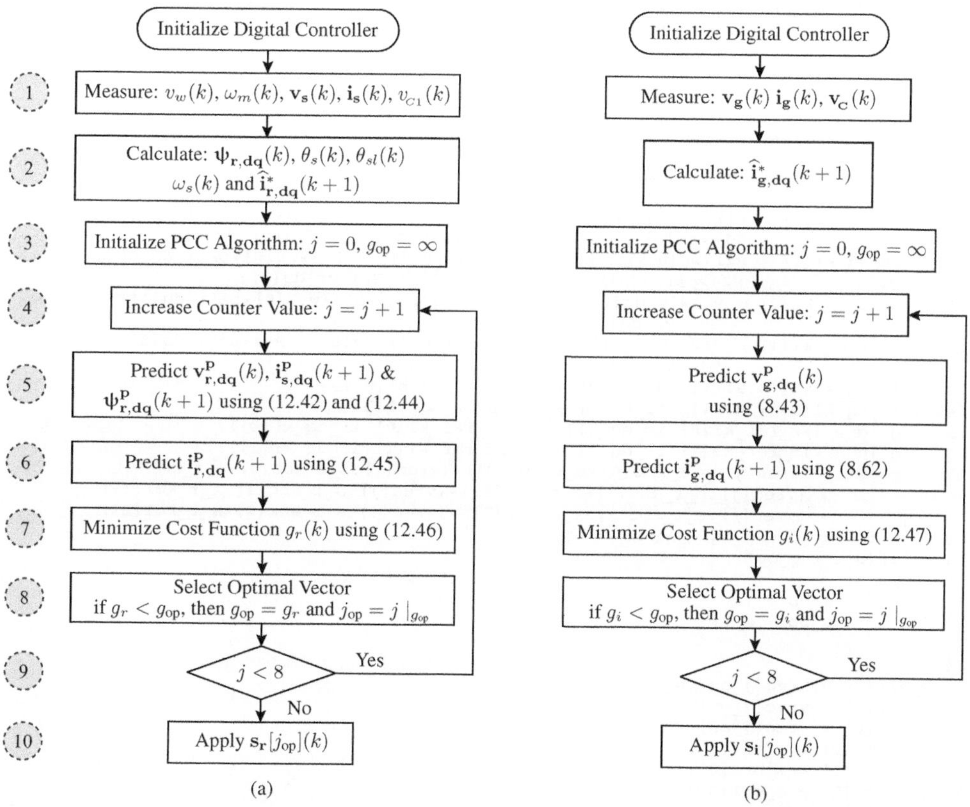

Figure 12.12 Flowchart of the dq-frame PCC algorithm for: (a) RSC and (b) GSC.

RSC and GSC output voltages are predicted in block ⑤. For the RSC algorithm, the stator currents and rotor flux linkages are additionally calculated. The rotor and grid currents are predicted with the help of block ⑥. Block ⑦ evaluates the RSC and GSC cost functions. An optimal vector j_{op} corresponding to the minimum (optimal) cost function value is stored in block ⑧. After the completion of an iterative loop for eight counter values, the switching signals that correspond to j_{op} are produced by block ⑩ and applied directly to the RSC and GSC.

S-Function Builder programming is conducted in `MATLAB` environment to realize the PCC scheme for RSC (Algorithm 12.1). The program lines are explained as follows: PCC initialization is shown in lines `01-02`; the `for` loop is initialized in lines `03-04`; RSC voltages are predicted in lines `05-08`; stator currents are predicted in lines `09-10`; rotor flux linkages are calculated in lines `11-12`; rotor currents are predicted in lines `13-14`, the cost function is defined in lines `15-18`; the optimal vector `j_op` number is selected in lines `19-23`; the `for` loop is terminated in line `24`; the optimal three-phase switching states are produced in lines `25-27` on the basis of the `j_op` value; the switching states are stored in lines `28-30` to be used in the next iteration.

ALGORITHM 12.1 Indirect PCC Algorithm for DFIG with RSC

```
\\ P11 = P22 = P33 = P44 = 1-(Ts/tau_sigma); P12 = -P21 = P34 = -P43 = Ts;
\\ P13 = P24 = (kr*Ts)/(sigma*Ls*tau_r); P14 = -P23 = (kr*Ts)/(sigma*Ls);
\\ P31 = P42 = (Lm*Ts)/tau_r; G11 = G22 = Ts/(sigma*Ls);
\\ G13 = G24 = -(kr*Ts)/(sigma*Ls); G33 = G44 = Ts;
\\ H11 = H22 = -Lm/Lr; H13 = H24 = 1/Lr;
01 j_op = 1000;
02 g_op = 1000000000;
03 for(j = 0; j < 8; j++)
04 {
05     vr_ab[0] = vc[0]*vrec[j].s_r_ab[0];
06     vr_ab[1] = vc[0]*vrec[j].s_r_ab[1];
07     vr_dq[0] = vr_ab[0]*cstheta_sl[0] + vr_ab[1]*cstheta_sl[1];
08     vr_dq[1] = -vr_ab[0]*cstheta_sl[1] + vr_ab[1]*cstheta_sl[0];
09     is_dq_k1[0] = P11*is_dq[0] + P12*ws[0]*is_dq[1] + P13*psir_dq[0]
                    + P14*wr[0]*psir_dq[1] + G11*vs_dq[0] + G13*vr_dq[0];
10     is_dq_k1[1] = P21*ws[0]*is_dq[0] + P22*is_dq[1] + P23*wr[0]*psir_dq[0]
                    + P24*psir_dq[1] + G22*vs_dq[1] + G24*vr_dq[1];
11     psir_dq_k1[0] = P31*is_dq[0] + P33*psir_dq[0] + P34*wsl[0]*psir_dq[1]
                    + G33*vr_dq[0];
12     psir_dq_k1[1] = P42*is_dq[1] + P43*wsl[0]*psir_dq[0] + P44*psir_dq[1]
                    + G44*vr_dq[1];
13     ir_dq_k1[0] = H11*is_dq_k1[0] + H13*psir_dq_k1[0];
14     ir_dq_k1[1] = H22*is_dq_k1[1] + H24*psir_dq_k1[1];
15     g_id = (ir_ref_k1[0]-ir_dq_k1[0])*(ir_ref_k1[0]-ir_dq_k1[0]);
16     g_iq = (ir_ref_k1[1]-ir_dq_k1[1])*(ir_ref_k1[1]-ir_dq_k1[1]);
17     g_swr = (vrec[j].s_r_abc[0]-s_r_km1[0])*(vrec[j].s_r_abc[0]-s_r_km1[0])
             + (vrec[j].s_r_abc[1]-s_r_km1[1])*(vrec[j].s_r_abc[1]-s_r_km1[1])
             + (vrec[j].s_r_abc[2]-s_r_km1[2])*(vrec[j].s_r_abc[2]-s_r_km1[2]);
18     g_r = g_id + g_iq + alpha[0]*g_swr;
19     if(g_r < g_op)
20         {
21             j_op = j;
22             g_op = g_r;
23         }
24 }
25 s_ar1[0] = vrec[j_op].s_r[0];
26 s_br1[0] = vrec[j_op].s_r[1];
27 s_cr1[0] = vrec[j_op].s_r[2];
28 s_r_km1[0] = s_ar1[0];
29 s_r_km1[1] = s_br1[0];
30 s_r_km1[2] = s_cr1[0];
```

CASE STUDY 12.1 Steady-State Analysis of DFIG WECS

Case Study Objective: To support the analysis of the indirect PCC scheme, the steady state performance of a DFIG WECS is analyzed during two operating conditions: (1) subsynchronous mode with 0.7-pu wind speed and (2) supersynchronous mode with 1.0-pu wind speed.

Parameters: 3.0-MW, 690-V, 60-Hz DFIG WECS parameters are given as follows: V_s = 398.4 V, V_r = 91.63 V, R_s = 1.443 mΩ, R_r = 1.125 mΩ, L_{ls} = 0.094 mH, L_{lr} = 0.085 mH, L_m = 0.802 mH, ψ_s = 1.5057 Wb (*peak*), ψ_r = 1.7576 Wb (*peak*), P_p = 2, and n_m = 2160 rpm (DFIG #2 in Table A.8). The GSC parameters, with respect rotor base values, are as follows: r_i = 0.34 mΩ (0.01 pu), L_i = 0.02 mH (0.2 pu), and v_{dc}^* = 281 V (3.062 pu). The reference stator reactive power is zero (Q_s^* = 0). The indirect PCC scheme is implemented with T_s = 100 μs.

Analysis:

Figure 12.13 shows the steady state waveforms of DFIG WECS during a subsynchronous mode of operation with v_w= 8.4 m/s (0.7 pu). The slip angle θ_{sl}, dq-axes and phase-a rotor currents, RSC phase voltage, phase-a stator voltage and current, phase-a grid (GSC) voltage and current, and phase-a output voltage and current are outlined in subplots (a) to (f), respectively. In the subsynchronous mode, the DFIG slip (s) is noted as 0.16.

The slip angle θ_{sl} monotonically increases from 0 to 2π, as explained earlier in Figure 12.6. The rotor electrical speed, mechanical input power, and electromagnetic torque with v_w = 0.7 pu are

$$\omega_r = \omega_{m,R} \times P_p \times v_{w,\,\text{pu}} = 2160 \times 2 \times (2 \times \pi/60) \times 0.7 = 316.7 \text{ rad/s} \quad (0.7 \text{ pu})$$

$$P_m = P_{m,R} \times v_{w,\,\text{pu}}^3 = -3000 \times 10^3 \times 0.7^3 = -1029 \text{ kW} \quad (0.343 \text{ pu})$$

$$T_e = \frac{P_m\,P_p}{\omega_r} = \frac{-1029 \times 10^3 \times 2}{316.7} = -6.499 \times 10^3 \text{ N.m} \quad (-0.49 \text{ pu}).$$

By using the models in Equations (12.21) and (12.23), the dq-axes rotor currents calculated from the electromagnetic torque T_e and stator reactive power Q_s are expressed as follows:

$$i_{dr} = -\frac{T_e\,2\,\omega_s\,L_s}{3\,P_p\,L_m\,v_{ds}} = 1619.5 \text{ A } (peak) \quad (0.428 \text{ pu})$$

$$i_{qr} = \left(\frac{2\,L_s}{3\,v_{ds}\,L_m}\right) Q_s - \left(\frac{1}{\omega_s\,L_m}\right) v_{ds} = -1863.4 \text{ A } (peak) \quad (-0.493 \text{ pu})$$

$$i_r = \sqrt{i_{dr}^2 + i_{qr}^2} = 2468.8 \text{ A } (peak) \quad (0.65 \text{ pu}).$$

The dq-axes rotor currents in Figure 12.13(b) match with the aforementioned theoretical analysis. The dq-axes rotor currents are well regulated at their reference values by the indirect PCC scheme with a 0.75% average tracking error. By regulating the dq-axes rotor currents, the electromagnetic torque and stator reactive power are maintained at their reference value, thus indicating that the control objectives of DFIG WECS are met. The total harmonic distortion (THD) of a phase-a rotor current is 1.41%, which is comparable to the 0.67% THD obtained with PMSG (shown in Figure 9.11(c)). The switching frequency ($f_{sw,r}$) of RSC is measured as 1960 Hz with the weighting factor $\lambda_{sw,r}$ = 0. The RSC phase voltage v_{ar} with respect to the generator neutral and its fundamental v_{ar1} is shown in Figure 12.13(c).

With the help of the model presented in Equation (12.23), the stator output power is calculated by

$$P_s = -\frac{i_{dr}\,3\,v_{ds}\,L_m}{2\,L_s} = -1225 \text{ kW} \quad (0.408 \text{ pu}).$$

Similar to PMSG and SCIG WECS, a negative sign for P_s indicates that the power extracted from wind flows from the generator to the WFCP. Given the subsynchronous mode of operation, the absolute value of stator output P_s in the above expression is higher than the absolute value of P_m. The difference between P_s and P_m corresponds to rotor power P_r, that is,

$$|P_r| = |P_s| - |P_m| = 1225 - 1029 = 196 \text{ kW} \quad (0.065 \text{ pu}).$$

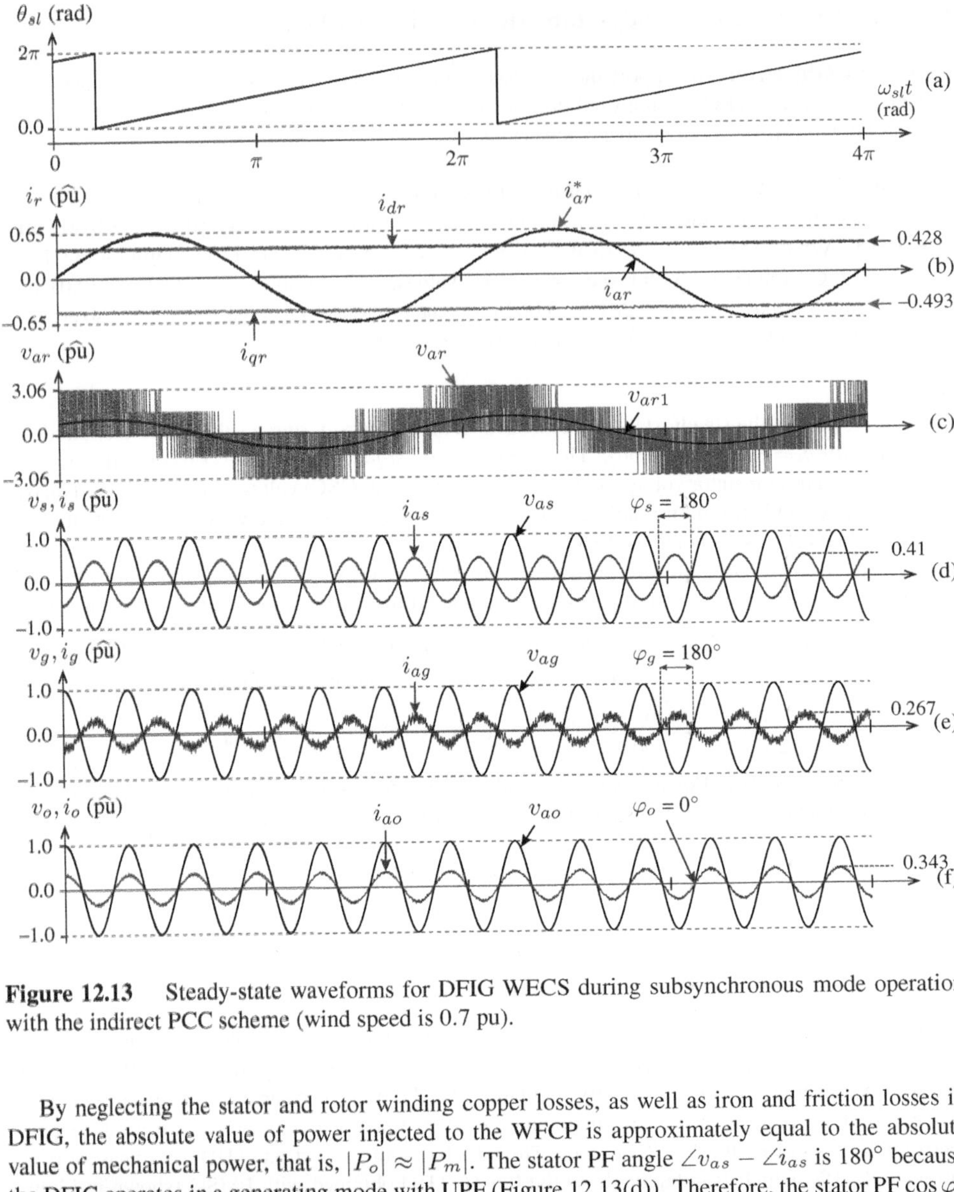

Figure 12.13 Steady-state waveforms for DFIG WECS during subsynchronous mode operation with the indirect PCC scheme (wind speed is 0.7 pu).

By neglecting the stator and rotor winding copper losses, as well as iron and friction losses in DFIG, the absolute value of power injected to the WFCP is approximately equal to the absolute value of mechanical power, that is, $|P_o| \approx |P_m|$. The stator PF angle $\angle v_{as} - \angle i_{as}$ is 180° because the DFIG operates in a generating mode with UPF (Figure 12.13(d)). Therefore, the stator PF $\cos\varphi_s$ is –1. Given the subsynchronous mode of operation, the current flows from the WFCP to the rotor circuit via the power converter. Therefore, the grid (GSC) currents are out of phase with the grid voltages, that is, $\angle v_{ag} - \angle i_{ag}$ is 180° (Figure 12.13(e)). The WFCP (WT output) voltage and current are in phase because of the UPF operation; therefore, $\angle v_{ao} - \angle i_{ao}$ is 0° and $\cos\varphi_o$ is 1 (Figure 12.13(f)). With the aforementioned computed powers and UPF, the stator, grid (GSC), and overall WT output currents become the following:

$$|i_s| = \sqrt{2} \times |I_s| = \frac{|P_s|}{3\,V_s} = \sqrt{2} \times \frac{1225 \times 10^3}{3 \times 398.4} = 1450 \text{ A } (peak) \quad (0.41\ \widehat{\text{pu}})$$

$$|i_g| = \sqrt{2} \times |I_g| = \frac{|P_r|}{3\,V_g} = \sqrt{2} \times \frac{196 \times 10^3}{3 \times 91.63} = 1008 \text{ A } (peak) \quad (0.267\ \widehat{\text{pu}})$$

$$|i_o| = \sqrt{2} \times |I_o| = \frac{|P_o|}{3\,V_o} = \sqrt{2} \times \frac{1209 \times 10^3}{3 \times 398.4} = 1218 \text{ A } (peak) \quad (0.343\ \widehat{\text{pu}}).$$

From the analysis, the pu value of i_o is equal to the pu value of P_m (0.343 pu). The PCC scheme for GSC ensures that the grid currents are thoroughly tracked to their references with an average tracking error of 6.18%. The GSC operates with an average switching frequency $f_{sw,i}$ of 1770 Hz. The THD of i_s and i_o is noted as 2.1% and 6.4%, respectively.

The aforementioned analysis is repeated with the rated wind speed (v_w = 1.0 pu) and the corresponding simulation results are given in Figure 12.14. The DFIG operates in a supersynchronous mode with a –0.2 slip. The rotor, stator, GSC, and output power and currents are calculated in a similar manner and are compared with the rated values in Table A.8. During supersynchronous mode, the rotor power expression is modified as $|P_r| = |P_m| - |P_s|$. The $f_{sw,r}$ = 1725 Hz, $f_{sw,i}$ = 1690 Hz, THD of i_{ar} = 0.95%, THD of i_{as} = 1.07%, THD of i_{ao} = 2.39%, and the tracking error of rotor and grid currents is 0.75% and 6.5%, respectively.

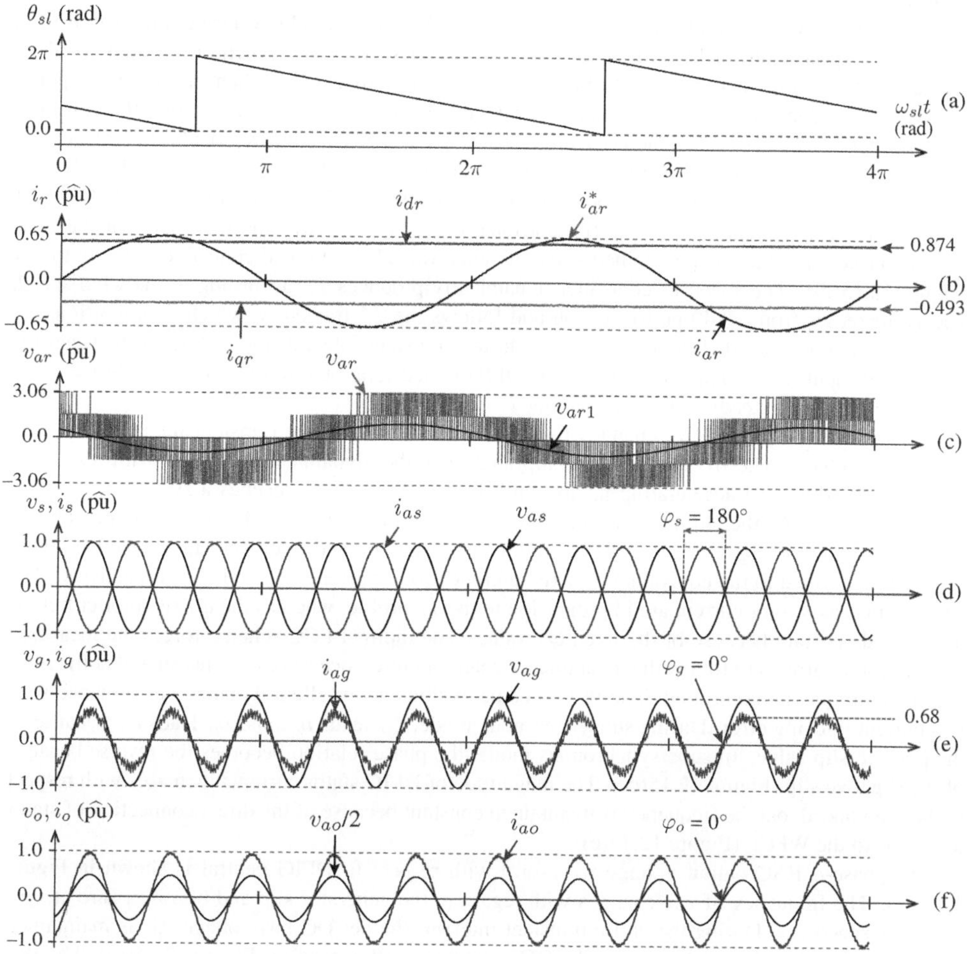

Figure 12.14 Steady-state waveforms for DFIG WECS during supersynchronous mode operation with the indirect PCC scheme (wind speed is 1.0 pu).

The numerically calculated stator-, rotor-, GSC-, and output-side variables match with the provided MATLAB simulation results. During the subsynchronous and supersynchronous modes of operation, the rotor and grid currents follow their respective references such that the overall control objectives of DFIG WECS are met. ■

CASE STUDY 12.2 Transient Analysis of DFIG WECS

Case Study Objective: As an extension to the previous case study, the dynamic performance of DFIG WECS during a transition from the subsynchronous mode to the supersynchronous mode is investigated in this section. A ramp change in wind speed is applied to increase the rotor speed. The DFIG stator-, rotor-, grid-, and WT output-side waveforms, as well as the DC link, are studied with respect to the varying wind speed.

Parameters: The 3.0-MW, 690-V, 60-Hz DFIG WECS parameters are similar to those in case study 12.1. To reduce the simulation runtime, the original moment of inertia J_m is changed to 10 kg.m^2.

Analysis:
Figure 12.15 shows the simulated waveforms for a 3.0-MW DFIG WECS during a transition from 0.7-pu rotor speed (subsynchronous mode) to 1.0-pu rotor speed (supersynchronous mode). Given the large moments of inertia in the turbine and generator shafts, the mechanical system response in MW WTs is considerably slower than the grid-side electrical response. Therefore, the generator speed cannot be changed instantaneously, although the wind speed changes suddenly. For this reason, the wind speed v_w profile is considered to gradually change from 8.4 m/s (0.7 pu) to 12 m/s (1.0) from t = 0.2 s to 1.2 s, such that the rotor speed increases from 0.7 pu to 1.0 pu. The OTSR MPPT produces a linear reference speed ω_m^* (proportional to v_w) for the RSC indirect PCC scheme. The measured rotor speed ω_m is maintained at its reference value by the speed control loop PI controller (Figure 12.15(a)). The speed PI controller dynamically produces T_e^* according to the wind speed. The T_e increases from –0.49 pu to –1.0 pu and follows the T_e^* trajectory, which is proportional to the square of rotor speed (Figure 12.15(b)). During transient interval, the difference between the mechanical input torque T_m and T_e helps the DFIG to accelerate the rotor speed ω_m. At t = 1.2 s, the system reaches a steady state, thus making T_e equal to T_m.

The DFIG slip dynamically changes from 0.16 to –0.2 during the transient interval and reach zero at t = 0.63 s. The frequency of the slip angle θ_{sl} also dynamically changes with respect to ω_m. The θ_{sl} shows an accelerating nature until the rotor speed ω_m reaches a synchronous speed w_s at t = 0.63 s. As the slip becomes negative, the θ_{sl} profile becomes the reverse and exhibits a decelerating nature (Figure 12.15(c)). The reference torque T_e^* and reference stator reactive power Q_s^* are transformed to the equivalent dq-axis reference rotor currents i_{dr}^* and i_{qr}^*, respectively. The changes in d-axis rotor current are in proportion to the T_e profile, whereas the q-axis rotor current is maintained constant because of the fixed Q_s^* value. The indirect PCC scheme forces the measured dq-axes rotor currents to follow the dynamically changing reference rotor currents such that the rotor speed and torque are controlled. The PCC scheme exhibits an excellent dynamic performance with less current tracking error. During subsynchronous mode, i_{ar} leads i_{br}, and i_{br} leads i_{cr} because of the positive slip value. In supersynchronous mode, the phase relation becomes the reverse because of the negative slip (Figure 12.15(d)). The magnitude of DFIG stator currents increases with respect to the rotor speed, but the frequency is maintained constant because of the direct connection of stator terminals to the WFCP (Figure 12.15(e)).

The phase-a RSC output voltage waveform with respect to DFIG neutral is shown in Figure 12.15(f). The frequency of v_{ar} changes with regard to the generator slip and becomes zero at synchronous speed ω_s. During the entire transient interval, the net DC-bus voltage v_{dc} is maintained constant at the v_{dc}^* value of 3.06 pu by the DC control loop PI controller (Figure 12.15(g)). The WT output active power P_o follows the ω_m spectrum and the WT output reactive power Q_o is maintained zero by the indirect PCC scheme employed for RSC (Figure 12.15(i)). In PMSG and SCIG WECS, the GSC controls the WECS reactive power but the RSC influences the reactive power in DFIG WECS. The WT output current magnitude increases in proportion to the active power extracted from wind. The output d-axis current i_{do} follows the envelop of three-phase output currents, whereas the q-axis output current i_{qo} is maintained zero because of the UPF operation (Figure 12.15(h)). This case study demonstrates the effectiveness of the indirect PCC scheme in achieving a high dynamic performance operation for DFIG WECS. ■

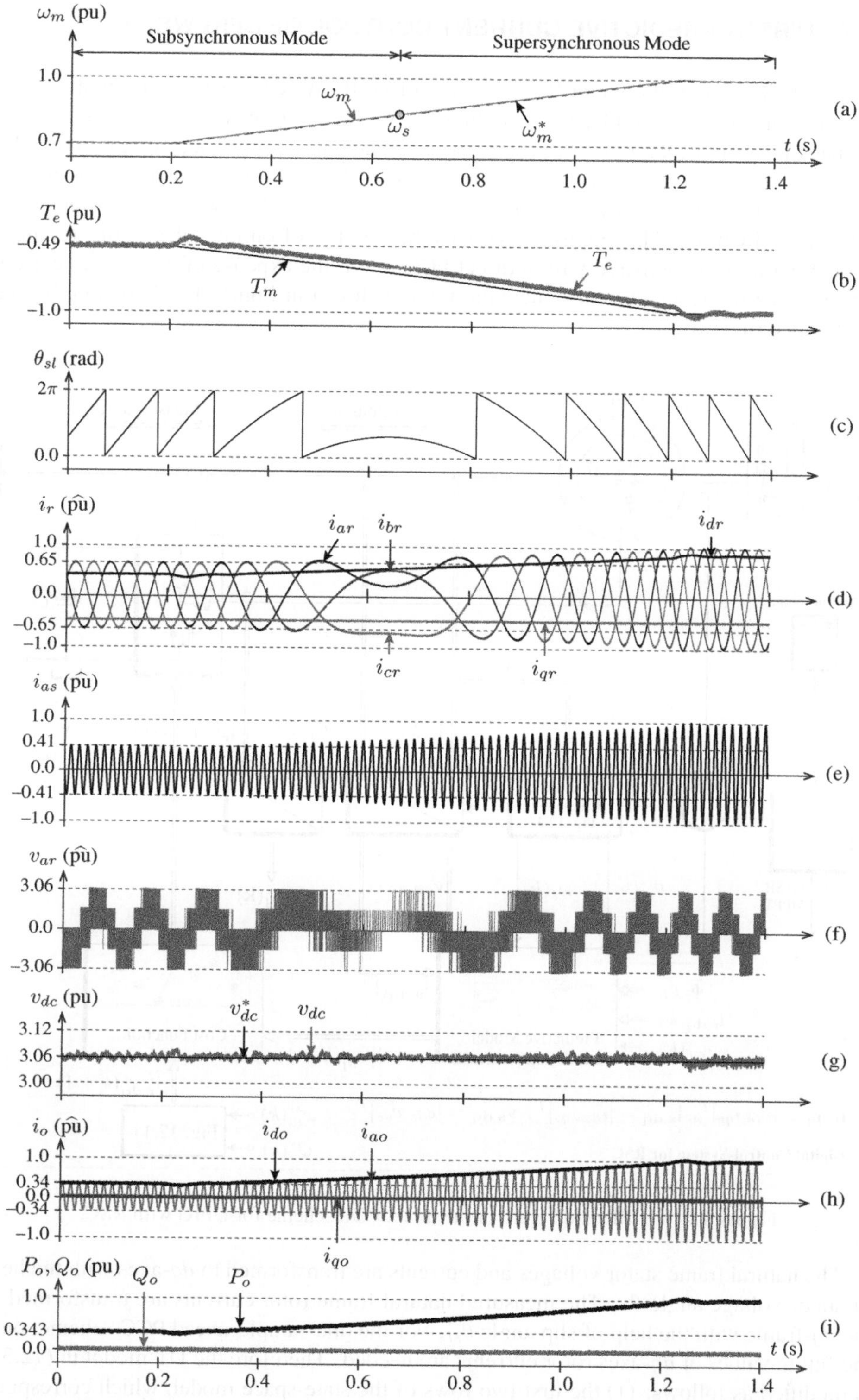

Figure 12.15 Transient waveforms for DFIG WECS during transition from subsynchronous to supersynchronous mode of operation.

12.7 DIRECT PREDICTIVE CURRENT CONTROL OF DFIG WECS

The block diagram of the direct PCC scheme for DFIG WECS is shown in Figure 12.16. This control scheme is similar to the indirect PCC discussed in Section 12.6 in terms of feedback measurements, reference rotor currents calculation, reference current extrapolation, and cost function design, except that the direct PCC requires feedback rotor current measurement and does not need additional calculations for the rotor flux linkages (ψ_{dr} and ψ_{qr}) estimation. This approach reduces the number of online calculations (computational burden) as compared with indirect PCC but at the expense of three extra current sensors. The reference currents calculation is not shown in Figure 12.16 to simplify the block diagram.

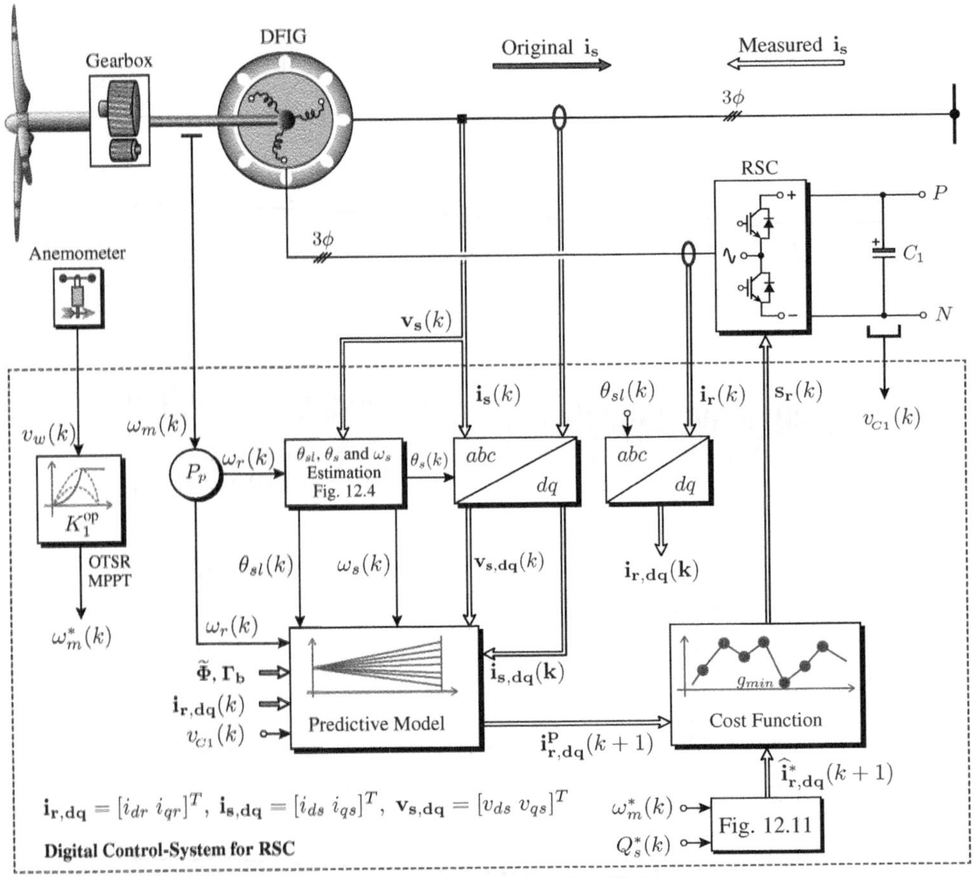

Figure 12.16 Block diagram of the direct PCC scheme for DFIG with RSC.

The natural frame stator voltages and currents are transformed to dq-axes with the help of stator voltage angle θ_s. The measured natural frame rotor currents are transformed to the dq-frame with the help of slip angle θ_{sl}. For the one-sample-ahead PCC scheme, only the future values of dq-axes rotor currents are needed. Therefore the DT model in (12.33) is modified as follows: (1) the first two rows of the state-space model, which correspond to the stator current prediction, are omitted; (2) rotor voltages v_{dr} and v_{qr} are represented in terms of dq-axes switching signals similar to Equation (12.42). The resultant sampled data model for the direct PCC scheme is obtained by the following:

$$\begin{bmatrix} i^p_{dr}(k+1) \\ i^p_{qr}(k+1) \end{bmatrix} = \underbrace{\begin{bmatrix} \phi_{31} & \phi_{32}(k) & \phi_{33} & \phi_{34}(k) \\ \phi_{41}(k) & \phi_{42} & \phi_{43}(k) & \phi_{44} \end{bmatrix}}_{\mathbf{\Phi}(k)} \begin{bmatrix} i_{ds}(k) \\ i_{qs}(k) \\ i_{dr}(k) \\ i_{qr}(k) \end{bmatrix} + \underbrace{\begin{bmatrix} \gamma_{31} & \gamma_{32} & \gamma_{33} & \gamma_{34} \\ \gamma_{41} & \gamma_{42} & \gamma_{43} & \gamma_{44} \end{bmatrix}}_{\mathbf{\Gamma_b}} \begin{bmatrix} v_{ds}(k) \\ v_{qs}(k) \\ v_{C1}(k)\, s^p_{dr1}(k) \\ v_{C1}(k)\, s^p_{qr1}(k) \end{bmatrix} \quad (12.48)$$

where $\mathbf{\Phi}(k)$ and $\mathbf{\Gamma_b}$ correspond to the modified DT matrices. ϕ_{31} to ϕ_{44} are the third- and fourth-row DT matrix elements of $\mathbf{\Phi}(k)$ in (12.34). Similarly, γ_{31} to γ_{44} are the third- and fourth-row DT matrix elements of $\mathbf{\Gamma_b}$ in (12.34). The predicted rotor currents are compared with the reference rotor currents by using a cost similar to the one defined with indirect PCC (Equation (12.46)). The switching state combination, which minimizes the cost function, is selected and applied to the RSC during $(k+1)$ sampling instant.

The flowcharts of the *dq*-frame direct PCC algorithms for RSC and GSC are shown in Figures 12.17(a) and 12.17(b), respectively. After the digital controller initialization and calculation of reference rotor and grid currents, the PCC algorithm enters an iterative

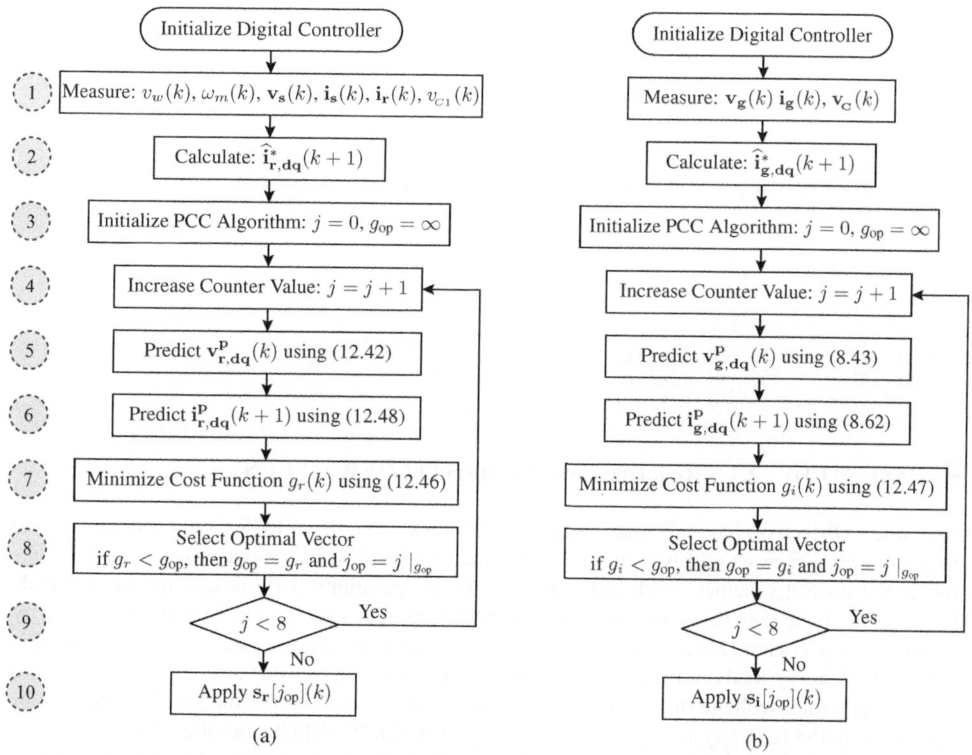

Figure 12.17 Flowchart of the *dq*-frame PCC algorithm for: (a) RSC and (b) GSC.

loop where the rotor and grid currents are predicted and the cost function minimization is performed. An optimal vector j_{op} and its corresponding switching signals that minimize the cost function are selected and directly applied to the RSC and GSC.

Algorithm 12.2 shows the MATLAB *S-Function Builder* programming for the *dq*-frame direct PCC for DFIG with RSC. The external function declarations for *S-Function* are the same for the indirect PCC algorithm. The description of the program lines is given as follows: the PCC is initialized in lines `01-02`; the `for` loop is initialized in lines `03-04`; RSC voltages are predicted in lines `05-08`; rotor currents are predicted in lines `09-10`; the cost function is defined in lines `11-13`; the optimal vector `j_op` number is selected in lines `14-18`; the `for` loop is terminated in Line `19`; optimal three-phase switching states are produced in lines `20-22` based on the `j_op` value. The indirect PCC program in lines `09-14` in Algorithm 12.1 are replaced by lines `09-10` in this section. As demonstrated through this programming, the direct PCC involves less number of online calculations and computational burden than the indirect PCC.

ALGORITHM 12.2 Direct PCC Algorithm for DFIG with RSC

```
\\ P31 = P42 = Rs*Lm*Ts; P32 = -P41 = Ls*Lm*Ts; P33 = P44 = 1-(Rr*Ls*Ts);
\\ P34 = -P43 = -Lr*Ls*Ts; k_phi = ws*sigma*Ls*Lr*Ts;
\\ G31 = G42 = -Lm*Ts; G33 = G44 = Ls*Ts;
01 j_op = 1000;
02 g_op = 1000000000;
03 for(j = 0; j < 8; j++)
04 {
05     vr_ab[0] = vc[0]*vrec[j].s_r_ab[0];
06     vr_ab[1] = vc[0]*vrec[j].s_r_ab[1];
07     vr_dq[0] = vr_ab[0]*cstheta_sl[0] + vr_ab[1]*cstheta_sl[1];
08     vr_dq[1] = -vr_ab[0]*cstheta_sl[1] + vr_ab[1]*cstheta_sl[0];
09     ir_dq_k1[0] = P31*is_dq[0] + P32*wr[0]*is_dq[1] + P33*ir_dq[0]
                   + P34*wr[0]*ir_dq[1] + k_phi*ir_dq[1]
                   + G31*vs_dq[0] + G33*vr_dq[0];
10     ir_dq_k1[1] = P41*wr[0]*is_dq[0] + P42*is_dq[1] + P43*wr[0]*ir_dq[0]
                   - k_phi*ir_dq[0] + P44*ir_dq[1]
                   + G42*vs_dq[1] + G44*vr_dq[1];
11     g_id = (ir_ref_k1[0]-ir_dq_k1[0])*(ir_ref_k1[0]-ir_dq_k1[0]);
12     g_iq = (ir_ref_k1[1]-ir_dq_k1[1])*(ir_ref_k1[1]-ir_dq_k1[1]);
13     g_r = g_id + g_iq;
14     if(g_r < g_op)
15         {
16             j_op = j;
17             g_op = g_r;
18         }
19 }
20 s_ar1[0] = vrec[j_op].s_r[0];
21 s_br1[0] = vrec[j_op].s_r[1];
22 s_cr1[0] = vrec[j_op].s_r[2];
```

CASE STUDY 12.3 Reactive Power Control of DFIG WECS

Case Study Objective: The grid code requirements indicate that the WECS should support the grid reactive power whenever requested by the grid operator regardless of the power conversion configuration and control schemes employed. During normal operations, UPF is maintained at WFCP. During grid voltage sags (low-voltage ride-through), a leading reactive power should be provided by the wind energy system [27, 28]. The voltage sags are caused by short circuit faults and the sudden connection of heavy loads in the power system. The lagging reactive power is needed during grid voltage swells (high-voltage ride-through), which are mainly caused by the sudden disconnection of heavy loads in the power system [1, 29]. The ability of the DFIG WECS and direct PCC scheme in supplying different reactive powers to the utility is analyzed in this case study. The dynamic performance of DFIG WECS is studied with step changes in reactive power reference.

Parameters: The DFIG WECS parameters in case study 12.1 are considered in this section. The DFIG is assumed to be operating in a steady state with 0.7-pu rotor speed (subsynchronous mode). The rated rotor current $i_{r,R}$ is 3780 A (*peak*).

Analysis:
The maximum stator reactive power reference Q_s^* limit, which can be applied to the RSC digital control system, depends on the operating condition. The q-axis rotor current i_{qr} corresponding to Q_s^* should be selected such that the rotor currents never exceed their rated value. For example, the d-axis rotor current i_{dr} is noted as 1619.5 A (*peak*) for a 0.7 pu rotor speed, as shown in case study 12.1. During this operating condition, the maximum value of i_{qr} which can be supplied by the RSC, is expressed as follows:

$$i_{qr}^{\max} = \pm\sqrt{i_{r,R}^2 - i_{dr}^2} = \sqrt{3780^2 - 1619.5^2} = \pm\,3415.5 \text{ A } (\textit{peak}) \quad (\pm\,0.904 \;\widehat{\text{pu}}).$$

In the DFIG operation, the q-axis rotor current is always negative. Therefore, the positive value in the aforementioned expression can be omitted. Thus, the q-axis rotor current is –3415.5 A (*peak*). The maximum Q_s^* which corresponds to the maximum i_{qr} is given by the following:

$$Q{}^{*\max}_{s} = \pm\left(i_{qr}^{\max} + \frac{v_{ds}}{\omega_s L_m}\right)\frac{3\,v_{ds}\,L_m}{2\,L_s} = \pm\,1174 \text{ kVAR} \quad (\pm\,0.391 \text{ pu}).$$

The aforementioned analysis indicates that the Q_s^* value should never exceed $\pm\,0.391$ pu when the DFIG operates at 0.7 pu. A similar analysis can be conducted with other rotor speed values. The simulated waveforms for a 3.0 MW DFIG WECS during a step change in Q_s^* from 0 pu to –0.391 pu are shown in Figure 12.18. The corresponding q-axis reference rotor current is calculated by the following:

$$\begin{aligned} i_{qr}^* &= \left(\frac{2\,L_s}{3\,v_{ds}\,L_m}\right)Q_s^* - \left(\frac{v_{ds}}{\omega_s\,L_m}\right) \\ &= -1863.4 \text{ A } (\textit{peak}) \quad (-0.493\;\widehat{\text{pu}}) \qquad \text{for } Q_s^* = 0 \text{ pu} \\ &= -3415.5 \text{ A } (\textit{peak}) \quad (-0.904\;\widehat{\text{pu}}) \qquad \text{for } Q_s^* = -0.391 \text{ pu.} \end{aligned}$$

From $\omega_{sl}t = 0$ to 2π, the stator reactive power reference Q_s^* is set to zero. During this condition, the DFIG WECS variables are noted as $i_{dr} = 0.428\;\widehat{\text{pu}}$, $i_{qr} = –0.493\;\widehat{\text{pu}}$, $i_r = 0.65\;\widehat{\text{pu}}$, $i_o = 0.343\;\widehat{\text{pu}}$, $P_s = 0.408$ pu, $P_o = 0.343$ pu, and $Q_o = 0$ pu (refer to case study 12.1 for details). At $\omega_{sl}t = 2\pi$, a step change in Q_s^* from 0 pu to –0.391 pu is applied. The q-axis rotor current tracks to its reference quickly without affecting the d-axis current control loop. The P_s and P_o are maintained at constant with their respective values. The WECS variables from $\omega_{sl}t = 2\pi$ to 4π are calculated as follows:

$$\begin{aligned} i_r &= \sqrt{i_{dr}^2 + i_{qr}^2} = \sqrt{1619.5^2 + (-3415.5)^2} = 3780 \text{ A } (\textit{peak}) \quad (1.0\;\widehat{\text{pu}}) \\ \varphi_s &= \tan^{-1}\frac{Q_s}{P_s} = \tan^{-1}\frac{-0.391}{0.408} = -43.8^\circ \\ \text{PFcos}\,\varphi_s &= \cos\,(-43.8^\circ) = 0.72 \text{ (lagging)} \\ Q_{o,\,\text{pu}} &= P_{o,\,\text{pu}} \tan\varphi_o = 0.343 \times \cos\,(43.8)^\circ = 0.33 \text{ pu.} \end{aligned}$$

This analysis implies that the output current i_{ao} lags the output voltage v_{ao} by 43.8°. In the preceding calculation, the sign of Q_o is opposite to the sign of Q_s because of the generator convention (stator currents are measured in opposite direction). Thus, a negative Q_s value produces a lagging PF at the WFCP. This is in opposite to the notation employed for GSC (refer to Figure 8.6). With $Q_s^* = –0.391$ pu, the direct PCC scheme-related parameters are observed as follows: $f_{sw,r}$ = 1875 Hz, $f_{sw,i}$ = 1740 Hz, THD of i_{ar} = 0.925%, THD of i_{as} = 1.62%, and THD of i_{ao} = 4.22%. The average tracking error of rotor and grid currents is 0.78% and 6.13%, respectively.

The performance of DFIG WECS during a step change in Q_s^* from 0 pu to 0.391 pu is analyzed in a similar way and the corresponding simulation results are given in Figure 12.19. The WECS

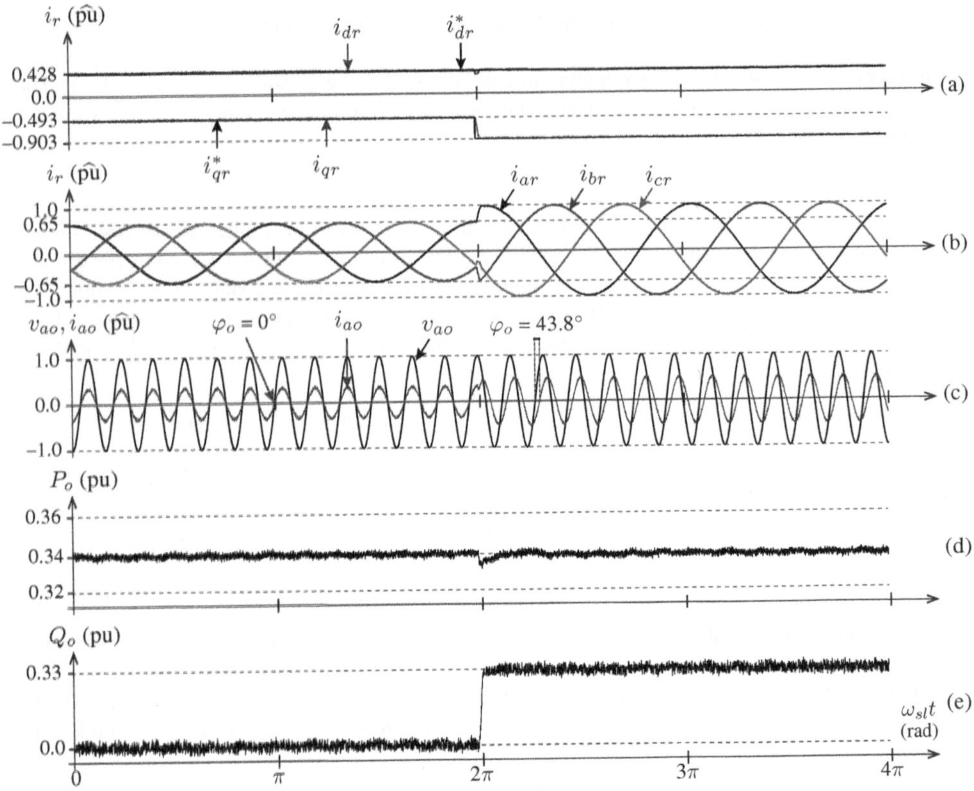

Figure 12.18 Dynamic response of the direct PCC scheme-driven DFIG WECS during a step change in Q_s^* from 0 pu to −0.391 pu.

variables corresponding to this new operating condition are calculated as follows:

$$i^*_{qr} = \left(\frac{2\,L_s}{3\,v_{ds}\,L_m}\right) Q_s^* - \left(\frac{v_{ds}}{\omega_s\,L_m}\right) = -312.6 \text{ A } (peak) \quad (-0.08\ \widehat{\text{pu}})$$

$$i_r = \sqrt{i^2_{dr} + i^2_{qr}} = \sqrt{1619.5^2 + (-312.6)^2} = 1650 \text{ A } (peak) \quad (0.44\ \widehat{\text{pu}})$$

$$\varphi_s = \tan^{-1}\frac{Q_s}{P_s} = \tan^{-1}\frac{0.391}{0.408} = 43.8°$$

$$\text{PFcos}\,\varphi_s = \cos(43.8°) = 0.72 \text{ (leading)}$$

$$Q_{o,\,\text{pu}} = P_{o,\,\text{pu}} \tan\varphi_o = -0.343 \times \cos(43.8)° = -0.33 \text{ pu.}$$

The i_{ao} leads the v_{ao} by 43.8°. With Q_s^* = 0.391 pu, the PCC scheme-related parameters are noted as $f_{sw,r}$ = 1985 Hz, $f_{sw,i}$ = 1755 Hz, THD of i_{ar} = 2.17%, THD of i_{as} = 1.62%, and THD of i_{ao} = 4.25%. The tracking error of rotor and grid currents is 0.76% and 5.89%, respectively.

The operating condition in Figure 12.18 with a negative Q_s^* is useful for supplying the lagging PF to WFCP during grid voltage swells. A positive Q_s^* value, which is similar to that in Figure 12.19, is used in supplying the leading PF to WFCP during grid voltage sags. During a step change in Q_s^*, only the magnitude change is noticed with three-phase rotor currents; however, both the magnitude and phase angle change are noticed for output currents. The rotor and output currents track their respective reference currents with fast dynamic responses and without any overshoot. This case study verifies that the direct PCC scheme and DFIG WECS offer a high control performance to meet the grid code requirements. ■

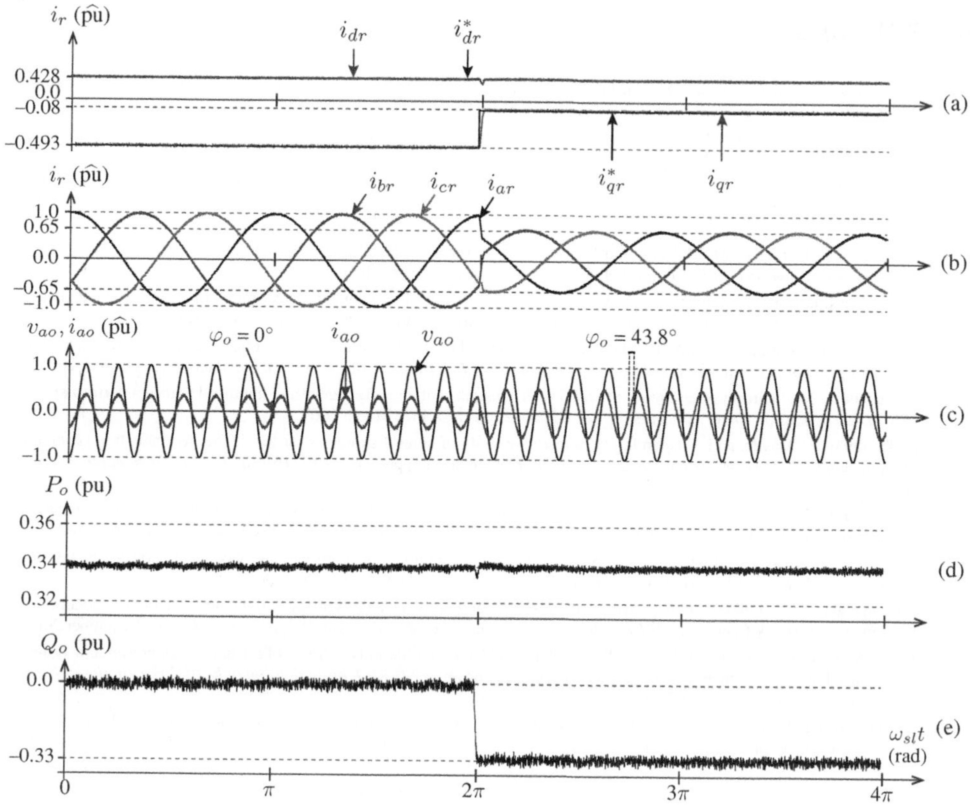

Figure 12.19 Dynamic response of the direct PCC scheme-driven DFIG WECS during a step change in Q_s^* from 0 pu to 0.391 pu.

12.8 CONCLUDING REMARKS

In this chapter, the modeling and control of BTB VSC-based DFIG WECS is presented as an extension to the SCIG WECS analysis presented in Chapter 11. The control of DFIG by FOC with SRF fixed to a stator flux or voltage angle is analyzed in detail. The equivalent circuit of DFIG, and the correlation between rotor currents and stator active and reactive powers are established. The indirect and direct rotor current dynamic models are presented in CT and DT frames. The digital control system development along with the calculation of SRF reference rotor currents is discussed to assist the design of the PCC scheme for DFIG WECS. The indirect and direct PCC schemes are analyzed with detailed design steps, flowcharts, MATLAB *S-Function Builder* programming, and case studies. The indirect PCC uses SCIG models to indirectly predict the future behavior of rotor currents without the need for feedback rotor currents. The direct PCC is simple and straightforward for digital implementation; however, it requires additional current sensors to measure the feedback rotor currents. The performance of DFIG WECS with PCC schemes is analyzed in detail during (1) the subsynchronous and supersynchronous modes of operation, (2) transition from subsynchronous to supersynchronous mode, and (3) unity, leading, and lagging PF operation. The observations indicate that MPC is an attractive alternative to the classical SVOC-FOC scheme. The discrete and nonlinear nature of VSCs is naturally handled by the MPC, thus leading to an excellent steady state and fast dynamic response.

REFERENCES

1. V. Yaramasu, B. Wu, P. C. Sen, S. Kouro, and M. Narimani, "High-power wind energy conversion systems: State-of-the-art and emerging technologies," *Proceedings of the IEEE*, vol. 103, no. 5, pp. 740–788, May 2015.
2. S. Muller, M. Deicke, and R. De Doncker, "Doubly fed induction generator systems for wind turbines," *IEEE Industry Applications Magazine*, vol. 8, no. 3, pp. 26–33, May 2002.
3. R. Datta and V. T. Ranganathan, "Variable-speed wind power generation using doubly fed wound rotor induction machine-a comparison with alternative schemes," *IEEE Transactions on Energy Conversion*, vol. 17, no. 3, pp. 414–421, September 2002.
4. M. Liserre, R. Cardenas, M. Molinas, and J. Rodríguez, "Overview of multi-MW wind turbines and wind parks," *IEEE Transactions on Industrial Electronics*, vol. 58, no. 4, pp. 1081–1095, April 2011.
5. R. Cardenas, R. Pena, S. Alepuz, and G. Asher, "Overview of control systems for the operation of DFIGs in wind energy applications," *IEEE Transactions on Industrial Electronics*, vol. 60, no. 7, pp. 2776–2798, July 2013.
6. J. Arbi, M.-B. Ghorbal, I. Slama-Belkhodja, and L. Charaabi, "Direct virtual torque control for doubly fed induction generator grid connection," *IEEE Transactions on Industrial Electronics*, vol. 56, no. 10, pp. 4163–4173, October 2009.
7. D. Zhi and L. Xu, "Direct power control of DFIG with constant switching frequency and improved transient performance," *IEEE Transactions on Energy Conversion*, vol. 22, no. 1, pp. 110–118, March 2007.
8. E. Tremblay, S. Atayde, and A. Chandra, "Comparative study of control strategies for the doubly fed induction generator in wind energy conversion systems: A DSP-based implementation approach," *IEEE Transactions on Sustainable Energy*, vol. 2, no. 3, pp. 288–299, July 2011.
9. R. Pena, J. Clare, and G. Asher, "Doubly fed induction generator using back-to-back PWM converters and its application to variable-speed wind-energy generation," *IET Electric Power Applications*, vol. 143, no. 3, pp. 231–241, May 1996.
10. B. H. Chowdhury and S. Chellapilla, "Double-fed induction generator control for variable speed wind power generation," *International Journal of Electric Power Systems Research*, vol. 76, no. 9-10, pp. 786–800, 2006.
11. J. Ekanayake and N. Jenkins, "Comparison of the response of doubly fed and fixed-speed induction generator wind turbines to changes in network frequency," *IEEE Transactions on Energy Conversion*, vol. 19, no. 4, pp. 800–802, December 2004.
12. R. Cardenas, R. Pena, G. Tobar, J. Clare, P. Wheeler, and G. Asher, "Stability analysis of a wind energy conversion system based on a doubly fed induction generator fed by a matrix converter," *IEEE Transactions on Industrial Electronics*, vol. 56, no. 10, pp. 4194–4206, October 2009.
13. R. Pena, R. Cardenas, E. Reyes, J. Clare, and P. Wheeler, "A topology for multiple generation system with doubly fed induction machines and indirect matrix converter," *IEEE Transactions on Industrial Electronics*, vol. 56, no. 10, pp. 4181–4193, October 2009.
14. L. Xu, D. Zhi, and B. Williams, "Predictive current control of doubly fed induction generators," *IEEE Transactions on Industrial Electronics*, vol. 56, no. 10, pp. 4143–4153, October 2009.
15. R. Teodorescu, M. Liserre, and P. Rodriguez, *Grid Converters for Photovoltaic and Wind Power Systems*. Chichester, UK: Wiley-IEEE Press, January 2011.
16. B. Wu, Y. Lang, N. Zargari, and S. Kouro, *Power Conversion and Control of Wind Energy Systems*, 1st ed., ser. IEEE Press Series on Power Engineering. Hoboken, NJ: Wiley-IEEE Press, July 2011.
17. L. Xu and P. Cartwright, "Direct active and reactive power control of DFIG for wind energy generation," *IEEE Transactions on Energy Conversion*, vol. 21, no. 3, pp. 750–758, September 2006.
18. Y. Tang and L. Xu, "A flexible active and reactive power control strategy for a variable speed constant frequency generating system," *IEEE Transactions on Power Electronics*, vol. 10, no. 4, pp. 472–478, July 1995.
19. J. Holtz, "The representation of AC machine dynamics by complex signal flow graphs," *IEEE Transactions on Industrial Electronics*, vol. 42, no. 3, pp. 263–271, June 1995.
20. J. Holtz, "Sensorless control of induction motor drives," *Proceedings of the IEEE*, vol. 90, no. 8, pp. 1359–1394, August 2002.
21. G. Abad, J. Lopez, M. Rodriguez, L. Marroyo, and G. Iwanski, *Doubly Fed Induction Machine: Modeling and Control for Wind Energy Generation Applications*, ser. IEEE Press Series on Power Engineering. Wiley-IEEE Press, 2011.
22. J. Elizondo, A. Olloqui, M. Rivera, M. Macias, O. Probst, O. Micheloud, and J. Rodriguez, "Model-based predictive rotor current control for grid synchronization of a DFIG driven by an indirect matrix converter," *IEEE Journal of Emerging and Selected Topics in Power Electronics*, vol. 2, no. 4, pp. 715–726, December 2014.
23. D. Zhi, L. Xu, and B. Williams, "Model-based predictive direct power control of doubly fed induction generators," *IEEE Transactions on Power Electronics*, vol. 25, no. 2, pp. 341–351, February 2010.
24. S. Bhowmik, R. Spee, and J. H. R. Enslin, "Performance optimization for doubly fed wind power generation systems," *IEEE Transactions on Industry Applications*, vol. 35, no. 4, pp. 949–958, July/August 1999.
25. A. Tapia, G. Tapia, J. Ostolaza, and J. Saenz, "Modeling and control of a wind turbine driven doubly fed induction generator," *IEEE Transactions on Energy Conversion*, vol. 18, no. 2, pp. 194–204, June 2003.
26. E. Koutroulis and K. Kalaitzakis, "Design of a maximum power tracking system for wind-energy-conversion applications," *IEEE Transactions on Industrial Electronics*, vol. 53, no. 2, pp. 486–494, April 2006.
27. J. Morren and S. W. H. De Haan, "Ridethrough of wind turbines with doubly-fed induction generator during a voltage dip," *IEEE Transactions on Energy Conversion*, vol. 20, no. 2, pp. 435–441, June 2005.
28. J. Lopez, P. Sanchis, X. Roboam, and L. Marroyo, "Dynamic behavior of the doubly fed induction generator during three-phase voltage dips," *IEEE Transactions on Energy Conversion*, vol. 22, no. 3, pp. 709–717, September 2007.
29. M. Tsili and S. Papathanassiou, "A review of grid code technical requirements for wind farms," *IET Renewable Power Generation*, vol. 3, no. 3, pp. 308–332, September 2009.

APPENDIX A

TURBINE AND GENERATOR PARAMETERS

This appendix provides the parameters of wind turbines, permanent magnet synchronous generators (PMSGs), squirrel-cage induction generators (SCIGs), doubly fed induction generators (DFIGs), and three-phase grid. The case studies and simulation results presented in this book have two distinct power ratings: 750 kW and 3.0 MW. These two power ratings correspond to the turbine mechanical output and not to the electrical power injected to grid. All wind turbine, generator, and converter parameters are designed for these power ratings. For 3.0-MW power rating, two voltage classes are used: 690 V (line-to-line) for low-voltage operation; 3000 V (line-to-line) for medium-voltage operation. The three possible wind energy conversion system (WECS) configurations are as follows:

- 750-kW, 690-V WECS: 750-kW wind turbine + 750-kW, 690-V wind generator + one channel of BTB 2L-VSC + 750-kVA, 690-V grid system.
- 3.0-MW, 690-V WECS: 3.0-MW wind turbine + 3.0-MW, 690-V wind generator + parallel BTB 2L-VSCs + 3.0-MVA, 690-V grid system.
- 3.0-MW, 3000-V WECS: 3.0-MW wind turbine + 3.0-MW, 3000-V wind generator + one channel of BTB 3L-VSC + 3.0-MVA, 3000-V grid system.

The PMSGs are broadly classified into the following three groups on the basis of generator shaft speed:

- Low-speed generators that do not use a gearbox between the turbine and generator.
- Medium-speed generators that use a single-stage gearbox.
- High-speed generators that use a conventional three-stage gearbox.

All of the aforementioned generators have been applied commercially. The parameters of these groups of generators are developed by considering surface-mount PMSG (SPMSG) and inset PMSG (IPMSG). The SPMSGs are popular for low-speed operations, whereas IPMSGs are primarily deployed in high-speed operations. However, SPMSG and IPMSG parameters are given for low-, medium-, and high-speed operations in order to support the theoretical analysis and comparisons. Low-speed SPMSG parameters are developed for passive generator-side converters that operate close to the unity power factor. For all generator and converter parameters, the corresponding base values are listed. All the wind generator parameters are referred to the stator side.

A.1 NOTATION OF GENERATOR VARIABLES

In this section, the notation of wind generator variables is briefly discussed. All parameters correspond to the generator mode operation. The variables common for the PMSG, SCIG, and DFIG are listed as follows:

P_m Generator shaft input power (W). This value is negative for all wind generators. If the losses in gearbox and drivetrain are ignored, then the turbine output power P_T becomes equal to P_m.

T_m Generator shaft input torque (N.m). This value is negative for the generator cases, consistent with the negative sign of shaft input power P_m. The value of generator torque is lower than the turbine output torque T_T if a gearbox is employed. T_m is formulated as $T_m = T_T / r_{gb}$, where r_{gb} is the gearbox ratio.

P_s Generator stator output active power (W). In the generator mode, the sign of P_s is negative and coherent with the sign of P_m and T_m. This value is slightly lower than P_m because of generator losses.

Q_s Generator stator output reactive power (VAR). A positive and negative signs for Q_s indicates leading and lagging reactive power, respectively. Q_s is positive for PMSG, negative for induction generators.

S_s Generator stator output apparent power (VA).

$V_{s,ll}$ Generator stator *rms* line-to-line voltage (V).

V_s Generator stator *rms* line-to-ground voltage (V).

I_s Generator stator *rms* line current (V).

f_s Generator stator frequency (Hz).

$cos\, \varphi_s$ Generator power factor. A negative value implies leading power factor. A positive value indicates lagging power factor.

P_p Generator number of pole pairs.

n_m Generator mechanical/shaft speed (rpm).

ω_m Generator mechanical/shaft angular speed (rad/s).

ω_r Generator electrical angular speed (rad/s).

H_m Combined turbine and generator shaft inertia constant (s).

J_m Combined turbine and generator shaft moment of inertia (kg.m^2).

The variables specific to the PMSG are listed as follows:

R_s Stator winding resistance (Ω). This value is the same in the abc-, $\alpha\beta$-, or dq-frame; hence, no specific subscript is added.

L_{ds} d-axis synchronous inductance (H).

L_{qs} q-axis synchronous inductance (H). This variable is equal to L_{ds} for SPMSG and higher than L_{ds} for IPMSG.

ψ_r Rotor flux linkage (Wb, *peak*). In PMSG, ψ_r is established by permanent magnets; in wound rotor synchronous generators, the current flowing through the field winding produces the rotor flux linkage.

θ_T Optimal torque angle (degrees). This variable is defined for IPMSG only. θ_T refers to the optimal angle of stator current with respect to the q-axis. θ_T provides useful information in the MTPA control scheme.

The variables related to induction generator are described as follows:

R_s Stator winding resistance (Ω).

R_r Rotor winding resistance (Ω).

L_{ls} Stator leakage inductance (H).

L_{lr} Rotor leakage inductance (H).

L_m Magnetizing inductance (H).

L_s Stator self-inductance (H). $L_s = L_{ls} + L_m$.

L_r Rotor self-inductance (H). $L_r = L_{lr} + L_m$.

s Slip of the induction generator. This variable is negative for the generator when the shaft speed is higher than the synchronous speed.

Ψ_s Stator flux linkage (Wb, *rms*).

Ψ_r Rotor flux linkage (Wb, *rms*).

ψ_s Stator flux linkage (Wb, *peak*).

ψ_r Rotor flux linkage (Wb, *peak*).

The additional variables related to the DFIG are given as follows:

P_r Rotor active power (W).

$V_{r,ll}$ Generator rotor *rms* line-to-line voltage (V).

V_r Generator rotor *rms* line-to-ground voltage (V).

I_r Generator rotor *rms* line current (V).

A.2 BASE VALUES

In this section, calculation of base values for wind generators and utility grid is discussed. The base values calculation for wind generators is given by,

S_{B_s} Generator base apparent power = S_s (VA).

V_{B_s} Generator *rms* base voltage = $V_s = V_{s,ll}/\sqrt{3}$ (V).

I_{B_s} Generator *rms* base current = $S_{B_s}/(3 \times V_{B_s})$ (A).

f_{B_s} Generator base frequency = f_s (rated) (Hz).

ω_{B_s} Generator base angular frequency = $2 \times \pi \times f_{B_s}$ (rad/s).

Z_{B_s} Generator base impedance = V_{B_s}/I_{B_s} (Ω).

L_{B_s} Generator base inductance = Z_{B_s}/ω_{B_s} (H).

C_{B_s} Generator base capacitance = $1/(Z_{B_s} \times \omega_{B_s})$ (F).

Ψ_{B_s} Generator *rms* base flux-linkage = V_{B_s}/ω_{B_s} (Wb).

v_{B_s} Generator *peak* base voltage = $V_{B_s} \times \sqrt{2}$ (V).

i_{B_s} Generator *peak* base current = $I_{B_s} \times \sqrt{2}$ (A).

ψ_{B_s} Generator *peak* base flux-linkage = v_{B_s}/ω_{B_s} (Wb).

The base values for utility grid in Type 4 are calculated as follows:

S_{Bg} Grid base apparent power = S_g (VA).

V_{Bg} Grid *rms* base voltage = $V_g = V_{g,ll}/\sqrt{3}$ (V).

I_{Bg} Grid *rms* base current = $S_{Bg}/(3 \times V_{Bg})$ (A).

f_{Bg} Grid base frequency = f_g (rated) (Hz).

ω_{Bg} Grid base angular frequency = $2 \times \pi \times f_{Bg}$ (rad/s).

Z_{Bg} Grid base impedance = V_{Bg}/I_{Bg} (Ω).

L_{Bg} Grid base inductance = Z_{Bg}/ω_{Bg} (H).

C_{Bg} Grid base capacitance = $1/(Z_{Bg} \times \omega_{Bg})$ (F).

v_{Bg} Grid *peak* base voltage = $V_{Bg} \times \sqrt{2}$ (V).

i_{Bg} Grid *peak* base current = $I_{Bg} \times \sqrt{2}$ (A).

The base values for utility grid in Type 3 are calculated as follows:

S_{Br} Grid base apparent power = $P_{r,R}$ (VA).

V_{Br} Grid *rms* base voltage = $V_r = V_{r,ll}/\sqrt{3}$ (V).

I_{Br} Grid *rms* base current = $S_{Br}/(3 \times V_{Br})$ (A).

f_{Br} Grid base frequency = f_s (rated) (Hz).

ω_{Br} Grid base angular frequency = $2 \times \pi \times f_{Br}$ (rad/s).

Z_{Br} Grid base impedance = V_{Br}/I_{Br} (Ω).

L_{Br} Grid base inductance = Z_{Br}/ω_{Br} (H).

C_{Br} Grid base capacitance = $1/(Z_{Br} \times \omega_{Br})$ (F).

v_{Br} Grid *peak* base voltage = $V_{Br} \times \sqrt{2}$ (V).

i_{Br} Grid *peak* base current = $I_{Br} \times \sqrt{2}$ (A).

A.3 PER-UNIT VALUES

In the theoretical analysis as well as in simulation results, the parameters, variables, and signals are represented in per-unit (pu) and peak-per-unit ($\widehat{\text{pu}}$) values. The pu values are obtained by dividing the parameter/variable with *rms* base quantity. Similarly, the $\widehat{\text{pu}}$ values are obtained by dividing the variable with *peak* base quantity. The common notation employed in this book is to represent the constant parameters, and *rms* and *average* values in pu value, and to represent *peak* values in $\widehat{\text{pu}}$ values. In this section, calculation of pu and $\widehat{\text{pu}}$ values for the wind turbine, wind generators, and utility grid is discussed. Subscript R corresponds to the rated value.

Wind Turbine

v_w (pu)	per-unit wind speed = $v_w/v_{w,R}$
P_T (pu)	per-unit turbine output power = $P_T/P_{T,R}$
ω_T (pu)	per-unit turbine angular speed = $\omega_T/\omega_{T,R}$

PMSG, SCIG, and DFIG

P_m (pu)	per-unit mechanical input power = $P_m/P_{m,R}$
T_m (pu)	per-unit mechanical torque = $T_m/\lvert T_{m,R}\rvert$
T_e (pu)	per-unit electromagnetic torque = $T_e/\lvert T_{m,R}\rvert$
P_s (pu)	per-unit stator output active power = P_s/S_{B_s}
$P_{cu,s}$ (pu)	per-unit stator copper losses = $P_{cu,s}/S_{B_s}$
Q_s (pu)	per-unit stator output reactive power = Q_s/S_{B_s}
S_s (pu)	per-unit stator output apparent power = S_s/S_{B_s}
V_s (pu)	per-unit stator phase voltage = V_s/V_{B_s}
I_s (pu)	per-unit stator line current = I_s/I_{B_s}
f_s (pu)	per-unit stator frequency = f_s/f_{B_s}
ω_s (pu)	per-unit stator angular frequency = ω_s/ω_{B_s}
n_m (pu)	per-unit mechanical speed (rpm) = $n_m/n_{m,R}$
ω_m (pu)	per-unit mechanical angular speed = $\omega_m/\omega_{m,R}$
ω_r (pu)	per-unit electrical angular speed = $\omega_r/\omega_{r,R}$
ω_{sl} (pu)	per-unit angular slip speed = $\omega_{sl}/\omega_{sl,R}$
R_s (pu)	per-unit stator winding resistance = R_s/Z_{B_s}
R_r (pu)	per-unit rotor winding resistance = R_r/Z_{B_s}
L_{ls} (pu)	per-unit stator leakage inductance = L_{ls}/L_{B_s}
L_{lr} (pu)	per-unit rotor leakage inductance = L_{lr}/L_{B_s}
L_m (pu)	per-unit magnetizing inductance = L_m/L_{B_s}
L_s (pu)	per-unit stator inductance = L_s/L_{B_s}
L_r (pu)	per-unit rotor inductance = L_r/L_{B_s}
L_{ds} (pu)	per-unit d-axis inductance = L_{ds}/L_{B_s}
L_{qs} (pu)	per-unit q-axis inductance = L_{qs}/L_{B_s}
s (pu)	per-unit slip of generator = s/s_R
Ψ_s (pu)	per-unit stator flux linkage = Ψ_s/Ψ_{B_s}
Ψ_r (pu)	per-unit rotor flux linkage = Ψ_r/Ψ_{B_s}
v_s ($\widehat{\text{pu}}$)	peak-per-unit stator phase voltage = v_s/v_{B_s}
v_{ds} ($\widehat{\text{pu}}$)	peak-per-unit d-axis stator voltage = v_{ds}/v_{B_s}
v_{qs} ($\widehat{\text{pu}}$)	peak-per-unit q-axis stator voltage = v_{qs}/v_{B_s}
i_s ($\widehat{\text{pu}}$)	peak-per-unit stator line current = i_s/i_{B_s}
i_{ds} ($\widehat{\text{pu}}$)	peak-per-unit d-axis stator current = i_{ds}/i_{B_s}
i_{qs} ($\widehat{\text{pu}}$)	peak-per-unit q-axis stator current = i_{qs}/i_{B_s}

ψ_s ($\widehat{\text{pu}}$)	peak-per-unit stator flux linkage = ψ_s/ψ_{Bs}
ψ_r ($\widehat{\text{pu}}$)	peak-per-unit rotor flux linkage = ψ_r/ψ_{Bs}

Utility Grid in Type 4 WECS

P_g (pu)	per-unit grid active power = P_g/S_{Bg}
Q_g (pu)	per-unit grid reactive power = Q_g/S_{Bg}
S_g (pu)	per-unit grid apparent power = S_g/S_{Bg}
V_g (pu)	per-unit grid phase voltage = V_g/V_{Bg}
I_g (pu)	per-unit grid line current = I_g/I_{Bg}
v_g ($\widehat{\text{pu}}$)	peak-per-unit grid phase voltage = v_g/v_{Bg}
v_{dg} ($\widehat{\text{pu}}$)	peak-per-unit d-axis grid voltage = v_{dg}/v_{Bg}
v_{qg} ($\widehat{\text{pu}}$)	peak-per-unit q-axis grid voltage = v_{qg}/v_{Bg}
i_g ($\widehat{\text{pu}}$)	peak-per-unit grid line current = i_g/i_{Bg}
i_{dg} ($\widehat{\text{pu}}$)	peak-per-unit d-axis grid current = i_{dg}/i_{Bg}
i_{qg} ($\widehat{\text{pu}}$)	peak-per-unit q-axis grid current = i_{qg}/i_{Bg}

Power Converters in Type 4 WECS

$v_{r,ab}$ (pu)	rectifier line-to-line voltage = $v_{r,ab}/v_{dc}^*$
$v_{i,ab}$ (pu)	inverter line-to-line voltage = $v_{i,ab}/v_{dc}^*$
v_{dc} (pu)	per-unit DC-link voltage = v_{dc}/V_{Bg}
r_i (pu)	per-unit GSC harmonic filter resistance = r_i/Z_{Bg}
L_i (pu)	per-unit GSC harmonic filter inductance = L_i/L_{Bg}
C_{dc} (pu)	per-unit DC-link capacitance = C_{dc}/C_{Bg}
P_{dc} (pu)	per-unit boost converter DC power = P_{dc}/S_{Bg}
v_{in} (pu)	per-unit boost converter input DC voltage = v_{in}/V_{Bg}
i_{dc} (pu)	per-unit boost converter DC inductor current = i_{dc}/I_{Bg}
r_{dc} (pu)	per-unit resistance of DC inductor = r_{dc}/Z_{Bg}
L_{dc} (pu)	per-unit inductance of DC inductor = L_{dc}/L_{Bg}
C_{in} (pu)	per-unit input DC capacitance = C_{in}/C_{Bg}

DFIG-Specific

P_r (pu)	per-unit rotor active power = P_r/S_{Bs}
Q_r (pu)	per-unit rotor reactive power = Q_r/S_{Bs}
V_r (pu)	per-unit rotor phase voltage = V_r/V_{Br}
I_r (pu)	per-unit rotor current = I_r/I_{Br}
v_r ($\widehat{\text{pu}}$)	peak-per-unit rotor phase voltage = v_r/v_{Br}
v_{dr} ($\widehat{\text{pu}}$)	peak-per-unit d-axis rotor voltage = v_{dr}/v_{Br}
v_{qr} ($\widehat{\text{pu}}$)	peak-per-unit q-axis rotor voltage = v_{qr}/v_{Br}
i_r ($\widehat{\text{pu}}$)	peak-per-unit rotor current = i_r/i_{Br}

i_{dr} ($\widehat{\text{pu}}$)	peak-per-unit d-axis rotor current = i_{dr}/i_{B_r}
i_{qr} ($\widehat{\text{pu}}$)	peak-per-unit q-axis rotor current = i_{qr}/i_{B_r}
V_g (pu)	per-unit grid phase voltage = V_g/V_{B_r}
I_g (pu)	per-unit grid line current = I_g/I_{B_r}
v_g ($\widehat{\text{pu}}$)	peak-per-unit grid phase voltage = v_g/v_{B_r}
v_{dg} ($\widehat{\text{pu}}$)	peak-per-unit d-axis grid voltage = v_{dg}/v_{B_r}
v_{qg} ($\widehat{\text{pu}}$)	peak-per-unit q-axis grid voltage = v_{qg}/v_{B_r}
i_g ($\widehat{\text{pu}}$)	peak-per-unit grid line current = i_g/i_{B_r}
i_{dg} ($\widehat{\text{pu}}$)	peak-per-unit d-axis grid current = i_{dg}/i_{B_r}
i_{qg} ($\widehat{\text{pu}}$)	peak-per-unit q-axis grid current = i_{qg}/i_{B_r}
v_{dc} (pu)	per-unit DC-link voltage = v_{dc}/V_{B_r}
r_i (pu)	per-unit GSC harmonic filter resistance = r_i/Z_{B_r}
L_i (pu)	per-unit GSC harmonic filter inductance = L_i/L_{B_r}
C_{dc} (pu)	per-unit DC-link capacitance = C_{dc}/C_{B_r}

EXAMPLE A.1 Calculation of Base and pu Values

Consider a three-phase SCIG #1 parameters in Table A.7. The rated values of SCIG are given as follows: S_s = 830.4 kVA, $V_{s,ll}$ = 690 V, f_s = 60 Hz, R_s = 3.996 mΩ, R_r = 4.529 mΩ, L_{ls} = 0.173 mH, L_m = 5.745 mH, ψ_s = 1.5037 (Wb, *peak*), ψ_r = 1.4258 (Wb, *peak*). The following is calculated: (a) base values for SCIG and (b) per-unit values of generator parameters.

Solution:

(a) The base values of SCIG are calculated as follows:

$$S_{B_s} = S_s = 830.4 \times 10^3 \text{ VA}, \qquad V_{B_s} = \frac{V_{s,ll}}{\sqrt{3}} = \frac{690}{\sqrt{3}} = 398.4 \text{ V},$$

$$I_{B_s} = \frac{S_{B_s}}{3 \times V_{B_s}} = \frac{830.4 \times 10^3}{3 \times 398.4} = 694.8 \text{ A}, \qquad \omega_{B_s} = 2 \times \pi \times f_{B_s} = 377 \text{ rad/s},$$

$$Z_{B_s} = \frac{V_{B_s}}{I_{B_s}} = 0.573 \text{ A}, \qquad L_{B_s} = \frac{Z_{B_s}}{\omega_{B_s}} = 1.52 \text{ mH},$$

$$C_{B_s} = \frac{1}{(Z_{B_s} \times \omega_{B_s})} = 4629.2\ \mu\text{F}, \qquad \Psi_{B_s} = \frac{V_{B_s}}{\omega_{B_s}} = 1.057 \text{ Wb}, \textit{rms},$$

$$v_{B_s} = V_{B_s} \times \sqrt{2} = 563.4 \text{ V}, \qquad i_{B_s} = I_{B_s} \times \sqrt{2} = 982.6 \text{ A}.$$

(b) The per unit values of the selected generator parameters are given below:

$$R_s \text{ (pu)} = \frac{R_s}{Z_{B_s}} = \frac{3.996 \times 10^{-3}}{0.573} = 0.007, \qquad R_r \text{ (pu)} = \frac{R_r}{Z_{B_s}} = \frac{4.529 \times 10^{-3}}{0.573} = 0.008,$$

$$L_{ls} \text{ (pu)} = \frac{L_{ls}}{L_{B_s}} = \frac{0.173 \times 10^{-3}}{1.521 \times 10^{-3}} = 0.11, \qquad L_m \text{ (pu)} = \frac{L_m}{L_{B_s}} = \frac{5.745 \times 10^{-3}}{1.521 \times 10^{-3}} = 3.78,$$

$$\psi_s\ (\widehat{\text{pu}}) = \frac{\psi_s}{\psi_{B_s}} = \frac{1.5037}{1.495} = 1.01, \qquad \psi_r\ (\widehat{\text{pu}}) = \frac{\psi_r}{\psi_{B_s}} = \frac{1.4258}{1.495} = 0.95.$$

♦

A.4 WIND TURBINE PARAMETERS

The parameters of 750-kW and 3.0-MW wind turbines are presented in Table A.1, along with the description of variables.

Table A.1 Rated parameters of variable-speed wind turbines

Rated Values	Turbine #1	Turbine #2
Rated Power, P_T (kW)	–750	–3000
Rated Torque, T_T (kN.m)	–318	–1273
Rated Speed, n_T (rpm)	22.5	22.5
Rotor Radius, r_T (m)	22.9	43.36
Swept Area, A_T (m^2)	1648	5905
Air Density, ρ (kg.m^2)	1.225	1.225
Rated Wind Speed $v_{w,R}$ (m/s)	12	12
Cut-in Wind Speed $v_{w,i}$ (m/s)	3	3
Cut-out Wind Speed $v_{w,o}$ (m/s)	25	25
Optimal Power Coefficient, C_p^{op}	0.43	0.48
Optimal Tip-Speed Ratio, λ_T^{op}	4.4971	8.5128
Intermittent Tip-Speed Ratio, λ_I^{op}	5.3371	12.1256
Turbine Constant, C_1	–0.6175	0.3915
Turbine Constant, C_2	116	116
Turbine Constant, C_3	0.4	0.4
Turbine Constant, C_4	0	0
Turbine Constant, C_5	5	5
Turbine Constant, C_6	21	21
Turbine Constant, C_7	0.1405	0.0192

The wind turbine output power is calculated as follows:

$$P_T = P_w \times C_p = \frac{1}{2}\rho A_T v_w^3 C_p \tag{A.1}$$

where ρ is air density (kg/m^3), A_T is the rotor swept area (m^2), r_T is the blade radius (m), and $v_{w,R}$ is the rated wind speed (m/s). C_p represents the power coefficient of the rotor blades. Practical wind turbines have C_p in the range of 0.32 to 0.52. C_p is defined below in terms of turbine coefficients C_1 to C_7:

$$C_p = C_1\left(\frac{C_2}{\lambda_I} - C_3\beta - C_4\beta^2 - C_5\right)e^{-\frac{C_6}{\lambda_I}} + C_7\lambda_T \tag{A.2}$$

where β is the pitch angle, which can be adjusted by the pitch control system. λ_T corresponds to the optimal tip-speed ratio (TSR). λ_T is defined by the following:

$$\lambda_T = \lambda_T^{\mathrm{op}} = \frac{\omega_T\, r_T}{v_{w,R}}. \tag{A.3}$$

Variable-speed wind turbines operate at an optimal TSR value during all wind speed conditions. In Equation (A.2), λ_I is intermittent TSR and is related to λ_T and β as demonstrated below:

$$\frac{1}{\lambda_I} = \frac{1}{\lambda_T + 0.08\beta} - \frac{0.035}{\beta^3 - 1}. \tag{A.4}$$

The gearbox ratio is not provided in Table A.1. This ratio is calculated according to the rated speed of wind turbine and generator, that is,

$$r_{gb} = \frac{n_m}{n_r}. \tag{A.5}$$

The gearbox ratio is equal to 1 for direct-drive turbine because n_m is equal to n_r. The gearbox ratio is 17.78 for medium-speed PMSGs, 80 for high-speed PMSGs, 80.62 for SCIGs, and 96 for DFIGs.

A.5 THREE-PHASE GRID PARAMETERS

The parameters of three types of utility grid and grid-side converters are listed in Table A.2. The base values of the grid are also provided. The inverter-side harmonic filter has 0.01 pu internal resistance and 0.2 pu inductance. The capacitance of DC-link capacitor(s) is 4.0 pu. The reference DC-bus voltage is 3.062 pu.

Table A.2 Rated parameters of three-phase grid and grid-side converter

Rated Values	Grid #1	Grid #2	Grid #3
Rated Power, P_g (kW)	750	3000	3000
Rated Line-to-Line Voltage, $V_{g,ll}$ (V, *rms*)	690	690	3000
Rated Phase Voltage, V_g (V, *rms*)	398.4	398.4	1732.1
Rated Current, I_g (A, *rms*)	627.6	2510.2	577.4
Rated Frequency, f_g (Hz)	60	60	60
Rated Frequency, ω_g (rad)	377	377	377
Resistance of Filter, r_i (mΩ)	6.35	1.59	30
Inductance of Filter, L_i (mH)	0.3368	0.084	1.592
DC Link Capacitance, C_{dc} (μF)	16714	66878	3537
Reference DC-bus Voltage, v_{dc}^* (V)	1220	1220	5304
S_{Bg} (kVA)	750	3000	3000
V_{Bg} (V)	398.4	398.4	1732.1
I_{Bg} (A)	627.6	2510.2	577.4
Z_{Bg} (Ω)	0.635	0.159	3.0
L_{Bg} (mH)	1.684	0.421	7.96
C_{Bg} (μF)	4178.6	16714.4	884.2
v_{Bg} (V)	563.4	563.4	2449.5
i_{Bg} (A)	887.5	3550	816.5

A.6 PERMANENT MAGNET SYNCHRONOUS GENERATOR PARAMETERS

The parameters of low-, medium-, and high-speed PMSGs are given in Tables A.3, A.4, and A.5, respectively. The parameters of low-speed SPMSG for passive generator-side converter are given Table A.6.

Table A.3 Parameters of low-speed (with no gearbox) PM synchronous generators

Rated Values	**SPMSG #1**	**IPMSG #1**	**SPMSG #2**	**IPMSG #2**	**SPMSG #3**	**IPMSG #3**
P_m (kW)	–750	–750	–3000	–3000	–3000	–3000
T_m (kN.m)	–318	–318	–1273	–1273	–1273	–1273
P_s (kW)	–741	–741	–2964	–2964	–2964	–2964
Q_s (kVAR)	324	324	1294	1294	1294	1294
S_s (kVA)	809	809	3234	3234	3234	3234
$V_{s,ll}$ (V, *rms*)	690	690	690	690	3000	3000
V_s (V, *rms*)	398	398	398	398	1732	1732
I_s (A, *rms*)	676	676	2706	2706	622	622
f_s (Hz)	9.75	9.75	9.75	9.75	9.75	9.75
$\cos\varphi_s$	–0.92	–0.92	–0.92	–0.92	–0.92	–0.92
P_p	26	26	26	26	26	26
n_m (rpm)	22.5	22.5	22.5	22.5	22.5	22.5
ω_m (rad/s)	2.36	2.36	2.36	2.36	2.36	2.36
ω_r (rad/s)	61.26	61.26	61.26	61.26	61.26	61.26
R_s (mΩ)	6.52	6.52	1.63	1.63	30.52	30.52
L_{ds} (mH)	3.85	3.74	0.96	0.94	18.17	17.68
L_{qs} (mH)	3.85	5.88	0.96	1.47	18.17	27.79
ψ_r (Wb, *peak*)	8.53	8.295	8.53	8.295	37.09	36.06
θ_T (degrees)	—	19.28	—	19.28	—	19.28
H_m (s)*	4.8	4.8	5.8	5.8	5.8	5.8
J_m (kg.m^2)*	1398×10^3	1398×10^3	6758×10^3	6758×10^3	6758×10^3	6758×10^3
H_m (s)†	0.4	0.4	0.4	0.4	0.4	0.4
J_m (kg.m^2)†	117×10^3	117×10^3	466×10^3	466×10^3	466×10^3	466×10^3
Base Values	**SPMSG #1**	**IPMSG #1**	**SPMSG #2**	**IPMSG #2**	**SPMSG #3**	**IPMSG #3**
S_{B_s} (kVA)	809	809	3234	3234	3234	3234
V_{B_s} (V)	398	398	398	398	1732	1732
I_{B_s} (A)	676	676	2706	2706	622	622
Z_{B_s} (Ω)	0.59	0.59	0.147	0.147	2.78	2.78
L_{B_s} (mH)	9.62	9.62	2.404	2.404	45.43	45.43
C_{B_s} (μF)	27710	27710	110840	110840	5865	5865
Ψ_{B_s} (Wb)	6.51	6.51	6.51	6.51	28.28	28.28
v_{B_s} (V)	563	563	563	563	2450	2450
i_{B_s} (A)	957	957	3827	3827	880	880
ψ_{B_s} (Wb)	9.20	9.20	9.20	9.20	39.99	39.99

For all the SPMSGs, the pu values are $R_s = 0.01$ pu, $L_{ds} = L_{qs} = L_s = 0.4$ pu, and $\psi_r = 0.928$ $\widehat{\text{pu}}$.
For all the IPMSGs, the pu values are $R_s = 0.01$ pu, $L_{ds} = 0.389$ pu, $L_{qs} = 0.612$ pu, and $\psi_r = 0.902$ $\widehat{\text{pu}}$.
* is original value, and † is recommended value to reduce the simulation runtime.

Table A.4 Parameters of medium-speed (with one-stage gearbox) PM synchronous generators

Rated Values	SPMSG #4	IPMSG #4	SPMSG #5	IPMSG #5	SPMSG #6	IPMSG #6
P_m (kW)	–750	–750	–3000	–3000	–3000	–3000
T_m (kN.m)	–17.91	–17.91	–71.62	–71.62	–71.62	–71.62
P_s (kW)	–741	–741	–2964	–2964	–2964	–2964
Q_s (kVAR)	324	324	1294	1294	1294	1294
S_s (kVA)	809	809	3234	3234	3234	3234
$V_{s,ll}$ (V, *rms*)	690	690	690	690	3000	3000
V_s (V, *rms*)	398	398	398	398	1732	1732
I_s (A, *rms*)	676	676	2706	2706	622	622
f_s (Hz)	40	40	40	40	40	40
$\cos\varphi_s$	–0.92	–0.92	–0.92	–0.92	–0.92	–0.92
P_p	6	6	6	6	6	6
n_m (rpm)	400	400	400	400	400	400
ω_m (rad/s)	41.89	41.89	41.89	41.89	41.89	41.89
ω_r (rad/s)	251.3	251.3	251.3	251.3	251.3	251.3
R_s (mΩ)	6.52	6.52	1.63	1.63	30.52	30.52
L_{ds} (mH)	0.938	0.912	0.235	0.228	4.43	4.31
L_{qs} (mH)	0.938	1.433	0.235	0.358	4.43	6.77
ψ_r (Wb, *peak*)	2.08	2.022	2.08	2.022	9.04	8.79
θ_T (degrees)	—	19.28	—	19.28	—	19.28
H_m (s)	4.8	4.8	5.8	5.8	5.8	5.8
J_m (kg.m^2)	4.42×10^3	4.42×10^3	21.38×10^3	21.38×10^3	21.38×10^3	21.38×10^3
Base Values	**SPMSG #1**	**IPMSG #1**	**SPMSG #2**	**IPMSG #2**	**SPMSG #3**	**IPMSG #3**
S_{B_s} (kVA)	809	809	3235	3235	3235	3235
V_{B_s} (V)	398	398	398	398	1732	1732
I_{B_s} (A)	676	676	2706	2706	622	622
Z_{B_s} (Ω)	0.59	0.59	0.147	0.147	2.78	2.78
L_{B_s} (mH)	2.34	2.34	0.59	0.59	11.07	11.07
C_{B_s} (μF)	6754	6754	27017	27017	1430	1430
Ψ_{B_s} (Wb)	1.59	1.59	1.59	1.59	6.89	6.89
v_{B_s} (V)	563	563	563	563	2450	2450
i_{B_s} (A)	957	957	3826	3826	880	880
ψ_{B_s} (Wb)	2.24	2.24	2.24	2.24	9.74	9.74

For all the SPMSGs, the pu values are $R_s = 0.01$ pu, $L_{ds} = L_{qs} = L_s = 0.4$ pu, and $\psi_r = 0.928\ \widehat{\text{pu}}$.
For all the IPMSGs, the pu values are $R_s = 0.01$ pu, $L_{ds} = 0.389$ pu, $L_{qs} = 0.612$ pu, and $\psi_r = 0.902\ \widehat{\text{pu}}$.

Table A.5 Parameters of high-speed (with three-stage gearbox) PM synchronous generators

Rated Values	SPMSG #7	IPMSG #7	SPMSG #8	IPMSG #8	SPMSG #9	IPMSG #9
P_m (kW)	–750	–750	–3000	–3000	–3000	–3000
T_m (kN.m)	–3.98	–3.98	–15.92	–15.92	–15.92	–15.92
P_s (kW)	–741	–741	–2964	–2964	–2964	–2964
Q_s (kVAR)	324	324	1294	1294	1294	1294
S_s (kVA)	809	809	3234	3234	3234	3234
$V_{s,ll}$ (V, *rms*)	690	690	690	690	3000	3000
V_s (V, *rms*)	398	398	398	398	1732	1732
I_s (A, *rms*)	676	676	2706	2706	622	622
f_s (Hz)	60	60	60	60	60	60
$\cos\varphi_s$	–0.92	–0.92	–0.92	–0.92	–0.92	–0.92
P_p	2	2	2	2	2	2
n_m (rpm)	1800	1800	1800	1800	1800	1800
ω_m (rad/s)	188.5	188.5	188.5	188.5	188.5	188.5
ω_r (rad/s)	377	377	377	377	377	377
R_s (mΩ)	6.52	6.52	1.63	1.63	30.52	30.52
L_{ds} (mH)	0.625	0.608	0.16	0.15	2.95	2.87
L_{qs} (mH)	0.625	0.955	0.16	0.24	2.95	4.52
ψ_r (Wb, *peak*)	1.39	1.35	1.39	1.35	6.03	5.86
θ_T (degrees)	—	19.28	—	19.28	—	19.28
H_m (s)	4.8	4.8	5.8	5.8	5.8	5.8
J_m (kg.m^2)	219	219	1060	1060	1060	1060
Base Values	**SPMSG #1**	**IPMSG #1**	**SPMSG #2**	**IPMSG #2**	**SPMSG #3**	**IPMSG #3**
S_{Bs} (kVA)	809	809	3234	3234	3234	3234
V_{Bs} (V)	398	398	398	398	1732	1732
I_{Bs} (A)	676	676	2706	2706	622	622
Z_{Bs} (Ω)	0.59	0.59	0.147	0.147	2.78	2.78
L_{Bs} (mH)	1.56	1.56	0.39	0.39	7.38	7.38
C_{Bs} (μF)	4504	4504	18011	18011	953	953
Ψ_{Bs} (Wb)	1.06	1.06	1.06	1.06	4.59	4.59
v_{Bs} (V)	563	563	563	563	2450	2450
i_{Bs} (A)	957	957	3827	3827	880	880
ψ_{Bs} (Wb)	1.49	1.49	1.49	1.49	6.49	6.49

For all the SPMSGs, the pu values are $R_s = 0.01$ pu, $L_{ds} = L_{qs} = L_s = 0.4$ pu, and $\psi_r = 0.928$ $\widehat{\text{pu}}$.
For all the IPMSGs, the pu values are $R_s = 0.01$ pu, $L_{ds} = 0.389$ pu, $L_{qs} = 0.612$ pu, and $\psi_r = 0.902$ $\widehat{\text{pu}}$.

Table A.6 Parameters of low-speed PMSG for passive generator-side converters

Rated Values	SPMSG #10	SPMSG #11	SPMSG #12
P_m (kW)	–750	–3000	–3000
T_m (kN.m)	–318	–1273	–1273
P_s (kW)	–740	–2962	–2962
Q_s (kVAR)	152	607	607
S_s (kVA)	756	3023	3023
$V_{s,ll}$ (V, *rms*)	690	690	3000
V_s (V, *rms*)	398	398	1732
I_s (A, *rms*)	632	2529	582
f_s (Hz)	9.75	9.75	9.75
$\cos\varphi_s$	–0.98	–0.98	–0.98
P_p	26	26	26
n_m (rpm)	22.5	22.5	22.5
ω_m (rad/s)	2.36	2.36	2.36
ω_r (rad/s)	61.26	61.26	61.26
R_s (mΩ)	8.02	2.01	37.52
L_{ds} (mH)	2.07	0.52	9.75
L_{qs} (mH)	2.07	0.52	9.75
ψ_r (Wb, *peak*)	9.13	9.13	39.68
H_m (s)*	4.8	5.8	5.8
J_m (kg.m^2)*	1307×10^3	6317×10^3	6317×10^3
H_m (s)†	0.4	0.4	0.4
J_m (kg.m^2)†	109×10^3	436×10^3	436×10^3
Base Values	**SPMSG #10**	**SPMSG #11**	**SPMSG #12**
S_{B_s} (kVA)	756	3023	3023
V_{B_s} (V)	398	398	1732
I_{B_s} (A)	632	2529	582
Z_{B_s} (Ω)	0.63	0.16	2.98
L_{B_s} (mH)	10.29	2.57	48.61
C_{B_s} (μF)	25901	103610	5482
Ψ_{B_s} (Wb)	6.51	6.51	28.28
v_{B_s} (V)	563	563	2450
i_{B_s} (A)	894	3577	823
ψ_{B_s} (Wb)	9.20	9.20	39.99

For all the SPMSGs, the pu values are $R_s = 0.013$ pu, $L_{ds} = L_{qs} = L_s = 0.2$ pu, and $\psi_r = 0.992$ $\widehat{\text{pu}}$.
* is original value, and † is recommended value to reduce the simulation runtime.

A.7 SQUIRREL-CAGE INDUCTION GENERATOR PARAMETERS

Three types of SCIG parameters are given in Table A.7.

Table A.7 Parameters of squirrel-cage induction generators

Rated Values	**SCIG #1**	**SCIG #2**	**SCIG #3**
P_m (kW)	–750	–3000	–3000
T_m (kN.m)	–3.948	–15.793	–15.793
P_s (kW)	–738.5	–2953.7	–2953.7
Q_s (kVAR)	–379.7	–1518.8	–1518.8
S_s (kVA)	830.4	3321.4	3321.4
$V_{s,ll}$ (V, *rms*)	690	690	3000
V_s (V, *rms*)	398.4	398.4	1732.1
I_s (A, *rms*)	694.8	2779.1	639.2
f_s (Hz)	60	60	60
$\cos \varphi_s$	–0.889	–0.889	–0.889
P_p	2	2	2
n_m (rpm)	1814	1814	1814
s	–0.0078	–0.0078	–0.0078
ω_m (rad/s)	189.96	189.96	189.96
ω_r (rad/s)	379.92	379.92	379.92
R_s (mΩ)	3.996	0.999	18.885
R_r (mΩ)	4.529	1.132	21.404
L_{ls} (mH)	0.173	0.0431	0.815
L_{lr} (mH)	0.173	0.0431	0.815
L_m (mH)	5.745	1.4372	27.168
ψ_s (Wb, *peak*)	1.5037	1.5037	6.538
ψ_r (Wb, *peak*)	1.4258	1.4258	6.199
H_m (s)	4.8	5.8	5.8
J_m (kg.m^2)	221	1068	1068
Base Values	**SCIG #1**	**SCIG #2**	**SCIG #3**
S_{B_s} (kVA)	830.4	3231.4	3231.4
V_{B_s} (V)	398.4	398.4	1732.1
I_{B_s} (A)	694.8	2779.1	639.2
Z_{B_s} (Ω)	0.573	0.143	2.709
L_{B_s} (mH)	1.521	0.380	7.188
C_{B_s} (μF)	4626.3	18504.8	978.9
Ψ_{B_s} (Wb, *rms*)	1.057	1.057	4.594
v_{B_s} (V)	563.4	563.4	2449.5
i_{B_s} (A)	982.6	3930.3	903.9
ψ_{B_s} (Wb, *rms*)	1.494	1.494	6.497

For all the SCIGs, the pu values are $R_s = 0.00697$ pu, $R_r = 0.00789$ pu, $L_{ls} = L_{lr} = 0.1134$ pu, $L_m = 3.779$ pu, $\psi_s = 1.006$ $\widehat{\text{pu}}$, and $\psi_r = 0.954$ $\widehat{\text{pu}}$.

A.8 DOUBLY FED INDUCTION GENERATOR PARAMETERS

The parameters of 750-kW and 3.0-MW DFIG are given in Table A.8.

Table A.8 Parameters of doubly-fed induction generators

Rated Values	DFIG #1	DFIG #2	DFIG #3
P_m (kW)	–750	–3000	–3000
T_m (kN.m)	–3.316	–13.263	–13.263
P_s (kW)	620.3	2481.3	2481.3
P_r (kW)	118.9	475.9	475.9
$V_{s,ll}$ (V, *rms*)	690	690	3000
V_s (V, *rms*)	398.4	398.4	1732.1
$V_{r,ll}$ (V, *rms*)	158.7	158.7	690
V_r (V, *rms*)	91.63	91.63	398.4
I_s (A, *rms*)	519.1	2076.2	477.5
I_r (A, *rms*)	668.3	2673.1	614.8
f_s (Hz)	60	60	60
P_p	2	2	2
n_m (rpm)	2160	2160	2160
s	–0.2	–0.2	–0.2
n_m range (rpm)	1350–2160	1350–2160	1350–2160
ω_m (rad/s)	226.2	226.2	226.2
ω_r (rad/s)	452.4	452.4	452.4
R_s (mΩ)	5.773	1.443	27.285
R_r (mΩ)	4.499	1.125	21.264
L_{ls} (mH)	0.038	0.094	1.785
L_{lr} (mH)	0.034	0.085	1.607
L_m (mH)	3.209	0.802	15.168
ψ_s (Wb, *peak*)	1.5057	1.5057	6.5464
ψ_r (Wb, *peak*)	1.7556	1.7556	7.6332
H_m (s)	4.8	5.8	5.8
J_m (kg.m^2)	170	680	680
Base Values	**DFIG #1**	**DFIG #2**	**DFIG #3**
S_{B_s} (kVA)	750	3000	3000
V_{B_s} (V)	398.4	398.4	1732.1
I_{B_s} (A)	627.6	2510.2	577.4
Z_{B_s} (Ω)	0.635	0.159	3.0
L_{B_s} (mH)	1.684	0.421	7.958
C_{B_s} (μF)	4178.6	16714	884.2
Ψ_{B_s} (Wb, *rms*)	1.057	1.057	4.594
v_{B_s} (V)	563.4	563.4	2449.5
i_{B_s} (A)	887.5	3549.9	816.5
ψ_{B_s} (Wb, *rms*)	1.494	1.494	6.497

For all the DFIGs, the pu values are $R_s = 0.0091$ pu, $R_r = 0.0071$ pu, $L_{ls} = 0.2243$ pu, $L_{lr} = 0.2019$ pu, $L_m = 1.9061$ pu, $\psi_s = 1.008$ $\widehat{\text{pu}}$, and $\psi_r = 1.175$ $\widehat{\text{pu}}$.

APPENDIX B
CHAPTER APPENDICES

B.1 APPENDIX FOR CHAPTER 4

To assess the performance of MPC schemes, the following parameters are defined by using the guidelines in [1–4]. The percentage mean absolute current reference tracking error is defined as an absolute difference between the reference and measured current for m number of samples with respect to the *rms* value of reference current:

$$\%\, e_{i_x} = \frac{\frac{1}{m}\sum_{k=0}^{m} |i_x^*(k) - i_x(k)|}{I_x^*} \times 100, \qquad x \in \{a,\, b,\, c,\, \alpha,\, \beta,\, d,\, q\} \tag{B.1}$$

where m represents the number of samples. In the simulation and experimental tests, 25,000 samples are considered corresponding to 15 fundamental cycles. The expression shown above can be used for three-phase currents and two-phase currents in $\alpha\beta$ and dq frames. For three-phase currents, overall tracking error is computed as an average of tracking error in each phase, that is,

$$\%\, e_i = \frac{\%\, e_{ia} + \%\, e_{ib} + \%\, e_{ic}}{3}. \tag{B.2}$$

The average tracking error in the $\alpha\beta$ frame can be calculated in a similar manner. The percentage mean absolute active and reactive power tracking errors are defined with respect to the base apparent power S as follows:

$$\%\, e_P = \frac{\frac{1}{m}\sum_{k=0}^{m} |P^*(k) - P(k)|}{S} \times 100 \tag{B.3}$$

$$\%\, e_Q = \frac{\frac{1}{m}\sum_{k=0}^{m} |Q^*(k) - Q(k)|}{S} \times 100. \tag{B.4}$$

Model Predictive Control of Wind Energy Conversion Systems, First Edition. Venkata Yaramasu and Bin Wu.

Companion Website: www.wiley.com/go/yaramasu_wu_mpc_wecs

The percentage mean absolute DC-link capacitors voltage deviation $\% e_{vc}$ is formulated for a three-level converter as follows:

$$\% e_{vc} = \frac{\frac{1}{m}\sum_{k=0}^{m} |v_{C1}(k) - v_{C2}(k)|}{v_{dc}^*} \times 100. \tag{B.5}$$

A similar approach can be used to calculate the voltage deviation of capacitors in other multilevel converters. The overall percentage of total harmonic distortion (THD) for three-phase currents is calculated as follows:

$$\%\,\mathrm{THD} = \frac{1}{3} \cdot \sum_{x=a,b,c} \frac{\sqrt{i_{2,x}^2 + i_{3,x}^2 + .. + i_{n,x}^2}}{i_{1,x}} \times 100 \tag{B.6}$$

where $i_{n,x}$ and $i_{1,x}$ are the nth order harmonic and fundamental components of phase-x measured currents, respectively.

The FCS-MPC leads to variable-switching frequency operation. The average device switching frequency obtained by the MPC scheme can be computed by measuring the cumulative number of switch changes and by dividing them with respect to the time period T considered for the measurement of switch changes. This approach is demonstrated for a two-level converter as follows:

$$f_{sw} = \frac{f_{sw,a} + f_{sw,b} + f_{sw,c}}{3} = \frac{n_{sw,a} + n_{sw,b} + n_{sw,c}}{3T} \tag{B.7}$$

where $f_{sw,a}$, $f_{sw,b}$, and $f_{sw,c}$ are the switching frequencies of semiconductor devices in phase-a, -b, and -c, respectively. $n_{sw,a}$, $n_{sw,b}$, and $n_{sw,c}$ are number of switch changes in phase-a, -b, and -c, respectively. Considering that the two switching devices in a two-level converter operate in a complementary manner, the overall switching frequency can be obtained by measuring only the number of switch changes in each upper-leg switch for 15 fundamental cycles.

Similarly, the average switching frequency of a 3L-converter is calculated as follows:

$$f_{sw} = \frac{f_{sw,a} + f_{sw,b} + f_{sw,c}}{3} = \frac{n_{sw,a1} + n_{sw,a2} + n_{sw,b1} + n_{sw,b2} + n_{sw,c1} + n_{sw,c2}}{6T}. \tag{B.8}$$

The switching frequencies of other multilevel converters can be calculated in a similar manner.

REFERENCES

1. V. Yaramasu, M. Rivera, B. Wu, and J. Rodriguez, "Model predictive current control of two-level four-leg inverters – Part I: Concept, algorithm and simulation analysis," *IEEE Transactions on Power Electronics*, vol. 28, no. 7, pp. 3459–3468, July 2013.
2. V. Yaramasu, "Predictive control of multilevel converters for megawatt wind energy conversion systems," Ph.D. dissertation, Ryerson University, Toronto, ON, Canada, 2014, available at: http://digital.library.ryerson.ca/islandora/object/RULA%3A3459.
3. H. Young, M. Perez, J. Rodriguez, and H. Abu-Rub, "Assessing finite-control-set model predictive control: A comparison with a linear current controller in two-level voltage source inverters," *IEEE Industrial Electronics Magazine*, vol. 8, no. 1, pp. 44–52, March 2014.
4. R. Vargas, P. Coftes, U. Ammann, J. Rodríguez, and J. Pontt, "Predictive control of a three-phase neutral-point-clamped inverter," *IEEE Transactions on Industrial Electronics*, vol. 54, no. 5, pp. 2697–2705, October 2007.

B.2 APPENDIX FOR CHAPTER 5

The summary of voltage vectors and corresponding switching vectors, switching signals, output voltages, and input DC branch currents are presented in this appendix for two-level voltage source converter in Table B.1, current source converter in Table B.2, three-level (3L) diode-clamped (neutral-point clamped) converter in Table B.3, 3L flying capacitor converter in Tables B.4 and B.5, and direct matrix converter in Table B.6.

Table B.1 Summary of 2L-VSC output voltages and input DC branch currents for all possible switching states

	Voltage Vector	Switching Vector	Upper-Leg Switching Signals			Output Phase Voltages			Input DC Currents	
No.	$\mathbf{V}$	$\mathbf{S}_{abc}$	s_{a1}	s_{b1}	s_{c1}	v_{aN}	v_{bN}	v_{cN}	i_P	i_N
1	$\mathbf{V}_0$	[000]	'0'	'0'	'0'	0	0	0	0	0
2	$\mathbf{V}_1$	[100]	'1'	'0'	'0'	v_{C1}	0	0	i_a	$i_b + i_c$
3	$\mathbf{V}_2$	[110]	'1'	'1'	'0'	v_{C1}	v_{C1}	0	$i_a + i_b$	i_c
4	$\mathbf{V}_3$	[010]	'0'	'1'	'0'	0	v_{C1}	0	i_b	$i_a + i_c$
5	$\mathbf{V}_4$	[011]	'0'	'1'	'1'	0	v_{C1}	v_{C1}	$i_b + i_c$	i_a
6	$\mathbf{V}_5$	[001]	'0'	'0'	'1'	0	0	v_{C1}	i_c	$i_a + i_b$
7	$\mathbf{V}_6$	[101]	'1'	'0'	'1'	v_{C1}	0	v_{C1}	$i_a + i_c$	i_b
8	$\mathbf{V}_7$	[111]	'1'	'1'	'1'	v_{C1}	v_{C1}	v_{C1}	0	0

Table B.2 Summary of CSC output currents and input DC voltage for all possible switching states

	Current Vector	Switching Vector	Upper-Leg Switching Signals			Lower-Leg Switching Signals			Output Currents			Input Voltage
No.	$\mathbf{I}$	$\mathbf{S}_{abc}$	s_{a1}	s_{b1}	s_{c1}	s_{a2}	s_{b2}	s_{c2}	i_a	i_b	i_c	v_{dc}
1	$\mathbf{I}_0$	[300]	'1'	'0'	'0'	'1'	'0'	'0'	0	0	0	0
2	$\mathbf{I}_0$	[030]	'0'	'1'	'0'	'0'	'1'	'0'	0	0	0	0
3	$\mathbf{I}_0$	[003]	'0'	'0'	'1'	'0'	'0'	'1'	0	0	0	0
4	$\mathbf{I}_1$	[210]	'1'	'0'	'0'	'0'	'1'	'0'	i_{dc}	$-i_{dc}$	0	$v_a - v_b$
5	$\mathbf{I}_2$	[201]	'1'	'0'	'0'	'0'	'0'	'1'	i_{dc}	0	$-i_{dc}$	$v_a - v_c$
6	$\mathbf{I}_3$	[021]	'0'	'1'	'0'	'0'	'0'	'1'	0	i_{dc}	$-i_{dc}$	$v_b - v_c$
7	$\mathbf{I}_4$	[120]	'0'	'1'	'0'	'1'	'0'	'0'	$-i_{dc}$	i_{dc}	0	$v_b - v_a$
8	$\mathbf{I}_5$	[102]	'0'	'0'	'1'	'1'	'0'	'0'	$-i_{dc}$	0	i_{dc}	$v_c - v_a$
9	$\mathbf{I}_6$	[012]	'0'	'0'	'1'	'0'	'1'	'0'	0	$-i_{dc}$	i_{dc}	$v_c - v_b$

Table B.3 Summary of 3L-VSC output voltages and input DC branch currents for all possible switching states

	Voltage Vector	Switching Vector	Upper-Leg Switching Signals						Output Phase Voltages			Input DC Currents		
No.	V	$\mathbf{S_{abc}}$	s_{a1}	s_{a2}	s_{b1}	s_{b2}	s_{c1}	s_{c2}	v_{aN}	v_{bN}	v_{cN}	i_P	i_Z	i_N
1	$\mathbf{V_0}$	[000]	‘0’	‘0’	‘0’	‘0’	‘0’	‘0’	0	0	0	0	0	0
2	$\mathbf{V_0}$	[111]	‘0’	‘1’	‘0’	‘1’	‘0’	‘1’	v_{C2}	v_{C2}	v_{C2}	0	0	0
3	$\mathbf{V_0}$	[222]	‘1’	‘1’	‘1’	‘1’	‘1’	‘1’	v_{dc}	v_{dc}	v_{dc}	0	0	0
4	$\mathbf{V_1}$	[100]	‘0’	‘1’	‘0’	‘0’	‘0’	‘0’	v_{C2}	0	0	0	i_a	$i_b + i_c$
5	$\mathbf{V_1}$	[211]	‘1’	‘1’	‘0’	‘1’	‘0’	‘1’	v_{dc}	v_{C2}	v_{C2}	i_a	$i_b + i_c$	0
6	$\mathbf{V_2}$	[110]	‘0’	‘1’	‘0’	‘1’	‘0’	‘0’	v_{C2}	v_{C2}	0	0	$i_a + i_b$	i_c
7	$\mathbf{V_2}$	[221]	‘1’	‘1’	‘1’	‘1’	‘0’	‘1’	v_{dc}	v_{dc}	v_{C2}	$i_a + i_b$	i_c	0
8	$\mathbf{V_3}$	[010]	‘0’	‘0’	‘0’	‘1’	‘0’	‘0’	0	v_{C2}	0	0	i_b	$i_a + i_c$
9	$\mathbf{V_3}$	[121]	‘0’	‘1’	‘1’	‘1’	‘0’	‘1’	v_{C2}	v_{dc}	v_{C2}	i_b	$i_a + i_c$	0
10	$\mathbf{V_4}$	[011]	‘0’	‘0’	‘0’	‘1’	‘0’	‘1’	0	v_{C2}	v_{C2}	0	$i_b + i_c$	i_a
11	$\mathbf{V_4}$	[122]	‘0’	‘1’	‘1’	‘1’	‘1’	‘1’	v_{C2}	v_{dc}	v_{dc}	$i_b + i_c$	i_a	0
12	$\mathbf{V_5}$	[001]	‘0’	‘0’	‘0’	‘0’	‘0’	‘1’	0	0	v_{C2}	0	i_c	$i_a + i_b$
13	$\mathbf{V_5}$	[112]	‘0’	‘1’	‘0’	‘1’	‘1’	‘1’	v_{C2}	v_{C2}	v_{dc}	i_c	$i_a + i_b$	0
14	$\mathbf{V_6}$	[101]	‘0’	‘1’	‘0’	‘0’	‘0’	‘1’	v_{C2}	0	v_{C2}	0	$i_a + i_c$	i_b
15	$\mathbf{V_6}$	[212]	‘1’	‘1’	‘0’	‘1’	‘1’	‘1’	v_{dc}	v_{C2}	v_{dc}	$i_a + i_c$	i_b	0
16	$\mathbf{V_7}$	[200]	‘1’	‘1’	‘0’	‘0’	‘0’	‘0’	v_{dc}	0	0	i_a	0	$i_b + i_c$
17	$\mathbf{V_8}$	[210]	‘1’	‘1’	‘0’	‘1’	‘0’	‘0’	v_{dc}	v_{C2}	0	i_a	i_b	i_c
18	$\mathbf{V_9}$	[220]	‘1’	‘1’	‘1’	‘1’	‘0’	‘0’	v_{dc}	v_{dc}	0	$i_a + i_b$	0	i_c
19	$\mathbf{V_{10}}$	[120]	‘0’	‘1’	‘1’	‘1’	‘0’	‘0’	v_{C2}	v_{dc}	0	i_b	i_a	i_c
20	$\mathbf{V_{11}}$	[020]	‘0’	‘0’	‘1’	‘1’	‘0’	‘0’	0	v_{dc}	0	i_b	0	$i_a + i_c$
21	$\mathbf{V_{12}}$	[021]	‘0’	‘0’	‘1’	‘1’	‘0’	‘1’	0	v_{dc}	v_{C2}	i_b	i_c	i_a
22	$\mathbf{V_{13}}$	[022]	‘0’	‘0’	‘1’	‘1’	‘1’	‘1’	0	v_{dc}	v_{dc}	$i_b + i_c$	0	i_a
23	$\mathbf{V_{14}}$	[012]	‘0’	‘0’	‘0’	‘1’	‘1’	‘1’	0	v_{C2}	v_{dc}	i_c	i_b	i_a
24	$\mathbf{V_{15}}$	[002]	‘0’	‘0’	‘0’	‘0’	‘1’	‘1’	0	0	v_{dc}	i_c	0	$i_a + i_b$
25	$\mathbf{V_{16}}$	[102]	‘0’	‘1’	‘0’	‘0’	‘1’	‘1’	v_{C2}	0	v_{dc}	i_c	i_a	i_b
26	$\mathbf{V_{17}}$	[202]	‘1’	‘1’	‘0’	‘0’	‘1’	‘1’	v_{dc}	0	v_{dc}	$i_a + i_c$	0	i_b
27	$\mathbf{V_{18}}$	[201]	‘1’	‘1’	‘0’	‘0’	‘0’	‘1’	v_{dc}	0	v_{C2}	i_a	i_c	i_b

v_{dc} is net dc bus voltage and is equal to $v_{C1} + v_{C2}$.
The three-phase currents are assumed to be balanced, i.e., $i_a + i_b + i_c = 0$.

Table B.4 (Part I) Summary of 3L-FCC output voltages and flying capacitors current for all possible switching states

	Voltage Vector	Switching Vector	Upper-Leg Switching Signals						Output Phase Voltages			FC Currents		
No.	**V**	$\mathbf{S_{abc}}$	s_{a1}	s_{a2}	s_{b1}	s_{b2}	s_{c1}	s_{c2}	v_{aN}	v_{bN}	v_{cN}	i_{af}	i_{bf}	i_{cf}
1	$\mathbf{V_0}$	[000]	'0'	'0'	'0'	'0'	'0'	'0'	0	0	0	0	0	0
2	$\mathbf{V_0}$	[1A1A1A]	'1'	'0'	'1'	'0'	'1'	'0'	$v_{dc} - v_{af}$	$v_{dc} - v_{bf}$	$v_{dc} - v_{cf}$	i_a	i_b	i_c
3	$\mathbf{V_0}$	[1A1A1B]	'1'	'0'	'1'	'0'	'0'	'1'	$v_{dc} - v_{af}$	$v_{dc} - v_{bf}$	v_{cf}	i_a	i_b	$-i_c$
4	$\mathbf{V_0}$	[1A1B1A]	'1'	'0'	'0'	'1'	'1'	'0'	$v_{dc} - v_{af}$	v_{bf}	$v_{dc} - v_{cf}$	i_a	$-i_b$	i_c
5	$\mathbf{V_0}$	[1A1B1B]	'1'	'0'	'0'	'1'	'0'	'1'	$v_{dc} - v_{af}$	v_{bf}	v_{cf}	i_a	$-i_b$	$-i_c$
6	$\mathbf{V_0}$	[1B1A1A]	'0'	'1'	'1'	'0'	'1'	'0'	v_{af}	$v_{dc} - v_{bf}$	$v_{dc} - v_{cf}$	$-i_a$	i_b	i_c
7	$\mathbf{V_0}$	[1B1A1B]	'0'	'1'	'1'	'0'	'0'	'1'	v_{af}	$v_{dc} - v_{bf}$	v_{cf}	$-i_a$	i_b	$-i_c$
8	$\mathbf{V_0}$	[1B1B1A]	'0'	'1'	'0'	'1'	'1'	'0'	v_{af}	v_{bf}	$v_{dc} - v_{cf}$	$-i_a$	$-i_b$	i_c
9	$\mathbf{V_0}$	[1B1B1B]	'0'	'1'	'0'	'1'	'0'	'1'	v_{af}	v_{bf}	v_{cf}	$-i_a$	$-i_b$	$-i_c$
10	$\mathbf{V_0}$	[222]	'1'	'1'	'1'	'1'	'1'	'1'	v_{dc}	v_{dc}	v_{dc}	0	0	0
11	$\mathbf{V_1}$	[1A00]	'1'	'0'	'0'	'0'	'0'	'0'	$v_{dc} - v_{af}$	0	0	i_a	0	0
12	$\mathbf{V_1}$	[1B00]	'0'	'1'	'0'	'0'	'0'	'0'	v_{af}	0	0	$-i_a$	0	0
13	$\mathbf{V_1}$	[21A1A]	'1'	'1'	'1'	'0'	'1'	'0'	v_{dc}	$v_{dc} - v_{bf}$	$v_{dc} - v_{cf}$	0	i_b	i_c
14	$\mathbf{V_1}$	[21A1B]	'1'	'1'	'1'	'0'	'0'	'1'	v_{dc}	$v_{dc} - v_{bf}$	v_{cf}	0	i_b	$-i_c$
15	$\mathbf{V_1}$	[21B1A]	'1'	'1'	'0'	'1'	'1'	'0'	v_{dc}	v_{bf}	$v_{dc} - v_{cf}$	0	$-i_b$	i_c
16	$\mathbf{V_1}$	[21B1B]	'1'	'1'	'0'	'1'	'0'	'1'	v_{dc}	v_{bf}	v_{cf}	0	$-i_b$	$-i_c$
17	$\mathbf{V_2}$	[1A1A0]	'1'	'0'	'1'	'0'	'0'	'0'	$v_{dc} - v_{af}$	$v_{dc} - v_{bf}$	0	i_a	i_b	0
18	$\mathbf{V_2}$	[1A1B0]	'1'	'0'	'0'	'1'	'0'	'0'	$v_{dc} - v_{af}$	v_{bf}	0	i_a	$-i_b$	0
19	$\mathbf{V_2}$	[1B1A0]	'0'	'1'	'1'	'0'	'0'	'0'	v_{af}	$v_{dc} - v_{bf}$	0	$-i_a$	i_b	0
20	$\mathbf{V_2}$	[1B1B0]	'0'	'1'	'0'	'1'	'0'	'0'	v_{af}	v_{bf}	0	$-i_a$	$-i_b$	0
21	$\mathbf{V_2}$	[221A]	'1'	'1'	'1'	'1'	'1'	'0'	v_{dc}	v_{dc}	$v_{dc} - v_{cf}$	0	0	i_c
22	$\mathbf{V_2}$	[221B]	'1'	'1'	'1'	'1'	'0'	'1'	v_{dc}	v_{dc}	v_{cf}	0	0	$-i_c$
23	$\mathbf{V_3}$	[01A0]	'0'	'0'	'1'	'0'	'0'	'0'	0	$v_{dc} - v_{bf}$	0	0	i_b	0
24	$\mathbf{V_3}$	[01B0]	'0'	'0'	'0'	'1'	'0'	'0'	0	v_{bf}	0	0	$-i_b$	0
25	$\mathbf{V_3}$	[1A21A]	'1'	'0'	'1'	'1'	'1'	'0'	$v_{dc} - v_{af}$	v_{dc}	$v_{dc} - v_{cf}$	i_a	0	i_c
26	$\mathbf{V_3}$	[1A21B]	'1'	'0'	'1'	'1'	'0'	'1'	$v_{dc} - v_{af}$	v_{dc}	v_{cf}	i_a	0	$-i_c$
27	$\mathbf{V_3}$	[1B21A]	'0'	'1'	'1'	'1'	'1'	'0'	v_{af}	v_{dc}	$v_{dc} - v_{cf}$	$-i_a$	0	i_c
28	$\mathbf{V_3}$	[1B21B]	'0'	'1'	'1'	'1'	'0'	'1'	v_{af}	v_{dc}	v_{cf}	$-i_a$	0	$-i_c$
29	$\mathbf{V_4}$	[01A1A]	'0'	'0'	'1'	'0'	'1'	'0'	0	$v_{dc} - v_{bf}$	$v_{dc} - v_{cf}$	0	i_b	i_c
30	$\mathbf{V_4}$	[01A1B]	'0'	'0'	'1'	'0'	'0'	'1'	0	$v_{dc} - v_{bf}$	v_{cf}	0	i_b	$-i_c$
31	$\mathbf{V_4}$	[01B1A]	'0'	'0'	'0'	'1'	'1'	'0'	0	v_{bf}	$v_{dc} - v_{cf}$	0	$-i_b$	i_c
32	$\mathbf{V_4}$	[01B1B]	'0'	'0'	'0'	'1'	'0'	'1'	0	v_{bf}	v_{cf}	0	$-i_b$	$-i_c$

v_{dc} is net dc bus voltage; v_{af}, v_{bf}, and v_{cf} are three-phase flying capacitors voltage.
The three-phase currents are assumed to be balanced, i.e., $i_a + i_b + i_c = 0$.

Table B.5 (Part II) Summary of 3L-FCC output voltages and flying capacitors current for all possible switching states

	Voltage Vector	Switching Vector	Upper-Leg Switching Signals						Output Phase Voltages			FC Currents		
No.	**V**	$\mathbf{S_{abc}}$	s_{a1}	s_{a2}	s_{b1}	s_{b2}	s_{c1}	s_{c2}	v_{aN}	v_{bN}	v_{cN}	i_{af}	i_{bf}	i_{cf}
33	$\mathbf{V_4}$	[1A22]	'1'	'0'	'1'	'1'	'1'	'1'	$v_{dc}-v_{af}$	v_{dc}	v_{dc}	i_a	0	0
34	$\mathbf{V_4}$	[1B22]	'0'	'1'	'1'	'1'	'1'	'1'	v_{af}	v_{dc}	v_{dc}	$-i_a$	0	0
35	$\mathbf{V_5}$	[001A]	'0'	'0'	'0'	'0'	'1'	'0'	0	0	$v_{dc}-v_{cf}$	0	0	i_c
36	$\mathbf{V_5}$	[001B]	'0'	'0'	'0'	'0'	'0'	'1'	0	0	v_{cf}	0	0	$-i_c$
37	$\mathbf{V_5}$	[1A1A2]	'1'	'0'	'1'	'0'	'1'	'1'	$v_{dc}-v_{af}$	$v_{dc}-v_{bf}$	v_{dc}	i_a	i_b	0
38	$\mathbf{V_5}$	[1A1B2]	'1'	'0'	'0'	'1'	'1'	'1'	$v_{dc}-v_{af}$	v_{bf}	v_{dc}	i_a	$-i_b$	0
39	$\mathbf{V_5}$	[1B1A2]	'0'	'1'	'1'	'0'	'1'	'1'	v_{af}	$v_{dc}-v_{bf}$	v_{dc}	$-i_a$	i_b	0
40	$\mathbf{V_5}$	[1B1B2]	'0'	'1'	'0'	'1'	'1'	'1'	v_{af}	v_{bf}	v_{dc}	$-i_a$	$-i_b$	0
41	$\mathbf{V_6}$	[1A01A]	'1'	'0'	'0'	'0'	'1'	'0'	$v_{dc}-v_{af}$	0	$v_{dc}-v_{cf}$	i_a	0	i_c
42	$\mathbf{V_6}$	[1A01B]	'1'	'0'	'0'	'0'	'0'	'1'	$v_{dc}-v_{af}$	0	v_{cf}	i_a	0	$-i_c$
43	$\mathbf{V_6}$	[1B01A]	'0'	'1'	'0'	'0'	'1'	'0'	v_{af}	0	$v_{dc}-v_{cf}$	$-i_a$	0	i_c
44	$\mathbf{V_6}$	[1B01B]	'0'	'1'	'0'	'0'	'0'	'1'	v_{af}	0	v_{cf}	$-i_a$	0	$-i_c$
45	$\mathbf{V_6}$	[21A2]	'1'	'1'	'1'	'0'	'1'	'1'	v_{dc}	$v_{dc}-v_{bf}$	v_{dc}	0	i_b	0
46	$\mathbf{V_6}$	[21B2]	'1'	'1'	'0'	'1'	'1'	'1'	v_{dc}	v_{bf}	v_{dc}	0	$-i_b$	0
47	$\mathbf{V_7}$	[200]	'1'	'1'	'0'	'0'	'0'	'0'	v_{dc}	0	0	0	0	0
48	$\mathbf{V_8}$	[21A0]	'1'	'1'	'1'	'0'	'0'	'0'	v_{dc}	$v_{dc}-v_{bf}$	0	0	i_b	0
49	$\mathbf{V_8}$	[21B0]	'1'	'1'	'0'	'1'	'0'	'0'	v_{dc}	v_{bf}	0	0	$-i_b$	0
50	$\mathbf{V_9}$	[220]	'1'	'1'	'1'	'1'	'0'	'0'	v_{dc}	v_{dc}	0	0	0	0
51	$\mathbf{V_{10}}$	[1A20]	'1'	'0'	'1'	'1'	'0'	'0'	$v_{dc}-v_{af}$	v_{dc}	0	i_a	0	0
52	$\mathbf{V_{10}}$	[1B20]	'0'	'1'	'1'	'1'	'0'	'0'	v_{af}	v_{dc}	0	$-i_a$	0	0
53	$\mathbf{V_{11}}$	[020]	'0'	'0'	'1'	'1'	'0'	'0'	0	v_{dc}	0	0	0	0
54	$\mathbf{V_{12}}$	[021A]	'0'	'0'	'1'	'1'	'1'	'0'	0	v_{dc}	$v_{dc}-v_{cf}$	0	0	i_c
55	$\mathbf{V_{12}}$	[021B]	'0'	'0'	'1'	'1'	'0'	'1'	0	v_{dc}	v_{cf}	0	0	$-i_c$
56	$\mathbf{V_{13}}$	[022]	'0'	'0'	'1'	'1'	'1'	'1'	0	v_{dc}	v_{dc}	0	0	0
57	$\mathbf{V_{14}}$	[01A2]	'0'	'0'	'1'	'0'	'1'	'1'	0	$v_{dc}-v_{bf}$	v_{dc}	0	i_b	0
58	$\mathbf{V_{14}}$	[01B2]	'0'	'0'	'0'	'1'	'1'	'1'	0	v_{bf}	v_{dc}	0	$-i_b$	0
59	$\mathbf{V_{15}}$	[002]	'0'	'0'	'0'	'0'	'1'	'1'	0	0	v_{dc}	0	0	0
60	$\mathbf{V_{16}}$	[1A02]	'1'	'0'	'0'	'0'	'1'	'1'	$v_{dc}-v_{af}$	0	v_{dc}	i_a	0	0
61	$\mathbf{V_{16}}$	[1B02]	'0'	'1'	'0'	'0'	'1'	'1'	v_{af}	0	v_{dc}	$-i_a$	0	0
62	$\mathbf{V_{17}}$	[202]	'1'	'1'	'0'	'0'	'1'	'1'	v_{dc}	0	v_{dc}	0	0	0
63	$\mathbf{V_{18}}$	[201A]	'1'	'1'	'0'	'0'	'1'	'0'	v_{dc}	0	$v_{dc}-v_{cf}$	0	0	i_c
64	$\mathbf{V_{18}}$	[201B]	'1'	'1'	'0'	'0'	'0'	'1'	v_{dc}	0	v_{cf}	0	0	$-i_c$

Table B.6 Summary of DMC source voltages and filter currents for all possible switching states

	Switching Vector	Switching Signals									Source Phase-to-Phase Voltages			Filter Currents		
No.	$\mathbf{S}_{123}$	s_{a1}	s_{b1}	s_{c1}	s_{a2}	s_{b2}	s_{c2}	s_{a3}	s_{b3}	s_{c3}	v_{abs}	v_{bcs}	v_{cas}	i_{ae}	i_{be}	i_{ce}
1	**[222]**	‘1’	‘0’	‘0’	‘1’	‘0’	‘0’	‘1’	‘0’	‘0’	0	0	0	0	0	0
2	**[111]**	‘0’	‘1’	‘0’	‘0’	‘1’	‘0’	‘0’	‘1’	‘0’	0	0	0	0	0	0
3	**[000]**	‘0’	‘0’	‘1’	‘0’	‘0’	‘1’	‘0’	‘0’	‘1’	0	0	0	0	0	0
4	**[200]**	‘1’	‘0’	‘0’	‘0’	‘0’	‘1’	‘0’	‘0’	‘1’	$-v_{cae}$	0	v_{cae}	i_{as}	0	$-i_{as}$
5	**[100]**	‘0’	‘1’	‘0’	‘0’	‘0’	‘1’	‘0’	‘0’	‘1’	v_{bce}	0	$-v_{bce}$	0	i_{as}	$-i_{as}$
6	**[122]**	‘0’	‘1’	‘0’	‘1’	‘0’	‘0’	‘1’	‘0’	‘0’	$-v_{abe}$	0	v_{abe}	$-i_{as}$	i_{as}	0
7	**[022]**	‘0’	‘0’	‘1’	‘1’	‘0’	‘0’	‘1’	‘0’	‘0’	v_{cae}	0	$-v_{cae}$	$-i_{as}$	0	i_{as}
8	**[011]**	‘0’	‘0’	‘1’	‘0’	‘1’	‘0’	‘0’	‘1’	‘0’	$-v_{bce}$	0	v_{bce}	0	$-i_{as}$	i_{as}
9	**[211]**	‘1’	‘0’	‘0’	‘0’	‘1’	‘0’	‘0’	‘1’	‘0’	v_{abe}	0	$-v_{abe}$	i_{as}	$-i_{as}$	0
10	**[020]**	‘0’	‘0’	‘1’	‘1’	‘0’	‘0’	‘0’	‘0’	‘1’	v_{cae}	$-v_{cae}$	0	i_{bs}	0	$-i_{bs}$
11	**[010]**	‘0’	‘0’	‘1’	‘0’	‘1’	‘0’	‘0’	‘0’	‘1’	$-v_{bce}$	v_{bce}	0	0	i_{bs}	$-i_{bs}$
12	**[212]**	‘1’	‘0’	‘0’	‘0’	‘1’	‘0’	‘1’	‘0’	‘0’	v_{abe}	$-v_{abe}$	0	$-i_{bs}$	i_{bs}	0
13	**[202]**	‘1’	‘0’	‘0’	‘0’	‘0’	‘1’	‘1’	‘0’	‘0’	$-v_{cae}$	v_{cae}	0	$-i_{bs}$	0	i_{bs}
14	**[101]**	‘0’	‘1’	‘0’	‘0’	‘0’	‘1’	‘0’	‘1’	‘0’	v_{bce}	$-v_{bce}$	0	0	$-i_{bs}$	i_{bs}
15	**[121]**	‘0’	‘1’	‘0’	‘1’	‘0’	‘0’	‘0’	‘1’	‘0’	$-v_{abe}$	v_{abe}	0	i_{bs}	$-i_{bs}$	0
16	**[002]**	‘0’	‘0’	‘1’	‘0’	‘0’	‘1’	‘1’	‘0’	‘0’	0	v_{cae}	$-v_{cae}$	i_{cs}	0	$-i_{cs}$
17	**[001]**	‘0’	‘0’	‘1’	‘0’	‘0’	‘1’	‘0’	‘1’	‘0’	0	$-v_{bce}$	v_{bce}	0	i_{cs}	$-i_{cs}$
18	**[221]**	‘1’	‘0’	‘0’	‘1’	‘0’	‘0’	‘0’	‘1’	‘0’	0	v_{abe}	$-v_{abe}$	$-i_{cs}$	i_{cs}	0
19	**[220]**	‘1’	‘0’	‘0’	‘1’	‘0’	‘0’	‘0’	‘0’	‘1’	0	$-v_{cae}$	v_{cae}	$-i_{cs}$	0	i_{cs}
20	**[110]**	‘0’	‘1’	‘0’	‘0’	‘1’	‘0’	‘0’	‘0’	‘1’	0	v_{bce}	$-v_{bce}$	0	$-i_{cs}$	i_{cs}
21	**[112]**	‘0’	‘1’	‘0’	‘0’	‘1’	‘0’	‘1’	‘0’	‘0’	0	$-v_{abe}$	v_{abe}	i_{cs}	$-i_{cs}$	0
22	**[210]**	‘1’	‘0’	‘0’	‘0’	‘1’	‘0’	‘0’	‘0’	‘1’	v_{abe}	v_{bce}	v_{cae}	i_{as}	i_{bs}	i_{cs}
23	**[201]**	‘1’	‘0’	‘0’	‘0’	‘0’	‘1’	‘0’	‘1’	‘0’	$-v_{cae}$	$-v_{bce}$	$-v_{abe}$	i_{as}	i_{cs}	i_{bs}
24	**[120]**	‘0’	‘1’	‘0’	‘1’	‘0’	‘0’	‘0’	‘0’	‘1’	$-v_{abe}$	$-v_{cae}$	$-v_{bce}$	i_{bs}	i_{as}	i_{cs}
25	**[102]**	‘0’	‘1’	‘0’	‘0’	‘0’	‘1’	‘1’	‘0’	‘0’	v_{bce}	v_{cae}	v_{abe}	i_{cs}	i_{as}	i_{bs}
26	**[021]**	‘0’	‘0’	‘1’	‘1’	‘0’	‘0’	‘0’	‘1’	‘0’	v_{cae}	v_{abe}	v_{bce}	i_{bs}	i_{cs}	i_{as}
27	**[012]**	‘0’	‘0’	‘1’	‘0’	‘1’	‘0’	‘1’	‘0’	‘0’	$-v_{bce}$	$-v_{abe}$	$-v_{cae}$	i_{cs}	i_{bs}	i_{as}

The source phase-to-phase voltages are: $v_{abs} = v_{as} - v_{bs}$, $v_{bcs} = v_{bs} - v_{cs}$, and $v_{cas} = v_{cs} - v_{as}$.
The filter phase-to-phase voltages are: $v_{abe} = v_{ae} - v_{be}$, $v_{bce} = v_{be} - v_{ce}$, and $v_{cae} = v_{ce} - v_{ae}$.
The three-phase source and filter currents are assumed to be balanced, i.e., $i_{as} + i_{bs} + i_{cs} = 0$ and $i_{ae} + i_{be} + i_{ce} = 0$.

APPENDIX C

MATLAB DEMO PROJECTS

To help the readers with the implementation of FCS-MPC scheme in `MATLAB/Simulink` environment, the following two demo projects are posted in the companion website of this book, along with detailed instructions:

- Stationary ($\alpha\beta$) frame PCC of 2L-VSC feeding an inductive-resistive load.
- Synchronous (dq) frame PCC of 2L-VSC feeding an inductive-resistive load.

The two demo projects have been developed by using `SimPowerSystems` toolbox in `MATLAB/Simulink` 2014a release software. The MPC algorithms are realized by *S-Function Builder* blocks, which can be used directly in the real-time implementation. The *S-Function Builder* programming given in each chapter can be combined with the demo files to develop MPC schemes for variable-speed WECS. The readers are expected to setup the `mex` before using the demo files. At the `MATLAB` command window, type the following command and follow on-screen instructions:

```
» mex -setup
```


Model Predictive Control of Wind Energy Conversion Systems, First Edition. Venkata Yaramasu and Bin Wu.

Companion Website: www.wiley.com/go/yaramasu_wu_mpc_wecs

INDEX

Model Predictive Control of Wind Energy Conversion Systems, First Edition. Venkata Yaramasu and Bin Wu.

Companion Website: www.wiley.com/go/yaramasu_wu_mpc_wecs

D

E

F

G

H

I

L

M

N

O

P

W

Y

Z

IEEE Press Series on Power Engineering

Series Editor: M. E. El-Hawary, Dalhousie University, Halifax, Nova Scotia, Canada

The mission of IEEE Press Series on Power Engineering is to publish leading-edge books that cover the broad spectrum of current and forward-looking technologies in this fast-moving area. The series attracts highly acclaimed authors from industry/academia to provide accessible coverage of current and emerging topics in power engineering and allied fields. Our target audience includes the power engineering professional who is interested in enhancing their knowledge and perspective in their areas of interest.

1. *Principles of Electric Machines with Power Electronic Applications, Second Edition*
M. E. El-Hawary

2. *Pulse Width Modulation for Power Converters: Principles and Practice*
D. Grahame Holmes and Thomas Lipo

3. *Analysis of Electric Machinery and Drive Systems, Second Edition*
Paul C. Krause, Oleg Wasynczuk, and Scott D. Sudhoff

4. *Risk Assessment for Power Systems: Models, Methods, and Applications*
Wenyuan Li

5. *Optimization Principles: Practical Applications to the Operations of Markets of the Electric Power Industry*
Narayan S. Rau

6. *Electric Economics: Regulation and Deregulation*
Geoffrey Rothwell and Tomas Gomez

7. *Electric Power Systems: Analysis and Control*
Fabio Saccomanno

8. *Electrical Insulation for Rotating Machines: Design, Evaluation, Aging, Testing, and Repair, Second Edition*
Greg Stone, Edward A. Boulter, Ian Culbert, and Hussein Dhirani

9. *Signal Processing of Power Quality Disturbances*
Math H. J. Bollen and Irene Y. H. Gu

10. *Instantaneous Power Theory and Applications to Power Conditioning*
Hirofumi Akagi, Edson H. Watanabe, and Mauricio Aredes

11. *Maintaining Mission Critical Systems in a 24/7 Environment*
Peter M. Curtis

12. *Elements of Tidal-Electric Engineering*
Robert H. Clark

13. *Handbook of Large Turbo-Generator Operation and Maintenance, Second Edition*
Geoff Klempner and Isidor Kerszenbaum

14. *Introduction to Electrical Power Systems*
Mohamed E. El-Hawary

15. *Modeling and Control of Fuel Cells: Distributed Generation Applications*
M. Hashem Nehrir and Caisheng Wang

16. *Power Distribution System Reliability: Practical Methods and Applications*
Ali A. Chowdhury and Don O. Koval

17. *Introduction to FACTS Controllers: Theory, Modeling, and Applications*
Kalyan K. Sen and Mey Ling Sen

18. *Economic Market Design and Planning for Electric Power Systems*
James Momoh and Lamine Mili

19. *Operation and Control of Electric Energy Processing Systems*
James Momoh and Lamine Mili

20. *Restructured Electric Power Systems: Analysis of Electricity Markets with Equilibrium Models*
Xiao-Ping Zhang

21. *An Introduction to Wavelet Modulated Inverters*
S.A. Saleh and M.A. Rahman

22. *Control of Electric Machine Drive Systems*
Seung-Ki Sul

23. *Probabilistic Transmission System Planning*
Wenyuan Li

24. *Electricity Power Generation: The Changing Dimensions*
Digambar M. Tigare

25. *Electric Distribution Systems*
Abdelhay A. Sallam and Om P. Malik

26. *Practical Lighting Design with LEDs*
Ron Lenk and Carol Lenk

27. *High Voltage and Electrical Insulation Engineering*
Ravindra Arora and Wolfgang Mosch

28. *Maintaining Mission Critical Systems in a 24/7 Environment, Second Edition*
Peter Curtis

29. *Power Conversion and Control of Wind Energy Systems*
Bin Wu, Yongqiang Lang, Navid Zargari, and Samir Kouro

30. *Integration of Distributed Generation in the Power System*
Math H. Bollen and Fainan Hassan

31. *Doubly Fed Induction Machine: Modeling and Control for Wind Energy Generation Applications*
Gonzalo Abad, Jesus Lopez, Miguel Rodrigues, Luis Marroyo, and Grzegorz Iwanski

32. *High Voltage Protection for Telecommunications*
Steven W. Blume

33. *Smart Grid: Fundamentals of Design and Analysis*
James Momoh

34. *Electromechanical Motion Devices, Second Edition*
Paul C. Krause, Oleg Wasynczuk, and Steven D. Pekarek

35. *Electrical Energy Conversion and Transport: An Interactive Computer-Based Approach, Second Edition*
George G. Karady and Keith E. Holbert

36. *ARC Flash Hazard and Analysis and Mitigation*
J. C. Das

37. *Handbook of Electrical Power System Dynamics: Modeling, Stability, and Control*
Mircea Eremia and Mohammad Shahidehpour

38. *Analysis of Electric Machinery and Drive Systems, Third Edition*
Paul Krause, Oleg Wasynczuk, S. D. Sudhoff, and Steven D. Pekarek

39. *Extruded Cables for High-Voltage Direct-Current Transmission: Advances in Research and Development*
Giovanni Mazzanti and Massimo Marzinotto

40. *Power Magnetic Devices: A Multi-Objective Design Approach*
S. D. Sudhoff

41. *Risk Assessment of Power Systems: Models, Methods, and Applications, Second Edition*
Wenyuan Li

42. *Practical Power System Operation*
Ebrahim Vaahedi

43. *The Selection Process of Biomass Materials for the Production of Bio-Fuels and Co-Firing*
Najib Altawell

44. *Electrical Insulation for Rotating Machines: Design, Evaluation, Aging, Testing, and Repair, Second Edition*
Greg C. Stone, Ian Culbert, Edward A. Boulter, and Hussein Dhirani

45. *Principles of Electrical Safety*
Peter E. Sutherland

46. *Advanced Power Electronics Converters: PWM Converters Processing AC Voltages*
Euzeli Cipriano dos Santos Jr. and Edison Roberto Cabral da Silva

47. *Optimization of Power System Operation, Second Edition*
Jizhong Zhu

48. *Power System Harmonics and Passive Filter Designs*
J. C. Das

49. *Digital Control of High-Frequency Switched-Mode Power Converters*
Luca Corradini, Dragan Maksimovic, Paolo Mattavelli, and Regan Zane

50. *Industrial Power Distribution, Second Edition*
Ralph E. Fehr, III

51. *HVDC Grids: For Offshore and Supergrid of the Future*
Dirk Van Hertem, Oriol Gomis-Bellmunt, and Jun Liang

52. *Advanced Solutions in Power Systems: HVDC, FACTS, and Artificial Intelligence*
Mircea Eremia, Chen-Ching Liu, and Abdel-Aty Edris

53. *Power Grid Operation in a Market Environment: Economic Efficiency and Risk Mitigation*
Hong Chen

54. *Current Signature Analysis for Condition Monitoring of Cage Induction Motors: Industrial Application and Case Histories*
William T. Thomson and Ian Culburt

55. *Model Predictive Control of Wind Energy Conversion Systems*
Venkata Yaramasu and Bin Wu